MW00396367

Springer Series in Optical Sciences   Volume 1

Founded by H. K. V. Lotsch

Springer

Berlin
Heidelberg
New York
Barcelona
Hong Kong
London
Milan
Paris
Singapore
Tokyo

Springer Series in Optical Sciences

Walter Koechner

Solid-State
Laser Engineering

Fifth Revised and Updated Edition

With 472 Figures and 55 Tables

 Springer

Dr. WALTER KOECHNER

Fibertek, Inc., 510 Herndon Parkway
Herndon, VA 20170, USA
Wkoechner@aol.com

Library of Congress Cataloging-in-Publication Data
Koechner, Walter, 1937– . Solid-state laser engineering / Walter Koechner. –
5th rev. and updated ed.
p. cm. – (Springer series in optical sciences; v.1)
Includes bibliographical references and index. ISBN 3-540-65064-4 (alk. paper)
1. Solid-state lasers. I. Title. II. Series. TA1705.K63 1999 621.36'61–dc21 99-15546 CIP

ISBN 3-540-65064-4 Springer-Verlag Berlin Heidelberg New York

ISBN 3-540-60237-2 4th edition Springer-Verlag Berlin Heidelberg New York

Typeset in TEX by K. Steingraeber, Heidelberg, using a Springer TEX macro-package
Cover concept by eStudio Calamar Steinen using a background picture from The Optics Project. Courtesy of John T. Foley, Professor, Department of Physics and Astronomy, Mississippi State University, USA
Cover production: *design & production* GmbH, Heidelberg

Printed on acid-free paper   SPIN 10695027   56/3144-5 4 3 2 1 0

Preface to the Fifth Edition

This book, written from an industrial vantage point, provides a detailed discussion of solid-state lasers, their characteristics, design and construction, and practical problems. The title *Solid-State Laser Engineering* has been chosen because the emphasis is placed on engineering and practical considerations of solid-state lasers. I have tried to enhance the description of the engineering aspects of laser construction and operation by including numerical and technical data, tables, and curves.

The book is mainly intended for the practicing scientist or engineer who is interested in the design or use of solid-state lasers, but the response from readers has shown that the comprehensive treatment of the subject makes the work useful also to students of laser physics who want to supplement their theoretical knowledge with the engineering aspects of lasers. Although not written in the form of a college textbook, the book might be used in an advanced college course on laser technology.

The aim was to present the subject as clearly as possible. Phenomenological descriptions using models were preferred to an abstract mathematical presentation, even though many simplifications had then to be accepted. Results are given in most cases without proof since I have tried to stress the application of the results rather than the derivation of the formulas. An extensive list of references is cited for each chapter to permit the interested reader to learn more about a particular subject.

Again, gratified by the wide acceptance of the previous edition of *Solid-State Laser Engineering*, I have updated and revised the fifth edition to include developments and concepts which have emerged during the last several years. Since the publication of the fourth edition, continued dramatic changes have taken place in the development of solid-state lasers. Today, systems range from tiny, diode-pumped micro-chip lasers to stadium-sized Nd:glass lasers under construction at the National Ignition Facility. The combination of diode-pump sources with innovative pump and resonator designs has dramatically improved beam quality obtainable from solid-state lasers. Spectral coverage, and output power at different wavelengths, have been considerably increased as a result of the emergence of improved nonlinear crystals. Also, table-top femtosecond laser sources have become a reality. At the high end of the power range, flashlamp-pumped Nd:YAG lasers up to the 5 kW level are employed for welding applications, and a number of diode-pumped lasers with outputs in the kW range have been demonstrated at various laboratories. For military and spaceborne systems, where compact packaging and low power consumption is of critical importance, diode-pumped lasers have become an enabling technology for many applications.

Like in previous editions, obsolete material has been deleted and new information has been added. In particluar, the following areas have been expanded:

- In the chapter on laser materials (Chap. 2), several crystals such as $Nd:YVO_4$, $Yb:YAG$ and $Tm:YAG$ are discussed in more detail. Although developed many years ago, only in combination with diode-pumping have these crystals become attractive for a number of applications.
- Recent developments of high-voltage and current semiconductor switches are incorporated in the chapter on pump sources (Chap. 6). These devices provide controllable on-off switches for flashlamps under certain operating conditions.
- The chapter on passive Q-switches (Chap. 8) has been rewritten because the emergence of crystals doped with absorbing ions or containing color centers has greatly improved the durability and reliability of passive Q-switches.
- The chapter on nonlinear devices (Chap. 10) has been expanded to include periodically-poled crystals such as $LiNbO_3$. Incorporation of these crystals in OPO's has considerably increased the output of mid-infrared lasers.
- The chapter on laser-induced damage to optical components (Chap. 11) has been completely rewritten to include the latest findings and results of coatings and bulk damage. Also, a section on self-focusing as it relates to high-brightness solid-state lasers has been added. Self-focusing, a major design consideration in large Nd:glass lasers, and an effect exploited in Kerr lens mode-locking, can be an issue in high brightness diode-pumped lasers.

The material presented in this book reflects the author's experience gained in directing solid-state laser R&D over a 30-year period. This book would not have been possible without the many contributions to the field of laser engineering that have appeared in the open literature and which have been used here as the basic source material. I apologize to any of my colleagues whose work has not been acknowledged or adequately represented in this book.

It is impossible to describe all the new materials, laser configurations, and pumping schemes which have been developed over the last several years. In doing so, the book would merely become a literature survey with commentaries. Readers interested in specific devices are referred to the original literature. Very good sources of information are the IEEE Journal of Quantum Electronics, Optics Letters, and the conference proceedings of the CLEO and Solid-State Laser Conferences.

My special thanks are due to Mindy Levenson, for typing the new material, and to the editor, Dr. H. Lotsch, for his support in preparing the new edition for printing.

None of the editions of this book could have been written without the encouragement, patience and support of my wife Renate.

Herndon, VA   *Walter Koechner*
May 1999

Contents

1. Introduction

In this introductory chapter we shall outline the basic ideas underlying the operation of solid-state lasers. In-depth treatments of laser physics can be found in a number of excellent textbooks [1.1].

1.1 Optical Amplification

To understand the operation of a laser we have to know some of the principles governing the interaction of radiation with matter.

Atomic systems such as atoms, ions, and molecules can exist only in discrete energy states. A change from one energy state to another, called a transition, is associated with either the emission or the absorption of a photon. The wavelength of the absorbed or emitted radiation is given by Bohr's frequency relation

$$E_2 - E_1 = h\nu_{21} , \tag{1.1}$$

where E_2 and E_1 are two discrete energy levels, ν_{21} is the frequency, and h is Planck's constant. An electromagnetic wave whose frequency ν_{21} corresponds to an energy gap of such an atomic system can interact with it. To the approximation required in this context, a solid-state material can be considered an ensemble of very many identical atomic systems. At thermal equilibrium, the lower energy states in the material are more heavily populated than the higher energy states. A wave interacting with the substance will raise the atoms or molecules from lower to higher energy levels and thereby experience absorption.

The operation of a laser requires that the energy equilibrium of a laser material be changed such that energy is stored in the atoms, ions, or molecules of this material. This is achieved by an external pump source which transfers electrons from a lower energy level to a higher one. The pump radiation thereby causes a "population inversion." An electromagnetic wave of appropriate frequency, incident on the "inverted" laser material, will be amplified because the incident photons cause the atoms in the higher level to drop to a lower level and thereby emit additional photons. As a result, energy is extracted from the atomic system and supplied to the radiation field. The release of the stored energy by interaction with an electromagnetic wave is based on stimulated or induced emission.

Stated very briefly, when a material is excited in such a way as to provide more atoms (or molecules) in a higher energy level than in some lower level, the material will be capable of amplifying radiation at the frequency corresponding to the energy level difference. The acronym "laser" derives its name from this process: "**L**ight **A**mplification by **S**timulated **E**mission of **R**adiation."

A quantum mechanical treatment of the interaction between radiation and matter demonstrates that the stimulated emission is, in fact, completely indistinguishable from the stimulating radiation field. This means that the stimulated radiation has the same directional properties, same polarization, same phase, and same spectral characteristics as the stimulating emission. These facts are responsible for the extremely high degree of coherence which characterizes the emission from lasers. The fundamental nature of the induced or stimulated emission process was already described by A. Einstein and M. Planck.

In solid-state lasers, the energy levels and the associated transition frequencies result from the different quantum energy levels or allowed quantum states of the electrons orbiting about the nuclei of atoms. In addition to the electronic transitions, multiatom molecules in gases exhibit energy levels that arise from the vibrational and rotational motions of the molecule as a whole.

1.2 Interaction of Radiation with Matter

Many of the properties of a laser may be readily discussed in terms of the absorption and emission processes which take place when an atomic system interacts with a radiation field. In the first decade of this century Planck described the spectral distribution of thermal radiation, and in the second decade Einstein, by combining Planck's law and Boltzmann statistics, formulated the concept of stimulated emission. Einstein's discovery of stimulated emission provided essentially all of the theory necessary to describe the physical principle of the laser.

1.2.1 Blackbody Radiation

When electromagnetic radiation in an isothermal enclosure, or cavity, is in thermal equilibrium at temperature T, the distribution of radiation density $\varrho(\nu)\,d\nu$, contained in a bandwidth $d\nu$, is given by Planck's law

$$\varrho(\nu)\,d\nu = \frac{8\pi\nu^2\,d\nu}{c^3}\,\frac{h\nu}{e^{h\nu/kT}-1}\,, \tag{1.2}$$

where $\varrho(\nu)$ is the radiation density per unit frequency [Js/cm^3], k is Boltzmann's constant, and c is the velocity of light. The spectral distribution of thermal radiation vanishes at $\nu = 0$ and $\nu \to \infty$, and has a peak which depends on the temperature.

The factor

$$\frac{8\pi\nu^2}{c^3} = p_n \tag{1.3}$$

in (1.2) gives the density of radiation modes per unit volume and unit frequency interval. The factor p_n can also be interpreted as the number of degrees of freedom associated with a radiation field, per unit volume, per unit frequency interval. The expression for the mode density p_n [modes s/cm^3] plays an important role in connecting the spontaneous and the induced transition probabilities.

For a uniform, isotropic radiation field, the following relationship is valid

$$W = \frac{\varrho(\nu)c}{4} , \tag{1.4}$$

where W is the blackbody radiation [W/cm^2] which will be emitted from an opening in the cavity of the blackbody. Many solids radiate like a blackbody. Therefore, the radiation emitted from the surface of a solid can be calculated from (1.4).

According to the Stefan-Boltzmann equation, the total black body radiation is

$$W = \sigma T^4 , \tag{1.5}$$

where $\sigma = 5.68 \times 10^{-12}$ W/cm^2 K^4. The emitted radiation W has a maximum which is obtained from Wien's displacement law

$$\frac{\lambda_{\max}}{\mu m} = \frac{2893}{T/K} . \tag{1.6}$$

For example, a blackbody at a temperature of 5200 K has its radiation peak at 5564 Å, which is about the center of the visible spectrum.

A good introduction to the fundamentals of radiation and its interaction with matter can be found in [1.2].

1.2.2 Boltzmann's Statistics

According to a basic principle of statistical mechanics, when a large collection of similar atoms is in thermal equilibrium at temperature T, the relative populations of any two energy levels E_1 and E_2, such as the ones shown in Fig. 1.1, must be related by the Boltzmann ratio

$$\frac{N_2}{N_1} = \exp\left(\frac{-(E_2 - E_1)}{kT}\right) , \tag{1.7}$$

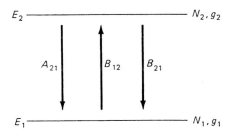

Fig. 1.1. Two energy levels with population N_1, N_2 and degeneracies g_1, g_2, respectively

where N_1 and N_2 are the number of atoms in the energy levels E_1 and E_2, respectively. For energy gaps large enough that $E_2 - E_1 = h\nu_{21} \gg kT$, the ratio is close to zero, and there will be very few atoms in the upper energy level at thermal equilibrium. The thermal energy kT at room temperature ($T \approx 300\,\text{K}$) corresponds to an energy gap $h\nu$ with $\nu \approx 6 \times 10^{12}\,\text{Hz}$, which is equivalent in wavelength to $\lambda \approx 50\,\mu\text{m}$. Therefore, for any energy gap whose transition frequency ν_{21} lies in the near-infrared or visible regions, the Boltzmann exponent will be very small at normal temperatures. The number of atoms in any upper level will then be very small compared to the lower levels. For example, in ruby the ground level E_1 and the upper laser level E_2 are separated by an energy gap corresponding to a wavelength of $\lambda \approx 0.69\,\mu\text{m}$. Since $h = 6.6 \times 10^{-34}\,\text{Ws}^2$, then $E_2 - E_1 = h\nu = 2.86 \times 10^{-19}\,\text{Ws}$. With $k = 1.38 \times 10^{-23}\,\text{Ws K}$ and $T = 300\,\text{K}$, it follows that $N_2/N_1 \approx \exp(-69)$. Therefore at thermal equilibrium virtually all the atoms will be in the ground level.

Equation (1.7) is valid for atomic systems having only non-degenerate levels. If there are g_i different states of the atom corresponding to the energy E_i, then g_i is defined as the degeneracy of the ith energy level.

We recall that atomic systems, such as atoms, ions, molecules, can exist only in certain stationary states, each of which corresponds to a definite value of energy and thus specifies an energy level. When two or more states have the same energy, the respective level is called degenerate, and the number of states with the same energy is the multiplicity of the level. All states of the same energy level will be equally populated, therefore the number of atoms in levels 1 and 2 is $N_1 = g_1 N_1'$ and $N_2 = g_2 N_2'$, where N_1' and N_2' refer to the population of any of the states in levels 1 and 2, respectively. It follows then from (1.7) that the populations of the energy levels 1 and 2 are related by the formula

$$\frac{N_2}{N_1} = \frac{g_2}{g_1} \frac{N_2'}{N_1'} = \frac{g_2}{g_1} \exp\left(\frac{-(E_2 - E_1)}{kT}\right) . \tag{1.8}$$

At absolute zero temperature, Boltzmann's statistics predicts that all atoms will be in the ground state. Thermal equilibrium at any temperature requires that a state with a lower energy be more densely populated than a state with a higher energy. Therefore N_2/N_1 is always less than unity for $E_2 > E_1$ and $T > 0$. This means that optical amplification is not possible in thermal equilibrium.

1.2.3 Einstein's Coefficients

We can most conveniently introduce the concept of Einstein's A and B coefficients by loosely following Einstein's original derivation. To simplify the discussion, let us consider an idealized material with just two nondegenerate energy levels, 1 and 2, having populations of N_1 and N_2, respectively. The total number of atoms in these two levels is assumed to be constant

$$N_1 + N_2 = N_{\text{tot}} . \tag{1.9}$$

Radiative transfer between the two energy levels which differ by $E_2 - E_1 = h\nu_{21}$ is allowed. The atom can transfer from state E_2 to the ground state E_1 by emitting energy; conversely, transition from state E_1 to E_2 is possible by absorbing energy. The energy removed or added to the atom appears as quanta of $h\nu_{21}$. We can identify three types of interaction between electromagnetic radiation and a simple two-level atomic system:

Absorption. If a quasimonochromatic electromagnetic wave of frequency ν_{21} passes through an atomic system with energy gap $h\nu_{21}$, then the population of the lower level will be depleted at a rate proportional both to the radiation density $\varrho(\nu)$ and to the population N_1 of that level

$$\frac{\partial N_1}{\partial t} = -B_{12}\varrho(\nu)N_1 , \tag{1.10}$$

where B_{12} is a constant of proportionality with dimensions cm^3/s^2 J.

The product $B_{12}\varrho(\nu)$ can be interpreted as the probability per unit frequency that transitions are induced by the effect of the field.

Spontaneous Emission. After an atom has been raised to the upper level by absorption, the population of the upper level 2 decays spontaneously to the lower level at a rate proportional to the upper level population.

$$\frac{\partial N_2}{\partial t} = -A_{21}N_2 , \tag{1.11}$$

where A_{21} is a constant of proportionality with the dimensions s^{-1}. The quantity A_{21}, being a characteristic of the pair of energy levels in question, is called the spontaneous transition probability because this coefficient gives the probability that an atom in level 2 will spontaneously change to a lower level 1 within a unit of time.

Spontaneous emission is a statistical function of space and time. With a large number of spontaneously emitting atoms there is no phase relationship between the individual emission processes; the quanta emitted are incoherent. Spontaneous emission is characterized by the lifetime of the electron in the excited state, after which it will spontaneously return to the lower state and radiate away the energy. This can occur without the presence of an electromagnetic field.

Equation (1.11) has a solution

$$N_2(t) = N_2(0) \exp\left(\frac{-t}{\tau_{21}}\right) , \tag{1.12}$$

where τ_{21} is the lifetime for spontaneous radiation of level 2. This radiation lifetime is equal to the reciprocal of the Einstein's coefficient,

$$\tau_{21} = A_{21}^{-1} . \tag{1.13}$$

In general, the reciprocal of the transition probability of a process is called its lifetime.

Stimulated Emission. Emission takes place not only spontaneously but also under stimulation by electromagnetic radiation of appropriate frequency. In this case, the atom gives up a quantum to the radiation field by "induced emission" according to

$$\frac{\partial N_2}{\partial t} = -B_{21}\varrho(\nu_{21})N_2 \, , \tag{1.14}$$

where B_{21} again is a constant of proportionality.

Radiation emitted from an atomic system in the presence of external radiation consists of two parts. The part whose intensity is proportional to A_{21} is the spontaneous radiation; its phase is independent of that of the external radiation. The part whose intensity is proportional to $\varrho(\nu)B_{21}$ is the stimulated radiation; its phase is the same as that of the stimulating external radiation.

The probability of induced transition is proportional to the energy density of external radiation in contrast to spontaneous emission. In the case of induced transition there is a firm phase relationship between the stimulating field and the atom. The quantum which is emitted to the field by the induced emission is coherent with it.

But we shall see later, the useful parameter for laser action is the B_{21} coefficient; the A_{21} coefficient represents a loss term and introduces into the system photons that are not phase-related to the incident photon flux of electric field. Thus the spontaneous process represents a noise source in a laser.

If we combine absorption, spontaneous, and stimulated emission, as expressed by (1.10, 11, and 14), we can write for the change of the upper and lower level populations in our two-level model

$$\frac{\partial N_1}{\partial t} = -\frac{\partial N_2}{\partial t} = B_{21}\varrho(\nu)N_2 - B_{12}\varrho(\nu)N_1 + A_{21}N_2 \, . \tag{1.15}$$

The relation

$$\frac{\partial N_1}{\partial t} = -\frac{\partial N_2}{\partial t} \tag{1.16}$$

follows from (1.9).

In thermal equilibrium, the number of transitions per unit time from E_1 to E_2 must be equal to the number of transitions from E_2 to E_1. Certainly, in thermal equilibrium

$$\frac{\partial N_1}{\partial t} = \frac{\partial N_2}{\partial t} = 0 \, . \tag{1.17}$$

Therefore we can write

$$\underset{\substack{\text{Spontaneous} \\ \text{emission}}}{N_2 A_{21}} \quad + \quad \underset{\substack{\text{Stimulated} \\ \text{emission}}}{N_2\varrho(\nu)B_{21}} \quad = \quad \underset{\text{Absorption}}{N_1\varrho(\nu)B_{12}} \, . \tag{1.18}$$

Using the Boltzmann equation (1.8) for the ratio N_2/N_1, we then write the above expression as

$$\varrho(\nu_{21}) = \frac{(A_{21}/B_{21})}{(g_1/g_2)(B_{12}/B_{21})\exp{(h\nu_{21}/kT)} - 1}. \tag{1.19}$$

Comparing this expression with the black body radiation law (1.2), we see that

$$\frac{A_{21}}{B_{21}} = \frac{8\pi\nu^2 h\nu}{c^3} \quad \text{and} \quad B_{21} = \frac{g_1 B_{12}}{g_2}. \tag{1.20}$$

The relations between the A's and B's are known as Einstein's relations. The factor $8\pi\nu^2/c^3$ in (1.20) is the mode density p_n given by (1.3).

In solids the speed of light is $c = c_0/n$, where n is the index of refraction and c_0 is the speed of light in vacuum.

For a simple system with no degeneracy, that is, one in which $g_1 = g_2$, we see that $B_{21} = B_{12}$. Thus, the Einstein coefficients for stimulated emission and absorption are equal. If the two levels have unequal degeneracy, the probability for stimulated absorption is no longer the same as that for stimulated emission.

1.2.4 Phase Coherence of Stimulated Emission

The stimulated emission provides a phase-coherent amplification mechanism for an applied signal. The signal extracts from the atoms a response that is directly proportional to, and phase-coherent with, the electric field of the stimulating signal. Thus the amplification process is phase-preserving. The stimulated emission is, in fact, completely indistinguishable from the stimulating radiation field. This means that the stimulated emission has the same directional properties, same polarization, same phase, and same spectral characteristics as the stimulating emission. These facts are responsible for the extremely high degree of coherence which characterizes the emission from lasers. The proof of this fact is beyond the scope of this elementary introduction, and requires a quantum mechanical treatment of the interaction between radiation and matter. However, the concept of induced transition, or the interaction between a signal and an atomic system, can be demonstrated, qualitatively, with the aid of the classical electron-oscillator model.

Electromagnetic radiation interacts with matter through the electric charges in the substance. Consider an electron which is elastically bound to a nucleus. One can think of electrons and ions held together by spring-type bonds which are capable of vibrating around equilibrium positions. An applied electric field will cause a relative displacement between electron and nucleus from their equilibrium position. They will execute an oscillatory motion about their equilibrium position. Therefore, the model exhibits an oscillatory or resonant behavior and a response to an applied field. Since the nucleus is so much heavier than the electron, we assume that only the electron moves. The most important model for understanding the interaction of light and matter is that of the harmonic oscillator. We take as our model a single electron, assumed to be bound to its equilibrium position by a linear restoring force. We may visualize the electron as a point of mass suspended by springs. Classical electromagnetic theory asserts that any oscillating electric charge will act as a miniature antenna or dipole and will continuously radiate away electromagnetic energy to its surroundings.

A detailed description of the electric dipole transition and the classical electron-oscillator model can be found in [1.3].

1.3 Absorption and Optical Gain

In this section we will develop the quantitative relations that govern absorption and amplification processes in substances. This requires that we increase the realism of our mathematical model by introducing the concept of atomic line-shapes. Therefore, the important features and the physical processes which lead to different atomic lineshapes will be considered first.

1.3.1 Atomic Lineshapes

In deriving Einstein's coefficients we have assumed a monochromatic wave with frequency ν_{21} acting on a two-level system with an infinitely sharp energy gap $h\nu_{21}$. We will now consider the interaction between an atomic system having a finite transition linewidth $\Delta\nu$ and a signal with a bandwidth $d\nu$.

Before we can obtain an expression for the transition rate for this case, it is necessary to introduce the concept of the atomic lineshape function $g(\nu, \nu_0)$. The distribution $g(\nu, \nu_0)$, centered at ν_0, is the equilibrium shape of the linewidth-broadened transitions. Suppose that N_2 is the total number of ions in the upper energy level considered previously. The spectral distribution of ions per unit frequency is then

$$N(\nu) = g(\nu, \nu_0)N_2 \ . \tag{1.21}$$

If we integrate both sides over all frequencies we have to obtain N_2 as a result:

$$\int_0^\infty N(\nu)d\nu = N_2 \int_0^\infty g(\nu, \nu_0)d\nu = N_2 \ . \tag{1.22}$$

Therefore the lineshape function must be normalized to unity:

$$\int_0^\infty g(\nu, \nu_0)d\nu = 1 \ . \tag{1.23}$$

If we know the function $g(\nu, \nu_0)$, we can calculate the number of atoms $N(\nu)d\nu$ in level 1 which are capable of absorbing in the frequency range ν to $\nu + d\nu$, or the number of atoms in level 2 which are capable of emitting in the same range.

From (1.21) we have

$$N(\nu)\,d\nu = g(\nu,\nu_0)\,d\nu\,N_2 \ . \tag{1.24}$$

From the foregoing it follows that $g(\nu,\nu_0)$ can be defined as the probability of emission or absorption per unit frequency. Therefore $g(\nu)\,d\nu$ is the probability that a given transition will result in an emission (or absorption) of a photon with energy between $h\nu$ and $h(\nu + d\nu)$. The probability that a transition will occur between $\nu = 0$ and $\nu = \infty$ has to be 1.

It is clear from the definition of $g(\nu,\nu_0)$ that we can, for example, rewrite (1.11) in the form

$$-\frac{\partial N_2}{\partial t} = A_{21}N_2 g(\nu,\nu_0)\,d\nu \ , \tag{1.25}$$

where N_2 is the total number of atoms in level 2, and $\partial N_2/\partial t$ is the number of photons spontaneously emitted per second between ν and $\nu + d\nu$.

The linewidth and lineshape of an atomic transition depends on the cause of line broadening. Optical frequency transitions in gases can be broadened by lifetime, collision, or Doppler broadening, whereas transitions in solids can be broadened by lifetime, dipolar or thermal broadening, or by random inhomogeneities. All these linewidth-broadening mechanisms lead to two distinctly different atomic lineshapes, the homogeneously and the inhomogeneously broadened line [1.4].

The Homogeneously Broadened Line

The essential feature of a homogeneously broadened atomic transition is that every atom has the same atomic lineshape and frequency response, so that a signal applied to the transition has exactly the same effect on all atoms in the collection. This means that within the linewidth of the energy level each atom has the same probability function for a transition.

Differences between homogeneously and inhomogeneously broadened transitions show up in the saturation behavior of these transitions. This has a major effect on the laser operation. The important point about a homogeneous lineshape is that the transition will saturate uniformly under the influence of a sufficiently strong signal applied anywhere within the atomic linewidth.

Mechanisms which result in a homogeneously broadened line are lifetime broadening, collision broadening, dipolar broadening, and thermal broadening.

Lifetime Broadening. This type of broadening is caused by the decay mechanisms of the atomic system. Spontaneous emission or fluorescence has a radiative lifetime. Broadening of the atomic transition due to this process is related to the fluorescence lifetime τ by $\Delta\omega_a\tau = 1$, where ω_a is the bandwidth.

Actually, physical situations in which the lineshape and linewidth are determined by the spontaneous emission process itself are vanishingly rare. Since the natural or intrinsic linewidth of an atomic line is extremely small, it is the linewidth that would be observed from atoms at rest without interaction with one another.

Collision Broadening. Collision of radiating particles (atoms or molecules) with one another and the consequent interruption of the radiative process in a random manner leads to broadening. As an atomic collision interrupts either the emission or the absorption of radiation, the long wave train which otherwise would be present becomes truncated. The atom restarts its motion after the collision with a completely random initial phase. After the collision the process is restarted without memory of the phase of the radiation prior to the collision. The result of frequent collisions is the presence of many truncated radiative or absorptive processes.

Since the spectrum of a wave train is inversely proportional to the length of the train, the linewidth of the radiation in the presence of collision is greater than that of an individual uninterrupted process.

Collision broadening is observed in gas lasers operated at higher pressures, hence the name pressure broadening. At higher pressures collisions between gas atoms limit their radiative lifetime. Collision broadening, therefore, is quite similar to lifetime broadening, in that the collisions interrupt the initial state of the atoms.

Dipolar Broadening. Dipolar broadening arises from interactions between the magnetic or electric dipolar fields of neighboring atoms. This interaction leads to results very similar to collision broadening, including a linewidth that increases with increasing density of atoms. Since dipolar broadening represents a kind of coupling between atoms, so that excitation applied to one atom is distributed or shared with other atoms, dipolar broadening is a homogeneous broadening mechanism.

Thermal Broadening. Thermal broadening is brought about by the effect of the thermal lattice vibrations on the atomic transition. The thermal vibrations of the lattice surrounding the active ions modulate the resonance frequency of each atom at a very high frequency. This frequency modulation represents a coupling mechanism between the atoms, therefore a homogeneous linewidth is obtained. Thermal broadening is the mechanism responsible for the linewidth of the ruby laser and Nd : YAG laser.

The lineshape of homogeneous broadening mechanisms lead to a Lorentzian lineshape for atomic response. For the normalized Lorentz distribution, the equation

$$g(\nu) = \left(\frac{\Delta\nu}{2\pi}\right)\left[(\nu - \nu_0)^2 + \left(\frac{\Delta\nu}{2}\right)^2\right]^{-1} \qquad (1.26)$$

is valid. Here, ν_0 is the center frequency, and $\Delta\nu$ is the width between the half-power points of the curve. The factor $\Delta\nu/2\pi$ assures normalization of the area under the curve according to (1.23). The peak value for the Lorentz curve is

$$g(\nu_0) = \frac{2}{\pi \Delta \nu} \; . \tag{1.27}$$

The Inhomogeneously Broadened Line

Mechanisms which cause inhomogeneous broadening tend to displace the center frequencies of individual atoms, thereby broadening the overall response of a collection without broadening the response of individual atoms. Different atoms have slightly different resonance frequencies on the same transition, for example, owing to Doppler shifts. As a result, the overall response of the collection is broadened. An applied signal at a given frequency within the overall linewidth interacts strongly only with those atoms whose shifted resonance frequencies lie close to the signal frequency. The applied signal does not have the same effect on all the atoms in an inhomogeneously broadened collection.

Since in an inhomogeneously broadened line interaction occurs only with those atoms whose resonance frequencies lie close to the applied signal frequency, a strong signal will eventually deplete the upper laser level in a very narrow frequency interval. The signal will eventually "burn a hole" in the atomic absorption curve. Examples of inhomogeneous frequency-shifting mechanisms include Doppler broadening and broadening due to crystal inhomogeneities.

Doppler Broadening. The apparent resonance frequencies of atoms undergoing random motions in a gas are shifted randomly so that the overall frequency response of the collection of atoms is broadened. A particular atom moving with a velocity component ν relative to an observer in the z direction will radiate at a frequency measured by the observer as $\nu_0(1 + v/c)$. When these velocities are averaged, the resulting lineshape is Gaussian. Doppler broadening is one form of inhomogeneous broadening, since each atom emits a different frequency rather than one atom having a probability distribution for emitting any frequency within the linewidth. In the actual physical situation, the Doppler line is best visualized as a packet of homogeneous lines of width $\Delta \nu_n$, which superimpose to give the observed Doppler shape. The He-Ne laser has a Doppler-broadened linewidth. Most visible and near-infrared gas laser transitions are inhomogeneously broadened by Doppler effects.

Line Broadening Due to Crystal Inhomogeneities. Solid-state lasers may be inhomogeneously broadened by crystalline defects. This happens only at low temperatures where the lattice vibrations are small. Random variations of dislocations, lattice strains, etc., may cause small shifts in the exact energy level spacings and transition frequencies from ion to ion. Like Doppler broadening, these variations do not broaden the response on an individual atom, but they do cause the exact resonance frequencies of different atoms to be slightly different. Thus random crystal imperfection can be a source of inhomogeneous broadening in a solid-state laser crystal.

A good example of an inhomogeneously broadened line occurs in the fluorescence of neodymium-doped glass. As a result of the so-called glassy state, there are variations, from rare earth site to rare earth site, in the relative atomic positions occupied by the surrounding lattice ions. This gives rise to a random distribution of static crystalline fields acting on the rare-earth ions. Since the line shifts corresponding to such crystal-field variations are larger, generally speaking, than the width contributed by other factors associated with the transition, an inhomogeneous line results.

The inhomogeneous-broadened linewidth can be represented by a Gaussian frequency distribution. For the normalized distribution, the equation

$$g(\nu) = \frac{2}{\Delta\nu} \left(\frac{\ln 2}{\pi} \right)^{1/2} \exp\left[-\left(\frac{\nu - \nu_0}{\Delta\nu/2} \right)^2 \ln 2 \right] \qquad (1.28)$$

is valid, where ν_0 is the frequency at the center of the line, and $\Delta\nu$ is the linewidth at which the amplitude falls to one-half. The peak value of the normalized Gaussian curve is

$$g(\nu_0) = \frac{2}{\Delta\nu} \left(\frac{\ln 2}{\pi} \right)^{1/2} . \qquad (1.29)$$

In Fig. 1.2 the normalized Gaussian and Lorentz lines are plotted for a common linewidth.

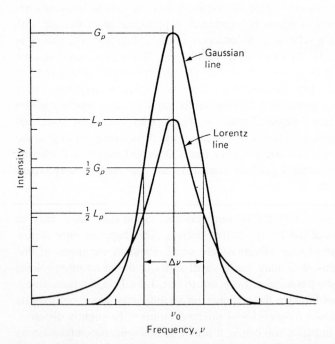

Fig. 1.2. Gaussian and Lorentz lines of common linewidth (G_p and L_p are the peak intensities)

1.3.2 Absorption by Stimulated Transitions

We assume a quasicollimated beam of energy density $\varrho(\nu)$ incident on a thin absorbing sample of thickness dx; as before, we consider the case of an optical system that operates between only two energy levels as illustrated schematically in Fig. 1.1. The populations of the two levels are N_1 and N_2, respectively. Level 1 is the ground level and level 2 is the excited level. We consider absorption of radiation in the material and emission from the stimulated processes but neglect the spontaneous emission. From (1.15 and 1.20) we obtain

$$-\frac{\partial N_1}{\partial t} = \varrho(\nu)B_{21}\left(\frac{g_2}{g_1}N_1 - N_2\right) . \tag{1.30}$$

As we recall, this relation was obtained by considering infinitely sharp energy levels separated by $h\nu_{21}$ and a monochromatic wave of frequency ν_{21}.

We will now consider the interaction between two linewidth-broadened energy levels with an energy separation centered at ν_0, and a half-width of $\Delta\nu$ characterized by $g(\nu,\nu_0)$ and a signal with center frequency ν_s and bandwidth $d\nu$. The situation is shown schematically in Fig. 1.3. The spectral width of the signal is narrow, as compared to the linewidth-broadened transition. If N_1 and N_2 are the total number of atoms in level 1 and level 2, then the number of atoms capable of interacting with a radiation of frequency ν_s and bandwidth $d\nu$ are

$$\left(\frac{g_2}{g_1}N_1 - N_2\right)g(\nu_s,\nu_0)d\nu . \tag{1.31}$$

The net change of atoms in energy level 1 can be expressed in terms of energy density $\varrho(\nu)d\nu$ by multiplying both sides of (1.30) with photon energy $h\nu$ and

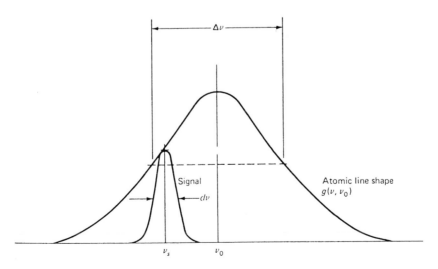

Fig. 1.3. Linewidth-broadened atomic transition line centered at ν_0 and narrow band signal centered at ν_s

dividing by the volume V. We will further express the populations N_1 and N_2 as population densities n_1 and n_2.

Equation (1.30) now becomes

$$-\frac{\partial}{\partial t}[\varrho(\nu_s)d\nu] = \varrho(\nu_s)d\nu \, B_{21}h\nu g(\nu_s, \nu_0)\left(\frac{g_2}{g_1}n_1 - n_2\right) . \tag{1.32}$$

This equation gives the net rate of absorbed energy in the frequency interval $d\nu$ centered around ν_s. In an actual laser system the wavelength of the emitted radiation, corresponding to the signal bandwidth $d\nu$ in our model, is very narrow as compared to the natural linewidth of the material. Ruby, for example, has a fluorescent linewidth of 5 Å, whereas the linewidth of the laser output is typically 0.1 to 0.01 Å. The operation of a laser, therefore, can be fairly accurately characterized as the interaction of linewidth-broadened energy levels with a monochromatic wave. The photon density of a monochromatic radiation of frequency ν_0 can then be represented by a delta function $\delta(\nu - \nu_0)$. After integrating (1.32) in the interval $d\nu$, we obtain, for a monochromatic signal of frequency ν_s and a linewidth-broadened transition,

$$-\frac{\partial\varrho(\nu_s)}{\partial t} = \varrho(\nu_s)B_{21}h\nu_s g(\nu_s, \nu_0)\left(\frac{g_2}{g_1}n_1 - n_2\right) . \tag{1.33}$$

The signal will travel through the material of thickness dx in the time $dt = dx/c = (n/c_0)dx$. Then, as the wave advances from x to $x + dx$, the decrease of energy in the beam is

$$-\frac{\partial\varrho(\nu_s)}{\partial x} = h\nu_s\varrho(\nu_s)g(\nu_s, \nu_0)B_{21}\left(\frac{g_2}{g_1}n_1 - n_2\right)\frac{1}{c} . \tag{1.34}$$

Integration of (1.34) gives

$$\frac{\varrho(\nu_s)}{\varrho_0(\nu_s)} = \exp\left[-h\nu_s g(\nu_s, \nu_0)B_{21}\left(\frac{g_2}{g_1}n_1 - n_2\right)\frac{x}{c}\right] . \tag{1.35}$$

If we introduce an absorption coefficient $\alpha(\nu_s)$,

$$\alpha(\nu_s) = \left(\frac{g_2}{g_1}n_1 - n_2\right)\sigma_{21}(\nu_s) , \qquad \text{where} \tag{1.36}$$

$$\sigma_{21}(\nu_s) = \frac{h\nu_s g(\nu_s, \nu_0)B_{21}}{c} . \tag{1.37}$$

Then we can write (1.35) as

$$\varrho(\nu_s) = \varrho_0(\nu_s)\exp[-\alpha(\nu_s)x] . \tag{1.38}$$

Equation (1.38) is the well-known exponential absorption equation for thermal equilibrium condition $n_1 g_2/g_1 > n_2$. The energy of the radiation decreases exponentially with the depth of penetration into the substance. The maximum possible absorption occurs when all atoms exist in the ground state n_1. For equal population of the energy states $n_1 = (g_1/g_2)n_2$, the absorption is eliminated and the material is transparent. The parameter σ_{21} is the cross section for the radiative transition $2 \to 1$. The cross section for stimulated emission σ_{21} is related to the absorption cross section σ_{12} by the ratio of the level degeneracies,

$$\frac{\sigma_{21}}{\sigma_{12}} = \frac{g_1}{g_2} . \tag{1.39}$$

The cross section is a very useful parameter to which we will refer in the following chapters. If we replace B_{21} by the Einstein relation (1.20), we obtain σ_{21} in a form which we will find most useful:

$$\sigma_{21}(\nu_s) = \frac{A_{21}\lambda_0^2}{8\pi n^2} g(\nu_s, \nu_0) . \tag{1.40}$$

As we will see later, the gain for the radiation building up in a laser resonator will be highest at the center of the atomic transitions. Therefore, in lasers we are mostly dealing with stimulated transitions which occur at the center of the linewidth.

If we assume $\nu \approx \nu_s \approx \nu_0$, we obtain, for the spectral stimulated emission cross section at the center of the atomic transition for a Lorentzian lineshape,

$$\sigma_{21} = \frac{A_{21}\lambda_0^2}{4\pi^2 n^2 \Delta\nu} , \tag{1.41}$$

and for a Gaussian lineshape,

$$\sigma_{21} = \frac{A_{21}\lambda_0^2}{4\pi n^2 \Delta\nu} \left(\frac{\ln 2}{\pi}\right)^{1/2} . \tag{1.42}$$

Here we have introduced into (1.40) the peak values of the lineshape function, as given in (1.27 and 1.29) for the Lorentzian and Gaussian curves respectively. For example, in the case of the R_1 line of ruby, where $\lambda_0 = 6.94 \times 10^{-5}$ cm, $n = 1.76$, $\tau_{21} = (1/A_{21}) = 3$ ms and $\Delta\nu = 11$ cm^{-1} one finds, according to (1.41), $\sigma_{21} = 4.0 \times 10^{-20}$ cm^2. In comparing this value with the data provided in Table 2.2, we have to distinguish between the spectroscopic cross section and the effective stimulated emission cross section. (This will be discussed in Sect. 2.3.1 for the case of Nd : YAG). The effective stimulated emission cross section is the spectroscopic cross section times the occupancy of the upper laser level relative to the entire manifold population. In ruby, the upper laser level is split into two sublevels, therefore the effective stimulated emission cross section is about half of the value calculated from (1.41).

1.3.3 Population Inversion

According to the Boltzmann distribution (1.7), in a collection of atoms at thermal equilibrium there are always fewer atoms in a higher-lying level E_2 than in a lower level E_1. Therefore the population difference $N_1 - N_2$ is always positive, which means that the absorption coefficient $\alpha(\nu_s)$ in (1.36) is positive and the incident radiation is absorbed (Fig. 1.4).

Suppose that it were possible to achieve a temporary situation such that there are more atoms in an upper energy level than in a lower energy level. The normally positive population difference on that transition then becomes negative, and the normal stimulated absorption as seen from an applied signal on that transition is correspondingly changed to stimulated emission, or amplification of the applied signal. That is, the applied signal gains energy as it interacts with the atoms and hence is amplified. The energy for this signal amplification is supplied by the atoms involved in the interaction process. This situation is characterized by a negative absorption coefficient $\alpha(\nu_s)$ according to (1.36). From (1.34) it follows that $\partial\varrho(\nu)/\partial x > 0$.

The essential condition for amplification is that there are more atoms in an upper energy level than in a lower energy level; i.e., for amplification,

$$N_2 > N_1 \quad \text{if} \quad E_2 > E_1 \,, \tag{1.43}$$

as illustrated in Fig. 1.5. The resulting negative sign of the population difference $(N_2 - g_2 N_1/g_1)$ on that transition is called a population inversion. Population inversion is clearly an abnormal situation; it is never observed at thermal equilibrium. The point at which the population of both states is equal is called the "inversion threshold".

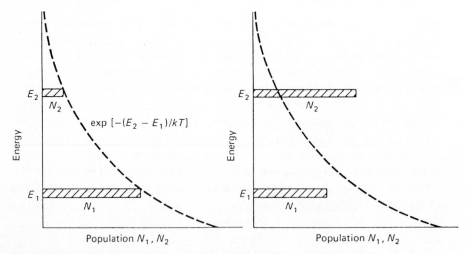

Fig. 1.4. Relative populations in two energy levels as given by the Boltzmann relation for thermal equilibrium

Fig. 1.5. Inverted population difference required for optical amplification

Stimulated absorption and emission processes always occur side by side independently of the population distribution among the levels. So long as the population of the higher energy level is smaller than that of the lower energy level, the number of absorption transitions is larger than that of the emission transitions, so that there is an overall attenuation of the radiation. When the numbers of atoms in both states are equal, the number of emissions becomes equal to the number of absorptions; the material is then transparent to the incident radiation. As soon as the population of the higher level becomes larger than that of the lower level, emission processes predominate and the radiation is enhanced collectively during passage through the material. To produce an inversion requires a source of energy to populate a specified energy level; we call this energy the pump energy.

In Sect. 1.4 we will discuss the type of energy level structure an atomic system must possess in order to make it possible to generate an inversion. Techniques by which the atoms of a solid-state laser can be raised or pumped into upper energy levels are discussed in Sect. 6.1. Depending on the atomic system involved, an inverted population condition may be obtainable only on a transient basis, yielding pulsed laser action; or it may be possible to maintain the population inversion on a steady-state basis, yielding continuous-wave (cw) laser action.

The total amount of energy which is supplied by the atoms to the light wave is

$$E = \Delta N h \nu , \tag{1.44}$$

where ΔN is the total number of atoms which are caused to drop from the upper to the lower energy level during the time the signal is applied. If laser action is to be maintained, the pumping process must continually replenish the supply of upper-state atoms. The size of the inverted population difference is reduced not only by the amplification process but also by spontaneous emission which always tends to return the energy level populations to their thermal equilibrium values.

1.4 Creation of a Population Inversion

We are concerned in this section with how the necessary population inversion for laser action is obtained in solid-state lasers. We can gain considerable understanding on how laser devices are pumped and how their population densities are inverted by studying some simplified but fairly realistic models.

The discussion up to this point has been based on a hypothetical $2 \leftrightarrow 1$ transition and has not been concerned with how the levels 2 and 1 fit into the energy level scheme of the atom. This detached point of view must be abandoned when one tries to understand how laser action takes place in a solid-state medium. As already noted, the operation of the laser depends on a material with narrow energy levels between which electrons can make transitions. Usually these levels are due to impurity atoms in a host crystal. The pumping and laser processes

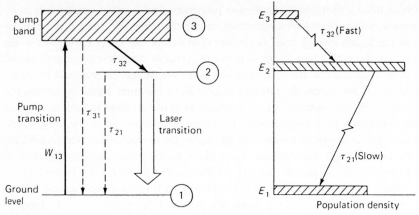

Fig. 1.6. Simplified energy level diagram of a three-level laser

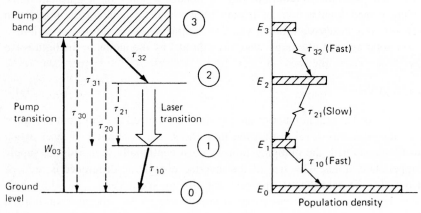

Fig. 1.7. Simplified energy level diagram of a four-level laser

in real laser systems typically involve a very large number of energy levels, with complex excitation processes and cascaded relaxation processes among all these levels. Operation of an actual laser material is properly described only by a many-level energy diagram. The main features can be understood, however, through the familiar three-level or four-level idealizations of Figs. 1.6 and 1.7. More detailed energy level diagrams of some of the most important solid-state laser materials are presented in Chap. 2.

1.4.1 The Three-Level System

Figure 1.6 shows a diagram which can be used to explain the operation of an optically pumped three-level laser, such as ruby. Initially, all atoms of the laser material are in the lowest level 1. Excitation is supplied to the solid by radiation of frequencies which produce absorption into the broad band 3. Thus, the pump light raises atoms from the ground state to the pump band, level 3. In general, the "pumping" band, level 3, is actually made up of a number of bands, so that the optical pumping can be accomplished over a broad spectral range. Most of the excited atoms are transferred by fast radiationless transitions into the intermediate sharp level 2. In this process the energy lost by the electron is transferred to the lattice. Finally, the electron returns to the ground level by the emission of a photon. It is this last transition that is responsible for the laser action. If pumping intensity is below laser threshold, atoms in level 2 predominantly return to the ground state by spontaneous emission. Ordinary fluorescence acts as a drain on the population of level 2. After the pump radiation is extinguished, level 2 is emptied by fluorescence at a rate that varies from material to material. In ruby, at room temperature, the lifetime of level 2 is 3 ms. When the pump intensity is above laser threshold, the decay from the fluorescent level consists of stimulated as well as spontaneous radiation; the stimulated radiation produces the laser output beam. Since the terminal level of the laser transition is the highly populated ground state, a very high population must be reached in the E_2 level before the $2 \to 1$ transition is inverted.

It is necessary, in general, that the rate of radiationless transfer from the uppermost level to the level at which the laser action begins be fast compared with the other spontaneous transition rates in a three-level laser. Therefore, the lifetime of the E_2 state should be large in comparison with the relaxation time of the $3 \to 2$ transition, i.e.,

$$\tau_{21} \gg \tau_{32} . \tag{1.45}$$

The number of atoms N_3 in level E_3 is then negligible compared with the number of atoms in the other two states, i.e., $N_3 \ll N_1, N_2$. Therefore,

$$N_1 + N_2 \approx N_{\text{tot}} . \tag{1.46}$$

A vital aspect of the three-level system is that the atoms are in effect pumped directly from level 1 into the metastable level 2 with only a momentary pause as they pass through level 3. With these conditions, we can calculate as if only two levels were present. In order that an equal population is achieved between the E_2 and E_1 levels, one-half of all atoms must be excited to the E_2 level:

$$N_2 = N_1 = \frac{N_{\text{tot}}}{2} . \tag{1.47}$$

In order to maintain a specified amplification, the population of the second level must be larger than that of the first level. In most cases which are of practical importance, however, the necessary inversion $(N_2 - N_1)$ is small compared with the total number of all atoms. The pump power necessary for maintaining this inversion is also small compared with the power necessary for achieving equal population of the levels.

The disadvantage of a three-level system is that more than half of the atoms in the ground state must be raised to the metastable level E_2. There are thus many atoms present to contribute to the spontaneous emission. Moreover, each of the atoms which participate in the pump cycle transfer energy into the lattice from the $E_3 \rightarrow E_2$ transition. This transition is normally radiationless, the energy being carried into the lattice by phonons.

1.4.2 The Four-Level System

The four-level laser system, which is characteristic of the rare earth ions in glass or crystalline host materials, is illustrated in Fig. 1.7. Note that a characteristic of the three-level laser material is that the laser transition takes place between the excited laser level 2 and the final ground state 1, the lowest energy level of the system. This leads to low efficiency. The four-level system avoids this disadvantage. The pump transition extends again from the ground state (now level E_0) to a wide absorption band E_3. As in the case of the three-level system, the atoms so excited will proceed rapidly to the sharply defined level E_2. The laser transition, however, proceeds now to a fourth, terminal level E_1, which is situated above the ground state E_0. From here the atom undergoes a rapid non-radiative transition to the ground level. In a true four-level system, the terminal laser level E_1 will be empty. To qualifiy as a four-level system a material must possess a relaxation time between the terminal laser level and the ground level which is fast compared to the fluorescent lifetime, i.e., $\tau_{10} \ll \tau_{21}$. In addition the terminal laser level must be far above the ground state so that its thermal population is small. The equilibrium population of the terminal laser level 1 is determined by the relation

$$\frac{N_1}{N_0} = \exp\left(\frac{-\Delta E}{kT}\right), \tag{1.48}$$

where ΔE is the energy separation between level 1 and the ground state, and T is the operating temperature of the laser material. If $\Delta E \gg kT$, then $N_1/N_0 \ll 1$, and the intermediate level will always be relatively empty. In some laser materials the energy gap between the lower laser level and the ground state is relatively small and, therefore, they must be cooled to function as four-level lasers. In a four-level system an inversion of the $2 \rightarrow 1$ transition can occur even with vanishingly small pump power, and the high pump rate, necessary to maintain equilibrium population in the aforementioned three-level system, is no longer needed. In the most favorable case, the relaxation times of the $3 \rightarrow 2$ and $1 \rightarrow 0$ transitions in the four-level system are short compared with the spontaneous emission lifetime of the laser transition τ_{21}. Hence we can also carry out the calculations as if only the E_1 and E_2 states were populated.

1.4.3 The Metastable Level

After this brief introduction to the energy level structure of solid-state lasers we can ask the question, "what energy level scheme must a solid possess to make it a useful laser?" As we have seen in the previous discussion, the existence of a metastable level is of paramount importance for laser action to occur. The relatively long lifetime of the metastable level provides a mechanism by which inverted population can be achieved. Most transitions of atoms show rapid non-radiative decay, because the coupling of the internal atomic oscillations to the surrounding lattice is strong. Radiative decay processes can occur readily, but most have short lifetimes and broad linewidths. Only a few transitions of selected atoms in solids turn out to be decoupled from the lattice vibrations. These transitions have a radiative decay which leads to relatively long lifetimes.

In typical laser systems with energy levels, such as illustrated by Fig. 1.6 and 7, the $3 \rightarrow 2$ transition frequencies, as well as the $1 \rightarrow 0$ transition frequencies, all fall within the frequency range of the vibration spectrum of the host crystal lattice. Therefore, all these transitions can relax extremely rapidly by direct nonradiative decay, i.e., by emitting a phonon to the lattice vibrations, with $\tau_{32}, \tau_{10} \approx 10^{-8}$ to 10^{-11} s. However, the larger $3 \rightarrow 0$, $3 \rightarrow 1$, $2 \rightarrow 0$, and $2 \rightarrow 1$ energy gaps in these atoms often correspond to transition frequencies that are higher than the highest possible vibration frequency of the crystal lattice. Such transitions cannot relax via simple single-phonon spontaneous emission, since the lattice simply cannot accept phonons at those high frequencies. These transitions must then relax either by radiative (photon) emission or by multiple-phonon processes. Since both these processes are relatively weak compared to direct single-phonon relaxation, the high-frequency transitions will have much slower relaxation rates ($\tau_{21} \approx 10^{-5}$ to 10^{-3} s in many cases). Therefore, the various levels lumped into level 3 will all relax mostly into level 2 while level 2 itself is metastable and long-lived because there are no other levels located close below it into which it can decay directly.

The existence of metastable levels follows from quantum mechanical considerations that will not be discussed here. However, for completeness we will at least explain the term "forbidden transition". As we have seen in Sect. 1.2.4, the mechanism by which energy exchange takes place between an atom and the electromagnetic fields is the dipole radiation. As a consequence of quantum-mechanical considerations and the ensuing selection rules, transfer between certain states cannot occur due to forbidden transitions. The term "forbidden" means that a transition among the states concerned does not take place as a result of the interaction of the electric dipole moment of the atom with the radiation field. As a result of the selection rules, an atom may get into an excited state from which it will have difficulty returning to the ground state. A state from which all dipole transitions to lower energy states are forbidden is metastable; an atom entering such a state will generally remain in that state much longer than it would in an ordinary excited state from which escape is comparatively easy.

In the absence of a metastable level, the atoms which become excited by pump radiation and are transferred to a higher energy level will return either directly to the ground state by spontaneous radiation or by cascading down on intermediate levels, or they may release energy by phonon interaction with the lattice. In order for the population to increase at the metastable laser level, several other conditions have to be met. Let us consider the more general case of a four-level system illustrated in Fig. 1.7. (Note that a three-level system can be thought of as a special case of a four-level scheme where level 1 and level 0 coincide). Pumping takes place between two levels and laser action takes place between two other levels. Energy from the pump band is transferred to the upper laser level by fast radiative transitions. Energy is removed from the lower laser level again by fast radiationless transitions.

For electrons in the pump band at level 3 to transfer to level 2 rather than return directly to the ground state, it is required that $\tau_{30} \gg \tau_{32}$. For population to build up, relaxation out of the lower level 1 has to be fast, $\tau_{21} \gg \tau_{10}$. Thus, as a first conclusion, we may say that if the right relaxation time ratio exists between any two levels (such as 3 and 2) in an energy level system, a population inversion should be possible. If so, then obtaining a large enough inversion for successful laser operation becomes primarily a matter of the right pumping method. The optical pumping method is generally feasible only in laser materials which combine a narrow laser emission line with a broad absorption transition, so that a broad-band intense light source can be used as the pump source. An exception is a solid-state laser which is pumped by another laser, such as a diode laser for example. In this case the requirement for a broad absorption range for the pump band can be relaxed.

Having achieved population inversion in a material by correct combination of relaxation times and the existence of broad pump bands, the linewidth of the laser transition becomes very important. In the following chapter we will see that the optical gain for a given population inversion is inversely proportional to linewidth. Therefore, the metastable level should have a sufficiently narrow linewidth.

1.5 Laser Rate Equations

The dynamic behavior of a laser can be described with reasonable precision by a set of coupled rate equations [1.5]. In their simplest forms, a pair of simultaneous differential equations describe the population inversion and the radiation density within a spatially uniform laser medium. We will describe the system in terms of the energy-level diagrams shown in Figs. 1.6 and 1.7. As we have seen in the preceding discussions, two energy levels are of prime importance in laser action: the excited upper laser level E_2 and the lower laser level E_1. Thus for many analyses of laser action an approximation of the three- and four-level systems by a two-level representation is very useful.

The rate-equation approach used in this section involves a number of sim-plifying assumptions; in using a single set of rate equations we are ignoring longitudinal and radial variations of the radiation within the laser medium. In spite of these limitations, the simple rate-equation approach remains a useful tool and, properly used, provides a great deal of insight into the behavior of real solid-state laser devices. We will derive from the rate equations the threshold condition for laser actions, and obtain a first-order approximation of the relaxation oscillations in a solid-state laser. Furthermore, in Chap. 4 we will use the rate equations to calculate the gain in a laser amplifier.

In general, the rate equations are useful in predicting the gross features of the laser output, such as average and peak power, Q-switched pulse-envelope shape, threshold condition, etc. On the other hand, many details of the nature of the laser emission are inaccessible from the point of view of a simple rate equation. These include detailed descriptions of the spectral, temporal, and spatial distributions of the laser emission. Fortunately, these details can often be accounted for independently.

In applying the rate equations to the various aspects of laser operation, we will find it more convenient to express the probability for stimulated emission $\varrho(\nu)B_{21}$ by the photon density ϕ and the stimulated emission cross section σ.

With (1.37) we can express the Einstein coefficient for stimulated emission B_{21} in terms of the stimulated emission cross section $\sigma(\nu)$,

$$B_{21} = \frac{c}{h\nu g(\nu)}\sigma_{21}(\nu) , \tag{1.49}$$

where $c = c_0/n$ is the speed of light in the medium. The energy density per unit frequency $\varrho(\nu)$ is expressed in terms of the lineshape factor $g(\nu)$, the energy $h\nu$, and the photon density ϕ [photons/cm^2] by

$$\varrho(\nu) = h\nu g(\nu)\phi. \tag{1.50}$$

From (1.49 and 50) we obtain

$$B_{21}\varrho(\nu) = c\sigma_{21}(\nu)\phi . \tag{1.51}$$

Three-Level System

In order to approximate the three-level system with a two-level scheme, we assume that the transition from the pump band to the upper laser level is so fast that $N_3 \approx 0$. Therefore pumping does not affect the other processes at all except to allow a mechanism of populating the upper level and thereby obtaining population inversion ($N_2 > N_1$).

Looking at Fig. 1.6, this assumption requires that the relaxation time ratio τ_{32}/τ_{21} be very small. In solid-state lasers $\tau_{32}/\tau_{21} \approx 0$ is a good approximation. Spontaneous losses from the pump band to the ground state can be expressed by the quantum efficiency η_Q. This parameter, defined as

$$\eta_Q = \left(1 + \frac{\tau_{32}}{\tau_{31}}\right)^{-1} \leq 1 , \tag{1.52}$$

specifies what fraction of the total atoms excited to level 3 drop from there to level 2, thus becoming potentially useful for laser action. A small η_Q obviously requires a correspondingly larger pump power.

The changes in the electron population densities in a three-level system, based on the assumption that essentially all of the laser ions are in either level 1 or level 2, are

$$\frac{\partial n_1}{\partial t} = \left(n_2 - \frac{g_2}{g_1} n_1 \right) c\phi\sigma + \frac{n_2}{\tau_{21}} - W_p n_1 \tag{1.53}$$

and

$$\frac{\partial n_2}{\partial t} = -\frac{\partial n_1}{\partial t} , \tag{1.54}$$

since

$$n_{\text{tot}} = n_1 + n_2 , \tag{1.55}$$

where W_p is the pumping rate $[\text{s}^{-1}]$.

The terms of the right-hand side of (1.53) express the net stimulated emission, the spontaneous emission, and the optical pumping.

The time variation of the population in both levels due to absorption, spontaneous, and stimulated emission is obtained from (1.15). Note that the populations N_1 and N_2 are now expressed in terms of population densities n_1 and n_2. To take into account the effect of pumping, we have added the term $W_p n_1$, which can be thought of as the rate of supply of atoms to the metastable level 2. More precisely, $W_p n_1$ is the number of atoms transferred from the ground level to the upper laser level per unit time per unit volume. The pump rate W_p is related to the pump parameter W_{13} in Fig. 1.6 by

$$W_p = \eta_Q W_{13} . \tag{1.56}$$

The negative sign in front of $W_p n_1$ in (1.53) indicates that the pump mechanism removes atoms from the ground level 1 and increases the population of level 2.

If we now define the inversion population density by

$$n = n_2 - \frac{g_2 n_1}{g_1} \tag{1.57}$$

we can combine (1.53, 54, and 57) to obtain

$$\frac{\partial n}{\partial t} = -\gamma n\phi\sigma c - \frac{n + n_{\text{tot}}(\gamma - 1)}{\tau_f} + W_p(n_{\text{tot}} - n), \tag{1.58}$$

where

$$\gamma = 1 + \frac{g_2}{g_1} \quad \text{and} \quad \tau_f = \tau_{21} . \tag{1.59}$$

In obtaining (1.58) we have used the relations

$$n_1 = \frac{n_{\text{tot}} - n}{1 + g_2/g_1} \quad \text{and} \quad n_2 = \frac{n + (g_2/g_1)n_{\text{tot}}}{1 + g_2/g_1} . \tag{1.60}$$

Another equation, usually regarded together with (1.58), describes the rate of change of the photon density within the laser resonator,

$$\frac{\partial \phi}{\partial t} = c\phi\sigma n - \frac{\phi}{\tau_c} + S, \tag{1.61}$$

where τ_c is the decay time for photons in the optical resonator and S is the rate at which spontaneous emission is added to the laser emission.

If we consider for the moment only the first term on the right, which is the increase of the photon density by stimulated emission, then (1.61) is identical to (1.33). However, for the time variation of the photon density in the laser resonator we must also take into account the decrease of radiation due to losses in the system and the increase of radiation due to a small amount of spontaneous emission which is added to the laser emission. Although very small, this term must be included because it provides the source of radiation which initiates laser emission.

An important consideration for initiation of laser oscillation is the total number p of resonant modes possible in the laser resonator volume V_R, since in general only a few of these modes are initiated into oscillations. This number is given by the familiar expression (1.3),

$$p = 8\pi\nu^2 \frac{\Delta\nu V_R}{c^3} , \tag{1.62}$$

where ν is the laser optical frequency, and $\Delta\nu$ is the bandwidth of spontaneous emission. Let p_L be the number of modes of the laser output. Then S can be expressed as the rate at which spontaneous emission contributes to stimulated emission, namely,

$$S = \frac{p_L n_2}{p\tau_{21}} . \tag{1.63}$$

The reader is referred to Chap. 3 for a more detailed description of the factor τ_c which appears in (1.61). For now we only need to know that τ_c represents all the losses in an optical resonator of a laser oscillator. Since τ_c has the dimension of time, the losses are expressed in terms of a relaxation time. The decay of the photon population in the cavity results from transmission and absorption at the end mirrors, "spillover" diffraction loss due to the finite apertures of the mirrors, scattering and absorptive losses in the laser material itself, etc. In the absence of the amplifying mechanism, (1.61) becomes

$$\frac{\partial \phi}{\partial t} = -\frac{\phi}{\tau_c} , \tag{1.64}$$

the solution of which is $\phi(\tau) = \phi_0 \exp(-t/\tau_c)$.

The importance of (1.61) should be emphasized by noting that the right-hand side of this equation describes the net gain per transit of an electromagnetic wave passing through a laser material.

Four-Level System

We will assume again that the transition from the pump band into the upper laser level occurs very rapidly. Therefore the population of the pump band is negligible, i.e., $n_3 \approx 0$. With this assumption the rate of change of the two laser levels in a four-level system is

$$\frac{dn_2}{dt} = W_p n_0 - \left(n_2 - \frac{g_2}{g_1} n_1\right)\sigma\phi c - \frac{n_2}{\tau_{21} + \tau_{20}} , \tag{1.65}$$

$$\frac{dn_1}{dt} = \left(n_2 - \frac{g_2}{g_1} n_1\right)\sigma\phi c + \frac{n_2}{\tau_{21}} - \frac{n_1}{\tau_{10}} , \tag{1.66}$$

$$n_{\text{tot}} = n_0 + n_1 + n_2 . \tag{1.67}$$

From (1.65) follows that the upper laser level population in a four-level system increases due to pumping and decreases due to stimulated emission and spontaneous emissions into level 1 and level 0. The lower level population increases due to stimulated and spontaneous emission and decreases by a radiationless relaxation process into the ground level. This process is characterized by the time constant τ_{10}. In an ideal four-level system the terminal level empties infinitely fast to the ground level. If we let $\tau_{10} \approx 0$, then it follows from (1.66) that $n_1 = 0$. In this case the entire population is divided between the ground level 0 and the upper level of the laser transition. The system appears to be pumping from a large source that is independent of the lower laser level. With $\tau_{10} = 0$ and $n_1 = 0$, we obtain the following rate equation for the ideal four-level system

$$n = n_2 \tag{1.68}$$

and

$$n_{\text{tot}} = n_0 + n_2 \approx n_0 \quad \text{since} \quad n_2 \ll n_0 . \tag{1.69}$$

Therefore, instead of (1.58), we have

$$\frac{\partial n_2}{\partial t} = -n_2\sigma\phi c - \frac{n_2}{\tau_f} + W_p(n_0 - n_2). \tag{1.70}$$

The fluorescence decay time τ_f of the upper laser level is given by

$$\frac{1}{\tau_f} = \frac{1}{\tau_{21}} + \frac{1}{\tau_{20}} , \tag{1.71}$$

where $\tau_{21} = A_{21}^{-1}$ is the effective radiative lifetime associated with the laser line. In the equation for the rate of change of the upper laser level we have again taken into account the fact that not all atoms pumped to level 3 will end up at the upper laser level. It is

$$W_p = \eta_Q W_{03} \,, \tag{1.72}$$

where the quantum efficiency η_Q depends on the branching ratios which are the relative relaxation rates for the atoms along the various possible downward paths,

$$\eta_Q = \left(1 + \frac{\tau_{32}}{\tau_{31}} + \frac{\tau_{32}}{\tau_{30}}\right)^{-1} \leq 1. \tag{1.73}$$

As already indicated in the case of a three-level system the quantum efficiency is the probability of an absorbed pump photon producing an active atom in the upper laser level. Some of the absorbed pump photons will not produce an active atom in the upper laser level. Some, for example, may decay to manifolds other than the manifold containing the upper laser level while others may decay to the ground level by radiationless transitions. The equation which describes the rate of change of the photon density within the laser resonator is the same as in the case of the three-level system.

Summary

The rate equation applicable to three- and four-level systems can be expressed by a single pair of equations, namely, (1.58 and 61), where $\gamma = 1 + g_2/g_1$ for a three-level system and $\gamma = 1$ for a four-level system. The factor γ can be thought of as an "inversion reduction factor" since it corresponds to the net reduction in the population inversion after the emission of a single photon. In a four-level system, see (1.70), we have $\gamma = 1$ since the population inversion density is only reduced by one for each photon emitted. For a three-level system, see (1.58), we have $\gamma = 2$ if we assume no degeneracy, i.e., $g_2/g_1 = 1$. This reflects the fact that in this case the population inversion is reduced by two for each stimulated emission of a photon because the emitting photon is not only lost to the upper laser level, but also increases the lower laser level by one. The parameters τ_f and W_p are defined by (1.56, 59, 72, and 73) for the three- and four-level systems. The factor S in (1.61), which represents the initial noise level of ϕ due to spontaneous emission at the laser frequency, is small and needs to be considered only for initial starting of the laser action. It will be dropped from this point on.

A more detailed analysis of the laser rate equations can be found in [1.1, 3].

2. Properties of Solid-State Laser Materials

Materials for laser operation must possess sharp fluorescent lines, strong absorption bands, and a reasonably high quantum efficiency for the fluorescent transition of interest. These characteristics are generally shown by solids (crystals or glass) which incorporate in small amounts elements in which optical transitions can occur between states of inner, incomplete electron shells. Thus the transition metals, the rare earth (lanthanide) series, and the actinide series are of interest in this connection. The sharp fluorescence lines in the spectra of crystals doped with these elements result from the fact that the electrons involved in transitions in the optical regime are shielded by the outer shells from the surrounding crystal lattice. The corresponding transitions are similar to those of the free ions. In addition to a sharp fluorescence emission line, a laser material should possess pump bands within the emission spectrum of readily available pump sources such as arc lamps and laser diode arrays.

The three principal elements leading to gain in a laser are:

- *The host material* with its macroscopic mechanical, thermal and optical properties, and its unique microscopic lattice properties.
- *The activator/sensitizer ions* with their distinctive charge states and free-ion electronic configurations.
- *The optical pump source* with its particular geometry, spectral irradiance, and temporal characteristic.

These elements are interactive and must be selectetd self-consistently to achieve a given system performance.

In this chapter we consider the properties of various host materials and activator/sensitizer combinations. Pump sources for solid-state lasers are treated in Chap. 6.

2.1 Overview

The conditions for laser action at optical frequencies were first described by *Schawlow* and *Townes* [2.1] in 1958. The first demonstration of laser action by *Maiman* [2.2] was achieved in 1960 using ruby ($Cr^{3+} : Al_2O_3$), a crystalline solid system. The next step in the development of solid-state lasers was the operation of trivalent uranium in CaF_2 and divalent samarium in CaF_2 by *Sorokin* and

Stevenson [2.3]. In 1961 *Snitzer* [2.4] demonstrated laser action in neodymium-doped glass. The first continuously operating crystal laser was reported in 1961 by *Johnson* and *Nassau* [2.5] using Nd^{3+} : $CaWO_4$. Since then laser action has been achieved from trivalent rare earths (Nd^{3+}, Er^{3+}, Ho^{3+}, Ce^{3+}, Tm^{3+}, Pr^{3+}, Gd^{3+}, Eu^{3+}, Yb^{3+}), divalent rare earths (Sm^{2+}, Dy^{2+}, Tm^{2+}), transition metals (Cr^{3+}, Ni^{2+}, Co^{2+}, Ti^{3+}, V^{2+}), and the actinide ion U^{3+} embedded in various host materials. Optically pumped laser action has been demonstrated in hundreds of ion-host crystal combinations covering a spectral range from the visible to the mid-infrared.

The exceptionally favorable characteristics of the trivalent neodymium ion for laser action were recognized at a relatively early stage in the search for solid-state laser materials. Thus, Nd^{3+} was known to exhibit a satisfactorily long fluorescence lifetime and narrow fluorescence linewidths in crystals with ordered structures, and to possess a terminal state for the laser transition sufficiently high above the ground state so that cw operation at room temperature was readily feasible. Therefore, this ion was incorporated as a dopant in a variety of host materials, i.e., glass, $CaWO_4$, $CaMoO_4$, CaF_2, LaF_3, etc., in an effort to make use of its great potential. However, most of these early hosts displayed undesirable shortcomings, either from the standpoint of their intrinsic physical properties or because of the way in which they interacted with the Nd^{3+} ions. Finally, yttrium aluminum garnet ("YAG") was explored by *Geusic* et al. [2.6] as a host for Nd^{3+} and its superiority to other host materials was quickly demonstrated. Nd : YAG lasers displayed the lowest thresholds for cw operation at room temperature of any known host-dopant combination.

2.1.1 Host Materials

Solid-state host materials may be broadly grouped into crystalline solids and glasses. The host must have good optical, mechanical and thermal properties to withstand the severe operating conditions of practical lasers. Desirable properties include hardness, chemical inertness, absence of internal strain and refractive index variations, resistance to radiation-induced color centers, and ease of fabrication.

Several interactions between the host crystal and the additive ion restrict the number of useful material combinations. These include size disparity, valence, and spectroscopic properties. Ideally the size and valence of the additive ion should match that of the host ion it replaces.

In selecting a crystal suitable for a laser ion host one must consider the following key criteria:

i) The crystal must possess favorable optical properties. Variations in the index of refraction lead to inhomogeneous propagation of light through the crystal which results in poor beam quality.

ii) The crystal must possess mechanical and thermal properties that will permit high-average-power operation. The most important parameters are thermal conductivity, hardness and fracture strength.

iii) The crystal must have lattice sites that can accept the dopant ions and that have local crystal fields of symmetry and strength needed to induce the desired spectroscopic properties. In general, ions placed in a crystal host should have long radiative lifetimes with cross sections near $10^{-20}\,\mathrm{cm^2}$.

iv) It must be possible to scale the growth of the impurity-doped crystal, while maintaining high optical quality and high yield. It appears that the greatest prospect for successful growth scaling is for crystals that melt congruently at temperatures below $1300°\mathrm{C}$. This relatively low melting temperature permits the use of a wide variety of crucible materials and growth techniques.

Glasses

Glasses form an important class of host materials for some of the rare earths, particularly Nd^{3+}. The outstanding practical advantage compared to crystalline materials is the tremendous size capability for high-energy applications. Rods up to 1 m in length and over 10 cm in diameter and disks up to 90 cm in diameter and several cm thick have been produced. The optical quality can be excellent, and beam angles approaching the diffraction limit can be achieved. Glass, of course, is easily fabricated and takes a good optical finish. Laser ions placed in glass generally show a larger fluorescent linewidth than in crystals as a result of the lack of a unique and well-defined crystalline surrounding for the individual active atom. Therefore, the laser thresholds for glass lasers have been found to run higher than their crystalline counterparts. Also, glass has a much lower thermal conductivity than most crystalline hosts. The latter factor leads to a large thermally induced birefringence and optical distortion in glass laser rods when they are operated at high average powers. Ions which have been made to lase in glass include Nd^{3+}, which will be discussed in detail in Sect. 2.3.2, Yb^{3+}, Er^{3+}, Tm^{3+}, and Ho^{3+}. Glass doped with erbium is of special importance, because its radiation of $1.55\,\mu\mathrm{m}$ does not penetrate the lens of the human eye, and therefore cannot destroy the retina. Because of the three-level behavior of erbium and the small absorption of pump light by Er^{3+}, multiple doping with neodymium and ytterbium is necessary to obtain satisfactory system efficiency. In this technique, called sensitization, the amplifying ion either absorbes radiation at wavelengths other than those of the laser ion and then radiates within the pump band of the laser ion or transfers its excitation energy directly to the laser ion. Because of its three-level operation, the Nd^{3+}-Yb^{3+}-Er^{3+} : glass is at least an order of magnitude less efficient than the Nd : glass lasers [2.7]. For specific details on laser glasses, the reader is referred to [2.8].

A large number of crystalline host materials have been investigated since the discovery of the ruby laser. Crystalline laser hosts generally offer as advantages over glasses their higher thermal conductivity, narrower fluorescence linewidths, and, in some cases, greater hardness. However, the optical quality and doping homogeneity of crystalline hosts are often poorer, and the absorption lines are generally narrower. For an overview of crystalline lasers, see [2.9].

Oxides

Sapphire. The first laser material to be discovered (ruby laser) employed sapphire as a host. The Al_2O_3 (sapphire) host is hard, with high thermal conductivity, and transition metals can readily be incorporated substitutionally for the Al. The Al site is too small for rare earths, and it is not possible to incorporate appreciable concentrations of these impurities into sapphire. Besides ruby which is still used today, Ti-doped sapphire has gained significance as a tunable-laser material. The properties of ruby and Ti-sapphire will be discussed in Sects. 2.2 and 2.5.2.

Garnets. Some of the most useful laser hosts are the synthetic garnets: yttrium aluminum, $Y_3Al_5O_{12}$ (YAG), gadolinium gallium garnet , $Gd_3Ga_5O_{12}$ (GGG) [2.6, 10], and gadolinium scandium aluminum garnet $Gd_3Sc_2Al_3O_{12}$ (GSGG) [2.6, 11]. These garnets have many properties that are desirable in a laser host material. They are stable, hard, optically isotropic, and have good thermal conductivities, which permits laser operation at high average power levels.

In particular, yttrium aluminum garnet doped with neodymium (Nd : YAG) has achieved a position of dominance among solid-state laser materials. YAG is a very hard, isotropic crystal, which can be grown and fabricated in a manner that yields rods of high optical quality. At the present time, it is the best commercially available crystalline laser host for Nd^{3+}, offering low threshold and high gain. The properties of Nd : YAG are discussed in more detail in Sect. 2.3.1. Besides Nd^{3+}, the host crystal YAG has been doped with Tm^{3+}, Er^{3+}, Ho^{3+}, and Yb^{3+}. Laser action in these materials were first reported in [2.12, 13].

In recent years, Nd : GSGG co-doped with Cr^{3+} has been employed in a number of laser systems. Cr^{3+} considerably increases the absorption of flash-lamp radiation and transfers the energy very efficiently to Nd. This laser will be discussed in more detail in Sect. 2.3.3.

Aluminate. In 1969 a crystal host derived from the same Y_2O_3-Al_2O_3 system as YAG was discovered [2.15, 17]. The crystal yttrium ortho aluminate (YAlO_3), termed YAlO or YAP is the 1:1 compound or perovskite phase, YAG is the 3:5 compound or garnet phase. Many physical properties of YAP such as hardness and thermal conductivity are similar to those of YAG. In contrast to YAG, which is cubic and isotropic, YAP is orthorhombic and anisotropic. The anisotropy of the spectral properties of YAlO_3 enables one to select crystallographic orientations of the laser rod which optimize particular performance characteristics. Thus rod orientations can be chosen for high gain and low thresholds or, alternatively, for low gain and high energy storage required for Q-switching operation.

The fluorescence in $Nd:YAlO_3$ occurs predominantly in three strong lines at 1.0645, 1.0725, and 1.0795 μm. For light propagating along the b axis of the crystal, the gain is maximum at 1.0795 μm and is comparable to that of the 1.064-μm line in $Nd:YAG$. For a beam propagating along the c axis of $Nd:YAlO_3$ the gain is maximum at 1.064 μm, but it is only about one-half that in $Nd:YAG$. Laser action in YAP has been achieved with ions including Nd^{3+}, Er^{3+}, Ho^{3+}, Tm^{3+} [2.15, 18].

Despite several major potential advantages of YAP over YAG, such as a polarized output, the capability of accepting a higher concentration of Nd, the possibility of varying gain by changing the rod axis, and lower costs because of faster crystal growth, the material disappeared from the market only two years after its introduction. YAP's disadvantages were a lower efficiency than expected and erratic performance. It was found, for example, that during the growth process Fe^{3+} impurities enter the single crystal [2.16], which increase the absorption loss at 1.06 μm and cause fluorescence quenching. However, over the years it has been possible to improve the optical quality of the YAP crystals [2.14] and the material is commercially available again.

Oxysulfide. The application of rare earth oxysulfides as laser host materials has also been explored. The entire oxysulfide series, from lanthanum oxysulfide through lutetium oxysulfide and yttrium oxysulfide, possess the same (uniaxial) crystal structure. Thus, solid solutions of any concentration of rare earth activator in any other rare earth oxysulfide host are possible. Host materials which are transparent from 0.35 to 7 μm include lanthanum, gadolinium, yttrium, and lutetium oxysulfide. Laser action was observed at 1.075 μm for Nd and La_2O_2S. The lasing transition cross section of $Nd:LOS$ was measured to be about one-third of that of $Nd:YAG$ [2.19]. The optical quality of LOS crystals is far below the quality of YAG crystals.

Phosphates and Silicates

Laser oscillations have been produced in crystals of Nd^{3+}-doped calcium fluorophosphate or $Ca_5(PO_4)_3F$. The host crystal has the mineral name fluorapatite, from which the name FAP was coined [2.20]. This material is unique in that the Nd^{3+} fluorescence spectrum is predominantly concentrated in a single narrow and intense line, whereas the absorption spectrum is relatively broad and intense. Fluorapatites have low oscillation thresholds and high slope efficiencies, but they are soft, are susceptible to the formation of color centers, and their low thermal conductivity leads to strong thermal distortions. Because of these latter properties, FAP did not become a popular laser material. In 1972 another apatite became available: silicate oxyapatite or CaLaSOAP. In contrast to FAP, CaLaSOAP is considerably harder; however, its thermal conductivity is nearly equal to FAP's and only one-ninth the thermal conductivity of YAG.

The energy storage in $Nd:SOAP$ is about five times that of YAG. In addition to SOAP's high energy storage, the material has the potential that large crystals can be grown inexpensively. Crystals 15 cm long with diameters as large as 1.2 cm have been grown at growth rates of 2 to 3 mm/h. It was reported that the optical quality of SOAP rods 7.5 by 0.6 cm was comparable to the quality of YAG rods with undetectable scattering and less than 0.5 fringe. The main disadvantages of SOAP are a low laser damage threshold and low thermal conductivity. Physical, spectroscopic, and laser characteristics of neodymium-doped silicate oxyapatite are reported in [2.21–23].

Laser action has been achieved in the neodymium pentaphosphates $YNdP_5O_{14}$, $LaNdP_5O_{14}$, and $ScNdP_5O_{14}$ [2.24–28]. An outstanding property of these materials is the high gain which can be achieved as a result of the favorable position of the $^4I_{15/2}$ manifold relative to the upper laser states. As a result of the high gain, cw oscillation at room temperature has been obtained in $ScNdP_5O_{14}$ with a pump power of only 4 mW using an argon laser as a pump source [2.28].

Tungstates, Molybdates, Vanadates, and Beryllates

$CaWO_4$ was the most popular material for Nd before YAG became commercially available. The rare earth substitutes for Ca, but only in the trivalent oxidation state, and hence charge compensation is needed. For optimum laser performance, substitution of Na^+ for Ca^{2+} was found to be best [2.29]. The material is very prone to fracture, even when well annealed, and thus considerable care is required when the boules are being fabricated into laser rods. The thermal conductivity is three to four times greater than that of the glasses. The absorption spectra of Nd^{3+} in this material consists of a large number of rather fine lines.

Sodium rare earth molybdates and tungstates have served as host materials for active ions. In these materials, which are similar to $CaWO_4$ and $CaMoO_4$, one-half of the calcium atoms are randomly replaced with sodium and the other half with rare earth. Laser action has been observed from Nd^{3+} in NaLa $(MoO_4)_2$ and $NaGd(WO_4)_2$ and from $NaNd(WO_4)_2$. Interest in Nd : NaLa $(MoO_4)_2$ stems from the fact that this material has a low gain, intermediate to Nd : glass and Nd : YAG and thus is capable of higher efficiencies than the former material and greater energy storage than the latter material [2.30]. The thermal conductivity of Nd : NaLa(MoO_4) is three times that of Nd : glass. However, the lower thermal conductivity and the higher thermal expansion coefficient of Nd : NaLa(MoO_4)$_2$ as compared to Nd : YAG results in a Nd : NaLa(MoO_4)$_2$ laser rod having much greater thermal stress than a Nd : YAG rod when both are subjected to the same average input power during lasing action.

Nd^{3+}-doped yttrium orthovanadate (YVO_4) has shown relatively low threshold at pulsed operation [2.31]. However, early studies of this crystal were hampered by severe crystal growth problems, and as a result YVO_4 was discarded as a host. With the emergence of diode pumping, Nd : YVO_4 has become an important solid-state laser material (Sect. 2.5.3), because it has very attractive features, such as a large stimulated emission cross section [2.32] and a high absorption of the pump wavelength, and the growth problem has been overcome for the small crystals required with this pump source.

A candidate of the beryllates is Nd^{3+}-doped lanthanum beryllate ($Nd:La_2Be_2O_5$). Since La^{3+} is the largest of the rare-earth ions, $La_2Be_2O_5$ (BEL) has large distribution coefficients for other trivalent rare-earth ions. For this reason boules of BEL with high concentrations of rare-earth ions are much more readily grown than are YAG and other hosts based on the yttrium ion. Since the thermal conductivity and the cross section are considerably lower than those of YAG the material has not found applications.

An interesting property was reported by *Chin* et al. [2.33] regarding this material. Nd:BEL is optically bi-axial, and has positive and negative thermal coefficients for the refractive index. Therefore, an optical path can be selected which minimizes thermal lensing, thus leading to athermal behavior.

Fluorides

The divalent fluorides are relatively soft, isotropic crystals. Rare earth-doped CaF_2 crystals have been studied extensively [2.3, 34], since this material was the host of many early solid-state lasers. The doping of trivalent rare earth into fluoride hosts requires charge compensation, which represents a major drawback. In recent years yttrium lithium fluoride ($YLiF_4$), a uniaxial crystal, has received attention as a host for Ho^{3+}, Er^{3+} [2.35, 36] and Nd^{3+} [2.37]. $YLiF_4$ is transparent to 1500 Å, therefore, high-current-density xenon flashlamps which emit strongly in the blue and near-ultraviolet can be used as pump sources without damage to the material. The linewidth of Er:YLF is only $10\,cm^{-1}$, indicating fairly high gain. In order to obtain efficient operation from a Ho:YLF laser, the material must be sensitized with Er-Tm.

Nd:YLF offers a reduction in thermal lensing and birefringence combined with improved energy storage relative to Nd:YAG. The thermomechanical properties of Nd:YLF, however, are not as good as those of Nd:YAG. Considerable development of Nd:YLF has taken place in the areas of crystal growth, spectroscopy, material characterization and laser physics. The Nd:YLF laser will be discussed in Sect. 2.3.4 since it is one of the most important laser materials.

Ceramics

Laser action has also been achieved by doping optical ceramics with Nd [2.38]. Advantages of this type of host material over glass or crystals include low cost and higher thermal conductivity, and better thermal shock resistance compared with glass. However, glass ceramics are plagued with high scattering losses.

2.1.2 Active Ions

Before proceeding to a discussion of the active laser ions, we will review briefly the nomenclature of atomic energy levels.

Different energy states of electrons are expressed by different quantum numbers. The electrons in an atom are characterized by a principal quantum number

n, an orbital angular momentum l, the orientation of the angular momentum vector m, and a spin quantum number s. A tradition from the early days of line-series allocation has established the following method of designating individual electronic orbits: a number followed by a letter symbolizes the principal quantum number n and the angular momentum number l, respectively. The letters s, p, d, f stand for $l = 0, 1, 2, 3$, respectively. For example a $3d$ electron is in an orbit with $n = 3$ and $l = 2$.

To designate an atomic energy term one uses by convention capital letters with a system of subscripts and superscripts. The symbol characterizing the term is of the form $^{2S+1}L_J$, where the orbital quantum numbers $L = 0, 1, 2, 3, 4$ are expressed by the capital letters S, P, D, F, G, H. A superscript to the left of the letter indicates the value $(2S + 1)$, i.e., the multiplicity of the term due to possible orientation of the resultant spin S. Thus a one-electron system $(S = \frac{1}{2})$ has a multiplicity 2. L and S can combine to a total angular momentum J, indicated by a subscript to the right of the letter. Thus the symbol $^2P_{3/2}$ shows an energy level with an orbital quantum number $L = 1$, a spin of $S = \frac{1}{2}$, and a total angular momentum of $J = \frac{3}{2}$. The complete term description must include the configuration of the excited electron, which precedes the letter symbol. Thus the ground state of Li has the symbol $2s\ ^2S_{1/2}$.

When an atom contains many electrons, the electrons that form a closed shell may be disregarded and the energy differences associated with transitions in the atom may be calculated by considering only the electrons outside the closed shell.

In describing the state of a multielectron atom, the orbital angular momenta and the spin angular momenta are added separately. The sum of the orbital angular momenta are designated by the letter L, and the total spin is characterized by S. The total angular momentum J of the atom may then be obtained by vector addition of L and S. The collection of energy states with common values of J, L, and S is called a term.

In the following section, a qualitative description is given of some of the prominent features of the most important rare earth, actinide, and transition metal ions.

Rare Earth Ions

The rare earth ions are natural candidates to serve as active ions in solid-state laser materials because they exhibit a wealth of sharp fluorescent transitions representing almost every region of the visible and near-infrared portions of the electromagnetic spectrum. It is a characteristic of these lines that they are very sharp, even in the presence of the strong local fields of crystals, as a result of the shielding effect of the outer electrons.

The ground state electronic configuration of the rare earth atom consists of a core which is identical to xenon, plus additional electrons in higher orbits. In xenon, the shells with quantum numbers $n = 1, 2, 3$ are completely filled. The shell $n = 4$ has its s, p and d subshells filled, whereas the 4f subshell capable of accommodating 14 electrons is completely empty. However, the $n = 5$ shell has acquired its first 8 electrons which fill the 5s and 5p orbits. The electronic configuration for xenon is:

$$1s^2 2s^2 2p^6 3s^2 3p^6 3d^{10} 4s^2 4p^6 4d^{10} 5s^2 5p^6 .$$

Elements beyond xenon, which has the atomic number 54, have this electronic structure and, in addition, have electrons in the 4f, 5d, 6s, etc. orbits. Cesium, barium and lanthanum are the elements between xenon and the rare earths. Cs has one, and Ba has two 6s electrons, and La has in addition one electron in the 5d orbit. Rare earth elements begin with the filling of the inner vacant 4f orbits. For example, the first rare earth element cerium has only one electron in the f-orbit:

$$\text{Ce}: \quad \ldots 4f 5s^2 5p^6 5d 6s^2$$

and the important rare earth neodymium has 4 electrons in the f-orbit

$$\text{Nd}: \quad \ldots 4f^4 5s^2 5p^6 6s^2 .$$

Since the first nine shells and subshells up to $4d^{10}$ are completely filled, only the outer electron configuration is indicated.

In crystals, rare earth ions are normally trivalent, but under appropriate conditions the valence state can also be divalent. When a trivalent ion is formed the atom gives up its outermost 6s electrons, the atom loses also its 5d electron if it has one, otherwise one of the 4f electrons is lost. For example, trivalent Ce has the electronic configuration

$$\text{Ce}^{3+}: \quad \ldots 4f 5s^2 5p^6$$

and trivalent Nd has the configuration

$$\text{Nd}^{3+}: \quad \ldots 4f^3 5s^2 5p^6 .$$

As one can see, the trivalent ions of rare earths have a simpler configuration than the corresponding atoms. Ions from the rare earths differ in electronic structure only by the number of electrons in the 4f shell as illustrated in Table 2.1. When a divalent rare earth ion is formed, the atom gives up its outermost 6s electrons.

The fluorescence spectra of rare earth ions arise from electronic transitions between levels of the partially filled 4f shell. Electrons present in the 4f shell can be raised by light absorption into unoccupied 4f levels. The 4f states are well shielded by the filled 5s and 5p outer shells. As a result, emission lines are relatively narrow and the energy level structure varies only slightly from one host to another. The effect of the crystal field is normally treated as a perturbation on the free-ion levels. The perturbation is small compared to spin-orbit and electrostatic interactions among the 4f electrons. The primary change in the energy levels is a splitting of each of the free-ion levels in many closely spaced levels caused by the Stark effect of the crystal field. In crystals the free-ion levels are then referred to as manifolds. Figure 2.5 provides a nice illustration of the splitting of the Nd^{3+} manifolds into sublevels as a result of the YAG crystal field.

Table 2.1. Electronic configuration of trivalent rare earths

Element number	Trivalent rare earth	Number of 4f electrons	Ground state
58	Cerium, Ce^{3+}	1	$^2F_{5/2}$
59	Praseodymium, Pr^{3+}	2	$^3H_4$
60	Neodymium, Nd^{3+}	3	$^4I_{9/2}$
61	Promethium, Pm^{3+}	4	$^5I_4$
62	Samarium, Sm^{3+}	5	$^6H_{5/2}$
63	Europium, Eu^{3+}	6	$^7F_0$
64	Gadolinium, Gd^{3+}	7	$^8S_{7/2}$
65	Terbium, Tb^{3+}	8	$^7F_6$
66	Dysprosium, Dy^{3+}	9	$^6H_{15/2}$
67	Holmium, Ho^{3+}	10	$^5I_8$
68	Erbium, Er^{3+}	11	$^4I_{15/2}$
69	Thulium, Tm^{3+}	12	$^3H_6$
70	Ytterbium, Yb^{3+}	13	$^2F_{7/2}$
71	Lutetium, Lu^{3+}	14	$^1S_0$

Neodymium. Nd^{3+} was the first of the trivalent rare earth ions to be used in a laser, and it remains by far the most important element in this group. Stimulated emission has been obtained with this ion incorporated in at least 100 different host materials, and a higher power level has been obtained from Nd lasers than from any other four-level material. The principal host materials are YAG and glass. In these hosts stimulated emission is obtained at a number of frequencies within three different groups of transitions centered at 0.9, 1.06, and 1.35 μm. Radiation at these wavelengths results from $^4F_{3/2} \rightarrow {}^4I_{9/2}, {}^4I_{11/2}, {}^4I_{13/2}$ transitions, respectively.

The nomenclature of the energy levels may be illustrated by a discussion of the Nd^{3+} ion. This ion has three electrons in the $4f$ subshell. In the ground state their orbits are so aligned that the orbital angular momentum adds up to $3 + 2 + 1 = 6$ atomic units. The total angular momentum $L = 6$ is expressed by the letter I. The spins of the three electrons are aligned parallel to each other, providing an additional $\frac{3}{2}$ units of angular momentum, which, when added antiparallel to the orbital angular momentum, gives a total angular momentum of $6 - \frac{3}{2} = \frac{9}{2}$ units. According to the quantum rules for the addition of angular momenta, the vector sum of an orbital angular momentum of 6 and a spin angular momentum of $\frac{3}{2}$ may result in the following four values of the total angular momentum: $\frac{9}{2}, \frac{11}{2}, \frac{13}{2}$, and $\frac{15}{2}$. The levels corresponding to these values are $^4I_{9/2}, {}^4I_{11/2}, {}^4I_{13/2}$, and $^4I_{15/2}$. The first of these, which has the lowest energy, is the ground state; the others are among the first few excited levels of Nd^{3+}. These levels are distinguished by the orientation of the spins with respect to the resultant orbital angular momentum. Other excited levels are obtained when another combination of the orbital angular momenta is chosen.

Erbium. Numerous studies of the absorption and fluorescence properties of erbium in various host materials have been conducted to determine its potential as an active laser ion. Laser oscillation was observed most frequently in the wavelength region 1.53 to 1.66 μm arising from transitions between the $^4I_{13/2}$ state and the $^4I_{15/2}$ ground state Er^{3+}. Stimulated emission in the vicinity of 1.6 μm is of interest, because the eye is less subject to retinal damage by laser radiation at these wavelengths due to the greatly reduced transmissivity of the ocular media.

Host materials of erbium have included YAG [2.12], YLF [2.35], YALO$_3$ [2.39], LaF$_3$ [2.40], CaWO$_4$ [2.41], CaF$_2$ [2.42], and various glasses [2.43, 44]. The terminal level in Er^{3+} is between 525 cm^{-1} for Er:YAG and 50 cm^{-1} for Er:glass. At room temperature all levels of the terminal $^4I_{15/2}$ manifold are populated to some degree, thus this transition forms a three-level laser scheme with a correspondingly high threshold. Laser action is generally achieved either by lowering the temperature to depopulate the higher-lying levels of the $^4I_{15/2}$ manifold, or by codoping the materials with trivalent ytterbium to improve the optical pumping efficiency via $Yb^{3+} \rightarrow Er^{3+}$ energy transfer.

Since YAG possesses the largest ground-state splitting of all host materials doped with Er and has other properties which make it the best available host for many applications, emphasis is placed mainly on the optimization of sensitized Er:YAG. Of particular interest is YAG, highly doped with Er which produces an output around 2.9 μm.

Erbium glass lasers are based on phosphate and silicate glass which is co-doped with neodymium, chromium or ytterbium. The Nd, Cr or Yb ions act as sensitizing agents by absorbing pump light in regions where the erbium is relatively transparent. Er:YAG and Er:glass lasers are discussed in greater detail in Sect. 2.4.

Holmium. Laser action in Ho^{3+} has been reported in many different host materials [2.45]. Because the terminal level is only about 250 cm^{-1} above ground level, the lower laser level has a relatively high thermal population at room temperature. While Ho:YAG and Ho:YLF have proven to be efficient lasers, operation has been limited in most cases to cryogenic temperatures, which will depopulate the lower laser level. Previous efforts in flashlamp pumped 2 μm lasers have concentrated on Er:Tm:Ho doped YAG and YLF.

It was discovered that Cr-sensitized Tm:Ho :YAG offers several advantages over the Er sensitized materials. In a Cr:Tm:Ho:YAG laser, a very efficient energy transfer process between the co-dopants takes place. Cr^{3+} acts to efficiently absorb flashlamp energy, which is then transferred to Tm with a transfer quantum efficiency approaching 2 (2 photons in the ir for each pump photon). From Tm the energy is efficiently transferred to Ho. Lasing occurs at the $^5I_7$-$^5I_8$ Ho transition at a wavelength of 2080 nm [2.46]. Laser diode pumping of a Tm:Ho:YAG laser via an absorption line in Tm^{3+} at 780 nm is also possible [2.47]. Chromium doping is not necessary in this case (see also Sect. 2.5.4).

Thulium. Efficient flashlamp and laser diode-pumped laser operation has been achieved in Tm^{3+} : YAG and Tm^{3+} : YLF co-doped either with Cr^{3+} or Ho^{3+}. The output wavelength for the $^3F_4 - ^3H_6$ transition is 2.014 μm. The thulium ion has an absorption at 785 nm which is useful for diode pumping. Diode-pumped Tm : YAG lasing at 2.01 μm and Tm : Ho : YAG lasing at 2.09 μm are discussed in Sect. 2.5.4. Flashlamp-pumped Cr : Tm : YAG lasers have achieved slope efficiencies of 4.5% and pulse energies exceeding 2 J. Cr-doping provided for efficient absorption of the flashlamp radiation [2.48]. Flashlamp-pumped Cr : Tm : YAG can achieve tunable output between 1.945 and 1.965 μm.

Praseodymium, Gadolinium, Europium, Ytterbium, Cerium. Laser action in all these triply ionized rare earths has been reported; however, only marginal performance was obtained in hosts containing these ions with the exception of ytterbium. Diode-pumped Yb : YAG has become an important laser which is described in Sect. 2.6.

Samarium, Dysprosium, Thulium. The divalent rare earths Sm^{2+}, Dy^{2+}, and Tm^{2+} have an additional electron in the $4f$ shell, which lowers the energy of the $5d$ configuration. Consequently, the allowed $4f$-$5d$ absorption bands fall in the visible region of the spectrum. These bands are particularly suitable for pumping the laser systems. Tm^{2+}, Dy^{2+}, and Sm^{2+} have been operated as lasers, all in a CaF_2 host. For laser operation, these crystals must be refrigerated to at least 77 K.

Actinide Ions

The actinides are similar to the rare earths in having the $5f$ electrons partially shielded by $6s$ and $6p$ electrons. Most of the actinide elements are radioactive, and only uranium in CaF_2 has been successfully used in a laser [2.3]. The host was doped with 0.05 % uranium. Laser action occurred at 2.6 μm between a metastable level and a terminal level some 515 cm^{-1} above the ground state.

Transition Metals

Important members of the transition metal group include the ruby (Cr^{3+} : Al_2O_3), alexandrite (Cr^{3+} : $BeAl_2O_4$) and Ti : sapphire (Ti^{3+} : Al_2O_3) laser, which are discussed in separate sections. Laser action has been observed in most other transition metals and particularly in Ni^{2+} and Co^{2+}. Considerable effort has gone into investigations of the Co : MgF_2 laser. Earlier work has been extended by *Moulton* et al. who have developed a room temperature Co : MgF_2 laser which is pumped by the 1.32 μm line of a Nd : YAG laser [2.49, 50]. This tunable laser will be further described in Sect. 2.5.

Summary

Compilations of useful materials, parameters and references on laser host and impurity ions can be found in [2.8, 9, 51–53].

2.2 Ruby

The ruby laser, although a three-level system, still remains in some use today for a few limited applications. From an application point of view, ruby is attractive because its output lies in the visible range, in contrast to most rare earth four-level lasers, whose outputs are in the near-infrared region. Spectroscopically, ruby possesses an unusually favorable combination of a relatively narrow linewidth, a long fluorescent lifetime, a high quantum efficiency, and broad and well-located pump absorption bands which make unusually efficient use of the pump radiation emitted by available flashlamps.

Physical Properties

Ruby chemically consists of sapphire (Al_2O_3) in which a small percentage of the Al^{3+} has been replaced by Cr^{3+}. This is done by adding small amounts of Cr_2O_3 to the melt of highly purified Al_2O_3. The pure single host crystal is uniaxial and possesses a rhombohedral or hexagonal unit cell, as shown in Fig. 2.1. The crystal has an axis of symmetry, the so-called c axis, which forms the major diagonal of the unit cell. Since the crystal is uniaxial, it has two indices of refraction, the ordinary ray having the E vector perpendicular to the c (optic) axis, and the extraordinary ray having the E vector parallel to the c axis.

As a laser host crystal, sapphire has many desirable physical and chemical properties. The crystal is a refractory material, hard and durable. It has good thermal conductivity, is chemically stable, and is capable of being grown to very high quality. Ruby is grown by the Czochralski method. In this procedure the solid crystal is slowly pulled from a liquid melt by initiation of growth on high-quality seed material. Iridium crucibles and rf heating are used to contain the melt and control the melt temperature, respectively. The crystal boules can be grown in the $0°$, $60°$, or $90°$ configuration, where the term refers to the angle between the growth axis and the crystallographic c axis. For laser-grade ruby the $60°$ type is commonly used.

As has already been stated, the active material in ruby is the Cr^{3+} ion. This ion has three d electrons in its unfilled shell; the ground state of the free ion is described by the spectroscopic symbol ^4A. The amount of doping is nominally 0.05 weight percent Cr_2O_3. However, in some applications it is desirable to lower the Cr^{3+} concentration to approximately 0.035 weight percent to obtain maximum beam quality in ruby oscillators.

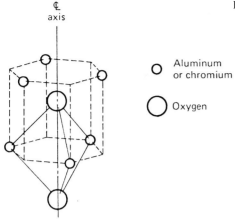

Fig. 2.1. Crystal structure of sapphire

¢
axis

○ Aluminum
   or chromium

◯ Oxygen

Laser Properties of Ruby

A simplified energy level diagram of ruby is given in Fig. 2.2. In ruby lasers, population inversion with respect to the so-called ^2E level is obtained by optically pumping Cr^{3+} ions from the $^4A_2$ ground state to the broad pump bands $^4F_2$ and $^4F_1$. The lifetime at the pump bands, which are each about 1000 Å wide, located in the green (18,000 cm^{-1}) and in the violet (25,000 cm^{-1}), is extremely short, with the ions returning to a metastable state ^2E. This metastable level is split into two sublevels with a separation of $\Delta E = 29$ cm^{-1}. The upper one is the $2\overline{A}$ and the lower one the $\overline{E}$ sublevel. The two transitions ($\overline{E} \to {}^4A_2$ and $2\overline{A} \to {}^4A_2$) are referred to as the R_1 and R_2 lines. Each is approximately 5 Å wide at room temperature, and the lines lie at the end of the visible, at 6943 and 6929 Å.

At thermal equilibrium the difference in population between the $\overline{E}$ and $2\overline{A}$ level is

$$\frac{n(2\overline{A})}{n(\overline{E})} = \exp\left(\frac{\Delta E}{kT}\right) = K \ . \tag{2.1}$$

At room temperature the Boltzmann factor is $K = 0.87$. The fluorescence in ruby consists of the R_1 and R_2 lines. However, laser action takes place only at the R_1 line, i.e., between the $\overline{E}$ and $^4A_2$ level. The R_1 line attains laser threshold before the R_2 line because of the higher inversion. Once laser action commences in the R_1 line, the $\overline{E}$ level becomes depleted and population transfer from the nearby $2\overline{A}$ level proceeds at such a fast rate that the threshold level is never reached for the R_2 line.

The relaxation time between the $2\overline{A}$ and $\overline{E}$ levels is very short, on the order of a nanosecond or less. For laser pulses which are long compared to this time constant the population ratio of the two states is kept unchanged, but since $\overline{E}$ decays much faster, almost the entire initial population of the two states decays through R_1 emission. If we compare the energy level diagram of ruby with our simplified scheme of a three-level system in Fig. 1.6, the levels $^4F_1$ and $^4F_2$ jointly

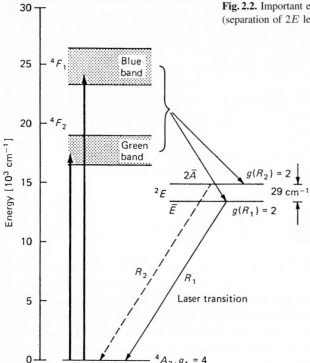

Fig. 2.2. Important energy levels of Cr^{3+} in ruby (separation of $2E$ levels not to scale)

constitute level 3, whereas the ^2E and $^4A_2$ states represent level 2 and level 1, respectively. We can write

$$n_2 = n_2(R_1) + n_2(R_2) \tag{2.2}$$

for the metastable level and n_1 for the ground level. Threshold and gain in ruby depends only on the population of level $n_2(R_1)$. However, in relating gain and threshold to the population of the ground level n_1 or to the total number of Cr^{3+} ions n_{tot}, one has to take the population of $n_2(R_2)$ into account. In ruby all levels are degenerate, that is,

$$g(n_1) = 4 , \quad g(R_1) = g(R_2) = 2 . \tag{2.3}$$

Because of the higher degeneracy of the ground state, amplification occurs when the R_1 level is at least one-half as densely populated as the ground state.

$$n_2(R_1) = \tfrac{1}{2} n_1 . \tag{2.4}$$

The two upper levels are related by

$$n_2(R_2) = K n_2(R_1) . \tag{2.5}$$

Since

$$n_2(R_1) + n_2(R_2) + n_1 = n_{tot} \,, \tag{2.6}$$

we have the following population at threshold (300 K)

$$n_2(R_1) = \frac{n_{tot}}{3 + K} = 0.26 n_{tot} \,,$$

$$n_2(R_2) = \frac{K n_{tot}}{3 + K} = 0.22 n_{tot} \,, \tag{2.7}$$

$$n_1 = \frac{2 n_{tot}}{3 + K} = 0.52 n_{tot} \,.$$

Thus we must have just under one-half of the atoms in the two upper levels in order to reach threshold. At complete inversion we have

$$n_2(R_1) = \frac{n_{tot}}{K + 1} = 0.53 n_{tot} \,,$$

$$n_2(R_2) = \frac{K n_{tot}}{K + 1} = 0.47 n_{tot} \,, \tag{2.8}$$

$$n_1 = 0 \,.$$

For normal laser operation the population densities of the various levels are between those given for threshold and total inversion. The amount of energy per unit volume which can be extracted from the inverted ruby depends on the population of levels $n_2(R_1)$ and $n_2(R_2)$, provided that the pulse duration is long enough that these two levels remain in thermal equilibrium. With a Cr^{3+} concentration of $n_{tot} = 1.58 \times 10^{19} \, cm^{-3}$ and a photon energy of $h\nu = 2.86 \times 10^{-19}$ Ws, we obtain for the maximum upper-state energy density, when all atoms exist in the excited energy states $n_2(R_1)$ and $n_2(R_2)$, a value of $E = n_{tot} h\nu = 4.52 \, J/cm^3$. The maximum energy which can be extracted, assuming complete inversion, is, according to (2.7, 8),

$$E_{ex\ max} = \frac{2 h\nu n_{tot}}{3 + K} = 2.35 \, J/cm^3 \,.$$

For pulses which are short compared to the relaxation time between levels $n_2(R_1)$ and $n_2(R_2)$, only energy stored in the level $n_2(R_1)$ can be extracted. Again, assuming an initial complete inversion of the material, we can extract a maximum energy per unit volume of

$$E'_{ex\ max} = \frac{2 h\nu n_{tot}}{3(1 + K)} = 1.6 \, J/cm^3 \,.$$

The upper-state energy density at the inversion level $E_{uth} = h\nu n_{tot}(1+K)/(3+K)$ has a numerical value of $E_{uth} = 2.18 \, J/cm^3$. The small-signal gain coefficient in ruby is

$$g_0 = \sigma_{21} \left[n_2(R_1) - \frac{g(R_1) n_1}{g(n_1)} \right] \,, \tag{2.9}$$

where σ_{21} is the stimulated emission cross section of the R_1 line and $n_2(R_1)$ is the population density of the E level. With (2.2, 5 and 6) we can write

$$g_0 = \sigma_{21} \left(\frac{3 + K}{2(1 + K)} n_2 - \frac{n_{\text{tot}}}{2} \right) . \tag{2.10}$$

It is customary to express the gain coefficient in terms of the absorption coefficient. Since $\sigma_{21} = \sigma_{12} g(n_1)/g(R_1)$, the gain coefficient can be expressed as

$$g_0 = \alpha_0 \left(\frac{(3 + K)n_2}{(1 + K)n_{\text{tot}}} - 1 \right) , \tag{2.11}$$

where $\alpha_0 = \sigma_{12} n_{\text{tot}}$ is the absorption coefficient of ruby. With the approximation $K \approx 1$ the expressions above can be simplified to

$$g_0 = \alpha_0 \left(\frac{2n_2}{n_{\text{tot}}} - 1 \right) = \sigma_{21} \left(n_2 - \frac{n_{\text{tot}}}{2} \right) = \sigma_{12}(2n_2 - n_{\text{tot}}) . \tag{2.12}$$

With all the chromium ions in the ground state ($n_2 = 0$), the gain of the unexcited ruby crystal is $g_0 = -\alpha_0$. The maximum gain achieved at total inversion ($n_2 = n_{\text{tot}}$) is $g_0 = 2\alpha_0/(1 + K) \approx \alpha_0$. Numerical values for α_0 and σ_{12} are obtained from the absorption data for ruby. Figure 2.3 shows the absorption coefficient and absorption cross section for the R lines of ruby as a function of wavelength [2.54]. As one can see from these curves, the absorption coefficient of the R lines (and, therefore, gain in the presence of inversion) for light having its E vector normal to the c axis is greater than that for light with E parallel to the c axis. This accounts for the polarization of the output from lasers employing ruby rods with c axis orientations away from the rod axis. The absorption cross section at the peak of the R_1 line is seven times higher for a beam polarized perpendicular to the c axis as for a beam polarized parallel to the c axis. From Fig. 2.3 follows an absorption cross section of $\sigma_{12} = 1.22 \times 10^{-20} \, \text{cm}^2$ and an absorption coefficient of $\alpha_0 = \sigma_{12} n_{\text{tot}} = 0.2 \, \text{cm}^{-1}$ for the peak of the R_1 line ($E \perp c$) at a doping concentration of $n_{\text{tot}} = 1.58 \times 10^{19} \, \text{cm}^{-3}$. Other published values are $\sigma_{12} = 1.7 \times 10^{-20} \, \text{cm}^2$ [2.55] and $\alpha_0 \approx 0.28 \, \text{cm}^{-1}$ [2.56]. Using Cronemeyer's data we obtain for the stimulated emission cross section in ruby a value of $\sigma_{21} = 2.5 \times 10^{-20} \, \text{cm}^2$. The maximum gain coefficient for complete inversion in a 0.05 % Cr-doped rod is then $g_{0(\text{max})} = 0.215 \, \text{cm}^{-1}$. The various parameters for maximum gain and absorption in ruby are related according to

$$g_{0(\text{max})} = \frac{\sigma_{21} n_{\text{tot}}}{K + 1} \approx \frac{2\sigma_{12} n_{\text{tot}}}{K + 1} \approx \frac{2\alpha_0}{K + 1} . \tag{2.13}$$

Despite the 20 % lower stimulated emission cross section and the lower population density, laser action at the R_2 line can be achieved in ruby either by employing dielectrically coated resonator mirrors which have a sufficiently higher transmission at R_1 than at the R_2 line, or by inserting a polarizer and a retardation plate inside the resonator.

The absorbed pump energy required to obtain threshold is

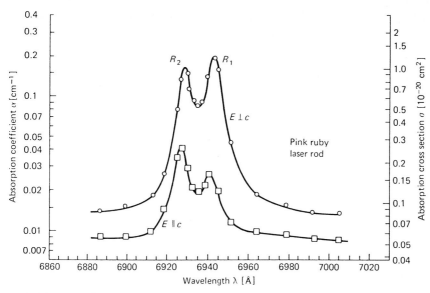

Fig. 2.3. Absorption coefficient and cross section of R_1 and R_2 lines in ruby as a function of wavelength, incident beam polarized parallel $(E \parallel c)$ and orthogonal $(E \perp c)$ to c axis of crystal. Cr^{3+} concentration is $1.58 \times 10^{19}\,cm^{-3}$ [2.54]

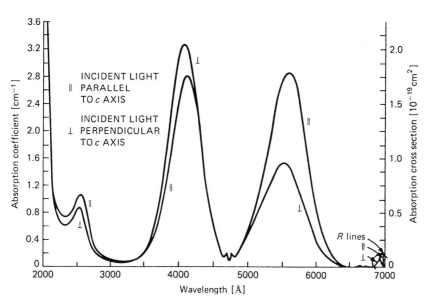

Fig. 2.4. Absorption coefficient and cross section of ruby pump bands for unpolarized light as a function of wavelength. Cr^{3+} concentration is $1.88 \times 10^{19}\,cm^{-3}$ [2.56]

$$E_{ab} \approx \frac{n_{tot}h\nu_p}{2\eta_Q} , \tag{2.14}$$

where $\nu_p = 6.25 \times 10^{14}\,\mathrm{s}^{-1}$ is the average frequency of the two main pump bands and η_1 is the quantum efficiency. Since the average quantum efficiency in ruby for the two main pump bands is $\eta_Q \approx 0.70$, one obtains $E_{ab} = 4.55\,\mathrm{J/cm}^3$.

The absorption spectrum of ruby is given in Fig. 2.4 [2.56]. The best-quality ruby is grown with the crystal c axis at $60°$ to the boule axis. From Fig. 2.4 follows that in this case the absorption spectrum is different for a light beam incident parallel or normal to the c axis of the crystal. When such a ruby is pumped in a diffuse reflecting pump cavity, the loci of constant inversion are elliptical as a result of the differences in absorption of the pump light in directions parallel and perpendicular to the c axis. This effect, which causes the beam pattern in the near and far field to be elliptical, can be observed experimentally. Important ruby laser parameters obtained from [2.54–56] are listed in Table 2.2.

Table 2.2. Optical and laser properties of ruby at room temperature

Property	Values and units
Cr_2O_3 doping	0.05 wt. %
Cr^{3+} concentration	1.58×10^{19} ions cm^3
Output wavelengths, $25°$ C	R_1, 14403 cm^{-1}, 6943 Å
	R_2, 14432 cm^{-1}, 6929 Å
Fluorescent lifetime	3.0 ms at 300 K
Spectral linewidth	11 cm^{-1}, 5.3 Å
Quantum efficiency	0.7
Separation of R_1 and R_2 lines	29 cm^{-1}, 870 GHz, 14 Å
Absorption coefficient and cross section	$\alpha_{R_1} = 0.2$ cm^{-1}
of laser line (R_1 level, $E \perp c$)	$\sigma_{R_1} = 1.22 \times 10^{-20}$ cm^2
Stimulated emission cross section	$\sigma_{21} = 2.5 \times 10^{-20}$ cm^2
Major pump bands	
Blue (4040 Å)	$\alpha_{\parallel} = 2.8$ cm^{-1}; $\alpha_{\perp} = 3.2$ cm^{-1}
Green (5540 Å)	$\alpha_{\parallel} = 2.8$ cm^{-1}; $\alpha_{\perp} = 1.4$ cm^{-1}
Refractive index at 6943 Å	1.763 ordinary ray $E \perp c$
	1.755 extraordinary ray $E \parallel c$

2.3 Nd : Lasers

From the large number of Nd : doped materials the lasers discussed in more detail in this chapter have gained prominence. Nd : YAG, because of its high gain and good thermal and mechanical properties, is by far the most important solid-state laser for scientific, medical, industrial and military applications. Nd : glass is important for laser fusion drivers because it can be produced in large sizes. Nd : Cr : GSGG has received attention because of the good spectral match between the flashlamp emission and the absorption of the Cr ions. An efficient energy transfer between the Cr and Nd ions results in a highly efficient Nd: laser. Nd: YLF is a good material for a number of applications, because the output is polarized, and the crystal exhibits lower thermal birefringence. Nd: YLF has a higher energy storage capability (due to its lower gain coefficient) compared to Nd: YAG and its output wavelength matches that of phosphate Nd: glass, therefore modelocked and Q-switched Nd : YLF lasers have become the standard oscillators for large glass lasers employed in fusion research. Nd: YVO_4 has become a very attractive material for small diode-pumped lasers because of its large emission cross section and strong absorption at 809 nm.

2.3.1 Nd: YAG

The Nd: YAG laser is by far the most commonly used type of solid-state laser. Neodymium-doped yttrium aluminum garnet (Nd: YAG) possesses a combination of properties uniquely favorable for laser operation. The YAG host is hard, of good optical quality, and has a high thermal conductivity. Furthermore, the cubic structure of YAG favors a narrow fluorescent linewidth, which results in high gain and low threshold for laser operation. In Nd: YAG, trivalent neodymium substitutes for trivalent yttrium, so charge compensation is not required.

Five years after *Geusic* et al. [2.6] reported the first successful lasing of Nd: YAG, rapid strides had been made in improving both the quality of the material and pumping technique so that available cw power outputs rose from the fractional watt level initially obtained to several hundred watts from a single laser rod [2.57]. On the other hand, in single-crystal Nd: YAG fiber lasers, threshold was achieved with absorbed pump powers as small as 1 mW [2.58]. Today, more than thirty-five years after its first operation, the Nd: YAG laser has emerged as the most versatile solid-state system in existence.

Physical Properties

In addition to the very favorable spectral and lasing characteristics displayed by Nd: YAG, the host lattice is noteworthy for its unusually attractive blend of physical, chemical, and mechanical properties. The YAG structure is stable from the lowest temperatures up to the melting point, and no transformations have been reported in the solid phase. The strength and hardness of YAG are lower than ruby but still high enough so that normal fabrication procedures do not produce any serious breakage problems.

Pure $Y_3Al_5O_{12}$ is a colorless, optically isotropic crystal which possesses a cubic structure characteristic of garnets. In Nd: YAG about 1 % of Y^{3+} is substituted by Nd^{3+}. The radii of the two rare earth ions differ by about 3 %. Therefore, with the addition of large amounts of neodymium, strained crystals are obtained – indicating that either the solubility limit of neodymium is exceeded or that the lattice of YAG is seriously distorted by the inclusion of neodymium. Some of the important physical properties of YAG are listed in Table 2.3, together with optical and laser parameters [2.59–67].

Laser Properties

The Nd:YAG laser is a four-level system as depicted by a simplified energy level diagram in Fig. 2.5. The laser transition, having a wavelength of 10641 Å, originates from the R_2 component of the $^4F_{3/2}$ level and terminates at the Y_3 component of the $^4I_{11/2}$ level. At room temperature only 40 % of the $^4F_{3/2}$ population is at level R_2; the remaining 60 % are at the lower sublevel R_1 according to Boltzmann's law. Lasing takes place only by R_2 ions whereby the R_2 level population is replenished from R_1 by thermal transitions. The ground level of Nd:YAG is the $^4I_{9/2}$ level. There are a number of relatively broad energy levels, which together may be viewed as comprising pump level 3. Of the main pump bands shown, the 0.81 and 0.75 μm bands are the strongest. The terminal laser level is 2111 cm^{-1} above the ground state and thus the population is a factor of $\exp(\Delta E/kT) \approx \exp(-10)$ of the ground-state density. Since the terminal level is not populated thermally, the threshold condition is easy to obtain.

Table 2.3. Physical and optical properties of Nd:YAG

Chemical formula	Nd:$Y_3Al_5O_{12}$
Weight % Nd	0.725
Atomic % Nd	1.0
Nd atoms/cm^3	1.38×10^{20}
Melting point	1970°C
Knoop hardness	1215
Density	4.56 g/cm^3
Rupture stress	1.3–2.6 $\times 10^6$ kg/cm^2
Modulus of elasticity	3×10^6 kg/cm^2
Thermal expansion coefficient	
[100] orientation	8.2×10^{-6} °C^{-1}, 0–250 °C
[110] orientation	7.7×10^{-6} °C^{-1}, 10–250 °C
[111] orientation	7.8×10^{-6} °C^{-1}, 0–250 °C
Linewidth	4.5 Å
Stimulated emission cross section	
$R_2 - Y_3$	$\sigma_{21} = 6.5 \times 10^{-19}$ cm^2
$4F_{3/2} - {}^4I_{11/2}$	$\sigma_{21} = 2.8 \times 10^{-19}$ cm^2
Fluorescence lifetime	230 μs
Photon energy at 1.06 μm	$h\nu = 1.86 \times 10^{-19}$ J
Index of refraction	1.82 (at 1.0 μm)
Scatter losses	$\alpha_{sc} \approx 0.002$ cm^{-1}

The upper laser level, $^4F_{3/2}$, has a fluorescence efficiency greater than 99.5 % [2.59] and a fluorescence lifetime of 230 μs. The branching ratio of emission from $^4F_{3/2}$ is as follows [2.61]: $^4F_{3/2} \to {}^4I_{9/2} = 0.25$, $^4F_{3/2} \to {}^4I_{11/2} = 0.60$, $^4F_{3/2} \to {}^4I_{13/2} = 0.14$, and $^4F_{3/2} \to {}^4I_{15/2} < 0.01$. This means that almost all the ions transferred from the ground level to the pump bands end up at the upper laser level, and 60 % of the ions at the upper laser level cause fluorescence output at the $^4I_{11/2}$ manifold.

At room temperature the main 1.06-μm line in Nd:YAG is homogeneously broadened by thermally activated lattice vibrations. The spectroscopic cross sec-

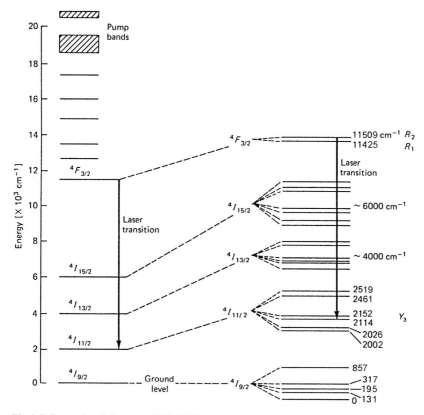

Fig. 2.5. Energy level diagram of Nd : YAG

tion for the individual transition between Stark sublevels has been measured [2.62] to be $\sigma(R_2 - Y_3) = 6.5 \times 10^{-19} \text{cm}^2$. At a temperature of 295 K, the Maxwell-Boltzmann fraction in the upper Stark sublevel is 0.427, implying an effective cross section for Nd : YAG of $\sigma^*(^4F_{3/2} - {}^4I_{11/2}) = 2.8 \times 10^{-19} \text{cm}^2$. The effective stimulated-emission cross section is the spectroscopic cross section times the occupancy of the upper laser level relative to the entire $^4F_{3/2}$ manifold population.

Table 2.4. Thermal properties of Nd : YAG

Property	Units	300 K	200 K	100 K
Thermal conductivity	W cm^{-1} K^{-1}	0.14	0.21	0.58
Specific heat	W s g^{-1} K^{-1}	0.59	0.43	0.13
Thermal diffusivity	cm^2 s^{-1}	0.046	0.10	0.92
Thermal expansion	K^{-1} × 10^{-6}	7.5	5.8	4.25
$\partial n/\partial T$	K^{-1}	7.3 × 10^{-6}	–	–

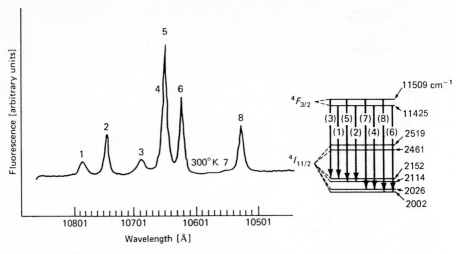

Fig. 2.6. Fluorescence spectrum of Nd^{3+} in YAG at 300 K in the region of 1.06 μm [2.67]

Figure 2.6 shows the fluorescence spectrum of Nd^{3+} in YAG near the region of the laser output with the corresponding energy levels for the various transitions. The absorption of Nd:YAG in the range 0.3 to 0.9 μm is given in Fig. 2.7. Thermal properties of Nd:YAG [2.68] are summarized in Table 2.4.

Nd:YAG Laser Rods

Commercially available laser crystals are grown exclusively by the Czochralski method. Growth rates, dopants, annealing procedures, and final size generally determine the manufacturing rate of each crystal. The boule axis or growth direction is customarily in the [111] direction. The high manufacturing costs of Nd:YAG are mainly caused by the very slow growth rate of Nd:YAG, which is of the order of 0.5 mm/h. Typical boules of 10 to 15 cm in length require a growth run of several weeks.

Boules grown from Nd:YAG typically contain very few optically observable scattering centers, and show negligible absorption at the lasing wavelength. However, all Nd:YAG crystals grown by Czochralski techniques show a bright core running along the length of the crystal when positioned between crossed polarizers. Strain flares are also visible, radiating from the core toward the surfaces of the crystal. Electron microprobe studies have revealed that in the core region the Nd concentration can run as much as twice as high as in the surrounding areas. The cores originate from the presence of facets on the growth interface which have a different distribution coefficient for neodymium than the surrounding growth surface. These compositional differences cause corresponding differences in thermal expansion coefficients which, in turn, give rise to the observed stress patterns during the cool down of the crystals from the growth temperature. Annealing does not seem to eliminate the cores and, thus far, no

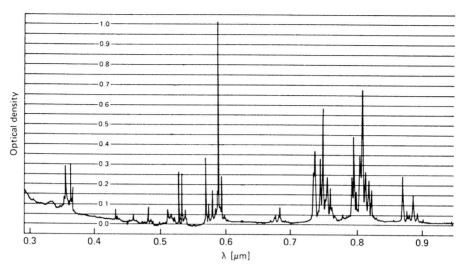

Fig. 2.7. Absorption spectrum of Nd : YAG at 300 K

way has been found of avoiding the formation of facets on the growth interface. However, by choosing the growth direction carefully and by maintaining as steep an interface angle as is practical, the stresses can be confined to a very localized region and high-optical quality rods can be fabricated from the surrounding material. This means, of course, that in order to provide rods of a given diameter, the crystal must be grown with a diameter that is somewhat more than twice as large. The boules are processed by quartering into sections. Rods can be fabricated with maximum diameters of about 15 mm and lengths of up to 150 mm. The optical quality of such rods is normally quite good and comparable to the best quality of Czochralski ruby or optical glass. For example, 6 mm by 100 mm rods cut from the outer sections of 20 mm by 150 mm boules typically may show only 1 to 2 fringes in a Twyman-Green interferometer.

Neodymium concentration by atom percent in YAG has been limited to 1.0 to 1.5 %. Higher doping levels tend to shorten the fluorescent lifetime, broaden the linewidth, and cause strain in the crystal, resulting in poor optical quality. The Nd doping level in YAG is sometimes expressed in different concentration units: A concentration of 1.0 % Nd atoms in the lattice is equivalent to 0.727 % Nd or 0.848 % Nd_2O_3 by weight, respectively. The concentration of Nd^{3+} sites in these cases is 1.386×10^{20} cm^{-3}.

In specifying Nd : YAG rods, the emphasis is on size, dimensional tolerance, doping level, and passive optical tests of rod quality. Cylindrical rods with flat ends are typically finished to the following specifications: end flat to $\lambda/10$, ends parallel to $\pm$ 4 arc seconds, perpendicularity to rod axis to ± 5 minutes, rod axis parallel to within $\pm 5°$ to [111] direction. Dimensional tolerances typically are ± 0.5 mm on length and ± 0.025 mm on diameter. Most suppliers furnish the laser crystals with a photograph showing the fringe pattern of the crystal as examined by a Twyman-Green interferometer. A double-pass Twyman-Green interferometer quickly reveals strained areas, small defects, or processing errors.

In a particular application the performance of a Nd : YAG laser can be somewhat improved by the choice of the optimum Nd concentration. As a general guideline, it can be said that a high doping concentration (approximately 1.2 %) is desirable for Q-switch operation because this will lead to high energy storage. For cw operation, a low doping concentration (0.6 to 0.8 %) is usually chosen to obtain good beam quality.

It is worth noting that in contrast to a liquid or a glass, a crystal host is not amenable to uniform dopant concentration. This problem arises as a result of the crystal-growth mechanism. In the substitution of the larger Nd^{3+} for a Y^{3+} in $Y_3Al_5O_{12}$, the neodymium is preferentially retained in the melt. The increase in concentration of Nd from the seed to the terminus of a 20-cm long boule is about 20 to 25 %. For a laser rod 3 to 8 cm long, this end-to-end variation may be 0.05 to 0.10 % of Nd_2O_3 by weight.

Different Laser Transitions in Nd : YAG

Under normal operating conditions the Nd : YAG laser oscillates at room temperature on the strongest $^4F_{3/2} \rightarrow ^4I_{11/2}$ transition at 1.0641 μm. It is possible, however, to obtain oscillation at other wavelengths by inserting etalons or dispersive prisms in the resonator, by utilizing a specially designed resonant reflector as an output mirror [2.69], or by employing highly selective dielectrically coated mirrors [2.70]. These elements suppress laser oscillation at the undesirable wavelength and provide optimum conditions at the wavelength desired. With this technique over 20 transitions have been made to laser in Nd : YAG [2.71–74]. The relative output of those transitions at which room-temperature cw operation has been achieved is listed in Table 2.5 [2.74]. The relative performance is compared to 1.064 μm emission for which 71.5 W output was obtained.

The 1.064- and 1.061-μm transitions, $^4F_{3/2} \rightarrow ^4I_{11/2}$, provide the lowest threshold laser lines in Nd : YAG. At room temperature the 1.064-μm line $R_2 \rightarrow Y_3$ is dominant, while at low temperatures the 1.061-μm line $R_1 \rightarrow Y_1$ has the lower threshold [2.75]. If the laser crystal is cooled, additional laser transitions are obtained, most notably the 1.839-μm line and the 0.946-μm line [2.76, 77].

The design and operation of a high-power Nd : YAG laser operating at 1.3 μm was reported in [2.78]. The laser achieved in the long-pulse mode an output in excess of 8 J per pulse at 5 Hz, and up to 165 W of laser output at 50 Hz.

2.3.2 Nd : Glass

There are a number of characteristics which distinguish glass from other solid-state laser host materials. Its properties are isotropic. It can be doped at very high concentrations with excellent uniformity, and it can be made in large pieces of diffraction-limited optical quality. In addition, glass lasers have been made, in a

Table 2.5. Main room-temperature transitions in Nd: YAG

Wavelength ([μm], air)	Transition	Relative Performance
1.05205	$R_2 \rightarrow Y_1$	46
1.06152	$R_1 \rightarrow Y_1$	92
1.06414	$R_2 \rightarrow Y_3$	100
1.0646	$R_1 \rightarrow Y_2$	≈ 50
1.0738	$R_1 \rightarrow Y_3$	65
1.0780	$R_1 \rightarrow Y_4$	34
1.1054	$R_2 \rightarrow Y_5$	9
1.1121	$R_2 \rightarrow Y_6$	49
1.1159	$R_1 \rightarrow Y_5$	46
1.12267	$R_1 \rightarrow Y_6$	40
1.3188	$R_2 \rightarrow X_1$	34
1.3200	$R_2 \rightarrow X_2$	9
1.3338	$R_1 \rightarrow X_1$	13
1.3350	$R_1 \rightarrow X_2$	15
1.3382	$R_2 \rightarrow X_3$	24
1.3410	$R_2 \rightarrow X_4$	9
1.3564	$R_1 \rightarrow X_4$	14
1.4140	$R_2 \rightarrow X_6$	1
1.4440	$R_1 \rightarrow X_7$	0.2

variety of shapes and sizes, from fibers a few micrometers in diameter to rods 2 m long and 7.5 cm in diameter and disks up to 90 cm in diameter and 5 cm thick.

There is a wide variety of Nd-doped laser glasses depending on the compositions of the glass network former and the network-modifying ions. Among various laser glasses only silicates and phosphates are commercially available with sufficient optical, mechanical and chemical properties. The Nd: phosphates are generally characterized by a large stimulated emission cross section ($\sigma = 3.7$–4.5×10^{-20} cm^2) and a relatively small nonlinear index ($n_2 = 0.91$–1.15×10^{-13} ESU). Typically, the cross section of a phosphate laser glass is 50 % higher than a comparable silicate. The Nd: phosphate glasses have been adopted in large laser systems employed for fusion research [2.79–83].

Since the advent of lasers, thousands of glasses have been formulated to investigate the effects of changes in glass network and network-modifier ions on the spectroscopic and lasing parameters of neodymium. The host glass has an important influence on the ability of the lasing ion to absorb light from the optical pumping source, to store this energy, and to release it to amplify the laser beam. Energy storage by the lasing ion is governed by its absorption properties, excited-state lifetimes, and quantum efficiency. For rare-earth laser glasses, the energy-storage capability varies only slightly with the host glass. The rate of energy extraction, on the other hand, is governed by the product of the intensity of the extracting beam and the stimulated-emission cross section σ of the lasing ions. Both of these factors are strongly influenced by the characteristics of the host glass. Hence, by appropriate choice of the host glass, one can produce lasers with wide varying performance.

The most common commercial optical glasses are oxide glasses, principally silicates and phosphates, i.e., SiO_2 and P_2O_5 based. Table 2.6 summarizes some important physical and optical properties of commercially available silicate and phosphate glasses. The 1053-nm gain cross sections of available phosphates range from 3.0×10^{-20} to 4.2×10^{-20} cm^2, and are generally larger than the 1064-nm cross sections of silicate glasses. Silicate and phosphate glasses have fluorescent decay times of around 300 μs at doping levels of 2×10^{20} Nd atoms/cm^3. Nonradiative processes account for 50–60 % of the excited decay.

In high-power lasers, with intense radiant fluxes such as employed in fusion research, nonlinear contributions to the refractive index and two-photon absorption become critical. It has been shown that those two nonlinear effects are minimized in fluoride glasses such as fluorphosphate and fluorberyllate.

Table 2.6. Physical and optical properties of Nd-doped glasses

Glass Type Spectroscopic Properties	$Q-246$ Silicate (Kigre)	$Q-88$ Phosphate (Kigre)	$LHG-5$ Phosphate (Hoya)	$LHG-8$ Phosphate (Hoya)	$LG-670$ Silicate (Schott)	$LG-760$ Phosphate (Schott)
Peak Wavelength [nm]	1062	1054	1054	1054	1061	1054
Cross Section [$\times 10^{-20}$ cm^2]	2.9	4.0	4.1	4.2	2.7	4.3
Fluorescent Lifetime [μs]	340	330	290	315	330	330
Linewidth FWHM [nm]	27.7	21.9	18.6	20.1	27.8	19.5
Density [g/cm^3]	2.55	2.71	2.68	2.83	2.54	2.60
Index of Refraction [Nd]	1.568	1.545	1.539	1.528	1.561	1.503
Nonlinear Index n_2 [10^{-13} esu]	1.4	1.1	1.28	1.13	1.41	1.04
dn/dt (20°–40°C [10^{-6}/°C]	2.9	−0.5	8.6	−5.3	2.9	−6.8
Thermal Coefficient of Optical Path (20°–40°C)[10^{-6}/°C]	+8.0	+2.7	+4.6	+0.6	8.0	–
Transformation Point [°C]	518	367	455	485	468	–
Thermal Expansion coeff. (20°–40° [10^{-7}/°C]	90	104	86	127	92.6	138
Thermal Conductivity [W/m °C]	1.30	0.84	1.19	–	1.35	0.67
Specific Heat [J/g °C]	0.93	0.81	0.71	0.75	0.92	0.57
Knoop Hardness	600	418	497	321	497	–
Young's Modulus [kg/mm^2]	8570	7123	6910	5109	6249	–
Poisson's Ratio	0.24	0.24	0.237	0.258	0.24	0.27

Laser Properties

There are two important differences between glass and crystal lasers. First, the thermal conductivity of glass is considerably lower than that of most crystal hosts. Second, the emission lines of ions in glasses are inherently broader than in crystals. A wider line increases the laser threshold value of amplification. Nevertheless, this broadening has an advantage. A broader line offers the possibility of obtaining and amplifying shorter light pulses, and, in addition, it permits the storage of larger amounts of energy in the amplifying medium for the same linear amplification coefficient. Thus, glass and crystalline lasers complement each other. For continuous or very high repetition-rate operation, crystalline materials provide higher gain and greater thermal conductivity. Glasses are more suitable for high-energy pulsed operation because of their large size, flexibility in their physical parameters, and the broadened fluorescent line.

Unlike many crystals, the concentration of the active ions can be very high in glass. The practical limit is determined by the fact that the fluorescence lifetime and, therefore the efficiency of stimulated emission, decreases with higher concentrations. In silicate glass, this decrease becomes noticeable at a concentration of 5 % Nd_2O_3.

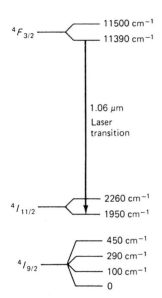

Fig. 2.8. Partial energy level diagram of Nd^{3+} in glass

Figure 2.8 shows a simplified energy level diagram of Nd: glass. The Nd^{3+} ion in glass represents a four-level system. The upper laser level indicated in Fig. 2.8 is the lower-lying component of the $^4F_{3/2}$ pair with a several-hundred microsecond spontaneous emission lifetime. The terminal laser level is the lower-lying level of the pair in the $^4I_{11/2}$ multiplet. The $^4I_{11/2}$ group empties spontaneously by a radiationless phonon transition to the $^4I_{9/2}$ ground state about 1950 cm^{-1} below. Published values for the $^4I_{11/2}$ level lifetime vary from 10 to 100 ns [2.84–88]. This lifetime is difficult to measure because the transition $^4I_{11/2} \rightarrow {}^4I_{9/2}$, at a wavelength of 5 μm, is absorbed by the glass host. The degeneracy of either member of the $^4F_{3/2}$ level is 1. Measurements performed by several researchers [2.86, 87] seem to indicate that the degeneracy of the terminal laser level is likely to be 1 or 2.

Due to the large separation of the terminal laser level from the ground state, even at elevated temperatures there is no significant terminal-state population and, therefore, no degradation of laser performance. In addition, the fluorescent linewidth of the neodymium ion in glass is quite insensitive to temperature variation; only a 10 % reduction is observed in going from room temperature to liquid nitrogen temperature. As a result of these two characteristics, it is possible to operate a neodymium-doped glass laser with little change in performance over a temperature range of $-100°$ to $+100°$ C.

Figure 2.9 shows the pump bands of Nd : glass. In comparing Fig. 2.9 with Fig. 2.7, one notes that the locations of the absorption peaks in Nd : YAG and Nd : glass are about the same; however, in Nd : glass the peaks are much wider and have less fine a structure compared to Nd : YAG.

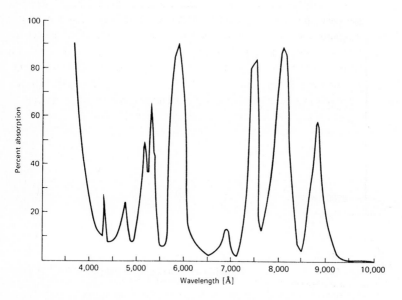

Fig. 2.9. Absorption versus wavelength of Nd : glass. (Material: ED-2; thickness: 6.3 mm)

Commercially Available Laser Rods. Glass is a mixture of oxides. Its main constituents are nonmetal oxides, such as SiO_2, B_2O_3, and P_2O_5. Different metal oxides alter the structure in various ways and make it possible to obtain a large variety of properties. The components are mixed before melting, with the laser activators also added to the batch. The mixture is heated in a heat-resistant crucible. The principal laser glass manufacturers use either platinium or ceramic crucibles, clay pots, or ceramic continuous tanks to contain the melt. When the melt has reached a high viscosity it is cast into a mold. Finally, the glass in the mold is placed into an annealing furnace where it is very slowly cooled down.

Glass laser rods are fabricated in a large variety of sizes. Typical rod sizes are between 10 and 50 cm in length, with diameters from 1 to 3 cm. However, rods up to 1 m in length and 10 cm in diameter are commercially available. Standard rod end configurations are the same as those mentioned for Nd : YAG.

Data Sources on Laser Glasses. Useful reviews on glass lasers can be found in [2.8, 9, 89, 90]. Pertinent materials parameters are listed in the data sheets of the major Nd : glass manufacturers.

2.3.3 Nd : Cr : GSGG

Soon after the invention of the Nd : YAG laser attempts were made to increase the efficiency of transferring radiation from the pump source to the laser crystal by utilizing a second dopant called a "sensitizer". A particularly attractive sensitizer is Cr^{3+} because the broad absorption bands of chromium can efficiently absorb light throughout the whole visible region of the spectrum. The concept of improving efficiency by co-doping a Nd laser crystal with Cr^{3+} ions is based on transferring excitation, absorbed by the broad Cr^{3+} absorption bands, over to the Nd^{3+} ions. No improvement was achieved with the host crystal YAG because all Cr^{3+} excitation was deposited in the ^2E level and the spin-forbidden nature of the $^2E \rightarrow {}^4A_2$ transition resulted in an inefficient transfer process.

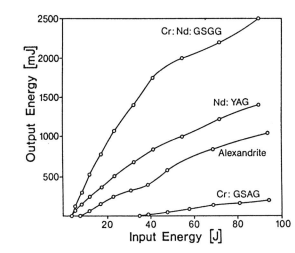

Fig. 2.10. Output energy versus input energy for Cr : Nd : GSGG, Nd : YAG, alexandrite, and Cr : GSAG [2.97]

However, researchers at the Lebedev Institute and the University of Hamburg [2.91–95] discovered that nearly 100 % transfer efficiency could be achieved in the co-doped garnet crystal GSGG. Unlike YAG, a large percentage of the Cr^{3+} excitation in GSGG appears in the $^4T_2$ state, nonradiative transfer to Nd^{3+} ions can occur via the $^4T_2 \rightarrow {}^4A_2$ transition, which is spin-allowed and has a good spectral overlap with the Nd^{3+} levels. Experiments performed in several laboratories showed, with flashlamp-pumped operation, nearly a factor-of-three improvement in slope efficiency for the doubly doped garnet compared to a Nd:YAG crystal. It was found that the pumping efficiency improvement is not a strong function of the Cr concentration for the range 1–2×10^{20}/cm^3 [2.96].

Figure 2.10 illustrates the high efficiency which can be achieved with this material by comparing it to Nd:YAG and to tunable lasers such as alexandrite and Cr:GSAG [2.97]. The experiments were carried out in a silver plated single-ellipse pump cavity using a 5×75 mm^2 flashlamp. The laser rods were 6×75 mm^2 in an identical laser-resonator configuration but with output coupling optimized for the particular laser. Another comparison between Nd:YAG and co-doped GSGG is shown in Fig. 2.11 [2.98]. The efficiency factors indicated on the curves are close to state-of-the art performance for both materials.

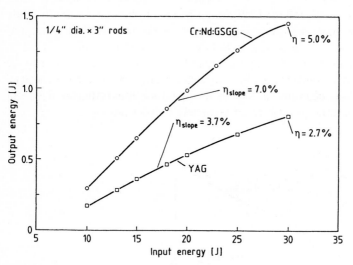

Fig. 2.11. Efficiency comparison of Nd:YAG and Cr:Nd:GSGG [2.98]

The higher pump efficiency of Nd:Cr:GSGG does not automatically translate into better system performance because Nd:Cr:GSGG does exhibit much stronger thermal focusing and stress birefringence, as compared to Nd:YAG. The absorption efficiency and the heat-deposition rate for the Cr:Nd:GSGG rod is almost three times those of Nd:YAG, a consequence of the broad red and blue absorption bands of the Cr^{3+} sensitizer [2.99]. As a consequence, thermal focusing power as a function of lamp input power has been reported to be several times larger in Cr:Nd:GSGG than in Nd:YAG [2.99, 100]. Therefore, if beam brightness is the criteria, rather than output energy, some of the advantage of GSGG is offset, particularly at high average powers. Measurements at the author's laboratory have shown that under single-shot and low-repetition-rate operation, a factor 2 higher beam brightness is achieved in a GSGG system as

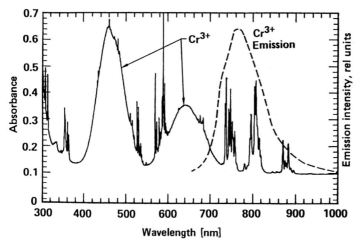

Fig. 2.12. Absorption spectra of Cr:Nd:GSGG [2.98]

compared to YAG. However, as the repetition rate is increased, this difference became smaller and beam brightness for a given input was about equal in both systems at 25 Hz for a Q-switched system at 100 mJ output.

Table 2.7. Material properties of GSGG

Property	Cr:Nd:GSGG
Thermal conductivity, κ	6.0 W/m°C
Heat capacity, C_P	0.40 J/g°C
Thermal expansion, α	$7.4 \times 10^{-6}/°C$
Young's modulus, E	212 MPa
Poisson's ratio, ν	0.28
Index of refraction, 1060 nm	1.942
Fluorescence lifetime, τ	281 μs
Stimulated emission	
Cross section, σ	3.1×10^{-19} cm^2
Density	6.44 g/cm^3
Elasto-optic coefficients	
$\quad P_{11}$	-0.012
$\quad P_{12}$	0.019
$\quad P_{44}$	-0.0665

The explanation for the high efficiency of co-doped GSGG follows from Fig. 2.12 which shows the absorption spectrum of this material. Two high-intensity Cr^{3+} absorption bands having maxima at 450 and 640 nm covering an appreciable part of the visible region, and narrow Nd lines can be observed. In contrast to Nd:Cr:YAG the transfer-time in the GdScGa-Garnet (Cr $\rightarrow$ Nd) = 10 μs is about 3 orders of magnitude faster. This is due to a favorable red shift of the $^4T_2$ energy band of Cr^{3+} and the high radiative transition probability of the $^4T_2 \rightarrow ^4A_2$ transition. Both the enhancement of energetic $3d$-$4f$ overlap and the increased $3d$-$3d$ oscillator strength result in the observed strong multipole coupling of the sensitizer Cr^{3+} and acceptor Nd^{3+}.

Material parameters of Cr : Nd : GSGG are listed in Table 2.7. A comparison with Nd : YAG reveals a lower thermal conductivity and a lower heat capacity for GSGG but, in general, the materials parameters are fairly close. The stimulated emission cross section, and therefore the gain of Cr : Nd : GSGG is about half of that in YAG. This can be an advantage for Q-switch operation since more energy can be stored before amplified spontaneous emission (ASE) starts to deplete the metastable level.

2.3.4 Nd : YLF

During the last 10 years, the crystal quality of the scheelite-structured host lithium yttrium fluoride has been dramatically improved and the material has gained a firm foothold for a number of applications.

The laser transition of Nd : LiYF$_4$ (Nd : YLF) at 1053 nm matches well the peak gain of Nd doped phosphate and fluorophosphate glasses. As a result, it is currently being used in master oscillators for amplifier chains using these glasses. Nd : YLF is also employed in medium energy Q-switched oscillator and oscillator-amplifiers because it requires fewer stages as compared to Nd : YAG lasers for the same output energy.

The material has also advantages for diode pumping since the fluorescence lifetime in Nd : YLF is twice as long as in Nd : YAG. Laser diodes are power limited, therefore, a larger pump time afforded by the longer fluorescence time, provides for twice the energy storage from the same number of diodes.

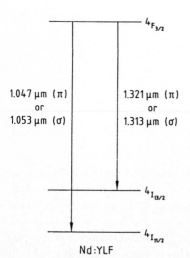

Fig. 2.13. Simplified energy level diagram of Nd : YLF

Figure 2.13 shows a simplified energy level diagram of Nd : YLF. Depending on the polarization, two lines each are obtained around 1.05 and 1.3 μm. For example, with an intracavity polarizer one can select either the 1047 nm (extraordinary) or 1053 (ordinary) transition. The same can be done for the two 1.3 μm transitions, however, in addition lasing at the 1.05 μm lines has to be suppressed. All lines originate on the same Stark split $^4F_{3/2}$ upper level.

Material properties of Nd : YLF are listed in Table 2.8. The relatively large thermal conductivity allows efficient heat extraction, and its natural birefringence overwhelms thermally induced birefringence eliminating the thermal depolarization problems of optically isotropic hosts like YAG. The cross-section for YLF is about a factor of 2 lower than YAG. For certain lasers requiring moderate Q-switch energies, the lower gain offers advantages in system architecture compared to the higher gain material Nd : YAG.

Table 2.8. Properties of Nd doped lithium yttrium fluoride (YLF)

Lasing wavelength [nm]	1053 (σ)
	1047 (π)
Index of Refraction	$n_0 = 1.4481$
$\lambda = 1.06 \mu$m	$n_e = 1.4704$
Fluorescent lifetime	480μs
Stimulated emission	$1.8 \times 10^{-19} (\pi)$
Cross-section [cm^2]	$1.2 \times 10^{-19} (\sigma)$
Density [g/cm^3]	3.99 (undoped)
Hardness [Mohs]	4–5
Elastic modulus [N/m^2]	7.5×10^{10}
Strength [N/m^2]	3.3×10^7
Poisson's ratio	0.33
Thermal conductivity [W/cm-K]	0.06
Thermal Expansion coefficient [$^\circ$C^{-1}]	a axis : 13×10^{-6}
	c axis : 8×10^{-6}
Melting point [$^\circ$C]	825

The energy storage in Q-switched operation of Nd oscillators and amplifiers is constrained by the onset of parasitic oscillations. To first order, the energy storage limit of two materials is inversely proportional to the ratio of the stimulated emission cross-section. Therefore, higher storage densities are obtained in the lower cross-section material Nd : YLF as compared to Nd : YAG. Figure 2.14 shows the performance of a Q-switched YLF oscillator, producing an output of 400 mJ [2.101]. This output level would require the addition of an amplifier in the case of a Nd : YAG laser. Single oscillators with up to 0.5 J Q-switched output, and oscillator-amplifier configurations with several Joules output usually result in less complicated structures if designed with Nd : YLF rods as compared to Nd : YAG [2.102, 103]. The weaker thermal lensing in Nd : YLF, coupled with

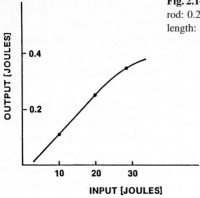

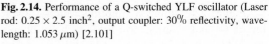

Fig. 2.14. Performance of a Q-switched YLF oscillator (Laser rod: 0.25×2.5 inch^2, output coupler: 30% reflectivity, wavelength: 1.053 μm) [2.101]

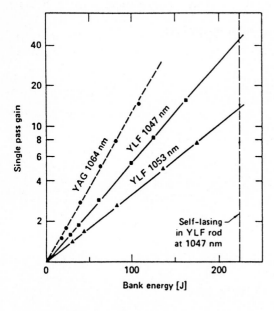

Fig. 2.15. Single pass gain versus pump energy for YLF and YAG [2.104]

its natural birefringence, provide additional advantages in certain applications compared to Nd:YAG [2.104, 105].

Figure 2.15 shows measured single-pass gain versus pump energy for the two transitions [2.104]. The measurements of the Nd:YLF amplifier were performed on a 10 mm diameter $\times$76 mm YLF rod with 1 % Nd doping. The crystal was pumped in a silver-coated double elliptical pumping head. The YAG data were obtained by replacing the Nd:YLF with a 10 mm diameter Nd:YAG rod in the same amplifier head.

2.3.5 Nd : YVO$_4$

Neodymium-doped yttrium vanadate has several spectroscopic properties that are particularly relevant to laser diode pumping. The two outstanding features are a large stimulated emission cross section which is five times higher than Nd : YAG, and a strong broadband absorption at 809 nm [2.106].

The potential of Nd : YVO$_4$ as an important laser material has been recognized already in 1966 [2.31]. However, crystals could not be grown free of scattering centers and absorbing color center defects. The unavailability of high-quality crystals of the size required for flashlamp pumping proved to be the major obstacle to further development. The high gain achievable in Nd : YVO$_4$ and the strong absorption of diode pump radiation require crystals only a few millimeters in length. With the constraint of large crystal size removed, and with further improvements in the crystal growth process, material of high optical quality became available.

The vanadate crystal is naturally birefringent and laser output is linearly polarized along the extraordinary π-direction. The polarized output has the advantage that it avoids undesirable thermally induced birefringence.

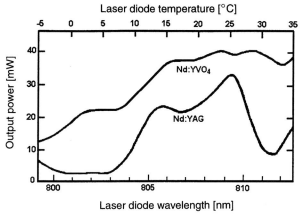

Fig. 2.16. Output from a Nd : YVO$_4$ and Nd : YAG laser as a function of diode pump temperature and wavelength [2.108]

Pump absorption in this uniaxial crystal is also polarization dependent. The strongest absorption occurs for pump light polarized in the same direction as the laser radiation. The absorption coefficient is about four times higher compared to Nd : YAG in the π-direction [2.107]. The sublevels at the $^4F_{5/2}$ pump band are more resolved in Nd : YAG, whereas in Nd : YVO$_4$ Stark splitting is smaller and the multiple transitions are more compacted. The result is a broader and less spiky absorption profile of Nd : YVO$_4$ compared to Nd : YAG around the pump wavelength of 809 nm [2.108]. Figure 2.16 gives an indication of the broader and smoother absorption profile of this material as compared to Nd : YAG. From this data follows also that Nd : YVO$_4$ laser performance is more tolerant to diode temperature variations because of the large pump bandwidth. If one defines this bandwidth as the wavelength range where at least 75% of the pump radiation is absorbed in a 5 mm thick crystal, then one obtains for Nd : YVO$_4$ a value of 15.7 nm, and 2.5 nm for Nd : YAG [2.106].

Table 2.9. Material parameters for Nd: YVO$_4$ [2.106, 107]

Laser cross section	15.6×10^{-19} cm^2
Laser wavelength	1064.3 nm
Linewidth	0.8 nm
Fluorescence lifetime	100 μs
Peak pump wavelength	808.5 nm
Peak absorption coefficient	37 (π polarization)
at 808 nm [cm^{-1}]	10 (σ polarization)
Nd doping	1% (atomic Nd)

Table 2.9 summarizes important material parameters for Nd : YVO$_4$. The material does have several drawbacks, the principal one is a shorter excited state lifetime than Nd : YAG. As expressed in (3.68) the pump input power to reach threshold for cw operation depends on the product of $\sigma\tau$. Therefore, the large cross section σ of Nd : YVO$_4$ is partially offset by its shorter fluorescence lifetime τ. Fluorescence lifetime is also a measure of the energy storage capability in Q-switched operation. Large energy storage requires a long fluorescence lifetime. As far as thermal conductivity is concerned, it is only half as high as Nd : YAG and somewhat lower than Nd : YLF.

The properties of Nd : YVO$_4$ can best be exploited in an end pumped configuration and a number of commercial laser systems are based on fiber-coupled diode arrays pumping a small vanadate crystal. Actually, Nd : YVO$_4$ is the material of choice for cw endpumped lasers in the 5 watt output region. These systems are often also internally frequency doubled to provide output at 532 nm. In endpumped systems the pump beam is usually highly focused and it is difficult to maintain a small beam waist over a distance of more than a few millimeters. In this case a material such as Nd : YVO$_4$ which has a high absorption coefficient combined with high gain is very advantageous.

Impressive overall efficiencies have been achieved in Nd : YVO$_4$ as illustrated in Figs. 3.47 and 6.63. Considerably higher output is obtained from Nd : YVO$_4$ than from Nd : YAG. Actually, the highest efficiency TEM$_{00}$ performance has been demonstrated in this material [2.108]. For cw-pumped, repetitively Q-switched operation, the relatively short upper-state lifetime of Nd : YVO$_4$ requires high pulse rates of (50–100) kHz in order to achieve average power close to cw performance. The high gain in combination with a short fluorescence lifetime produces relatively short Q-switch pulses compared to Nd : YAG and Nd : YLF. Pulsewidth in repetitively Q-switched systems is typically around 10 ns.

Various end-pumped Nd:YVO$_4$ laser systems were described in [2.107–112] and the results of experiments with small side pumped slabs were summarized in [2.113, 114].

2.4 Er:Lasers

Erbium-laser performance is not very impressive in terms of efficiency or energy output. However, erbium has attracted attention because of two particular wavelengths of interest. A crystal, such as YAG, highly doped with erbium produces an output around 2.9 μm, and Er-doped phosphate glass generates an output at 1.54 μm. Both of these wavelengths are absorbed by water, which leads to interesting medical applications in the case of the 2.9 μm lasers, and to eye safe military rangefinders in the case of the shorter wavelength.

2.4.1 Er:YAG

In 1975 Soviet researchers [2.115–117] discovered laser operation at room temperature in highly doped Er:YAG, i.e. erbium concentrations around 50 %. Laser emission occurs at 2940 nm. The spectroscopy of Er^{3+} in YAG indicated that the upper laser level of the 2940 nm transition is pumped by light at wavelengths less than 600 nm. Therefore, the pump efficiency is not very high in this material. Also, due to a very long, lower-level lifetime the Er:YAG laser cannot be Q-switched. The 2940 nm lasing transition is between the $^4I_{11/2}$ and $^4I_{13/2}$ states of the Er^{3+} ion. The lower level has a much longer lifetime than the upper level (2 ms as compared to 0.1 ms) and so this transition can be blocked by the accumulation of population in the $^4I_{13/2}$ state. Despite these above mentioned drawbacks, the laser is of particular interest due to its wavelength which coincides with a water absorption line. Optical absorption by the water in tissue is extremely large (> 3000 cm^{-1}), making potential medical applications such as plastic surgery very attractive.

Figure 2.17 shows the output vs. input energy of an oscillator containing a 75 mm long by 6.25 mm diameter 50% Er:YAG rod. The pump pulse duration was 170 μs FWHM. The pump cavity was a double ellipse, silver coated and containing two flashlamps with 7 mm bore diameters. The output increases linearly to over 600 mJ per pulse for a 360 J pump pulse. Repetition rate was 1 Hz. At higher repetition rates the effects of thermal lensing caused a reduction in efficiency and distorted the beam shape [2.118].

Er:YAG can also be pumped with laser diode arrays. Close to 1 W output was obtained in a 50% Er:YAG crystal pumped with a InGaAs array at 963 nm [2.119]. The laser operated at a repetition rate of 100 Hz and a pulse duration of 400 μs. The laser was tunable over a range of 6 nm around the line center at 2936 nm. Tuning was accomplished by tilting an etalon.

Besides Er:YAG, other erbium lasers such as Er:YALO$_3$, Er:YLF and Er:Cr:YSGG have been investigated [2.120, 121]. These lasers provide addi-

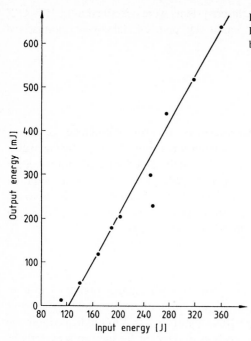

Fig. 2.17. Output versus input energy for 50 % Er : YAG pumped in a double elliptical cavity by two xenon filled flashlamps [2.122]

tional laser lines between 2.71 to 2.92 μm, or at 1.7 μm or they have the potential for improved performance. Also laser emission from Er : YLF pumped by a laser diode array for both pulsed and cw modes has been reported [2.122].

2.4.2 Er : Glass

Erbium has been made to lase at 1.54 μm in both silicate and phosphate glasses [2.123]. Due to the three-level behavior of erbium, and the weak absorption of pump radiation, co-doping with other rare earth ions is necessary to obtain satisfactory system efficiency. In fact, the erbium must be sensitized with ytterbium if the glass is to lase at all at room temperature. The Yb acts as sensitizing agent by absorbing pump light in regions where the erbium ion is relatively transparent. The principal absorption band is the Yb^{3+} band at 1 μm.

The flash-lamp current density required to match this band is about 1000–2000 A/cm^2. Such current densities require pulse times of the order of several milliseconds which are, however, compatible with the long fluorescent transition lifetimes of the dopant ions. Xenon flashlamps have the spectral output which best matches this pumping requirement of the erbium laser. The energy level scheme relevant to the Er^{3+} is between the $^4I_{13/2}$ multiplet in the region of 6500 cm^{-1} (1.54 μm) above the ground state, and the ground state $^4I_{15/2}$ levels (Fig. 2.18).

The optimum erbium concentration is usually found to fall in the range of 0.2 to 0.5 wt. % Er_2O_3 (or about 2 to 5 × 10^{19} Er^{3+} ions per cubic centimeter). This optimum is a balance between several factors. The Er^{3+} concentration must

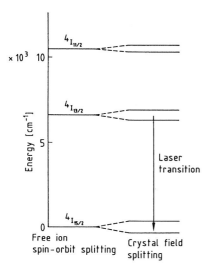

Fig. 2.18. Er^{3+} energy level scheme

be sufficiently low to produce a reasonable threshold for lasing. The Er^{3+} ion is a three-level laser and about 60% of the ions must be excited before threshold is reached. However, the Er^{3+} concentration must also be sufficiently high to produce a high energy-transfer efficiency from the Yb^{3+} to the Er^{3+}.

Yb:Er:glass has a relatively low absorption in the regime between 400–900 nm. The only strong absorption is between 0.9–1 μm due to a transition in Yb^{3+}. As a result, the absorption of Yb:Er:glass does not match the emission from flashlamps very well. The efficiency of flashlamp pumping can be improved by adding Cr^{3+} as a sensitizer to the Yb:glass [2.126]. Cr^{3+} absorbs flashlamp radiation in two broad bands centered at 450 and 640 nm, and emits radiation in a broad band centered at 760 nm. This allows energy to be transferred from Cr^{3+} to the $^4I_{9/2}$ and $^4I_{11/2}$ states of Er^{3+} and the $^2F_{5/2}$ state of Yb^{3+}. Material properties of Er:Glass are listed in Table 2.10.

Q-switched performance from an erbium phosphate laser glass is illustrated in Fig. 2.19. The laser contains a 3×50 mm glass-rod, and a rotating prism as Q-switch. Flashlamp pulse length is 3 ms. This laser is typical for eye-safe range finders. Small, handheld rangefinders are the major application for Q–switched Er:glass lasers [2.124]. High repetition rate, or high-average-power eye-safe military lasers, usually employ a Nd:YAG laser which is wavelength-shifted to 1.5 μm by means of a Raman cell or parametric oscillator.

Table 2.10. Material properties of an erbium doped phosphate glass

Emission wavelength	1.54 μm
Fluorescent lifetime	8 ms
Index of refraction at 1.54 μm	1.531
dn/dT	$63 \times 10^{-7}/°C$
Thermal expansion (α)	$124 \times 10^{-7}/°C$
Thermo-optic coefficient (W)	$-3 \times 10^{-7}/°C$

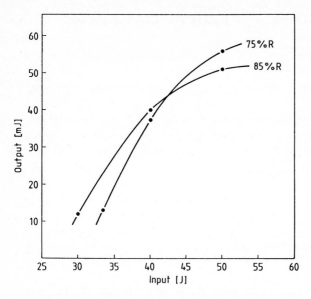

Fig. 2.19. Q-switched performance from an erbium phosphate glass oscillator [KIGRE, Inc.]

Employing an athermal phosphate glass, a high repetition-rate Er : glass laser has been built and operated at 30 Hz [2.125] for fiber-fault location in communications fiber networks.

2.5 Tunable Lasers

In most lasers, all of the energy released via stimulated emission by the excited medium is in the form of photons. Tunability of the emission in solid-state lasers is achieved when the stimulated emission of photons is intimately coupled to the emission of vibrational quanta (phonons) in a crystal lattice. In these "vibronic" lasers, the total energy of the lasing transition is fixed, but can be partitioned between photons and phonons in a continuous fashion. The result is broad wavelength tunability of the laser output. In other words, the existence of tunable solid-state lasers is due to the subtle interplay between the Coulomb field of the lasing ion, the crystal field of the host lattice, and electron-phonon coupling permitting broad-band absorption and emission. Therefore, the gain in vibronic lasers depends on transitions between coupled vibrational and electronic states; that is, a phonon is either emitted or absorbed with each electronic transition.

The history of tunable solid-state lasers can be traced back to work performed over 35 years ago at Bell Laboratories. In 1963, Bell-Labs researchers reported the first vibronic laser, a nickel-doped magnesium fluoride (Ni : MgF_2) device [2.127]. The same group later built a series of vibronic lasers using nickel, cobalt or vanadium as the dopant and MnF_2, MgO, MgF_2, ZnF_2, or $KMgF_3$ as the host crystal [2.128–130]. Tunable in the 1.12 to 2.17 μm range, these flashlamp-pumped lasers had a serious drawback: they operated only when cooled to cryogenic temperatures. Despite these early results, this field lay dormant for more than a decade, probably because of the need for cryogenic cooling of the laser material, the emergence of Nd : YAG and the concentration on dye lasers and color-center lasers [2.131]. The first room-temperature vibronic laser, reported by Bell Labs in 1974, was a flashlamp-pumped Ho : BaY_2F_8 device emitting at 2.17 μm [2.132].

In 1977, the crystal alexandrite (chromium doped $BeAl_2O_4$) was made to lase on a vibronic transition [2.133–135]. The crystal field strength at the chromium site in alexandrite is sufficiently low to permit thermal activation of broad $^4T_2$ fluorescence. The result is a low-gain, four-level laser tunable in the range of 0.7 to 0.8 μm.

During the mid-1970s, research started on a variety of divalent transition-metal doped crystals such as Ni^{2+}, V^{2+} and Co^{2+} at MIT Lincoln Laboratory [2.136]. Most of the materials investigated, such as V : MgF_2 for example, suffered from excited state absorption which reduces the net gain to an unacceptably low level. A notable exception is Co^{2+} : MgF_2 which turned out to be a useful laser.

If pumped by a 1.3 μm Nd laser, the cobalt laser is tunable from 1750–2500 nm. This earlier work resulted also in the discovery of the Ti : sapphire laser by *Moulton* in 1982 [2.137]. In the sapphire host, Ti^{2+} exhibits no appreciable excited-state absorption, and therefore has a tuning range that spans virtually its entire fluorescence emission wavelength range.

The spectral ranges of typical tunable lasers are shown in Fig. 2.20.

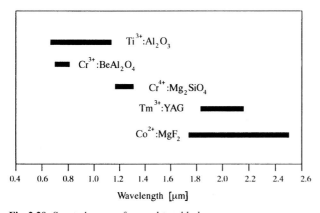

Fig. 2.20. Spectral range of several tunable lasers

Chromium Laser

Chromium has enjoyed considerable success as a tunable-laser ion due to the special nature of its trivalent d^3 electronic configuration. Particular attributes of Cr over other transition metals are its chemical stability, broad pump bands, large energy-level splitting, and reduced excited state absorption (ESA). These advantages have led to at least a dozen crystals being demonstrated as good laser hosts for Cr. Chromium has a laser tuning range of about 100 nm, with the center wavelength shifted by the particular crystal host.

Alexandrite. Cr in chrysoberyl ($BeAl_2O_4$) occupies the 700–800 nm range and was the first tunable ion solid-state laser considered for practical use. Details of this commercially available laser material are given in Sect. 2.5.1.

Cr : GdScGa-Garnet. Cr^{3+}-doped Ga garnets present a class of transition metal ion lasers which can operate pulsed as well as cw at room temperature [2.138–141]. Due to a low-crystal field at the Cr site, the $^4A_2$-$^4T_2$ splitting is nearly equal to the $^4A_2$-$^4T_2$ separation. Thus, in contrast to ruby or alexandrite, the fluorescence is totally dominated by the broadband four-level $^4T_2 \rightarrow {}^4A_2$ transition.

Garnet crystals are described by the formula $C_3A_2D_3O_{12}$, where one chooses large ions A = Ga, Sc, Lu for the octahedral site, D = Ga for the tetrahedral site, and C = Y, Gd, La, Lu for the dodecahedral site. Broadband four-level fluorescence has been obtained in the 700–950 nm spectral range from Cr-doped YGG, YScGG (YSGG), GGG, GdScGG(GSGG), and LaLuGG(LLGG) with lifetimes ranging from 240 to 70 μs, respectively. The crystals are grown by the standard Czochralski technique at $\approx 1750°$ C.

One compound, Cr^{3+}-doped GdScGa-garnet, has received particular attention. In Cr^{3+}-doped GdScGa-garnet, the R-line fluorescence is totally absent and due to a large 2400 cm^{-1} Stokes shift of the $^4T_2$ level, the whole fluorescence is channelled into the broad band (4 level) $^4T_2 \rightarrow {}^4A_2$ transition. Pulsed and cw laser action has been obtained from Cr : GSGG over a tuning range from approximately 700 to 900 nm. The development of the alexandrite and Ti : sapphire lasers, which overlap the tuning range of Cr : GSGG have overshadowed this material, since the former can be flashlamp pumped, and the latter has a wider tuning range.

Cr : KZnF$_3$. This fluorine perovskite was the second tunable solid-state laser material (after alexandrite) to be offered commercially [2.142]. Cr : KZnF$_3$ operates from 785 to 865 nm, with good slope efficiency when laser pumped between 650 to 700 nm. Similarly to Cr : GSGG, this material has been displaced by the more versatile Ti : sapphire laser.

Cr : Forsterite

Both pulsed and cw laser operation has been achieved in Cr^{4+} : Mg_2SiO_4 with Nd : YAG lasers at 1.06 μm and 532 nm as pump sources [2.143–145]. This laser is of interest because the tuning range covers the spectral region from 1167 to 1345 nm which is not accessible with other lasers. A distinguishing feature of laser actions in Cr : Mg_2SiO_4 is that the lasing ion is not trivalent chromium (Cr^{3+}) as is the case with other chromium-based lasers but the active ion in this crystal is tetravalent chromium (Cr^{4+}) which substitutes for silicon (Si^{4+}) in a tetrahedral site. Single crystals of Cr-doped forsterite are grown by the Czochralski method. Forsterite has a fluorescence lifetime of 25 μs and therefore has to be pumped by another laser. Stimulated emission cross-section is 1.44×10^{-19} cm^2, about one-fourth of Nd : YAG.

Cr : LiSAF

Cr^{3+}-doped $LiSrAlF_6$ has a tuning range from 780–920 nm and can be flashlamp and diode pumped. Very large crystals of high optical quality can be grown from this material. Cr : LiSAF lasers are particularly important for femtosecond pulse generation and will be discussed in Sect. 2.5.3 and Chap. 9.

Co : MgF₂

Interest in mid-IR lasers is stimulated by remote sensing and medical applications. Room temperature operation of Co, Tm, Ho and Er doped materials has made this spectral region accessible to practical applications. Co : MgF₂ has a particularly wide tuning range which extends from 1750 to 2500 nm [2.49, 50]. A commercial version of this laser employs a 1.32 μm Nd : YAG laser as a pump source. The laser achieves an output of 980 mJ at 10 Hz repetition rate [2.146]. Q-switching has also been achieved at about 20 mJ per pulse. Figure 2.21 shows the tuning range of the Co : MgF₂ laser, which requires different sets of optics to span the whole range.

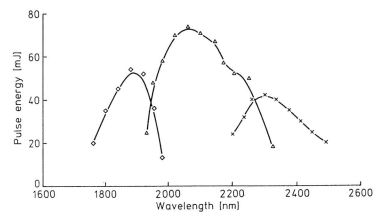

Fig. 2.21. Tuning range of the Co : MgF₂ laser [2.146]

Titanium Laser

The Ti : sapphire laser has an exceptionally wide tuning range and a large gain cross-section, i.e., about 50 % of Nd : YAG. Also, large, high quality crystals are commercially available. These attributes have led to the use of this material for many laser applications. Further details will be discussed in Sect. 2.5.2.

Rare-Earth Tunable Lasers

Tm and Ho doped YAG, YLF and YAP lasers are primarily employed as laser sources in the $2\,\mu$m region, a wavelength of interest for certain medical applications and coherent radar systems. However, limited tunability can also be achieved with these lasers. In the previous examples, tunability was afforded by the 3d electron of the transition metal ions Cr^{3+} and Ti^{3+}. In Tm^{3+}:YAG, for example, the considerable phonon broadening and high multiplicity of the Stark levels of the 4f electron provides tunability from 1.87 to $2.16\,\mu$m (Sect. 2.5.4). Similarly, holmium lasers have some tunability around $2.067\,\mu$m. For example, a diode pumped Ho:Tm:YLF laser was tuned with an etalon between 2.06 and $2.07\,\mu$m [2.147]. Holmium lasers are usually co-doped with Tm to provide increased absorption for the pump source. Although not particularly considered a tunable laser, Yb:YAG has tunability from 1.018 to $1.053\,\mu$m [2.148]. This laser is mainly of interest for high power generation.

Tm, Ho and Yb lasers have been known for years, but flashlamp pumping is not particularly efficient in these 3-level lasers. The fact that these crystals can be pumped with laser diodes has generated renewed interest in these lasers materials. Tm:YAG is discussed in Sect. 2.5.4 and Yb:YAG in Sect. 2.6.

2.5.1 Alexandrite Laser

Alexandrite ($BeAl_2O_4$:Cr^{3+}) developed by *Walling* et al. [2.149] is the common name for chromium-doped chrysoberyl – one of the olivine family of crystals – with four units of $BeAl_2O_4$ forming an orthorhombic structure. The crystal is grown in large boules by the Czochralski method much like ruby and YAG. Laser rods up to 1 cm in diameter and 10 cm long with a nominal 2-fringe total optical distortion are commercially available. The chromium concentration of alexandrite is expressed in terms of the percentage of aluminum ions in the crystal which have been replaced by chromium ions. The Cr^{3+} dopant concentration, occupying the Al^{3+} sites, can be as high as 0.4 atomic percent and still yield crystals of good optical quality. A concentration of 0.1 atomic percent represents 3.51×10^{19} chromium ions per cubic centimeter.

Alexandrite is optically and mechanically similar to ruby, and possesses many of the physical and chemical properties of a good laser host. Hardness, strength, chemical stability and high thermal conductivity (two-thirds that of ruby and twice that of YAG) enables alexandrite rods to be pumped at high average powers without thermal fracture. Alexandrite has a thermal fracture limit which is 60 % that of ruby and five times higher than YAG. Table 2.11 lists the chrysoberyl material properties [2.150].

Table 2.11. Material parameters of alexandrite

Laser wavelength [nm]	700–818
Stimulated emission cross-section [cm^2]	1.0×10^{-20}
Spontaneous lifetime [μs]	260 ($T = 298$ K)
Doping density [at. %]	0.05–0.3
Fluorescent linewidth [Å]	1000
Inversion for 1% gain per cm [cm^{-3}]	$2–10 \times 10^{17}$
Stored energy for 1% gain per cm [J/cm^3]	0.05–0.26
Gain coefficient for 1 J/cm^3 stored energy [cm^{-1}]	0.038–0.19
Index of refraction: (750 nm)	$E \parallel a$ 1.7367
	$E \parallel b$ 1.7421
	$E \parallel c$ 1.7346
Thermal Expansion:	$\parallel a$ 5.9 $\times 10^{-6}$/K
	$\parallel b$ 6.1
	$\parallel c$ 6.7
Thermal conductivity:	0.23 W/cm-K
Melting point:	1870°C
Hardness:	2000 kg/mm^2

Due to its orthorhombic structure, alexandrite is biaxial with the principal axes of the index ellipsoid along the crystallographic axes. Light emitted from the laser is polarized with the E vector parallel to the b axis. The gain in the $E \parallel b$ polarization is 10 times that in the alternate polarizations. Alexandrite lases at room temperature with flashlamp pumping throughout the range 701 to 818 nm. The alexandrite absorption bands are very similar to those of ruby, and span the region from about 380 to 630 nm with peaks occurring at 410 and 590 nm. Figure 2.22 shows the absorption bands of alexandrite. The laser gain cross-section increases from 7×10^{-21} cm^2 at 300 K to 2×10^{-20} cm^2 at 475 K which results in improved laser performance at elevated temperature. The

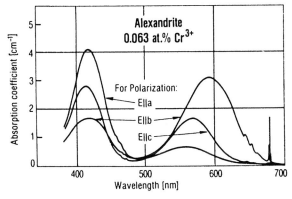

Fig. 2.22. Absorption spectrum of alexandrite [Allied Corps. Data Sheet]

$260\,\mu s$, room-temperature fluorescence lifetime permits effective energy storage and Q-switched operation.

Laser action has also been demonstrated on the R line at 680.4 nm. This three-level mode is analogous to the lasing in ruby except that the stimulated emission cross-section in alexandrite $(3 \times 10^{-19}\,\mathrm{cm}^2)$ is ten times larger than for ruby.

Kinetics. As noted above, alexandrite can operate both as a four-level vibronic laser and as a three-level system analogous to ruby. As a three-level laser, it has a high threshold, fixed output wavelength (680.4 nm at room temperature) and relatively low efficiency. Obviously, the primary interest of alexandrite lies in its vibronic nature.

The basic physics of the 4-level alexandrite laser can be discussed with reference to the energy level diagram (Fig. 2.23). The $^4A_2$ level is the ground state, and $^4T_2$ is the absorption state continuum. Vibronic lasing is due to emission from the $^4T_2$ state to excited vibronic states within $^4A_2$. Subsequent phonon emission returns the system to equilibrium. Since alexandrite is an intermediate crystal field material $(E \approx 800\,\mathrm{cm}^{-1})$ there is coupling between the ^2E state and the $^4T_2$. The lifetimes of each of these states is 1.5 ms and 6.6 μs, respectively. The two phosphorescent R lines emitted from ^2E occur in the vicinity of 680 nm, as for ruby. The terminal laser level is a set of vibrational states well above the ground state. The initial laser level is a level $800\,\mathrm{cm}^{-1}$ above a long-lived storage level and in thermal equilibrium with it.

Due to the vibronic nature of the alexandrite laser, the emission of a photon is accompanied by the emission of phonons. These phonons contribute to thermalization of the ground-state vibrational levels. The laser wavelength depends on which vibrationally excited terminal level acts as the transition terminus; any energy not released by the laser photon will then be carried off by a vibrational phonon, leaving the chromium ion at its ground state.

Of fundamental importance in the kinetics of alexandrite is the cross-sectional probability σ_{2a} that the excited chromium ions will themselves absorb laser

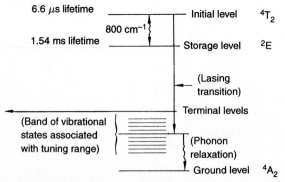

Fig. 2.23. Energy level diagram for chromium ions in alexandrite. The variable partition in de-excitation energy between photons and phonons leads to wavelength tunability

photons circulating in the cavity. This probability must be small compared to the effective emission cross section, otherwise the excited ions will absorb a significant amount of the emitted laser power before it escapes into the resonator. The excited Cr^{3+} ions which absorb the laser emission decay immediately back to their original excited state distribution, so that there is no net loss of excited state population due to excited state absorption. The intracavity flux, of course, suffers a loss due to this absorption contributing to additional heating of the laser rod. In alexandrite the excited-ion absorption band has a deep broad minimum just where the laser emission gain is maximum. At the band center, σ_{2a} is less than 10 % of σ. If σ_{2a} were greater than σ then lasing could not occur at all. In fact, the latter is responsible for the long-wavelength tuning limit in alexandrite [2.151].

Temperature Effects in Alexandrite. As the temperature increases, the gain of alexandrite increases, the gain peak shifts to a longer wavelength and the fluorescence lifetime decreases. The four-level model can be used to predict the temperature dependence of the laser performance. The ^2E state acts as a storage level for $^4T_2$. Thus, as the temperature of alexandrite increases, the vibronic continua in $^4T_2$ are successively populated from ^2E in accordance with the Boltzmann distribution, and the stimulated emission cross-section increases.

However, raising the temperature also tends to populate the terminal levels – especially those which lie closest to the ground level and which therefore correspond to the highest-energy (shortest-wavelength) photons. Since laser performance is highest with a maximally populated initial level and a minimally populated terminal level, it can be seen that increasing the temperature has two conflicting effects. The result is that performance is positively affected by temperature increases only for wavelengths above 730 nm.

Another adverse effect of higher temperature on laser performance is the reduction of the fluorescent decay time. The fluorescent lifetime is $260 \mu s$ at room temperature and $130 \mu s$ at $100°$ C. The total radiative quantum efficiency is nearly constant and equal to unity in the regime of interest. As the temperature is increased, the initial laser level $(^4T_2)$ has an increased share of the excited population; since this level has a much higher decay rate than the storage level ^2E the overall fluorescence lifetime of the upper level (the combined storage and initial levels) is reduced. Therefore, at some higher temperature the storage time becomes much shorter than the flashlamp pulse duration and much energy is lost in fluorescence. This situation limits the advantage derived from the increasing population of the initial level.

Guch [2.152] evaluated the performance of a flashlamp pumped alexandrite laser at temperatures from ambient to $310°$ C. As illustrated in Fig. 2.24a, the variation in laser output over the $34°–310°$ C range is dramatic. The threshold performance indicates that laser gain rises significantly, as temperature increases. Figure 2.24b illustrates laser output energy as a function of temperature for a fixed input to the flashlamp. The ability of alexandrite lasers to sustain high gain

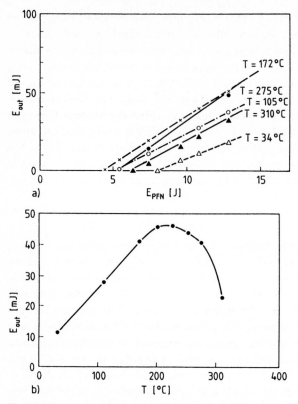

Fig. 2.24. Temperature dependence of alexandrite **(a)** Alexandrite laser energy as a function of pump energy at temperatures from 34 to 310° C. **(b)** Alexandrite laser output energy as a function of temperature for fixed 10.8-J flashlamp output [2.152]

and efficiency at temperatures above those normally encountered in solid-state lasers is particularly striking.

Alexandrite Laser Performance. The development of the alexandrite laser has reached maturity after almost 20 years of efforts. Its high average-power performance is about 100 W if operated at 250 Hz [2.153]. Overall efficiency is close to 0.5 %. Tunability over the range of approximately 700 to 818 nm has been demonstrated with tuning accomplished in a manner similar to dye lasers: a combination of etalons and birefringent filters. With these standard spectral control devices 0.5 cm^{-1} linewidths and tunability over 150 nm has been achieved. Alexandrite has been lased in pulsed and cw modes; it has been Q-switched and mode locked.

A rod 10 cm long and 0.63 cm in diameter, when lased in a stable resonator, yields over 5 J long-pulsed, and as much as 2 J with pulse duration less than 30 ns when Q-switched [2.152]. The reason for such high output energies is the fact that alexandrite is a low-gain medium. Highly stable frequency locking (without loss of bandwidth control) was achieved with an injection power ten orders of magnitude smaller than the oscillator output [2.154].

2.5.2 Ti : Sapphire

Since laser action was first reported [2.137], the Ti : Al$_2$O$_3$ laser has been the subject of extensive investigations and today it is the most widely used tunable solid-state laser. The Ti : sapphire laser combines a broad tuning range of about 400 nm with a relatively large gain cross-section which is half of Nd : YAG at the peak of its tuning range. The energy level structure of the Ti^{3+} ion is unique among transition-metal laser ions in that there are no d-state energy levels above the upper laser level. The simple energy-level structure ($3d^1$ configuration) eliminates the possibility of excited-state absorption of the laser radiation, an effect which has limited the tuning range and reduced the efficiency of other transition-metal-doped lasers [2.155].

In this material, a Ti^{3+} ion is substituted for an Al^{3+} ion in Al$_2$O$_3$. Laser crystals, grown by the Czochralski method, consist of sapphire doped with 0.1% Ti^{3+} by weight. Crystals of Ti : Al$_2$O$_3$ exhibit a broad absorption band, located in the blue-green region of the visible spectrum with a peak around 490 nm. A relatively weak absorption band is observed in the ir region which has been shown to be due Ti^{3+} - Ti^{4+} pairs. This residual ir absorption interferes with efficient laser operation, particularly in the case of flashlamp pumping. Optimized crystal growth techniques and additional annealing processes have drastically reduced this absorption band compared to earlier crystals. The great interest in this material arises from the broad vibronic fluorescence band which allows tunable laser output between 670–1070 nm, with the peak of the gain curve around 800 nm.

The absorption and fluorescence spectra for Ti : Al$_2$O$_3$ are shown in Fig. 2.25. The broad, widely separated absorption and fluorescence bands are caused by the strong coupling between the ion and host lattice, and are the key to broadly tunable laser operation. The laser parameters of Ti : Al$_2$O$_3$ are listed in Table 2.12.

Besides having favorable spectroscopic and lasing properties, one other advantage of Ti : Al$_2$O$_3$ are the material properties of the sapphire host itself, namely very high thermal conductivity, exceptional chemical inertness and mechanical rigidity. Titanium sapphire is available from commercial vendors in sizes of 3.5 cm diameter by 15 cm long and, due to the well-developed growth technology for sapphire, of good optical quality.

Ti : sapphire lasers have been pumped with a number of sources such as argon and copper vapor lasers, frequency doubled Nd : YAG and Nd : YLF lasers, as well as flashlamps.

Flashlamp pumping is very difficult to achieve in Ti : sapphire because a very high pump flux is required. The reason for that is the short fluorescence lifetime of 3.2 μm which results in a small product of stimulated emission cross-section times fluorescence lifetime (σt_f). The population inversion in a laser required to

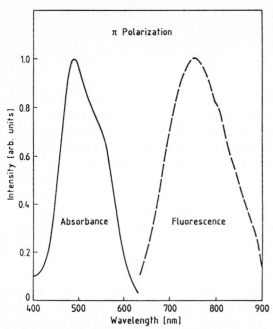

Fig. 2.25. Absorption and fluorescence spectra of the Ti^{3+} ion in Al_2O_3 (sapphire) [2.155]

achieve threshold is inversely proportional to σt_f as will be shown in Sect. 3.1. However, improvements in crystal quality which lead to the removal of residual absorption bands, in combination with special flashlamps have resulted in output energies of 3 J per pulse at 2% efficiency. The flashlamps designed for high wall loadings were spectrally enhanced by using a dye surrounding the laser rod to convert near-uv light from the flashlamp into blue-green fluorescence which is within the absorption band of Ti : sapphire [2.156]. With design improvements in the lamp-discharge circuit and the pump cavity, an average power of 220 W at an efficiency of 2.2% has been achieved recently [2.157].

Table 2.12. Laser parameters of Ti : Al_2O_3

Index of refraction	$n = 1.76$
Fluorescent lifetime	$\tau = 3.2\,\mu s$
Fluorescent linewidth (FWHM)	$\Delta\lambda \sim 180\,nm$
Peak emission wavelength	$\lambda_p \sim 790\,nm$
Peak stimulated emission cross section	
parallel to c axis	$\sigma_{p\parallel} \sim 4.1 \times 10^{-19}\,cm^2$
perpendicular to c axis	$\sigma_{p\perp} \sim 2.0 \times 10^{-19}\,cm^2$
Stimulated emission cross section	
at $0.795\,\mu m$ ($\parallel c$ axis)	$\sigma_\parallel = 2.8 \times 10^{-19}\,cm^2$
Quantum efficiency of converting a $0.53\,\mu m$ pump photon into an inverted site	$\eta_Q \approx 1$
Saturation fluence at $0.795\,\mu m$	$E_{sat} = 0.9\,J/cm^2$

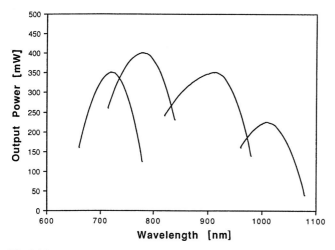

Fig. 2.26. Tuning range of a Ti : sapphire laser pumped by a Nd : YLF laser at 1 kHz [2.158]

Commercial Ti : sapphire lasers are pumped by argon lasers to obtain cw output, and by frequency-doubled Nd : YAG or Nd : YLF lasers for pulsed operation. In the cw mode, typical performance is close to 1 W output with a 5 W argon pump. Tuning ranges from about 700 nm to 1050 nm require several sets of cavity mirrors. For pulsed solid-state lasers as pump source, output energies range from a few mJ at repetition rates of around 1 kHz, to 100 mJ per pulse at 20 pps.

Figure 2.26 displays the oputput of a Ti : sapphire laser pumped at 1 kHz by a frequency-doubled Nd : YLF laser which had an average output of 1.7 W at 527 nm. The pump pulses had a duration between 200–300 ns, whereas the Ti : sapphire laser yielded output pulses of 10–12 ns duration. This pulse shortening is characteristic of gain switched operation. Recently, all solid-state Ti : sapphire lasers have been designed which employ diode-pumped, frequency-doubled Nd : YLF and Nd : YAG lasers as the pump source [2.159–161].

A very important application of Ti : sapphire lasers is the generation and amplification of femtosecond mode-locked pulses. Kerr-lens mode locking and chirped pulse amplification with Ti : sapphire lasers is discussed in Chap. 9.

2.5.3 Cr : LiSAF

A relatively new chromium host is Cr^{3+} : $LiSrAlF_6$ (LiSAF) which has a tuning range from 780–920 nm and an excited lifetime of 67 μs [2.162]. This material is very similar to Cr^{3+} : $LiCaAlF_6$ (LiCAF) which has a tuning range from 720–840 nm and a lifetime of 170 μs [2.163]. Since the peak emission of LiSAF is four times larger than LiCAF, it generally performs better, and most of the recent laser work has concentrated on LiSAF.

The LiSAF host crystal is uniaxial and the Cr^{3+} emission has been shown to be strongly π-polarized (E$\|$c). The absorption and emission spectra are displayed

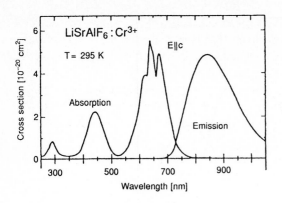

Fig. 2.27. Absorption and emission spectra of Cr:LiSAF [2.162]

Table 2.13. Comparison of relevant laser parameters for Cr:LiSAF and Ti:Sapphire

	Cr:LiSAF	Ti:Sapphire
Peak wavelength [nm]	850	790
Linewidth [nm]	180	230
Emission cross section [10^{-19} cm^2]	0.5	4.1
Fluorescence lifetime [μs]	67	3.2
Refractive Index	1.41	1.76
Scattering loss [cm^{-1}]	0.002	0

in Fig. 2.27 [2.162]. The peak of the $^4T_2 \rightarrow \, ^4A_2$ emission occurs at 830 nm and has a cross-section of 4.8×10^{-20} cm^2. Very large laser rods with diameters up to 25 mm have been fabricated from crystals grown by the Czochralski technique.

A comparison of important laser parameters of Cr:LiSAF with those of Ti:sapphire is shown in Table 2.13 [2.162, 163]. Peak emission of LiSAF is at a slightly longer wavelength as compared to Ti:sapphire, but there is a good overlap between the spectra. Actually, the output from mode-locked Ti:sapphire oscillators has been amplified by Cr:LiSAF amplifiers [2.164]. The gain bandwidth is narrower, but still comparable, to Ti:sapphire. The major differences between the two crystals is the emission cross-section, fluorescent lifetime and the thermal and mechanical properties. Whereas the gain of Cr:LiSAF is approximately one order of magnitude lower than that of Ti:sapphire, it has a long enough lifetime of 67 μs to permit efficient flashlamp pumping.

LiSAF is a rather soft and mechanically weak crystal with properties more related to glass than the far superior Ti:sapphire crystal. A comparison of the thermal and physical properties of LiSAF and glass is provided in Table 2.14 [2.165]. An important design consideration in using LiSAF crystals is the maintenance of the proper pH value of the cooling water for liquid-immersed laser rods. For example, at a pH value of 5, the material has a dissolution rate which is 100 times higher than YLF. However, at a pH value around 7, the dissolution rate is on the same order as YLF and phosphate glass. Unlike Ti : sapphire, flashlamp pumping of Cr : LiSAF is fairly straight-forward, due to the relatively long lifetime of the upper state level and the excellent overlap of the absorption bands with the emission of flashlamps. LiSAF can also be pumped with AlGaInP diodes at 670 nm [2.166, 167]. Laser action has even been achieved by pumping highly doped crystals at the wings of the absorption profile at 752 nm with AlGaAs diodes [2.168].

Table 2.14. Comparison of thermal and physical properties of LiSAF and glass

	Cr:LiSAF	Glass
Thermal shock resistance [W/m$^{1/2}$]	≈ 0.4	≈ 0.4
Fracture strength [kg/mm^2]	3.9	5
Thermal expansion coefficient ($\times 10^{-6}$/°C)	22	11.4
Young's modulus [Gpa]	100	50
Microhardness [kg/mm^2]	197	≈ 500
Fracture toughness [MPam$^{1/2}$]	0.4	0.45
Thermal conductivity [Wm^{-1}K^{-1}]	3.09	0.62

Cr : LiSAF has found applications as a flashlamp or diode-pumped laser source with tunable output around 850 nm [2.169]. The broadband emissions of Cr : LiSAF makes this crystal attractive for the generation and amplification of femtosecond mode-locked pulses. Of particular interest is a diode-pumped, all solid-state tunable source for femtosecond pulse generation. As will be discussed in Chap. 9, systems have been developed which range from small diode-pumped Cr : LiSAF mode-locked oscillators [2.170], to very large flashlamp-pumped amplifier stages with rod diameters up to 25 mm [2.164].

2.5.4 Tm : YAG

Thulium and thulium sensitized holmium lasers have outputs in the 2 μm region, a wavelength of interest for coherent radar systems, remote sensing and medical applications. The possibility of pumping Tm-doped crystals with readily available powerful GaAlAs laser diodes at 785 nm, has stimulated interest in these materials. Representative crystals of this class of three-level lasers are Tm : YAG [2.171, 173–176], Tm : YAP [2.172], Ho : Tm : YAG [2.47] and Ho : Tm : YLF [2.178–180]. In the singly-doped crystals, laser radiation is produced by the Tm^{3+} ion at 2 μm. In the doubly-doped crystals, pump radiation is absorbed by the thulium ion, and by a cross-relaxation process, energy is transferred to the holmium ion. Laser emission occurs at 2.1 μm between the $^5I_7$ and $^5I_8$ levels in holmium.

As an example of these tunable lasers in the 2 μm region, we will consider diode-pumped Tm : YAG. It seems to be the preferred crystal for a number of applications. Although Ho : Tm : YAG has a larger cross-section as compared to Tm : YAG, it generally has poorer output performance because of increased losses

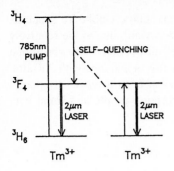

Fig. 2.28. Energy levels and pumping diagram for Tm : YAG [2.175]

introduced by up conversion from the upper laser level [2.71, 172, 174]. However, quite recently, impressive performance levels at room temperature have been obtained in Ho : Tm : YLF with output energies up to 125 mJ at single frequency [2.178]. This crystal has lower up conversion losses as compared to Ho : Tm : YAG. Also the extremely long lifetime of the upper laser level of 14 ms provides high energy storage capability for Q-switch operation at low repetition rates.

In the previous section, tunability of solid-state lasers was achieved by the 3d electron of the transition metal ions Cr^{3+} and Ti^{3+}. In Tm^{3+} : YAG, the considerable phonon broadening and high multiplicity of the Stark levels of the 4f electron provides relative broad tunability around $2\,\mu$m.

The relevant energy levels and transitions of Tm : YAG are shown in Fig. 2.28. Pump radiation at 785 nm transfers Tm^{3+} ions from the $^3H_6$ ground state into the $^3H_4$ level of Tm. From the $^3H_4$ pump level the ions relax down to the upper laser level $^3F_4$. Laser action takes place between level $^3F_4$ and the lower laser level, which lies within the $^3H_6$ ground-state manifold. Output radiation is around $2\,\mu$m.

At high Tm concentrations, cross-relaxation between adjacent ions can occur, whereby the excitation is undergoing the transition $^3H_4 \rightarrow {}^3F_4$, while an unexcited ion simultaneously undergoes the transition $^3H_6 \rightarrow {}^3F_4$. This process leaves two Tm^{3+} ions in the upper laser level $^3F_4$ for every ion originally excited. In other words, two ions in the upper laser level are generated for every pump photon absorbed. At typical concentration levels of (3–12)% Tm^{3+} the quantum efficiency of pumping into the $^3F_4$ state is nearly two.

The fluorescence spectrum of Tm : YAG is shown in Fig. 2.29. The peak emission is at $2.02\,\mu$m and the bandwidth of fluorescence is about 400 nm. Tunability is typically from 1.87 to $2.16\,\mu$m. The absorption spectrum of Tm : YAG is

Table 2.15. Material parameters of Tm : YAG [2.175, 177]

Pump wavelength	780–785 nm
Peak laser wavelength	$2.02\,\mu$m
Effective cross section at 25°C	$2 \times 10^{-21}\,\mathrm{cm}^2$
Fluorescence lifetime	10 ms
Tunability	$1.87–2.16\,\mu$m

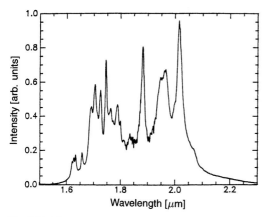

Fig. 2.29. Fluorescence spectrum for Tm : YAG [2.175]

centered at 785 nm and has a linewidth of 4 nm which makes pumping attractive with GaAlAs diodes. Pertinent laser parameters are summarized in Table 2.15 [2.177]. Compared to Nd : YAG, the disadvantages of Tm : YAG are the higher threshold to overcome the residual population in the ground level and the two orders of magnitude smaller gain cross-section. Since the lower laser level is in the ground-state manifold and therefore, partially populated, quasi three-level lasers require stronger pump intensities or they require cooling to reduce the population in the lower laser level.

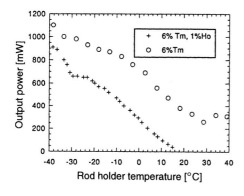

Fig. 2.30. Output power of Tm : YAG and Tm : Ho : YAG as a function of temperature [2.177]

The rather dramatic improvement in laser performance when the Tm : YAG crystal is operated at reduced temperature is illustrated in Fig. 2.30. Also shown is the output dependence of Ho : Tm : YAG on temperature. Because of the more complete three-level nature of this material, the temperature dependence is even stronger than in Tm : YAG.

Instead of pumping Tm : YAG at the peak absorption line, sometimes "wing pumping" at 805 nm is employed. The lower absorption coefficient at that wavelength provides a longer pathlength in the crystal which is advantageous in high power lasers for better heat removal and reduced thermal loading. Also, pump diode performance and reliability is improved because at the longer wavelength the aluminum concentration in GaAlAs is lower. In an end-pumped 3 mm diameter by 55 mm long Tm:YAG rod, a cw output of 115 W was achieved by pumping at 805 nm with 460 W of cw pump power [2.173]. The output was achieved by maintaining the crystal temperature at 3°C.

2.6 Yb:YAG

Although Yb:YAG has been known for decades, interest was not very high in this material since it lacks any pump bands in the visible. The crystal has only a single absorption feature around 942 nm. With conventional pump sources such as flashlamps, the threshold is very high which eliminated this material from any serious considerations. With the emergence of powerful InGaAs laser diodes which emit at 942 nm several researchers recognized the potential of Yb:YAG [2.181]. The relevant energy level diagram of Yb:YAG is very simple and consists of the $^2F_{7/2}$ ground state and $^2F_{5/2}$ excited state manifolds separated by about 10,000 cm^{-1}. The laser wavelength is at 1.03 μm, a representative emission spectrum for a 5.5 atomic percent doped sample is shown in Fig. 2.31.

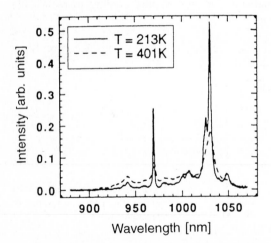

Fig. 2.31. Emission spectra of Yb:YAG with 5.5 at % doping [2.182]

The laser transition is $^2F_{5/2}$ to $^2F_{7/2}$ with the terminal level 623 cm^{-1} above the ground state. The thermal energy at room temperature is 200 cm^{-1} therefore the terminal state is thermally populated which makes Yb:YAG a quasi three level system. By comparison, the terminal laser level in Nd:YAG is about 2000 cm^{-1} above the ground state. Being a quasi three-level laser Yb:YAG absorbs at 1.03 μm unless it is pumped to inversion. At room temperature the thermal population of the lower laser level is about 5.5%. The absorbed pump power per volume needed to achieve and maintain transparency at the laser wavelength is $I = f_a n_t h\nu_p / \tau_f$ where f_a is the fraction of the total ion density n_t occupying the lower laser level, $h\nu_p$ is the energy per pump photon, τ_f is the lifetime of the upper level. With $f_a = 0.055$, $n_t = 1.38 \times 10^{20}\,\text{cm}^{-3}$ at 1% doping, $h\nu_p = 2.11 \times 10^{-19}\,\text{J}$ and $\tau_f = 0.95\,\text{ms}$ the absorbed pump power needed to reach inversion is $1.7\,\text{kW/cm}^3$. Of course, a higher power density is required to overcome optical losses and reach laser threshold, and for an efficient operation the laser has to be pumped about (5–6) times above threshold. Typically, in this laser, small volumes of material are pumped on the order of $10\,\text{kW/cm}^3$.

Yb : YAG performance is strongly dependent on temperature and can be improved by cooling the crystal which reduces the thermal population and increases the stimulated emission cross-section from $2.1 \times 10^{-20}\,\text{cm}^2$ at room temperature to twice this value at 220 K [2.182].

Figure 2.32 illustrates the dependence of optical efficiency, i.e. the conversion of pump light into laser output, on temperature [2.183]. In this laser a thin disc, about 0.4 mm thick, is pumped with fiber-coupled laser diodes in a multiple-path arrangement. The Yb : YAG crystal is mounted on a heat sink which could be cooled to low temperature. Two different pump sources were evaluated. With the smaller diode array the efficiency increased from 45% at room temperature to 64% at $-74°$C. With the larger pump source the efficiency changes from 38% at room temperature to 55% at $-65°$C. At this low temperature the output was 100 W for the 183 W pump. At the more typical operating temperature of $-9°$C for this laser, a pump power of 225 W was needed to generate the same output. Clearly, laser efficiency and cost of the pump source can be traded for overall system efficiency which includes power and cost for refrigeration. Quite noticeable is the high optical-to-optical efficiency which can be achieved in these lasers.

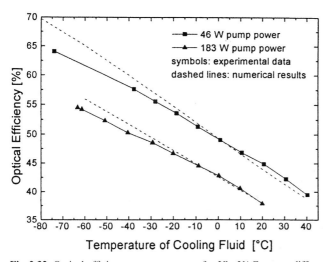

Fig. 2.32. Optical efficiency vs. temperature for Yb : YAG at two different pump powers [2.183]

In most of the large systems built to date the temperature of the crystal is maintained between $-10°C$ and $20°C$ depending on the particular design and mode of operation. To maintain the laser crystal at such a temperature under intensive pump radiation requires, as a minimum, some kind of refrigeration.

Despite these obvious drawbacks, Yb:YAG has a number of redeeming features which motivated the development of very powerful systems.

Pumping of Yb:YAG with an InGaAs pump source produces the smallest amount of crystal heating compared to any other major laser system. Actually the pump radiation in this material generates only about one-third of the heat compared to Nd:YAG. The fractional thermal loading, i.e. the ratio of heat generated to absorbed energy is around 11% for Yb:YAG pumped at 943 nm and 32% for Nd:YAG pumped at 808 nm [2.184]. This substantially reduced thermal dissipation is the result of a very small energy difference between the photons of the pump and laser radiation. This quantum defect or Stokes shift, is 9% in Yb:YAG vs. 24% in Nd:YAG. The thermal load generated in a laser medium is of primary concern for high-power applications. The reduced thermal heat load can potentially lead to higher-average-power systems with better beam quality than possible with Nd:YAG. Other unique advantages of Yb:YAG are an absorption bandwidth of 18 nm [2.185] for diode pumping which is about 10 times broader than the 808-μm absorption in Nd:YAG. This significantly relaxes temperature control needed for the diode pump source. Yb:YAG has a long lifetime of 951 μs [2.182], which reduces the number of quasi-cw pump diodes required for a given energy per pulse output. As will be discussed in Chap. 6, laser diodes are power limited, a long fluorescence time permits the application of a long pump pulse which, in turn, generates a high energy output. Important materials parameters of Yb:YAG are listed in Table 2.16.

Table 2.16. Pertinent material parameters for Yb:YAG [2.185]

Laser wavelength	1030 nm
Radiative lifetime at room temperature	951 μs
Peak emision cross section	2.1×10^{-20} cm^2
Peak absorption wavelength	942 nm
Pump bandwidth at 942 nm	18 nm
Pump cross section at 942 nm	7.7×10^{-21} cm^2
Doping density (1% at.)	1.38×10^{20} cm^3

The presence of only one excited-state manifold eliminates problems associated with excited-state absorption and up-conversion processes. Despite the quasi three-level behavior of Yb:YAG, the obvious advantages of very low fractional heating, broad absorption at the InGaAs wavelength, long lifetime combined with high conductivity and tensile strength of the host crystal have raised the prospect of high power generation with good beam quality.

The major interest lies in large systems at the kW level for industrial applications. A number of crystals have been doped with Yb, in particular apatite crystals, but YAG is favored for power scaling because it has excellent thermomechanical properties. The unique properties of Yb : YAG have resulted in a number of different pumping schemes, such as end and side pumping of a small rod, or face pumping of a thin disc, which have all generated outputs in the multi-hundred watt range.

In the end-pumped configuration, the output radiation from a stack of diode bars is collimated and funneled into the front end of a 2 mm diameter and 60 mm long rod [2.186]. The crystal is doped at a low concentration of 0.44% such that the absorption path is long and heat removal can take place along the barrel of the rod. The crystal temperature was maintained at 0°C with a water–alcohol mixture. Figure 2.33 illustrates the performance of the system, in which up to 434 W in cw mode was achieved.

In the side pumped approach a 2 mm diameter and 20 mm long crystal doped at 1% Yb produced up to 600 W output in a quasi-cw mode, i.e. 460 mJ pulses with a 0.8 ms pulse duration [2.187, 188]. The rod temperature was held at 18°C by liquid cooling of the rod surfaces.

Extremely small volumes, on the order of a few mm^3, are pumped in the thin-disc approach [2.183, 189, 190]. In this design, discs with a thickness of typically (0.3–0.4) mm are pumped from the front in a multiple pass active mirror design. The discs, mounted on a cold finger, are cooled from the backside. The pump beam is typically (2.5–7) mm in diameter. The discs are heavily doped up to 11%. At a temperature of 1°C an output power of 264 W was measured for one particular set-up. Electrical input power to the diodes was 2060 W which gives an electrical to optical efficiency of 12.8%.

In general, the efficiency of quasi three level Yb : YAG lasers is comparable to the overall efficiency obtained in large Nd : YAG systems, however, the much reduced heat load in Yb : YAG permits generation of comparable outputs in much smaller laser crystals.

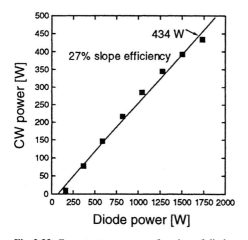

Fig. 2.33. Cw output power as a function of diode power of an end-pumped Yb : YAG [2.186]

3. Laser Oscillator

In Chap. 1 we studied the processes which lead to optical amplification in laser materials. The regenerative laser oscillator is essentially a combination of two basic components: an optical amplifier, and an optical resonator. The optical resonator, comprised of two opposing plane-parallel or curved mirrors at right angles to the axis of the active material, performs the function of a highly selective feedback element by coupling back in phase a portion of the signal emerging from the amplifying medium.

Figure 3.1 shows the basic elements of a laser oscillator. The pump lamp inverts the electron population in the laser material, leading to energy storage in the upper laser level. If this energy is released to the optical beam by stimulated emission, amplification takes place. Having been triggered by some spontaneous radiation emitted along the axis of the laser, the system starts to oscillate if the feedback is sufficiently large to compensate for the internal losses of the system. The amount of feedback is determined by the reflectivity of the mirrors. Lowering the reflectivity of the mirrors is equivalent to decreasing the feedback factor. The mirror at the output end of the laser must be partially transparent for a fraction of the radiation to "leak out" or emerge from the oscillator.

An optical structure composed of two plane-parallel mirrors is called a Fabry-Pérot resonator. In Chap. 5 we will discuss the temporal and spatial mode structures which can exist in such a resonator. For the purpose of this discussion it is sufficient to know that the role of the resonator is to maintain an electromagnetic field configuration whose losses are replenished by the amplifying medium through induced emission. Thus, the resonator defines the spectral, directional,

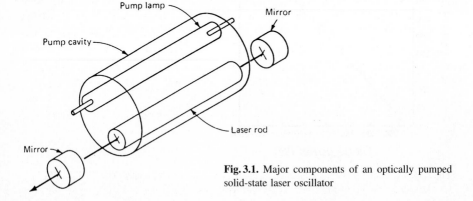

Fig. 3.1. Major components of an optically pumped solid-state laser oscillator

and spatial characteristics of the laser radiation, and the amplifying medium serves as the energy source.

In this chapter we will develop an analytical model of a laser oscillator that is based mainly on laser system parameters.

3.1. Operation at Threshold

We will calculate the threshold condition of a laser oscillator composed of two mirrors having the reflectivities R_1 and R_2, and an active material of length l. We assume a gain coefficient per unit length of g in the inverted laser material. In each passage through the material the intensity gains by a factor of $\exp(gl)$. At each reflection a fraction $1 - R_1$ or $1 - R_2$ of the energy is lost. Starting at one point, the radiation will suffer two reflections before it can pass the same point in the original direction. The threshold condition is established by requiring that the photon density – after the radiation has traversed the laser material, been reflected by mirror with R_1, and returned through the material to be reflected by mirror with R_2 – be equal to the initial photon density. Then on every complete two-way passage of the light through the laser the loss will just equal the gain. We can express the threshold condition by

$$R_1 R_2 \exp(2gl) = 1 \ . \tag{3.1}$$

The regenerative amplifier becomes unstable when the amplification per transit exceeds the losses. In this case oscillations will build up, starting from a small disturbance. Clearly, if the round-trip gain

$$G = R_1 R_2 \exp(2gl) \tag{3.2}$$

is larger than 1, radiation of the proper frequency will build up rapidly until it becomes so large that the stimulated transitions will deplete the upper level and reduce the value of g. The condition of steady state is reached if the gain per pass exactly balances the internal and external losses. This process, called gain saturation, will be discussed in Sect. 3.2. In an oscillator a number of loss mechanisms are responsible for attenuating the beam; the most important ones are reflection, scattering, and absorption losses in the mirrors, the amplifying medium, and all other elements in the resonator, and diffraction losses.

We will find it convenient to lump all of the dissipative losses into a single parameter, the absorption coefficient per unit length α. The condition for oscillation is then

$$R_1 R_2 \exp(g - \alpha)2l = 1 \ . \tag{3.3}$$

In Sect. 1.4 we characterized all the loss mechanisms by a single parameter τ_c which is equal to the decay time constant of the radiation in a passive resonator.

90   3. Laser Oscillator

Resonators are characterized by the quality factor Q, which is defined as the ratio of energy stored in the resonator to power dissipated from the resonator per unit angular frequency ω_0. The resonator Q defined in this way is equal to

$$Q = 2\pi \left[1 - \exp\left(\frac{-T_0}{\tau_c}\right)\right]^{-1} \approx \frac{2\pi\tau_c}{T_0} = 2\pi\nu_0\tau_c ,$$
(3.4)

where $\omega_0 = 2\pi\nu_0 = 2\pi/T_0$.

The loss mechanism, besides limiting the lifetime of the oscillation, causes a broadening of the resonance frequency. The width $\Delta\nu$ of the resonance curve at which the intensity has fallen off to half the maximum value is

$$\Delta\nu = (2\pi\tau_c)^{-1} .$$
(3.5)

If we introduce this expression into (3.4) we obtain for the Q value

$$Q = \frac{\nu_0}{\Delta\nu} .$$
(3.6)

The decay time constant of the radition τ_c can also be defined as the average lifetime of the photons in the resonator. A photon in the resonator will have some average lifetime before being scattered or emitted or lost in other ways to the optical system. If we relate τ_c to the fractional power loss ε per round trip, we obtain

$$\varepsilon = \frac{t_R}{\tau_c} ,$$
(3.7)

where $t_R = 2l'/c$ is the round-trip time of a photon in a resonator having an optical length l'. Rearranging (3.3) yields

$$2gl = -\ln R_1 R_2 + 2\alpha l .$$
(3.8)

The expression on the right is the total fractional power loss per round trip. Since $2gl = \varepsilon = t_R/\tau_c$, we obtain

$$\tau_c = \frac{2l'}{c}(-\ln R_1 R_2 + 2\alpha l)^{-1} .$$
(3.9)

Miscellaneous losses, such as absorption and scattering at the mirrors and diffraction losses of the resonator, can be thought of as leakage from the rear mirror. Hence the reduced reflectivity R_2 of the rear mirror $R_2 = 1 - L_M$ takes into account the miscellaneous losses. In practice, L_M does not exceed a few percent. With the approximation

$$\ln(1 - L_M) \approx -L_M ,$$
(3.10)

one can combine the optical losses in the resonator with the losses in the crystal:

$$L = 2\alpha l + L_{\mathrm{M}} \;. \tag{3.11}$$

where L is the two-way loss in the resonator.

With this approximation (3.9) reduces to

$$\tau_{\mathrm{c}} = \frac{2l'}{c(L - \ln R_1)} \;. \tag{3.12}$$

In a typical pulsed laser, the transmission of the output mirror is around 50 % and the combined losses are around 10 %. If we assume a typical resonator length of 50 cm, we obtain a photon lifetime of $\tau_{\mathrm{c}} = 5.5$ ns. In a continuous Nd : YAG laser, transmission of the output mirror is typically 90 %, therefore $\tau_{\mathrm{c}} \approx 17$ ns, all other parameters being the same.

With the aid of (3.10, 11) we can express the threshold condition (3.3) in the following form:

$$2gl = L - \ln R_1 \approx T + L \;. \tag{3.13}$$

The approximation $-\ln R_1 \approx T$ is valid only for values of R_1 close to one.

We turn now to the rate equation (1.61), which gives the photon density in the amplifying medium. It is clear from this equation that for onset of laser emission the rate of change of the photon density must be equal to or greater than zero. Thus at laser threshold for sustained oscillation the condition

$$\frac{\partial \phi}{\partial t} \geq 0 \tag{3.14}$$

must be fulfilled, which enables us to obtain from (1.61) the required inversion density at threshold,

$$n \geq \frac{1}{c\sigma \tau_{\mathrm{c}}} \;. \tag{3.15}$$

In deriving this expression we have ignored the factor S, which denotes the small contribution from spontaneous emission to the induced emission. The reader should note that by introducing $g(\nu_{\mathrm{s}}) = -\alpha(\nu_{\mathrm{s}}) = n\sigma_{21}(\nu_{\mathrm{s}})$ according to (1.36) and expressing τ_{c} by (3.9), this equation is identical to the threshold condition (3.3).

We may write the threshold condition in terms of the fundamental laser parameters. Upon substitution of $\sigma_{21}(\nu_{\mathrm{s}})$ from (1.40), we obtain

$$n_2 - \frac{g_2 n_1}{g_1} > \frac{\tau_{21} 8\pi \nu^2}{\tau_{\mathrm{c}} c^3 g(\nu_{\mathrm{s}}, \nu_0)} \;. \tag{3.16}$$

The lineshape factor $g(\nu_s, \nu_0)$ and therefore the stimulated emission cross section σ are largest at the center of the atomic line. Thus from (3.16) we can see qualitatively how the linewidth of the laser output is related to the linewidth of the atomic system. Self-sustained oscillation which develops from noise will occur in the neighborhood of the resonant frequency, because only at a narrow spectral range at the peak will the amplification be large enough to offset losses. Consequently, the output of the laser will be sharply peaked, and its linewidth will be much narrower than the atomic linewidth.

It is also obvious from this equation that an increase of the inversion by stronger pumping will increase the laser linewidth because the threshold condition can now be met for values of $g(\nu_s, \nu_0)$ farther away from the center. As we will see in Chap. 5 the linewidth of an actual laser system is related to the linewidth of the active material, the level of pump power, and the properties of the optical resonator. The threshold condition at the center of the atomic line is obtained by introducing the peak values of the amplification curve into (3.16). If $g(\nu_s, \nu_0)$ has a Lorentzian shape with full width at half-maximum of $\Delta\nu$ centered about ν_s, then $g(\nu_0) = 2/\pi\Delta\nu$ and

$$n_2 - \frac{g_2 n_1}{g_1} > \frac{\tau_{21} 4\pi^2 \Delta\nu\nu_0^2}{\tau_c c^3} . \tag{3.17}$$

For a Gaussian lineshape, $g(\nu_0) = 2(\pi \ln 2)^{1/2}/\pi\Delta\nu$, and the start-oscillation condition is still given by (3.17) with $\Delta\nu$ replaced by $\Delta\nu/(\pi \ln 2)^{1/2}$. Again, we have assumed that the laser threshold will be reached first by a resonator mode whose resonant frequency lies closest to the center of the atomic line.

From (3.17) we can infer those factors favoring high gain and low threshold for a laser oscillator. In order to achieve a low threshold inversion, the atomic linewidth $\Delta\nu$ of the laser material should be narrow. Furthermore, the incidental losses in the laser cavity and crystal should be minimized to increase the photon lifetime τ_c. It is to be noted that the critical inversion density for threshold depends only on a single resonator parameter, namely τ_c. A high reflectivity of the output mirror will increase τ_c and therefore decrease the laser threshold. However, this will also decrease the useful radiation coupled out from the laser. We will address the question of optimum output coupling in Sect. 3.4.

We will now calculate the pumping rate W_p which is required to maintain the oscillator at threshold. For operation at or near threshold the photon density ϕ is very small and can be ignored. Setting $\phi = 0$ in the rate equation (1.58) and assuming a steady-state condition of the inversion, $\partial n/\partial t = 0$, as is the case in a conventional operation of the laser oscillator, we obtain for a three-level system

$$\frac{n}{n_{\text{tot}}} = \frac{W_p \tau_{21} - g_2/g_1}{W_p \tau_{21} + 1} , \tag{3.18}$$

and for a four-level system from (1.70)

$$\frac{n_2}{n_0} = \frac{W_p \tau_f}{W_p \tau_f + 1} \approx W_p \tau_f \,. \tag{3.19}$$

Other factors being equal, four-level laser systems have lower pump-power thresholds than three-level systems. In a four-level system an inversion is achieved for any finite pumping rate W_p. In a three-level system we have the requirement that the pumping rate W_p exceeds a minimum or threshold value given by

$$W_{p(th)} = \frac{g_2}{\tau_{21} g_1} \tag{3.20}$$

before any inversion at all can be obtained. Whereas for a four-level material the spontaneous lifetime has no effect on obtaining threshold inversion, in a three-level material the pump rate required to reach threshold is inversely proportional to τ_{21}. Thus, for three-level oscillators only materials with long fluorescence lifetimes are of interest.

The reader is reminded again that (3.18, 19) are valid only for a negligible photon flux ϕ. This situation occurs at operation near threshold; it will later be characterized as the regime of small-signal amplification. We will now calculate the minimum pump power which has to be absorbed in the pump bands of the crystal to maintain the threshold inversion. This will be accomplished by first calculating the fluorescence power at threshold, since near above threshold almost all the pump power supplied to the active material goes into spontaneous emission. The fluorescence power per unit volume of the laser transition in a four-level system is

$$P_f = \frac{h\nu n_{th}}{\tau_f} \,. \tag{3.21}$$

where $n_2 = n_{th}$ is the inversion at threshold.

In a three-level system at threshold, $n_2 \approx n_1 \approx n_{tot}/2$ and

$$P_f \approx \frac{h\nu n_{tot}}{2\tau_{21}} \,. \tag{3.22}$$

In order that the critical inversion is maintained, the loss by fluorescence from the upper laser level must be supplied by the pump energy. As a result, we obtain for the absorbed pump power P_{ab} needed to compensate for population loss of the laser level by spontaneous emission

$$P_{ab} = \frac{\nu_p P_f}{\nu_L \eta_Q} = \frac{P_f}{\eta_u} \,. \tag{3.23}$$

The factor ν_p/ν_L represents the ratio of the photon energy at the pump-band and the laser wavelength and η_Q is the quantum efficiency as defined in (1.56

and 73). The difference between the pump power and the fluorescence power represents the thermal power which is released to the lattice of the crystal.

3.2 Gain Saturation

In the previous section we considered the conditions for laser threshold. Threshold was characterized by a steady-state population inversion, i.e., $\partial n/\partial t = 0$ in the rate equations. In doing this we neglected the effect of stimulated emission by setting $\phi = 0$. This is a good assumption at threshold, where the induced transitions are small compared with the number of spontaneous processes.

As the threshold is exceeded, however, stimulated emission and photon density in the resonator build up. Far above threshold we have to consider a large photon density in the resonator. From (1.58) we can see that $\partial n/\partial t$ decreases for increasing photon density. Steady state is reached when the population inversion stabilizes at a point where the upward transitions supplied by the pump source equal the downward transitions caused by stimulated and spontaneous emission. With $\partial n/\partial t = 0$ one obtains for the steady-state inversion population in the presence of a strong photon density ϕ

$$n = n_{\text{tot}} \left(W_{\text{p}} - \frac{\gamma - 1}{\tau_{\text{f}}} \right) \left(\gamma c \sigma \phi + W_{\text{p}} + \frac{1}{\tau_{\text{f}}} \right)^{-1}. \tag{3.24}$$

The photon density ϕ is given by the sum of two beams travelling in opposite directions through the laser material.

We will now express (3.24) in terms of operating parameters. From Sect. 1.3 we recall that the gain coefficient $g = -\alpha$ is defined by the product of stimulated emission and inversion population. Furthermore, we will define a gain coefficient which the system would have at a certain pump level in the absence of stimulated emission. Setting $\phi = 0$ in (3.24), we obtain the small-signal gain coefficient

$$g_0 = \sigma_{21} n_{\text{tot}} [W_{\text{p}} \tau_{\text{f}} - (\gamma - 1)](W_{\text{p}} \tau_{\text{f}} + 1)^{-1} \tag{3.25}$$

which an active material has when pumped at a level above threshold and when lasing action is inhibited by blocking the optical beam or by removing one or both of the resonator mirrors. If feedback is restored, the photon density in the resonator will increase exponentially at the onset with g_0. As soon as the photon density becomes appreciable, the gain of the system is reduced according to

$$g = g_0 \left(1 + \frac{\gamma c \sigma_{21} \phi}{W_{\text{p}} + (1/\tau_{\text{f}})} \right)^{-1} \tag{3.26}$$

where g is the saturated gain coefficient. Equation (3.26) was obtained by introducing (3.25) into (3.24) and using $g = \sigma_{21} n$.

We can express ϕ by the power density I in the system. With $I = c\phi h\nu$ we obtain

$$g = \frac{g_0}{1 + I/I_s} ,\tag{3.27}$$

where

$$I_s = \left(W_p + \frac{1}{\tau_f}\right) \frac{h\nu}{\gamma\sigma_{21}} .\tag{3.28}$$

The parameter I_s defines a flux in the active material at which the small-signal gain coefficient g_0 is reduced by one-half [3.1].

In a four-level system $W_p \ll 1/\tau_f$ and $\gamma = 1$, so (3.28) reduces to

$$I_s = \frac{h\nu}{\sigma_{21}\tau_f} .\tag{3.29}$$

For a three-level system the saturation flux is

$$I_s = \frac{h\nu[W_p + (1/\tau_{21})]}{\sigma_{21}[1 + (g_2/g_1)]} .\tag{3.30}$$

As we can see from (3.25), the small-signal gain depends only on the material parameters and the amount of pumping power delivered to the active material. The large-signal or saturated gain depends in addition on the power density in the resonator.

In a four-level system, a very interesting relationship between the stimulated emission lifetime τ_{st} and the saturation power density I_s can be obtained. Since the total number of downward transitions per second depends on the power density, the lifetime of decay for the excitation of the upper level will show a similar dependence. We may write for the total number of downward transitions per second

$$\frac{\partial n_2}{\partial t} = n_2 \left(\frac{1}{\tau_f} + \frac{1}{\tau_{st}}\right) ,\tag{3.31}$$

where τ_f is the fluorescence decay time of the upper laser level, and τ_{st} is the stimulated emission lifetime.

From (1.10, 51) it follows

$$\tau_{st} = \frac{h\nu}{\sigma_{21}I} ,\tag{3.32}$$

where I is the power density in the active material. Thus, as the excitation power is increased beyond the oscillation threshold, the portion of power going into stimulated emission increases. Comparing (3.29) and (3.32), we obtain

$$\tau_{st} = \tau_f \left(\frac{I_s}{I} \right) . \tag{3.33}$$

For a power density in the laser material which equals the saturation power density, the stimulated emission lifetime equals the fluorescence decay time of the upper laser level.

Gain saturation as a function of steady-state radiation intensity must be analyzed for lasers with homogeneous and inhomogeneous line broadening. Equation (3.27) is valid only for the former case, in which the gain decreases proportionately over the entire transition line. As we have seen in Chap. 2, a ruby laser has a homogeneously broadened bandwidth, whereas in Nd : glass the interaction of the active ion with the electrostatic field of the host leads to an inhomogeneous line. However, in solid-state materials such as Nd : glass, the cross-relaxation rate is very fast. The latter is associated with any process characteristic of the laser medium that affects the transfer of excitation within the atomic spectral line so as to prevent or minimize the departure of this line from the equilibrium distribution. It has been shown that in the case of a very fast cross-relaxation within the inhomogeneous line, the saturated gain is in agreement with that of a homogeneously broadened bandwidth [3.2].

3.3 Circulating Power

For a single-pass laser amplifier we can write the equation for energy density at each point (x) in the material:

$$\frac{dI(x)}{dx} = \frac{g_0 I(x)}{1 + I(x)/I_s} - \alpha I(x). \tag{3.34}$$

If we assume unsaturated operation, that is, $I(x) \ll I_s$, the single-pass gain is

$$G = \frac{I(l)}{I(0)} = \exp(g_0 - \alpha)l \tag{3.35}$$

for a laser amplifier of length l. Due to the loss coefficient α, the gain equation (3.34) becomes transcendental and hence only solvable numerically. This is a case which we will study in Chap. 4.

Assume that mirrors are placed at the ends of the laser rod. There are now two waves propagating through the amplifier in opposite directions, as shown in Fig. 3.2. The gain saturation is a function of the total power density in the medium. The calculations of the circulating power density involves the solution of two coupled differential equations for the two waves propagating through the gain medium. The result of this calculation is presented in [3.1, 3]. For the case where the transmission T of the output mirror is small, the results can be simplified. In this case

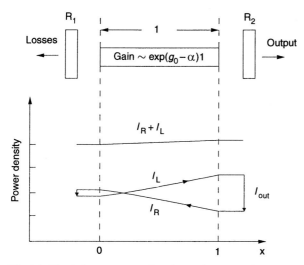

Fig. 3.2. Circulating power traveling from left to right (I_L) and right to left (I_R) in a laser oscillator

$$I_{\text{cir}} \approx I_L(x) \approx I_R(x) \ , \tag{3.36}$$

where I_L and I_R is the power density of the wave propagating to the left and right, respectively.

The power density in the gain medium, or the intracavity power density I_{int} is therefore

$$I_{\text{int}} = I_L(x) + I_R(x) \approx 2I_{\text{cir}} \ . \tag{3.37}$$

The saturated gain coefficient g is obtained by substituting I in (3.27) with I_{int}. Laser output and intracavity power density are related according to

$$I_{\text{int}} = \left(\frac{1+R}{1-R} \right) I_{\text{out}} \tag{3.38}$$

and the output power from the oscillator is

$$P_{\text{out}} = A \left(\frac{1-R}{1+R} \right) I_{\text{int}} \tag{3.39}$$

where A is the cross-section of the laser rod, and R is the reflectivity of the output coupler.

3.4 Oscillator Performance Model

In this section we will develop a model for the laser oscillator. First we will discuss the various steps involved in the conversion process of electrical input to laser output. After that, we will relate these energy transfer mechanisms to parameters which are accessible to external measurements of the laser oscillator. The purpose of this section is to gain insight into the energy conversion mechanisms and therefore provide an understanding of the dependency and inter relationship of the various design parameters which may help in the optimization of the overall laser efficiency. In almost all applications of lasers, it is a major goal of the laser designer to achieve the desired output performance with the maximum system efficiency.

3.4.1 Energy-Transfer Mechanisms

The flow of energy from electrical input to laser output radiation is illustrated schematically in Fig. 3.3. Also listed are the principal factors and design issues which influence the energy conversion process. There are different ways of partitioning this chain of transfer processes. This approach was chosen from an engineering point of view which devides the conversion process into steps related to individual system components. As shown in Fig. 3.3, the energy transfer from electrical input to laser output can conveniently be expressed as a four-step process:

Conversion of Electrical Input Delivered to the Pump Source to Useful Pump Radiation. We define as useful radiation, the emission from the pump source which falls into the absorption bands of the laser medium. The pump source efficiency η_P is therefore the fraction of electrical input power which is emitted as optical radiation within the absorption region of the gain medium. The output of a laser diode or a diode array represents all useful pump radiation provided the spectral output is matched to the absorption band of the gain medium. We will find it convenient to express the pump radiation of a laser diode by the slope efficiency η_P and threshold input energy E_{THD}. The electrical device efficiency

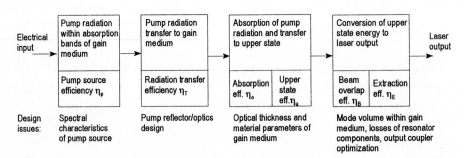

Fig. 3.3. Energy flow in a solid state laser system

of the laser diode will be expressed by η_{PS}. Typical values for commercially available cw and quasi-cw diode arrays are $\eta_P = 0.40$–0.50 and $\eta_{PS} = 0.30$–0.40. For flashlamp or cw arc lamp pumped systems, the pump source efficiency may be defined as

$$\eta_P = \int_{\lambda_1}^{\lambda_2} P_\lambda d\lambda / P_{EL} \tag{3.40}$$

where P_λ is the radiative power per unit wavelength emitted by the lamp, and the integral is taken over the wavelength range λ_1 to λ_2 which is useful for pumping the upper laser level. The output characteristics of arc lamps and their dependency on operating parameters will be discussed in Sect. 6.1. The measurement of η_P for broad-band sources is somewhat involved and requires either a calorimetric measurement of the power absorbed in a sample of the laser material, or integration over the source emission spectrum and the absorption spectrum of the gain material. Typical values are $\eta_P = 0.04$–0.08.

Transfer of the Useful Pump Radiation Emitted by the Pump Source to the Gain Medium. The transfer of flashlamp pump radiation to the laser medium is accomplished by means of a completely enclosed reflective chamber or pump cavity. The radiation transfer efficiency η_T can be defined as

$$P_T = \eta_T P_P \tag{3.41}$$

where P_P is the useful pump radiation emitted by the source, and P_T is the fraction of this radiation transferred into the laser material. The factor η_T is a combination of the capture efficiency, defined by the fraction of rays leaving the source and intersecting the laser rod, and the transmission efficiency. The former is based on the geometrical shape of the pump cavity, diameter and separation of the pump source and laser rod. The latter is a function of the reflectivity of the walls of the pump cavity, reflection losses at the rod surface and coolant jacket, absorption losses in the coolant fluid, and radiation losses through the clearance holes at the side walls of the pump cavity. Expressions and numerical values for η_T derived from ray trace analysis or Monte-Carlo techniques for a large number of pump cavity configurations are given in Sect. 6.3. For closed coupled cavities, typical values are $\eta_T = 0.3$–0.6.

In diode-pumped lasers the radiation transfer is much simpler. In so-called end-pumped lasers, the transfer system usually consists of lenses for the collection and focusing of diode radiation into the laser crystal. Furthermore, in side-pumped systems, the laser diodes are mounted in close proximity to the laser crystal without the use of any intervening optics. If we express reflection losses and spill-over losses at the optics or active medium by the parameter r, we can write

$$\eta_T = (1 - r) \,. \tag{3.42}$$

Since the laser crystal and optical components are all anti-reflection coated, the radiation transfer losses are very small in these systems. Values for the radiation transfer efficiency are typically $\eta_T = 0.85$–0.98.

Absorption of Pump Radiation by the Gain Medium and Transfer of Energy to the Upper Laser Level. This energy transfer can be divided into two processes. The first is the absorption of useful pump radiation by the gain medium expressed by η_a and the second is the efficient transfer from the ground state to the upper laser level expressed by η_u.

The absorption efficiency is the ratio of power P_a absorbed to power P_T entering the laser medium.

$$\eta_a = \frac{P_a}{\int_{\lambda_1}^{\lambda_2} P_T d\lambda} . \tag{3.43}$$

The quantity η_a is a function of the absorption coefficient α_λ at the spectral region of the pump source and the path length in the gain medium. Calculation of η_a requires integration of the product of α_λ and $P_r(\lambda)$ over the spectral absorption region and volume of the laser medium. If the pump radiation is totally diffuse inside the laser rod, as is the case when the lateral surface of the rod is rough ground, a good approximation is

$$\eta_a = 2\alpha_{av} R \exp(-\alpha_{av} R) \tag{3.44}$$

where R is the radius of the laser rod, and α_{av} is the average absorption coefficient taken over the lamp spectral distribution $P_r(\lambda)$ [3.4]. For diode pumped lasers, the absorption efficiency can be approximated by

$$\eta_a = 1 - \exp(-\alpha_D l) \tag{3.45}$$

where α_D is the absorption coefficient of the laser crystal at the wavelength emitted by the laser diode, and l is the path length in the crystal. Detailed data on η_a for both flashlamp and laser diode radiation can be found in Chap. 6.

The upper state efficiency may be defined as the ratio of the power emitted at the laser transition to the power absorbed into the pump bands. This efficiency is the product of two contributing factors

$$\eta_u = \eta_Q \eta_S \tag{3.46}$$

where η_Q is the quantum efficiency, which is defined as the number of photons contributing to laser emission divided by the number of pump photons, and η_S is the Stokes factor which represents the ratio of the photon energy emitted at the laser transition $h\nu_L$ to the energy of a pump photon $h\nu_p$, i.e.,

$$\eta_S = \left(\frac{h\nu_L}{h\nu_p} \right) = \frac{\lambda_P}{\lambda_L} \tag{3.47}$$

where λ_P and λ_L is the wavelength of the pump transition and the laser wavelength, respectively. The quantum efficiency is close to 1 for most common laser materials. For example, for Nd : YAG emitting at 1064 nm which is pumped by a laser diode array at 808 nm, we obtain $\eta_Q = 0.95$, $\eta_S = 0.76$, and $\eta_u = 0.72$.

In a flashlamp-pumped system, the value of η_S is an average value derived from considering the whole absorption spectrum of the laser.

Conversion of the Upper State Energy to Laser Output. The efficiency of this process can be divided into the fractional spatial overlap of the resonator modes with the pumped region of the laser medium and the fraction of the energy stored in the upper laser level which can be extracted as output.

The beam overlap efficiency η_B describes the spatial overlap between the resonator modes and the pump power, or gain distribution of the laser medium. In an amplifier, η_B is a measure of the spatial overlap of the input beam with the pump or gain distribution in the laser material. This subject usually does not receive a lot of attention in laser literature, but a poor overlap of the gain region of the laser with the laser-beam profile is often the main reason why a particular laser performs below expectations. For example, the generally disappointing performance of slab lasers can often be traced to an insufficient utilization of the rectangular gain cross-section by the laser beam. Likewise, the low overall efficiency of lasers with a TEM_{00} mode output is the result of a mode volume which occupies only a small fraction of the gain region of the laser rod. On the other hand, so-called end-pumped lasers, where the output from a laser diode pump is focused into the gain medium, achieve near perfect overlap.

Instead of comparing the pump with the mode volume, it is often sufficient to compare the cross-sections, if we assume that the radial distribution of the resonator mode does not change appreciably inside the laser rod. For example, in a 50 cm long resonator, with two curved mirrors of 5 m radius at each end, the TEM_{00} mode changes only 5% over the length of the resonator (Sect. 5.1.3). In this case, it is convenient to express the beam fill factor as the spatial overlap of the transverse mode structure and radial pump power distribution.

The beam overlap efficiency η_B can be defined by an overlap integral between the pump and resonator mode distribution [3.5, 6]. This quantity can be calculated from

$$\eta_B = \int g_r(r) I_B(r) 2\pi r \, dr \Big/ \int I_B^2(r) 2\pi r \, dr \qquad (3.48a)$$

where $g(r)$ is the gain profile and I_B is beam profile. If we assume Gaussian profiles with spot sizes of w_g and w_B for the gain and beam profiles, respectively, we can derive a beam overlap efficiency

$$\eta_B = \frac{2w_B^2}{w_g^2 + w_B^2} \quad \text{for} \quad w_g > w_B \qquad (3.48b)$$

and

$$\eta_B = 1 \quad \text{for} \quad w_g \leq w_B . \qquad (3.48c)$$

For a multimode laser, the pump and cavity modes have a high spatial overlap and the extraction efficiency is close to unity. In the case of side-pumped lasers with a fundamental-mode output, the spatial overlap can be low if the pump- and cavity-mode distributions are not adjusted for good overlap.

Values for η_B can range from as low as 0.1, for example, for a laser rod of 5 mm diameter operated inside a large radius mirror resonator containing a 1.5 mm aperture for fundamental mode control, to 0.95 for an end-pumped laser operating at TEM_{00}. Innovative resonator designs, employing unstable resonators, internal lenses, variable reflectivity mirrors, etc. can achieve TEM_{00} mode operation typically at $\eta_B = 0.3$–0.5. Multimode lasers typically achieve $\eta_B = 0.8$–0.9. The overlap of a Gaussian pump and laser beam, such as can exist in an end-pumped solid-state laser, has been treated analytically in the literature [3.6]. If a laser oscillator is followed by an amplifier, a telescope is usually inserted between these two stages in order to match the oscillator beam to the diameter of the amplifier and thereby optimize η_B.

The circulating power in an optical resonator is diminished by internal losses described by the round trip loss L (Sect. 3.2) and by radiation coupled out of the resonator. For reasons that will become apparent in the next section, we will define an extraction efficiency η_E which describes the fraction of total available upper state energy or power which appears at the output of the laser

$$\eta_E = P_{out}/P_{avail} .   \tag{3.49}$$

Expressions for η_E will be given in Sect. 3.4.2, Chaps. 4 and 8 for the cw oscillator, laser amplifier and Q-switch oscillator, respectively.

An indication of the reduction of available output power due to losses in resonator can be obtained from the coupling efficiency

$$\eta_c = T/L + T .   \tag{3.50}$$

As will be explained in the next section, the slope of the output vs input curve of a laser is directly proportional to this factor, whereas the overall system efficiency of a laser is directly proportional to η_E.

The conversion processes described so far are equally applicable to cw and pulsed lasers, provided that the power terms are replaced with energy and integration over the pulse length is carried out where appropriate. The energy flow depicted in Fig. 3.3 can also be extended to laser amplifiers and Q-switched systems because the discussion of pump source efficiency, radiation transfer, etc. is equally applicable to these systems. Even the definition for the extraction efficiency remains the same, however the analytical expressions for η_E are different as will be discussed in Chap. 4 for the laser amplifier.

If the laser oscillator is Q-switched, additional loss mechanisms come into play which are associated with energy storage at the upper laser level and with the transient behavior of the system during the switching process. The Q-switch process will be described in detail in Chap. 8. However, for completeness of the discussion on energy transfer mechanisms in a laser oscillator, we will briefly discuss the loss mechanisms associated with Q-switch operations.

In a Q-switched laser, a large upper state population is created in the laser medium, and stimulated emission is prevented during the pump cycle by introduction of a high loss in the resonator. At the end of the pump pulse, the loss is removed (the resonator is switched to a high Q) and the stored energy in the gain medium is converted to optical radiation in the resonator from which it is coupled out by the output mirror.

There are losses prior to opening of the Q-switch, such as fluorescence losses and Amplified Spontaneous Emission (ASE) losses which will depopulate the upper state stored energy. Also, not all of the stored energy available at the time of Q-switching is converted to optical radiation. For a Q-switched laser, the extraction efficiency η_E in Fig. 3.3 can be expressed as

$$\eta_E = \eta_{St}\eta_{ASE}\eta_{EQ} \tag{3.51}$$

where η_{St} and η_{ASE} account for the fluorescence and ASE losses prior to the opening of the Q-switch, and η_{EQ} is the extraction efficiency of the Q-switch process.

Assuming a square pump pulse of duration t_p, the maximum upper state population reached at the end of the pump cycle is given by [3.3]

$$n_2(t_p) = n_0 W_p \tau_f \left[1 - \exp(-t_p/\tau_f)\right] . \tag{3.52}$$

Since the total number of atoms raised to the upper level during the pump pulse is $n_0 W_p t_p$, the fraction available at the time of Q-switching ($t = t_p$) is

$$\eta_{St} = \frac{[1 - \exp(-t_p/\tau_f)]}{t_p/\tau_f} . \tag{3.53}$$

The storage efficiency η_{St} is therefore the ratio of the energy stored in the upper laser level at the time of Q-switching to the total energy deposited in the upper laser level. From the expression for η_{St} follows that for a pump pulse equal to the fluorescence lifetime ($t_p = \tau_f$) the storage efficiency is 0.63. Clearly, a short pump pulse increases the overall efficiency of a Q-switched laser. However, a shorter pump pulse puts an extra burden on the pump source because the pump has to operate at a higher peak power to deliver the same energy to the laser medium.

If the inversion reaches a critical value, the gain can be so high such that spontaneous emission after amplification across the gain medium may be large enough to deplete the laser inversion. Furthermore, reflections from internal surfaces can increase the path length or allow multiple passes inside the gain section which will make it easier for this unwanted radiation to build up. In high gain oscillators, multi-stage lasers, or in laser systems having large gain regions, ASE coupled with parasitic oscillations present the limiting factor for energy storage.

We can define η_{ASE} as the fractional loss of the stored energy density to ASE and parasitic oscillations

$$\eta_{ASE} = 1 - E_{ASE}/E_{ST} . \tag{3.54}$$

Minimizing reflections internal to the laser medium by AR coatings, and providing a highly scattering, absorbing, or low reflection (index matching) surface of the laser rod, coupled with good isolation between amplifier stages will minimize ASE and parasitic losses. The occurence of ASE can often be recognized in a laser oscillator or amplifier as a saturation in the laser output as the lamp input is increased (see also Sect. 4.4.1).

The fraction of the stored energy E_{ST} available at the time of Q-switching to energy E_{EX} extracted by the Q-switch can be expressed as the Q-switch extraction efficiency

$$\eta_{EQ} = E_{EX}/E_{ST} .$$

(3.55)

The fraction of initial inversion remaining in the gain medium after emission of a Q-switched pulse is a function of the initial threshold and final population inversion densities. These parameters are related via a transcendental equation, as shown in Chap. 8.

3.4.2 Laser Output

In this subsection, we will describe the basic relationships between externally measurable quantities, such as laser output, threshold and slope efficiency, and internal systems and materials parameters.

After the pump source in a laser oscillator is turned on, the radiation flux in the resonator which builds up from noise will increase rapidly. As a result of the increasing flux, the gain coefficient decreases according to (3.27) and finally stabilizes at a value determined by (3.13). A fraction of the intracavity power is coupled out of the resonator and appears as useful laser output according to (3.39). If we combine (3.13, 27, 39), the laser output takes the form

$$P_{out} = A \left(\frac{1 - R}{1 + R} \right) I_S \left(\frac{2g_0 l}{L - \ln R} - 1 \right) .$$

(3.56)

In this equation, I_S is a materials parameter, A and l are the cross-section and length of the laser rod, respectively, and R is the reflectivity of the output coupler. These quantities are usually known, whereas the unsaturated gain coefficient g_0 and the resonator losses L are not known. We will now relate g_0 to system parameters, and describe methods for the measurement of g_0 and the losses L in an oscillator.

The Four-Level System. The population inversion in a four-level system as a function of pump rate is given by (3.19). Making the assumption that $W_P \tau_f \ll 1$ and multiplying both sides of this equation by the stimulated emission cross section yields

$$g_0 = \sigma_{21} n_0 W_P \tau_f .$$

(3.57)

Now we recall from Chap. 1, that $W_P n_0$ gives the number of atoms transferred from the ground level to the upper laser level per unit time and volume, i.e.,

$$W_P n_0 = \eta_Q W_{03} n_0 = \eta_Q P'_{ab}/h\nu_p V = \eta_Q \eta_S P'_{ab}/h\nu_L V \qquad (3.58)$$

where $P'_{ab}/h\nu_p V$ are the number of atoms transferred to the pump band per unit time and volume, and η_Q and η_S are the quantum efficiency and Stokes factor defined earlier, respectively. If we introduce (3.58) into (3.57), we can express the small signal gain coefficient in terms of absorbed pump power

$$g_0 = \sigma_{21} \tau_f \eta_Q \eta_S P'_{ab}/h\nu_L V = \eta_Q \eta_S \eta_B P_{ab}/I_S V \qquad (3.59)$$

where I_S is the saturation flux defined in (3.29), and the absorbed pump power P'_{ab} in the gain region is related to the total absorbed pump power P_{ab} in the laser rod by $P'_{ab} = \eta_B P_{ab}$. The beam overlap efficiency η_B was defined in (3.48). With (3.46, 56, 59), we can express the laser output in terms of absorbed pump power

$$P_{out} = \left(\frac{1-R}{1+R}\right) A I_S \left(\frac{2\eta_u \eta_B P_{ab}}{(L - \ln R) A I_S} - 1\right). \qquad (3.60)$$

The absorbed pump power in the laser material is related to the electrical input to the pump source by

$$P_{ab} = \eta_P \eta_T \eta_a P_{in} . \qquad (3.61)$$

With (3.59 and 61) we can establish a simple relationship between the small signal, single pass gain and lamp input power

$$\ln G_0 = g_0 l = K P_{in} \qquad (3.62)$$

where for convenience, we have combined all the terms on the right-hand side into a single conversion factor

$$K = \eta_P \eta_T \eta_a \eta_u \eta_B / A I_S . \qquad (3.63)$$

With the value of K either calculated or measured, one can plot the small-signal, single-pass gain as a function of lamp input power. At the end of this section, we will describe an experimental method of determining K. It is important to note that $G_0 = \exp(g_0 l)$ is the one-way gain for a given pump input P_{in} that would be reached in the absence of saturation effects. In the literature, the term is referred to as small signal, or unsaturated, single pass gain.

If we introduce (3.62) into (3.56), the output of the laser can be expressed as

$$P_{out} = \sigma_S (P_{in} - P_{th}) \qquad (3.64)$$

where σ_S is the slope efficiency of the output vs input curve, as shown in Fig. 3.4,

$$\sigma_S = \frac{2(1-R)}{(1+R)(L - \ln R)} \eta_P \eta_T \eta_a \eta_u \eta_B , \qquad (3.65)$$

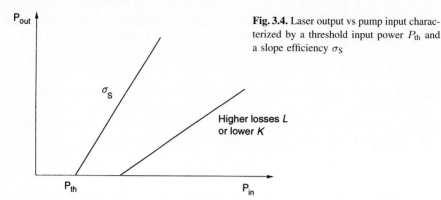

Fig. 3.4. Laser output vs pump input characterized by a threshold input power P_{th} and
a slope efficiency σ_S

and P_{th} is the input power at threshold

$$P_{th} = \frac{(L - \ln R)AI_S}{2\eta_P \eta_T \eta_a \eta_u \eta_B} . \tag{3.66}$$

If we introduce the materials parameter for I_S from (3.29) into (3.66), and relate
the energy absorbed to energy emitted at the laser transition $h\nu_L = \eta_u h\nu_p$, and we
further combine the remaining efficiency factors into an overall pump efficiency
η_{pe} where

$$\eta_{pe} = \eta_p \eta_t \eta_a \eta_B , \tag{3.67}$$

then we can express the laser-threshold condition in a form often presented in
the literature

$$P_{th} = (L - \ln R)\frac{Ah\nu_p}{2\sigma\tau_f\eta_{pe}} \tag{3.68}$$

where ν_p is the energy of the pump photons. From (3.68) it follows that a
laser material with a large product of the stimulated emission cross-section and
fluorescence lifetime ($\sigma\tau_f$) will have a low laser threshold. The slope efficiency
σ_S is simply the product of all the efficiency factors discussed in Sect. 3.4.1.
The input power P_{th} required to achieve laser threshold is inversely proportional
to the same efficiency factors. Therefore, a decrease of any of the η-terms will
decrease the slope efficiency and increase the threshold, as shown in Fig. 3.4. In
the expressions for σ_S and P_{th} we have left R and L in explicit form, because
these parameters are subject to optimization in a laser resonator. The factor K,
on the other hand, combines system parameters which are related to the pump
process; they are more difficult to change in a completed system. As expected,
higher optical losses L, caused by reflection, scattering, or absorption, increase
the threshold input power and decrease the slope efficiency.

From (3.39, 64) we can calculate the laser output power and the total flux
inside the resonator for a given input power as a function of output coupling.
The general shape of P_{out} and I as a function of T are shown in Fig. 3.5. The
output is zero for $T = 0$, and also for a very large transmission where the laser

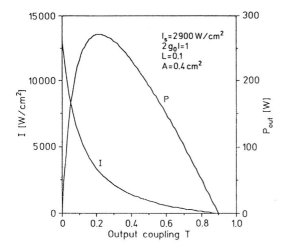

Fig. 3.5. Laser output power P_{out} and total flux I inside the resonator as a function of output coupling T

just reaches threshold for the specified input power, i.e. $T_m = 2g_0l - L$. The flux inside the resonator is also zero for $T = T_m$. However, for $T = 0$, the flux reaches a high value in order to drive the gain down to a value which equals the internal losses, i.e., $2gl = L$. Therefore, $I_{MAX} = [(g_0/g) - 1]I_S$ and the maximum flux in the resonator is determined by the ratio of unsaturated to saturated gain, or by the factor the system is operated above threshold.

As shown in Fig. 3.5, the output reaches a maximum for a specific value of T. Equation (3.56) can be readily differentiated with respect to T in order to determine the output coupling which maximizes P_{out}, i.e.,

$$T_{opt} = \left(\sqrt{2g_0l/L} - 1 \right) L . \tag{3.69}$$

As we can see from this expression, the transmission of the output mirror must be increased if the small signal gain (or input power) is increased. In pulsed solid-state lasers, the optimum transmission of the output mirror is typically 50–70%. In cw-pumped systems, the optimum output mirror transmission is usually between 2% and 20%. The large difference in the output mirror transmission for the two modes of operation is due to the fact that pulsed systems are operated at much higher input powers P_{in}. Pulsed systems therefore have a correspondingly higher gain. For example, flashlamp input powers for pulsed systems range typically from 100 kW to 10 MW. The smaller value would be realized in a typical military-type high-repetition-rate Nd : YAG system pumped by a flashlamp pulse of 20 J energy and 200 μs duration. A large Nd : glass oscillator would typically be operated at 500 J input at a duration of the flashlamp pulse of approximately 500 μs. In contrast to these high flashlamp peak powers, continuous lasers such as Nd : YAG lasers are normally driven at input powers between 1 and 12 kW.

Introducing the expression for T_{opt} into (3.56) gives the laser output at the optimum output coupling

$$P_{opt} = g_0l I_S A \left(1 - \sqrt{L/2g_0l} \right)^2 . \tag{3.70}$$

If we use the definition for g_0 and I_S given before, we can readily see that $g_0 I_S = n_2 h\nu / \tau_f$, which represents the total excited state power per unit volume. (A similar expression will be derived in Chap. 4 for the available energy in a laser amplifier.) Therefore, the maximum available power from the oscillator is

$$P_{\text{avail}} = g_0 l I_S A = \eta_P \eta_T \eta_a \eta_u \eta_B P_{\text{in}} \qquad (3.71)$$

where the second expression on the right hand side is obtained from (3.62). The optimum power output can be expressed as

$$P_{\text{opt}} = \eta_E P_{\text{avail}} , \qquad (3.72)$$

where

$$\eta_E = \left(1 - \sqrt{L/2g_0 l} \right)^2 \qquad (3.73a)$$

is the extraction efficiency already mentioned in Sect. 3.4.1. The behavior of η_E as a function of the loss-to-gain ratio $L/2g_0 l$ is depicted in Fig. 3.6. The detrimental effect of even a very small internal loss on the extraction efficiency is quite apparent. For example, in order to extract at least 50% of the power available in the laser material, the internal loss has to be less than 10% of the unsaturated gain. Achievement of a high extraction efficiency is particularly difficult in cw systems because the gain is relatively small and unavoidable resonator losses can represent a significant fraction of the gain. In Fig. 3.7 the extraction efficiency for the optimized resonator is plotted for different values of L and $g_0 l$.

The overall system efficiency of a solid-state laser is directly proportional to the extraction efficiency

$$\eta_{\text{sys}} = \eta_E \eta_P \eta_T \eta_a \eta_u \eta_B . \qquad (3.73b)$$

If one compares (3.73b) with the expression for the slope efficiency of a laser given by (3.65), then the first term is replaced by the extraction efficiency η_E.

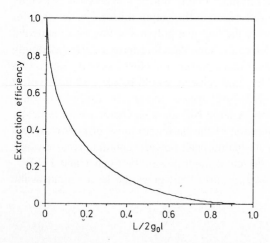

Fig. 3.6. Extraction efficiency η_E as a function of loss-to-gain ratios $L/2g_0 l$

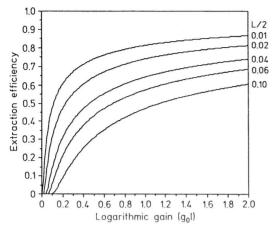

Fig. 3.7. Extraction efficiency for the optimized resonator as a function of the single-pass logarithmic gain. Parameter is the one-way resonator loss

As we recall the first term in (3.65) in the coupling efficiency η_c as defined in (3.50) for values of R close to one. In case of a diode pump source, the diode slope efficiency η_P has to be replaced by the electrical diode efficiency η_{PS}.

The sensitivity of the laser output to values of T which are either above or below T_{opt} is illustrated in Fig. 3.8. For resonators which are either over- or under-coupled, the reduction of power compared to that available at T_{opt} depends on how far the system is operated above threshold. For oscillators far above threshold, the curve has a broad maximum and excursions of $\pm 20\%$ from T_{opt} do not reduce the output by more than a few percent.

Detailed discussions of the optimization of the laser output and extraction efficiency have also been given in [3.3, 7].

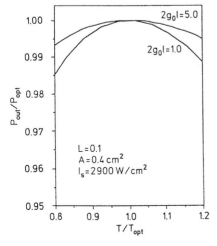

Fig. 3.8. Sensitivity of laser output for non-optimal output coupling

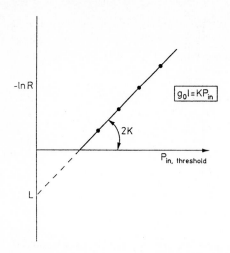

Fig. 3.9. Measurement of the resonator losses as well as the product of all the efficiency factors involved in the energy transfer mechanism of a laser

As we have seen, the resonator losses and the gain in the laser material play an important part in the optimization process of a laser system. Following a method first proposed by *Findlay* and *Clay* [3.8], the resonator losses can be determined by using output mirrors with different reflectivities and determining threshold power for lasing for each mirror. According to (3.13) and with (3.62), we can write

$$-\ln R = 2K P_{th} - L \tag{3.74}$$

where R is the reflectivity of the output mirror and P_{th} is the input at threshold.

Extrapolation of the straight-line plot of $-\ln R$ versus P_{th}, at $P_{th} = 0$, yields the round-trip resonator loss L, as shown in Fig. 3.9. The slope of the straight line is $2K$. With K measured, the product of all the η-factors is known and the unsaturated gain as a function of lamp input can be plotted. From the knowledge of the product of the η-terms, adjustments to individual terms can be made to fit the measured data to the calculated or estimated values.

For example, in a diode-pumped laser η_P and η_u are known very accurately from the beginning. The factors η_T and η_a can be calculated with reasonable accuracy. However, the value of the beam overlap efficiency η_B is usually associated with the greatest uncertainty.

The Three-Level System. The population inversion in a three-level system as a function of the pump rate is given by (3.18). If we multiply both sides of this equation by the stimulated emission cross section σ_{21}, we obtain

$$g_0 = \alpha_0 \frac{W_P \tau_{21} - 1}{W_P \tau_{21} + 1} \tag{3.75}$$

where $\alpha_0 = \sigma_{21} n_{tot}$ is the absorption coefficient of the material when all atoms exist in the ground state. In the absence of pumping, (3.75) simply becomes $g_0 = -\alpha_0$. In order to simplify our analysis, we assumed $g_2 = g_1$. The saturation flux for a three-level system is given in (3.30). If we multiply this equation with n_{tot} and introduce from (3.22) the expression for the fluorescence power output P_F' per unit volume at the population inversion we obtain

$$I_S = (W_P \tau_{21} + 1)P_F'/\alpha_0 . \tag{3.76}$$

From (3.75, 76) in conjunction with the expression for laser threshold (3.13) and the saturated gain coefficient and output coupling (3.38, 39) we can derive the output equation for a three-level laser

$$P_{out} = \frac{(1 - L_N)(1 - R)}{(L - \ln R)} P_F \left(W_P \tau_{21} - \frac{1 + L_N}{1 - L_N} \right) \tag{3.77}$$

where

$$L_N = (L - \ln R)/2\alpha_0 l \tag{3.78}$$

and $P_F = lAP_F'$ is the total fluorescence power at inversion.

We assume now that the pump rate W_P is a linear function of lamp input P_{in}:

$$W_P \tau_{21} = K P_{in} . \tag{3.79}$$

In a four-level system, the unsaturated gain is directly proportional to the pump rate and therefore to the input power of the pump source. In a three-level system, we have instead

$$g_0 = \alpha_0 \frac{K P_{in} - 1}{K P_{in} + 1} . \tag{3.80}$$

At inversion $W_P \tau_{21} = 1$, and the pump input is converted to fluorescence output through the energy transfer mechanism expressed by the η-terms discussed before. Therefore,

$$P_{in} = \frac{1}{K} = \frac{P_F}{\eta_P \eta_T \eta_a \eta_u \eta_B} . \tag{3.81}$$

If we introduce this expression into (3.77), we obtain for the slope efficiency

$$\sigma_S = \frac{(1 - L_N)(1 - R)}{(L - \ln R)} \eta_P \eta_T \eta_a \eta_u \eta_B = \frac{(1 - L_N)(1 - R)}{(L - \ln R)} K P_F \tag{3.82}$$

and for the threshold input power

$$P_{th} = \frac{(1 + L_N)P_F}{(1 - L_N)\eta_P \eta_T \eta_a \eta_u \eta_B} = \frac{(1 + L_N)}{(1 - L_N)K} . \tag{3.83}$$

In these expressions σ_S, P_{th}, R and l are measurable parameters, whereas P_f and α_0 must be calculated from the basic materials parameters. For example, for ruby $\alpha_0 = 0.2\,\mathrm{cm}^{-1}$ and $P_f = 727\,\mathrm{W/cm}^3$. The factor K and, therefore, the gain in the system as well as the losses L, can be calculated from the threshold input power and slope efficiency.

If we compare these expressions with the equations derived for a four-level laser, some of the basic differences between these systems can be illustrated. For example, in a four-level system which has no absorption losses ($L = 0$) and no coupling losses ($R = 1$), the threshold is $P_{th} = 0$. In the three-level system the threshold is $P_{th} = P_f/\eta_P\eta_T\eta_a\eta_U\eta_B$, which is the input power required to achieve inversion. It is also quite apparent that changes in reflectivity R affect the threshold in a three-level system much less than in a four-level system. For example, if we use output mirror reflectivities of $R = 0.93$, 0.53, 0.29, and 0.18 in a laser oscillator and assume a loss of $L = 0.20$, then according to (3.83) in a ruby oscillator with a 20-cm-long crystal ($2l\alpha_0 = 8$) the relative change in threshold is $1 : 1.1 : 1.3 : 1.5$, whereas in a four-level system changes in threshold are $1 : 2.7 : 4.8 : 6.3$.

The calculations carried out so far are valid only for a three-level system operated at steady state, i.e., a system which is operated either cw or pulsed with a pump pulse long compared to the fluorescence time τ_{21}. In practice, a three-level system such as ruby is normally pumped with a 1-ms-long pump pulse, which is short compared to the 3-ms spontaneous emission time. For this case we will replace the fluorescence power at inversion P_f by the energy E_{ui} which is stored in the upper level at inversion. Furthermore, the lamp input power P_{in} will be substituted by the input energy E_{in}. Equation (3.80) is plotted in Fig. 3.10 together with the expression $g_0 = (K/l)E_{in}$ which is valid for a four-level system. The numerical parameters were chosen to be $K = 10^{-3}\,\text{J}^{-1}$, $l = 15\,\text{cm}$, $\alpha = 0.2\,\text{cm}^{-1}$.

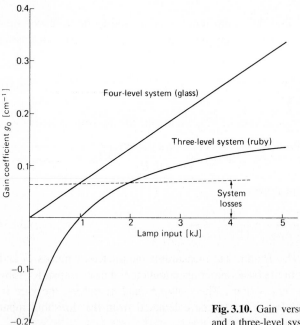

Fig. 3.10. Gain versus lamp input for a four-level and a three-level system

In performing these kinds of model comparisons it has to be pointed out that in a free running oscillator the output consists of a series of random spikes rather than a smooth pulse. Therefore, the parameters calculated above are averaged values obtained by integrating over the whole pulse length.

3.5 Relaxation Oscillations

So far in this chapter we have considered only the steady-state behavior of the laser oscillator. Let us now consider some aspects of transient or dynamic behavior. Relaxation oscillations are by far the most predominant mechanisms causing fluctuations in the output of a solid-state laser. Instead of being a smooth pulse, the output of a pumped laser is comprised of characteristic spikes. In cw-pumped solid-state lasers the relaxation oscillations, rather than causing spiking of the output, manifest themselves as damped, sinusoidal oscillations with a well-defined decay time.

3.5.1 Theory

In many solid-state lasers the output is a highly irregular function of time. The output consists of individual bursts with random amplitude, duration, and separation (see, for example, Fig. 3.16). These lasers typically exhibit what is termed "spiking" in their output. We will explain the phenomena of the spike formation with the aid of Fig. 3.11. When the laser pump source is first turned on there are a negligible number of photons in the cavity at the appropriate frequency. The pump radiation causes a linear buildup of excited atoms and the population is inverted.

Although under steady-state oscillation conditions N_2 can never exceed $N_{2,th}$, under transient conditions the pump can raise N_2 above the threshold level, because no laser oscillation has yet been built up and no radiation yet exists in the cavity to pull N_2 back down by means of stimulated emission.

The laser oscillation does not begin to build up, in fact, until after N_2 passes $N_{2,th}$, so that the net round-trip gain in the laser exceeds unity. Then, however, because N_2 is considerably in excess of $N_{2,th}$, the oscillation level will actually build up very rapidly to a value of the photon flux ϕ substantially in excess of the steady-state value for the particular pumping level.

But, when $\phi(t)$ becomes very large, the rate of depletion of the upper-level atoms due to stimulated emission becomes correspondingly large, in fact considerably larger than the pumping rate W_p. As a result, the upper-level population $N_2(t)$ passes through a maximum and begins to decrease rapidly, driven downward by the large radiation density. The population $N_2(t)$ is driven back below the threshold level $N_{2,th}$; the net gain in the laser cavity becomes less than unity, and so the existing oscillation in the laser cavity begins to die out.

To complete the cycle of this relaxation process, once the radiation level has decreased below the proper steady-state level, the stimulated emission rate

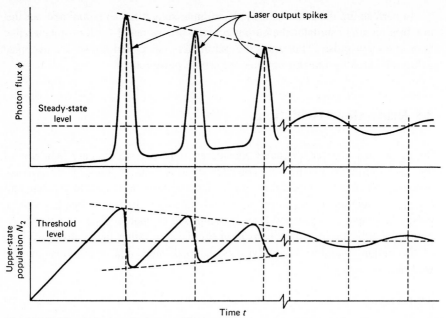

Fig. 3.11. Spiking behavior of a laser oscillator

again becomes small. At this point the pumping process can begin to build the population level N_2 back up toward and through the threshold value again. This causes the generation of another burst of laser action, and the system can again go through a repeat performance of the same or a very similar cycle.

Turning now to the rate equation, we can interpret these curves as follows: At the beginning of the pump pulse we can assume that the induced emission is negligible because of the low photon density. During this time we may neglect the term containing ϕ in (1.53) and write

$$\frac{dn}{dt} = W_p n_{tot} . \tag{3.84}$$

The population inversion therefore increases linearly with time before the development of a large spiking pulse. As the photon density builds up, the stimulated emission terms become important and for the short duration of one pulse the effect of the pumping can be neglected. Therefore, during the actual spiking pulse the rate equations can be written by neglecting both the pumping rate for the excess population and the cavity loss rate in (1.58, 61):

$$\frac{dn}{dt} = -\gamma c\sigma n\phi , \quad \frac{d\phi}{dt} = +c\sigma n\phi . \tag{3.85}$$

The photon density thus grows with time and the population inversion decreases with time. The photon density reaches a peak when the decreasing inversion reaches the threshold value n_{th}. The inversion reaches a minimum at

$\gamma n c \sigma \phi \approx W_p n_{tot}$. The cycle repeats itself, forming another spike. The inversion fluctuates in a zigzag fashion around the threshold value n_{th}. As time passes, the peaks become smaller and the curve becomes damped sinusoidal.

Solutions of the laser rate equations predict a train of regular and damped spikes at the output of the laser [3.9–11]. Most lasers, however, show completely irregular, undamped spikes. This discrepancy between theory and experiment is due to the fact that the spiking behavior dies out very slowly in most solid-state lasers and therefore persists over the complete pump cycle. Furthermore, mechanical and thermal shocks and disturbances present in real lasers act to continually reexcite the spiking behavior and keep it from damping out. Hence many lasers, especially the ruby laser, spike continuously without ever damping down to the steady state. Depending on the system parameters such as mode structure, resonator design, pump level, etc., the spiking may be highly irregular in appearance or it may be regular. Conditions for regular spiking in pulsed ruby lasers are summarized in [3.11, 12].

In cw-pumped lasers, such as Nd : YAG, the relaxation oscillations are much weaker and usually consist of damped sinusoidal oscillations around the steady-state value. These oscillations may be treated as perturbations of the steady-state population inversion and photon density given in the rate equations (1.58, 61). Compared to the fluorescence time τ_f, the relaxation oscillations have a much shorter period, therefore the term including τ_f in the rate equations can be ignored (that is, $\tau_f \to \infty$).

We now introduce a small perturbation Δn into the steady-state value of the population inversion n; similarly, a perturbation $\Delta \phi$ is introduced into the steady state of the photon density ϕ. Thus we may write

$$\tilde{n} = n + \Delta n \quad \text{and} \quad \tilde{\phi} = \phi + \Delta \phi \ . \tag{3.86}$$

We now proceed to eliminate the population inversion n from (1.61). This is done by first differentiating the equation and then substituting $\partial n / \partial t$ from (1.58). The differential equation is then linearized by introducing n and ϕ from (3.86). Neglecting products of ($\Delta n \Delta \phi$), we finally obtain

$$\frac{d^2(\Delta \phi)}{dt^2} + c \sigma \phi \frac{d(\Delta \phi)}{dt} + (\sigma c)^2 \phi n (\Delta \phi) = 0 \ . \tag{3.87}$$

The solution of this equation gives the time variation of the photon density

$$\Delta \phi \approx \exp\left(-\frac{\sigma c \phi}{2}\right) t \sin[\sigma c (\phi n)^{1/2} t] \ . \tag{3.88}$$

The frequency $\omega_s = \sigma c (\phi n)^{1/2}$ and the decay time constant $\tau_R = 2/\sigma c \phi$ of this oscillation can be expressed in terms of laser parameters by noting that $I = c \phi h \nu$ and $n = 1/c \sigma \tau_c$. The latter expression follows from (1.61) for the steady-state condition, i.e. $\partial \phi / \partial t = 0$, and ignoring the initial noise level S. With the introduction of the intracavity power density I and the photon decay time τ_c we obtain:

$$\omega_s = \sqrt{\frac{\sigma I}{\tau_c h\nu}} \quad \text{and} \quad \tau_R = \frac{2h\nu}{I\sigma} \,. \tag{3.89}$$

These expressions can be further simplified for the case of a four-level system by introducing the saturation power density I_s, leading to

$$\omega_s = \sqrt{\frac{I}{I_s \tau_f \tau_c}} \quad \text{and} \quad \tau_R = 2\tau_f \left(\frac{I_s}{I}\right) \,. \tag{3.90}$$

Note, that the greater the power density I and therefore the output power from the laser, the higher the oscillation frequency. The decay time τ_R will decrease for higher output power.

From these equations it follows that the damping time is proportional to the spontaneous lifetime. This is the reason that relaxation oscillations are observed mainly in solid-state lasers where the upper-state lifetime is relatively long. Oscilloscope traces of the relaxation oscillations of a ruby and a cw-pumped Nd : YAG laser are presented in Fig. 3.15 and 3.26a, respectively. Frequency spectra of the relaxation oscillations of cw-pumped Nd : YAG oscillators are depicted in Fig. 3.26b and 5.45. Investigations of the output fluctuations of cw-pumped Nd : YAG lasers can be found in [3.22–29] (see also Sects. 3.6 and 5.3.1).

3.5.2 Spike Suppression in Solid-State Lasers

Relaxation oscillations are a fundamental property of the laser and are produced by the dynamic interaction between the radiation within the resonator and the energy stored in the active laser medium. Subsequently, relaxation oscillations can be stimulated by a variation in the optical losses of a particular resonator mode, or by a change of the gain in the laser medium. The former includes mechanical vibration of the resonator components, and temporal and spatial variations of the optical path in the laser rod brought about by changes of the cooling conditions of the rod surface. The latter includes rapid variations and fluctuations in the output of arc lamps and lamp instabilities due to arc wander.

The task of reducing relaxation oscillations requires eliminating the sources which start these oscillations, and providing electronic feedback to attenuate the amplitude of the oscillations once they do occur. Stabilization of laser oscillations using electronic feedback dates back as far as 1962 [3.13].

An active feed back system typically consists of a beam splitter and photodiode at the output of the laser, an intracavity modulator and an electronic drive circuit. The photodiode samples a fraction of the laser output and provides a feedback signal to the modulator via the electronic drive circuit. As the output of the laser increases above a certain value due to the onset of a spike, the loss in the resonator is increased by the feedback circuit. Such an electronic feedback circuit can convert the spiky output from a solid-state laser into a fairly smooth output. Obtaining a smooth pulse is offset by the disadvantage of a reduced output. As intracavity modulators, Kerr cells [3.13, 15], Pockels cells [3.14, 27] and acousto-optic modulators [3.16] have been employed. In order to obtain a reasonably smooth output, the delay time between the signal derived from the photodiode and the change in transmission of the internal modulator has to be short compared to the duration of an individual spike. Since the individual spikes have a duration of approximately 0.1–1 μs, this requires a bandwidth of at least 20–50 MHz for the feedback system. Electronic feedback loops which change the transmission of an intracavity modulator have been designed for a number of solid state lasers such as ruby, Nd : YAG, Nd : glass and Nd : YLF [3.10, 3.13–16]. A typical feedback control system for the reduction of output fluctuations in a diode-pumped Nd : YAG laser is depicted in Fig. 5.44.

In water cooled, arc lamp pumped lasers it is very difficult to reduce output fluctuations below a certain level. An enabling technology for designing low noise oscillators is the use of diode-laser pump sources, in combination with conduction cooling and a monolithic resonator/gain medium design. Instabilities and fluctuations caused by the arc discharge are eliminated. Conductive cooling removes mechanical vibrations introduced by the water pump, and temporal optical disturbances in the laser rod by the turbulent water flow are eliminated. A monolithic design drastically reduces mechanical instabilities and the laser operates in a single transverse and longitudinal mode. Extremely low noise oscillators, comprised of conductively cooled, diode-pumped monolithic lasers with an active feedback loop that suppresses relaxation oscillations were described in [3.17–19]. In diode-pumped lasers, the feedback loop controls the drive current to the laser-diode pump source.

3.5.3 Gain Switching

In the previous subsection, we described techniques to suppress relaxation oscillations; in gain switching – the subject treated here – the phenomena of spiking is exploited to produce a high peak power pulse. If a solid state laser is pumped by another laser, it is possible to pump at such a fast pump rate that the population inversion and gain reach a level considerably above threshold before the laser oscillation has time to build-up in the resonator. As the radiation increases in time, it will then deplete the upper state population. The response of a laser to a very fast pump pulse which drives the inversion far above threshold is in the form of a relaxation oscillation depicted in Fig. 3.11 [3.20, 21, 30]. If the pump pulse is not only fast but also shorter than the width of the first peak of the relaxation oscillation, the radiation emitted from the system will consist of only the first spike. The population inversion will have been driven to a value below laser threshold after the first spike and it will not grow because there is no pump energy to replenish the population.

The use of a gain-switching method to enhance the peak power of a diode-pumped monolithic Nd : YAG laser has been described in [3.21]. A single-longitudinal mode operation is established by pumping the oscillator at its threshold pump level of 1.5 mW for about 500 μs. The device is then gain switched by increasing the diode drive current for several microseconds which increases the power to 50 mW. This results in the emission of a relaxation oscillation from the

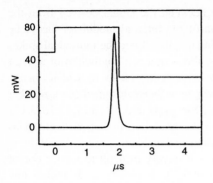

Fig. 3.12. Diode drive current (upper trace) and Nd : YAG laser output (lower trace) of a gain switched system [3.21]

Nd : YAG crystal. The gain-switching pulse is turned off at the end of the first relaxation oscillation. The emission of further relaxation pulses is thus avoided. This is sketched in Fig. 3.12. The drive current at $\tau = 0$ is switched from a value just above the threshold condition to a high value. After $2\,\mu s$ the first relaxation oscillation has developed. At that point, the drive current to the pump source is turned off.

Both gain switching and Q-switching allow the generation of short pulses. Q-switched pulses are generated by storing energy in the upper state of the active medium by a pump pulse having a duration on the order of the upper-state lifetime. After the inversion and gain have been established, the loss in the resonator is suddenly switched off. In gain switching, energy is deposited very quickly in the upper state, i.e., gain is switched on before the radiation in the resonator has time to build up from noise.

From this discussion, it is clear why gain switching as compared to Q-switching is employed only in special cases in solid-state lasers. A Q-switch permits the transformation of a pump pulse of relatively low power and long duration into the emission of a very short pulse of high peak power. For example, in a Nd : YAG laser the flashlamp pump pulse is typically $200\,\mu s$, and the Q-switch output pulse is on the order of 20 ns, which results in a time compression of 4 orders of magnitude. In gain switching, peak power and pulse width of the pump pulse and laser output are on the same order.

As an example of a gain-switched laser, we consider the pumping of Ti : sapphire with a frequency-doubled Q-switched Nd : YAG laser. Ti : sapphire is difficult to pump with a flashlamp because the short upper state lifetime of $3.2\,\mu s$ requires a pump pulse of similar pulse width. Therefore, a common pump source for a pulsed Ti : sapphire laser is a frequency-doubled, Q-switched Nd : YAG laser. In the specific example considered [3.30], the pulse width of 10 ns from the Nd : YAG laser pump is significantly shorter than the $3.2\,\mu s$ upper state lifetime in Ti : sapphire, and the pump pulse is also significantly shorter than the power build-up time which may be on the order of 50–200 ns. Due to the short pump pulse, the inversion is initially driven far above threshold and the Ti : sapphire laser responds with the emission of a gain-switched pulse with a pulse width on the order of 10–40 ns depending on the intensity of the pump source.

3.6 Examples of Regenerative Oscillators

In this section we will relate the performance characteristics of typical solid state lasers such as ruby, Nd : glass, Nd : YAG and alexandrite to the oscillator model described earlier in this chapter. Some of the systems such as ruby, Nd : glass and alexandrite are generally pumped by flashlamps, and Nd : YAG, the most versatile of all lasers, is either pumped by flashlamps, cw arc lamps, or by laser diode-arrays. An example for each type of these lasers will be provided.

3.6.1 Ruby

Figure 3.13 shows a photograph of the head of a typical ruby laser employed extensively in the 1960s and early 1970s. This configuration, developed by Korad Corp., was the first commercially available solid-state laser. Ruby is treated here because of its historical importance and as an example of a 3-level laser.

The primary components of the laser head are the laser rod, a helical flash-lamp, and a reflector. The laser rod is surrounded by a close-coupled helical flashlamp. A cylindrical reflector around the flashlamp aids in efficiently direct-ing the flashlamp pump light into the laser rod. The entire laser housing, including the region occupied by the flashlamp, is filled with cooling water. Laser heads contain ruby rods ranging in size from 7.5 cm by 1 cm to 20 cm by 2 cm. De-pending on the size of the flashlamp, input energies are between 3 and 20 kJ. In addition to the laser head, shown in Fig. 3.13, a complete laser system contains the following modules: an optical rail which supports the laser head, resonator mirrors, and other optical components; a power supply; a flashlamp trigger unit; an energy-storage capacitor bank; and a water cooler.

We will now examine the performance of an oscillator composed of a ruby rod (0.05% doping) which is 10.4 cm long and 0.95 cm in diameter. The optical resonator consists of two flat mirrors with a separation of 71 cm. The performance of the system around threshold as a function of output coupling is shown in Fig. 3.14. As one can see, a mirror reflectivity around 50% is about optimum for this system. A reflectivity considerably higher than this value results in an output versus input curve which has a very low slope efficiency. On the other hand, a reflectivity which is appreciably lower than 50% results in a very high threshold. The curvature in the curves near the threshold is caused by nonuniform pumping. More pump light is absorbed in the areas immediately below the surfaces than in the center of the rod. Therefore the rod starts to lase with a ring-like structure. At a somewhat higher energy input the full cross section of the rod will start to lase. The curve for $R = 0.92$ is straight even at threshold, which indicates that at this high reflectivity, pumping uniformity is insignificant and the whole rod starts to lase almost uniformly. From these curves we can calculate the parameters K and L according to (3.82, 83). Taking, for example, the curve with the 92% mirror, we have $R_1 = 0.92$. The slope efficiency of this curve is $\sigma_s = 3.6 \times 10^{-3}$ and the rod has a length of 10 cm and a volume of $V = 7.8$ cm^3. In Chap. 2 we calculated the upper-level energy density at inversion $E_{uth} = 2.18$ J/cm^3, therefore the total

Fig. 3.13. Cutaway of a ruby laser head

stored energy is $E_{ui} = 17\,\text{J}$. We need also the absorption coefficient of ruby, which is $\alpha_0 = 0.2\,\text{cm}^{-1}$. Introducing these values into (3.82, 83) yields $L = 10\%$ and $K = 4.7 \times 10^{-4}\,\text{J}^{-1}$. Therefore the gain in this rod as a function of input energy can be expressed as

$$g_0 = 0.2 \frac{0.47 E_{\text{in}}\,[\text{kJ}] - 1}{0.47 E_{\text{in}}\,[\text{kJ}] + 1}\ \ [\text{cm}^{-1}] \,.$$

From the value for K it follows that inversion in this rod occurs at the flashlamp input energy of $E = 1/K = 2.1\,\text{kJ}$.

The performance of the same oscillator operated at higher input energies and with the optimum output coupler was measured.

The output vs input of a 10.4-cm by 1.4-cm rod doped with 0.03 % Cr^{3+} can be approximated by

$$E_{\text{out}} = 0.013(E_{\text{in}} - 2.5\,\text{kJ}) \,.$$

For the 10.4-cm by 0.95-cm rods doped with 0.05% Cr^{3+}, a typical output versus input curve can be described by

$$E_{\text{out}} = 0.0075(E_{\text{in}} - 2.2\,\text{kJ}) \,.$$

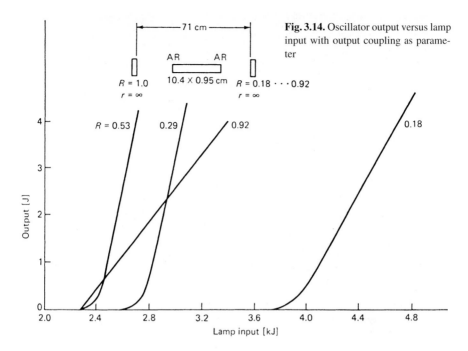

Fig. 3.14. Oscillator output versus lamp input with output coupling as parameter

If we introduce the values of slope efficiency $\sigma_s = 7.5 \times 10^{-3}$ and threshold energy $E_{th} = 2.2 \times 10^3$ J together with the other pertinent rod parameters in (3.82, 83), we obtain $L = 0.27$ and $K = 7.3 \times 10^{-4}$ J^{-1}. The most important factors which contribute to losses L are the scattering losses, the ruby absorption losses caused by the rod holders which shadow the ends of the rods from the pump radiation, and possible losses in the coatings. The gain as a function of input energy is

$$g_0 = 0.2 \frac{0.73 E_{in}[\text{kJ}] - 1}{0.73 E_{in}[\text{kJ}] + 1} [\text{cm}^{-1}] .$$

At maximum input energy of 5 kJ, this rod possesses a small-signal gain coefficient of $g_0 = 0.11$ cm^{-1} and a single-pass gain of $G_0 = \exp(g_0 l) = 3.3$.

The typical time dependence of the oscillator output pulses is shown in Fig. 3.15.

In a diffusely reflecting cavity a large-diameter rod is able to capture a larger fraction of the pump light than a smaller rod. Furthermore, a large-diameter rod provides a longer absorption path for the pump light and can therefore absorb a larger fraction of the pump radiation. In ruby we assume an average absorption coefficient in the pump bands of $\alpha = 1.5$ cm^{-1} for a 0.05%-doped rod and $\alpha = 1.0$ cm^{-1} for a 0.03% Cr^{3+} concentration. The absorbed pump radiation as a function of rod diameter was given in (3.44).

A rod doped with 0.03% which is 1 cm in diameter will absorb 60% of the incident pump radiation; for a 2-cm-diameter rod doped at 0.05%, this fraction

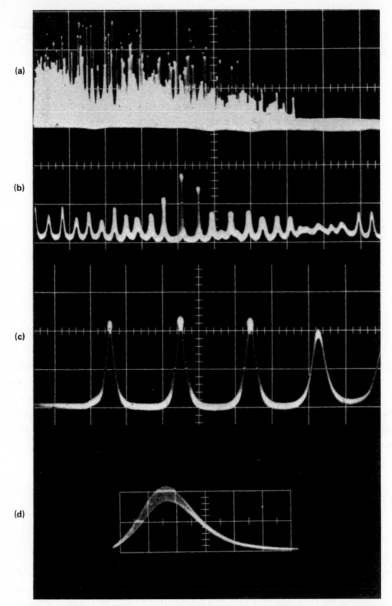

Fig. 3.15a–d. Temporal structure of the ruby laser output. Oscillator is comprised of a 10-cm by 1-cm ruby rod pumped at 4.75 kJ and two flat mirrors separated by 60 cm. Output is 25 J. The upper trace (**a**) shows the complete laser output. The lower traces (**b–d**) represent portions of the output pulse taken at an expanded scale. The time scale is (**a**) 200 μs, (**b**) 5 μs, (**c**) 1 μs, and (**d**) 100 ns/div. The energy of the individual spike shown in (**d**) is 7.5 mJ

increases to 67%. Therefore, η_a will be larger for large-diameter and/or higher-doped crystals. The factor η_C depends mainly on the optical losses in the system. For example, a combined loss of $L = 0.10$ and an output reflector of $R_1 = 0.45$ for the case of a 10-cm-long ruby crystal results in a coupling efficiency of $\eta_C = 0.48$. The longer rods will have higher scattering and absorption losses, which will decrease η_C. The efficiency factors can be related to the energy levels per cubic centimeter of ruby crystal. For a ruby crystal at population inversion we can write

Stored energy in upper level: $E_{u,th} = 2.18 \, \mathrm{J/cm^3}$

Minimum absorbed energy: $E_{ab} = E_{u,th}/\eta_U = 4.4 \, \mathrm{J/cm^3}$

Lamp input at inversion: $E_{in} = E_{ab}/\eta_P \eta_T \eta_a \eta_B = 70\text{–}190 \, \mathrm{J/cm^3}$

Lamp input at threshold: $E'_{in} = E_{in}/\eta_C = 100\text{–}320 \, \mathrm{J/cm^3}$

We will conclude this subsection on ruby oscillators by briefly summarizing the performance of a cw-pumped oscillator. Since ruby operates as a three-level system, about one-half of the Cr^{3+} ions must be excited into the upper laser level before inversion is obtained. At population inversion the fluorescence power is, according to Sect. 3.4, $P_f = 726 \, \mathrm{W/cm^3}$. In order to sustain population inversion, at least $P_{ab} = P_f/\eta_U = 1.4 \, \mathrm{kW/cm^3}$ of pump power has to be absorbed by the crystal. Assuming the same efficiency factors η_P, η_T, η_a as we had in the pulsed case, the electrical input power required for the lamp is about $70 \, \mathrm{kW/cm^3}$ of active material. Since it is not possible to concentrate the power of an incoherent light source into a column smaller than that of the source itself, the necessary pump radiation must be produced in a light source having an extremely high power density. Continuous operation can be expected, therefore, only in very small volumes of ruby. All ruby lasers which have been operated continuously used mercury arc lamps as pump sources [3.31, 32]. In a capillary mercury arc lamp which has typically a 1-mm bore, a wall-stabilized arc contracts toward the center of the tube at an internal pressure of 200 atm and forms an extremely brilliant filamentary light source of 0.3–0.6 mm effective diameter.

A cw-pumped ruby laser, which used a rod 2 mm in diameter and 50 mm in length, generated an output of 1.3 W at an input of 2.9 kW [3.31]. Only a small part of the crystal's cross section was excited by the filament arc, and lasing action occurred only in the small volume of $6 \times 10^{-3} \, \mathrm{cm^3}$. Using this value, the lamp input power per unit volume of active material required to obtain threshold is approximately $230 \, \mathrm{kW/cm^3}$. The individual efficiency factors of the system were estimated to be $\eta_P = 0.25$, $\eta_a = 0.04$, $\eta_T = 0.80$, $\eta_U = 0.50$, and $\eta_C = 0.30$. The main reason for the poor overall efficiency was the low absorption of useful pump light by the small lasing volume.

3.6.2 Nd : Glass

We shall now examine the performance of a Nd : glass oscillator. The laser rod, 15 cm long and 1 cm in diameter, is pumped by two linear flashlamps in a highly polished double elliptical cylinder. The flashlamps, matching the laser rod in size, generate a $600\,\mu$s-long pulse with a total energy of up to 1 kJ. The oscillator starts to lase at a threshold input energy of 300 J and produces an output of 10 J at an input of 1 kJ. In order to determine the gain and the inherent losses of the oscillator, threshold was measured as a function of mirror reflectivity, as shown in Fig. 3.16. The measurement reveals a total loss in the resonator of $L = 0.21$ and a value of $1.1 \times 10^{-3}\,\mathrm{J}^{-1}$ for the parameter K. With K measured, we can express the small-signal coefficient as a function of input energy according to $g_0 = K E_{\mathrm{in}}$. For the 15-cm-long laser rod we obtain $g_0 = 7 \times 10^{-5} E_{\mathrm{in}}$ [cm^{-1}]. For example, for a lamp input of 1 kJ, the gain coefficient is $g_0 = 0.07\,\mathrm{cm}^{-1}$ and the single-pass gain is $G_0 = 2.9$.

Typical performance data obtained from water-cooled laser heads containing helical flashlamps and ED-2 glass rods are displayed in Fig. 3.17. The various curves obtained by testing eight different glass rods illustrate the spread in performance depending on the glass rod quality. The average rod performances can be characterized by a threshold of 1.5 KJ, a slope efficiency of 2%, and a total efficiency at maximum input of 1.4%, that is,

$$E_{\mathrm{out}} = 0.02(E_{\mathrm{in}} - 1.5\,\mathrm{kJ})\ .$$

Given the distribution of performance data shown in Fig. 3.17, the laser manufacturer's minimum guaranteed performance is 55 J, allowing a margin for component variations and system degradation. The insert shows a measurement of

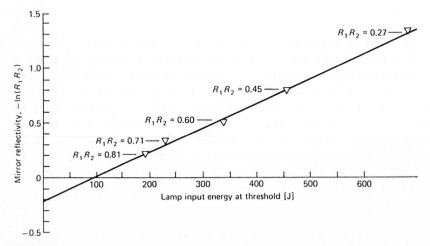

Fig. 3.16. Lamp input energy for laser threshold as a function of mirror reflectivity (15-cm by 1-cm ED-2 glass rod pumped by two linear flashlamps)

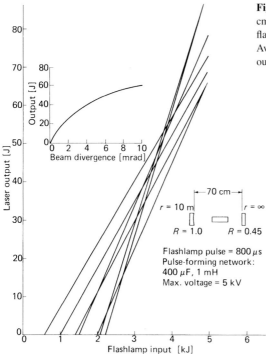

Fig. 3.17. Performance of 15-cm by 1.2-cm Nd:glass rods pumped by a helical flashlamp (pump length 10 cm). Insert: Average beam divergence versus energy output

beam divergence at maximum lamp input energy. For most rods, half of the output energy is emitted in a beam with a 3–4-mrad beam divergence. From larger Nd:glass rods pumped at higher input energies, correspondingly larger outputs are obtainable, as is illustrated in Fig. 3.18. For example, from an ED-2 laser rod 25 cm long and 2 cm in diameter pumped at 20,000 J, an output of 200 J can be achieved.

Before we leave this section on glass oscillators we will consider a few examples of very large systems. Figure 3.19 shows the performance of a system composed of a 94-cm-long and 1-cm-diameter ED-2 glass rod, a single linear flashlamp (EG&G FX-77-35), and silver foil wrapped closely around the rod and the lamp.

The highest output energy from a single-element laser oscillator was obtained from a 95-cm-long and 3.8-cm-diameter glass rod pumped by four 1-m linear flashlamps in a close-wrap configuration. The output from the system was 5 kJ at an input of 180 kJ [3.33].

3.6.3 Nd:YAG

To illustrate the application of the equations which we derived in this chapter, let us consider a cw-pumped Nd:YAG laser. The system contains a laser rod 75 mm in length and 6.2 mm in diameter. The laser rod is pumped by two krypton arc lamps capable of a maximum electrical input power of 12 kW. The laser

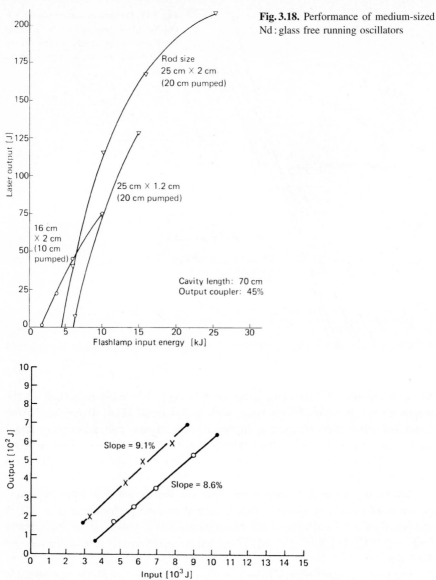

Fig. 3.18. Performance of medium-sized Nd : glass free running oscillators

Fig. 3.19. Output versus input of a Nd : glass rod (94 cm by 1 cm) pumped by a single linear flashlamp. (o) 5 ms pulse width; (×) 2.2 ms pulse width. Output reflection 28 %

rod and the pump lamps are contained in a highly polished, gold-plated, double-elliptical cylinder. The major and minor axes of a single ellipse are 38 and 33 mm, respectively. Cooling of the rod and lamps is accomplished by circulating water in flowtubes which surround the crystal and lamps. The interior of the reflec-tor is also water-cooled. The optical resonator is composed of two dielectrically coated mirrors which have a separation of 40 cm. The rear mirror has a concave

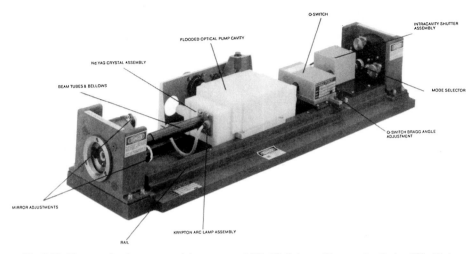

O-SWITCH

INTRACAVITY SHUTTER ASSEMBLY

FLOODED OPTICAL PUMP CAVITY

Nd:YAG CRYSTAL ASSEMBLY

BEAM TUBES & BELLOWS

MODE SELECTOR

O-SWITCH BRAGG ANGLE ADJUSTMENT

MIRROR ADJUSTMENTS

KRYPTON ARC LAMP ASSEMBLY

RAIL

Fig. 3.20. Photograph of a commercial cw-pumped Nd : YAG laser (Quantronix, Series 100, High power laser systems)

curvature of 2 m, whereas the front mirror is flat. The krypton arc lamps, having a cold-fill pressure of 2 atm, operate at a maximum current level of 55 A. Phase-controlled SCRs in the power supply permit operation of the lamps at any desired input level.

Figure 3.20 shows a photograph of a typical, commercially available, cw-pumped Nd : YAG laser. The laser head contains a single arc lamp. The mechanical shutter is used to stop laser oscillations for short periods of time without having to turn off the arc lamp. The purpose of the mode selector and Q-switch will be discussed in Chaps. 5 and 8. Bellows are employed between each optical element in the laser head in order to seal out dust and dirt particles from the optical surfaces. The laser head cover is sealed with a gasket in order to further reduce environmental contamination.

In Fig. 3.21 the output power of the two lamp systems mentioned above is plotted against lamp input power for different Nd : YAG crystals and front-mirror reflectivities. The highest output from a 7.5-cm-long crystal was achieved with a front mirror of 85% reflectivity. The output versus input curve for this mirror shows a slope efficiency of σ_s = 0.026 and an extrapolated threshold of P_{th} = 2.8 kW. The nonlinear portion of the curve close to threshold is due to the focusing action of the mirror-finished elliptical cylinder. At first, only the center of the rod lases. In Fig. 3.22 the lamp input power required to achieve laser threshold was measured for different mirror reflectivities. If one plots $\ln(1/R_1)$ rather than R_1, one obtains a linear function according to (3.74). From this measurement follows a value of the pumping coefficient of $K = 72 \times 10^{-6}\,\mathrm{W}^{-1}$ and a combined loss of L = 0.075. With these two values known, it is possible to plot the small-signal, single-pass rod gain as a function of lamp input power, as shown in Fig. 3.23. Assuming a ground-state population density of $6.0 \times 10^{19}\,\mathrm{cm}^{-3}$ and a stimulated emission cross section of $2.8 \times 10^{-19}\,\mathrm{cm}^2$, it follows

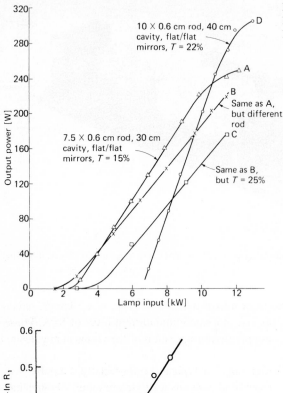

Fig. 3.21. Continuous output versus lamp input of a powerful Nd : YAG laser

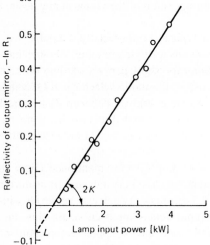

Fig. 3.22. Threshold power input as a function of mirror reflectivity

from Fig. 3.23 that at maximum lamp input, only 0.6% of the total neodymium ion concentration is inverted. At threshold, which occurs at around 2 kW input, the small-signal gain coefficient is about $0.01 \, \text{cm}^{-1}$, which corresponds to only a 0.06% inversion of the total ground-state population.

Using (3.21) we can calculate the total fluorescence output of the laser at threshold. With $V = 2.3 \, \text{cm}^3$, $\tau_\text{f} = 230 \, \mu\text{s}$, $h\nu = 1.86 \times 10^{-19} \, \text{Ws}$, and $n_\text{th} = 1.1 \times 10^{16} \, \text{cm}^{-3}$, one obtains $P_\text{f} = 20 \, \text{W}$.

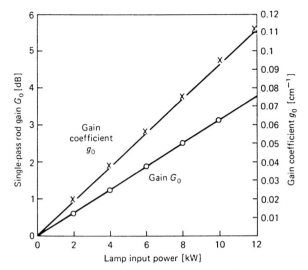

Fig. 3.23. Small-signal, single-pass rod gain and gain coefficient as a function of lamp input

The numerical values for K, I_s, A, and L were introduced into (3.63–65), and the output power as a function of the front mirror reflectivity was calculated for different lamp input powers. The result of the calculations is shown in Fig. 3.24. As can be seen, relatively broad maxima of the optimum mirror reflectivities are obtained. The mirrors which give the highest output power for the different input powers are located along the dashed curve. This curve can also be obtained from (3.69). Also indicated in Fig. 3.24 are the experimentally determined reflectivities which gave the highest output. Figure 3.25 shows the intracavity power density as a function of laser output power for a fixed lamp input power of 12 kW. The parameter is the reflectivity of the front mirror. This curve is obtained from Fig. 3.24 (uppermost curve) and using (3.39). As we can see from this figure, the circulating power density in the resonator increases for the higher reflectivities despite the reduction of output power.

Like most solid-state lasers, Nd:YAG exhibits relaxation oscillation. Figure 3.26 exhibits oscilloscope traces of the relaxation oscillations of a small cw-pumped Nd:YAG laser. This laser has a 3-by-63-mm crystal pumped by two 1000-W halogen-cycle tungsten filament lamps in a double-elliptical reflector configuration. The maximum output which can be achieved with this system is about 10 W at an input of 2 kW. The optical resonator is formed by two 5-m concave mirrors which are 35 cm apart.

In practice, relaxation oscillations in solid-state lasers have amplitudes and intervals between oscillations which appear to be almost random. Most of the irregularities are caused by multimode operation, mechanical instabilities of the resonator, and pumping nonuniformities. Under very carefully controlled conditions and single-mode operation, regular periodic relaxation oscillations can be observed. Figure 3.26a shows an oscilloscope trace of a relaxation oscillation.

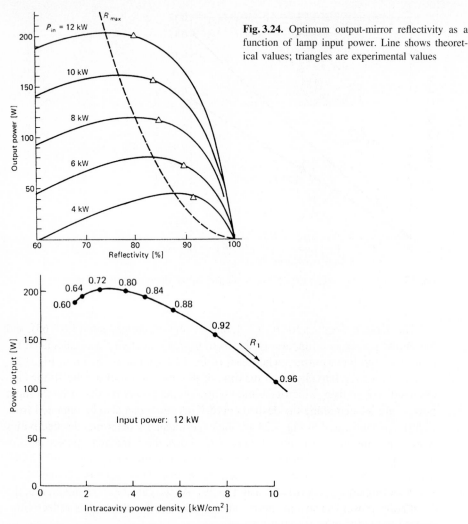

Fig. 3.24. Optimum output-mirror reflectivity as a function of lamp input power. Line shows theoretical values; triangles are experimental values

Fig. 3.25. Intracavity power density as a function of laser power output. Parameter is the reflectivity of the output coupler

The oscillation is a damped sine wave with only a small content of harmonics. Figure 3.26b displays the spectrum of the relaxation oscillations, as obtained by a spectrum analyzer. From (3.90), it follows that the resonant frequency is proportional to $(P_{out})^{1/2}$ of the laser. Figure 3.26b illustrates this dependence. With (3.90) we can calculate the center frequency f_s and the time constant τ_R of the relaxation oscillations. With the laser operated at 1 W output, a mirror transmission of $T = 0.05$, and a beam diameter of 0.12 cm, one obtains $I = 4.0 \times 10^3$ W/cm^2; $l = 35$ cm is the length of the cavity, and $L = 0.03$ are the combined cavity losses. With these values, and $I_s = 2900$ W/cm^2 and $\tau_f = 230\,\mu$s, it follows that $\tau_R = 333\,\mu$s and $f_s = 72$ kHz. The measured center frequency of the relaxation

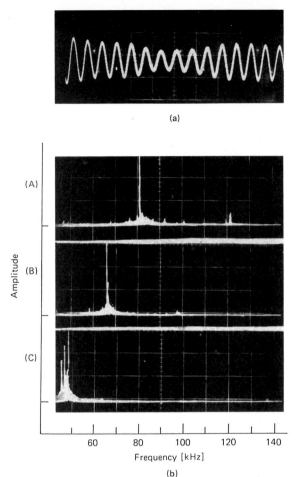

(a)

(b)

Fig. 3.26a,b. Relaxation oscillation of a cw-pumped Nd:YAG laser. (a) Oscilloscope trace showing the temporal behavior of a relaxation oscillation: time scale: $20\,\mu s$/div. (b) Frequency spectrum of relaxation oscillations at different output power levels: (A) 1.3 W, (B) 1.0 W, (C) 0.25 W

oscillation is according to Fig. 3.26b about 68 kHz. In cw-pumped systems one can reduce the amplitude fluctuations caused by relaxation oscillations by orders of magnitude by proper design procedures, which we will discuss in Chap. 5.

The largest Nd:YAG crystals which are readily available are about 15 cm long and 1 cm in diameter. The output from a 10-cm-long crystal pumped by a helical flashlamp is shown in Fig. 3.27. The overall efficiency of the system at maximum input is between 0.8 and 1% depending on the particular crystal. The output energy as a function of beam angle is plotted in the insert of Fig. 3.27. As can be seen, for most crystals half of the total output energy is contained in a beam angle between 3 and 4 mrad, and 90% of the energy is contained in a 6- to 8-mrad angle.

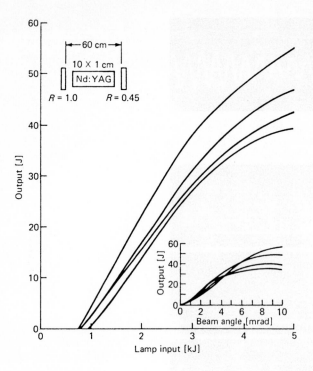

Fig. 3.27. Performance of a pumped pulsed Nd : YAG laser oscillator. Pulse width is 1 ms. Major illustration shows output versus input. Insert shows beam divergence versus output

3.6.4 Alexandrite

Figure 3.28 shows the output performance of a 9.5 mm diameter by 76 mm long alexandrite rod contained in a double-elliptical pump cavity and pumped by two flashlamps. The resonator length was 42 cm. The figure shows the dependence of output pulse energy on input energy with various rod coolant temperatures and

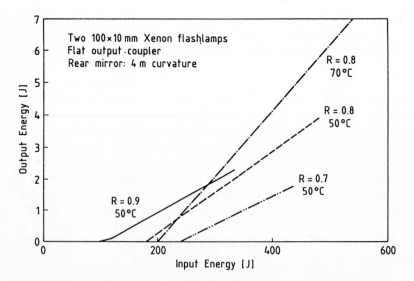

Fig. 3.28. Output energy versus input energy for alexandrite laser at elevated temperatures [3.34]

output mirror reflectivities. With a mirror reflectivity of 80% and a rod coolant temperature of 70° C, the threshold energy was 200 J and the slope efficiency was 2%. The system was operated at a repetition rate of 15 Hz. The temperature dependence of the emission and excited state absorption cross-section of alexandrite is illustrated in Fig. 3.29. As part of an effort to compare theoretical predictions with experimental results, computer predictions and measured output vs. input data are plotted in Fig. 3.30 for a single elliptical pump cavity system. The experimental conditions are listed in Fig. 3.30, the predictions were modelled based on the equations given in Sect. 3.4.

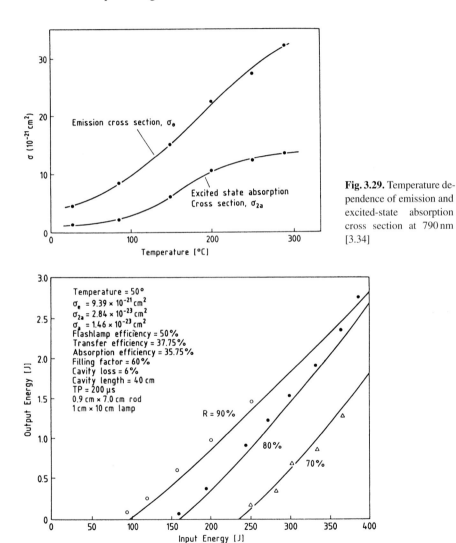

Fig. 3.29. Temperature dependence of emission and excited-state absorption cross section at 790 nm [3.34]

Fig. 3.30. Energy output vs input energy for an alexandrite laser as function of output coupling [3.34]

3.6.5 Laser-Diode-Pumped Systems

With the rapid advances in laser-diode technology, there are now single diodes, 1-cm diode bars and stacked diode bars commercially available at power levels which make these devices very attractive as pump sources for solid-state lasers. Diodes fabricated from GaAlAs with different percentages of Ga and Al emit at wavelengths in the vicinity of 800 nm ranging from approximately 700 to 900 nm. These emission wavelengths coincide very well with strong absorption bands of several lasing ions. In particular, high power laser diode arrays can be fabricated to give a 3–4 nm wide spectral output which can be temperature tuned to coincide with the $^4F_{5/2}$ absorption bands of neodymium in various host materials.

Because of this excellent spectral match, the output from the diode pump is very efficiently utilized in producing a population inversion in the laser material. Since the output of a diode array is directional, the pump radiation can be transferred to the laser material with little loss, either in a close coupled configuration, or by directing the radiation into the gain medium via optical elements. Moreover, because the diode pumps are partially coherent devices, their output beams may be tightly focused or otherwise adjusted to spatially match the resonator modes of the solid-state laser. In the following examples, it will become apparent that the excellent spectral and spatial characteristics of diode pumps as compared to flashlamps, translate into a high pump source efficiency, a high pump radiation transfer efficiency and high spatial overlap between the pump region and resonator modes.

As will be discussed in Chap. 6, diode-pumped systems can be divided into side pumped and end-pumped systems. In the side-pumped geometry, the diode arrays are placed along the length of the laser rod or slab and pump the active material perpendicular to the direction of propagation of the laser resonator mode.

In the end-pumped geometry, the pump radiation is collimated and focused longitudinally into the laser material collinear with the resonator mode. These configurations take full advantage of the spectral as well as spatial properties of laser diodes. In this section, we will discuss examples of both classes of systems.

Side-Pumped Nd : Laser Oscillators

As an example of a side-pumped Nd : laser oscillator, we will consider first a simple laboratory set-up as illustrated in Fig. 3.31. A number of different laser materials were each side-pumped with a diode array close-coupled to the polished rod barrels without intervening optics. The pump array consisted of a stack of five 1 cm pulsed GaAlAs laser bars each containing 40 laser subarrays. The diode array produced 50 mJ/pulse output with a 200 μs pulse width at the maximum current rating of 80 A per pulse. At that current rating, the array had a linewidth of about 4 nm. Figure 3.32 shows the diode output as a function of electrical input. The diode-array output wavelength was temperature tuned with a thermo-electric cooler between 805 and 812 nm. A translation stage was used to optimize

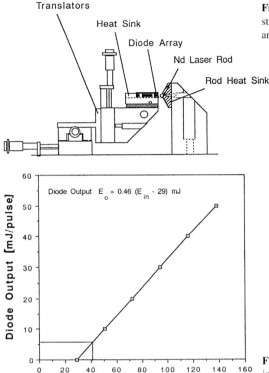

Translators

Heat Sink

Diode Array

Nd Laser Rod

Rod Heat Sink

Fig. 3.31. Experimental set-up of a solid-state laser side-pumped with a laser diode-array

Diode Output $E_o = 0.46 (E_{in} - 29)$ mJ

Fig. 3.32. Output energy vs. electrical input energy for a one cm by 5-bar laser diode array

the pump distribution in the laser rods which were mounted on a highly polished and silver-coated heat sink. The pump radiation emitted from the diode array has a divergence of 40° by 10° full width half maximum.

The polished cylindrical surface of the rod focuses the radiation towards the center, and radiation passing through the active material is reflected back also towards the center by the highly reflective back-surface. The highest output was achieved with the smallest separation between the laser rods and the diode array, i.e., the front facet of the array was almost touching the laser rod. Larger separations produce high Fresnel losses due to the very large angles of incidence at the laser-rod surface.

The laser materials utilized in these experiments carried out in the author's lab, see also [3.35], were a 3.5 mm diameter by 20 mm long, 1.1% doped Nd : YAG rod, a 3.5 mm diameter by 20 mm long, 1% doped Nd : BeL rod, and a 3 mm diameter by 20 mm long, 6% doped Nd : Phosphate Glass (Kigre Q-98) rod. The birefringent Nd : BeL rod was cut along the "y" axis to produce laser output at 1.07 μm.

The 12 cm long laser resonator consisted of a 50 cm-curvature concave total reflector, and a flat partially reflecting output coupler. Output from the lasers was multimode and matched the shape of the pumped area.

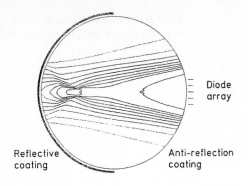

Fig. 3.33. Pump intensity distribution inside a Nd: YAG rod

Diode array

Reflective coating

Anti-reflection coating

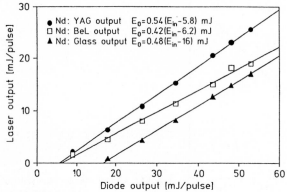

Fig. 3.34. Laser output energy vs. pump pulse energy

Figure 3.33 shows the calculated pump radiation distribution inside the Nd: YAG laser rod obtained from a ray trace analysis which takes into account the spectral and spatial properties of the source and the spectral absorption by the laser material. The laser output beam had an almost rectangular shape, filling the area between the diode array and the back reflector. The hot spot located approximately between the center of the rod and the back reflector is the result of the focusing action of the cylindrical front and back surface. The pumped area of the 3.5 mm diameter laser rod was about 70% of the total cross-section or $A = 0.07\ cm^2$.

The output from the diode-array-pumped lasers is exhibited in Figs. 3.34 and 35. The laser output followed the current input to the diode with less than $10\,\mu s$ build-up time and a small amount of relaxation oscillator. Therefore, for the purpose of analysis, the lasers were assumed to be in a steady-state regime.

Figure 3.34 shows the output energy from the lasers as a function of diode-array pump energy at $0.8\,\mu m$. The optical slope efficiencies are 54% for Nd: YAG, 42% for Nd: BeL, and 48% for Nd: Glass.

Figure 3.35 depicts the output energy from the lasers as a function of electrical energy input to the diode array. The electrical slope efficiencies for the lasers

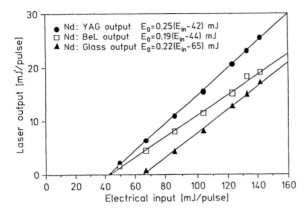

Fig. 3.35. Laser output energy vs. electrical input energy

are 25% for Nd : YAG, 19% for Nd : BeL, and 22% for Nd : Glass. The overall electrical "wallplug" efficiencies at maximum output for the three lasers are 18%, 14%, and 12%, respectively.

The electrical slope efficiencies of these curves are obviously the product of the slope efficiencies of the curves in Figs. 3.32 and 34. Furthermore, the threshold values in Fig. 3.35 are the sum of the diode laser threshold (which occurs at 29 mJ electrical input energy) and the solid-state laser thresholds. The optimum output coupling for each laser was experimentally determined and the results are shown in Fig. 3.36. From the three materials, Nd : YAG has the largest stimulated emission cross-section and therefore the highest gain coefficient. Therefore, for comparable resonator internal losses one would expect the highest output coupling for the Nd : YAG material according to (3.69).

From the experimental data presented in these figures, we can obtain a fairly accurate picture of the various steps involved in the conversion of electrical input power to laser output. Table 3.1 lists the numerical values for the various efficiencies for the three laser materials.

Table 3.1. Energy transfer efficiencies of several laser oscillators

Transfer Process		Nd : YAG	Nd : BeL	Nd : Glass
Diode Slope Efficiency:	η_P	0.46	0.46	0.46
Transfer Efficiency:	η_T	0.99	0.89	0.95
Absorption Efficiency:	η_a	0.91	0.94	0.94
Stokes Shift:	η_S	0.76	0.76	0.76
Quantum Efficiency:	η_Q	0.95	0.82	0.85
Coupling Efficiency:	η_C	0.87	0.77	0.86
Beam Overlap Efficiency:	η_B	0.95	0.95	0.95
Optical Slope Efficiency:	σ_S	0.54	0.42	0.48
Electrical Slope Efficiency:	σ_S	0.25	0.19	0.22

The diode slope efficiency η_P is experimentally determined by measuring the output with a power meter or calorimeter as function of electrical input. The transfer efficiency η_T is strictly a function of Fresnel losses of the cylindrical surface of the rods. In the case of the Nd : YAG rod, one half of the barrel which is facing the diode array is anti-reflection coated. The coating has less than 0.25% loss at 810 nm for normal incidence, but reflection losses are higher at larger angles. The 1% reflection loss is an average value obtained by considering all angles. The other two materials were not coated; therefore the reflection losses are higher.

The absorption efficiency η_a was calculated by employing a computer code specifically designed for diode pumping of materials ([3.36], see also Chap. 6 for details). In this work, the absorption of the laser diode radiation for a number of lasers is calculated vs. path length in the laser material. From this reference follows that for the Nd : YAG laser, for example, about 70% of the incident radiation is absorbed in a 3 mm path length. This is about the average distance between the front surface and the back reflector of the 3.5 mm diameter crystal. From the radiation incident on the back reflector, 30% was estimated to be lost due to absorption and scattering and radiation being reflected into areas of the rod which are outside the gain region. Nd : BeL and Nd : Glass have a higher absorption coefficient for the diode radiation; therefore η_a is somewhat higher for these materials. Absorption efficiencies as a function of material thickness and laser pump wavelength for a number of Nd laser materials can be found in [3.37].

The Stokes shift is simply the ratio of the pump and laser wavelength, and the quantum efficiency is a materials parameter. The coupling efficiency η_C was determined by measuring the resonator losses according to the method described in Sect. 3.4.2. For Nd : YAG, the resonator round trip loss was 1.5%, and for Nd : BeL and Nd : Glass, the loss was 1.8% and 1%, respectively. The output coupling for these three materials was 10% for Nd : YAG, and 6% for Nd : BeL and Nd : Glass. The beam overlap efficiency takes into account pump radiation in areas which have not reached lasing threshold. It is the least known of all the parameters and was adjusted such that the product of the η-terms agreed with the measured slope efficiencies. The optical slope efficiency $\sigma = \eta_T \eta_a \eta_S \eta_Q \eta_C \eta_B$ is the slope of the curves in Fig. 3.34 representing laser output vs. optical pump power, whereas the electrical slope efficiency is the slope of the laser output vs. electrical input curves which includes η_P.

We will now calculate a number of laser parameters from this data by using the Nd : YAG laser as an example. From the values in Table 3.1, one can calculate the gain of the laser as a function of input power. From (3.62) and with $A = 0.07 \, \text{cm}^2$, $I_S = 2.9 \times 10^3 \, \text{W/cm}^2$ and $\Delta t_P = 200 \, \mu s$ follows

$$g_0 l = 15 \times 10^{-3} E_D / \text{mJ}$$

or if we want to express the gain in terms of electrical input power we have to introduce into (3.62) the diode characteristics $E_D = \eta_P(E_{in} - E_{THD})$. The laser gain vs. electrical input energy

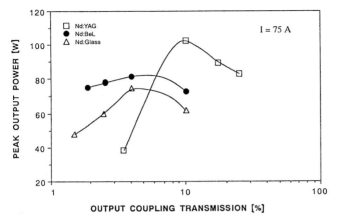

Fig. 3.36. Laser output vs. output coupling

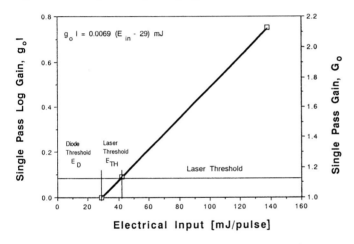

Fig. 3.37. Single-pass logarithmic gain as a function of electrical input energy

$$g_0 l = 6.9 \times 10^{-3}(E_{\text{in}} - 29 \,\text{mJ})$$

is plotted in Fig. 3.37. Also indicated in this figure are the thresholds for the diode laser and solid-state laser. The maximum single pass gain for the laser is $G_0 = \exp(g_0 l) = 2.1$. The laser output at this gain level is $E_{\text{out}} = 24 \,\text{mJ}$, the available energy at the upper level follows from (3.71), with the values above one obtains $E_{\text{avail}} = 30 \,\text{mJ}$. Therefore the extraction efficiency of the laser is $\eta_E = 0.80$. This value can also be obtained from (3.73). The optimum transmission of the laser is according to (3.69) $T_{\text{opt}} = 0.13$. Putting this value into (3.66) gives a threshold input energy of $E_{\text{TH}} = 11 \,\text{mJ}$. The values agree reasonably well with the measured data. The reader is reminded that these performance parameters are based on average values of the highly non-uniform, and multimode beam profile.

The next example illustrates the performance of a relatively large diode array side-pumped oscillator. The system produces an energy per pulse of about

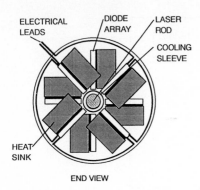

Fig. 3.38. Cross-section (*left*) and photograph (*right*) of diode-pumped Nd : YAG laser head

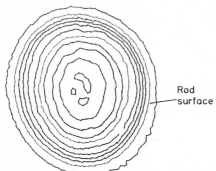

Rod surface

Fig. 3.39. Pump distribution of a 16-diode array side pumped Nd : YAG crystal (each line represents a 10% change in intensity)

0.5 J at a repetition rate of 40 Hz. Critical design issues for this laser include heat removal from the diode arrays and laser rod, and the overlapping of the pump and resonator mode volumes. In side-pumped configurations, laser-diode arrays are not required to be coherent, and pump power can be easily scaled with multiple arrays around the outside of the rod or along its axis. Instead of one diode-array pumping the laser crystal, this particular laser employs 16 diode-arrays located symmetrically around the rod. As shown in Fig. 3.38, the diode pumps are arranged in four rings, each consisting of four arrays. Since each array is one cm long, the total pumped length of the 6.6 cm × 0.63 cm Nd : YAG crystal is 4 cm. This arrangement permits the incorporation of large water-cooled heat sinks required for heat dissipation, and it also provides for a very symmetrical pump profile. An eight-fold symmetry is produced by rotating adjacent rings of diodes by 45°. A photograph of the extremely compact design is also shown in Fig. 3.38. The symmetrical arrangement of the pump sources around the rod produces a very uniform pump distribution, as illustrated in Fig. 3.39. The intensity profile displays the fluorescence output of the rod taken with a CCD camera. In Fig. 3.40, the output vs. optical pump input is plotted for long pulse multimode and TEM_{00} mode operation. Shown also is the output for Q-switch TEM_{00} mode operation. The resonator configuration for the long pulse, multimode operation is

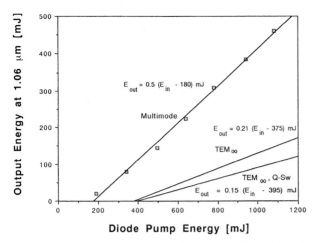

Fig. 3.40. Output vs. input energy for diode-pumped Nd : YAG oscillator

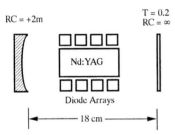

Fig. 3.41. Resonator configuration for multimode operation

depicted in Fig. 3.41. The TEM$_{00}$ mode performance was achieved with a variable reflectivity mirror and a concave-convex resonator structure which will be described in Chap. 5.

The multimode laser output can be expressed by

$$E_{\text{out}} = 0.5 \, (E_{\text{opt}} - 180 \, \text{mJ})$$

where E_{opt} is the optical pump energy from the 16 diode arrays. The electrical input energy E_{in} required to achieve E_{opt} is

$$E_{\text{opt}} = 0.5 \, (E_{\text{in}} - 640 \, \text{mJ}) \, .$$

Combining these two output-input curves relates the laser output with the electrical input energy

$$E_{\text{out}} = 0.25 \, (E_{\text{in}} - 1000 \, \text{mJ}) \, .$$

The slope efficiency of the laser is 25% and the overall electrical efficiency at the maximum output of 460 mJ per pulse is 16%. The optimum output coupling was experimentally determined. Figure 3.42 shows a plot of the laser output for different values of the reflectivity. The different efficiency factors of the system

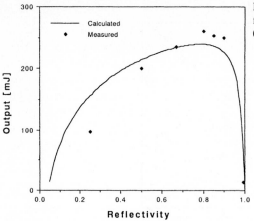

Fig. 3.42. Oscillator output vs. mirror reflectivity (calculated curve based on $T_{opt} = 0.2$, $g_0 l = 1.4$, $L = 0.03$)

are listed in Table 3.2. The slope efficiency of the diode array was measured with a power meter. The coupling of radiation to the rod is lower than in the previous example because in this system a water jacket and coolant are in the optical path of the pump radiation. The sapphire water jacket is AR coated for the pump radiation. The absorbed pump radiation in the 6.3 mm diameter crystal was calculated as before from a computer code.

Table 3.2. Energy transfer efficiencies of a large diode-pumped Nd:YAG oscillator

Transfer Process		Multimode	Single Mode
Diode Slope Efficiency:	η_P	0.50	0.50
Transfer Efficiency:	η_T	0.95	0.95
Absorption Efficiency:	η_a	0.90	0.90
Stokes Shift:	η_S	0.76	0.76
Quantum Efficiency:	η_Q	0.95	0.95
Coupling Efficiency:	η_C	0.90	0.90
Beam Overlap Efficiency:	η_B	0.90	0.38
Electrical Slope Efficiency:	σ_S	0.25	0.10

The output coupling follows from $T_{opt} = 0.20$ and the measured round trip loss of $L = 2.2\%$. The gain/mode overlap efficiency η_B was estimated by comparing the beam profile with the pump distribution in the laser rod. The final value was adjusted for the product of the η-terms to agree with the measured slope efficiency of 25%. For TEM$_{00}$ mode operation, the major difference is a substantially reduced value for η_B as a result of a smaller beam. All other parameters remain unchanged.

End-Pumped Laser Oscillators

In this so-called end-pumped configuration, the radiation from a single laser-diode or diode bar is focused to a small spot on the end of a laser rod. With a suitable choice of focusing optics, the diode pump radiation can be adjusted to coincide with the diameter of the TEM_{00} resonator mode. The end pumping configuration thus allows the maximum use of the energy from the laser diodes.

Using this longitudinal pumping scheme, the fraction of the active laser volume excited by the diode laser can be matched quite well to the TEM_{00} lasing volume. A solid-state laser pumped in this manner operates naturally in the fundamental spatial mode without intracavity apertures.

Many excellent systems with extremely compact packaging and high efficiency are commercially available which utilize the mode matching pump profile. By way of illustrating the concept, we will consider the configuration illustrated in Fig. 3.43 which was originally proposed by *Sipes* [3.38]. The output from a diode array with 200 mW output at 808 nm is collimated and focused into a 1 cm long × 0.5 cm diameter 1% Nd : YAG sample. The resonator configuration is plano-concave, with the pumped end of the Nd : YAG rod being coated for high reflection at 1.06 μm, and with an output coupler having a 5 cm radius of curvature and a reflectivity at 1.06 μm of 95%.

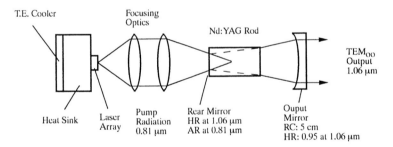

Fig. 3.43. End-pumped laser oscillator

Figure 3.44 displays electrical input power versus 1.06 μm output power for the configuration illustrated in Fig. 3.43. We see that for approximately 1 W of electrical input power, 80 mW of Nd : YAG output is measured.

The electrical slope efficiency of the laser is 13%, and the overall efficiency at 80 mW output is 8%. Also shown in Fig. 3.44 is the performance of the diode array which has a slope efficiency of 34% and an overall efficiency of 22% at 220 mW output. The energy transfer steps of the laser are listed in Table 3.3. The slope efficiency η_P of the diode laser output is a measured quantity. The transfer efficiency η_T includes the collection of diode radiation by the lens system and reflection losses at the surfaces. The pump radiation is completely absorbed in the 1-cm long crystal, i.e., $\eta_a = 1$. The coupling efficiency η_C follows from the measured 1% one-way loss and an output coupling of 5%. As stated by the author,

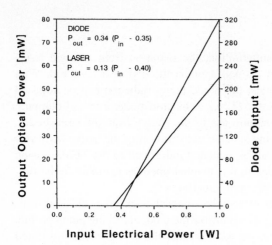

Fig. 3.44. Nd : YAG laser and diode output vs. electrical input power

the large ellipticity of the diode array beams made it difficult to focus the entire pump beam into the laser resonator mode. The value of η_B is therefore lower than can be achieved with an optimized system. The measured slope efficiency of the laser is the product of the η-terms listed in Table 3.3.

Table 3.3. Energy transfer efficiencies of an end-pumped Nd : YAG oscillator

Transfer Process		
Diode Slope Efficiency:	η_P	0.34
Transfer Efficiency:	η_T	0.90
Absorption Efficiency:	η_a	1.00
Stokes Shift:	η_S	0.76
Quantum Efficiency:	η_Q	0.95
Coupling Efficiency:	η_C	0.71
Beam Overlap Efficiency:	η_B	0.85
Electrical Slope Efficiency:	σ_S	0.13

This example of an end-pumped laser demonstrates the high overall performance at TEM$_{00}$ mode which can be achieved with this type of system.

The attractive features of end-pumped lasers are a very compact design combined with high beam quality and efficiency as a result of the good overlap between the pumped region and the TEM$_{00}$ laser mode. Details of mode matching of a Gaussian pump beam with a Gaussian TEM$_{00}$ laser beam can be found in [3.39].

Figure 3.45 displays a photograph of a typical end-pumped laser. The optical components shown separately, are the laser diode array, two cylindrical and two spherical lenses, Nd : YAG crystal, polarizer, Q-switch, and output coupler. The end mirror of the resonator is coated onto the laser crystal. The diode array is

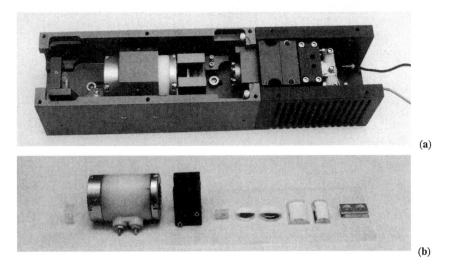

(a)

(b)

Fig. 3.45. Photograph of assembled end-pumped laser (**a**), and disassembled optical components (**b**) [Fibertek, Inc.]

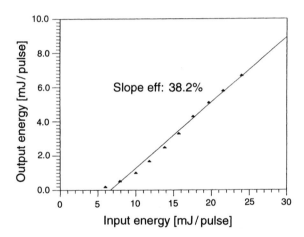

Slope eff: 38.2%

Fig. 3.46. Laser output as a function of pump energy at the laser crystal [3.40]

comprised of three quasi-cw 1-cm bars. The three-bar stack had a maximum output energy of 27 mJ in a 200 μs pulse. At that pulse energy, the laser output was 6.7 mJ in the TEM_{00} mode and 8.4 mJ multimode (Fig. 3.46). The pulse repetition was 50 pps. The resonator loss was determined by measuring the pump threshold as a function of the reflectivities of the output coupler, as described in Sect. 3.4.2. The measurement yielded a value of 2.4% for the resonator round-trip loss. The maximum output was obtained for an output coupler reflectivity of 95%. The optical and mechanical design of this laser is depicted in Figs. 6.61 and 6.62.

Laser-diode end-pumped cw operation has been extended to a number of laser materials. Figure 3.47 compares the output from Nd : BeL, Nd : YAG and

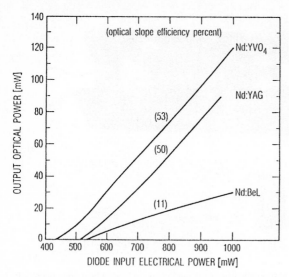

Fig. 3.47. Laser output as a function of pump laser diode electrical input. Optical slope efficiency percents are in parentheses. (YAG and YVO$_4$ at 97 % and BeL at 99 % output couplers) [3.43]

Nd : YVO$_4$ [3.43]. The spectral differences of the materials studied have a considerable impact on laser threshold and slope efficiency. The comparative power curves of Fig. 3.47 were compiled based on the optimized value of output coupler and diode-pump wavelength for each crystal. Nd : YVO$_4$ clearly exhibits the lowest threshold despite its larger optical losses and short fluorescence lifetime (95 μs).

In recent years, end-pumped lasers have been constructed with many different Nd-doped laser materials, as well as Ho-, Tm-, Yb- and Cr-doped crystals. Information regarding the latest developments can be found in the proceedings and technical digests of the various annual laser conferences. The performance of various diode-pumped Nd lasers has been summarized in [3.44]. The following examples demonstrate the large range of pulse shapes and wavelengths, which can be generated with end-pumped solid-state lasers. Picosecond pulses have been generated in Nd : YAG [3.45] and Nd : YLF [3.46], and femtosecond pulses in Cr : LiSAF [3.47, 48]. Endpumped Nd : YAG frequency doubled to 0.532 μm [3.41], Yb : YAG at 1.03 μm [3.42], and Tm : YAG at 2.0 μm [3.49] oscillators have all generated output powers in excess of 100 W. High-efficiency operation with optical-to-optical conversion up to 50% has been achieved in Nd : YVO$_4$ pumped with stacks of diode bars [3.50].

In the diode-pumped lasers discussed so far, the active element is a thin slab or rod a few millimeters long. An extreme case of an end-pumped laser is the fiber laser. The development of low-loss rare-earth-doped fibers has led to the construction of a number of single-mode fiber lasers [3.51–55]. Due to the small volume of the active core, low threshold and efficient operation has been achieved.

The pumping of solid-state laser materials with single diodes or diode bars is a very rapidly emerging technology and new lasers and pumping configurations are reported at a rapid pace.

3.7 Travelling-Wave Oscillator

The oscillators which we have discussed so far are characterized by standing waves in the resonator. In an oscillator consisting of a ring-like resonator utilizing 3 or 4 mirrors and a nonreciprocal optical gate, a travelling wave can be generated. The optical gate provides a high loss for one of the two counter-circulating travelling waves. The wave with the high loss is suppressed, and a unidirectional output from the laser is obtained. A typical laser cavity configuration of a travelling-wave oscillator is shown in Fig. 3.48. The system consists of a four-mirror rectangular resonator, a Brewster-ended laser rod, a $\lambda/2$ plate, and a Faraday rotator. Three mirrors are coated for maximum reflectivity at the laser wavelength and the fourth mirror is partially transparent. The unidirectional optical gate is formed by a half-wave plate and a Faraday rotator, consisting of a glass rod located within a solenoid-generated axial magnetic field.

The Faraday element, which should possess a high Verdet constant, rotates the plane of polarization (defined by the Brewster-ended rod) of the two circulating beams by a small angle $\pm \Theta$, the sign being dependent on the propagation direction and polarity of the magnetic field. The half-wave plate is orientated with one of its axes at an angle $\beta/2$ with respect to the polarization of the beams. The magnetic field is adjusted such that for the wave travelling in the clockwise direction, the half-wave plate rotates the plane of polarization by an angle β and the Faraday cell by an angle $+\Theta$ so that the total rotation is $\beta + \Theta$. The differential loss $\Delta\alpha$ between the counterrotating waves is then

$$\Delta\alpha = \sin^2(\beta + \Theta) - \sin^2(\beta - \Theta) . \tag{3.91}$$

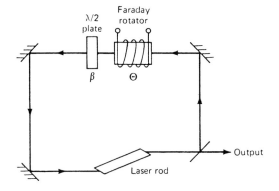

Fig. 3.48. Optical schematic of a travelling-wave oscillator [3.56]

For maximum power output, the half-wave plate and the intensity of the magnetic field on the Faraday rotator are adjusted so that $\beta = \Theta$. Then the entire loss, proportional to $4\Theta^2$, is experienced only in the clockwise direction, leaving the wave in the counterclockwise direction unattenuated. As a result the laser will oscillate in a unidirectional wave.

Instead of the rectangular four-mirror cavity, a three-mirror cavity can be employed equally well. Also, the Brewster-ended rod can be replaced with a polarizer and a flat-ended rod. In some designs, rather than using a Faraday rotator, the travelling mode is produced by means of an asymmetric resonator [3.57]. Travelling-wave oscillators have generated interest mainly as a way to eliminate "spatial hole burning" (Sect. 5.2) caused by the standing-wave distribution of the intensity in a conventional oscillator. Since travelling-wave oscillators of the type depicted in Fig. 3.48 are more complicated to construct and require more optical components compared to standing-wave oscillators, these systems have not found any real applications in the past.

However, the interest in travelling-wave lasers has dramatically increased with the emergence of laser diodes as practical pumps for Nd:YAG lasers. The compact designs made possible with end-pumped diode-laser geometries have resulted in monolithic ring lasers where the functions of the elements shown in Fig. 3.48 are performed by a single Nd:YAG crystal. Unidirectional ring lasers contain three essential elements: a polarizer, a half-wave plate, and a Faraday rotator. The polarizer, half-wave plate equivalent, and the Faraday rotator are all embodied in the nonplanar ring Nd:YAG laser first proposed by *Kane* et al. [3.58], and illustrated in Fig. 3.49. With a magnetic field H present in the direction shown, the YAG crystal itself acts as the Faraday rotator, the out-of-plane total internal reflection bounces (labelled A and C) act as the half-wave plate, and the output coupler (mirror D) acts as a partial polarizer. Polarization selection results from non-normal incidence at the dielectrically coated output mirror.

The basic idea of a monolithic, diode-pumped uni-directional ring laser is to provide the equivalent of a discrete element design of a half-wave plate with a fast axis rotation angle which is half of the Faraday rotation angle. Since the Faraday rotation is small, the equivalent wave-plate rotation angle is also made small. The design of non-planar ring oscillators incorporating the above concepts has matured considerably over the past years. A number of modified versions of the original monolithic ring laser have also been reported [3.59–62]. Compactness, single mode operation combined with excellent frequency and amplitude stability due to the monolithic structure are the main practical advantages of this design.

As we will discuss in Sect. 5.2.3 these devices are important as seed lasers for injection-seeded Nd:YAG oscillators.

The architecture of the monolithic ring laser precludes the use of intracavity elements such as those required for efficient second harmonic generation, Q-switching or tuning of the single-frequency output. For these cases, a laser-diode pumped ring laser, comprised of discrete elements can be constructed [3.63, 64].

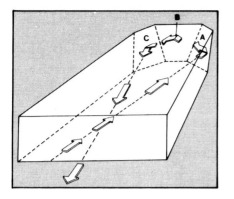

Fig. 3.49. Design of a monolithic ring Nd : YAG laser [3.58]

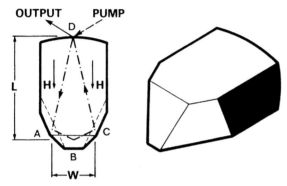

Figure 3.50 depicts a ring-configuration resonator for a cw Ti : sapphire laser. The pump beam generated by an argon laser provides end-pumping of the Ti : sapphire crystal. The Faraday rotator and $\lambda/2$-waveplate combination assure unidirectional operation. The remaining resonator components, a birefringent filter and an etalon are for wavelength selection and linewidth narrowing, respectively.

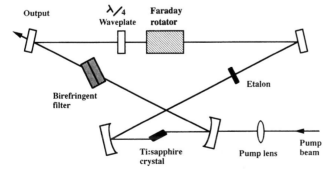

Fig. 3.50. The ring configuration for a cw Ti : sapphire laser cavity (courtesy of Schwartz Electro-Optics)

4. Laser Amplifier

In this chapter we will discuss the gain in energy for a laser beam passing through an optically active material. The use of lasers as pulse amplifiers is of great interest in the design of high-energy, high-brightness light sources. The generation of high-energy pulses is based on the combination of a master oscillator and multistage power amplifier. For the purpose of illustrating the amplifier concept and principles we assume a straightforward system, as shown in Fig. 4.1. In this scheme an amplifier is driven by an oscillator which generates an initial laser pulse of moderate power and energy. In the power amplifier with a large volume of active material the pulse power can grow, in extreme cases, up to 100 times.

In an oscillator-amplifier system, pulse width, beam divergence, and spectral width are primarily determined by the oscillator, whereas pulse energy and power are determined by the amplifier. Operating an oscillator at relatively low energy levels reduces beam divergence and spectral width. Therefore, from an oscillator-amplifier combination one can obtain, either a higher energy than is achievable from an oscillator alone, or the same energy in a beam which has a smaller beam divergence and narrower linewidth. Generally speaking, the purpose of adding an amplifier to a laser oscillator is to increase the brightness $B[W \, \mathrm{cm}^{-2}\mathrm{sr}^{-1}]$ of the output beam

$$B = \frac{P}{A\Omega} \, , \qquad (4.1)$$

where P is the power of the output beam emitted from the area A, and Ω is the solid-angle divergence of the beam. Multiple-stage amplifier systems can be built

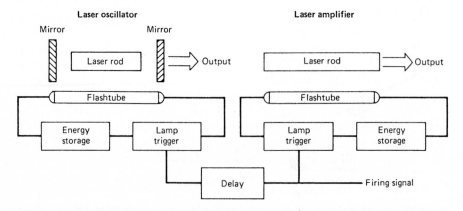

Fig. 4.1. Schematic diagram of a laser oscillator-amplifier configuration

if higher amplifications are required. In extreme cases Nd : glass laser systems have been constructed in which a 1-mJ output from an oscillator is amplified to a 1-kJ beam.

In the design of laser amplifiers the following aspects must be considered:

– Gain and energy extraction.
– Wavefront and pulse-shape distortions introduced by the amplifier.
– Energy and power densities at the optical elements of the amplifier system.
– Feedback in the amplifier which may lead to superradiance or prelasing.

Of primary interest in the design of amplifiers is the gain which can be achieved and the energy which can be extracted from the amplifier. The rod length in an amplifier is determined primarily by the desired gain, while the rod diameter, set by damage threshold considerations, is dependent on the output energy. We shall see in the following sections that the gain of an amplifier pumped at a certain inversion level depends on the intensity and duration of the input pulse.

To a first approximation we can assume the growth of input energy to be exponential, since the stimulated emission is proportional to the exciting photon flux. It will be seen, however, that exponential amplification will occur only at low photon-flux levels. A deviation from the exponential gain regime in an amplifier occurs when an optical pulse traveling in the inverted medium becomes strong enough to change the population of the laser levels appreciably. The optical amplifier will exhibit saturation effects as a result of depletion of the inversion density by the driving signal. Taking an extreme case, we can see that if a high-intensity light pulse is incident on a laser rod, the stimulated emission can completely deplete the stored energy as it progresses. Then the gain can be expected to be linear with the length of the rod rather than exponential.

Let us assume that a low-level signal is incident on a long amplifier which is uniformly inverted. At the beginning the signal increases exponentially with distance, then after a transition region the signal amplitude grows linearly with length, and finally the signal no longer increases at all. This happens when the gain per unit length just balances the losses per unit length in the rod.

So far we have discussed the effect of the input signal level on the gain of the amplifier. We will also have to consider the pulse length and its influence on the amplification mechanism. In an amplifier where the input pulse is considerably shorter than the fluorescence lifetime of the medium, such as a Q-switch pulse or a mode-locked pulse, energy is extracted from the amplifier which was stored in the amplifying medium prior to the arrival of the pulse. If the signal is long compared to the fluorescent time, such as the output from a free running oscillator or a cw laser, a steady-state gain best characterizes the amplification mechanism. Hence the effect of spontaneous emission and pumping rate on the population inversion cannot be neglected.

The two regimes of light amplification – short pulse and steady state or long pulse – will be discussed in Sects. 4.1 and 4.2. The phenomenon of gain saturation is common to both time domains. We will see, however, that in the case of pulses which are short compared to the fluorescence lifetime of the material the amplification depends on the energy density, whereas in the case of long pulses or cw mode of operation the gain depends on power density.

Any amplification process is associated with some kind of distortion. In Sect. 4.3 we will summarize wavefront and pulse-shape distortions associated with the amplification of optical pulses. In multiple-stage amplifier systems the stability of the system is of prime concern to the laser designer. In Sect. 4.4 the conditions for stable operation of amplifiers will be discussed.

4.1 Pulse Amplification

The events during the amplifier action are assumed to be fast compared with the pumping rate W_p and the spontaneous emission time τ_f. Therefore $t_p \ll \tau_f, W_p^{-1}$, t_p being the width of the pulse which passes through the amplifying medium.

Thus the amplification process is based on the energy stored in the upper laser level prior to the arrival of the input signal. As the input pulse passes through the amplifier, the atoms are stimulated to release the stored energy. The amplification process can be described by the rate equations (1.58,61). If we ignore the effect of fluorescence and pumping during the pulse duration, we obtain for the population inversion

$$\frac{\partial n}{\partial t} = -\gamma nc\sigma\phi . \tag{4.2}$$

The growth of a pulse traversing a medium with an inverted population is described by the nonlinear, time-dependent photon-transport equation, which accounts for the effect of the radiation on the active medium and vice versa,

$$\frac{\partial \phi}{\partial t} = cn\sigma\phi - \frac{\partial \phi}{\partial x} c . \tag{4.3}$$

The rate at which the photon density changes in a small volume of material is equal to the net difference between the generation of photons by the stimulated emission process and the flux of photons which flows out from that region. The latter process is described by the second term on the right of (4.3). This term which characterizes a traveling-wave process is absent in (1.61).

Consider the one-dimensional case of a beam of monochromatic radiation incident on the front surface of an amplifier rod of length L. The point at which the beam enters the gain medium is designated the reference point, $x = 0$. The two differential equations (4.2, 3) must be solved for the inverted electron population n and the photon flux ϕ. *Frantz* and *Nodvik* [4.1] and others [4.2, 3] solved these nonlinear equations for various types of input pulse shapes.

If we take for the input to the amplifier a square pulse of duration t_p and initial photon density ϕ_0, the solution for the photon density is

$$\frac{\phi(x,t)}{\phi_0} = \left\{1 - [1 - \exp(-\sigma n x)] \exp\left[-\gamma\sigma\phi_0\left(c(t - \frac{x}{c})\right)\right]\right\}^{-1}, \tag{4.4}$$

where n is the inverted population density, assumed to be uniform throughout the laser material at $t = 0$. The energy gain for a light beam passing through a laser amplifier of length $x = l$ is given by

$$G = \frac{1}{\phi_0 t_p} \int_{-\infty}^{+\infty} \phi(l,t)dt . \tag{4.5}$$

After introducing (4.4) into (4.5) and integrating, we obtain

$$G = \frac{1}{c\gamma\sigma\phi_0 t_p} \ln\{1 + [\exp(\gamma\sigma\phi_0\tau_0 c) - 1]e^{n\sigma l}\} . \tag{4.6}$$

We shall cast this equation in a different form such that it contains directly measurable laser parameters. The input energy per unit area can be expressed as

$$E_{in} = c\phi_0 t_p h\nu . \tag{4.7}$$

A saturation fluence E_s can be defined by

$$E_s = \frac{h\nu}{\gamma\sigma} = \frac{E_{st}}{\gamma g_0} \tag{4.8}$$

where $E_{st} = h\nu n$ is the stored energy per volume, and $g_0 = n\sigma$ is the small-signal gain coefficient.

In a four-level system $\gamma = 1$, and the total stored energy per unit volume in the amplifier is

$$E_{st} = g_0 E_s . \tag{4.9}$$

The extraction efficiency η_E is the energy extracted from the amplifier divided by the stored energy in the upper laser level at the time of pulse arrival. With this definition, we can write

$$\eta_E = \frac{E_{out} - E_{in}}{g_0 \ell E_s} . \tag{4.10}$$

In this expression E_{out}, E_{in} is the amplifier signal output and input fluence, respectively. In a four-level system, all the stored energy can theoretically be extracted by a signal. In a three-level system $\gamma = 1 + g_2/g_1$, and only a fraction of the stored energy will be released because as the upper laser level is depleted, the lower-level density is building up.

Introducing (4.7, 8) into (4.6), one obtains

$$G = \frac{E_s}{E_{in}} \ln\left\{1 + \left[\exp\left(\frac{E_{in}}{E_s}\right) - 1\right]G_0\right\} . \tag{4.11}$$

This expression represents a unique relationship between the gain G, the input pulse energy density E_{in}, the saturation parameter E_s, and the small-signal, single-pass gain $G_0 = \exp(g_0 l)$.

Equation (4.11), which is valid for rectangular input pulses, encompasses the regime from small-signal gain to complete saturation of the amplifier. The equation can be simplified for these extreme cases. Consider a low-input signal E_{in} such that $E_{in}/E_s \ll 1$, and furthermore $G_0 E_{in}/E_s \ll 1$; then (4.11) can be approximated to

$$G \approx G_0 \equiv \exp(g_0 l) . \tag{4.12}$$

In this case, the "low-level gain" is exponential with rod length, and no saturation effects occur. This, of course, holds only for rod lengths up to a value where the output energy density $G_0 E_{in}$ is small compared to E_s.

For high-level energy densities such that $E_{in}/E_s \gg 1$, (4.11) becomes

$$G \simeq 1 + \left(\frac{E_s}{E_{in}} \right) g_0 l . \tag{4.13}$$

Thus, the energy gain is linear with the length of the gain medium, implying that every excited state contributes its stimulated emission to the beam. Such a condition obviously represents the most efficient conversion of stored energy to beam energy, and for this reason amplifier designs which operate in saturation are used wherever practical, with the major limitation being the laser damage threshold.

We will now recast (4.11) into a form which makes it convenient to model the energy output and extraction efficiency for single and multiple amplifier stages operated either in a single or double pass configuration. With the notation indicated in Fig. 4.2, E_0 is now the input to the amplifier and E_1 is the output fluence which are related by

$$E_1 = E_S \ln \left\{ 1 + \left[\exp\left(\frac{E_0}{E_S} \right) - 1 \right] \exp(g_0 l) \right\} . \tag{4.14}$$

The extraction efficiency is according to (4.10)

$$\eta_1 = (E_1 - E_0)/g_0 l E_S . \tag{4.15}$$

In a laser system which has multiple stages, these equations can be applied successively, whereby the output of one stage becomes the input for the next stage.

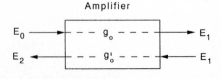

Amplifier

Fig. 4.2. Notation for the calculation of energy output fluence and extraction efficiency for one- and two-pass single or multiple amplifier stages

As already mentioned, efficient energy extraction from an amplifier requires that the input fluence is comparable to the saturation fluence of the laser transition. For this reason, amplifiers are often operated in a double-pass configuration; a mirror at the output returns the radiation a second time through the amplifier. A $\lambda/4$ waveplate is usually inserted between the amplifier and the mirror; this causes a 90° rotation of the polarization of the return beam. A polarizer in front of the amplifier separates the input from the output signal. In some situations, as shall be discussed in Sect. 10.4, the simple reflective mirror may be replaced by a phase conjugate mirror, in which case optical distortions in the amplifier chain will be reduced.

The output fluence E_2 from a two-pass amplifier can be calculated as follows:

$$E_2 = E_S \ln\left\{1 + \left[\exp(E_1/E_S) - 1\right] \exp(g_0'l)\right\} . \tag{4.16}$$

The input for the return pass is now E_1 which is obtained from (4.14) as the output of the first pass. The gain for the return pass is now lower because energy has been extracted from the gain medium on the first pass

$$g_0' = (1 - \eta_1)g_0 . \tag{4.17}$$

The extraction efficiency of the double-pass amplifier is

$$\eta_2 = (E_2 - E_0)/g_0 l E_S . \tag{4.18}$$

The extraction efficiency calculated from (4.14–18) for one- and two-pass amplifiers, for different values of $g_0 l$ and normalized input fluences, are plotted in Fig. 4.3. The results show the increase in extraction efficiency with higher input energies, and the considerable improvement one can achieve with double-pass amplifiers. Equations (4.14–18) can be readily applied to multistage systems, by writing a simple computer program which sequentially applies these equations

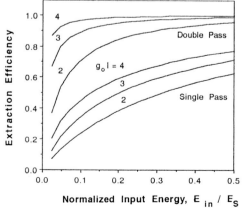

Fig. 4.3. Extraction efficiency for a one- and two-pass amplifier as a function of input intensity E_{in} and small signal logarithmic gain $g_0 l$. Input is normalized to saturation fluence E_S

to the different amplifier stages. In Sect. 4.1.3, we will illustrate the results of such a modeling effort for a 4-stage double pass amplifier chain.

It should be noted that the above equations assume a uniform gain coefficient and beam intensity profile. In most systems, both quantities will have a radially dependent profile. In this case, an effective gain coefficient can be calculated according to

$$g_{\text{eff}} = \int g_0(r) I_B(r) 2\pi r dr \Big/ \int I_B(r) 2\pi r dr \qquad (4.19)$$

where $g_0(r)$ is the radial gain distribution, and $I_B(r)$ is the radial intensity profile of the beam.

In laser amplifier technology the small-signal gain coefficient $g_0 = n\sigma_{21}$ is often expressed as

$$g_0 = \beta E_{\text{st}} , \qquad (4.20)$$

where E_{st} is the previously discussed stored energy per unit volume and

$$\beta = \frac{\sigma_{21}}{h\nu} \qquad (4.21)$$

is a parameter relating the gain to the stored energy. Equation (4.11) can be rearranged to take the form

$$E_{\text{out}} = E_s \ln\left\{ 1 + \left[\exp\left(\frac{E_{\text{in}}}{E_s}\right) - 1 \right] \exp(\beta E_{\text{st}} l) \right\} . \qquad (4.22)$$

Equations (4.11, 22) permit one to calculate the gain of an amplifier as a function of the input energy density, provided that the small-signal gain or the energy stored in the amplifier is known. These parameters, which depend on the input energy, the volume of the active material, the efficiency of the pump structure, etc., normally must be estimated. Later in this section we will list typical performance data for ruby, Nd : glass, and Nd : YAG amplifiers which can be used as guidelines.

One of the significant points of (4.11) for the design of laser amplifiers lies in the fact that if one data point of an existing amplifier is known, the performance of the amplifier under different operating conditions can be calculated. Also, the effect on the performance of changes in design, such as amplifier length or diameter or the use of multiple stages in a system, can be studied with the aid of this equation. Before we discuss practical examples, it should be noted that in deriving (4.11) two assumptions were made:

1) It was assumed that the pulse shape of the incident pulse was rectangular. However, it should be mentioned that despite this assumption, the above analysis holds, to a good approximation, for a symmetrical triangular-shaped pulse. This approximation becomes less valid in cases where amplified pulse

shapes differ significantly from incident pulse shapes as a result of the higher gain experienced by the leading edge of the pulse. In such cases, more accurate gain equations, given in [4.4, 5], should be used.

2) We have assumed a lossless amplifier. In real solid-state laser amplifiers there inevitably exist linear losses of radiation as a result of absorption and scattering caused by defects and impurities in the active medium. A linear loss limits the energy growth in the saturation regime. There the energy tends to grow linearly as a result of the amplification and to decrease exponentially as a result of loss.

Avizonis and *Grotbeck* [4.6] have derived an expression which describes the gain process in an amplifier without approximation. They obtained

$$\frac{dE(x)}{dx} = E_s g_0 \left[1 - \exp\left(\frac{-E(x)}{E_s} \right) \right] - \alpha E(x) , \tag{4.23}$$

where $E(x)$ is the pulse energy at point x, α is the loss coefficient per unit length, and x is the amplifier length coordinate.

Equation (4.23) can be solved analytically only if a zero loss is assumed ($\alpha = 0$). In this case the result is identical to (4.11). From (4.23) follows that, in the presence of losses, gain in the saturating regime occurs only for pulse energies below a limiting value E_{max}. For $\alpha \ll g_0$ this limiting value is

$$E_{max} = \frac{g_0 E_s}{\alpha} . \tag{4.24}$$

Most laser materials such as Nd : glass, and Nd : YAG have sufficiently small losses that they can be neglected in most cases in the design of amplifiers. Loss coefficients for these materials are typically less than $0.001 \, \text{cm}^{-1}$. For values of E_s between 5–$10 \, \text{J/cm}^2$ and g_0 between 0.1 and $0.5 \, \text{cm}^{-1}$, the theoretical value of the limiting energy E_{max} is in the range of several hundreds of joules per square centimeter. In practice, the realizable pulse output energy is limited to significantly lower values because of damage in the active medium. Therefore, only in optically very poor host materials, where scattering or absorption losses are high, need one take α into account. References treating laser amplifiers including losses can be found in [4.7, 8].

The amplification efficiency (i.e., the ratio of the energy extracted by the signal to the energy stored in the active medium) in laser amplifiers can be increased by passing the signal twice through the active medium. The double-pass technique yields a considerable increase in the small-signal gain. This advantage must be weighted against the added complexity of a multiple-pass amplifier brought about by the addition of beam-deflecting optics. Double-pass amplifiers are described in [4.9, 10]. In order to illustrate the usefulness of (4.22), we will calculate the gain for several amplifiers and compare the calculations with measured data.

4.1.1 Ruby Amplifiers

As mentioned in Sect. 2.2, laser action in ruby occurs on the R_1 emission line. However, since energy is stored in both the $\overline{E}$ and $2\overline{A}$ levels, a transfer of energy from the $2\overline{A}$ to the $\overline{E}$ level can take place during the amplification process. Whether energy transfer between the two levels actually occurs depends on whether the relaxation time between the $\overline{E}$ and $2\overline{A}$ levels is short or long compared with the length of the amplified pulse. The relaxation time between the two excited levels in ruby is on the order of 1 ns or less.

Amplification of Q-Switched Pulses

First we will consider the amplification mechanisms for pulses which are longer than 1 ns. In this case, the two upper levels (i.e., E and $2A$ levels) remain in thermal equilibrium and energy can be extracted from both levels. The stored energy in the combined upper levels is

$$E_{st} = h\nu n = h\nu(2n_2 - n_{tot}) . \tag{4.25}$$

The maximum energy which one may hope to extract from ruby is

$$E_{ex} = E_{st}/2 = E_{st}(\overline{E}) . \tag{4.26}$$

The extraction efficiency in ruby is therefore

$$\eta_E = E_{ex}/E_{st}(\overline{E}) \tag{4.27}$$

and from (2.12 and 4.20, 25), we obtain for ruby

$$g_0 = \beta E_{st}(\overline{E}) = \beta' E_{st} \quad \text{with} \quad \beta' = \beta/2 . \tag{4.28}$$

The saturation fluence defined as the ratio of extractable energy to the small-signal gain coefficient, is

$$E_s = \frac{h\nu}{\sigma_{21}} . \tag{4.29}$$

If we introduce the parameters listed in Table 2.2 into (4.28, 29), we obtain $E_s = 11.0 \, \text{J/cm}^2$ and $\beta' = 0.044 \, \text{cm}^2/\text{J}$. The expressions which relate the upper-state population density to the gain coefficient, stored energy, extractable energy, and total upper-state energy density are presented in graphical form in Fig. 4.4.

Having summarized the pertinent laser parameters for ruby, we turn now to (4.22). Introducing the materials parameter E_s and β' into this equation and assuming values for E_{st} and E_{in}, one can plot the output energy density of a ruby amplifier as a function of amplifier length. The curves in Fig. 4.5 were obtained by assuming a value of $E_{in} = 0.1 \, \text{J/cm}^2$ and three different values of the stored energy $E_{st} = 3.0$, 4.0, and 4.5 J/cm^3 for total inversion.

The practical significance of these curves lie in the fact that they can be used for any input energy and amplifier length. For example, assume that there is a

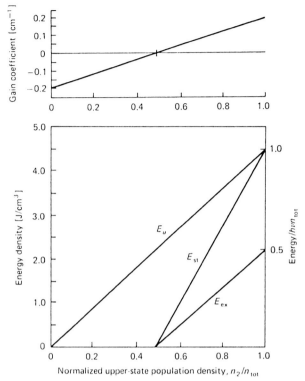

Fig. 4.4. Gain coefficient and upper-state energy in optically pumped ruby. E_u is the upper-state energy density, E_{st} is the stored energy, E_{ex} is the maximum extractable energy density, n_{tot} is the total Cr^{3+} concentration in ruby, and n_2 is the upper-state population density

total of $1\,J/cm^2$ being delivered by the oscillator to a 10-cm-long amplifier with a stored energy of $E_{st} = 4\,J/cm^3$. A line drawn through the ordinate at $1\,J/cm^2$ intersects the curve at A, adding to that the amplifier length, we obtain point B. Point C, which has the same abscissa as point B, intersects the curve at $4\,J/cm^2$, which is the output from the amplifier.

The only parameter that is difficult to determine is the rod energy storage E_{st} per unit volume. This parameter is dependent on the flashlamp input energy and pulse width, the design and efficiency of the pumping geometry, and the laser rod geometry. From the rate equations (1.53) we can obtain a relationship between the pump pulse intensity and the stored energy. Since $\phi = 0$ prior to the arrival of the signal, (1.53) has the solution

$$\frac{n_1}{n_{tot}} = \left\{1 + W_p\tau_f \exp\left[-\left(W_p + \frac{1}{\tau_f}\right)t_p\right]\right\}(1 + W_p\tau_f)^{-1} . \tag{4.30}$$

The pump rate W_p can be expressed by the lamp input energy per cubic centimeter of rod volume E_p, the pulse duration t_p, and an adjustable parameter f which is essentially the pump efficiency [4.11–13].

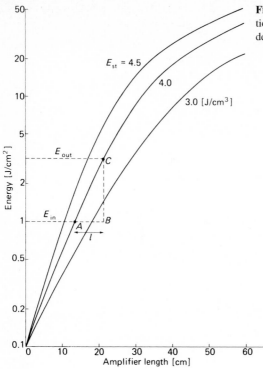

Fig. 4.5. Ruby amplifier gain as a function of amplifier length and stored energy density

$$W_{\mathrm{p}} = \frac{fE_{\mathrm{p}}}{t_{\mathrm{p}}} \ . \tag{4.31}$$

From (4.30 and 31) we obtain an expression of the population inversion as a function of flashlamp input energy per cubic centimeter of rod volume:

$$n = n_{\mathrm{tot}} \left[1 - 2 \left(\frac{1 + (fE_{\mathrm{p}}\tau_{\mathrm{f}}/t_{\mathrm{p}})\exp(-fE_{\mathrm{p}} - t_{\mathrm{p}}/\tau_{\mathrm{f}})}{1 + fE_{\mathrm{p}}\tau_f/t_{\mathrm{p}}} \right) \right] \ . \tag{4.32}$$

The expression $g_0 = \sigma_{12}n$ normalized to the absorption coefficient α_0 is plotted in Fig. 4.6 as a function of fE_{p} with t_{p} as a parameter. As illustrated, the gain which can be achieved for a given lamp input energy depends on the ratio of pulse width to fluorescence lifetime. For a longer pump pulse, gain is reduced because of the depletion of the upper level by spontaneous emission. For an infinitely short pump pulse, $t_{\mathrm{p}} = 0$ and the gain coefficient becomes

$$g_0 = \alpha_0[1 - 2\exp(-fE_{\mathrm{p}})] \ . \tag{4.33}$$

The parameter f depends only on the efficiency of the pump geometry and the spectral output of the lamp. In Fig. 4.7 the stored energy E_{st} as a function of lamp input energy per cubic centimeter of rod volume is plotted for a practical pulse length of $t_{\mathrm{p}} = 1$ ms. The parameter is the pumping coefficient f.

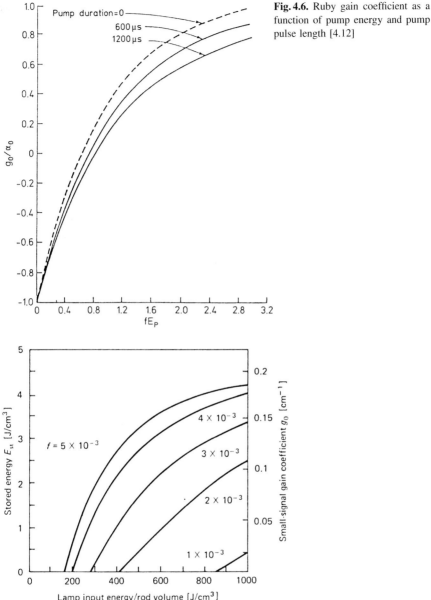

Fig. 4.6. Ruby gain coefficient as a function of pump energy and pump pulse length [4.12]

Fig. 4.7. Stored energy density in ruby as a function of lamp input energy per cubic centimeter of rod volume. Parameter is the pump efficiency factor f [cm^3/J]

Figures 4.5 and 7 completely describe the performance of a ruby amplifier in the Q-switch regime. Starting with the total lamp input power and the rod volume, the parameter E_p can be calculated. For a given E_p, there follows from Fig. 4.7 a value for the upper-state stored energy E_{st}. With this value known, from Fig. 4.5 the output of the amplifier for any input energy density can be obtained. The only assumption which has to be made is the numerical value of the pump coefficient f. A good estimate for most helical lamp geometries is $f = (4 - 8) \times 10^{-3}$ [cm^3/J] and $f = (6 - 8) \times 10^{-3}$[cm^3/J] for focusing pump cavities.

Once an amplifier has been built and one data point ($E_{\text{in}}, E_{\text{out}}, E_{\text{p}}, t_{\text{p}}$) has been measured the amplifier can, in most cases, be completely described for a variety of different operating conditions using the equations and graphs presented in this section. For example, if E_{out} is measured for a single input power E_{in}, then we can calculate the small-signal single-pass gain according to (4.11)

$$G_0 = \frac{\exp(E_{\text{out}}/E_{\text{s}}) - 1}{\exp(E_{\text{in}}/E_{\text{s}}) - 1} , \qquad (4.34)$$

and with G_0 known, the small-signal gain coefficient g_0 and the stored energy E_{st}. If the flashlamp input energy and pulse duration was recorded during the gain measurement, the pump parameter f can be determinded from (4.33). We will now compare the theoretical results with measurements performed on actual ruby amplifiers. In Fig. 4.8, the small-signal gain coefficient g_0 is plotted as a function of flashlamp input per cubic centimeter of rod volume for a number of different ruby laser amplifiers. The systems, varying in ruby rod size, are all pumped by helical flashlamps. The measured values can be approximated by (4.33) if one assumes a pump pulse of 0.8 ms and a pump parameter of $f = 8 \times 10^{-3}$[cm^3/J].

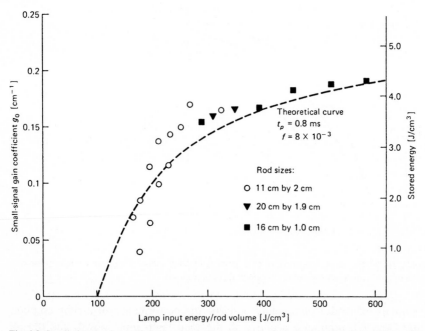

Fig. 4.8. Small-signal gain coefficient and stored energy versus lamp input energy per cubic centimeter of rod for several ruby laser heads

The flashlamp pulse in these systems is typically 0.8–1.0 ms wide at the 50 % power points. The signal input is normally delayed by about the same time with respect to the triggering of the flashlamp. In accordance with the calculated value, the gain curve bends over for high lamp input energies as a result of depletion of the ground level. Measurements on about 20 helical flashlamp pumped ruby laser amplifiers have shown that energy storages of $E_{st} = 3.5 - 3.8 \, \text{J/cm}^3$ are typical for a flashlamp input of 350 J/cm^3 of active material. The small-signal gain coefficient under these conditions is, therefore, approximately $g_0 = 0.15 - 0.17 \, \text{cm}^{-1}$ for rods doped at 0.05 % Cr^{3+}. Since the small-signal gain coefficient is proportional to the doping concentration, see (2.13), variations in the peformance of different ruby rods can often be traced to differences in doping concentration.

Amplification of Mode-Locked Pulses

For input pulses which are so short ($t_p \ll 1$ ns) that no thermalization between the $\overline{E}$ and $2\overline{A}$ level can take place during the amplification process, the energy which can be extracted from the ruby crystal depends on the population of the $\overline{E}$ level only. In Sect. 2.2, for the maximum energy which can be extracted from ruby in a subnanosecond pulse if prior to the arrival of the pulse complete inversion exists, we obtained

$$E'_{ex}, \max \approx \frac{n_{tot} h\nu}{3} .$$

(4.35)

Since the small-signal coefficient is the same as for the longer pulse, we obtain after introducing (4.35) and (2.13) into (4.8),

$$E'_s = \frac{2h\nu}{3\sigma_{21}} .$$

(4.36)

The numerical value for the saturation density is $E_s = 7.7 \, \text{J/cm}^2$. Gain saturation in connection with the amplification of mode-locked pulses is only of academic interest, because energy densities attainable in practice are considerably below E_s as a result of the limitation in peak power imposed by material damage. The performance of ruby amplifiers employed to amplify picosecond pulses have been described in [4.14].

4.1.2 Nd : Glass Amplifiers

An enormous data basis exists regarding the design of Nd : glass amplifiers since these systems have become the lasers of choice for inertial confinement fusion research. Motivated by requirements to drive inertial confinement fusion targets at ever higher powers and energies, very large Nd : glass laser systems have been designed, built, and operated at a number of laboratories throughout the world over the past thirty years.

During the 1960's and into the early 1970's, large glass laser systems consisted of pulsed oscillators followed by rod amplifiers. The introduction of face-pumped disk amplifiers relieved the inherent aperture constraint of rod amplifiers.

Initially the systems employed silicate glass as the host material for neodymium. The lower nonlinear refractive index and higher gain coefficient of phosphate glasses has provided a powerful incentive to build systems based upon these materials.

The architectural design of these master oscillator-pulsed amplifier (MOPA) systems is determined by the nonlinear aspects of propagation, namely self-focusing and gain saturation. For example, the component layout of a laser MOPA chain is shown schematically in Fig. 4.9. A MOPA chain consists of a master oscillator, which generates a well-controlled low-energy pulse for amplification, and a series of power amplifiers to increase the beam energy. The clear apertures of the power amplifiers increase stepwise down the chain to avoid optical damage as the beam energy grows.

Spatial filters are important elements in a high-peak-power laser system and are required to serve three purposes: removal of small-scale spatial irregularities from the beam before they grow exponentially to significant power levels; reduction of the self-induced phase front distortion in the spatial envelope of the beam; and expansion of the beam to match the beam profile to amplifiers of different apertures.

Laser oscillation in the chain is prevented by the appropriate placement along the chain of Faraday rotators and polarizer plates. To construct laser systems that provide more energy than can be obtained from a single MOPA chain, or to provide for multibeam irradiation geometries, the ouptut from a small MOPA chain is split into the desired number of beams, and each of these beams is used to drive a full MOPA chain. The system shown in Fig. 4.9 is an example of such a design. Shown is one 10 KJ beam line of the NOVA glass laser system built at Lawrence Livermore Laboratory. The complete system has 10 such identical beam lines producing a total output energy in excess of 100 KJ.

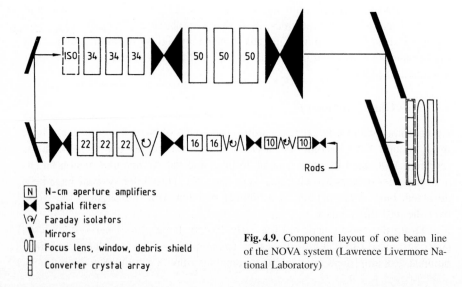

N N-cm aperture amplifiers
◄► Spatial filters
\◊/ Faraday isolators
\ Mirrors
◐◖ Focus lens, window, debris shield
▤ Converter crystal array

Fig. 4.9. Component layout of one beam line of the NOVA system (Lawrence Livermore National Laboratory)

An alternative system concept that uses components of MOPA chains consists of the division of the initial beam into two beams, followed by amplification, splitting, amplification, etc., until the desired number of beams and total energy are obtained.

It is outside the scope of this book to address the complex issues surrounding the design, development and operation of glass lasers employed in fusion target irradiation facilities. The reader interested in the design, modeling, and materials aspects of these lasers is referred to the extensive literature which covers the whole aspect of lasers for inertial confinement fusion. The monograph by *Brown* [4.15], the special issue on lasers for fusion of the IEEE Journal of Quantum Electronics [4.16], the annual reports issued by LLNL [4.17] as well as several overview papers on fusion glass lasers [4.18, 19] are recommended as an introduction to this area of laser technology.

In this section we will proceed to provide several basic guidelines which are important in the design of Nd : glass oscillator-amplifier systems.

In order to calculate gain and energy extraction from a glass amplifier for a given input energy we have to know the saturation density E_s and the small signal gain G_0 according to (4.11).

The saturation fluence E_s of a laser glass depends inversely on the gain cross-section σ and can be written as

$$E_s = (h\nu/\sigma)k \ . \tag{4.37}$$

The 1053 nm gain cross-section of phosphate glasses range from 3.0 to 4.2$\times$ 10^{-20} cm^2 and k is a parameter which is dependent on the output fluence and the duration of the amplified laser pulse. Saturation fluence can depend on the pulse duration if it is less than the lifetime of the lower laser level (less than 1 ns for most glasses). The dependence of E_s on output fluence has been attributed to a hole-burning mechanism [4.20]. It is believed that the saturating pulse couples more strongly to one fraction of the inverted ions in the glass.

Figure 4.10 shows representative values of saturation fluence versus output fluences for several phosphate glasses [4.20, 21]. The increase of saturation fluence for a higher output fluence is quite pronounced. Pulsewidths for these measurements ranged from 1 to 50 ns.

The next step in the design of a glass amplifier is the calculation of the small signal gain. According to (4.20) we need to determine the stored energy density in the rod, which depends on the flashlamp operating parameters and energy output as well as on the pump cavity design and transfer efficiency.

Figure 4.11a shows the axial small-signal gains of rod amplifiers with diameters of 25 mm (RA) and 50 mm (RB), respectively. The 380 mm long glass rods are pumped by six 15 mm I.D. xenon flashlamps. The glass material is Hoya LHG-7 with Nd_2O_3 concentration of either 1.2 (RA) or 0.6 wt%(RB), respectively. The rod is in contact with the cooling liquid which flows in a Pyrex jacket

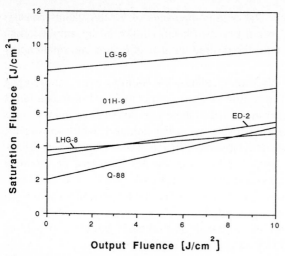

Fig. 4.10. Saturation versus output fluence for a number of silicate glasses (01H-9, LG-56, ED-2) and phosphate glasses (LHG-8, Q-88) [4.20a]

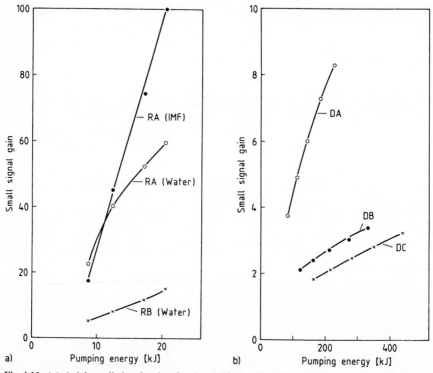

Fig. 4.11. (a) Axial small-signal gain of rod amplifiers with 2.5 cm (RA) and 5 cm clear aperture (Hoya laser glass LHG-7). The small rod was cooled either by water or by an index matching fluid (IMF) [4.22]. **(b)** Small-signal gain of several disk amplifiers (Hoya, LHG-7) with clear apertures of 10 cm (DA), 15 cm (DB) and 20 cm (DC) [4.22]

that surrounds the laser rod. The gain of the amplifer RA depends on the cooling liquid which suppresses the parasitic oscillation. The axial gain at the pumping energy of 22.5 kJ is 59 with water as the coolant and 100 with the index-matching fluid (aqueous solution of $ZnCl_2$ and $SmCl_3$). The axial gain of the amplifier RB is 15.5 at the pumping energy of 22.5 kJ. Gain coefficient and stored energies are given in Table 4.1.

For comparison, Table 4.1 and Fig. 4.11 show also data on these three types of disk amplifiers having clear apertures ranging from 10 to 20 cm. The numbers given in Table 4.1 for the stored energy density, and pumping efficiency E_s/E_p (energy stored in laser glass divided by the pumping energy can be used as guidelines for the design of glass amplifiers). For a critically damped lamp pulse of about 300 μs, one can expect typically a 0.5 to 1.5 % conversion of pump input energy per unit rod volume to energy stored per unit rod volume. The numerical value for the stored energy normally ranges from 0.2 to 0.8 J/cm^3.

The spread in the ratio of E_{st}/E_p and in the numerical value of E_{st} is due mainly to the following:

1) The conversion efficiency E_{st}/E_p increases for large-diameter amplifiers because they are optically thicker. However, because of the large rod volume, the pump energy density and the stored energy per volume decrease for large amplifiers, i.e., large amplifiers have low gain but high storage.

2) As the amplifying medium becomes large, fluorescence depumping effects such as amplified spontaneous emission and parasitic oscillations tend to decrease the pumping efficiency.

The effect of fluorescence losses is demonstrated rather dramatically in Fig. 4.11a. The use of an index-matching fluid eliminates the onset of radial

Table 4.1. System and performance parameters of rod (RA, RB) and disk (DA, DB, DC) Nd : phosphate glass amplifiers [4.22]

		Clear aperture [cm]	Laser glass dimension [cm]	Nd_2O_3 concent. [wt · %]	Flashlamp arc length [cm]	Pump energy [KJ]	Small signal gain	Gain coefficient $[cm^{-1}]$	Stored energy $[J/cm^3]$	Efficiency[a] [%]
ROD	RA	2.5	38 × 2.5	1.2	30 (6 lamps)	22.5	59	0.136	0.66	0.5
	RB	5.0	38 × 5.0	0.6	30 (6 lamps)	22.5	15.5	0.091	0.44	1.2
DISK	DA	10	21.4 × 11.4 × 2.4 (6 disks)	2.9	127 (16 lamps)	220	8.5	0.124	0.60	0.75
	DB	15	32.5 × 18.4 × 3 (4 disks)	2.0	127 (24 lamps)	330	3.4	0.085	0.41	0.70
	DC	20	40 × 21.4 × 3.2 (3 disks)	1.9	127 (32 lamps)	440	3.2	0.101	0.49	0.72

[a] Stored energy/pump energy

parasitic modes thereby allowing a much higher energy storage compared to a rod with a glass/water interface. Fluorescence losses are also the reason for the lower efficiency of the large 20 cm disk compared to the small 10 cm disk amplifier in Table 4.1.

3) Flashlamp pulse-width and -shape is critical in glass amplifiers because the lamp pulse is comparable to, or larger than, the fluorescence time of the gain storage medium. As a result, considerable depletion of the upper level can take place as a result of fluorescence.

The fluorescence time in Nd : glasses is between 200 and 300 μs. It is very difficult to obtain flashlamp pulses which are shorter than 300 μs without sacrificing lamp life to an intolerable value.

4) The main absorption bands for glass are between 0.6 and 0.9 μm. If on increases the electrical input to a flashlamp, then the higher current density will cause a blue shift of the radiation. As a result, the pump source becomes less efficient and one observes a gain foldover in the amplifier if one plots gain as a function of lamp input [4.7, 23, 24].

5) Of course, in addition to the factors mentioned above, the design of the pump chamber, flashlamp configuration, reflectivity of the walls etc. will have a major influence on peak energy storage and gain uniformity.

Equation (4.11) can be graphically represented, as shown in Fig. 4.12. The energy output is plotted as a function of small signal gain G_0, with the saturation

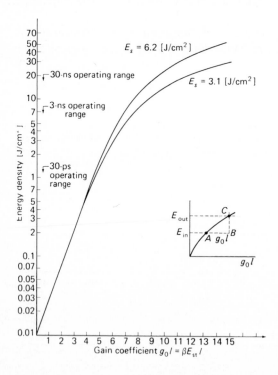

Fig. 4.12. Energy output from a Nd : glass amplifier as a function of small-signal gain coefficient g_0 and amplifier length

density as parameter. Depending on the glass type and desired output energy the saturation fluence can be obtained from Fig. 4.10. The small-signal gain G_0 follows from the stored energy density and rod length.

To obtain the gain or the output energy for a given input, the graph is used in the following manner. As shown in the insert, the input energy density determines the point A on the curve. The value for the gain factor $g_0 l$ for the amplifier must then be added to obtain point B. The abscissa of B intersects the curve at C. Also indicated on this graph are the maximum permissible output energy densities in Nd : glass amplifiers for several typical operating ranges. These energy densities are determined by the damage threshold of Nd : glass (Chap. 11). Although these energy levels are not strongly defined, most Nd : glass laser manufacturers will not recommend operation of systems beyond these limits. The maximum power densities for the indicated pulse widths of 30 ps, 3 ns, and 30 ns are 50 GW/cm^2, 2 GW/cm^2, and 0.7 GW/cm^2, respectively.

The two curves of Fig. 4.12 are for two different values of E_s, corresponding to a typical value of a phosphate glass such as Q-88 at an output fluence of around 3 J/cm^2, whereas the high value of E_s is more typical for a silicate glass such as 01H9 at the same output fluence.

4.1.3 Nd : YAG Amplifiers

Amplified spontaneous emission (ASE) and parasitic oscillations due to the high gain of Nd : YAG effectively limit the energy storage density and therefore the useful energy which can be extracted from a given rod [4.25, 26]. The small-signal gain coefficient g_0 and stored energy are related according to $g_0 = \beta E_{st}$. With the materials parameters listed in Table 2.4, we obtain for Nd : YAG, $\beta = 4.73$ cm^2/J, a value 30 times higher than that for Nd : glass. If we want to extract 500 mJ from a Nd : YAG rod 6.3 mm in diameter and 7.5 cm long, the minimum stored energy density has to be $E_{st} = 0.21$ J/cm^3. The small-signal single-pass gain in the rod will be $G_0 = \exp(\beta E_{st} l) = 1720$. A Nd : glass rod of the same dimensions would have a gain of 1.3.

As a result of the high gain in Nd : YAG, only small inversion levels can be achieved. Once the gain reaches a certain level, amplification of spontaneous emission effectively depletes the upper level. Furthermore, small reflections from the end of the rod or other components in the optical path can lead to oscillations. These loss mechanisms, which will be discussed in more detail in Sect. 4.4, lead to a leveling off of the output energy versus pump input energy curve in a Nd : YAG amplifier. Figure 4.13 shows plots of energy extracted from a Nd : YAG amplifier versus lamp input energy [4.25]. As can be seen from these curves, the maximum energy which can be extracted from the different rods reaches a saturation level. The data show that the saturation limit of the energy output density from a Nd : YAG amplifier tends to be fairly independent of rod length for rods longer than 50 mm. This contrasts with the strong dependence of energy storage density on rod diameter. In fact, rods with lengths longer than 50 mm tend to decrease the saturation limit, as shown in Fig. 4.13. A thin, long rod will provide a long path and therefore high gain for the spontaneous emission to build up, whereas

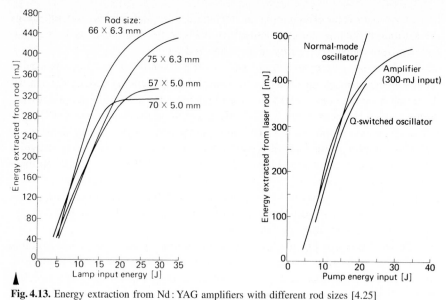

Fig. 4.13. Energy extraction from Nd : YAG amplifiers with different rod sizes [4.25]

Fig. 4.14. Energy extraction from Nd : YAG operated as a normal mode and Q-switched oscillator, and amplifier of Q-switched pulses [4.25]

in a relatively short rod of large diameter more total energy can be stored for the same total gain. Measurements have shown that for rods 5, 6.3, and 9 mm in diameter the maximum extractable energy is 0.3, 0.5, and 0.9 J, respectively. Increasing the temperature of a Nd : YAG rod will reduce its gain, and therefore more energy can be stored. For example, the extracted energy from an amplifier was increased from 770 to 926 mJ/cm^2 by raising the temperature from 26 to 96°C.

The relative performances of a laser rod when used as a normal-mode oscillator, as a Q-switched oscillator, and as a single-pass amplifier are displayed in Fig. 4.14 [4.25]. A 0.63- by 6.6-cm Nd : YAG laser rod in a silver-plated, single-ellipse, single-lamp pump cavity was used in all modes of operation. Normal-mode performance was achieved with two plane-parallel dielectric-coated mirrors. The pump pulse had a duration of approximately 100 μs at the half-power points. The Q-switched performance was obtained with a rotating prism switch in which the prism speed and mirror reflectivity had been optimized for maximum efficiency in the output range of 100 to 400 mJ. The single-pass amplifier performance represents the energy extracted from the rod with a 300-mJ input from an oscillator. These data show that all modes of operation are approximately equivalent until the 320-mJ output level is reached. Above this level the modes of operation which require energy storage become much less efficient. The Q-switched oscillator operates at a slightly lower efficieny than does the same rod in the amplifier mode.

If the Nd : YAG crystal is pumped with an input of 14 J, a total of 300 mJ can be extracted from the amplifier. Since the signal input is 300 mJ, the amplifier has a saturated gain of $G_s = 2$ and a total output of 600 mJ. In order to extract 300 mJ from this rod, at least $0.15 \, \text{J/cm}^3$ must be stored in the upper level. This corresponds to a small-signal gain of $G_0 = 108$. Using (4.13) we can compare the measured saturated gain with the theoretical gain. For an amplifier operated in the saturation regime $E_{in} \ll E_s$, the theoretical gain is $G = 1 + (E_s/E_{in})g_0 l$. where the saturation energy for Nd : YAG is $E_s = 1/\beta = 0.2 \, \text{J/cm}^2$. With $E_{in} = 1 \, \text{J/cm}^2$, $g_0 = 0.71 \, \text{cm}^{-1}$, and $l = 6.6 \, \text{cm}$ for this amplifier, we obtain $G = 2$, in agreement with the measurement.

As was mentioned in the beginning of this chapter, the fraction of stored energy which can be extracted from a signal depends on the energy density of the incoming signal. In this particular case the amplifier is completely saturated, since $E_{in} = 5E_s$. According to our previous discussion, this should result in a high extraction efficiency. The expression for the saturated gain can be written $G_s = 1 + (E_s/E_{in})\beta E_{st} l$. As we have seen, if we substitute the stored energy E_{st} in this equation with the extracted energy $E_{ex} = 300 \, \text{mJ}$, we obtain exactly the measured gain of the amplifier. This suggests that the extraction efficiency of this amplifier, $\eta_e = E_{ex}/E_{st}$, is close to 100 %. Figure 4.15 shows the measured extraction efficiency of this amplifier as a function of signal input.

The transition region in which the amplifier output as a function of signal input changes from an exponential (small-signal) relationship to a linear (saturated) relationship occurs for input energy densities of approximately $0.2 \, \text{J/cm}^2$, in accordance with the theoretical value. The pulse width for the data presented in Fig. 4.15 ranged from 15 to 22 ns. The curve in this figure shows that for efficient energy extraction a Nd : YAG amplifier should be operated with an input signal of around $1 \, \text{J/cm}^2$.

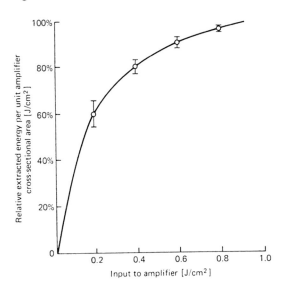

Fig. 4.15. Amplifier energy extraction as a function of oscillator energy density [4.25]

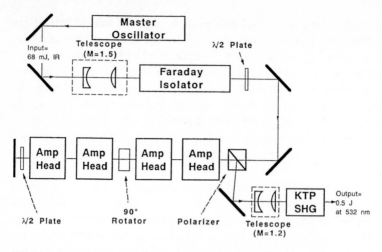

Fig. 4.16. Optical schematic of a high power multi-stage Nd : YAG laser

In our last example, we will describe a modern, multi-stage Nd : YAG master-oscillator power-amplifier (MOPA) design, as depicted in Fig. 4.16. The laser produces an output in the TEM$_{00}$ mode of 750 mJ at 1.064 μm at a repetition rate of 40 Hz. A harmonic generator converts this output to 532 nm with 65% conversion efficiency [4.27].

The system features an oscillator and 4 amplifiers in a double-pass configuration. The linearly polarized output from the oscillator is expanded by a telescope to match the amplifier rods. A Faraday rotator and $\lambda/2$ wave plate act as a one-way valve for the radiation, thereby isolating the oscillator from laser radiation and amplified spontaneous emission reflected back by the amplifier chain. The output from the oscillator passes through the 4 amplifiers and after reflection by a mirror, the radiation passes through the amplifiers a second time. A quarter-wave plate introduces a 90° rotation of the polarized beam after reflection by the rear mirror; this allows the radiation to be coupled out by the polarizer located at the input of the amplifier chain. After a slight expansion, the output beam is passed through a KTP crystal for second-harmonic generation. Located between the 2 pairs of amplifiers is a 90° rotator which serves the purpose of minimizing thermally-induced birefringence losses, as will be explained in Sect. 7.1.1. The changes in the polarization of the beam as it travels back and forth through the amplifier chain is illustrated in Fig. 4.17.

Each amplifier contains 16 linear diode arrays for side pumping of the Nd : YAG crystal. The optical pump energy for each amplifier is 900 mJ at 808 nm, or 4.5 kw at the pump pulse width of 200 μs. In each amplifier, the arrays are arranged in an eight-fold symmetrical pattern around the 7.6 mm diameter and 6.5 cm long laser rod in order to produce a uniform excitation. The active length of the rod pumped by the arrays is 4 cm. The small-signal, single pass gain of one amplifier as a function of pump energy is plotted in Fig. 4.18.

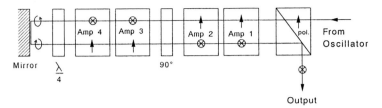

Mirror $\dfrac{\lambda}{4}$ 90°

Output

Fig. 4.17. Two-pass amplifier chain with polarization output coupling and birefringence compensation

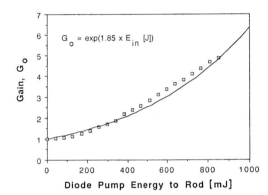

Fig. 4.18. Small signal, single pass gain as a function of optical pump energy

For low input signals, the gain increases exponentially with g_0l according to (4.11). The logarithmic gain g_0l is proportional to the pump energy, as derived in (3.62),

$$g_0l = KE_p \tag{4.38}$$

where E_p is the optical pump energy from the diode arrays, and the η-terms for an amplifier operated in the energy storage mode are

$$K = \eta_T\eta_a\eta_S\eta_Q\eta_B\eta_{ST}\eta_{ASE}/AE_S . \tag{4.39a}$$

For this particular amplifier design, the numerical values are $\eta_T = 0.88$ for the transfer efficiency, $\eta_a = 0.85$ for the absorption of pump radiation in the 7.6 mm diameter Nd : YAG crystal, $\eta_S = 0.76$ and $\eta_Q = 0.90$ for the Stokes shift and quantum efficiency, $\eta_B = 0.62$ for the overlap between the beam and gain region of the rod, $\eta_{ASE} = 0.90$ and $\eta_{ST} = 0.68$ for the storage efficiency. The latter ist calculated from (3.53) for a pump pulse length of $t_p = 200\,\mu s$ and a fluorescence lifetime of $T_F = 230\,\mu s$ for Nd : YAG. With $A = 0.25\,\mathrm{cm}^2$ and $E_S = 0.44\,\mathrm{J/cm}^2$, one obtains $K = 1.85$. The curve in Fig. 4.18 is based on this value of K, i.e.

$$E_{out}/E_{in} = \exp[1.85E_p(J)] . \tag{4.39b}$$

The energy output as a function of signal input of one amplifier stage single pass, and two amplifiers in a double-pass configuration is plotted in Fig. 4.19. The amplifiers are operated at a fixed pump energy of 900 mJ each. Also plotted

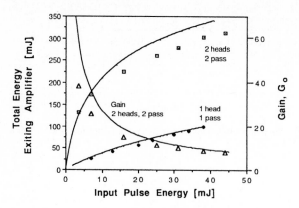

Fig. 4.19. Energy output as a function of signal input for a single amplifier stage and a two-amplifier, double pass configuration. Shown is also the gain for the latter configuration

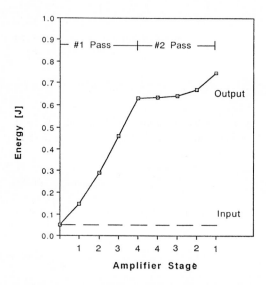

Fig. 4.20. Energy levels at successive stages in the amplifier chain. Points are measured values, and curve is obtained from computer model

in this figure is the gain for the two-amplifier double pass configuration. The amplifiers are highly saturated as can be seen from the nonlinear shape of the E_{out} vs. E_{in} curve and the drop in gain at the higher input levels.

The increase in energy as the signal pulse travels forward and backward through the amplifier chain is plotted in Fig. 4.20. Shown are the measured data points and a curve representing the values calculated from (4.14–18) with $E_{IN} = 50$ mJ, $A = 0.25\,\text{cm}^2$, $G_0 = 4.8$, $E_S = 0.44\,\text{J/cm}^2$. Energy extraction from each stage for both passes is depicted in Fig. 4.21. The data indicate that the amplifiers are totally saturated and all the stored energy within the beam is extracted. As the input beam travels the first time through the amplifiers, successively more energy is extracted from each stage since the ratio of E_{IN}/E_S increases. On the return pass, very little energy is removed from the last two stages because stored energy has already been depleted.

The logarithmic gain for the four stages having a 16 cm active length is $g_0 l = 6.0$. The double pass configuration increases the extraction efficiency from 0.8 to 1.0. It should be noted that the small-signal gain obtained from Fig. 4.18 is measured over the cross-section of the beam. The gain and beam profile are both centrally peaked in this design. Both the beam profile and the gain profile have a Gaussian shape with spot sizes of 2.85 and 4.25 mm, respectively. In order to avoid diffraction effects, provide for adequate beam alignment tolerance, and accommodate a slightly expanding beam, the beam cross-section at the $1/e^2$ intensity points is $A = 0.25 \, \text{cm}^2$, whereas the rod cross-section is $0.45 \, \text{cm}^2$.

Although the extraction is complete within the beam, stored energy is left at the outer regions of the rod. The beam fill factor takes into account this fact. Since both gain and beam profile are radially dependent, the beam fill factor cannot be calculated from the ratio of the areas, as would be the case with a uniform gain and beam profile. The beam overlap efficiency can be calculated provided the gain and beam profiles are known. These distributions were obtained from images taken with a CCD camera which recorded the profiles of the fluorescence and laser beam output. With the radial spot sizes of the Gaussian approximations given above, one obtains from (3.48b) a value of $\eta_B = 0.62$ for the beam overlap efficiency.

According to (4.9) the total stored energy at the upper laser level is $E_{ST} = g_0 l E_S A$. With $g_0 l = 1.55$, $E_S = 0.44 \, \text{J/cm}^2$ and $A = 0.25 \, \text{cm}^2$, one obtains $E_{St} = 170 \, \text{mJ}$ of stored energy within the volume addressed by the beam. The stored energy in the full cross-section of the rod is $E'_{St} = E_{St}/\eta_B = 260 \, \text{mJ}$. The values for E_{St} and E'_{St} are indicated in Fig. 4.21.

The electrical system efficiency η_{sys} of the amplifier chain is the product of the laser diode efficiency η_p, the conversion efficiency of optical pump power to the upper laser level at the time of energy extraction, and the extraction efficiency of the stored energy into laser output, i.e.

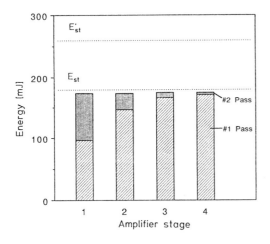

Fig. 4.21. Total energy extraction from each amplifier stage

$$\eta_{sys} = \frac{E_{out} - E_{in}}{E_{EL}} = \eta_p \eta_T \eta_a \eta_U \eta_B \eta_E \qquad (4.40)$$

where

$$\eta_U = \eta_S \eta_Q \quad \text{and} \quad \eta_E = \eta_{ST} \eta_{ASE} \eta_{EQ} . \qquad (4.41)$$

Table 4.2 lists the individual efficiencies of the amplifier system.

Table 4.2. Energy conversion efficiency of the two-pass multistage Nd : YAG amplifier

Laser diode array electrical efficiency η_P		0.35
Conversion of optical pump energy to upper state energy		
transfer efficiency	$\eta_T = 0.88$	
absorption efficiency	$\eta_a = 0.85$	0.51
Stokes efficiency	$\eta_S = 0.75$	
Quantum efficiency	$\eta_0 = 0.90$	
Conversion of upper state energy to laser output		
beam overlap efficiency	$\eta_B = 0.62$	
storage efficiency	$\eta_{st} = 0.68$	0.38
fractional loss	$\eta_{ASE} = 0.90$	
gain extraction efficiency	$\eta_{EG} = 1.00$	
Amplifier efficiency η_{sys}		0.068

4.2 Steady-State Amplification

If the pulse duration is long compared to the fluorescent lifetime τ_f, the population inversion and gain coefficient follow the intensity in a quasi-stationary way. This situation is realized, for example, with the amplification of laser pulses from a free running oscillator. The nonlinear gain characteristics for laser amplifiers as a function of steady-state radiation have been analyzed for media with homogeneous and inhomogeneous line broadening [4.28–30]. In the former case, the gain decreases proportionately over the entire transition line. In the case of an inhomogeneous line, spectral hole burning can take place if the driving signal is a narrow-band signal. However, in solid-state lasers, materials with an inhomogeneous line, such as Nd : glass, have a cross-relaxation which is fast compared to the long pulses discussed in this section. For example, the cross-relaxation rate is of the order of nanoseconds in Nd : glass. The following analysis is applicable to systems with homogeneous line broadening, such as in Nd : YAG, and with inhomogeneous line broadening, such as in Nd : glass, provided in the latter case that the driving signal is broad band and therefore taps the entire inversion density, or in the case of a narrow-band signal it is assumed that cross-relaxation is infinitely fast.

In Sect. 3.2 we derived an expression for the saturated gain in an oscillator. In the case of an amplifier, the gain depends on the coordinate z measured in the direction of the axis of the amplifier.

$$g(z) = \frac{g_0}{1 + I(z)/I_s} \; , \tag{4.42}$$

where g_0 is the small-signal gain, $g(z)$ is the gain at position z in the amplifier obtained at a signal level $I(z)$, and I_s is the saturation density defined as the signal power, which reduces the small-signal gain by a factor of one-half.

Assuming that the inversion n does not change much over a photon transit time in the amplifier, the power density $I(z)$ builds up along z in the following manner

$$\frac{dI}{dz} = \sigma n(z)I(z) = g(z)I(z) \quad . \tag{4.43}$$

We have ignored losses in this expression. Substituting g from (4.42) into (4.43) and then integrating over the power density I results, after some algebra, in [4.29, 30]

$$\frac{I_{in}}{I_s} = \frac{\ln(G_0/G)}{G - 1} \; , \tag{4.44}$$

where $G_0 = \exp(g_0 l)$ is the small-signal, single-pass gain of the amplifier of length l and $G = I_{out}/I_{in}$ is the effective gain. Equation (4.44), which is plotted in Fig. 4.22, permits one to calculate the power gain of solid-state lasers in the steady-state regime. For cases where $G \gg 1$, we can simplify (4.44) and write

$$\frac{I_{out}}{I_s} = \ln\left(\frac{G_0}{G}\right) \quad . \tag{4.45}$$

The gain drops to one-half for $I_{out} = 0.69 I_s$ and to $1/e$ for $I_{out} = I_s$.

For small-signal levels $G I_{in} \ll I_s$ we obtain $G = G_0$, or

$$I_{out} = I_{in} \exp(g_0 l) \quad . \tag{4.46}$$

Accordingly, the small-signal gain is identical for short-pulse and steady-state amplification. At large signal levels $I_{in} \gg I_s$, one obtains

$$I_{out} = I_{in} + I_s g_0 l \quad . \tag{4.47}$$

In this energy regime the output increases linearly with length.

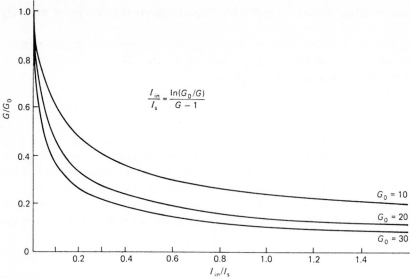

Fig. 4.22. Steady-state gain versus input power density in laser amplifiers. Parameters are the small-signal gain G_0 and the saturation power density I_s

$$\frac{I_{out}}{I_s} \approx \frac{g_0}{\alpha} \quad , \tag{4.48}$$

where α is the loss coefficient in the amplifying medium. For higher power densities the overall gain no longer increases with amplifier length.

4.2.1 Ruby Amplifier

The calculation of the steady-state gain in a ruby amplifier is somewhat complicated because the saturation density I_s depends not only on fundamental material parameters but also on the pump rate W_p. In principle, this is true for a four-level system as well. As we have seen in Chap. 3, in a four-level system $W_p \tau_f \ll 1$, and the effect of the pump rate on gain saturation can be ignored. In a three-level system $W_p \tau \geq 1$, and the pump rate must be considered in the term for saturation. In the pulse regime we defined a saturation energy E_s, in the case of steady-state amplification a power density I_s will be used to describe saturation.

If we introduce (4.8) into (3.30, 75) we can write

$$I_s = \frac{2E_s}{\tau_f(1 - g_0/\alpha_0)} \quad . \tag{4.49}$$

From this equation the gain as a function of signal input power can be obtained for ruby lasers, provided that the small-signal gain G_0 or the gain coefficient g_0 is known.

In order to illustrate the application of these calculations, we will consider the input versus output characteristics of a 20-cm by 2-cm ruby rod pumped by a helical flashlamp. The small-signal gain for this ruby laser amplifier in the pulsed mode was measured to be 29 for an input energy of 20 kJ. In the long-pulse mode the small-signal gain was $G_0 = 20$ for the same flashlamp input. Figure 4.23 shows the measured amplifier output as a function of input. The parameter is the flashlamp pump energy. For an input of 24 kJ, the amplifier can be described by (4.44) if one uses $G_0 = 30$ and $I_s = 50\,\mathrm{kW/cm^2}$, which follows from (4.49) for a gain coefficient of $0.17\,\mathrm{cm^{-1}}$. A best fit through the measured data for $G_0 = 20$ and $G_0 = 50$ is obtained if one uses also $I_s = 50\,\mathrm{kW/cm^2}$ for these gain curves. It turns out that the saturation parameter might have to be fitted to experimental data. As mentioned before, the equations derived in this subsection are based on the condition that the signal duration is long compared to the fluorescence time. In the typical pulse-width regime of about 1 ms, this condition is not met.

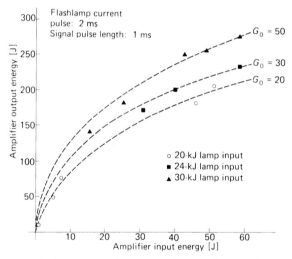

Fig. 4.23. Output versus input energy for a steady-state ruby amplifier. Signal pulse length was 1 ms and flashlamp current pulse had a duration of 2 ms

4.2.2 Nd : Glass Amplifier

With the materials parameter of Nd : glass $h\nu = 1.89 \times 10^{-19}\,\mathrm{Ws}$, $\gamma = 1$, $\sigma_{21} = 3.5 \times 10^{-20}\,\mathrm{cm^2}$, and $\tau_f = 300\,\mu\mathrm{s}$, we obtain $I_s = 20\,\mathrm{kW/cm^2}$ from (3.29). Introducing this parameter into (4.44), we can calculate the normalized gain as a function of the normalized input power for the amplification of long pulses. The gain as a function of input power level can also be obtained from Fig. 4.22. The small-signal gain must be known for a particular amplifier. If no experimental data is available, the small-signal gain can be estimated using the graphs and formulas derived in Sect. 4.1.

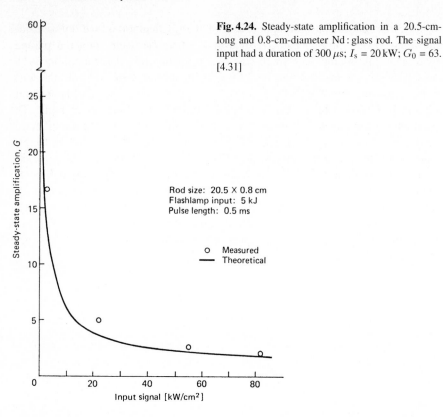

Fig. 4.24. Steady-state amplification in a 20.5-cm-long and 0.8-cm-diameter Nd : glass rod. The signal input had a duration of 300 μs; $I_s = 20$ kW; $G_0 = 63$. [4.31]

The performance of a Nd : glass amplifier [4.31] employed in the amplification of long pulses is shown in Fig. 4.24. The unsaturated gain for the amplifier was measured to be 63. At a maximum driving power density of 82.5 kW/cm², the gain decreased to 2.2. As seen from Fig. 4.24, there is a reasonably good agreement between theory and experiment.

4.3 Signal Distortion

As an optical signal propagates through a laser amplifier, distortions will arise as the result of a number of physical processes. We can distinguish between spatial and temporal distortions.

4.3.1 Spatial Distortions

The development of high-power laser oscillator-amplifier systems has led to considerable interest in the quality of the output beam attainable from these devices. Beam distortions produced during amplification may lead either to an increase in divergence or to localized high energy densities which can cause laser rod damage. We will consider the main phenomena producing a spatical distortion, which is analogous to a wavefront or intensity distortion.

Nonuniform Pumping Due to the exponential absorption of pump light, the center of the rod is pumped less than the edges. Figure 4.25 shows the stored energy profile of several Nd:glass amplifier stages of the Omega system built at the University of Rochester [4.32]. The curves were obtained by measuring the small-signal gain with a narrow probe beam across the rod on five amplifier stages of increasing diameter. In order to provide a uniform gain profile as much as possible, the Nd concentration is decreased for increasing rod diameters.

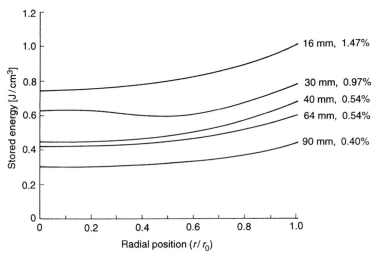

Fig. 4.25. Stored energy distribution in different Nd:glass rods [4.32]. Rod diameters and doping levels (wt.%) are indicated on the curves

Nonuniformities in the Active Material Even laser rods of excellent optical quality contain a small amount of inherent stress, index of refraction variations, gradients in the active ion concentration, contaminants, inclusions, etc. These nonuniformities will significantly modify the energy distribution of an incoming signal. An initially smooth energy distribution will contain ripples after passing through a laser amplifier.

Gain Saturation A beam propagating through an amplifier suffers distortion because of the saturation-induced change in the distribution of gain. The weaker portions of the signal are amplified relatively more than the stronger portions because they saturate the medium to a lesser degree. The spreading of a Gaussian input beam due to homogeneous gain saturation obtained by integrating (4.42) numerically step by step in the z direction of the amplifier is shown in Fig. 4.26 [4.33]. The figure shows the change in the radial intensity distribution for a

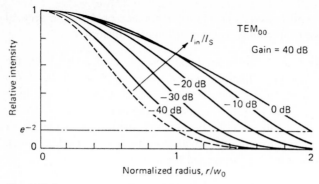

Fig. 4.26. Spreading of a TEM_{00} input beam due to homogenous gain saturation [4.33]

TEM_{00} beam (dashed line) entering a homogeneously broadened laser amplifier with a radially uniform small-signal gain of 40 dB. Each curve corresponds to a specific value of the normalized signal parameter I_{in}/I_s. The output distribution has been normalized to its value on the amplifier axis to make the distortion more evident. Figure 4.26 shows that even for input signals 10^{-4} below the saturation flux (i.e., a signal that begins to saturate only near the amplifier output), there is about a 10 % expansion of the e^{-2} radius. When the input has increased to a value that begins to saturate the amplifier at its input ($I_{in}/I_s = 1$), the e^{-2} radius has just about doubled. Calculations presented in [4.33] reveal that for a further increase in input signal the beam expansion falls below the $I_{in}/I_s = 1$ curve. This is to be expected, since for very large input signals the amplifier is so highly saturated that it adds very little to the signal passing through and thus cannot distort it. Hence, it would seem that the maximum distortion occurs when the input signal is about equal to the saturation flux. For either extreme, $I_{in} \ll I_s$ and $I_{in} \gg I_s$, the pulse-shape distortion is minimized.

Diffraction Effects An amplifier rod represents a limiting aperture for an incoming beam. Diffraction effects at the edges of the amplifier rod will give rise to Fresnel rings, which can strongly disturb the beam uniformity. One method to minimize the effect of diffraction is to utilize a Gaussian-shaped beam. If the beam size is chosen properly with respect to the limiting aperture, the energy density on the edges will be low enough to produce a negligible spatial modulation across the beam. *Campbell* and *DeShazer* [4.34] calculated the near-field beam pattern for Gaussian beams truncated by circular apertures for various combinations of aperture radius a and beam radius w. (For $w = a$ the amplitude of the beam is reduced to $1/e$ at the edge of the aperture, and 86.5 % of the energy is transmitted through the aperture). It was shown that truncation of a Gaussian beam by a circular aperture introduced structure in the near-field beam profile. However, if the ratio of aperture radius to beam radius is 2 or greater ($a \geq 2w$), then the structure is negligible and the transverse-beam profile remains very nearly Gaussian. For $a = 2w$ the power at the edges of the aperture is $-33\,\mathrm{dB}$ as compared to the maximum intensity.

Trenholme [4.35] has demonstrated empirically that if the beam is apertured so as to cut off portions of intensities below K, where K is a specified fraction of the beam's maximum intensity, then the peak-to-peak magnitude of the diffraction ripple will be on the order of $0.40\,K^{1/2}$.

Methods to reduce diffraction effects, in addition to tailoring the Gaussian beam profile to the limiting aperture, consist of using spatial filtering. A spatial filter consists of two positive lenses and a pinhole at the common focal plane. Spatial filtering in a amplifier chain is obtained if we match an aperture diameter to the first dark ring of the far-field pattern produced by the preceeding stage. No modulation will apear in the regime of the Airy disc. In a more general sense, low-pass spatial filters in the amplifier chain filter the beam noise resulting from inperfection of the optics and diffraction. By choosing the pinhole diameter one can control which spatial frequencies are passed. The use of spatial filters in high peak power systems leads to the requirement that they be operated in vacuum, because the high intensities reached at the focal spot lead to optical breakdown in air.

Figure 4.27 illustrates spatial modulation of a high power beam due to diffraction and ripples induced by dust and scattering particles. (Shiva beam line operated at $700\,\mathrm{J}$ in $600\,\mathrm{ps}$ [4.36]). Figure 4.27b shows the relative intensity of one Shiva beam line before and after installation of a spatial filter at the output [4.36]. As one can see from these densitometer scans, the intensity of the high frequency spike was reduced by a factor 4 by the low-pass filter. Besides the removal of unwanted ripples in the beam, spatial filters can at the same time be utilized for imaging or beam relaying and for magnification to match different diameter amplifiers [4.37].

Thermal Distortions The nonuniform pumping which leads to a higher gain coefficient on the edges of a laser rod also causes a nonuniform temperature profile across the rod. The absorbed pump power raises the temperature of the surface of the rod above the temperature of the rod center. As a result, a negative thermal lens is created, which distorts the wavefront of the beam. Associated with the thermal gradients are stresses in the active material which lead to birefringence. Therefore the thermal effect leads to strong optical aberrations which can be only partially compensated for by insertion of correcting elements. Thermal effects in laser rods are discussed in detail in Chap. 7.

Index Nonlinearity When a powerful laser pulse propagates through an isotropic material, it induces changes in the refractive index described to first order by the equation

$$n = n_0 + n_2 I \quad , \tag{4.50}$$

where n_0 is the linear and n_2 is the nonlinear index of refraction, and I is

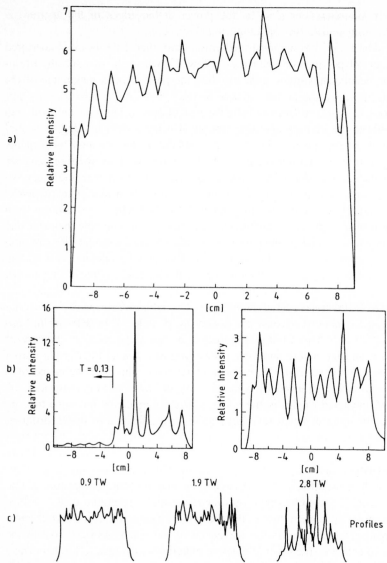

Fig. 4.27. (a) Relative intensity of the output beam of one arm of the Shiva system at 700 J in 600 ps [4.36]. **(b)** Relative intensity of one beam line of the SHIVA system at 1.5 TW, before and after installation of the output spatial filter [4.36]. **(c)** Beam profiles of a large Nd : glass laser showing progressive modulation growth (small scale beam breakup) as the power is increased [4.50]

the beam intensity. From (4.50) it follows that a laser beam propagating in a transparent medium induces an increase in the index of refraction by an amount proportional to the beam intensity. This effect which causes a self-induced aber-ration of the beam is known as whole-beam phase distortion [4.38, 39]. The index nonlinearity gives also rise to so-called small-scale ripple growth, in which case the amplitude of small perturbations grows exponentially with propagation distance [4.40–43].

The principal limit on the performance of a high-power glass laser, such as employed in fusion research, is the nonlinear refractive index [4.44–49]. In the presence of an intense light wave, with a power density of gigawatts per centimer squared or more, the refractive index of a transparent dielectric is increased. This intensity-induced index change is small, on the order of 5 ppm at 10^{10} W/cm^2, but it has profound effects.

The first problem caused by the nonlinear index is the accumulation of self-induced phase front distortions on the spatial envelope of the pulse (whole-beam phase distortion) which alter the focusing properties of the system output. In order to maintain the optical quality of the beam the integrated nonlinear index along the optical path given by the parameter

$$B = \frac{2\pi}{\lambda_0} \int \frac{\Delta n}{n} dl \tag{4.51}$$

must be kept below a critical value. In (4.51), where n is the refractive index, Δn is the nonlinear index change, and λ_0 is the wavelength of the light in vacuum, the integral is taken along any ray passing through the laser. The value of B at any position across the beam gives the radians of phase delay which that portion has undergone when compared to a low-intensity beam. For large systems developed for inertial confinement fusion, values of B at the beam center above about 9 (1.5 waves of delay) are unacceptable.

The second problem caused by the nonlinear index is small-scale ripple growth. In this process, a small region of higher intensity in the beam raises the index locally, which tends to focus light toward it and raises the intensity even more. It has been shown that ripples grow from their original amplitude by a factor of

$$G = e^B \tag{4.52}$$

which involves the same B integral as the whole-beam distortions. Experimentally, it is found that fusion-quality beams can be produced only when B is less than 4 or 5, because of the small-scale growth.

Figure 4.27c illustrates the progressive growth of beam modulation as the power is increased in a high power glass laser [4.50].

High-power solid-state laser systems must not only be carefully designed and built to minimize beam irregularities caused by dirt particles, material imperfections, diffraction from apertures, and other "noise" sources; some provision must also be made for removing the remaining small-scale structure, before it becomes large enough to cause severe degradation of the laser's focusing properties.

As was already mentioned, spatial filters accomplish the task of removal of unwanted ripples, imaging and beam magnification [4.49–51]. The exponential growth of ripples in a high-peak-power system if left uncontrolled will quickly lead to component damage and beam degradation. As already mentioned, a spatial filter is a low-pass filter, and the pin-hole diameter controls which spatial frequencies up to a cut-off are passed. The spatial filter is designed to remove the most damaging ripples while passing most of the energy. Therefore ripples, rather than growing exponentially, are reset at each spatial filter. The imaging property of a spatial filter also significantly reduces the effect of whole beam self-focusing. The existence of an intensity-dependent index of refraction leads to a steepening of gradients of an essentially smooth profile. A spatial filter prevents peaks from developing and results therefore in a more constant beam profile.

4.3.2 Temporal Distortions

Pulse-shape Distortion. For a square pulse traversing an amplifier, the leading pulse edge sees a larger inverted population than does the trailing edge. This occurs simply because the leading edge stimulates the release of some of the stored energy and decreases the population inversion for the trailing edge. Thus less energy is added to the final portions of a pulse than to the leading regions. The change of an arbitrary input pulse shape after propagation is described in [4.1, 2]. If we consider for simplicity a square pulse, we can write for the power of the pulse

$$P(x,t) = P_{\mathrm{in}} \left\{ 1 - [1 - \exp(-g_0 x)] \exp\left[- \left(\frac{t - x/c}{t_{\mathrm{p}}} \right) \right] \frac{E_{\mathrm{in}}}{E_{\mathrm{s}}} \right\}^{-1}, \qquad (4.53)$$

where g_0 is the small-signal gain coefficient, E_{in} and E_{s} are the input and saturation energy densities, respectively, and t_{p} is the length of the pulse. Note that the power is dependent on time and location x within the amplifying medium and also on the input energy $E_{\mathrm{in}} = P_{\mathrm{in}} t_{\mathrm{p}}$.

At the input of the amplifying medium ($x = 0$) we have $P(0, t) = P_{\mathrm{in}}$. For the square-pulse case considered here we note that for any value of input, the gain at the leading edge of the pulse ($x = ct$) is exponential with distance through the amplifier. The trailing edge of the pulse is characterized by $t_{\mathrm{p}} = t - x/c$. In Fig. 4.28 the change of a rectangular input pulse is shown. The pulse undergoes a sharpening of the leading edge and a reduction of its duration. Pulse-shape distortions for input pulses which have profiles other than rectangular are treated in [4.4, 5]. The pulse-shape evolution depends appreciably on the rise time and shape of the leading edge of the input pulse. In general, one observes a forward shift of the peak as the pulse transverses the amplifying medium.

Frequency Modulation. The intensity-induced index changes, are not only responsible for the catastrophic self-focusing problems in high-power glass laser amplifiers, but also cause a frequency shift of the pulse in the amplifying medium. In mode-locked pulses, this frequency modulation causes a broadening of the pulse width [4.41, 42, 52–57]. The frequency shift $\Delta\nu$ impressed on a wave go-

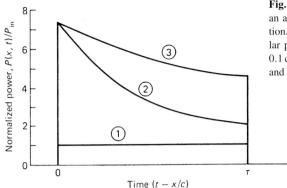

Fig. 4.28. Change of the pulse shape in an amplifier as a result of gain saturation. Curve 1 is the initially rectangular pulse, curves 2 and 3 are for $g_0 = 0.1\,\mathrm{cm}^{-1}$, $L = 20\,\mathrm{cm}$, $E_{\mathrm{in}}/E_{\mathrm{s}} = 0.5$, and $E_{\mathrm{in}}/E_{\mathrm{s}} = 0.1$, respectively

ing through a medium of length l, whose refractive index changes at the rate $\delta n/\delta t$, is given by

$$\Delta \nu = -\frac{l}{\lambda}\frac{\delta n}{\delta t} , \qquad (4.54)$$

where $\delta n = n_2|E|^2$, λ is the wavelength in vacuum, and δt is the rise time of the pulse. An experimental demonstration of this Doppler shift is given in [4.52]. The δn term is approximated for Nd:glass by

$$\delta n \approx (10^{-15}\,\mathrm{cm}^2/\mathrm{W}) \times I ,$$

where I is the intensity in watts per square centimeter. Thus a mode-locked pulse with a 5-ps rise time and a peak power of $3\,\mathrm{GW/cm}^2$ will experience a frequency shift of $\Delta\nu/\nu = 10^{-3}$ or $10\,\text{Å}$ in a 50-cm-long Nd:glass amplifier. *Duguay et al.* [4.56] measured a mean spectral broadening of $5\,\mathrm{cm}^{-1}$ per round trip in a mode-locked glass amplifier.

4.4 Gain Limitation and Amplifier Stability

One of the main considerations in the design of an amplifier or amplifier chain is its stability under the expected operating conditions. In a laser operating as an amplifier or as a Q-switched oscillator the active material acts as an energy storage device, and to a large extent its utility is determined by the amount of population inversion which can be achieved. As the inversion is increased in the active material, a number of different mechanisms begin to depopulate the upper level.

Depopulation can be caused by amplified stimulated emission (ASE), also sometimes called superfluorescence, a process which will be enhanced by radiation from the pump source falling within the wavelength of the laser transition, or by an increase of the pathlength in the gain medium either by internal or external reflections.

Also, if sufficient feedback exists in an amplifier due to reflections from optical surfaces, the amplifier will start to lase prior to the arrival of the signal pulse. If the reflections are caused by surfaces which are normal to the beam direction, the amplifier will simply become an oscillator and prelasing will occur prior to Q-switching or the arrival of a signal pulse. Reflections which include the cylindrical surfaces of a rod will lead to parasitic modes, propagating at oblique angles with respect to the optical axis of the laser.

4.4.1 Amplified Spontaneous Emission

The level of population inversion which can be achieved in an amplifier is usually limited by depopulation losses which are caused by amplified spontaneous emission. The favorable condition for strong ASE is a high gain combined with a long path length in the active material. In these situations, spontaneous decay occurring near one end of a laser rod can be amplified to a significant level before leaving the active material at the other end. A threshold for ASE does not exist, however, the power emitted as fluorescence increases rapidly with gain. Therefore, as the pump power increases, ASE becomes the dominant decay mechanism for the laser amplifier. At that point, an intense emission within a solid angle Ω around the axis of the active material is observed from each end of the rod [4.58]:

$$\Omega = A/L^2 \tag{4.55}$$

where L and A are the rod length and cross-sectional area, respectively. As a result of refraction at the end faces, the geometrical aperture angle of the rod is increased by n^2, as shown in Fig. 4.29. No mirrors are required for ASE to occur, however, reflections from a mirror or internal reflections from the cylindrical surfaces of the rod will increase the path length for amplified spontaneous emission which will lead to a further increase in intensity. This aspect will be treated in Sect. 4.4.2.

In high-gain, multistage amplifier systems ASE may become large enough to deplete the upper state inversion. ASE is particularly important in large Nd:glass systems employed in fusion experiments. Therefore, the subject of ASE in Nd:glass laser discs and slabs received a great deal of attention [4.59–65]. *Linford* et al. [4.66] provided an analytical expression for the fluorescence flux I_{ASE} from a laser rod as a function of small signal gain, which has been

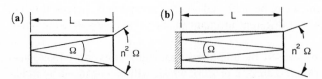

Fig. 4.29. Directionality and maximum pathlength for ASE in a laser rod without a reflector (**a**), and with a reflector on one end (**b**)

found very useful in estimating the severity of ASE. From this reference follows with the approximation $G_0 > 1$

$$\frac{I_{ASE}}{I_S} = \frac{\Omega}{4} \frac{G_0}{(\ln G_0)^{1/2}} \tag{4.56}$$

where I_S is the saturation flux and G_0 is the small signal gain of the active medium. We will apply this formula to a multistage Nd:YAG amplifier and compare the calculated with the measured values of ASE.

In Fig. 4.30, the measured ASE from a four stage, double-pass Nd:YAG amplifier chain is plotted as a function of diode-pump input. At an input of about 500 mJ into each amplifier ASE starts to become noticeable, and quickly increases in intensity to reach 75 mJ at an input of 900 mJ per amplifier. The detrimental effect of ASE can be seen in Fig. 4.31 which shows the output from the amplifier chain as a function of pump input. As the amplified spontaneous emission increases, the slope of output vs. input decreases and the difference can be accounted for by the loss due to ASE.

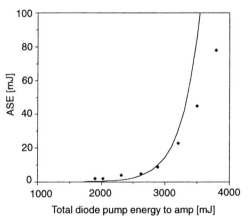

Fig. 4.30. ASE from a 4-stage double pass Nd:YAG amplifier chain. [Dots are measured values, and solid line represents calculated values from (4.56)]

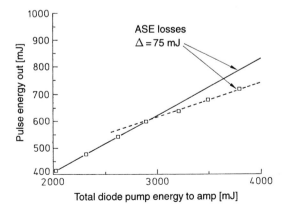

Fig. 4.31. Signal output vs. pump input from a multistage Nd:YAG amplifier chain. (Onset of ASE shown in Fig. 4.30 account for drop in output energy)

From (4.56) in conjunction with Fig. 4.18 which shows the gain vs. pump input for one amplifier head, we can calculate the ASE from the amplifier chain. We have to recall that $\Omega = A/4L^2$ since the path length for ASE is doubled in a double-pass system, and $G_0 = (G_0')^8$, where G_0' is the small signal gain in one amplifier. With $A = 0.45\,\text{cm}^2$, $L = 50\,\text{cm}$, $I_S = 2.9 \times 10^3\,\text{w/cm}^2$ and $t_P = 200\,\mu\text{s}$, one obtains the curve in Fig. 4.30 which closely matches the observed ASE output.

Figure 4.32 illustrates the case of amplified spontaneous emission from an uncoated Nd : glass rod, 1.25 cm in diameter and 25 cm long, pumped by a helical flashlamp. At a flashlamp input of 5 kJ, an output of 60 mJ in each direction is obtained. As can be seen from Fig. 4.32, the calculated full beam angle $\Theta = Dn/L$, which is $\Theta \approx 4.5°$ for the parameters mentioned above, contains 75% of the total energy.

ASE has been used in a few specialized applications as the source of output energy from a laser. Because of the high gain obtainable in long Nd : glass laser rods, several mirrorless oscillators, also called optical avalanche lasers, have been built. In these systems the spontaneous emission was amplified in one or two passes to generate a very powerful beam. In one such system [4.67], a photon avalanche was created by optically pumping six laser rods to a high inversion

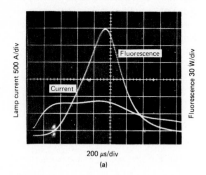

200 μs/div

(a)

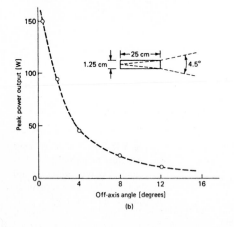

Off-axis angle [degrees]

(b)

Fig. 4.32. (a) Flashlamp current pulse and ASE output from a 25-cm Nd : glass rod pumped at 5 kJ. (b) Directionality of ASE

and then suddenly connecting them in series. The cumulative 155-dB gain was sufficient to generate a 1-GW, 70-ns photon avalanche pulse at the output. Because of the absence of any resonant structure, the output from a mirrorless laser is spectrally and temporally smooth, i.e., longitudinal and transverse-mode structures as well as spiking of the output do not occur in these systems. In contrast to spontaneous emission, ASE has some temporal coherence since each spontaneously emitted light avalanche is itself coherent.

So far we discussed the amplification of fluorescence generated by the active material itself. It has been observed, that flashlamp pump radiation which is within the spectral region of the laser transition will also be amplified and can lead to a reduction of stored energy in the laser material.

Gain saturation resulting from amplification of pump radiation in the active medium has been observed mainly in Nd:glass and Nd:YAG amplifiers [4.26, 68]. In addition to the depumping process caused by the lamp radiation, fluorescence being emitted from the laser rod escaping out into the laser pump cavity can be reflected back into the laser rod, and thereby stimulating further off-axis emission. Thus the presence of pump-cavity walls which have a high effective reflectivity at the fluorescence wavelength causes a transverse depumping action which depletes the energy available for on-axis stimulated emission. Elimination of these effects may require the use of optical filters in the pump cavity, cladding of the laser rod with a material which absorbs at the laser wavelength, or the addition of chemicals to the cooling fluid which serve the same purpose.

The influence of coolant properties upon the performance of a Nd:YAG laser amplifier is shown in Fig. 4.33. Two properties of the cooling fluid are of interest. These are the refractive index and the 1.06-μm absorption. The data is shown for a cored 2-cm-diameter by 10-cm-long Nd:YAG rod operated with water, FC-104, and air as cooling media [4.25]. FC-104 is a fluorcarbon flashlamp coolant made by the 3M Company. All fluids were used with and without a 1.06-μm absorbing filter (samarium-doped glass) in the pump cavity which was located between the rod and lamp. The addition of a 1.06-μm absorbing filter when water was used as a coolant made little difference in amplifier performance. Water possesses an appreciable intrinsic absorption at 1.06 μm. The performance obtained with FC-104, which is highly transparent at 1.06 μm, was greatly affected by the use of 1.06-μm absorbing filters. The performance obtained with air cooling was also affected by the use of a 1.06-μm absorbing filter, but the change was not nearly as marked as the change noted with FC-104. The primary difference between air and FC-104 in this experiment is their respective refractive indices. FC-104 has a low refractive index of approximately 1.28, but this is still significantly higher than that of air. With air cooling, the reflection of fluorescence at the rod surface is so high that it exceeds the return from the pump reflector, thereby limiting the effectiveness of the samarium filter. The dependence of the saturation level on absorption in the pump cavity at 1.06 μm, and on the refractive index of the material surrounding the rod, is attributed to the presence of amplified spontaneous emission from

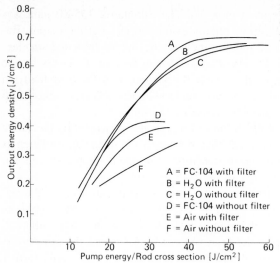

Fig. 4.33. Energy extraction density of YAG : Nd as a function of coolant properties [4.25]

A = FC-104 with filter
B = H_2O with filter
C = H_2O without filter
D = FC-104 without filter
E = Air with filter
F = Air without filter

the rod, as well as the presence of significant reflections of $1.06\,\mu m$ radiation form both the rod surfaces and pump reflectors. This energy storage limiting mechanism, although most pronounced in Nd : YAG rods because of the high gain obtained in this material, is not unique to Nd : YAG lasers.

Gain limitations in Nd : glass rods, resulting from amplified pump radiation and reflection of spontaneous emission from the pump cavity walls back into the rod, have led to the development of samarium-clad Nd : glass rods. Cladding of the rod also suppresses internal reflection from the rod walls when the refractive indices of the rod and cladding are matched.

4.4.2 Prelasing and Parasitic Modes

Prelasing and parasitic modes are consequences of the interaction of an inverted material with highly reflecting surroundings and are susceptible to design variation.

Prelasing. Laser action, occurring during the pump phase in an amplifier, results from the residual feedback of the various interfaces in the optical path. The condition for stable operation can be derived by considering the gain in an amplifier rod of length L pumped to an inversion level characterized by a gain coefficient g per unit length. Let the end faces of the rod exhibit reflectivities of r_1 and r_2. Then spontaneous emission emanating from any given location in the rod, traveling normal to the rod end faces, will exhibit a loop gain after one round trip in the rod of approximately $r_1 r_2 \exp(2gL)$. If this loop gain is greater than unity, oscillations will build up until the usable rod-stored energy is depleted. The requirement for an amplifier is thus

$$r_1 r_2 \exp(2gL) < 1 \,, \tag{4.57}$$

which sets an upper limit on rod length.

Parasitic Oscillations. Internal reflections at the boundaries of the active medium can drastically enhance the onset of ASE, particularly if these reflections lead to a closed path, i.e., a ray that is reflected upon itself. In this case, we have a strong feedback mechanism and as soon as the gain in the laser medium exceeds the reflection losses parasitic oscillations will set in. For example, in rods with polished walls there can exist internal modes completely contained by total internal reflection. These are particularly troublesome.

One type of internal mode, the whisper mode, propagates circumferentially and has no longitudinal component. This mode can penetrate into the rod as far as $r = R/n$, where n is the index of refraction [4.12, 68]. Another type of internal mode, the light-pipe mode, propagates down the rod in a zigzag fashion. The gain limitations due to internal radial modes are considerably improved in immersed rods particularly in case of an index matching fluid, or rods with rough surfaces, grooves or anti-reflection coatings [4.73, 74]. Longitudinal modes can be suppressed if the faces of the rod are either wedged or anti-reflection coated.

Several researchers have discussed parasitic processes which take place in laser disks [4.60, 69–71]. As the size of the disk becomes large, such processes begin to become significant. The threshold of the parasitic oscillation depends on the edge-coating material, which reduces the feedback by absorbing fluorescence.

The threshold for parasitic oscillation sets an upper limit on the energy storage attainable in Nd : glass laser discs. It has been shown [4.69] that the lowest-loss path lies in a plane across the diameter of the disc. The only losses on such a path come at the edge reflections. At the oscillation threshold this edge loss must just cancel the path gain. For very thin discs the length for this path is approximately the disc refractive index n times the diameter D (assuming that the disc is in air). At the oscillation threshold we have

$$R\,e^{nDg} = 1 \,, \tag{4.58}$$

where g is the gain coefficient in the laser material and R is the edge reflectivity. For typical design parameters, such as $D = 15\,\text{cm}$, $g = 0.2\,\text{cm}^{-1}$, and $n = 1.56$, the reflectivity must be less than 1 % to avoid oscillation. Cladding the edge of the disks with an index-matched absorbing glass results in a dramatic reduction of parasitic losses. Besides bulk parasitic modes there are surface modes. These occur primarily across the disk face because of the larger gain coefficient at the surface due to pump light absorption [4.65, 68].

Amplified spontaneous emission (ASE) strongly influences the performance of large aperture Nd : glass disk amplifiers. Even with perfectly absorbing edge claddings on the disks, depumping caused by single pass ASE increases rapidly with aperture, reducing peak gain and storage efficiency. ASE becomes the principal depumping process in large, edge-clad disks when the product of the gain coefficient and disk major axis gD becomes large (greater than 3) [4.60, 72].

Techniques Used to Prevent or Reduce Feedback in Amplifiers. In order to minimize prelasing, rod ends are cut at an oblique angle to the rod axis, often at the Brewster angle, or they are coated with antireflection layers. Reflections causing off-axis spontaneous emission from the cylindrical rod surfaces are minimized by use of index-matching fluids around the laser rod or by rough grinding of the cylindrical walls. If more than one stage of amplification is required, the separate stages are in part decoupled by one or more of the following measures. The distance between amplifier stages may be made as large as possible. Since ASE occurs in relatively large beam angles, a large distance between stages very effectively reduces feedback between stages, or even more effective are spatial filters. If the small-signal gain of an amplifier chain is very high (30 dB or more), optical isolators based on the Faraday effect, are placed between amplifier stages.

5. Optical Resonator

The light emitted by most lasers contains several discrete optical frequencies, separated from each other by frequency differences which can be associated with different modes of the optical resonator. It is common practice to distinguish two types of resonator modes: "Longitudinal" modes differ from one another only in their oscillation frequency; "transverse" modes differ from one another not only in their oscillation frequency, but also in their field distribution in a plane perpendicular to the direction of propagation. Corresponding to a given transverse mode are a number of longitudinal modes which have the same field distribution as the given transverse mode but which differ in frequency.

To describe the electromagnetic field variations inside optical resonators, the symbols TEM_{mnq} or TEM_{plq} are used. The capital letters stand for "transverse electromagnetic waves" and the first two indices identify a particular transverse mode, whereas q describes a longitudinal mode. Because resonators that are used for typical lasers are long compared to the laser wavelength, they will, in general, have a large number of longitudinal modes. Therefore, the index q which specifies the number of modes along the axis of the cavity will be very high. The indices for the transverse modes, which specify the field variations in the plane normal to the axis, are very much lower and sometimes may be only the first few integers.

The spectral characteristics of a laser, such as linewidth and coherence length, are primarily determined by the longitudinal modes; whereas beam divergence, beam diameter, and energy distribution are governed by the transverse modes. In general, lasers are multimode oscillators unless specific efforts are made to limit the number of oscillating modes. The reason for this lies in the fact, that a very large number of longitudinal resonator modes fall within the bandwidth exhibited by the laser transition, and a large number of transverse resonator modes can occupy the cross section of the active material.

5.1 Transverse Modes

The theory of modes in optical resonators has been treated in [5.1–3]; comprehensive reviews of the subject can also be found in [5.4, 5].

5.1.1 Intensity Distribution of Transverse Modes

In an optical resonator electromagnetic fields can exist whose distribution of amplitudes and phases reproduce themselves upon repeated reflections between the mirrors. These particular field configurations comprise the transverse electromagnetic modes of a passive resonator.

Transverse modes are defined by the designation TEM_{mn} for Cartesian coordinates. The integers m and n represent the number of nodes of zeros of intensity transverse to the beam axis in the vertical and horizontal directions. In cylindrical coordinates the modes are labeled TEM_{pl} and are characterized by the number of radial nodes p and angular nodes l. The higher the values of m, n, p, and l, the higher the mode order. The lowest-order mode is the TEM_{00} mode, which has a Gaussian intensity profile with its maximum on the beam axis. For modes with subscripts of 1 or more, intensity maxima occur that are off-axis in a symmetrical pattern. To determine the location and amplitudes of the peaks and nodes of the oscillation modes, it is necessary to employ higher-order equations which either involve Hermite (H) or Laguerre (L) polynomials. The Hermite polynomials are used when working with rectangular coordinates, while Laguerre polynomials are more convenient when working with cylindrical coordinates.

In cylindrical coordinates, the radial intensity distribution of allowable circularly symmetric TEM_{pl} modes is given by the expression

$$I_{pl}(r, \phi, z) = I_0 \varrho^l [L_p^l \varrho]^2 (\cos^2 l\phi) \exp(-\varrho) \tag{5.1}$$

with $\varrho = 2r^2(z)/w^2(z)$, where z is the propagation direction of the beam, and r, ϕ are the polar coordinates in a plane transverse to the beam direction. The radial intensity distributions are normalized to the spot size of a Gaussian profile; that is, $w(z)$ is the spot size of the Gaussian beam, defined as the radius at which the intensity of the TEM_{00} mode is $1/e^2$ of its peak value on the axis. L_p is the generalized Laguerre polynomial of order p and index l.

The intensity distribution given in (5.1) is the product of a radial part and an angular part. For modes with $l = 0$ (i.e., TEM_{p0}), the angular dependence drops out and the mode pattern contains p dark concentric rings, each ring corresponding to a zero of $L_p^0(\varrho)$. The radial intensity distribution decays due to the factor $\exp(-\varrho)$. The center of a pl mode will be bright if $l = 0$, but dark otherwise because of the factor ϱ^l. These modes, besides having p zeros in the radial direction, also have $2l$ nodes in azimuth.

The only change in a (pl) mode distribution comes through the dependence of the spot size $w(z)$ on the axial position z. However, the modes preserve the general shape of their electric field distributions for all values of z. As w increases with z, the transverse dimensions increase so that the sizes of the mode patterns stay in constant ratio to each other.

From (5.1) it is possible to determine any beam mode profile. Figure 5.1a depicts various cylindrical transverse intensity patterns as they would appear in the output beam of a laser. Note that the area occupied by a mode increases with the mode number. A mode designation accompanied by an asterisk indicates a mode which is a linear superposition of two like modes, one rotated 90° about the axis relative to the other. For example, the TEM mode designated 01* is made up of two TEM_{01} modes.

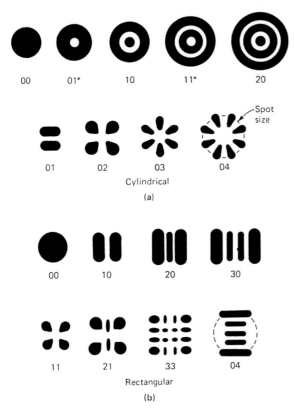

Fig. 5.1. Examples of (**a**) cylindrical and (**b**) rectangular transverse mode patterns. For cylindrical modes, the first subscript indicates the number of dark rings, whereas the second subscript indicates the number of dark bars across the pattern. For rectangular patterns, the two subscripts give the number of dark bars in the x and y directions [5.6]

The intensity distribution of the modes shown in Fig. 5.1a can be calculated if we introduce the appropriate Laguerre polynomials into (5.1), i.e.,

$$L_0^l(\varrho) = 1 , \quad L_1^0(\varrho) = 1 - \varrho ; \quad L_2^0(\varrho) = 1 - 2\varrho + \tfrac{1}{2}\varrho^2 .$$

A plot of the intensity distributions of the lowest-order mode and the next two higher-order transverse modes, i.e., TEM_{00}, TEM_{01*} and TEM_{10}, is illustrated in Fig. 5.2.

In rectangular coordinates the intensity distributions of a (m, n) mode is given by

$$I_{mn}(x, y, z) = I_0 \left[H_m \left(\frac{x(2)^{1/2}}{w(z)} \right) \exp \left(\frac{-x^2}{w^2(z)} \right) \right]^2$$

$$\times \left[H_n \left(\frac{y(2)^{1/2}}{w(z)} \right) \exp \left(\frac{-y^2}{w^2(z)} \right) \right]^2 . \tag{5.2}$$

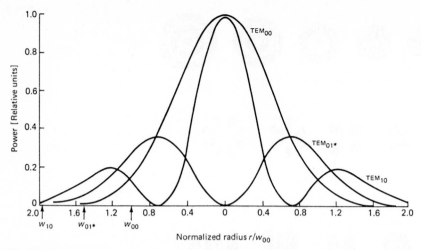

Fig. 5.2. Radial intensity distribution for TEM$_{00}$, TEM$_{01*}$, and TEM$_{10}$ modes. The radii are normalized to the beam radius w_{00} of the fundamental mode [5.7]

As before, $w(z)$ is the spot size at which the transverse intensity decreases to $1/e^2$ of the peak intensity of the lowest-order mode. The function $H_m(s)$ is the mth-order Hermite polynomial, for example, $H_0(s) = 1, H_1(s) = 2s$, and $H_2(s) = 4s^2 - 2$. At a given axial position z, the intensity distribution consists of the product of a function of x alone and a function of y alone. The intensity patterns of rectangular transverse modes are sketched in Fig. 5.1b. The m, n values of a single spatial mode can be determined by counting the number of dark bars crossing the pattern in the x and y directions. Note that the fundamental mode ($m = n = 0$) in this geometry is identical with the fundamental mode in cylindrical geometry.

The transverse modes shown in Fig. 5.1 can exist as linearly polarized beams, as illustrated by Fig. 5.3. By combining two orthogonally polarized modes of the same order, it is possible to synthesize other polarization configurations; this is shown in Fig. 5.4 for the TEM$_{01}$ mode.

5.1.2 Characteristics of a Gaussian Beam

A light beam emitted from a laser with a Gaussian intensity profile is called the "fundamental mode" or TEM$_{00}$ mode. Because of its importance it is discussed here in greater detail. The decrease of the field amplitude with distance r from the axis in a Gaussian beam is described by

$$E(r) = E_0 \exp\left(\frac{-r^2}{w^2}\right) . \tag{5.3}$$

Thus, the distribution of power density is

$$I(r) = I_0 \exp\left(\frac{-2r^2}{w^2}\right) . \tag{5.4}$$

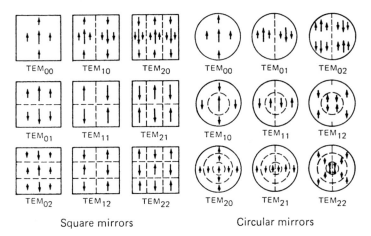

Square mirrors   Circular mirrors

Fig. 5.3. Linearly polarized resonator mode configurations for square and circular mirrors [5.1]

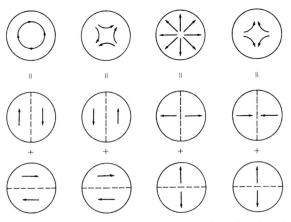

Fig. 5.4. Synthesis of different polarization configurations from the linearly polarized TEM_{01} mode

The quantity w is the radial distance at which the field amplitude drops to $1/e$ of its value on the axis, and the power density is decreased to $1/e^2$ of its axial value. The parameter w is often called the beam radius or "spot size," and $2w$, the beam diameter. The fraction of the total power of a Gaussian beam which is contained in a radial aperture of $r = w$, $r = 1.5w$, and $r = 2w$ is 86.5 %, 98.9 %, and 99.9 %. If a Gaussian beam is passed through a radial aperture of $3w$, then only 10^{-6} % of the beam power is lost due to the obstruction. For our subsequent discussion, an "infinite aperture" will mean a radial aperture in excess of three spot sizes.

If we consider now a propagating Gaussian beam, we note that although the intensity distribution is Gaussian in every beam cross section, the width of the intensity profile changes along the axis. The Gaussian beam contracts to a minimum diameter $2w_0$ at the beam waist where the phase front is planar. If one

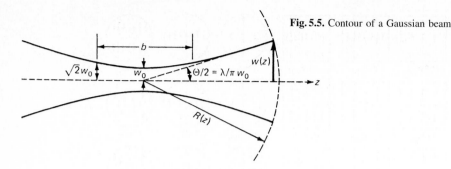

Fig. 5.5. Contour of a Gaussian beam

measures z from this waist, the expansion laws for the beam assume a simple form. The spot size a distance z from the beam waist expands as a hyperbola, which has the form

$$w(z) = w_0 \left[1 + \left(\frac{\lambda z}{\pi w_0^2} \right)^2 \right]^{1/2} .$$

(5.5)

Its asymptote is inclined at an angle $\Theta/2$ with the axis, as shown in Fig. 5.5, and defines the far-field divergence angle of the emerging beam. The full divergence angle for the fundamental mode is given by

$$\Theta = \lim_{z \to \infty} \frac{2w(z)}{z} = \frac{2\lambda}{\pi w_0} = 1.27 \frac{\lambda}{(2w_0)} .$$

(5.6)

From these considerations it follows that at large distances, the spot size increases linearly with z, and the beam diverges at a constant cone angle Θ. A most interesting point here is, that the smaller the spot size w_0 at the beam waist, the greater the divergence.

At sufficiently large distances from the beam waist, the wave has a spherical wavefront appearing to emanate from a point on the beam axis at the waist. If $R(z)$ is the radius of curvature of the wavefront that intersects the axis at z, then

$$R(z) = z \left[1 + \left(\frac{\pi w_0^2}{\lambda z} \right)^2 \right] .$$

(5.7)

It is important to note that in a Gaussian beam the wavefront has the same phase across its entire surface.

Sometimes the properties of a TEM_{00} mode beam are described by specifying a confocal parameter

$$b = \frac{2\pi w_0^2}{\lambda} ,$$

(5.8)

where b is the distance between the points at each side of the beam waist for which $w(z) = (2)^{1/2} w_0$ (Fig. 5.5).

Before we leave the subject of Gaussian beams, we will briefly point out the difference in the definition of beam divergence for Gaussian beams and plane waves. A laser operating at the TEM_{00} mode will have a beam divergence according to (5.6). For a plane wavefront incident upon a circular aperture of diameter D, the full cone angle of the central (Airy) disc, defined at the first minimum of the Fraunhofer diffraction pattern, is given by

$$\Theta_p = \frac{2.44\lambda}{D} .$$
(5.9)

The energy contained within this angle is about 84 % of the total energy transmitted by the aperture.

Equations (5.6, 9) are often confused, because various conventions have been adopted by different authors with the equation

$$\Theta_R = \frac{1.22\lambda}{D} ,$$
(5.10)

which represents the *half-* cone angle of the Fraunhofer diffraction pattern and also happens to be the "Rayleigh criterion" for the angular resolution of an optical instrument.

In laboratory work, a beam size is often obtained by measuring the diameter of the illuminated spot with a scale. This is not the spot size $2w_0$ as defined by (5.5). There is no obvious visual cue to the magnitude of the spot size in the appearance of the illuminated spot. Thus, "spot size" and "size of the illuminated spot" are totally different concepts. The former is a property of the laser cavity; the latter is a subjective estimate. To measure the spot size, the illuminated spot should be scanned with a photodetector behind a small pinhole. The resulting curve of intensity versus position of the pinhole will yield a Gaussian curve from which the spot size w_0 can be calculated.

Much faster results are obtained with electronic beam diagnostic instrumentation. These instruments consist of a CCD array camera, which samples the beam at the focal plane of a lens, a computer and beam analysis software. These instruments permit rapid 2-D or 3-D visualization of the beam profile and calculation of beam parameters.

5.1.3 Resonator Configurations

The most commonly used laser resonators are composed of two spherical or flat mirrors facing each other. We will first consider the generation of the lowest-order mode by such a resonant structure. Once the parameters of the TEM_{00} mode are known, all higher-order modes simply scale from it in a known manner. Diffraction effects due to the finite size of the mirrors will be neglected in this section.

The Gaussian beam indicated in Fig. 5.6 has a wavefront curvature of R_1 at a distance t_1 from the beam waist. If we put a mirror at t_1 whose radius of

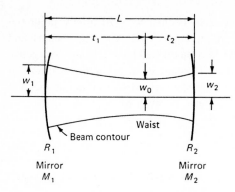

Fig. 5.6. Mode parameters of interest for a resonator with mirrors of unequal curvature

curvature equals that of the wavefront, then the mode shape has not been altered. To proceed further, we can go along the z axis to another point t_2 where the TEM_{00} has a radius of curvature R_2, and place there a mirror whose radius of curvature R_2 equals that of the spherical wavefront at t_2. Again the mode shape remains unaltered.

Therefore, to make a resonator, we simply insert two reflectors which match two of the spherical surfaces defined by (5.7). Alternatively, given two mirrors separated by a distance L, if the position of the plane $z = 0$ and the value of the parameter w_0 can be adjusted so that the mirror curvatures coincide with the wavefront surfaces, we will have found the resonator mode.

We will now list formulas, derived by *Kogelnik* and *Li* [5.4], which relate the mode parameters w_1, w_2, w_0, t_1, and t_2 to the resonator parameters R_1, R_2, and L. As illustrated in Fig. 5.6, w_1 and w_2 are the spot radii at mirrors M_1 and M_2, respectively; t_1 and t_2 are the distances of the beam waist described by w_0 from mirrors M_1 and M_2, respectively; and R_1 and R_2 are the curvatures of mirrors M_1 and M_2 which are separated a distance L. Labeling conventions are that concave curvatures are positive.

The beam radii at the mirrors are given by

$$w_1^4 = \left(\frac{\lambda R_1}{\pi}\right)^2 \frac{R_2 - L}{R_1 - L} \left(\frac{L}{R_1 + R_2 - L}\right) ,$$

$$w_2^4 = \left(\frac{\lambda R_2}{\pi}\right)^2 \frac{R_1 - L}{R_2 - L} \left(\frac{L}{R_1 + R_2 - L}\right) . \tag{5.11}$$

The radius of the beam waist, which is formed either inside or outside the resonator, is given by

$$w_0^4 = \left(\frac{\lambda}{\pi}\right)^2 \frac{L(R_1 - L)(R_2 - L)(R_1 + R_2 - L)}{(R_1 + R_2 - 2L)^2} . \tag{5.12}$$

The distances t_1 and t_2 between the waist and the mirrors, measured positive (Fig. 5.6), are

$$t_1 = \frac{L(R_2 - L)}{R_1 + R_2 - 2L} , \quad t_2 = \frac{L(R_1 - L)}{R_1 + R_2 - 2L} . \tag{5.13}$$

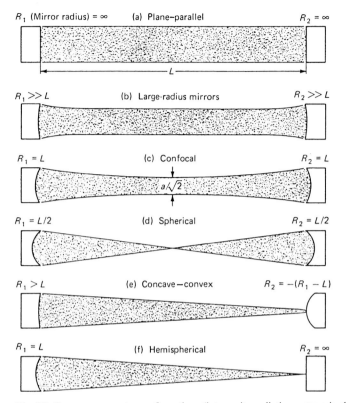

Fig. 5.7. Common resonator configurations (intracavity radiation pattern is shaded)

These equations treat the most general case of a resonator. There are many optical resonator configurations for which (5.11–13) are greatly simplified. Figure 5.7 shows some of the most commonly used geometries.

Mirrors of Equal Curvature

With $R_1 = R_2 = R$ we obtain from (5.11)

$$w_{1,2}^2 = \frac{\lambda R}{\pi} \left(\frac{L}{2R - L} \right)^{1/2} . \tag{5.14}$$

The beam waist which occurs at the center of the resonator $t_1 = t_2 = R/2$ is

$$w_0^2 = \frac{\lambda}{2\pi} [L(2R - L)]^{1/2} . \tag{5.15}$$

If we further assume that the mirror radii are large compared to the resonator length $R \gg L$, the above formula simplifies to

$$w_{1,2}^2 = w_0^2 = \left(\frac{\lambda}{\pi} \right) \left(\frac{RL}{2} \right)^{1/2} . \tag{5.16}$$

As follows from (5.16), in a resonator comprised of large-radius mirrors, the beam diameter changes very little as a function of distance.

A resonator comprised of mirrors having a radius of curvature on the order of 2 to 10 m, i.e., several times longer than the length of the resonator, is one of the most commonly employed configurations. Such a large-radius mirror resonator has a reasonable alignment stability and a good utilization of the active medium.

A special case of a symmetrical configuration is the spherical resonator which consists of two mirrors separated by twice their radius, that is, $R = L/2$. The corresponding beam consists of a mode whose dimensions are fairly large at each mirror and which focus down to a diffraction-limited point at the center of the resonator. A spherical resonator is rather sensitive to misalignment and the small spot can lead to optical damage.

Another very important special case of a resonator with mirrors of equal curvature is the confocal resonator. For this resonator the mirror separation equals the curvature of the identical mirrors, that is, $R = L$. From (5.14, 15) we obtain the simplified relation

$$w_{1,2} = \left(\frac{\lambda R}{\pi}\right)^{1/2} \quad \text{and} \quad w_0 = \frac{w_{1,2}}{(2)^{1/2}} . \tag{5.17}$$

The confocal configuration gives the smallest possible mode dimension for a resonator of given length. For this reason, confocal resonators are not often employed since they do not make efficient use of the active material.

Plano-Concave Resonator

For a resonator with one flat mirror ($R_1 = \infty$) and one curved mirror we obtain

$$w_1^2 = w_0^2 = \left(\frac{\lambda}{\pi}\right)[L(R_2 - L)]^{1/2} \quad \text{and} \quad w_2^2 = \left(\frac{\lambda}{\pi}\right)R_2\left(\frac{L}{R_2 - L}\right)^{1/2} . \tag{5.18}$$

The beam waist w_0 occurs at the flat mirror (that is $t_1 = 0$ and $t_2 = L$). A special case of this resonator configuration is the hemispherical resonator. The hemispherical resonator consists of one spherical mirror and one flat mirror placed approximately at the center of curvature of the sphere. The resultant mode has a relatively large diameter at the spherical mirror and focuses to a diffraction-limited point at the plane mirror. In practice, one makes the mirror separation L slightly less than R_2 so that a value of w_1 is obtained that gives reasonably small diffraction losses.

In solid-state lasers, the small spot size can lead to optical damage at the mirror. A near hemispherical resonator has the best alignment stability of any configuration, therefore it is often employed in low power lasers such as HeNe lasers.

Concave-Convex Resonator

The pertinent beam parameters for concave-convex resonators can be calculated if we introduce a negative radius $(-R_2)$ for the convex mirror into (5.11–13). A small-radius convex mirror in conjunction with a large-radius concave or plane mirror is a very common resonator in high-average-power solid-state lasers. As follows from the discussion in the next section, as a passive resonator such a configuration is unstable. However, in a resonator which contains a laser crystal, this configuration can be stable since the diverging properties of the concave mirror are counteracted by the focusing action of the laser rod. Since the concave mirror partially compensates for thermal lensing, a large mode volume can be achieved as will be shown in an example later in this chapter.

Plane-Parallel Resonator

The plane-paralles resonator, which can be considered a special case of a large-radius mirror configuration if $R_1 = R_2 = \infty$, has been thoroughly analyzed by *Fox* and *Li* [5.1] who showed that the output of the plane-parallel resonator does not consist of a plane-parallel wavefront. Instead, owing to fixed and rather large diffraction losses around the edges, there is a phase lag of approximately $30°$ near the edges which gives the wavefront a slight curvature.

Plane-parallel mirrors make good use of the volume of the active material, however they are very sensitive to even the slightest misalignment.

In calculating the mode size in a resonator, it must be noted that in most cases the resonator is formed by spherical mirrors with the reflecting surfaces deposited on plane-concave mirror blanks of index n. The front mirror of such a resonator acts as a negative lens and will change the characteristics of the emerging beam. The beam appears to have a different waist diameter and location. Taking the negative lens effect into account, a beam emerging from a resonator of equal mirror radii R can be described as having a beam waist of

$$w'_0 = w_0 \left(1 + \frac{L(n^2 - 1)}{2R}\right)^{-1/2} , \qquad (5.19)$$

with the waist located at

$$t_2 = \frac{L}{2} \frac{nR}{R + L(n^2 - 1)/2} . \qquad (5.20)$$

The parameter w_0 refers to the beam waist calculated previously for the resonator of equal mirror radii. As we can see from (5.19), the beam leaving the resonator appears to have a smaller beam waist, which indicates larger divergence, and the location of the waist appears to be moved toward the front mirror. The negative lens effect of the output mirror disappears if the outer surface of the output mirror has a radius of curvature given by

$$R'_1 = \frac{R_1(n - 1)}{n} . \qquad (5.21)$$

5.1.4 Stability of Laser Resonators

For certain combinations of R_1, R_2, and L the equations summarized in the previous sub-section give nonphysical solutions (i.e., imaginary spot sizes). This is the region where low-loss modes do not exist in the resonator.

Light rays that bounce back and forth between the spherical mirrors of a laser resonator experience a periodic focusing action. The effect on the rays is the same as in a periodic sequence of lenses [5.8, 9]. Rays passing through a stable sequence of lenses are periodically refocused. For unstable systems the rays become more and more dispersed the further they pass through the sequence. In an optical resonator operated in the stable region, the waves propagate between reflectors without spreading appreciably. This fact can be expressed by a stability criterion [5.10]

$$0 < \left(1 - \frac{L}{R_1}\right)\left(1 - \frac{L}{R_2}\right) < 1 . \tag{5.22}$$

To show graphically which type of resonator is stable and which is unstable, it is useful to plot a stability diagram on which each particular resonator geometry is represented by a point. This is shown in Fig. 5.8, where the parameters

$$g_1 = 1 - \frac{L}{R_1} , \quad g_2 = 1 - \frac{L}{R_2} \tag{5.23}$$

are drawn as the coordinate axes.

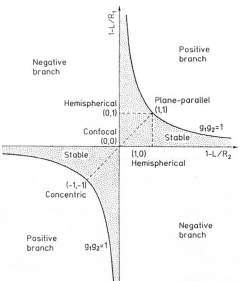

Fig. 5.8. Stability diagram for the passive laser resonator

All cavity configurations are unstable unless they correspond to points located in the area enclosed by a branch of the hyperbola $g_1 g_2 = 1$ and the coordinate axes. The origin of the diagram represents the confocal system.

The resonators located along the dashed line oriented at $45°$ with respect to the coordinate axis, are symmetric configurations, i.e., they have mirrors with the same radius of curvature.

The diagram is divided into positive and negative branches defining quadrants for which $g_1 g_2$ is either positive or negative. The reason for this classification becomes clear when we discuss unstable resonators.

5.1.5 Diffraction Losses

In any real laser resonator some part of the laser beam will be lost either by spillover at the mirrors or by limiting apertures, such as the lateral boundaries of the active material. These losses will depend on the diameter of the laser beam in the plane of the aperture and the aperture radius. If we take a finite aperture of radius a within the resonator into account, the diffraction losses depend on four parameters, R_1, R_2, L, and a, which describe the resonator; and on three parameters λ, m, and n, characterizing the particular optical beam present in the resonator. Fortunately, the losses depend only on certain combinations of these parameters. These combinations are the so-called Fresnel number,

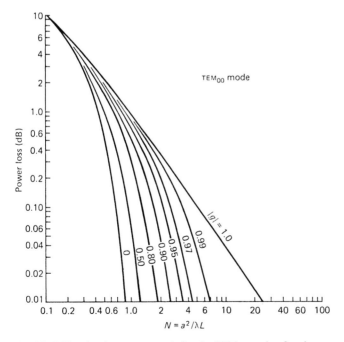

Fig. 5.9. Diffraction losses per transit for the TEM_{00} mode of various symmetrical and stable resonators [5.10]

$$N = \frac{a^2}{\lambda L} ,$$

(5.24)

and the quantities g_1 and g_2 which were defined in (5.23). The parameter N can be thought of as the ratio of the acceptance angle (a/L) of one mirror as viewed from the center of the opposing mirror to the diffraction angle (λ/a) of the beam. Therefore, when N is small, especially if $N<1$, the loss factor will be high because only a portion of the beam will be intercepted by the mirrors. When N is large, the losses will be low for the stable resonator configurations and large for the unstable resonators. If two resonators have the same values of N, g_1, and g_2, then they have the same diffraction loss, the same resonant frequency, and the same mode patterns. The fractional energy loss per transit due to diffraction effects for the lowest-order mode of a resonator with identical mirrors ($g_1 = g_2 = g$) is given in Fig. 5.9. For the plane-parallel resonator with circular aperture, an analytical expression of the diffraction losses for large Fresnel numbers ($F \geq 1$)

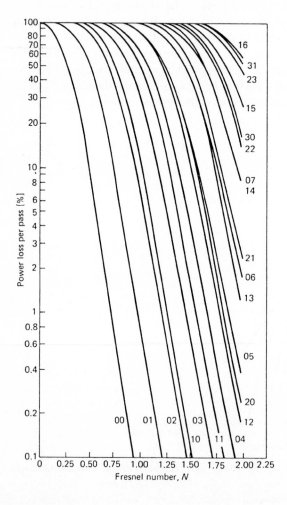

Fig. 5.10. Diffraction loss for various resonator modes as a function of the Fresnel number [5.12]

has been derived in [5.11], namely

$$\alpha = 8\kappa_{pl}^2 \frac{\delta(M + \delta)}{[(M + \delta)^2 + \delta^2]^2} , \tag{5.25}$$

where $\delta = 0.824$, $M = (8\pi N)^{1/2}$, and k_{pl} is the $(p + l)$th zero of the Bessel function of order l. The diffraction losses in a confocal resonator for a number of low-order modes are plotted in Fig. 5.10 [5.12]. The mode designations are in the cylindrical coordinate system. Note that all modes have very large losses for small N and that the losses drop with increasing N. However, the losses for the higher-order modes drop less rapidly than the losses for the lower-order modes.

5.1.6 Higher-Order Modes

The mode radius and the divergence angle for modes of cylindrical symmetry are usually defined in terms of the $1/e^2$ intensity points of the outermost lobe. As indicated in Fig. 5.2, the radius of each mode increases with increasing mode number. The ratio of the radii of different modes with respect to that of the lowest-order mode is designated c_{pl}. From (5.1) follows that the ratios of different mode diameters remain constant at any plane inside or outside of the cavity, in the near or far field. Therefore, the mode radius and the beam divergence of a higher-order mode can be related to the fundamental mode parameters as follows [5.7, 13]:

$$W_{pl}(z) = C_{pl}w_0 \quad \Theta_{pl} = C_{pl}\Theta , \tag{5.26}$$

where Θ is obtained from (5.6) and w_0 can be calculated for a particular resonator from the formulas summarized in Sect. 5.1.3. Numerical values for C_{pl}, as obtained from (5.1) for several low-order modes are $C_{00} \equiv 1$; $C_{01*} = 1.5$; $C_{10} = 1.9$ $C_{11*} = 2.15$; $C_{20} = 2.42$; $C_{21*} = 2.63$.

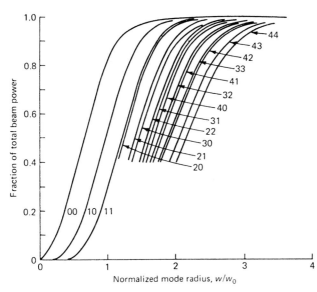

Fig. 5.11. Fraction of total power for a particular mode TEM$_{mn}$ within a circular section of radius R. The number on each curve corresponds to the individual modes [5.6]

For modes of rectangular symmetry, defining the mode radius and the divergence angle in terms of the $1/e^2$ intensity points of the outermost lobe is not adequate, particularly for those rectangular modes whose pattern is far from square. A possible description of rectangular mode sizes is in terms of a radius which contains something like 90 % of the beam power. This is a highly functional definition from an applications standpoint. The relative size of some rectangular modes have been calculated by *Kruger* [5.6, 14]. Figure 5.11 shows the fraction of the total beam power for a particular rectangular mode (TEM$_{mn}$) within a circular cross section of the beam. From Fig. 5.11 it is also possible to determine the relative beam divergence for any mode. If we define the mode size at the 90 % power points, the divergence angle for a particular mode is the TEM$_{00}$ divergence angle times the normalized radius at the 90% level. For example, since the cross section containing 90% of the power for the TEM$_{44}$ mode has a normalized radius $(w/w_0) = 2.74$, the divergence angle for TEM$_{44}$ is 2.74 times the TEM$_{00}$ divergence angle.

5.1.7 Active Resonator

So far we have discussed the modes in a passive resonator consisting of a pair of mirrors. Introducing an active element into the resonator, such as a laser crystal, in addition to altering the optical length of the cavity, will perturb the mode configuration, since the active material possesses a saturable, nonuniform gain and exhibits thermal lensing and birefringence. In high-gain, giant-pulse lasers, gain saturation at the center of a TEM$_{00}$ mode can lead to a flattening of the intensity profile [5.15, 16]. Also, pump nonuniformities leading to a nonuniform gain distribution across the beam will lead to non-Gaussian output intensity profiles. Theoretical and experimental investigations have shown that in solid-state lasers the governing mechanisms which distort the mode structure in the resonator are the thermal effects of the laser rod. As will be discussed in more detail in Chap. 7, optical pumping leads to a radial temperature gradient in the laser rod. As a result, in cw and high average power systems, the rod is acting like a positive thick lens of an effective focal length f, which is inversely proportional to the pump power.

The theory necessary to analyze resonators that contain optical elements other than the end mirrors has been developed by *Kogelnik* [5.17]. We will apply this theory to the case of a resonator containing an internal thin lens. To a first approximation, this lens can be thought of as representing the thermal lensing introduced by the laser rod. The more complex case of a distributed thick lens which more adequately describes thermal lensing has been treated in [5.17–20].

Beam properties of resonators containing internal optical elements are described in terms of an equivalent resonator composed of only two mirrors. The pertinent parameters of a resonator equivalent to one with an internal thin lens are

$$g_1 = 1 - \frac{L_2}{f} - \frac{L_0}{R_1} ; \quad g_2 = 1 - \frac{L_1}{f} - \frac{L_0}{R_2} , \tag{5.27}$$

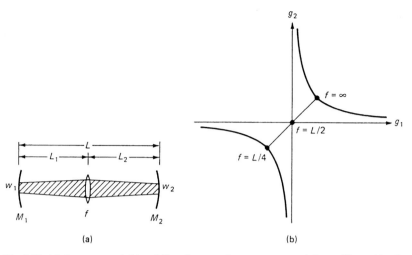

(a)   (b)

Fig. 5.12. (a) Geometry and (b) stability diagram of a resonator containing a thin positive lens

where $L_0 = L_1 + L_2 - (L_1 L_2 / f)$ and f is the focal lenth of the internal lens; L_1 and L_2 are the spacings between mirrors M_1, M_2 and the lens, as shown in Fig. 5.12a.

In any resonator, the TEM_{00} mode spot size at one mirror can be expressed as a function of the resonator parameters

$$w_1^2 = \frac{\lambda L}{\pi} \left(\frac{g_2}{g_1 (1 - g_1 g_2)} \right)^{1/2} . \tag{5.28}$$

The ratio of the spot sizes at the two mirrors is

$$\frac{w_1^2}{w_2^2} = \frac{g_2}{g_1} . \tag{5.29}$$

The stability condition (5.22) remains unchanged.

As an example we will consider a resonator with flat mirrors ($R_1 = R_2 = \infty$) and a thin lens in the center ($L_1 = L_2 = L/2$). From (5.27) we obtain

$$g = g_1 = g_2 = 1 - \frac{L}{2f} , \quad w_1^2 = w_2^2 = \left(\frac{\lambda L}{\pi} \right) (1 - g^2)^{-1/2} . \tag{5.30}$$

For $f = \infty$ the resonator configuration is plane-parallel; for $f = L/2$ we obtain the equivalent of a confocal resonator; and for $f = L/4$ the resonator corresponds to a spherical configuration.

The mode size in the resonator will grow to infinity as the mirror separation approaches four times the focal length of the laser rod. Figure 5.12b shows the location of a plane-parallel resonator with an internal lens of variable focal length in the stability diagram. A very detailed analysis of a resonator consisting of a flat and a curved mirror containing an internal lens has been presented in [5.19].

The discussion of resonators has so far been restricted to devices having axial symmetry with respect to the beam axis. Resonators that contain inclined surfaces, such as Brewster-ended rods, polarizers, prisms, etc., lack axial symmetry. The effect of these asymmetric devices is to produce astigmatic beams, i.e., beams that have different spot sizes, wavefront curvatures, and beam waist positions in two orthogonal directions. Axially asymmetric laser cavities have been analyzed in [5.21, 22].

5.1.8 Resonator Sensitivity

There are two contexts in which the term "stability" is used. First, laser resonators are said to be optically stable or unstable depending on the value of g_1 and g_2. Second, the mode size and position are sensitive to mechanical and optical perturbations of the optical elements. We will refer to the stability of the mode against these perturbations as the resonator sensitivity.

When designing resonators for an optimum mode size, it will be of the utmost importance to consider the resonator sensitivity to these mechanical and optical perturbations. Usually one is interested in the sensitivity of the resonator to two common types of perturbations: first, a time-varying thermal lensing effect caused by the laser rod, and second, misalignments of the resonator mirrors. The former perturbation leads mainly to a change in mode size and beam divergence, whereas the latter perturbation leads to a lateral displacement and angular tilt of the output beam which causes an increase of the diffraction losses and, therefore, a reduction of output power. First-order effects on the modes as a function of cavity perturbation are usually obtained by evaluation of a sensitivity matrix [5.19, 20].

Considering first the resonator's sensitivity to lensing effects, we note from (5.29, 30) that a resonator is insensitive to axial perturbations if the spot size w_1 is insensitive to changes of g_1 and g_2. A calculation of the relative sensitivities of various resonators to small changes in mirror radii of curvature has been carried out by *Chesler* and *Maydan* [5.23] and *Steffen* et al. [5.18]. A perturbation within the resonator producing changes in mirror radii is equivalent to the introduction of a lens of some focal length f. In order for the resonator to have a low sensitivity to axial perturbations, i.e., small spot size changes for large changes of g_1 and g_2, it is necessary that $dw_1/df = 0$. This condition is met for resonator geometries which satisfy the following equation [5.18]

$$2g_2g_1 - 1 + \left(\frac{g_1}{g_2}\right)\left(\frac{L_1}{L_2}\right)^2 + \frac{2g_1L_1}{L_2} = 0 . \tag{5.31}$$

One particular resonator satisfying (5.31) is determined by

$$g_1g_2 = \tfrac{1}{2} \quad L_1 = 0 . \tag{5.32}$$

In this case the internal lens is either absent or located at the surface of mirror R_1. Figure 5.13 presents the resonator stability diagram discussed in Sect. 5.1.4

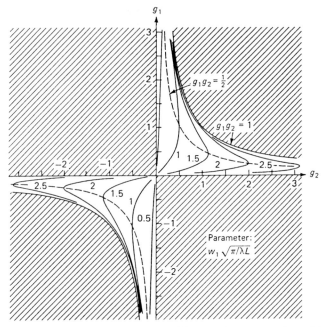

Fig. 5.13. Resonator stability diagram with curves for constant values of the normalized TEM_{00} mode spot size $w_1(\pi/\lambda L)^{1/2}$ at one mirror [5.18]

with curves of constant TEM_{00} mode spot sizes [5.18]. The curves, which are obtained from (5.28), reveal that the spot size is fairly insensitive to variations of g_1 and g_2 for resonator configurations which can be represented by points on the hyperbola $g_1 g_2 = 0.5$. Note that large spot sizes w_1 are obtained for resonators with large g_2 values. From (5.23) follows that in order for $g_2 > 1$, the radius of curvature of mirror R_2 has to become negative, which indicates a convex mirror according to our labeling convention.

The resonator sensitivity to mirror misalignment is related to the fact that the mode axis must be normal to each of the two mirrors. This can be satisfied only if the ray is incident along a line which passes through the center of curvature of the mirrors. Considering a typical resonator of mirror radii R_1 and R_2 and separation L, as shown in Fig. 5.14, a rotation of mirror M_1 through an angle Θ

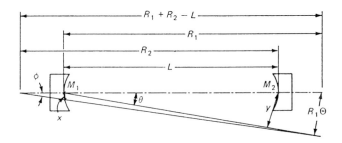

Fig. 5.14. Mirror alignment parameters

rotates the line joining the centers of curvatures of the two mirrors through an angle ϕ and causes a linear displacement x and y. Small-angle approximations are valid, and we have

$$x = \frac{R_1(R_2 - L)\Theta}{R_1 + R_2 - L} \, , \quad y = \frac{R_1 R_2 \Theta}{R_1 + R_2 - L} \, . \tag{5.33}$$

For a resonator with large radius mirrors of equal radii $R_1 = R_2 = R$ we obtain, from (5.33),

$$x = y = \frac{R\Theta}{2} \, . \tag{5.34}$$

Note that, if one of the mirrors is slightly tilted, the entire mode is displaced parallel to the resonator axis.

For a confocal resonator ($R_1 = R_2 = L$) we have

$$x = 0 \quad \text{and} \quad y = L\Theta \, . \tag{5.35}$$

In this case the mirror being tilted represents the pivot point for the mode axis.

If the flat mirror of a hemispherical resonator ($R_1 \approx \infty$, $R_2 \approx L$) is tilted, we obtain

$$x = (R_2 - L)\Theta \approx 0 \, , \quad y = L\Theta \, , \tag{5.36}$$

which is similar to that of a confocal resonator. On the other hand, if the spherical mirror is tilted ($R_1 \approx L$, $R_2 \approx \infty$), then

$$x \approx y \approx L\Theta \, . \tag{5.37}$$

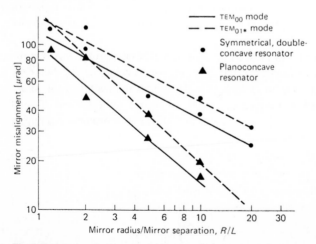

Fig. 5.15. Mirror misalignment in an argon laser which will produce a 10 % drop in output power versus the normalized mirror radius. The curves show the characteristics for a symmetrical resonator with two curved mirrors and for a plano-concave resonator. Sensitivity is plotted for the TEM_{00} and TEM_{01*} mode [5.7]

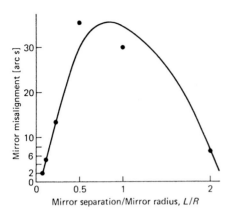

Fig. 5.16. Alignment tolerance of various types of resonators. Curve shows a reduction of output power by 50% of a He-Ne laser operated in a single transverse mode [5.26]

Comparing the sensitivity to angular tilt of the various resonator configurations, we note that, for example, a large-radius mirror resonator with $R = 10L$ is five times more sensitive to tilt than the confocal and hemispherical resonators.

In accordance with this theory, measurements performed on various lasers have shown that the alignment tolerances of a resonator with relatively short-radius mirrors is less stringent than those imposed with long-radius mirrors. Furthermore, a plano-concave resonator is more sensitive to misalignment than a resonator with two curved mirrors. Also, alignment tolerances become progressively less stringent for higher-order modes. Figure 5.15 gives experimental results obtained with an argon laser [5.25]. Clearly, from these curves it fol-

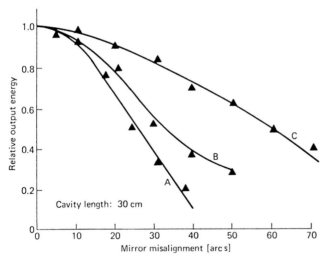

Fig. 5.17. Mirror misalignment sensitivity versus relative output power of a Nd:YAG laser. (**A**: Two flat mirrors, **B**: One flat, one 10-m-radius mirror, **C**: Two 10-m-radius mirrors). The mirror misalignment sensitivity to reduce the energy output by 10% was 12, 15, and 21 arc s, respectively. The laser, which was operated conventional mode, comprised a 5-mm by 50-mm Nd:YAG rod pumped by a single flashlamp. The cavity length was 30 cm and the front mirror was 70% reflective [5.27]

lows that the higher-order mode is less sensitive to mirror misalignment and, furthermore, that the confocal resonator has the highest tolerance in terms of misalignment. Figure 5.16 shows the alignment tolerance of various types of resonators for a HeNe laser operating single mode [5.26]. Again, the confocal resonator is far more forgiving for mirror misalignment than the other types. As is apparent from this figure, the alignment tolerances for a concentric-type resonator ($L/R = 2$) and a resonator having large-radius mirrors is about the same.

Figure 5.17 shows the drop in output power of a pulsed Nd : YAG laser as a function of mirror misalignment for different mirror combinations [5.27]. As is to be expected, the resonator containing two curved mirrors is least sensitive to misalignment.

5.1.9 Mode-Selecting Techniques

Many applications of solid-state lasers require operation of the laser at the TEM_{00} mode since this mode produces the smallest beam divergence, the highest power density, and, hence, the highest brightness. Focusing a fundamental-mode beam by an optical system will produce a diffraction-limited spot of maximum power per unit area. Generally speaking, in many applications it is a high brightness (power/unit area/solid angle) rather than large total emitted power that is desired from the laser.

Furthermore, the radial profile of the TEM_{00} mode is smooth. This property is particularly important at higher power levels, since multimode operation leads to the random occurrence of local maxima in intensity [5.36], so-called hot spots, which can exceed the damage threshold of the optical components in the resonator.

Transverse mode selection generally restricts the area of the laser cross section over which oscillation occurs, thus decreasing the total output power. However, mode selection reduces the beam divergence so that the overall effect of mode selection is an increase in the brightness of the laser. For example, the beam diameter and beam divergence for a TEM_{pl} mode increases with the factor C_{pl} introduced in Sect. 5.1.6, which means that for the same output power the brightness decreases by a factor of $(C_{pl})^{-4}$ for the higher-order modes.

Most lasers tend to oscillate not only in higher-order transverse modes, but in many such modes at once. Because of the fact that higher-order transverse modes have a larger spatial extent than the fundamental mode, a given size aperture will preferentially discriminate against higher-order modes in a laser resonator. As a result, the question of whether or not a laser will operate only in the lowest-order mode depends on the size of this mode and the diameter of the smallest aperture in the resonator. If the aperture is much smaller than the TEM_{00} mode size, large diffraction losses will occur which will prevent the laser from oscillating. If the aperture is much larger than the TEM_{00} mode size, then higher-order modes will have sufficiently small diffraction losses to be able to oscillate.

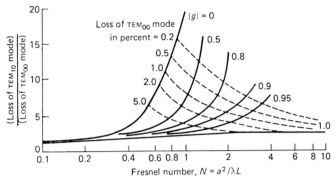

Fig. 5.18. Ratio of the losses per transit of the two lowest-order modes for the symmetrical resonator. The dashed curves are contours of constant loss for the TEM$_{00}$ mode [5.10]

The diffraction losses caused by a given aperture and the transverse mode selectivity achievable with an aperture of radius a is illustrated in Fig. 5.18 [5.10]. In this figure the ratio of the loss of the TEM$_{10}$ mode to the loss of the TEM$_{00}$ mode is plotted as a function of the Fresnel number for a symmetrical resonator. Note that the mode selectivity is strongly dependent on the resonator geometry, and is greatest for a confocal resonator and smallest for the plane-parallel resonator. From Fig. 5.18 it follows that the resonators of lasers operating in the TEM$_{00}$ mode will have Fresnel numbers on the order of approximately 0.5 to 2.0. For Fresnel numbers much smaller than these, the diffraction losses will become prohibitively high, and for much larger values of N mode discrimination will be insufficient.

These predictions are in agreement with the experimental observations. For example, typical Nd:YAG lasers have cavity lengths of 50 to 100 cm and TEM$_{00}$ operation typically requires the insertion of an aperture in the cavity with a diameter between 1 and 2 mm. Without an aperture, a 50-cm-long resonator with a 0.62-cm-diameter Nd:YAG rod as the limiting aperture will have a Fresnel number of 19. In Nd:glass lasers, where oscillator rods of 15-mm diameter are not uncommon, the Fresnel number would be 160 for the same resonator length.

We will now discuss typical resonator configurations which are used in the generation of fundamental mode output from solid-state lasers. Because the TEM$_{00}$ mode has the smallest beam diameter of all the resonator modes, a number of techniques have been developed to increase the TEM$_{00}$ mode volume in the active material, which is normally considerably larger in diameter than the mode size. We will find that a resonator designed for TEM$_{00}$ mode operation will represent a compromise between the conflicting goals of large mode radius, insensitivity to perturbation, good mode discrimination, and compact resonator length.

Large-Radius Mirror Configuration. The most common technique to produce TEM$_{00}$ mode output is the use of a nearly plane-parallel resonator with an internal aperture for mode selection. A typical example of this type of resonator

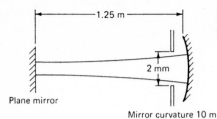

Fig. 5.19. Resonator employed to produce TEM_{00} mode from a ruby laser

is sketched in Fig. 5.19 [5.28]. The resonator has half the diffraction losses of a resonator which is 2.5 m long and has two 10-m-radius mirrors. The relevant parameters for the latter resonator operated at the ruby wavelength are $N = 0.58$ and $g = 0.75$. With N and g known, the diffraction losses for the symmetrical resonator can be obtained from Fig. 5.10. We find for the resonator illustrated in Fig. 5.19 a single-pass diffraction loss of 20 % for the TEM_{00} mode and 50 % for the TEM_{10} mode.

Resonators with Internal Beam Focusing. Spherical and hemispherical resonators and systems containing internal lenses have in common that they support large mode size differences in the resonator due to their focusing action. For example, in a hemispherical cavity the spot size in the limit can theoretically become zero at the flat mirror and grow to infinity for $L = R$. Location of the laser rod close to the curved mirror permits utilization of a large active volume. An example of this type of resonator is indicated in Fig. 5.20 [5.29]. Mode selection in this resonator, which was employed in a cw Nd:YAG laser, is achieved by axially moving the laser rod until it becomes the limiting aperture for TEM_{00} operation.

A simple resonator scheme, which can be used in cw experiments, is to operate the laser with two flat mirrors which are symmetrically moved farther apart until the TEM_{00} mode power is optimized. The thermal lensing of the rod makes this resonator equivalent to a symmetrical system with strongly curved mirrors. As was discussed in Sect. 5.1.8, theoretically the mode size in the crystal will grow to infinity as the mirror separation approaches four times the focal length of the laser rod.

Resonators with strong internal focusing suffer from several disadvantages which make them unattractive for most applications. In particular, since it is necessary to operate quite close to the edge of the optically stable region, the configurations are extremely sensitive to mechanical and optical perturbations.

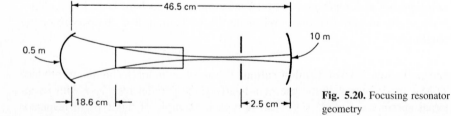

Fig. 5.20. Focusing resonator geometry

Convex-Concave Resonators. *Chesler* and *Maydan* [5.23] have described in some detail the use of convex-concave resonators for efficient and stable generation of TEM$_{00}$ power in solid-state lasers. With these resonators, one can make $g_2/g_1 > 1$ with the resultant increased size of w_1, and independently set $g_1 g_2 = 0.5$ so that the axial sensitivity to thermal focusing is minimized. Operation at the positive branch ($g_2 > 1$) rather than the negative branch ($g_2 < 1$) is usually preferred because the cone angle of the mode is smaller and allows a higher degree of overlap with the cylindrically shaped laser rod.

Optical resonators with one convex and one concave mirror offer significant advantages for TEM$_{00}$ mode operation of solid-state lasers. High efficiency and low sensitivity to perturbations and compactness have been obtained in comparison to other resonator types.

As an example we will calculate the parameters for a concave-convex resonator which has been used for a high-repetition-rate Nd:YAG laser [5.18] (Fig. 5.21a). The laser rod, with a diameter of 5 mm, was measured to have a focal length of 6 m for a particular input. The optical length of the cavity was restricted to 0.8 m. The optimum value of w_1 has been found empirically to be equal to one-half the laser rod radius. With the rod as close as possible to the output mirror, the following design parameters are obtained for the equivalent resonator (Fig. 5.21b).

$$w_1 = 1.25 \, \text{mm}, \quad f = 6 \, \text{m}, \quad L_1 = 0.1 \, \text{m}, \quad L_2 = 0.7 \, \text{m} \, .$$

Introducing the stability criterion $g_1 g_2 = 0.5$ into (5.29) yields an expression for the mode size as a function of g_1

$$w_1 = \left(\frac{\lambda L}{\pi g_1} \right)^{1/2} . \tag{5.38}$$

Introducing the values for L and w_1 into this equation, one obtains $g_1 = 0.16$, and from (5.32), $g_2 = 3.12$. From (5.27) it follows that $R_1 = 1.1 \, \text{m}$, $R_2 = -0.36 \, \text{m}$;

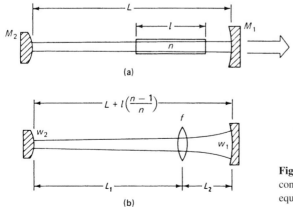

Fig. 5.21. Convex-concave resonator containing (**a**) a laser rod and (**b**) an equivalent passive resonator

and from (5.29) we obtain $w_2 = 0.28$ mm. The mirror M_1 and the rod surface can be combined by grinding a curvature $R_1' = nR_1$ onto the end of the rod.

Thermally Compensated Resonators. As we will discuss in more detail in Chap. 7, a cylindrical laser rod with uniform internal heating and surface cooling assumes a radial parabolic temperature gradient. This gradient and the associated thermal stresses lead to a radial variation of the index of refraction which transforms the laser rod into a bifocal positive lens with large aberrations and stress birefringence. The optical power of the crystal varies directly with pump power.

A scaling of the TEM_{00} output to higher powers is hindered by these strong thermal effects which occur in the laser rod. The principle problem is the degradation of the wavefront due to the highly aberrated nature of these distortions. A thermally stressed rod can be represented to a first order by a bifocal lens. The focal length of the lens is different for radial and tangential field components. In the ideal case of a perfectly uniform pumped rod, the thermal lens is spherical. In most cases the thermal lens is aspheric because pumping is stronger in the center of the rod compared to the outer regions. Also depending on the arrangement of the pump sources around the laser rod the thermal lens may not be radially symmetric. Since the thermally induced optical distortions of the TEM_{00} wavefront are the main reason for limiting the output, the major objective in obtaining a high-quality beam is to minimize the thermal load of the rod and to compensate existing distortions in the resonator.

In order to get an idea about the amount of compensation needed, usually a collimated HeNe laser beam is passed through the pumped rod, and the average focal length is measured as a function of pump power. The objective of a first-order compensation is to negate the positive lensing of the rod with a diverging optical component. The most common approaches are the insertion of a negative lens in the resonator, or a resonator featuring an internal telescope, or a design where one of the mirrors has a convex surface. The latter two approaches are discussed in Sect. 5.1.10. Higher-order phase aberrations have also been compensated with aspheric lenses [5.30].

A higher degree of thermal compensation can be achieved if the pump head of an oscillator is separated into two units and a $90°$ polarization rotator is placed between the laser rods. This eliminates bifocusing because the radial and tangential polarization of the beam are interchanged in the two crystals. Effective bi-focusing compensation requires uniformity of the thermal lensing within the two laser rods.

Figure 5.22 shows an example of a thermally compensated resonator designed to achieve high TEM_{00} mode output in a linearly polarized beam [5.33]. The $90°$ quartz rotator eliminates bifocusing and the convex rear mirror of the resonator compensates for the positive thermal lensing of the Nd:YAG crystal. Instead of using a convex resonator mirror, a negative lens is sometimes inserted between the two rods together with the quartz rotator. The resonator depicted in Fig. 5.22 contains also a Brewster plate to linearly polarize the laser output. Without the quartz rotator, the insertion of the polarizer generates a large de-

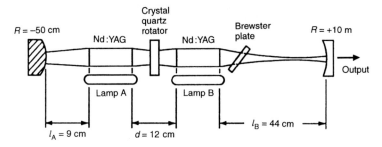

Fig. 5.22. Schematic diagram of a birefringence-compensated laser

polarization loss for the TEM_{00} mode. Due to stress birefringence the linearly polarized beam transmitted by the polarizer will be divided into two orthogonally polarized beams in the laser crystal. On its return path, the stress-induced orthogonally polarized portion of the beam will be rejected by the polarizer. The introduction of a 90° phase rotator between the two rods minimizes these losses because the birefringence induced by one rod is canceled in the other (see Sect. 7.1.1 for a detailed explanation). Clearly the approaches mentioned above rely on a radially symmetric wavefront distortion. Therefore uniform pumping helps in compensating phase aberrations.

Other measures for achieving high TEM_{00} mode output are more fundamental and address the origin of thermal distortions.

Changing from an arc lamp pumped system to diode pumping drastically reduces the thermal distortions in the laser crystal. For the same output power, a diode pump source generates only about 1/3 the amount of heat in a Nd : YAG crystal compared to an arc lamp. Also instead of using a cylindrical rod, a rectangular slab with an internal zig-zag path minimizes wavefront distortions in the resonator. As explained in Sect. 7.3, such a structure provides a high degree of self-compensation of optical beam distortons. If the TEM_{00} output beam has to be linearly polarized, a naturally birefringent crystal such as Nd : YLF can be advantageous in particular situations. The natural birefringence in Nd : YLF is much larger than the thermally induced birefringence.

Performance of Diffraction-Limited Systems. Before leaving the subject of mode selection, it should be pointed out that besides the mode selecting techniques which are discussed here, it is important to have a perfectly aligned system with clean and damage-free optical surfaces. Slight misalignments, tilts, or imperfections (dust particles) of laser reflectors can cause changes in the mode character of the output, favoring higher-order modes. For example, a particle or a damage spot located at the center of the beam can prevent oscillation in the TEM_{00} mode and cause oscillation at the TEM_{01^*} mode (for a theoretical discussion of this effect see, for example, [5.35]).

To illustrate the effect of mode selection on the performance of a laser system, Fig. 5.23 presents mode patterns and radial intensity distributions from a cw-pumped Nd : YAG laser. Shown are the first six modes from the laser, which

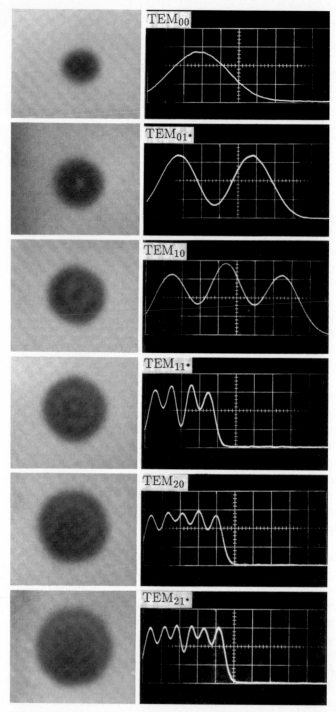

Fig. 5.23. Mode patterns and scans of the far-field beam intensity distribution of a cw-pumped Nd : YAG laser

were obtained by successively increasing the mode-selecting aperture. The mode patterns were taken in the far field (20 m away from the laser) by photographing the light striking a ground-glass surface through an infrared image intensifier. The radial intensity distribution of the transverse mode patterns was observed by sweeping the laser beam across a small-aperture detector using a rotating plane mirror and displaying the detector signal on an oscilloscope. Knowing the distance from the rotating mirror to the detector and the rate of rotation of the mirror, the absolute diameter of the beam can be accurately determined. The laser uses a 3-mm by 63-mm Nd:YAG rod which is pumped by two tungsten-filament lamps. The resonator comprises two mirrors with 4-m curvatures separated by 50 cm. With an input of 2 kW, a flat-ended rod was measured to have a focal length of 2 m. In this particular system the rod has a concave surface on each end with a radius of (-5) m. With $n = 1.78$ for Nd:YAG, it follows from $f = R/2(n - 1)$ that the focal length $f = -3.2$ m for the unpumped rod. The combination of thermal lensing and negative curvature results in an effective focal length of 5.3 m at 2 kW input, at which the beam-divergence data were taken. The system delivered a maximun of 8.5 W of multimode power. Mode selection was accomplished by insertion of different-sized apertures into the cavity. The various aperture sizes and the corresponding output powers, beam divergences, and mode structures are summarized in Table 5.1. From this table follows in accordance with theory that the brightness for higher-order modes decreases despite an increase in total output power.

Table 5.1. Performance of a cw-pumped Nd:YAG laser at different transverse modes

	TEM_{00}	TEM_{01^*}	TEM_{10}	TEM_{11^*}	TEM_{20}	TEM_{21^*}
Aperture size [mm]	1.4	1.6	1.8	2.0	2.2	2.3
Laser output power [W]	1.5	2.4	3.5	4.5	5.5	6.0
Beam divergence [mrad] ($1/e^2$ points)	1.9	2.3	2.8	3.1	3.4	3.6
Brightness [MW/cm^{2}sr]	28	18	20	18	15	15

Employing thermal compensation schemes, such as concave-convex resonators, cw-pumped Nd:YAG lasers using a 4-mm by 75-mm Nd:YAG rod, pumped by two krypton arc lamps in a double-elliptical cavity, will produce between 20 and 25 W TEM_{00} mode power at an input power of 5.5 kW. In a high-repetition-rate Nd:YAG system operating at 20 pps, a Q-switched energy of 60 mJ was achieved in the fundamental mode [5.19]. The resonator, which was 42 cm long, consisted of a rear convex mirror and a flat front mirror which was actually one end of the Nd:YAG rod. The flat-flat laser rod was 5 mm in diameter and 50 mm long; a 30 % reflective coating was used on the output end. The rod was pumped at 42 J input by a xenon flashlamp in a silver-plated single-elliptical cavity. The same system produced a non-Q-switched output of 500 mJ in a 100 μs-long-pulse.

5.1.10 Examples of Advanced Stable-Resonator Designs

A large amount of research has been devoted to the design of optical resonator configurations, which maximize energy extraction from solid state lasers. In fact, operation with stable cavities in the TEM_{00} mode, while producing a beam with a smooth and well-controlled spatial profile, in general results in a poor filling of the active volume and hence in a large waste of the stored energy. Improvements can be divided into the following optical designs: stable telescopic resonators and concave-convex resonators. Both schemes have been extensively studied and experimentally tested and also found commercial applications, exhibiting somewhat competing characteristics.

In stable telescopic resonators a magnifying telescope is added to a conventionals stable cavity to expand the mode cross section in the arm of the cavity where the field interacts with the active medium. In the concave-convex resonator, the same effect is achieved by the particular choice of mirror curvature and resonator length. In both cases, the beam quality remains good, but the mode volume is still limited, for the TEM_{00} mode. Furthermore, at the highest intensities, damage problems arise for the optical elements in the resonator section where the beam gets its smaller dimension. For solid-state lasers with low gain, or for systems where a Gaussian profile in the near field is required, the stable resonator is the only choice. Almost all laser applications require a small beam divergence, either to obtain a small spot size at a large range, or a high power density at the focal plane of a lens. Therefore, the challenge in designing a stable resonator is to maximize low-order mode power extraction. More specifically we can establish the following design criteria:

- The diameter of the TEM_{00} mode should be limited by the active material.
- The resonator should be dynamically stable, i.e., insensitive to pump-induced fluctuations of the rod's focal length.
- The resonator modes should be fairly insensitive to mechanical misalignments.

Lasers operating in the fundamental mode usually require the insertion of an aperture in the resonator to prevent oscillations of higher-order modes. In this case, the efficiency of the laser is generally lower, compared with multimode operation, due to the small volume of active material involved in the laser action. Large-diameter TEM_{00} modes can be obtained using special resonator configurations, but, if proper design criteria are not applied, the resonator becomes quite sensitive to small perturbations. Also, in solid-state lasers, thermal focusing of the rod greatly modifies the modes and the pump-induced fluctuations of the focal length may strongly perturb the laser output, even preventing any practical or reliable use of the laser. Efficient exploitation of the rod volume of a solid-state laser operating in the fundamental mode requires the solution of two conflicting problems: The mode volume in the rod has to be maximized, but the resonator should remain as insensitive as possible to focal length and alignment perturbations. Early solutions proposed compensation of the thermal lens by a convex mirror, or by negative lenses ground at the ends of the rod, that exactly eliminate the focusing effect of the rod. With these methods high power in a singlemode beam can be obtained; the compensation, however, is effective only for one particular value of the focal length. Large fundamental mode volume and good stability against thermal lens fluctuations have been achieved by a particular choice of mirror curvatures or by insertion of a telescope in the resonator.

In the following subsections we will discuss these two approaches which lead to the design of convex-concave and telescopic resonators.

In addition, at the end of this subsection, two resonator configurations are described, which are not new in concept, but have gained great significance in certain applications. The astigmatically compensated folded mirror resonator is important in the design of Ti : sapphire lasers and for Kerr-lens mode locking. The other configuration is the crossed porro prism, polarization-coupled resonator which has proven to be the most robust design for military lasers.

The Concave-Convex Resonator

The design procedure for resonators known as dynamic stable, in which the fluctuation of the mode volume in the rod is kept under control by an appropriate choice of mirror curvatures, has been described in (5.18, 23, 31, 32, 37). This earlier work on dynamic stable concave-convex resonators has been expanded considerably in [5.38–40]. *Magni* [5.39] has shown that, for a given resonator, as the focal length of the laser rod varies, there are always two zones of stability. These two zones have the same width in terms of the dynamic optical power of the rod. The fundamental mode volume in the laser rod is inversely proportional to this width. As far as tolerance to mirror alignment is concerned, the two zones have different misalignment sensitivities.

In the following sub-section, we will summarize some of the results presented in [5.39]. A number of new variables are introduced to describe the resonator parameters

$$u_1 = L_1 \left(1 - \frac{L_1}{R_1} \right) , \quad u_2 = L_2 \left(1 - \frac{L_2}{R_2} \right) , \quad x = \frac{1}{f} - \frac{1}{L_1} - \frac{1}{L_2} .$$

$$(5.39a)$$

Using the above equations, g_1 and g_2 can be expressed as

$$g_1 = -\frac{L_2}{L_1}(1 + xu_1) , \quad g_2 = -\frac{L_1}{L_2}(1 + xu_2) .$$

$$(5.39b)$$

By eliminating x in (5.39b) a linear relationship between g_1 and g_2 is derived which describes a given resonator configuration for changes in focal length.

$$g_2 = \left(\frac{L_1}{L_2} \right)^2 \frac{u_2}{u_1} g_1 + \frac{L_1}{L_2} \left(\frac{u_2}{u_1} - 1 \right) .$$

$$(5.39c)$$

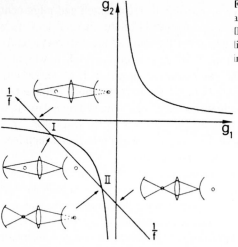

Fig. 5.24. Stability diagram and mode profiles at the edges of the stability zones (marked I and II) of a concave-convex resonator. The straight line $1/f$ represents a resonator with a variable internal lens [5.39]

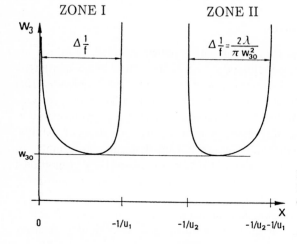

Fig. 5.25. Spot size w_3 inside the laser rod as a function of the optical power $1/f$ of the rod [5.39]

The intersections of this line with the axes and with the hyperbola $g_1 g_2 = 1$ defines two distinct stability zones indicated, as I and II in Fig. 5.24. The two zones have the same width in terms of x, as illustrated in Fig. 5.25. The parameter w_{30} is the spot size at the principal planes of the laser rod and is the same in both stability zones. The value of w_{30} is given by

$$w_{30}^2 = \frac{2\lambda}{\pi} |u_1|, \quad |u_1| > |u_2|. \tag{5.40a}$$

The relationship between the width of the stability zones and the minimum spot sizes on the lens is given by

$$w_{30}^2 = \frac{2\lambda}{\pi} \frac{1}{|\Delta \frac{1}{f}|}. \tag{5.40b}$$

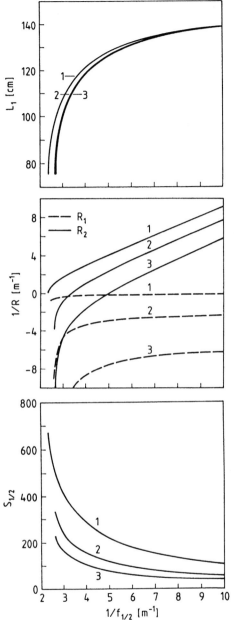

Fig. 5.26. Distance L_1, mirror curvatures $1/R_1$ and $1/R_2$, and misalignment sensitivity $S_{1/2}$ vs the optical power $1/f_{1/2}$ for a specific concave-convex resonator. Resonator length fixed at $L = 150$ cm wavelength $\lambda = 1.064\,\mu$m. The number on the curves is the spot size $w_{30} = 1,\ 2,\ 3$ mm [5.39]

The TEM$_{00}$ mode volume in the laser rod is inversely proportional to the width of the stability zone and hence to the range of input power to the lamp for which the resonator remains stable. The centers of the two stability zones in Fig. 5.25 are at

$$x_I = -\frac{1}{2u_1} \quad \text{and} \quad x_{II} = -\frac{1}{u_2} - \frac{1}{2u_1} .$$

(5.40c)

At the edges of the stable zones, defined by $g_1 g_2 = 0$ and $g_1 g_2 = 1$, the spot size w_3 goes to infinity.

Magni [5.39] treated in his analysis not only the stability of the resonator to small fluctuations of the focal length of the laser material, but also considered the mechanical stability of the resonator in terms of a misalignment sensitivity.

For example, the misalignment sensitivity $S_{1/2}$ in the middle of zone I can be expressed as

$$S_{1/2} = \frac{1}{w_{30}} \frac{2L_1 L_2}{(4L_1^2 + L_2^2)^{1/2}} . \tag{5.41}$$

To a very rough approximation, it may be assumed that $1/S_{1/2}$ is the tilt angle for about a 10% increase in resonator losses [5.41,42]. It was found, that the tolerance to mirror tilt is higher in zone I as compared to zone II.

The design of a resonator described in [5.39] proceeds by specifying a focal length $f_{1/2}$ of the laser rod, a spot size w_{30} inside the laser material and a total resonator length L. With these parameters given, the mirror curvatures R_1 and R_2, the separation of the mirrors from the principal planes of the rod L_1 and L_2, and the sensitivity parameter $S_{1/2}$ can be calculated. Please note that $L = L_1 + L_2$, the physical length is different, since it also includes the distance between the principal planes of the rod. In Fig. 5.26 the values of L_1, R_1, R_2, and $S_{1/2}$ for optimized resonators are plotted vs $f_{1/2}$ for a few values of w_{30}. From this figure the values of the curvature radii and the position of the rod can be readily obtained for a given focal length and for a given spot size of the mode in the rod.

Figure 5.27 illustrates a resonator design based on the results presented in Fig. 5.26. We assume a mode radius of $w_{30} = 3$ mm in the active medium. This is the optimum mode radius for a 6 mm diameter rod. The length of the resonator is given as 150 cm.

For a focal length of $f = 17$ cm ($1/f = 6$ m^{-1}) we obtain $R_1 = -14$ cm, $R_2 = 55$ cm and $L_1 = 130$ cm. The resonator has a misalignment sensitivity of about $S_{1/2} = 80$, which would indicate that for a tilt angle of about 12 mrad, the losses will increase by 10%. In Fig. 5.27 the distance from the end of the rod to the principal plane is $h = l/2n$, where l is the length of the laser rod.

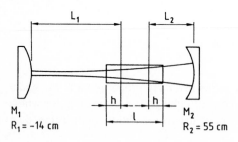

Fig. 5.27. Dynamically stable resonator for a strongly focusing rod

The Telescopic Resonator

Hanna et al. [5.43, 44] and *Sarkies* [5.45], reported on the use of a telescope in an Nd:YAG resonator (Fig. 5.28). An attractive feature of the telescope is that it allows easily controllable adjustment to compensate thermal lensing under varying pumping conditions. In addition, the telescopic resonator avoids the very small spot on the convex mirror of the convex-concave mirror design. This is particularly important at the high power levels typical for Q-switched Nd:YAG lasers.

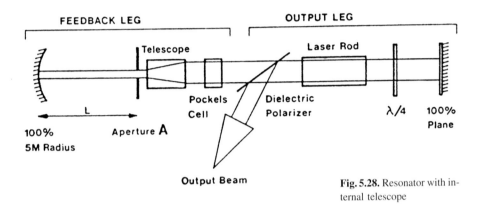

Fig. 5.28. Resonator with internal telescope

By introducing a suitably adjusted telescope into a Q-switched Nd:YAG laser resonator, it was possible to obtain reliable operation with a large-volume TEM$_{00}$ mode. The basic principle behind the resonator design is that of choosing a telescope adjustment which compensates the thermal lensing in the laser rod (thus permitting a large spot size) and at the same time ensuring that the spot size is insensitive to fluctuations in focal length of the thermal lens. The telescope performs two distinctly separate functions. Firstly, it reduces the size of the beam at the output end which increases the diffraction losses for higher-order modes; whereas on the input side the telescope magnification is chosen such that the beam matches the rod diameter. Secondly, the telescope is an element of variable focal length. It can therefore be adjusted to place the resonator anywhere on the stability diagram. Because the ratio of the diffraction losses of the higher-order modes to the lower-order ones increases as the telescope output beam decreases, the telescope can be adjusted to ensure that modes above a certain order do not reach threshold. Thus the mode selection process is controlled by two telescope parameters, the magnification M and the focal length f.

The telescope adjustment is chosen to minimize the effect of focal-length variations in the laser rod and at the same time ensures the optimum mode-selection properties of a confocal resonator. *Hanna* et al. [5.43] performed a detailed analysis of the telescopic resonator. The analysis provided simple design equations relating the spot size, resonator length, telescope magnification, and defocusing and diffraction losses. A short summary of the key design parameters is given below.

One can best understand the role of the telescope by considering, for simplicity, a short telescope of magnification M (where $f_2 = -Mf_1$) located close to the laser rod characterized by a focal length f_R, and a resonator mirror with an equivalent focal length f_M, as shown in Fig. 5.29a.

It can be shown that small defocusing of the telescope has two effects: it changes the spot size and it changes the wavefront curvature. Thus the telescope can be adjusted to achieve compensation of the thermal lens f_R of the laser rod by making

$$-\frac{1}{f_T} = \frac{1}{f_R} + \frac{1}{f_M} \tag{5.42}$$

where

$$\frac{1}{f_T} = \frac{\delta}{f_2^2} \tag{5.43}$$

is the focal length of the telescope for small defocusing δ. The expression on the right-hand side of (5.42) is the optical power of the laser rod/mirror combination.

The magnification M of the telescope has to be selected for maximum insensitivity of spot size to variations of f_R. This is achieved for

$$\frac{1}{2M^2L} = \frac{1}{f_T} + \frac{1}{f_R} + \frac{1}{f_M} \ . \tag{5.44}$$

The spot size in the laser rod is given by

$$w_1 = M(2L\lambda/\pi)^{1/2} \ . \tag{5.45}$$

Thus, introducing the correctly adjusted telescope allows the same large mode volume in the laser rod to be maintained but with a reduction of cavity length by M^2. The main limitation of this approach is that it exposes components in the reduced beam to higher intensity and thus greater damage risk.

Hanna et al. [5.43] have presented some of their key findings in graphical form. Figure 5.29b shows the spot sized w_1 (in laser rod) and w_2 (on left-hand mirror) versus telescope defocusing δ, with f_R fixed (5 m), or equivalently versus f_R with δ fixed at -6.2 mm. The laser parameters are as follows: $A = 0.55$ m, $B = 0.37$ m, $C = 0.16$ m, $f_1 = -0.05$ m, $f_2 = 0.20$ m, $M = 4$.

The main feature is the broad minimum of spot size in the laser rod (upper curve), implying insensitivity of the spot size to δ. For a fixed mirror curvature and telescope setting (i.e., f_M and f_T constant) it follows from (5.44, 45) that Fig. 5.29b also represents a plot of spot size versus f_R. The minimum of the upper curve therefore implies insensitivity to fluctuations in f_R. The desired operating point is at the bottom of this minimum and the telescope must therefore be defocused by the correct amount to ensure this.

In arriving at a resonator design, the main parameters to be chosen are spot size w_1 in the laser rod, resonator length L and magnification M. It is assumed

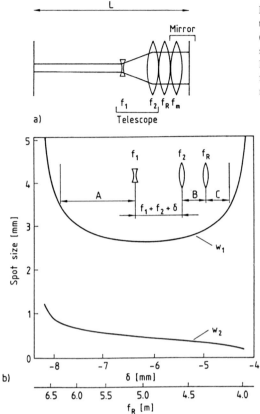

Fig. 5.29. (a) Resonator with internal telescope (f_1, f_2), focusing laser rod (f_R), and curved mirror (f_M). **(b)** Spot size w_1 in the laser rod, and w_2 on the left mirror, as a function of telescope defocusing δ or focusing f_R of the laser rod [5.44]

that f_R is known, this parameter is usually determined by passing a He-Ne laser beam through the laser rod and measuring the beam waist at the desired pump level.

With w_1 chosen, the choice of values for L and M is made according to (5.45) to give an acceptable compromise between a small M and hence an inconveniently large L, or small L and hence large M which may then lead to excessive intensity in the contracted beam. When L (and hence M) have been chosen, and a mirror has been selected (which determines f_M) then the telescope focal length f_T is obtained from (5.44) and finally the amount of defocusing from (5.43).

A circular aperture to select the TEM_{00} mode is inserted and centered. In practice, it is found that the aperture diameter should be ≈ 1.5 times the calculated spot diameter at the point of insertion to ensure suppression of the TEM_{01} mode.

Astigmatically Compensated Folded-Mirror Resonator

In the design of Ti:sapphire lasers, Kerr-lens mode locking, or cavity dumping with a fast acousto-optic switch, or in intra cavity frequency doubling, it is

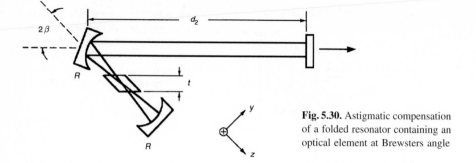

Fig. 5.30. Astigmatic compensation of a folded resonator containing an optical element at Brewsters angle

often necessary to have a highly concentrated beam inside the resonator. The requirement of a long resonator and a small beam waist are best satisfied by a folded resonator commonly employed in dye lasers. A resonator configuration containing an internal focus can be designed with internal lenses. An equivalent resonator is a three-mirror design depicted in Fig. 5.30, which is preferable over a lens system because reflection losses can be kept much smaller.

Such a folded resonator, because of the oblique angle of incidence of the center mirror, introduces astigmatism. Astigmatic beams have different spot sizes and beam-waist positions in two orthogonal directions. In highly focused resonators of this type, the adjustment of mirror spacing is critical as the cavity is stable only over a small range of values. In a highly astigmatic resonator, a condition can exist where the resonator is stable in only one coordinate unless steps are taken to minimize astigmatism.

The purpose of astigmatic compensation is to produce mode characteristics that are equal in x and y. For the case where an optical element between the two mirrors is at Brewsters angle, as shown in Fig. 5.30, the astigmatism of the mirror can be offset by those of the Brewster ended element.

With the angle of incidence chosen to achieve astigmatic compensation, the beam-waist diameters in the x and y directions are approximately equal, and focal points for x and y are at the same location.

A mirror used at an oblique angle of incidence has two focal points [5.111].

$$f_x = (R/2)/\cos\Theta \quad \text{and} \quad f_y = (R/2)\cos\Theta \tag{5.46}$$

because it focuses sagittal (xz) and tangential (yz) ray bundles at different locations.

A Brewster-angle laser rod has two different effective distances l_x and l_y, which the rays have to traverse [5.111, 112]

$$l_x = t\sqrt{n^2 + 1}/n^2 \quad \text{and} \quad l_y = t\sqrt{n^2 + 1}/n^4 \tag{5.47}$$

where t is the length, and n is the refractive index of the laser rod.

It has been shown [5.111] that the astigmatism of the oblique angle of incidence mirror can be compensated for by the astigmatism of the Brewster-angle laser crystal at an included angle 2Θ by

$$\sin\Theta \, \tan\Theta = \frac{2t(n^2 - 1)(n^2 + 1)^{1/2}}{Rn^4} \tag{5.48}$$

where R is the radius of curvature of the mirror.

In practice, the laser rod, mode locking or nonlinear crystal, as well as the mirror curvatures, are chosen to achieve the proper operating conditions for the laser. With the parameters t, n, and R chosen, compensation is achieved by adjusting the tilt angle of the mirror. As an example, we consider a diode end-pumped Nd : glass laser with the pump beam directed through one of the curved mirrors. With a rod length of $t = 3$ mm, an index of refraction of $n = 1.503$, and a mirror with a radius of curvature of $R = 125$ mm, the center mirror has to be tilted by $\Theta = 8.35°$ in order to achieve astigmatic compensation.

Instead of a three-mirror configuration, the folded mirrors are often used inside four-mirror designs, as illustrated in Figs. 3.50, 9.12, and 9.16.

Crossed-Porro Prisms, Polarization-Coupled Resonator

Lasers employed in military systems, such as range finders and target designators, have to operate in a totally different environment compared to commmercial or laboratory lasers. The design is dictated by the requirement that these lasers are insensitive to shock, vibration and large temperature excursions. For these reasons, reliable operation under severe environmental conditions is more important than optimum system performance.

A temperature and alignment insensitive optical resonator is a key feature of a military laser. As a result of decades of experience gained from fielded systems, a particular resonator-design concept and architecture has evolved, which is implemented in most military lasers.

The key features of this design are as follows:

- Beam turning is accomplished with prisms rather than mirrors.
- The resonator is folded to reduce the overall length of the system.
- Alignment is performed by rotation of several refractive wedges.
- The output beam is obtained through polarization coupling.

The optical design of a typical military Q-switched laser which incorporates these features is exhibited in Fig. 5.31.

Mirrors attached to adjustable mirror mounts have proven not to be stable enough from a mechanical and thermal point of view for military applications. Instead, as shown in Fig. 5.31, fixed mounted porro prisms and corner cubes are employed in the optical train. The self-aligning features of retro-reflectors provide the highest stability in the presence of harsh environmental conditions. Important for good beam quality are optically high-quality porro prisms with less than 5 μm edge chips at the roof line, and an angular tolerance of no more than

±5 arc sec from the 90° roof angle. The prisms and corner cube are aligned in a test rig during assembly and either glued or mechanically locked in place.

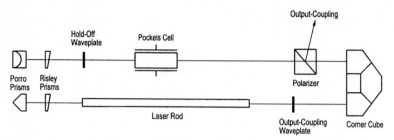

Fig. 5.31. Resonator using cross-porro prisms and polarization output coupling

The use of crossed porro prisms for the resonator requires a different approach from standard designs for beam adjustment and output coupling. Fine adjustment of the beam is accomplished by rotation of refractive alignment wedges. The wedges consist typically of 5–10 mm diameter and 3–5 mm thick fused silica with a 1° wedge angle. A so-called *Risley pair* permits beam adjustment in the x-y directions. In resonators which use two fixed mounted reflective mirrors, one Risley pair is required in front of each mirror. Also, for boresighting the beam outside the laser, a Risley pair is often used. Aligning a laser with a Risley pair requires some practice. The best way to start is by orienting the pair with a HeNe laser such that the wedges are at 90° with respect to each other. If the prisms are then installed in the laser, a slight rotation of one prism around this starting position will move the beam in an arc which is orthogonal to the arc of the other prism. In this case, the alignment procedure is somewhat similar to a more conventional x-y mirror adjustment. Once alignment of the cavity is completed, the mounting rings of the prisms are locked in place by set screws. It was found that refractive alignment prisms, mounted in stainless steel holders and clamped down by set screws, are the most reliable method of achieving long-term alignment stability of the resonator in the presence of severe environmental conditions.

In the design illustrated in Fig. 5.31 which utilizes Porro prisms instead of reflective mirrors, only one wedge at each end of the resonator is required because of the self-aligning feature of these prisms in the other direction. Output coupling is achieved by a waveplate/polarizer combination. The amount of output coupling can be adjusted by rotation of the $\lambda/2$ waveplate. In principle, the waveplate is not necessary because the optimum output coupling could also be achieved by a rotation of the Porro prism located behind the laser rod. A Porro prism introduces a phase shift between the s and p components of light because of the total internal reflections. The prism-induced phase shift depends on the refractive index. An additional phase shift can be induced by changing the prism azimuth angle with respect to the plane of polarization. Therefore, in a Porro prism resonator the phase shift required for output coupling can be achieved with the proper combination of prism induced phase shift and prism angle rotation. However, in most cases it is desirable to maintain the Porro prisms in a crossed fixed position to improve the alignment stability of the laser resonator.

Q-switching is accomplished with the Q-switch placed between the polarizer and waveplate-Porro prism combination. If, for example, the Q-switch is operated in the pulse-on mode, it needs to be located between two crossed polarizers. In this case, the waveplate-Porro prism combination is adjusted to transmit only radiation which is polarized orthogonally with respect to the transmission of the output coupling polarizer. A high Q in the cavity is established if a $V_{1/2}$ voltage is applied to the Pockels cell which rotates the polarization $90°$ in the Q-switch. The phase retardation introduced by the Porro prism has to be compensated for by a rotation of the waveplate in such a way, that the combination of both elements together with the output coupling polarizer, produces the maximum extinction ratio in the resonator.

It should also be noted that in the design depicted in Fig. 5.31, the Q-switch is subjected to a lower flux compared to other parts of the resonator, because a fraction of the radiation is coupled out before it reaches the Q-switch. This is important, because $LiNbO_3$, a common Q-switch crystal, has the lowest damage threshold of the optical components employed in the resonator.

In order to keep the laser systems short and compact, most military lasers are folded (Fig. 5.31). For the reasons discussed above, a corner cube rather than mirrors, is used for bending the beam. In addition to the specific optical components discussed so far, a military laser requires the use of expansion matched materials for all optical mounts, a hermetically sealed enclosure to protect the optical components, and kinematic mounting of critical components to avoid stress-induced distortions.

Optical Modeling

Besides the fundamental calculations and considerations presented in this subsection it is often necessary in the design of a complicated laser system to employ an optics analysis code. This is particularly the case in lasers which contain many optical elements and folded beam paths. We are normally interested in identifying potential problem areas which could lead to optical damage in our high-peak power systems. Therefore, of particular interest is the beam profile within the resonator and throughout the rest of the system, diffraction effects from limiting apertures, self-focusing caused by nonlinear refractive index changes in glass and crystals, back reflection from optical elements, etc.

For our work we found particularly useful a general purpose optical modeling code originally developed by the University of Arizona and Los Alamos National Laboratories. The software, called General Laser Analysis Design (GLAD) is a three-dimensional code that models the transverse electric field amplitude and phase by a two-dimensional array and the axial dimension by successive calculations [5.34]. The code permits state-of-the-art laser analysis of the near and far-field propagation through nonlinear gain media, lenses, mirrors, etc. including resonator modeling in the presence of aberration effects. GLAD, as well as similar physical-optics codes, employs a modular building block approach which models each component in sequence as the beam progresses through the system. These codes treat complex laser systems, in much the way that lens design programs are applied to geometrical optics.

5.2 Longitudinal Modes

5.2.1 Fabry-Perot Resonators

The Fabry-Perot resonator is not only an essential element of the laser, it is also used for mode selection and as an instrument to measure laser linewidth. Formulas needed to calculate the salient features of the different kinds of resonators employed in solid-state lasers are summarized below.

Basic Equations

A Fabry-Perot resonator consists of two plane-parallel optical surfaces; in classical optics this arrangement is known as a Fabry-Perot interferometer or as an etalon [5.46–48] (see below). Multiple reflections which occur between the surfaces cause individual components of the wave to interfere at M_1 and M_2 (Fig. 5.32). Constructive interference occurs at M_2 if all components leaving M_2 add in phase. For such preferred directions, the components reflected from M_1 destructively interfere and all the incident energy is transmitted by the etalon. For each member of either the reflected or the transmitted set of waves, the phase of the wave function differs from that of the preceding member by an amount which corresponds to a double traversal of the plate. This phase difference is

$$\delta = \left(\frac{2\pi}{\lambda} \right) 2nd \cos \Theta \; , \tag{5.49}$$

where nd is the optical thickness between the two reflecting surfaces, $\Theta = \Theta'/n$ is the beam angle in the material, and λ is the wavelength. The transmission of the Fabry-Perot resonator is

$$T = \left[1 + \frac{4r}{(1-r)^2} \sin^2 \left(\frac{\delta}{2} \right) \right]^{-1} \tag{5.50}$$

where r is the reflectivity of each of the two surfaces. The maximum value of the transmission, $T_{\max} = 1$, occurs when the path length differences between the transmitted beams are multiple numbers of the wavelength:

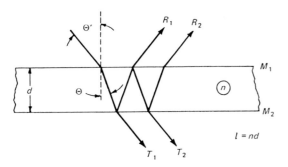

Fig. 5.32. Interference of a plane wave in a plane-parallel plate

$$2nd \cos \Theta = m\lambda : \quad m = 1, 2, 3, \ldots . \tag{5.51}$$

The reflectivity of the resonator can be expressed by

$$R = \left(1 + \frac{(1-r)^2}{4r \sin^2(\delta/2)}\right)^{-1} . \tag{5.52}$$

The maximum value of the reflectivity

$$R_{\max} = \frac{4r}{(1+r)^2} \tag{5.53}$$

is obtained when the path length difference of the light beam equals multiples of half-wavelength

$$2nd \cos \Theta = \frac{m\lambda}{2} \quad m = 1, 3, 5, \ldots . \tag{5.54}$$

Figure 5.33 illustrates the transmission and reflection properties of the etalon. In the absence of absorption losses, the transmission and reflectance of an etalon are complementary in the sense that $R+T = 1$. We define the ratio of the spacing between two adjacent passbands and the passband width as finesse $F = \Delta\lambda/\delta\lambda$:

$$F = \pi \left[2 \arcsin\left(\frac{2+4r}{(1-r)^2}\right)^{-1/2}\right]^{-1} \approx \frac{\pi(r)^{1/2}}{(1-r)} . \tag{5.55}$$

The approximation can be used if $r > 0.5$. For small values of r the finesse approaches the value $F \approx 2$ and (5.52) is reduced to

$$R = R_{\max} \sin^2 \frac{\delta}{2} . \tag{5.56}$$

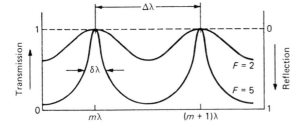

Fig. 5.33. Intensity of the reflected and transmitted beams as a function of phase difference

We see from (5.51) that the resonance condition of the etalon depends on the wavelenth λ, the optical thickness nd, and the beam angle Θ. A variation of any of these quantities will shift the etalon passband. The wavelength difference between two passbands is called the free spectral range of the etalon. The following equation gives the variations which are required to move a passband by one spectral range:

$$\Delta\lambda = \frac{\lambda_0^2}{2nd \cos\Theta} , \quad \Delta\nu = \frac{c}{2nd \cos\Theta} ,$$

$$\Delta(nd) = \frac{\lambda_0}{2 \cos\Theta} \quad \Delta\Theta = \left(\frac{\lambda}{nd}\right)^{1/2} \quad \text{at } \Theta = 0 . \tag{5.57}$$

The variations required to shift the transmission or reflection spectrum by the width of one passband are

$$\delta\lambda = \frac{\Delta\lambda}{F} ; \quad \delta l = \frac{\Delta l}{F} ; \quad \delta\Theta = \frac{\Delta\Theta}{F} . \tag{5.58}$$

Laser Resonator

If we consider for the moment only a single transverse mode, then according to (5.57) the separation of the longitudinal modes in a laser cavity is given by

$$\Delta\lambda = \frac{\lambda_0^2}{2L} \quad \text{or} \quad \Delta\nu = \frac{c}{2L} \tag{5.59}$$

where L is the optical length of the resonator.

For a resonator with $L = 75\,\text{cm}$ and $\lambda_0 = 6943\,\text{Å}$ one obtains $\Delta\lambda = 0.003\,\text{Å}$. A ruby laser has a linewidth of about $0.5\,\text{Å}$, which means that there are approximately 160 longitudinal modes within this linewidth. If the reflectivities of the two surfaces of a resonator are different, as in a laser cavity, the reflectivity r in (5.55) is the geometric mean reflectivity of the two mirrors: $r = (R_1 R_2)^{1/2}$. Typical values for a laser cavity are $R_1 = 0.5$ and $R_2 = 1.0$. Introducing these numbers into (5.55, 58), one obtains $F = 8.5$ and $\delta\lambda = 0.0004\,\text{Å}$ for the finesse and the spectral width of the empty resonator (Fig. 5.34).

In the presence of several transverse modes, additional resonant frequencies occur in the laser cavity. The frequency separation of different TEM_{plq} modes in a laser resonator is given by [5.4]

$$\Delta\nu = \left(\frac{c}{2L}\right) \left[\Delta q + \left(\frac{1}{\pi}\right) \Delta(2p+l)\,\text{arc}\,\cos\left(1 - \frac{L}{R}\right)\right] . \tag{5.60}$$

The term on the right containing Δq gives the frequency spacing of the axial modes which belong to a single transverse mode. The term with $\Delta(2p+l)$ describes the separation of the resonance frequencies of different transverse TEM_{pl} modes. Note that the resonant frequencies depend on $(2p+l)$ and not on p and l separately. Therefore, frequency degeneracies arise when $(2p+l)$ is equivalent for different modes. By replacing $(2p+l)$ with $(m+n)$, the cavity frequencies for transverse modes expressed in Cartesian coordinates are obtained.

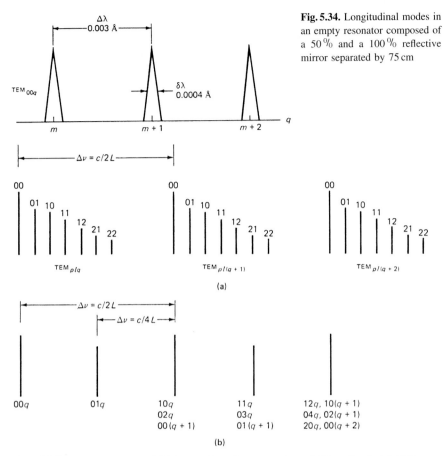

Fig. 5.35. Resonance spectrum of **(a)** long-radius mirror resonator and **(b)** confocal resonator

From (5.60) follows that the frequency spacing between transverse modes is not only a function of mirror separation, as is the case with axial modes, but depends also on the curvature of the mirrors.

For a near plane-parallel or long-radius resonator ($L \ll R$), the second term within the brackets of (5.60) becomes small compared to Δq. In this case the resonant-mode spectrum is composed of the relatively large axial mode spacing $c/2L$, with each axial mode surrounded by a set of transverse-mode resonances, as shown in Fig. 5.35a. For example, the 75-cm-long resonator mentioned before, if terminated by two mirrors with 10-m curvatures, will have a resonance spectrum of $\Delta\nu/\text{MHZ} = 200\Delta q + 28(2p+l)$. If the mirror radii are decreased, starting from the plane-parallel configuration, the transverse-mode frequency spacing increases while the axial-mode frequency intervals remain the same, provided that the mirror separation is constant. The extreme condition is reached for the confocal resonator, where (5.60) reduces to

$$\Delta \nu = \left(\frac{c}{2L} \right) \left[\Delta q + \left(\frac{1}{2} \right) \Delta(2p + l) \right] . \tag{5.61}$$

In the confocal resonator the resonance frequencies of the transverse modes resulting from changing p and l, either coincide, or fall halfway between the axial-mode frequencies (Fig. 5.35b). As the mirror curvature increases still more, the frequency interval of the transverse modes decreases and becomes zero for a concentric resonator. The frequency spacing $\Delta \nu$ expressed by (5.60) can be measured if a spectrum analyzer and photodetector of sufficiently fast response are available. The output from the detector will contain a beat-frequency signal corresponding to $\Delta \nu$ if more than one mode is oscillating.

Two techniques are widely used for obtaining a spectrally narrow output from solid-state lasers. They involve the use of either a resonant reflector or an intracavity tilted etalon. Both devices will be discussed next.

Resonant Reflectors

Fabry-Perot resonators of fixed spacing are generally referred to as etalons. If an etalon is employed in place of an output mirror in a laser cavity, it is referred to as a resonant reflector.

The normal operating region of a resonant reflector is at its maximum reflection. The value of the peak reflectivity is obtained from (5.53). If an uncoated etalon is used, the reflectivity

$$r = \left(\frac{n - 1}{n + 1} \right)^2 \tag{5.62}$$

must be introduced into (5.53).

As an example we will consider the properties of a single sapphire etalon employed as an output reflector in a ruby laser. If we assume a thickness $d = 3.2$ mm, an index of refraction $n = 1.76$, and an operating wavelength $\lambda = 6943$ Å, then we obtain from the preceding equations a maximum reflectivity of $R_{max} = 0.27$, a separation of the peaks of $\Delta \lambda = 0.43$ Å, and a half-width of $\delta \lambda = 0.22$ Å. As a comparison, an ordinary glass etalon with index of refraction 1.45 has a maximum reflectance of $R_{max} \approx 0.17$. The spectral characteristics of a 2.5-mm-thick quartz etalon is shown in Fig. 5.36 (top).

The resonance peaks of multiple-resonant reflectors are sharper and more separated than is the case in single-element devices. The maximum reflectivity of a multielement resonant reflector is [5.49]

$$R_{max} = \left(\frac{n^N - 1}{n^N + 1} \right)^2 , \tag{5.63}$$

where n is the refractive index of the plates and N is the number of reflecting surfaces.

It will be shown that the simple theory outlined in the preceding subsection, which is applicable to single-plate etalons, can be used to predict the main features of the spectral curve of a multielement etalon. More important, the simple theory provides a quick way to see how changes in the material, plate thickness, coatings, etc., will alter the reflectivity of a given device.

As an example, we will consider a three-plate etalon. The unit consists of three quartz plates with a thickness of 2.5 mm, separated by two spacers each 25 mm in length. Figure 5.36 (bottom) shows the calculated reflectivity as a function of wavelength for this device. As can be seen from this figure, the resonant reflector has a peak reflectivity of 65 % and the main peak has a half-width of 0.038 Å. The envelope of the individual resonance peaks is repeated after every 0.67 Å. There are six different resonance effects which occur in this device. With $n = 1.45$, $d_Q = 2.5$ mm, and $d_A = 25$ mm, one obtains a resonance due to: a single quartz plate $l = 3.62$ mm, $\Delta\lambda_1 = 0.66$ Å; one plate plus one air gap $l = 28.6$ mm, $\Delta\lambda_2 = 0.084$ Å; air gap alone $l = 25$ mm, $\Delta\lambda_3 = 0.096$ Å; two plates plus air gaps $l = 32.25$ mm, $\Delta\lambda_4 = 0.074$ Å; two plates plus both air gaps $l = 57.25$ mm, $\Delta\lambda_5 = 0.042$ Å; three plates plus both air gaps $l = 60.9$ mm, $\Delta\lambda_6 = 0.040$ Å.

Comparing these results with the actually calculated curve reveals that the resonance in the single plates determines the period of the whole device. The main peaks are formed by the resonance in the air space, whereas the minor peaks are caused by resonances including both air spaces and at least two plates. If one uses sapphire instead of quartz, then with $n = 1.79$ and $N = 4$ one obtains $R_{\max} = 0.67$ for a two-plate device.

Tilted Etalon

Consider an etalon of length d, refractive index n, with the surface normal inclined to the incident light beam at a small angle Θ'. The transmission of the tilted etalon is given by (5.50). The shift in the resonance wavelength due to a change in tilt angle Θ' is [5.50]

$$\Delta\lambda = \frac{-\lambda(\Delta\Theta')^2}{2n^2} \, . \tag{5.64}$$

Fabry-Perot Interferometer

The Fabry-Perot interferometer is commonly used to measure the linewidth of the laser emission. The arrangement normally used to display the familiar Fabry-Perot rings consists of a short-focal-length negative lens, an etalon with spacings of 5 to 40 mm, a long-focal-length positive lens, and a photographic plate or screen. The first lens converts the parallel beam from the laser into a diverging cone of light.

The resonances of the etalon cause an angularly selective transmission. The light transmitted by the interferometer is collected by a lens. Light beams leaving the etalon inclined at the same angle with respect to the optical axis of the etalon form a ring in the focal plane of the lens. The corresponding number of Fabry-Perot rings are displayed by a screen or they can be photographed. In order to

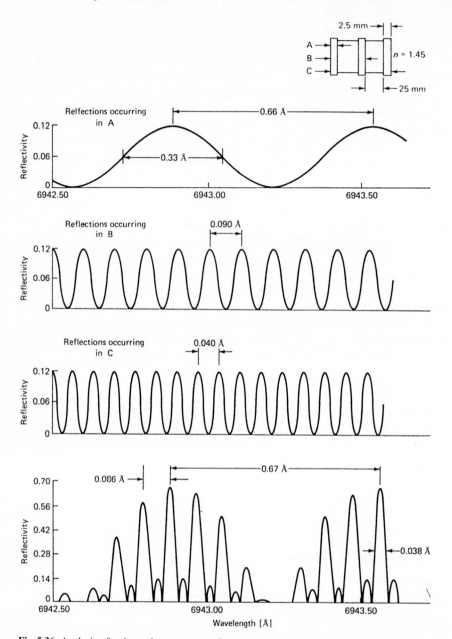

Fig. 5.36. Analysis of a three-plate resonant reflector

increase the diameter of the rings, a telescope is sometimes inserted between the interferometer and the imaging lens.

If the instrument is illuminated by a perfect monochromatic light source, one obtains a set of rings which correspond to directions for which constructive interference occurs ($2l \cos \Theta = m\lambda$). Since these interferometers are employed to measure the spectral characteristics of light beams, one is interested in the wavelength shift required to move the pattern from one ring to the next ring. This wavelength interval is called the free spectral range of the interferometer.

For an interferometer which consists of two dielectrically coated mirrors with a reflectivity of 90 % and separated by 40 mm, we obtain the following performance: The free spectral range is $\Delta\lambda_s = 0.062$ Å and the finesse is $F = 30$. The theoretical resolution of the instrument obtained from (5.58) is $\delta\lambda' = 0.002$ Å. However, the practical resolution is limited by the photographic film. As a rule of thumb, one usually assumes that the resolution is about one-tenth of the free spectral range. This means that a pattern showing rings which are not wider than 10 % of the ring separation indicates a laser linewidth of less than $\delta\lambda = 0.006$ Å.

5.2.2 Spectral Characteristics of the Laser Output

If a laser is operated without any mode-selecting elements in the resonator, then the spectral output will consist of a large number of discrete frequencies determined by the transverse and axial modes. The linewidth of the laser transition limits the number of modes that have sufficient gain to oscillate. The situation is diagrammed schematically in Fig. 5.37, which shows the resonance frequencies of an optical resonator and the fluorescence line of the active material. Laser emission occurs at those wavelengths at which the product of the gain of the laser transition and the reflectivity of the mirrors exceeds unity. In the idealized example shown, the laser would oscillate at seven axial modes.

Depending on the pumping level, for ruby and Nd : YAG, one finds a linewidth of approximately 0.3 to 0.5 Å for the laser emission in the absence of mode selection. In Nd : glass, because of the wide fluorescence curve, the laser linewidth is two orders of magnitude broader.

The spectral characteristics of a laser are quite frequently described in terms of bandwidth, linewidth, number of axial modes, and coherence length. We will now outline the relationships among these quantities.

If the laser emits K axial modes, the bandwidth between the two extreme modes is

$$\Delta\nu = \frac{(K-1)c}{2L} \quad \text{or} \quad \Delta\lambda = \frac{(K-1)\lambda^2}{2L} . \tag{5.65}$$

The beam emitted from a laser which emits a discrete number of integrally related wavelengths is strongly modulated. The situation can be illustrated by writing down the simplest case of two superimposed traveling waves whose wavelengths are specified by adjacent axial modes.

This situation is shown schematically in Fig. 5.38. The two waves interfere with one another, and produce traveling nodes which are found to be separated

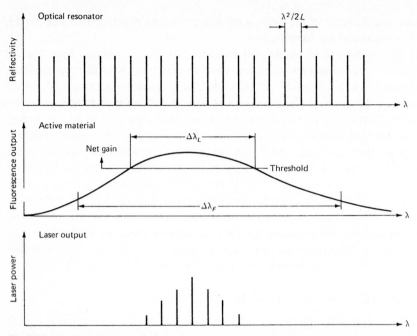

Fig. 5.37. Schematic diagram of spectral output of a laser without mode selection

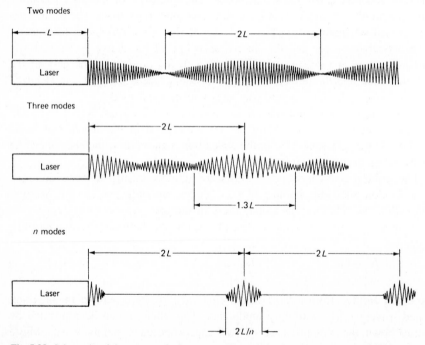

Fig. 5.38. Schematic of the output of a laser operating at two, three, and n longitudinal modes

from one another in time by twice the cavity separation. That is, the output of such a laser is modulated at a frequency of twice the end-mirror separation $(\nu_m = c/2L)$.

When lines at three integrally related frequencies are emitted, the output becomes more modulated; however, the maxima are still separated from one another by a distance of twice the mirror separation. As the number of integrally related modes increases, the region of constructive interference – which is inversely proportional to the number of oscillating modes – becomes narrower.

For Q-switched lasers which have been developed for holography, the coherence length has become an important parameter in specifying the spectral characteristics of the output beam.

The temporal coherence of any spectral source is defined as the path-length difference over which the radiation can still interfere with itself. Common techniques for measuring spectral bandwidth or coherence length of lasers include Fabry-Perot and Michelson interferometers, observation of beat frequencies, measurement of hologram brightness as a function of path length differences, and measurement of the fringe-free depth of field in a hologram.

In optics textbooks [5.46, 47], coherence length is defined as the path length difference for which fringe visibility in a Michelson interferometer is reduced to $1/(2)^{1/2}$. (Laser manufacturers sometimes use the 1/2 or $1/e^2$ points.)

The fringe visibility of an interferometer is defined as [5.46]

$$V = \frac{I_{\max} - I_{\min}}{I_{\max} + I_{\min}} . \tag{5.66}$$

In our subsequent discussion we will assume that the intensities of the axial modes in the laser output are equal. In this case the coherence length is directly related to the observed visibility of the interference fringes, and a simple relation between a set of longitudinal modes and temporal coherence can be obtained [5.51].

Single Axial-Mode Operation

The linewidth of the output of a laser operated in a single axial mode is usually many orders of magnitude narrower than the linewidth of the empty or passive resonator. The theoretical limit of the laser linewidth is determined by spontaneously emitted radiation which mixes with the wave already present in the resonator and produces phase fluctuations. The photons in the resonator all correspond to in-phase waves except for the spontaneous emission, which occurs at the same frequency but with a random phase. It is this phase jitter that causes the finite laser linewidth $\Delta\nu_L$ which is given by the Schalow-Townes limit [5.52]:

$$\Delta\nu_L = \frac{2\pi h\nu(\Delta\nu_c)^2}{P_{OUT}} \tag{5.67}$$

where $\Delta\nu_c$ is the linewidth of the passive resonator, $h\nu$ is the photon energy and P_{OUT} is the laser output.

According to (3.5 and 12) the passive resonator linewidth ($\Delta\nu_c$) is related to the photon lifetime τ_c in the resonator, resonator length l', round-trip loss L and output transmission T as follows:

$$\tau_c = \frac{1}{2\pi(\Delta\nu_c)} = \frac{2l'}{c(L+T)} \qquad (5.68)$$

where c is the speed of light. With this expression, the linewidth limit can also be expressed in the following form

$$\Delta\nu_L = \frac{h\nu}{2\pi\tau_c^2 P_{OUT}} . \qquad (5.69)$$

Since the laser output power P_{OUT} equals the number of photons n in the resonator times the energy per photon divided by the photon lifetime τ_c

$$P_{OUT} = \frac{nh\nu}{\tau_c} \qquad (5.70)$$

we find that the laser linewidth equals the empty resonator linewidth divided by the number of photons in the resonator, i.e.,

$$\Delta\nu_L = \frac{\Delta\nu_c}{n} . \qquad (5.71)$$

The most coherent solid state source is the diode pumped monolithic ring Nd : YAG laser discussed in Chap. 3. The monolithic design provides a short and stable resonator design which separates axial modes sufficiently to obtain single line operation. Diode pumping eliminates pump induced instabilities present in flashlamp lasers, and the ring configuration prevents spatial hole burning.

We will now calculate the fundamental linewidth limits for such a laser and compare it with the values achieved in actual systems. The physical length of these mini-lasers is typically 5 mm and the refractive index is 1.82, therefore the optical length of the resonator is $2l' = 1.82$ cm. The output coupling and internal losses for such a monolithic laser are $L + T = 0.01$. These values result in a lifetime of the photons in the resonator of $\tau_c = 6$ ns or a linewidth of the empty resonator of $\Delta\nu_c = 26$ MHz. The theoretical limit for the laser linewidth at a typical output power of 1 mW is then according to (5.69), $\Delta\nu_L = 1$ Hz. In practice, the laser linewidth is determined by temperature fluctuations and mechanical vibrations and instabilities which produce rapid changes in frequency. For example, the temperature-tuning coefficient for a monolithic Nd : YAG laser operating at 1064 nm is 3.1 GHz/°C. The major contribution is the temperature dependence of the refractive index. The design of narrow linewidth laser oscillators requires extremely good temperature control. A state-of-the-art diode-pumped monolithic Nd : YAG laser has achieved a linewidth of 300 Hz during a 5 ms observation time, 10 kHz in time intervals of 300 ms, and the laser exhibited a long term drift of 300 kHz/minute [5.53]. The laser did not feature an active stabilization feedback loop which would further reduce linewidth drift. Interest in lasers having a very narrow linewidth is for applications such as coherent radar systems, spectroscopy and holography.

Flashlamp- and coolant-induced instabilities made it extremely difficult in the past to achieve linewidths on the order of tens of MHz. Now with the emergence of miniature, diode-pumped monolithic Nd : YAG lasers, it is possible to achieve linewidths in the kHz regime.

Pulsed Nd : YAG lasers can be designed with a long coherence length by employing the cw monolithic Nd : YAG devices as seed lasers. An injection seeded Q-switched slave oscillator can produce a transform-limited linewidth as will be explained later in this chapter. In such a system, the linewidth is determined by the pulselength. The FWHM linewidth of a Gaussian shaped pulse with no frequence chirp is

$$\Delta \nu = \frac{2 \ln 2}{\pi t_P} \tag{5.72}$$

where t_P is the full width at the half-power points. A 20 ns transform limited Q-switch pulse with a Gaussian temporal profile will have a linewidth of 22 MHz. In a Q-switched ruby system, the output pulse has a strong frequency chirp which prevents transform-limited performance. The frequency chirp is caused by a change of the refractive index which, in turn, is the result of a change in population inversion.

When a pulse having a linewidth of $\Delta \nu$ interacts with another identical pulse delayed by a time τ, the fringe visibility can be expressed as

$$V = \exp\left(-\frac{(\pi \tau \Delta \nu)^2}{4 \ln 2} \right). \tag{5.73}$$

Since the transit time difference can be expressed in terms of an optical path-length difference $\Delta l = c\tau$ where c is the speed of light, the product $\tau(\Delta \nu)$ can be thought to be a normalized path length difference. In Fig. 5.39, fringe visibility is plotted as a function of $\tau(\Delta \nu)$. The coherence length l_c for which frige visibility has dropped to 0.71 is

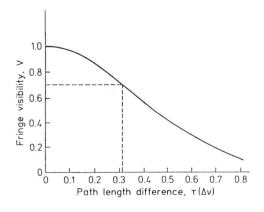

Fig. 5.39. Fringe visibility in a Michelson interferometer of a beam with a Gaussian spectral profile

$$l_c = \frac{0.64c}{\Delta\nu} \ . \tag{5.74}$$

Operation of the Laser at Two Longitudinal Lines. The linewidth of a single line is assumed to be very narrow compared to the mode separation. Therefore, the power spectrum of the laser is represented by two δ functions. The fringe visibility in this case is

$$V = \left| \cos\left(\frac{\pi\Delta l}{2L}\right) \right| \ . \tag{5.75}$$

This function, which has a periodicity of $2L$, is plotted in Fig. 5.40. The condition $V \geq 1/(2)^{1/2}$ is satisfied for

$$l_c = L \ . \tag{5.76}$$

Laser Emission at N Longitudinal Modes. The power spectrum of the laser is developed into a series of δ functions with the assumption that the modes have equal intensity. One obtains

$$V = \left| \frac{\sin(N\pi\,\Delta l/2L)}{N\,\sin(\pi\,\Delta l/2L)} \right| \ . \tag{5.77}$$

This function is also plotted in Fig. 5.40 for the case of a laser oscillating in three and four modes. The fringe visibility function is periodic in integer multiples of path difference $\Delta l = 2L$. The fringe visibility is equal or larger than $1/(2)^{1/2}$ for path length differences of

$$l_c = \frac{2L}{N} \ , \quad N \geq 2 \ . \tag{5.78}$$

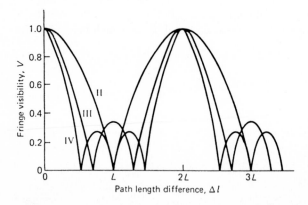

Fig. 5.40. Fringe visibility as a function of path length difference for a laser operating in two (II), three (III), and four (IV) longitudinal modes. L is the length of the laser resonator

5.2.3 Axial Mode Control

A typical solid-state laser will oscillate in a band of discrete frequencies which have an overall width of about 10^{-4} of the laser frequency. Although this is a rather monochromatic light source, there are still many applications for which greater spectral purity is required.

In one of the earliest attempts to narrow the spectral width of a laser, tilted Fabry-Perot etalons were employed as mode-selecting elements [5.54]. Also the concept of axial mode selection based on an analysis of the modes of a multiple-surface resonator was introduced [5.55, 56]. In these earlier works it was shown that it should be possible to discriminate against most of the modes of a conventional resonator by adding additional reflecting surfaces. Since then, many techniques have been developed to provide a narrow spectral linewidth from solid-state lasers.

In our discussion we will distinguish between three mode-selecting techniques:

- Interferometric mode selection
  A Fabry-Perot-type reflector is inserted between the two mirrors of the optical resonator. This will cause a strong amplitude modulation of the closely spaced reflectivity peaks of the basic laser resonator and thereby prevent most modes from reaching threshold.
- Enhancement of axial mode selection
  In this case, an inherent mode-selection process in the resonator is further enhanced by changing certain system parameters, such as shortening the resonator, removal of spatial hole burning or lengthening of the Q-switch build-up time.
- Injection seeding
  This technique takes advantage of the fact that stable, single longitudinal mode operation can readily be achieved in a very small crystal located within a short, traveling wave resonator. These devices which are end-pumped by a laser diode array, are by themselves not powerful enough for most applications, unless the output of such a device is coupled into a large slave oscillator for amplification [5.57].

Interferometric Mode Selection

The role of the resonant devices employed in interferometric mode selection is to provide high feedback for a single wavelength near the center of the fluorescence line, while at the same time discriminating against nearby wavelengths.

Resonant Reflectors. By replacing the standard dielectrically coated front mirror with a resonant reflector, the number of oscillating modes may be greatly reduced. As an example, Fig. 5.41 shows the output spectrum of a ruby laser operated without any mode-selecting elements in the cavity. The laser, operated with two dielectrically coated mirrors, emits a beam with a linewidth of about 0.5 Å at

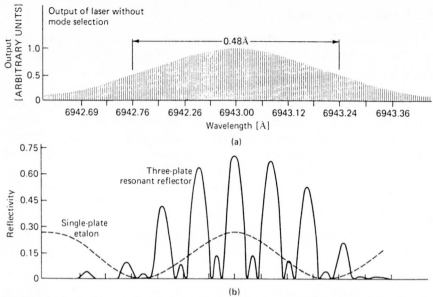

Fig. 5.41. (a) Spectral output of a ruby laser without mode selection. (b) Resonance curves of mode-selecting elements

the 50 % power points. The fluorescence linewidth of ruby is 5.5 Å at room temperature; however, above laser threshold the linewidth is reduced by a factor of 10. The envelope of the 0.5-Å-wide laser line contains approximately 160 longitudinal modes for a 75-cm-long resonator. The lines are separated from each other by 0.003 Å.

The dashed line in Fig. 5.41b shows the reflectivity versus wavelength of a single sapphire etalon of 3.2-mm thickness. The curve is drawn so that the peak reflectivity concides with the maximum in the gain profile. In this case the laser will emit a beam which has a linewidth of about 0.06 Å (20 modes). Actually, if the single sapphire etalon is not temperature-controlled, the reflectivity peak can be located anywhere with respect to the gain curve. The most undesirable location would be such that the reflectivity has a minimum at the peak of the laser line. In this case, one obtains a laser output composed of two groups of lines separated by 0.45 Å.

Also shown in Fig. 5.41b is the reflectivity curve of a three-plate resonant reflector. The reflectivity peaks of this device are much narrower as compared to a single sapphire etalon; this makes such a unit a better mode selector. Experiments have shown that this type of mode selector can reduce the number of axial modes to between one and three under proper operating conditions.

Resonant reflectors featuring one, two, or three etalons are in use. The etalons are usually fabricated from quartz or sapphire. Both materials have high damage thresholds. The advantage of sapphire over quartz is that higher peak reflectivity can be achieved for the same number of surfaces. The peak reflectivities for single-, double-, and triple-plate resonant reflectors are 0.13, 0.40, and 0.66 if quartz is used; and 0.25, 0.66, and 0.87 in the case of sapphire. Etalon thickness is typically 2 to 3 mm, which assures a sufficiently large spectral separation of the reflectivity maxima within the fluorescence curve so that lasing can occur on only one peak. In multiple-plate resonators, the spacing between the etalons is 20 to 25 mm in order to achieve a narrow width of the main peak.

The design of multiple-plate resonators can be optimized with the aid of computer programs which take into account the desired reflectivity, width, and separation of the resonance peaks [5.49]. It is important that the difference in thickness of the different plates in a multiple element resonator be held to less than one-tenth of wavelength. The parallelism of the spacers between the plates should also be held to this tolerance. Because of the relatively low peak reflectivity of resonant reflectors, these devices are used primarily with Q-switched solid-state lasers [5.58–66].

Optimum mode selection from a resonant reflector is achieved only when the reflection maximum of the device is centered at the peak of the fluorescence curve of the active material. Ideally, the gain at the adjacent reflection maxima should be insufficient to produce oscillation. Temperature tuning is the normal means of shifting the reflectivity peaks of the reflectors relative to the laser linewidth.

Table 5.2. Properties of materials frequently used in the design of resonant reflectors

Quantity	Dimension	Sapphire	Quartz	BK7	Stainless steel	Air
dn/dT	$10^{-6}/C$	12.6	10.3	1.86	–	−0.78
$(1/l)(dl/dT)$	$10^{-6}/C$	5.8	0.55	7.0	0.9	–
$n(6943\,\text{Å})$	–	1.76	1.455	1.51	–	1.0

Let us consider the changes which take place when the temperature of either the reflector or the laser or both is varied. A change in temperature will change the thickness and the index of refraction of an etalon. These effects combine to shift the wavelength of each resonant peak by an amount

$$\frac{d\lambda}{dT} = \lambda_0 \left(\frac{1}{l}\frac{dl}{dT} + \frac{1}{n}\frac{dn}{dT} \right) , \tag{5.79}$$

where n and l are the index of refraction and the thickness of the etalon, respectively. The first term on the right-hand side is the linear coefficient of expansion, and the second term is the thermal coefficient of the refractive index.

The temperature change necessary to cause an etalon to scan through one spectral range can be written in the form [5.59]

$$\Delta T = \frac{\lambda_0^2}{2nl(d\lambda/dT)} . \tag{5.80}$$

Using the materials parameter listed in Table 5.2, we obtain for a 3-mm sapphire etalon employed in a ruby laser: $d\lambda/dT = 0.09\,\text{Å/C}$ and $\Delta T = 5.0\ $C.

These numbers reveal that for a given wavelength the peak reflectivity of the etalon changes drastically with temperature, therefore temperature control of the resonant reflector is a necessity for stable and reproducible laser performance.

In most commercial lasers the temperature of the active material and the resonant reflector are both temperature-controlled by the cooling water circulated through the laser head and reflector housing. Ruby has a wavelength tuning rate of 0.067Å/C. If we operate a ruby system with the above-mentioned sapphire etalon, then a 1 C rise in water temperature will shift the ruby fluorescence peak by 0.067Å and the reflectivity peak of the etalon by 0.09Å toward longer wavelength. The relative shift between the two peaks is 0.023Å/C. The temperature dependence of the reflectivity of the mode selector can also be used to thermally tune the wavelength of the laser [5.60, 63].

Intracavity Resonances. Mode selection can also be achieved by using the laser rod itself as an etalon, by carefully aligning the flat and uncoated end of the laser rod to a flat-cavity mirror [5.65], or by adding additional mirrors which form three- or four-mirror laser resonators [5.67].

Intracavity Tilted Etalon. The stability, simplicity, low loss, and wide adaptability of the bandwidth and free spectral range makes the tilted etalon a versatile mode selector.

The etalon is inserted at a small angle in the laser resonator. The tilt effectively decouples the internal transmission etalon from the resonator; i.e., no other resonances will be formed with other surfaces in the main resonator. If the etalon is sufficiently misaligned, it acts simply as a bandpass transmission filter. The tilted etalon has no reflection loss for frequencies corresponding to its Fabry-Perot transmission maxima. At other frequencies the reflections from this mode selector are lost from the cavity, and thus constitute a frequency-dependent loss mechanism.

A narrow region of high transmission can be obtained by using sufficiently high-reflectivity coatings on the etalon. The fact, that for efficient mode selection tilted etalons must be dielectrically coated is a disadvantage in a high-power system, because of the possible damage, especially since the power density inside the etalon is approximately a factor of $(1 - R)^{-1}$ higher than the power density in the main resonator. Higher resistance to damage is the main reason why resonant reflectors are preferred as mode-selecting elements in pumped-pulsed, Q-switched lasers. On the other hand, resonant reflectors cannot be used for cw-pumped Nd : YAG lasers because the maximum reflectivity of these devices is too low. For the latter systems, tilted etalons are used extensively as mode selectors. The application of tilted etalons for the mode control of ruby, Nd : glass, and Nd : YAG systems can be found in [5.54, 65, 68–75].

The transmission maxima can be tuned into the central region of the gain curve by changing the tilt angle or by adjusting the etalon temperature. A requirement for mode selection is that the etalon transmission drops off sharply enough to allow only one mode to oscillate. The transmission T can be expanded around its maximum, yielding

$$T(\nu) = \left[1 + \frac{4R}{(1-R)^2}\left(\frac{2\pi nd\,d\nu}{c}\right)^2\right]^{-1}, \tag{5.81}$$

where $d\nu$ is the frequency deviation from the transmission maximum and R is the reflectivity of the etalon surfaces. If we assume that one mode oscillates at the center of the etalon's transmission curve ($d\nu = 0$), the mode with the second largest T will be spaced at approximately $d\nu = c/2L$.

For single-mode operation, the tilted Fabry-Perot etalon must have adequate selectivity to suppress axial modes adjacent to the desired one, and its free spectral range must be large enough so that the gain of the laser is sufficiently reduced at its next resonance. The difficulty which arises from these two requirements in choosing the optimum etalon thickness can be circumvented by employing two etalons of different thickness in the resonator [5.71]. Etalons have to be temperature-tuned to the center of the gain curve and stabilized to within 0.01°C [5.75].

Enhancement of Longitudinal Mode Selection

Here we will discuss the fact that single-mode operation is easier to obtain for a narrow laser linewidth, short resonator length, a large number of round trips in a Q-switched laser, and phase modulation of the standing wave inside the active material.

Cooling of the Active Medium. The linewidth of solid-state lasers, in particular ruby, decreases for decreasing temperature. This reduces the number of axial modes which have sufficient gain to oscillate. Cooling of a ruby crystal as a means of obtaining single axial-mode operation in combination with other mode-selecting techniques was discussed in [5.65].

Shortening of the Optical Resonator. The spectral separation of adjacent axial modes is inversely proportional to the length of the resonator. Thus, in a short resonator it is easier to discriminate against unwanted modes. For example, reliable single axial mode operation was achieved with an end-pumped Nd:YAG laser which had a length of 5 mm [5.76]. This short resonator is possible due to a monolithic design, where the coatings are directly applied to the rod. The axial mode spacing for such a short resonator is 16.7 GHz. Due to this large spacing (about 10% of the gain profile) adjacent modes have insufficient gain to reach laser threshold.

Lengthening of the Pulse Buildup Time. It was observed very early that passive Q-switches tend to act as mode selectors [5.77–81].

The mode-selection property of the Q-switching dyes was explained by *Sooy* [5.82] as a result of natural selection. Longitudinal-mode selection in the laser takes place while the pulse is building up from noise. During this buildup time, modes which have a higher gain or a lower loss will increase in amplitude more rapidly than the other modes. Besides differences in gain or losses between the modes, there is one other important parameter which determines the spectral output of the laser. This parameter is the number of round trips it takes for the pulse to build up from noise. The difference in amplitude between two modes becomes larger if the number of round trips is increased. Therefore, for a given loss difference between the modes it is important for good mode selection to allow as many round trips as possible.

Since a passive Q-switch requires more round trips for the development of a pulse as compared to a Pockels cell Q-switch, the former has better mode selection properties.

The growth with time t of the power P_n in mode n is given by [5.82]

$$P_n(t) = P_{\text{on}} \exp[k_n(t - t_n)^2] , \tag{5.82}$$

where P_{on} is the noise power in the mode at the start of buildup, t_n is the time at which the net gain for mode n reaches unity, k_n is equal to $(1/2T)(dg_n/dt)$, where T is the round trip time for the resonator, and g_n is the gain coefficient for mode n. Second, the ratio of the powers P_m, P_n in modes m and n, respectively, after q double passes of buildup, is given to a good approximation by

$$\frac{P_n}{P_m} = \left(\frac{1 - L_n}{1 - L_m}\right)^q (1 - L_n)^{q[(g_m/g_n)-1]} , \tag{5.83}$$

where L_m, L_n are the losses per double pass for modes m and n, respectively, and g_m, g_n are the gain coefficients for these modes. The first factor in (5.83) corresponds to loss discrimination and the second factor to gain discrimination.

In most lasers, gain differences between adjacent modes are too small to play a significant role in the mode-selecting process. For mode discrimination due to different values of the reflectivity of the mode selector we obtain, from (5.83)

$$\frac{P_n}{P_m} = \left(\frac{R_n}{R_m}\right)^q , \tag{5.84}$$

where $R_n = 1 - L_n, R_m = 1 - L_m$.

In Fig. 5.42 the output power ratio of two modes as a function of number of round trips is plotted. Parameter is the difference in reflectivity which these modes experience at the mode selector. The reflectivity peak of a mode selector can be approximated by

$$R = R_{\text{max}} \cos^2\left(\frac{\lambda}{\Delta\lambda_m} \frac{\pi}{2}\right) , \tag{5.85}$$

where $\Delta\lambda_m$ is the FWHM of the reflectivity curve and λ is the wavelength taken from the center of the peak.

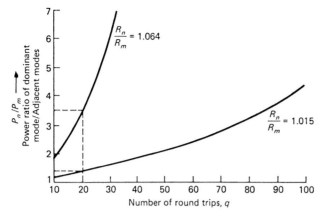

Fig. 5.42. Mode selection during the buildup of the pulse. The parameter R_n/R_m expresses the differences of the reflectivity which these two modes experience at the mode selector during one round trip

As an example, we will consider a 75-cm-long resonator and the three-plate etalon described in Sect. 5.2.1. Assuming the dominant mode right at the center of the reflecting peak, then the two adjacent modes will be shifted in wavelength by $\Delta\lambda = 0.003$ Å. With $\Delta\lambda_m = 0.038$ Å for the three-plate etalon, one obtains, from (5.85), $R/R_{\max} = 0.985$, which means that the difference in reflectivity for the two modes adjacent to the center mode is 1.5 %, or $R_n/R_m = 1.015$. In order to obtain single-mode operation from this system, we would need at least 155 round trips according to (5.84) if we use as a criterion for single-mode operation that the dominant mode should be at least ten times greater in peak power than any other mode. For the cavity length chosen we require, therefore, a pulse buildup time of 775 ns.

A logical step following the foregoing analysis is to use an active Q-switch, but to switch it in a way that ensures a large buildup time. This has been done by *Hanna* [5.84, 85], who operated a Pockels-cell Q-switch in a manner analogous to a saturable absorber. In this technique, initially the Pockels cell is only partially open and, therefore, presents a loss which is analogous to the low-level absorption loss of a saturable absorber. As a result of pumping, the net gain increases until it exceeds unity, and buildup of the giant pulse then starts. This buildup is monitored by a photodiode and its signal is used to trigger the Pockels cell to open completely when a preset signal level is reached. This is analogous to the bleaching of a saturable absorber at a particular intensity. Thus a long buildup time is achieved by an active Q-switch used in a way which is very closely analogous to a "passive" saturable absorber Q-switch. Utilizing this technique in conjunction with a resonant reflector, reliable single-mode operation in a Nd:CaWO$_4$ laser was obtained [5.86].

Lengthening of the pulse build-up is usually combined with intracavity mode selection. As an example, we will describe a Pockels cell Q-switched ruby oscillator containing a resonant reflector. The rear reflector is a multi-plate resonant reflector of the type shown in Fig. 5.36. However a fourth plate is added to increase the peak reflectivity of the device. The front reflector is a 2.5-mm-thick single sapphire etalon. The main function of this device is to provide an optimum output coupling of 25 % and to prevent oscillation at satellite peaks (spaced 0.086 Å from the main peak) of the multiplate resonator. Both mode-selecting elements are independently temperature controlled to within ±0.2 C. The laser head contains a 10-cm-long by 0.96-cm-diameter ruby rod pumped by a helical flashlamp. Operation at the TEM_{00} mode is accomplished by inserting a 1.5-mm aperture into the 75-cm-long resonator. Single-axial-mode operation is achieved by drastically decreasing the rise time of the Pockels-cell voltage. In a standard Pockels-cell Q-switch, the rise time of the voltage pulse on the crystal is typically 20 ns, and the laser output appears normally after 50 to 100 ns. In a 75-cm-long resonator this time delay amounts to about 10 to 20 round trips for the energy to build up. In this particular case the rise time is reduced to 1 μs, and the Q-switch pulse is emitted at the end of this time period; this suggests that about 200 round trips occurred before the pulse was emitted.

The system is operated only slightly above threshold. Single-transverse-mode operation is obtained at output energies below 50 mJ from a single oscillator. Single-transverse- and -longitudinal-mode operation is achieved at output levels between 10 and 15 mJ. Several amplifiers increased the energy to 10 J which was sufficient to obtain holograms with a scene depth of 5 m.

Reduction of Spatial Hole Burning. Limitations in obtaining a larger single-mode output are commonly attributed to spatial hole burning. Atoms located in the vicinity of the nodal planes of one axial mode will preferentially contribute to other modes. In gas lasers, spatial hole burning is all but prevented by the thermal motions of the atoms. In crystals, however, the amplitudes of the lattice vibrations are small compared to the light wavelength, and spatial averaging cannot take place. The efficiency of axial mode selection can be greatly increased by providing relative motion between the atoms in the active material and the electric field of the resonator.

In principle, spatial hole burning can be eliminated with traveling-wave ring structures [5.87–94], by generating circularly polarized light in the rod [5.96], with mechanical motion [5.97], or with electro-optic phase modulation [5.98].

Traveling-wave unidirectional ring lasers have been particularly successful in achieving narrow line width from a number of solid-state lasers. As was described in Sect. 3.2, a unidirectional ring laser contains three essential elements, a polarizer, a half-wave plate and a Faraday rotator. Ring lasers can be constructed with discrete elements [5.87–93], or monolithic versions of traveling-wave ring oscillators have been developed pumped by laser diodes [5.94, 99]. The first such device was the Monolithic Isolated Single mode End pumped Ring laser (MISER) developed by *Kane* et al. [5.94]. Diode pumping permits the design of very short resonators with large spacing between modes. In addition, a diode array provides a very stable pump source, with almost no amplitude fluctuations, and very little heat deposition into the lasing medium, all factors which are conducive to stable, single-line operation.

Most of the research on single-frequency, diode-laser-pumped Nd lasers has concentrated on monolithic devices where the optical cavity is formed by coated surfaces on the laser medium. While monolithic devices have significant advantages in terms of mechanical stability, they do not allow Q-switching or rapid tuning of the laser wavelength. Therefore, diode-pumped single-frequency lasers employing external mirrors in a ring geometry have been developed [5.93, 95].

Injection-Seeded Oscillator

The development of the injection-seeded oscillator, made possible by the availability of tiny seed lasers, has brought a revolutionary change to the technology of temporal mode control in solid-state lasers [5.57, 76, 83, 100, 101, 104, 113, 114]. This technique has shown to be simple to implement, requiring only a small diode-pumped single-frequency Nd laser to supply the seed beam. The development of the single-frequency injection laser grew out of the difficulty experienced in obtaining reliable stable single frequency operation in Q-switched Nd: YAG lasers.

Injection seeding is accomplished by introducing radiation from a low power (usually cw) stable single-frequency master oscillator into the high-power Q-switched laser cavity during the pulse buildup period. Both the injected single-mode radiation and spontaneous emission from the slave laser will be regeneratively amplified in the slave cavity. If the injected signal has enough power on a slave cavity resonance, the corresponding single axial mode will eventually saturate the homogeneously broadened gain medium and, in the absence of spatial hole burning, prevent development of any other axial modes from spontaneous emission.

The cw master oscillator provides a stable single-frequency TEM_{00} output for injection seeding. The seed laser then becomes the analog of the crystal oscillator in an RF transmitter. In [5.113], an injection-seeding theory has been developed which treats the threshold for an injection seeding and spectral purity of the laser output, as well as effects of position and angular alignment and frequency mismatch between the seed oscillator and power amplifier.

The most common master oscillator for injection seeding is a diode-pumped monolithic ring laser with a non-planar optical cavity [5.94]. Such a device produces an extremely stable single frequency with a cw output of a few milliwatts.

The diode-pumped monolithic ring laser has all the characteristics for achieving stable axial mode operation. The laser resonator is only a few millimeters long, therefore axial modes are spaced sufficiently apart for axial mode control. The ring laser configuration avoids spatial hole burning, and end-pumping provides a small gain volume for efficient and stable TEM_{00} mode operation. Furthermore, the monolithic design assures good mechanical stability of the resonator.

Figure 5.43 illustrates the injection seeding technique which can convert the single line output from a cw seed laser into a powerful Q-switched pulse which is transform-limited (linewidth is limited only by the temporal width of the laser pulse) on a single shot basis. The long term frequency drift is less than 50 MHz/h.

The injection seeding beam is introduced into the resonator cavity through reflection off the polarizing beam splitter. A Faraday rotator and polarizer combination provides the necessary isolation between the master laser and the Q-switched slave laser, both to protect the master laser from possible damage and to prevent destabilizing feedback into the master oscillator.

With the oscillator cavity of the slave laser in the low-Q condition, the injected pulse makes one round trip in the cavity and is then rejected by the polarizer. As the Q-switch opens, the injected pulse builds up rapidly in the cavity, extracting energy from the gain medium. If the original injected signal is of sufficient intensity, the laser mode that builds up will correspond to that of the injection source. Power gains of 10^9 can typically be achieved between the injection laser and the seeded oscillator output. A few milliwatt output of the injection laser is sufficient to obtain complete locking if the cavity resonance frequency is within 10% of the master oscillator frequency.

Active stabilization is required to ensure that the frequency detuning between the master and slave oscillators is within the tolerance for successful injection seeding. The frequency of the two cavities is maintained in coincidence by a feedback loop monitoring the build-up time of the oscillator pulse. If the master oscillator is properly injection-seeded, the Q-switch build-up time is considerably shorter as compared to a nonseeded resonator where the pulse has to build-up from noise. Frequency adjustments between the two lasers is performed by temperature tuning of the monolithic diode-pumped laser crystal in the master oscillator. Frequency sensitivity to temperature is about 3.1 GHz/K, and temperature stability of 0.01 K has been obtained using a proportional controller.

For single-frequency operation, it is necessary to avoid spatial hole-burning in the laser rod. This is especially true in a high-gain system where unextracted energy gives rise to "post lasing" following the main Q-switched pulse. In order to avoid spatial hole burning, it is necessary to avoid the formation of a standing wave in the laser rod. To accomplish this, quarter-wave plates are inserted in the resonator at each end of the laser rod, as shown in Fig. 5.43.

The quarter-wave plates convert the forward and backward traveling waves in the rod to circular polarization. Superposition of the two circularly polarized waves which are spatially displaced with respect to each other by $\lambda/4$, produces an electric field pattern in the rod which has the shape of a twisted ribbon with a spatial period of one optical wavelength [5.103]. Each polarization component of the electric field forms a standing wave, but the 90° phase shift between the components assures that the total energy density in a mode is uniform along the rod. Thus, hole burning in the rod is prevented.

The resonator shown in Fig. 5.43 is an unstable resonator with a Gaussian output mirror. Proper injection seeding requires strong transverse mode discrim-

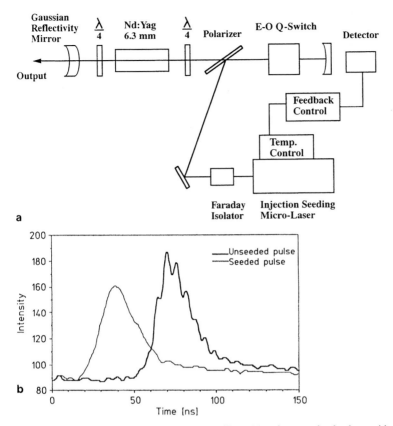

Fig. 5.43. Schematic of injection seeded slave oscillator (**a**) and temporal pulseshape with and without seeding (**b**)

ination in the slave oscillator [5.102]. As will be discussed in Sect. 5.5, unstable resonators exhibit this characteristic.

The temporal shape of the output pulse of the laser with and without injection seeding is shown in Fig. 5.43b. The output of the laser converted from a pulse characterized by beating between cavity modes in the uninjected condition, to an output characterized by a single frequency and smooth temporal profile with the injection laser on. Pulse-to-pulse amplitude jitter is also virtually eliminated in the injection-seeded system.

The output from the diode-array pumped slave oscillator is 50 mJ in a pulse with a smooth Gaussian time profile and a transform limited linewidth. The output from this oscillator was amplified by a 4-stage double-pass amplifier chain to 750 mJ/pulse.

5.3 Temporal and Spectral Stability

5.3.1 Amplitude Fluctuations

Fluctuations may be broadly categorized as short-term or long-term. Long-term fluctuations with periods greater than 1 s can be attributed to changes of parameters such as alignment and cleanliness of optical elements or degradation of the pump source.

Here we are concerned with short-term fluctuations with periods of less than a second. The sources of amplitude modulation of the laser output are mechanical vibrations of the optical components, thermal instabilities in the active material, variations in the pump lamp intensity, mode beating and mode hopping, relaxation oscillations, and quantum noise. The latter two phenomena are inherent properties of the laser oscillator. Which of the fluctuations mentioned above are dominant in a particular situation depends on the operating mode of the laser and the frequency interval of interest.

Conventional-Mode Operation. As has been discussed in Chap. 3, in this regime relaxation oscillations dominate all other noise mechanisms.

Q-Switched Pulsed-Pumped Laser. The envelope of a Q-switched pulse can show amplitude modulation due to mode beating of axial modes. The envelope is smooth for operation at a single axial mode and appears to be smooth for a very large number of axial modes, in which case amplitude modulation due to mode beating is reduced as a result of averaging effects.

Continuous-Pumped Lasers. All the noise mechanisms listed above can be studied in a cw-pumped Nd:YAG laser. Output fluctuations at low frequencies (below 1 kHz) can be traced to mechanical and thermal instabilities and fluctuations in the pump lamp intensity. The mechanical vibrations can be caused by the environment and transmitted to the laser, or they can be generated in the laser head itself by turbulence in the cooling flow [5.105]. Any vibrations of optical components will change the losses in the resonator; as a result, the output power will fluctuate. Besides vibrations transmitted to the laser head through its support structure, vibrations from the water pump transmitted to the laser head by the cooling water and hoses, as well as turbulence or air bubbles in the water surrounding the laser rod, can cause mechanical vibrations of the resonator mirrors, and laser rod [5.106]. It was found, that if these mechanisms are dominant, laser performance can be stabilized by acoustically decoupling the water cooler from the laser head and reducing the flow rate.

Changes in the intensity of the pump lamps will modulate the output of the laser. In krypton arc lamps, intensity variations are caused by arc wandering, plasma instabilities, and voltage ripple. The first two instabilities will produce random spatial and temporal variations of rod gain. Low-frequency modulation in arc lamps can be reduced by the addition of an active feedback loop to control lamp current.

Turbulence of hot air inside the laser head or dust particles in the beam can cause the optical beam to fluctuate. A bellows system which encloses the optical train is, therefore, a requirement for a low-noise laser.

Air bubbles in the cooling path of the lamps or laser rod can cause modulation of the intensity of the pump lamp. Modulation of the refractive index of the laser rod by the coolant can cause the output to fluctuate. In a typical cooling system, the laser rod is cooled by a turbulent flow of water around it. Therefore, the heat transfer between the rod and coolant changes constantly in time and space. The temperature variations within the rod create optical distortion through the dependence of the index of refraction of Nd:YAG on temperature. This mechanism becomes important if the laser rod is used as the limiting aperture for TEM_{00} operation. The use of a quartz sleeve around the laser rod to damp thermal disturbances [5.107] or an increase in the flow rate to create a more stable dynamic flow characteristic [5.108] have been suggested to overcome these problems. However, it was found [5.106] that reducing the beam diameter by means of an aperture has the same effect. The thermal inertia of the periphery of the rod provided sufficient damping to eliminate this problem.

In the 50- to 500-kHz range, depending on the particular system, output fluctuations are dominated by relaxation oscillations. Any small perturbation having a frequency component at the vicinity of the resonance frequency of the relaxation oscillation will cause large oscillations in the output power. A potential source for the excitation of relaxation oscillations is mechanical vibrations. Obviously, in order to reduce the modulation caused by relaxation oscillations, the mechanical laser structure must be extremely stable [5.109, 110].

At the lower end of the megahertz region, parasitic noise is caused by mode competition of transverse modes, whereas mode hopping of longitudinal modes causes amplitude modulation at frequencies of several hundred megahertz. Different transverse modes occupy different regions of the laser rod; as a result of pumping nonuniformities, the gain varies for each transverse mode. Different axial modes are located at different points on the laser gain curve. Any disturbances of the laser system will encourage or discourage individual modes. Therefore, the mode pattern changes constantly. Mode selection eliminates output fluctuations associated with mode competition.

In diode-pumped solid-state lasers, instabilities and fluctuations caused by the lamp discharge are eliminated. Furthermore, in conduction-cooled diode-pumped lasers, vibrations introduced by the cooling system are eliminated also. Actually, the diode-pumped monolithic ring laser is the building block for the design of ultra-low intensity noise lasers [5.94]. The monolithic resonator design drastically reduces mechanical- and vibrational-induced instabilities, and the laser operates in a single transverse and longitudinal mode. This laser produces an output with a relatively small amplitude fluctuation except for the frequency spectrum in which relaxation oscillations occur. By adding an electronic feedback loop, residual intensity fluctuations, as well as the relaxation oscillations, can be reduced to a level close to the shot-noise-limited performance [5.115–117].

The quantum or shot noise generated in a detector is a fundamental property and is due to the discrete nature of photon-electron generation. The value of shot noise for a detector irradiated by a Poisson-distributed stream of photons is given by

$$i_{\mathrm{RMS}} = (2e\Delta fI)^{1/2} \tag{5.86}$$

where I is the dc photo current in the detector, Δf is the electrical bandwidth, and e is the electron charge. Equation (5.86) expresses the limit of accuracy with which a current I can be measured at a fixed bandwidth Δf. Shot noise produces a white uniform noise floor at all frequencies.

Noise fluctuations in lasers can be reduced by a feedback loop using an electro-optic modulator in the resonator [5.118]. In diode-pumped solid-state lasers, intensity noise suppression can be achieved with a feedback system which modulates the drive current to the diodes.

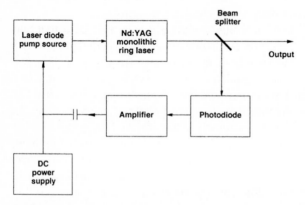

Fig. 5.44. Feedback control for the reduction of intensity noise in the laser output

Figure 5.44 displays the schematic of such a feedback control system. A fraction of the radiation emitted by the laser is sampled by a photo diode, after amplification and phase shifting, the detected signal is used to change the drive current of the diode pump source. This causes a modulation of the laser radiation. The servo loop is ac coupled to the power supply to avoid changes in the dc operation of the diode pump. If the feedback loop has the proper phase and gain with respect to the detected laser radiation, intensity fluctuations can be compensated. The feedback system has to be carefully designed in order to avoid noise amplification and to keep the control loop stable.

Figure 5.45 illustrates the reduction of intensity noise one can achieve with a carefully designed feedback control loop. Shown is the photo current noise power, as measured by a spectrum analyzer with a 10 kHz bandwidth, of a

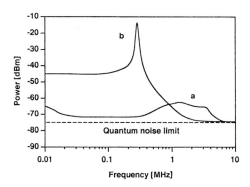

Fig. 5.45. Noise output of a monolithic cw Nd : YAG ring oscillator with *a* and without *b* feedback control [5.117]

diode-pumped Nd : YAG nonplanar ring laser in the free running mode and with feedback control [5.117]. The peak in the noise spectrum at about 300 kHz of the free-running laser is caused by relaxation oscillations. It was observed that frequencies below the relaxation oscillations, intensity noise in the laser output was mainly caused by noise in the pump-diode output. Radiation emitted from each of the individual stripes of the laser-diode beat together which results in the observed noise floor. At frequencies greater than 5 MHz, the output reaches the quantum-noise level.

With feedback control, there is a large reduction in the noise level for frequencies below 1 MHz. However, the intensity noise is increased between 1 and 5 MHz due to the excess noise generated by the feedback loop. Above 5 MHz, the laser noise is unaltered by the feedback loop. Also shown in Fig. 5.45, is the quantum-noise limit. Under feedback control, the spectral density of the output noise $\Delta P/P/\sqrt{\text{Hz}}$ was measured to be $3 \times 10^{-8}\sqrt{\text{Hz}}$ between 10 and 300 kHz which is 6.1 dB above the quantum-noise limit.

5.3.2 Frequency Control

Frequency control requires a certain amount of tunability of the laser emissions, either to lock the frequency to a reference or to an incoming signal for coherent detection, or to provide an output signal at a specific wavelength for spectroscopic studies.

Depending on the range of tunability, we can distinguish between:
- *Broadband tunability*: This is achieved with tunable lasers (Sect. 2.5) in combination with wavelength-sensitive elements such as birefringent tuning plates, dispersive prisms and gratings, as will be discussed in Sect. 5.6. The tuning range can be as large as 100 THz in the case of Ti : sapphire.
- *Tuning over the gain bandwidth of a narrow-band laser*: Small angular changes of an intra cavity etalon, or cavity length adjustments with a PZT mounted resonator mirror provide some means of tuning the wavelength of Nd lasers. For example, rapid tuning within the gain spectrum of Nd : YAG, which is about 180 GHz, has been achieved with a combination of tilt-angle adjustment of an intra-cavity etalon and cavity-length adjustment via a PZT

controlled mirror. The etalon was mounted on a galvanometer which allowed a very precise and repeatable angular rotation. The tuning range, covering many longitudinal modes, was on the order of 50 GHz [5.119].

- *Tuning within a single longitudinal line of a resonator*: This is the topic of the present subsection. As we shall see, depending on the length of the resonator, and therefore the separation of longitudinal modes and the method of introducing a frequency shift, tuning has been achieved up into the GHz regime.

In certain applications, such as coherent LADAR systems, a stabilized single-frequency laser source is required, both as local oscillator for the detection system, and as injection seeding source for a pulsed transmitter.

Miniature diode-pumped solid-state lasers, such as monolithic or microchip devices, are particularly attractive for this application because of their single-frequency operation. Furthermore, the short resonator of these devices supports a relatively large tuning range because longitudinal modes are widely separated. For example, a 2 mm long monolithic resonator has a mode spacing of about 41 GHz in Nd : YAG. In monolithic lasers, the frequency can be tuned by thermal expansion or by mechanical stress applied to the crystal. Discrete-element miniature lasers can be tuned by insertion of an electro-optic modulator into the cavity or by cavity-length adjustments.

In monolithic lasers such as the microchip laser, or nonplanar ring laser, the resonator mirrors are coated directly on the Nd : YAG or Nd : YLF crystal, which precludes the introduction of traditional intra-cavity elements. However, varying the temperature of the crystal will change the resonator length, and therefore frequency tune the laser emission. The sensitivity of the laser frequency to thermal changes can be calculated as follows. We assume a microchip laser consisting of a small disk of Nd : YAG polished flat and parallel on two sides which is end-pumped by a laser diode.

The resonant frequency of the laser cavity changes with cavity length according to

$$\frac{d\nu}{\nu} = -\frac{dl_R}{l_R} \tag{5.87}$$

where ν is the lasing frequency and $l_R = nl$ is the optical length of the resonator.

Temperature changes lead to a thermal expansion of the resonator and to thermal and stress-induced changes of the refractive index. To a first order, we can neglect stress-related index changes. Thermally-induced length and refractive-index changes of the microchip laser resonator can be expressed by

$$dl_R = nl \left(\alpha_e + \frac{1}{n} \frac{\delta n}{\delta T} \right) dT_{av} . \tag{5.88}$$

Introducing (5.88) into (5.87) yields

$$\frac{d\nu}{dT} = - \left(\alpha_e + \frac{1}{n} \frac{\delta n}{\delta T} \right) \nu \tag{5.89}$$

with the materials parameters for Nd : YAG from Tables 2.4 and 2.5, one obtains $d\nu/dT = -3.25\,\mathrm{GHz}/°\mathrm{C}$.

The actual tuning rate is the difference between the tuning rate of the resonator and the small temperature shift of the laser transition in Nd : YAG. Actual measured data of the thermal tuning coefficient in a monolithic Nd : YAG laser gave a value of $-2.5\,\mathrm{GHz}/°\mathrm{C}$.

Temperature tuning results in a large tuning coefficient, but thermal control of the laser frequency is relatively slow with a time constant on the order of one second. Thermal tuning was employed to control two diode-pumped monolithic Nd : YAG ring oscillators to within 40 kHz [5.120].

From the standpoint of frequency control, the large thermal tuning coefficient is actually a disadvantage, because in order to maintain good frequency stability, these lasers require extremely precise temperature control. This is usually achieved by mounting these devices on TE coolers and environmental shielding from temperature fluctuations.

Instead of changing the temperature of the laser crystal by means of a TE cooler, pump power modulation has been successfully employed to cause rapid frequency tuning in miniature lasers. As the pump power increases, more thermal energy is deposited in the gain medium, raising its temperature and therefore changing the resonant frequency of the laser cavity. Modulation of the pump power has the undesirable effect of changing the laser output power. In implementing this technique, TE coolers maintain a fixed average temperature of the crystal. Around this temperature point, rapid thermal changes are induced by means of current modulation of the pump-laser diode. The thermal response is sufficiently fast, to permit phase locking of the laser.

Experiments have been described in which two 1 mm long Nd : YAG microchip lasers operating at $1.32\,\mu\mathrm{m}$ have been frequency locked to another [5.121, 161]. In the free-running mode, a frequency jitter of 250 kHz over a 15 s period was observed, which was reduced to 50 Hz in the frequency-locked case. At a modulation frequency of 100 Hz, pump power changes of 1 mW produced an optical frequency change of 5.25 MHz.

Another technique for precisely tuning the laser frequency of a monolithic laser is based on stress-induced changes in the resonator. In these designs, stress is applied to the laser crystal by using a piezoelectric transducer. This results in a stress-induced birefringence of the refractive index, and a strain-related distortion or elongation of the resonator. In a disk shaped Nd : YAG monolithic laser with a thickness of 1 mm, a frequency range of 76.5 GHz could be scanned by applying stress transverse to the beam axis [5.162]. Bonding a thin plate of piezoelectric material to a monolithic Nd : YAG ring laser, permitted piezoelectric tuning over a range of 100 MHz with a speed fast enough to achieve phase locking of two lasers [5.163]. Tuning could be accomplished over several tens of MHz in a few microseconds. Frequency stability over 1 millisecond was better than 500 Hz.

In the case of discrete-element micro-cavity lasers, frequency tuning has been achieved by cavity length modulation or by incorporating an electro-optic modulator into the resonator. In the former design, an air gap between the two pieces forming a ring resonator was adjusted by a piezoelectric transducer. Cavity length tuning was achieved over a frequency range of 13.5 GHz at a speed on the order of milliseconds [5.164].

Intra-cavity electro-optic modulation of diode-pumped linear Nd:YAG and Nd:YVO$_4$ micro cavity lasers have been demonstrated [5.165–167]. Figure 5.46 illustrates a Nd:YAG microchip laser with a rapid frequency tuning capability. A composite laser cavity is formed by bonding a lithium tantalate wafer to a Nd:YAG crystal. Both wafers have a combined thickness of 1 mm. A high voltage applied parallel to the c-axis of the LiTaO$_3$ crystal modulated the laser output over a tuning range of 1.5 GHz [5.166].

The most stringent requirement for absolute frequency control is imposed on scientific lasers developed for gravitational-wave detectors. Absolute frequency stability of a laser can be achieved by locking the output to a high finesse reference Fabry-Perot cavity [5.168–170]. Employing a high-finesse reference cavity as frequency discriminator, the beat note of two diode-pumped Nd:YAG lasers can be reduced to fractions of one Hz.

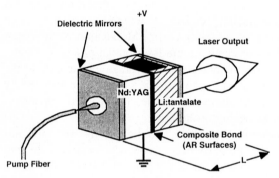

Fig. 5.46. Nd:YAG micro chip laser with intracavity electro-optic modulator [5.166]

5.4 Hardware Design

In the construction of a resonator for a laser oscillator, the important considerations are the mechanical stability and the quality of the optical components, protection of the optical train from the environment, and protection of the operator from radiation hazards.

Laser Support Structure. The key requirement of a laser resonator is that all optical components be accurately and rigidly fixed relative to each other, that the structure be insensitive to temperature variations, and that there be a provision for making angular adjustments of the beam.

General research involving lasers, is usually carried out with the components mounted on an optical table, since this provides maximum flexibility for the experimental set-up. The optical components of commercial lasers are usually mounted on a baseplate or on support structures made from aluminum extrusions. In the latter case, adjustment of the axial position of optical components can be provided by sliding carriages. Commercial lasers are also fully enclosed by a housing for protection against the environment and for eye-safety reasons.

In most cases the optical surfaces are additionally protected by enclosing the whole optical train with tubes or bellows. Enclosures may range from a mere safety cover to a rubber-sealed dust-proof cover, all the way to a hermetically sealed cover which will maintain a positive pressure on the inside.

In military systems weight and size constraints lead to densely packaged structures with aluminum as the material of choice. Often, the whole support structure is machined from one solid block of metal and the optical beam is folded to achieve a compact design.

For complex designs such as the laser shown, for example in Fig. 7.39a, a solid model based mechanical design and analysis capability is indispensable. The most widely used software is Pro/Engineer, a 3-D parametric solid modeling CAD system. It allows laser systems to be virtually built, assembled, and analyzed prior to fabrication. This capability produces an accurate virtual representation of the laser system allowing the designer to focus on technical intents and requirements rather than drawing accuracy. Such virtual representations provide the foundation for both engineering and manufacturing reviews prior to fabrication. Maintaining focus on product integrity is partially enhanced by the automated drafting process of the Pro/Engineer system. Detail drawing views, dimensions and bill-of-materials are automatically created directly from the model eliminating the traditional drafting errors inherent in 2D CAD systems. Changes in the model are automatically reflected in the detail drawings as well. Substantial quality enhancements, as well as error reduction, is achieved by the automated component interference checks of assemblies. The model checks for interfering components and identifies those requiring modification prior to fabrication. This solid modeling capability can be integrated with an analysis capability. Interactive analysis capabilities include; stress, deflection, thermal and modal. Analysis capabilities also include mass properties.

Mirror Mounts. The requirements for a good mirror mount are an independent, orthogonal, and backlash-free tilt with sufficient resolution and good thermal and mechanical stability. Commercial mirror mounts meeting these requirements usually employ a two-axis gimbal suspension, a three-point suspension using a torsion spring, two independent hinge points formed by leaf springs, or a metal diaphragm. Rotation is achieved by micrometers or differential screws.

Adjustable mirror mounts are, in general, not used in military laser systems because these components do not provide the required long term stability in the presence of temperature excursions, shock and vibration. Instead of mirrors, these systems employ crossed Porro prisms or retro reflectors. Beam alignment is accomplished by insertion of two thin prisms (Risley prism) into the beam. The prisms which can be independently rotated, will deflect the beam over a small cone angle. After alignment is completed, the prisms are locked in place. Output coupling from such a resonator is by means of a polarizer and Q-switch. An example of such a resonator has been given in Sect. 5.1.10.

Optical Surfaces. The polished surfaces of components, such as mirrors, laser rod, windows, lenses, doubling crystal, etc., used in a laser resonator must meet certain requirements in terms of surface quality, flatness, parallelism, and curvature. The quality of an optical surface is specified by the scratch and dig standards (MIL-13830A). According to this standard, surface quality is expressed by two numbers. The first number gives the apparent width of a scratch in microns and the second number indicates the maximum bubble and dig diameter in 10-μs steps. Components used in optical resonators should have an optical surface quality of about 15/5. As a comparison, commercially available lenses of standard quality have an 80/50 surface.

Mirror blanks for the resonator are either plano-concave, plano-plano, or in special cases plano-convex, made from schlierengrade fused quartz or BSC 2. Standard curvatures are flat, 10, 5, 3, 2, 1, 0.5, and 0.25 m radius of curvature. The surface finish should be $\lambda/10$ or better. Standard mirror blanks have diameters of 25, 12.5, and 6.2 mm, and thicknesses of 6.2 and 10 mm. A parallelism of 1 arc s or better, and a flatness of at least $\lambda/20$ is required for plane-parallel blanks to be used for etalons.

Coatings. Evaporated dielectric films are essential parts of the resonator optics, and their properties are important for the performance of the laser. A typical laser requires a highly reflective and a partially reflective mirror coating and several antireflection coatings on the laser rod, Q-switch crystal, windows, lenses, etc.

Dielectric thin films are produced by evaporation and condensation of transparent materials in a vacuum. The dielectric thin film must adhere well to the substrate. It should be hard, impervious to cleaning, have low loss, and should exhibit a high damage threshold. Commercially available coatings are either hard coatings which will meet MIL-C-675A or semihard coatings which will pass the Scotch-tape adherence test per MIL-M-13508B.

The simplest antireflection coating is a single quarter-wave film. The reflectivity of a single dielectric layer of index n_1, which has an optical thickness of $\lambda/4$, is, at normal incidence,

$$R = \left(\frac{n_s - n_1^2}{n_s + n_1^2} \right)^2 , \tag{5.90}$$

where n_s is the refractive index of the substrate. Zero reflection is achieved when $n_1 = (n_s)^{1/2}$.

Table 5.3. Substrate characteristics

	Index of refraction	Reflectivity percent	Reflectivity with single-layer MgF$_2$
Fused quartz (0.63 μm)	1.46	3.5	1.5
BSC-2 (0.63 μm)	1.52	4.2	1.4
Ruby (0.69 μm)	1.76	7.6	0.15
Nd : YAG (1.06 μm)	1.82	8.4	0.1

The lowest refractive index available as stable film is MgF$_2$ with $n_1 = 1.38$, a value which results in a perfect antireflection coating for a substrate with $n_s = 1.90$. For $n_s = 1, 80, 1.70$, and 1.5 the residual reflectance from a $\lambda/4$ MgF$_2$ coating is $\approx 0.1, 0.3$, and 1.4%, respectively. From Table 5.3 it follows that MgF$_2$ is an excellent match for ruby and Nd : YAG. In cases where the reflection from a single-layer antireflection coating is still too high, two or more dielectric-layers must be applied.

A substrate having two layers with index values of n_1 and n_2 of optical thickness $\lambda/4$ will have a total reflectance of

$$R = \left(\frac{n_2^2 - n_1^2 n_s}{n_2^2 + n_1^2 n_s} \right)^2 . \tag{5.91}$$

Zero reflection can be obtained if $(n_2/n_1)^2 = n_s$. If coating materials with the proper ratio n_2/n_1 are not available for a particular substrate, thicknesses which deviate from $\lambda/4$ must be used to achieve zero reflection from glass. Because the region of low reflectance of this type of coating is rather small, the coating is sometimes called "V"-coating. A very hard and durable two-layer coating frequently employed on glass substrates is the system ZrO$_2$-MgF$_2$.

High reflectivity, multilayer dielectric films can be tailored to give specific reflectance versus wavelength characteristics by the appropriate choice of the number of layers, layer thickness, and index of refraction of the materials. The simplest design of a multilayer coating is a stack of alternating films of equal optical thickness, corresponding to $\lambda/4$, but of two different refractive indices. It is most efficient to start and end with a high-index layer so that the structure will have an odd number of layers. We obtain

$$R_{\text{max}} = \left(\frac{n_1^{l+1} - n_2^{l-1} n_s}{n_1^{l+1} + n_2^{l-1} n_s} \right)^2 , \tag{5.92}$$

where n_1 is for the high-index material, n_2 is for the low-index material, and l is the odd number of $\lambda/4$ films.

Table 5.4 lists some of the most common thin-film materials used for solid-state laser optics. Titanium oxide, cerium oxide, and zinc sulfide are used mainly

Table 5.4. Thin-film materials

Chemical formula	Material	Index of refraction
MgF_2	Magnesium fluoride	1.38
ThF_4	Thorium fluoride	1.50
CeO_2	Cerium oxide	2.3
ZrO_2	Zirconium oxide	2.1
SiO_2	Quartz	1.45
ZnS	Zinc sulfide	2.35
TiO_2	Titanium oxide	2.28

as the high-index materials; magnesium fluoride and silicon dioxide are used for the low-index materials.

As follows from (5.92), the reflectance increases with increasing number of $\lambda/4$ films in discrete steps. The number of layers employed depends on the material combination and the desired reflectance value. Normally, it does not exceed 21 because scatter and absorption losses increase with the number of layers.

For commercially available mirror coatings, the maximum reflectance ranges between 99.5 and 99.8 %. Scattering and absorption losses combined are in the order of 0.5 to 0.2 %. Typical thin-film combination are ZnS-MgF_2, ZnS-ThF_4, CeO_2-MgF_2, TiO_2-SiO_2, and ZrO_2-SiO_2. The first system is very soft and must be handled with great care; the second coating is more durable and can be cleaned with alcohol and lens tissue; the last two coatings are extremely hard and durable.

A summary of optical coatings and thin-film techniques can be found in [5.122]. Optical properties of multilayer stacks and designs have been presented in the books by *Baumeister* [5.123], *Heavens* [5.124], *Macleod* [5.125], and *Vasicek* [5.126].

5.5 Unstable Resonators

As was discussed in Sect. 5.1, in a stable resonator the light rays are confined between the surfaces of the resonator mirrors and do not walk out past their edges. In order to produce a diffraction-limited output beam from a stable resonator, the Fresnel number must be on the order of unity or smaller (Sect. 5.1.5), otherwise sufficient discirimination against higher-order modes cannot be achieved. For practical resonator lengths, this usually limits the diameter of the TEM_{00} mode to a few millimeters or less.

If the resonator parameters g_1 and g_2 lie outside the shaded regions defined in Fig. 5.8, one obtains an unstable resonator. In these configurations, the light beam is no longer confined between the mirrors. A light beam in an "unconfining" or unstable resonator diverges away from the axis, as shown in Fig. 5.47, eventually radiation will spill around the edges of one or both mirrors. This fact

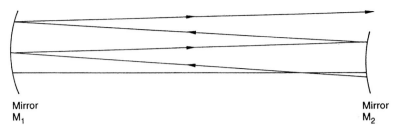

Mirror
M_1

Mirror
M_2

Fig. 5.47. Light ray in an unstable resonator

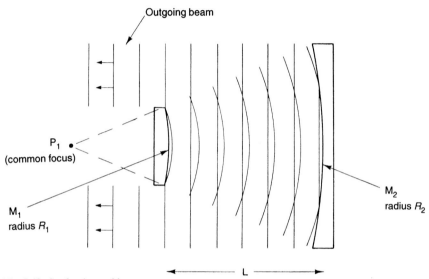

Outgoing beam

P_1
(common focus)

M_1
radius R_1

M_2
radius R_2

L

Fig. 5.48. Confocal unstable resonator

can, however, be used to advantage if these walk-off losses are turned into useful output coupling.

Figure 5.48 illustrates the spherical waves bouncing between the mirrors of an unstable resonator. In this so-called confocal configuration, the concave mirror's focal point coincides with the back focal point of the smaller convex mirror. Thus we see that P_1 is the common focus. Light traveling to the left is collimated and may conveniently be coupled out by providing an unobstructed aperture around mirror M_1.

The unstable resonator corresponds to a divergent periodic focusing system, whereby the beam expands on repeated bounces to fill the entire cross-section of at least one of the laser mirrors, however large it may be. The unstable resonator has a much higher-order mode discrimination, as compared to its stable counterpart, therefore a nearly diffraction-limited output beam from a large diameter gain medium can be obtained. In the near field, however, the output from an unstable resonator usually has an annular intensity pattern. As we have seen, stable resonator modes are well approximated by Gaussian-Laguerre or Hermite-Gaussian functions. Unstable resonator modes do not have a simple functional form; their intensity distributions must be generated numerically.

The unstable resonator first described by *Siegman* [5.127, 128] has been studied extensively both theoretically and experimentally. Excellent reviews are found in [5.129, 130]. The most useful property of an unstable resonator is the attainment of a large fundamental mode volume and good spatial mode selection at high Fresnel numbers. In other words, unstable resonators can produce output beams of low divergence in a short resonator structure which has a large cross-section.

Immediately following its invention, the significance of the unstable resonator was recognized for the extraction of diffraction-limited energy from large volume gas lasers. However, it took many years for unstable resonators to be applied to solid-state laser system. There are a number of reasons for this slow acceptance. The laser medium must be of high optical quality, for an unstable resonator to be effective. This requirement has limited applications of unstable resonators primarily to gas lasers, because the time- and power-dependent thermal distortions occurring in solid-state lasers made this type of resonator unattractive. In addition, the output coupler of an unstable resonator, having the dimensions of a few millimeters in typical solid-state lasers, is much more expensive and difficult to fabricate in comparison to a partially transmitting mirror required for a stable resonator.

Furthermore, the misalignment tolerance of an unstable resonator is smaller compared to its stable counterpart and the advantage of a large mode volume is achieved at a sacrifice of mode quality because of aperture-generated Fresnel fringes. The output from a solid-state laser is often passed through amplifier stages, or the oscillator may be followed by a harmonic generator. The near-field beam pattern of an unstable resonator which consists of a doughnut shaped beam with diffraction rings and a hot spot in the center is not very attractive in these applications.

About 10 years after its discovery, *Byer* and co-workers [5.131, 133] applied the unstable resonator concept for the first time to a Q-switched Nd:YAG oscillator/amplifier system. They did achieve a marked improvement in Nd:YAG output energy in a diffraction-limited mode.

However, only recently has the unstable resonator found applications in commercial lasers, mainly as a result of the availability of the variable reflectivity output mirror which provides a smooth and uniform output beam from an unstable resonator. Output coupling via such a mirror provides an elegant solution in overcoming the beam profile issue. Whether an unstable resonator has an advantage over a stable resonator for a particular system depends on the gain of the laser. As we will discuss later, gain, mode volume and sensitivity to misalignment are closely related in unstable resonators. (In a stable resonator, only mode volume and alignment stability are directly related as we have seen from the discussions in Sect. 5.1.10.) Generally speaking, only in a high-gain laser can a large mode size be realized in an unstable resonator at a reasonable misalignment tolerance.

The most useful form of an unstable resonator is the confocal unstable resonator [5.132, 134]. A primary advantage of this configuration is that it automatically produces a collimated output beam; this also means that the final pass of the beam through the gain medium is collimated.

Confocal configurations can be divided into positive-branch resonators which correspond to the case $g_1 g_2 > 1$, and negative-branch resonators for which $g_1 g_2 < 0$. The quadrants for the location of these resonators are labelled in the $g_1 - g_2$ plane of Fig. 5.8. Confocal unstable resonators of the positive or negative branches are shown in Fig. 5.49. These configurations are defined by the following relationships:

$$2L = R_1 + R_2 \tag{5.93}$$

or

$$L = f_1 + f_2 \; ,$$

where L is the optical length of the resonator, and R_1, R_2, f_1, f_2 are the radii and focal lengths of the two mirrors, respectively. For concave mirrors, the sign for R and f is positive, and for convex mirrors it is negative. A diverging output beam will result if the mirrors of a negative- or positive-branch confocal resonator are moved closer together, and a converging beam will result if the mirrors are moved farther apart. The light rays do not cross the optic axis in the positive-branch resonator, whereas the negative-branch resonator has an intracavity focus.

5.5.1 Confocal Positive-Branch Unstable Resonator

The confocal positive-branch unstable resonator is the most widely used form of the unstable resonator for solid-state lasers.

Referring to Fig. 5.49 the annular output beam has an outer diameter of D and inner diameter d, where d is also the diameter of the output coupler.

The resonator magnification

$$M = D/d \tag{5.94}$$

is the amount that the feedback beam is magnified when it travels one round trip in the resonator and becomes the output beam.

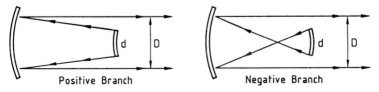

Fig. 5.49. Positive- and negative-branch confocal unstable resonators

The magnification in the transverse beam dimensions results in a decrease of intensity by a factor $1/M^2$ in each round trip, and radiation spilling out around the edges of the output mirror. A beam which is just contained within the mirrors will, after one round-trip, lose a fraction T of its energy.

The geometrical output coupling is related to the magnification M by

$$T = 1 - \frac{1}{M^2} \, . \tag{5.95}$$

If we insert a gain medium between the mirrors, the loss in energy has to be made up by the gain of the laser. According to the laser threshold condition (3.13) we obtain

$$2gl = L + \ln M^2 \, . \tag{5.96}$$

Ignoring internal resonator losses L for a moment, the round-trip gain of the laser has to be

$$G = \exp(2gl) \geq M^2 \, . \tag{5.97}$$

For a confocal resonator, the mirror radii are given by

$$R_1 = \frac{-2L}{M - 1} \, , \quad \text{and} \quad R_2 = \frac{2ML}{M - 1} \, , \tag{5.98}$$

where L is the length of the resonator and R_1 and R_2 are the output and back-cavity mirror curvatures. Note that the output mirror has a negative curvature and thus is convex, while the high-reflectivity mirror has positive curvature and is concave.

Siegman [5.129] has investigated the relationship between M and the output coupling T for confocal unstable resonators and has shown that T is less than the geometrically predicted value $(1 - 1/M^2)$. This is because the intensity distribution, according to wave optics, tends to be more concentrated towards the beam axis than that predicted by geometrical optics.

The equivalent Fresnel number of a resonator characterizes the destructive or constructive interference of the mode at the center of the feedback mirror due to the outcoupling aperture. For a positive-branch, confocal resonator, the equivalent Fresnel number is

$$N_{\text{eq}} = \frac{(M - 1)(d/2)^2}{2L\lambda} \tag{5.99}$$

where d is the diameter of the output mirror, and L is the resonator length. Plots of the mode eigenvalue magnitudes versus N_{eq}, show local maxima near half-integer N_{eq}'s.

Physically, the half-integer equivalent Fresnel numbers correspond to Fresnel diffraction peaks centered on the output coupler which leads to increased feedback into the resonator. Resonators should be designed to operate at half-integer equivalent Fresnel numbers (N_{eq}) to obtain best mode selectivity.

As a final step one has to take into account the effect of the laser-rod focal length f. One usually chooses an available mirror curvature R and calculates the rod focal length at the desired lamp input power required to achieve an effective mirror curvature R_{eff}. If the mirror to rod distance is less than the rod focal length, then

$$\frac{1}{f} + \frac{1}{R_{\text{eff}}} = \frac{1}{R} .$$ (5.100)

Essentially the focusing effect of the laser rod is compensated by increasing the radius of curvature of the mirror.

Equation 5.100 is only a first-order approximation, a more rigorous treatment is found in [5.141].

The design of such a resonator is an iterative process. It starts with some knowledge of the optimum output coupling T or the saturated gain coefficient g obtained from operating the laser with a stable resonator. From either of these two parameters, we obtain from (5.95 or 96) the magnification M. Selecting a value of 0.5, 1.5, 2.5, ... for N_{eq} from (5.99) provides a relationship between d and L. With (5.93, 94, 98) the pertinent resonator design parameters such as d, D, L, R_1 and R_2 can be calculated. The effect of thermal lensing is then accounted for by adjusting the radius of curvature according to (5.100).

In analogy to high-energy lasers utilizing unstable resonators, the output coupling can be accomplished by means of a scraper mirror or edge coupler. Figure 5.50 shows relevant adaptions for solid-state lasers.

Figure 5.50a illustrates an unstable resonator with an output scraper mirror. It is inclined at an angle of $45°$ to the resonator axis and has a hole in its

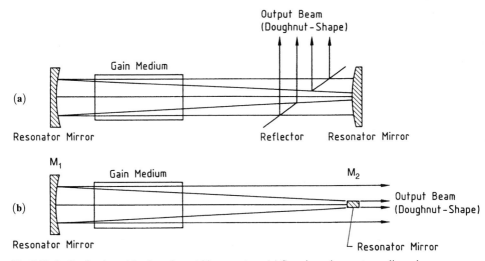

Fig. 5.50a,b. Confocal, positive branch unstable resonators. (**a**) Scraping mirror out-coupling scheme. (**b**) Small mirror out-coupling scheme

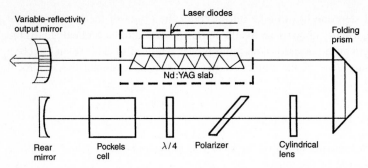

Fig. 5.51. Diode pumped Nd:YAG slab laser with positive branch unstable resonator and variable reflectivity output coupler [5.139]

center which allows light to pass through it and be fed back into the resonator to sustain the lasing. Because the end mirrors and scraper are oversize, this hole determines the size and shape of the beam outcoupled from the resonator. The design depicted in Fig. 5.50b consists of a concave mirror M_1 and a convex output mirror M_2, both of which are totally reflecting. The dot mirror M_2 is a small circular dielectrically coated area of radius d centered on a glass substrate. The output beam is collimated as it exits the resonator around the edges of the dot mirror.

While these two output coupling techniques are borrowed from high-energy gas lasers, the design shown in Fig. 5.51 is unique to solid-state lasers. Illustrated is a diode-pumped Nd:YAG slab laser with a positive-branch, confocal unstable resonator in a folded configuration. The output coupler is a dielectrically coated mirror with a variable reflectivity profile. This is the type of output coupler used in state-of-the-art unstable resonators for solid-state lasers. As will be discussed in Sect. 5.5.3, a semitransparent mirror with a radially variable reflectivity profile does not produce diffraction rings or a hot spot in the output beam, as one obtains from the output couplers depicted in Fig. 5.50. The resonator in Fig. 5.51 has a magnification $M = 1.38$ and a cavity length of 58 cm. The laser generated 100 mJ of Q-switched output at a repetition rate of 100 Hz with a near diffraction limited beam [5.139].

5.5.2 Negative-Branch Unstable Resonators

Due to the presence of an intra-cavity focal point, the negative-branch resonator has been neglected in practical laser applications. Despite the potential problem of air breakdown this resonator merits consideration due to its unique feature of relatively large misalignment tolerances.

It was found [5.142] that a Q-switched Nd:YAG laser featuring a negative-branch unstable resonator, was not significantly degraded with a mirror misalignment angle of as much as a few milliradians. Also system air breakdown was not experienced for Q-switched pulses on the order of 170 mJ and 12 ns pulse length. Therefore, for small solid-state lasers, typical of range finders and target

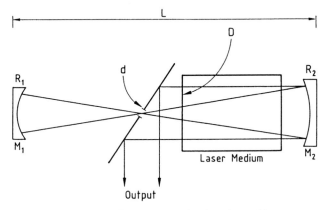

Fig. 5.52. Arrangement of a typical negative-branch unstable resonator

designators with peak powers not exceeding 5–10 MW, it is conceivable that a negative-branch unstable resonator could be employed.

The design parameters for a negative-branch resonator of the type shown in Fig. 5.52 are

$$R_1 = 2L/(M+1) \quad \text{and} \quad R_2 = 2ML/(M+1) ,$$
(5.101)

where L is the confocal resonator length and M is the optical magnification defined before.

A variation of the negative-branch unstable resonator was described by *Gobbi* et al. [5.143]. The design is based on the proper choice for the size of the field-limiting aperture d located at the common focal plane of the mirror. If the aperture is chosen such that a plane wave incident on it is focused by mirror M_1 to an Airy disk having the same diameter d, then this results in the removal of the hot spot inside the cavity and in a smoothing of the spatial profile.

If the aperture diameter d is chosen such that

$$d = 2(0.61\lambda f_1)^{1/2}$$
(5.102)

where $f_1 = R_1/2$, then only the Airy disk is allowed to propagate beyond the aperture, and on reflection from the mirror M_2, it is magnified, collimated, and ready to start another similar cycle.

The radius at which the beam has zero amplitude is not determined by the geometrical magnification M, but an effective magnification imposed by diffraction.

$$M_{\text{eff}} = 1.5M = -1.5f_2/f_1 \quad f_2 = R_2/2 .$$
(5.103)

The diameter of the collimated beam passing through the laser is

$$D = |M_{\text{eff}}|d .$$
(5.104)

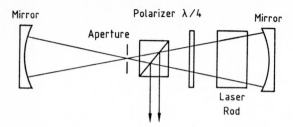

Fig. 5.53. Negative-branch unstable resonator with polarization output coupling

By adding the constraint on the aperture size that it match the Airy disk, the usual hot spot in the focal plane is completely removed by diffraction. Actually the combination of aperture d and mirror M_1 acts on the resonator field as a low-pass spatial filter. This accounts for the smoothness of the field profile.

The disadvantage of this design is the limited value of D which can be achieved in practical systems. In order to fill a large active volume with diameter D, such for example a slab laser, either M_{eff} or d has to be large (5.104). In order for d to be large, it follows from (5.102), that f_1 has to be large which in turn leads to a long resonator. A large M_{eff} requires a very high gain material, for example, a Q-switched Nd: YAG oscillator. A beam diameter of $D = 4.8$ mm inside the laser rod requires a resonator of 125 cm in length and a magnification of $M = 4$. In order to achieve the high gain required for this design the laser was pumped 2.5 times above threshold.

Another interesting feature of this design is the beam extraction from the resonator. Instead of a tilted scraper mirror, as shown in Fig. 5.52, beam extraction was achieved by means of a polarization coupling scheme, employing a polarizer and a quarter wave plate (Fig. 5.53).

5.5.3 Variable Reflectivity Output Couplers

One of the major disadvantages of an unstable resonator, namely the generation of an annular output beam containing diffraction rings and a hot spot in the center, can be eliminated by employing a partially transparent output coupler with a radially variable reflectivity profile.

In such a mirror, reflectivity decreases radially from a peak in the center down to zero over a distance comparable to the diameter of the laser rod. Such a resonator is, in principle at least, capable of sustaining a single transverse mode of a very large volume and with a smooth, uniform spatial profile. Whether a large volume can be realized in a practical system depends on the gain of the system. As will be discussed at the end of the section, only in a high-gain laser can a large mode size be utilized. Otherwise alignment stability becomes a very critical issue and the laser has only a very small tolerance for mirror misalignment caused by environmental conditions.

It clearly seems that if an unstable resonator is to be used in a solid-state laser, a Gaussian-type reflectivity output coupler is a very elegant solution in

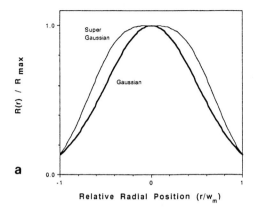

Fig. 5.54. Mirror with a Gaussian and super-Gaussian reflectivity profile (**a**) and unstable resonator employing such a mirror (**b**)

overcoming the beam profile issue. The merits of such a design in providing a smooth output profile in an unstable resonator has been recognized as far back as 1965 [5.135]. Various solutions have been proposed for the fabrication of variable reflectivity profiles. Among these have been birefringent elements [5.136–138] and a radially variable Fabry-Perot interferometer [5.140]. However, advances in thin film manufacturing technology have made it possible to produce mirrors with a Gaussian reflectivity profile by deposition of multilayer dielectric films of variable thickness on a transparent substrate [5.144–153]. This represents a real breakthrough, in terms of a practical implementation of unstable resonators and a number of commercially available Nd:YAG lasers incorporate such mirrors.

The reflectivity of a mirror having a Gaussian type reflectivity profile such as shown in Fig. 5.54a can be described as follows:

$$R(r) = R_{max} \exp[-2(r/w_m)^n] \tag{5.105}$$

where r is the radial coordinate, R_{max} is the peak reflectivity at the center, w_m is the mirror spot size or radial distance at which the reflectivity falls to $1/e^2$ of its peak value, and n is the order of the Gaussian profile. Most mirrors have a Gaussian profile for which $n = 2$; super-Gaussian profiles are obtained for $n > 2$. By adjusting the maximum reflectivity (R_{max}), the width (w_m), and the order of the Gaussian (n), the output energy for a given oscillator design with a specific magnification can be optimized, while maintaining a smooth output beam profile.

In the following discussion, we will address the design issues concerning Gaussian mirrors of order 2 ($n = 2$). Super Gaussian mirrors which can provide a higher beam fill factor as compared to Gaussian mirrors have been treated in [5.154–157]. Figure 5.54b illustrates the concept of the variable reflectivity unstable resonator.

Upon reflection at a Gaussian mirror, a Gaussian beam is transformed into another Gaussian beam with a smaller spot size

$$\frac{1}{(w_{\mathrm{r}})^2} = \frac{1}{w^2} + \frac{1}{w_{\mathrm{m}}^2} \qquad (5.106)$$

where w is the spot size of the incident beam, and w_{r} is the spot size of the reflected beam.

For an unstable resonator containing a variable reflectivity output coupler, the magnification M is the ratio between the beam sizes of the incident and reflected beam, given by

$$M = \frac{w}{w_{\mathrm{r}}} = [1 + (w/w_{\mathrm{m}})^2]^{1/2} \ . \qquad (5.107)$$

The effective output coupling from a resonator containing a variable reflectivity mirror depends on the average reflectivity of the mirror, which is a function of the profile of the incident beam. For a Gaussian beam (TEM$_{00}$ mode) the effective average reflectivity is

$$R_{\mathrm{eff},00} = R_{\mathrm{max}}/M^2 \ . \qquad (5.108)$$

For the next-higher-order mode (TEM$_{01}$), the effective reflectivity can be shown to be

$$R_{\mathrm{eff},01} = R_{\mathrm{max}}/M^4 \ . \qquad (5.109)$$

Mode discrimination is achieved because the effective mirror reflectivity for the TEM$_{01}$ mode is reduced by $(1/M^2)$ as compared to the TEM$_{00}$ mode.

Though the reflected beam is Gaussian, the same is not true for the transmitted beam I_{out}. The spatial profile of the transmitted beam is the product of the incident Gaussian beam and the transmission profile of the mirror

$$I_{\mathrm{out}}(r) = \left\{ 1 - R_{\mathrm{max}} \exp\left[-2(r/w_{\mathrm{m}})^2\right] \right\} I_0 \exp[-2(r/w)^2] \ . \qquad (5.110)$$

Figure 5.55 shows a plot of the profile of the output beam for various peak reflectivities of the mirror. As the curve illustrates, if R_{max} is greater than a certain value, a dip occurs in the center of the output beam. In these situations, the mirror transmission increases with radius faster than the beam intensity decreases. The condition for which the central dip of the mirror just begins to occur is given by

$$R_{\mathrm{max}} \le 1/M^2 \ . \qquad (5.111)$$

From (5.108 and 111) we obtain the condition that in order to have a filled-in output beam without a dip in it, the effective reflectivity must be less than

$$R_{\mathrm{eff}} = 1/M^4 \ . \qquad (5.112)$$

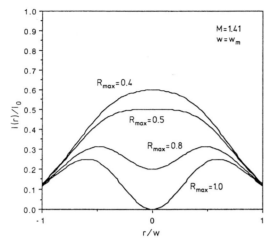

Fig. 5.55. Profile of the output beam for various peak reflectivities of the mirror

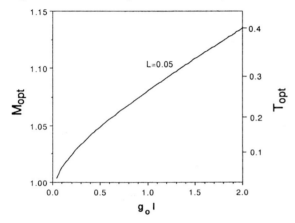

Fig. 5.56. Relationship of logarithmic gain $g_0 l$, T_{opt} and M_{opt}. Parameter is the resonator loss L

The output coupling for this resonator is then

$$T = 1 - 1/M^4 . \tag{5.113}$$

Since the optimum output coupling T_{opt} of an oscillator is related to the logarithmic gain and losses according to (3.67) we can plot M_{opt} as a function of these quantities as shown in Fig. 5.56.

Design Procedure. We assume that the oscillator was operated with a stable resonator and T_{opt} was determined. Alternatively, the laser can be operated as an amplifier and with a small probe beam, the gain and losses can be measured; in this case T_{opt} has to be calculated from (3.67). With T_{opt} known, we obtain from (5.113) the value of M. The spot size radius w of the Gaussian beam is usually selected to be half the radius of the laser rod r_{d} in order to avoid diffraction effects, i.e. $w/r_d = 0.50$. With this assumption, the mirror spot size w_M follows then from (5.107). Therefore, the mirror parameters R_{max} from (5.111), and w_M have been specified. The mirror radii and mirror separation follow from (5.93, 98) for the confocal unstable resonator. Thermal lensing of the laser rod can be accounted for by changing the curvature of one of the mirrors according to (5.100).

As an example, we use a diode-array pumped Nd:YAG laser rod for which the following parameters have been measured with a 1.06 μm probe beam: logarithmic gain $g_0 l = 1.5$, internal losses $L = 0.05$, thermal lensing $f = 6$ m. From these parameters follows, based on the discussion above: $T_{opt} = 0.34$, $R_{eff} = 0.66$, $M = 1.11$ and $R_{max} = 0.81$. The laser rod has a diameter of 5 mm. For good extraction efficiency, we select a diameter of the fundamental mode of half the rod diameter, or $w = 1.25$ mm. The spot radius for the mirror is therefore from (5.107) $w_M = w/(M^2 - 1)^{1/2} = 2.6$ mm. The focal length of the uniform reflectivity mirror was chosen $f_1 = 5$ m, which determines the focal length of the Gaussian mirror, $f_2 = -4.5$ m. Resonator length is $L = 0.5$ m. The focal length of mirror M_1 is increased according to (5.100) to compensate for thermal lensing, one obtains: $f_1' = 7.7$ m.

The operation of multi-hundred watt Nd:YAG lasers which incorporated positive- or negative-branch unstable resonators was first reported in [5.171–173]. Two unstable-resonator designs have been described which are particularly suitable for high-power solid-state lasers, because in these designs, the influence of thermal-rod lensing on the resonator properties is minimized. Both configurations, the rod-imaging negative branch [5.174], and the near-concentric positive [5.175, 176] branch unstable resonator, employ graded reflectivity mirrors.

As already mentioned, a negative-branch unstable resonator contains a focal point inside the cavity. A special case is the rod-imaging resonator shown in Fig. 5.57a. In this resonator, the rear mirror images the principal plane of the rod onto itself. In this case, it can be shown that the g parameter of the rear mirror does not depend on the focal length of the rod thermal lens.

If both principal planes of the thermal lens are imaged onto themselves by the mirrors, one obtains a stable concentric resonator. A near-concentric unstable resonator is formed if both mirrors are moved a small distance further apart from this position (Fig. 5.57b). It has been shown that this resonator is less sensitive to rod focal length changes as compared to other configurations. Theoretical and experimental investigations of these resonator concepts, as well as design procedures, can be found in [5.176].

In our laboratory, we tested a high-power diode-pumped, Nd:YAG laser with both of these unstable resonator configurations in an attempt to optimize system performance [5.177]. The dual-head laser contained a Nd:YAG rod of 10 mm diameter and a length of 69 mm. Each head is pumped by a five-fold symmetric stack of laser diodes emitting at 810 nm. The heads are pumped with 250 μs long pulses at a repetition rate of 200 Hz at a combined optical average pump power of 3.2 kW. With a close-coupled flat/flat resonator and an 80% reflective output coupler, a maximum of 425 W average output was obtained from the dual-head

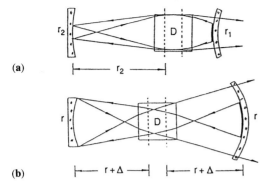

(a)

(b)

Fig. 5.57. Rod imaging (**a**) and near concentric (**b**) unstable resonator containing a laser rod with optical power D [5.176]

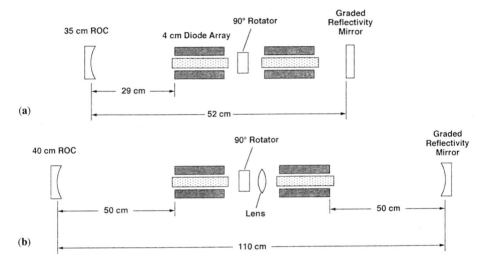

(a)

(b)

Fig. 5.58. Rod imaging (**a**) and symmetric near concentric (**b**) unstable resonator

oscillator. The laser heads were placed close together in all experiments. At a combined maximum input pump power of 3.2 kW, this arrangement behaves as one rod with a refractive power of 3.3 diopters.

The rod-imaging unstable resonator was formed by a concave rear mirror and a flat graded reflectivity output coupler. As shown in Fig. 5.58a, a 90° quartz rotator is placed between the laser rods to compensate for birefringence. The best result from this set-up was an output power of 250 W at a beam parameter product of 9.6 mm-mrad [defined as beam diameter times full angle). In a symmetric near-concentric unstable resonator configuration with a 10 mm separation from the concentric position, the best performance was at an output of 220 W at a beam divergence of 5.5 mm-mrad. The super Gaussian mirror had a central reflectivity of 87%, a spot radius of 2.5 mm, and a super Gaussian exponent of 9. In order to match the desired beam diameter to the available variable reflectivity mirrors, the refractive power of the rods was increased by placing a lens between the rods, as shown in Fig. 5.58b. For a given mirror spot size, this leads to a larger intra cavity beam inside the rod and to a beam cross-section reduction at the mirrors.

Based upon these experiments, we came to the following conclusions: The laser output power is about the same for both resonator configurations. The near-concentric unstable resonator was about twice as long in comparison to the rod-imaging design, but it produced a lower beam divergence. Consistent with theory, the near-concentric resonator was more tolerant to changes in thermal lensing of the rod (i.e., pump input power), as compared to the rod imaging resonator. Both designs are considerably less sensitive to pump input changes compared to a positive-branch confocal configuration. The advantages of lower beam divergence and higher tolerance to pump input changes of the near-concentric unstable resonator has to be traded-off against the much shorter rod-imaging resonator.

5.5.4 Gain, Mode Size and Alignment Sensitivity

Whether a large mode size in a resonator can be realized in practical situations depends upon the resonator's sensitivity to mirror tilt and thermal lensing. The former determines the laser's susceptibility to the effects of temperature variations, structural changes and shock and vibration, whereas the latter determines the laser's tolerance to changes in the operating parameters such as repetition rate, or laser output. For stable resonators, these issues have been addressed in Sects. 5.1.8, 10. Here we will discuss the sensitivity of the unstable resonator to mirror tilt. The principal effect of a small tilt of a resonator mirror is to steer the output beam. We consider the case of a tilted primary mirror because its sensitivity is greater than the sensitivity to feedback-mirror tilts. If the primary mirror is tilted by θ with respect to its aligned position, the propagation of the output beam is then tilted by an angle ϕ. The relationship between the mirror tilt angle θ and the beam angle ϕ has been treated in [5.134] for the confocal resonator using a hard aperture output coupler.

For the positive branch confocal resonator, the alignment sensitivity is

$$\frac{\phi}{\theta} = \frac{2M}{M-1} \tag{5.114}$$

and for the negative branch confocal resonator

$$\frac{\phi}{\theta} = \frac{2M}{M+1} . \tag{5.115}$$

The major conclusions we can draw from these results are as follows:

a) A confocal positive-branch resonator is more sensitive to mirror misalignment compared to a negative-branch resonator; this is particularly the case for values of M less than 2;

b) In a low-gain laser, the value of M will be small, therefore in such lasers, the unstable resonator will be very sensitive to mirror misalignment.

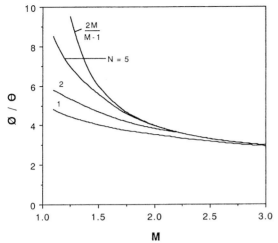

Fig. 5.59. Ratio of the fundamental mode beam-steering angle ϕ to the mirror misalignment angle θ for different values of the magnification M. Parameter is the equivalent Fresnel number N [5.144]

We will now turn our attention to the Gaussian mirror unstable resonator. *McCarthy* et al. [5.144] have presented an analytical solution to the problem of the misaligned optical resonator with a Gaussian mirror. The misalignment is introduced by tilting the uniform mirror by an angle θ. The Gaussian beam propagates then along an axis making an angle ϕ relative to its original propagation direction. As before, the quantity ϕ/θ characterizes the resonator sensitivity to misalignment.

The ratio ϕ/θ as a function of the resonator magnification M is plotted in Fig. 5.59. Parameter is the equivalent Fresnel number of the Gaussian mirror which is defined as

$$N = w_M^2/\lambda L . \tag{5.116}$$

For large values of N and magnifications larger than 2, the alignment sensitivity approaches the value given in (5.114) which is based on purely geometrical considerations. As we recall from our previous discussion, the gain of the laser determines the magnification M, whereas high extraction of the energy stored in the volume of the laser rod requires a large beam size, and therefore a large w_M and N.

It is apparent from the curves in Fig. 5.59 that for large Fresnel numbers in combination with small values of M, the resonator becomes extremely sensitive to misalignment. Unfortunately, most solid-state lasers have a small gain which results in a rather low M value. For the example chosen in the last section, one obtains $w_M = 2.6$ mm, and $L = 0.5$ m which gives $N = 13$ for the Nd wavelength. Based on the data present in Fig. 5.59, this is not a very robust resonator to say the least. This was verified in actual experiments, alignment was difficult to obtain and could be maintained only under laboratory conditions.

Actually, the analysis presented in [5.144] concludes that for values of M much less than 2, the unstable resonator has about the same misalignment sensitivity as a plane-plane resonator. Therefore, only in high-gain materials such as Nd : YAG with the pump beam tightly focused to achieve high inversion, will it be possible to achieve magnifications high enough such that a large beam diameter can be utilized.

Also in certain gas lasers, where the gain is high, and therefore $M > 2$, the resonator sensitivity is independent of mode size. In this regime, one can take full advantage of the unstable resonator in creating a large fundamental mode volume.

In low-gain systems, an unstable resonator has to be designed with a Fresnel number close to one for good alignment stability. This, of course, defeats the purpose of using an unstable resonator since with such a low Fresnel number, a stable resonator can provide the same performance.

5.6 Wavelength Selection

Vibronic lasers such as alexandrite, Ti : sapphire, Cr : GSGG have very broad gain curves, in the operation of these lasers it is necessary to use a wavelength selection technique to: (a) restrict laser action to a specified wavelength; and (b) tune the laser output. Several different methods are available (in principle) for providing the wavelength selection and tuning. These include (a) use of a prism inside the laser, (b) utilization of an adjustable optical grating within the laser, (c) use of intracavity etalons, or (d) use of one or more thin birefringent plates within the laser that are tilted at Brewster's angle.

The technique most commonly employed for the wavelength selection of tunable lasers is the birefringent filter [5.158]. In its simplest form, the birefringent filter consists of a single thin birefringent crystal located inside the laser (Fig. 5.60). For simplicity, we assume that the birefringent axes lie in the plane of the crystal, and that the crystal is tilted at Brewster's angle.

The crystal and birefringent axes, and the incident light are shown in more detail in Fig. 5.61. In this figure, α is the incidence angle between the incoming

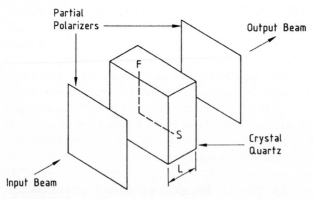

Fig. 5.60. Single-crystal wavelength selector

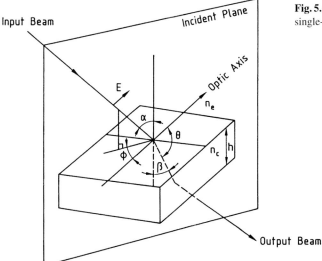

Fig. 5.61. Schematic of typical single-plate birefringent filter

laser beam and the normal to crystal plane, and ϕ is the angle between the plane of incidence and the crystal optic axis. The phase difference δ between the ordinary and extraordinary rays emerging from the crystal is given by

$$\delta = \frac{2\pi h(n_0 - n_e)\sin^2\theta}{\lambda \sin\alpha} \tag{5.117}$$

where λ is the laser wavelength in vacuum.

An alternate representation of the single stage tuning element is given in Fig. 5.62. The input and output partial polarizers represent the front and back surfaces of the quartz crystal situated at Brewster's angle. In addition, the crystal is assumed to be rotated to $\theta = 45$ degrees in this figure.

Wavelength selection occurs with the birefringent filter because of the two different crystal indices of refraction. When the laser light has a wavelength corresponding to an integral number of full-wave retardations, the laser operates as if the filter were not present. At any other wavelength, however, the laser mode polarization is modified by the filter and suffers losses at the Brewster surfaces.

Tunability of the laser is achieved by rotating the birefringent crystal in its own plane. This changes the included angle ϕ between crystal optic axis and the laser axes and, hence, the effective principal refractive indices of the crystal.

The amplitude transmittance of the single-stage filter of Fig. 5.60 has been calculated in [5.145]. For a quartz crystal rotated to $\phi = 45$ degrees and tilted to Brewster's angle, the transmittance at unwanted wavelengths is about 82 %. This may or may not provide adequate suppression of unwanted wavelengths for certain lasers.

One way to lower the filter transmittance in the rejection band is to use a stack of identical crystal plates that are similarly aligned, as shown in Fig. 5.63

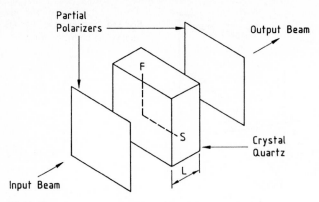

Fig. 5.62. Alternate representation of single-stage crystal filter

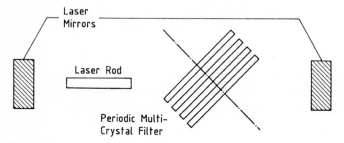

Fig. 5.63. Four-crystal wavelength selector

[5.160]. If one uses a stack of ten quartz plates, the resulting transmittance in the rejection band is about 15 %, which is certainly small enough to suppress unwanted laser frequencies.

Another technique for lowering the filter transmittance in the rejection band is to include more Brewster's angle surfaces in the laser. This reduces the value of T_0 for the filter at the cost of increasing the width of the central passband slightly.

Still another approach for narrowing the width of the central passband of the filter is to use several crystals in series whose thicknesses vary by integer ratios. One such filter, consisting of three crystals whose thicknesses are in the ratio 1:4:16, has been analzyed in [5.159]. The central passband of this filter is considerably narrower than those of the previously discussed designs. The disadvantage of this approach, however, is that numerous (unwanted) transmission spikes are now present, the largest of which has an amplitude of about 75 %.

It is obvious from these results that there are virtually an unlimited number of designs that can be tried, with corresponding tradeoffs in central passband width, stopband transmittance, presence of spikes in the stopband, and complexity of the birefringent filter. For all of these birefringent filter designs, tuning is continuous and easily implemented through rotation of the multiple crystals.

6. Optical Pump Systems

The pump source of a solid-state laser emits radiation in a spectral region that falls within the absorption bands of the lasing medium. Electrical energy, either continuous or pulsed is supplied to the pump source and converted into optical radiation. A highly reflective enclosure, or an optical system, couples the radiation from the pump source to the laser material. In this chapter we will discuss the salient features of the three components which together comprise the optical pump system of a solid-state laser: the pump source, the power supply, and the pump cavity or radiation transfer optics.

6.1 Pump Sources

In the application of light sources for pumping lasers, the primary objective is to convert electrical energy to radiation efficiently and to generate high-radiation fluxes in given spectral bands. The most efficient laser pump will produce maximum emission at wavelengths which excite fluorescence in the laser material, and produce minimal emission in all spectral regions outside of the useful absorption bands.

The various light sources which have been employed over the years to pump solid-state lasers are listed in Table 6.1. Today, only flashlamps, cw arc lamps, and laser diodes are of practical interest.

In the past, tungsten-halogen lamps have been a very popular source for many Nd:YAG lasers because of their low cost and simple power supply design. However, because of the low efficiency, this pump source has been replaced by cw arc lamps or laser diodes. In a vortex lamp the arc is stabilized by a gas jet which required a closed-cycle gas recirculation system. Several early Nd:YAG lasers have been pumped with this very powerful radiation source. Mercury arc lamps have been employed to pump small cw ruby lasers. Before the emergence of laser diodes, alkali-metal lamps, as well as the sun, have been considered as pump sources for space-based systems. LED's are the precursors to laser-diode pumping and several laboratory devices have been built with these sources. Chemical energy stored in photo flash bulbs has been utilized in the design of single shot Nd:YAG lasers. In classified programs, the use of radiation from explosive devices has been pursued as a pump source for solid-state lasers. The reader interested in details of the earlier pump sources is referred to previous editions of this book.

Table 6.1. Optical pump sources employed in solid-state lasers

```
Line                Line & blackbody           Blackbody
emitters            emitters                   sources
   |                   |                          |
Semiconductor        Arc lamps          Filament      Non-electric
diodes                                  lamps         sources
   |                   |                   |             |
Laser    LED      Noble    Metal    Tungsten-          Sun
                  gas      vapor    Halogen
   |                 |        |                          |
CW            Flashlamps   Mercury                    Flashbulb
   |                 |        |                          |
Quasi-CW      CW Arc       Alkali                     Explosives
              lamps        metal
                 |
              Vortex
```

Table 6.2. Brightness temperatures of typical pump sources

Filament lamps	2,400–3,400 K
Sun	5,800 K
Pulsed arc	5,000–15,000 K
CW arc	4,000–5,500 K

If we compare the spectral characteristics of the pump sources listed in Table 6.1, we find that on one side we have monochromatic pump sources, such as laser diodes, and at the other extreme black-body radiators, such as filament lamps. The discharge lamps operated at low current densities represent a compromise between a monochromatic source and a black-body radiator. Radiation from an arc discharge lamp is made up of both line and continuum components. The output spectra of arc lamps are complicated and difficult to describe in simple terms. Nevertheless, it is useful to speak of an effective brightness temperature to relate the radiant flux density at a given wavelength to that of an ideal black body at that temperature. Brightness temperatures for typical pump sources are given in Table 6.2, and spectral distributions of black-body radiators at temperatures which are achievable with typical pump sources are plotted in Fig. 6.1.

The dramatic improvement in laser efficiency which can be achieved in replacing a black-body pump source with a more efficient emitter can be illustrated in the following example: A 6 mm thick sample of Nd : YAG absorbs about 3% of the radiation from a black-body source at 2800 K. Therefore, a standard 1 kW tungsten-iodine filament lamp generates about 30 W of useful radiation for pumping a 6 mm diameter Nd : YAG laser rod. A krypton arc lamp has a radiation efficiency of 45% and a fractional utilization of the radiation by Nd : YAG of

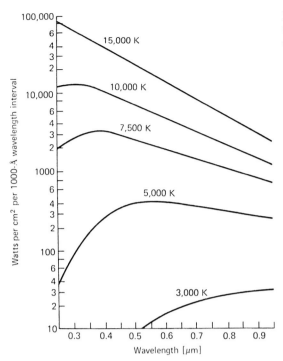

Fig. 6.1. Emission from black-body radiators at temperatures which are achievable with typical pump sources

20% for the same sample thickness as above. Therefore, 9% of electrical input is available as potential pump radiation. This reduces the required electrical input to 330 W for the above case. A cw laser diode array requiring 120 W of electrical input will generate 30 W output, all of which is within the absorption bands of Nd : YAG. A factor of eight improvement in efficiency, and in case of quasi-cw diodes which are more efficient compared to their cw counterparts, the improvement is over a factor of 10. A typical quasi cw array will produce 30 W average power at 808 nm at an electrical input of around 70 W.

This comparison addresses only the spectral characteristics of the various pump sources. Coherent pump sources, such as laser diodes, because of their spatial characteristics, permit concentration of pump radiation into the active medium with little losses. Whereas only a fraction of the omnidirectional radiation from a lamp pump source can be directed into the active medium. The result is an overall electrical efficiency for a tungsten lamp pumped laser of about 0.5%, which increases for arc lamp pumped systems to about (2–3)%, and diode-pumped lasers achieve around (10–15)%.

Pump sources for solid-state lasers range in size and power from tiny laser diodes at the 1 W level, pumping a microchip laser, to flashlamp systems employed in inertial confinement fusion which occupy large buildings and provide tens of megajoules of pump radiation. Table 6.3 lists some of the largest pump sources assembled for individual solid-state lasers. The largest solid-state laser built to date is the NOVA system at Lawrence Livermore National Laboratory.

Thousands of flashlamps pump Nd : glass slabs in 10 parallel beam lines. The most powerful commercial systems are cw arc lamp pumped Nd : YAG lasers built for industrial applications. Diode pump sources, comprised of many quasi-cw or cw 1-cm bars, at the multi-kilowatt level have been built at several facilities as part of R&D efforts. At the present time, the technology enhancement afforded by these powerful coherent pump sources does not outweigh the high cost for commercial applications.

Table 6.3. Summary of powerful pump sources utilized in solid-state lasers

Pump source	Number of individual emitters	Total pump energy/power	Laser output	Application	Laser
Flashlamps	5000	30 MJ	125 kJ (per pulse)	Laser fusion research	Nd : Glass
cw Arc lamps	16	140 kW	5 kW	industrial	Nd : YAG
Quasi cw diodes	1440*	14.4 kW (average)	3.6 kW (average)	R& D effort	Nd : YAG
cw diodes	220*	4.0 kW	1 kW	R& D effort	Nd : YAG

* 1-cm arrays

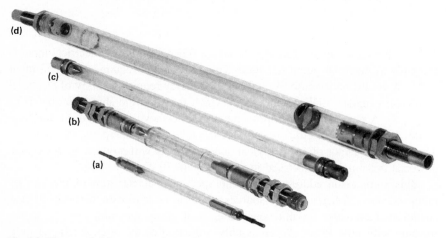

Fig. 6.2. Water-cooled krypton arc lamps for cw operation (**a**) lamp with tungsten rod seal (**b–d**) lamps and water jackets of solder seal lamps

Photographs of typical cw krypton arc lamps and xenon filled flashlamps are shown in Figs. 6.2 and 6.3. Packaging configurations of laser diode arrays are depicted in Figs. 6.17, 20, 25.

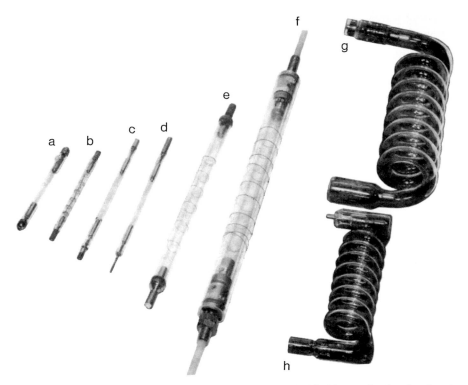

Fig. 6.3. Air-cooled (**a–c**) and water-cooled (**d–h**) linear and helical flashlamps of various length and bore diameter

6.1.1 Flashlamps

Lamp Design and Construction

Flashlamps used for laser pumping are essentially long arc devices designed so that the plasma completely fills the tube. A flashlamp consists of a linear or helical quartz tube, two electrodes which are sealed into the envelope, and a gas fill. Standard linear lamps have straight discharge tubes with wall thicknesses of 1 to 2 mm, bore diameters between 3 mm and 19 mm, and lengths from 5 cm to 1 m. Helical lamps, which have been used in the past, mainly in high-energy ruby and Nd : glass lasers, are constructed by wrapping heated quartz tubing around mandrels.

Flashlamps are typically filled with gas at a fill pressure of 300 to 700 torr at room temperature. Xenon is generally chosen as the gas fill for flashlamps because it yields a higher radiation output for a given electrical input energy than other gases. However, in a few special cases, such as small, low-energy lamps employed to pump Nd : YAG lasers, the low-atomic-weight gas krypton provides a better spectral match to the absorption bands of Nd : YAG.

The anode in flashlamps consists of pure tungsten, or often thoriated tungsten because the alloy is easier to machine. The cathode is comprised of a compressed pellet of porous tungsten impregnated with barium calcium aluminate. This pellet is attached onto a tungsten heat sink. The surface area of the tip must be large enough to handle the peak current while the shape of the tip positions the arc during the trigger pulse. During the lamp operation, barium is transported to the cathode surface where it forms a monolayer with a work function of about $2\,eV$ versus $4.5\,eV$ for pure tungsten. A lower work function improves electron emission for a given temperature. For the same reason, a cathode material with a low work function makes it easier to trigger the lamp. Since in a standard lamp the cathode is more emissive than the anode, flashlamps are polarized and will pass current in only one direction without damage.

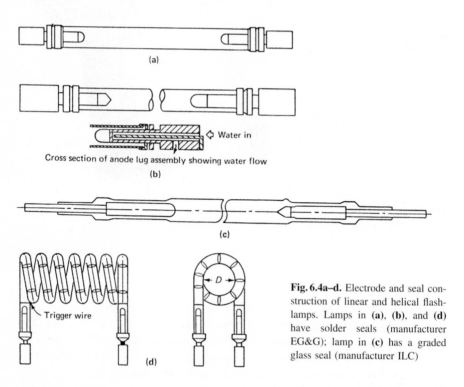

(a)

Cross section of anode lug assembly showing water flow

(b)

(c)

Trigger wire

(d)

Fig. 6.4a–d. Electrode and seal construction of linear and helical flashlamps. Lamps in **(a)**, **(b)**, and **(d)** have solder seals (manufacturer EG&G); lamp in **(c)** has a graded glass seal (manufacturer ILC)

There are two types of construction techniques in commercial use for sealing the electrode into the body of the flashlamp. One type, the tungsten-rod seal, uses an intermediate highly doped borsilica glass to seal the electrode to the fused quartz envelope. The glass balances the thermal stresses between fused silica, which has an extremely low coefficient of thermal expansion, and tungsten, which has a large coefficient. The second type, the solder seal, uses a copper rod, one end of which is brazed to an electrode and the other end welded to a nickel cup. The seal is made with a low-temperature indium solder between the copper-plated nickel cup and the platinum-coated end of the quartz envelope. The essential features of both types of lamps are shown in Fig. 6.4. The solder seal must be operated below the softening point of the solder, which is about 180° C. A unique feature of this type of quartz-to-metal seal is that the diameter of this seal joint can be as large as the quartz tubing itself. This permits a large cross section for the electrodes and, thus, an excellent path for dissipation of heat from the electrodes. In some cases the copper stem is water-cooled internally. The lamp is filled at room temperature through the hollow copper stem. When the filling is finished, the copper tube is crushed to seal in the gas. Thus, there is no "tip-off" on this lamp and little residual stress due to formation of the seal.

The tungsten-rod seal, on the other hand, can be operated at much higher temperatures. Since the heat removal through the small-diameter tungsten stem is not very large, the quartz tubing is shrunk into close proximity to the electrode to aid cooling of the electrode in radial directions. An advantage of the rod-seal construction is the fact that during evacuation the lamps can be baked out at 1000° C, and during the filling process the electrodes can be induction-heated to a bright red for their final outgassing. The lamps are then backfilled to the appropriate gas pressure and tipped off. The tungsten-rod seal permits the fabrication of high-purity contaminant-free lamps having long shelf lives.

The envelope and the rod-seal area of flashtubes are normally cooled by free or forced air, by pressurized nitrogen, or by a liquid such as water, water-alcohol mixtures, or a fluorinated hydrocarbon. Liquid cooling permits operation of the lamps at a maximum inner-tube wall surface loading of $300\,\mathrm{W/cm^2}$ of average power. Free-air convection cooling is limited to handling about $5\,\mathrm{W/cm^2}$ of dissipation: forcing air across the flashtube envelope enables one to dissipate up to $40\,\mathrm{W/cm^2}$.

Liquid-cooled linear flashlamps are available with outer quartz jackets which permit cooling of the lamps with a highly turbulent flow. The water jacket, which is attached to the flashlamp by means of O-ring seals, is reusable after the lamp has reached its end-of-life.

Optical Characteristics

The radiation output of a gas discharge lamp is composed of several different components, each corresponding to a different light-emission mechanism. The relative importance of each of these mechanisms depends strongly on the power density in the lamp, and so the low-power and the high-power optical output spectra are markedly different. The total radiation is made up of both line and continuum components. The line radiation corresponds to discrete transitions between the bound energy states of the gas atoms and ions (bound-bound transitions). The continuum is made up primarily of recombination radiation from gas ions capturing electrons into bound states (free-bound transitions) and of bremsstrahlung radiation from electrons accelerated during collisions with ions (free-free transitions). The spectral distribution of the emitted light depends in complex ways on electron and ion densities and temperatures.

In the continuous-power, wall-stabilized noble gas arc, current densities are such that there is a high number of bound-bound transitions and therefore the

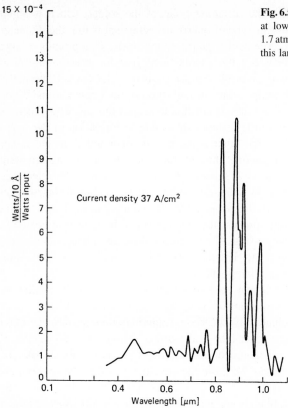

Fig. 6.5. Spectral emission from xenon at low current densities. Fill pressure 1.7 atm of Xe. Color temperature for this lamp is 5200 K [6.1]

radiation spectrum is characteristic of the fill gas and is broadened by increases in pressure.

In high-current-density, pulsed laser applications, the spectral output of the lamp is dominated by continuum radiation, and the line structure is seen as a relatively minor element. Between these two cases, a pulsed-power region can exist where the pulsed power level is such that discrete line radiation is still emitted and is superimposed on a strong background continuum. Spectral distributions over the range 0.35 to 1.1 μm were measured by Goncz and Newell [6.1] for representative pulsed and continuous xenon arc lamps. Figure 6.5 shows the output spectrum for a continuous xenon arc lamp. As can be seen, this lamp produces a very strong line structure in the infrared. In Fig. 6.6 the spectral emission of a xenon flashtube is plotted for two current densities. As a result of the high current densities, the line structure is in this case masked by a strong continuum. From Fig. 6.6 it also follows that a high current density shifts the spectral output toward the shorter wavelengths.

As we discussed at the beginning of this chapter, one can relate the radiation characteristics of a flashlamp $R(\lambda, T)$ to the characteristics of a black-body. The departure of any practical light source from the black-body characteristic is accounted for in the emissivity, $\varepsilon(\lambda, T)$, which varies between zero and one, and

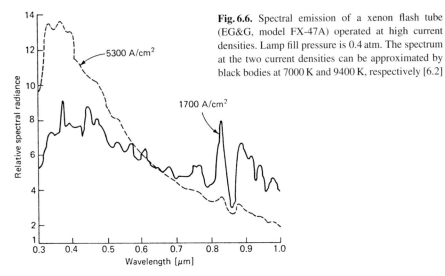

Fig. 6.6. Spectral emission of a xenon flash tube (EG&G, model FX-47A) operated at high current densities. Lamp fill pressure is 0.4 atm. The spectrum at the two current densities can be approximated by black bodies at 7000 K and 9400 K, respectively [6.2]

is both wavelength and temperature dependent; thus $R(\lambda, T) = \varepsilon(\lambda, T)B(\lambda, T)$. In a flashlamp, $\varepsilon(\lambda, T)$ depends strongly on temperature and wavelength. If we assume local thermodynamic equilibrium in the plasma, the black-body characteristic $B(\lambda, T)$ becomes the black-body curve corresponding to the electron temperature in the plasma. In general, the emissivity of the flashlamp radiation at longer wavelengths is greater than the emissivity at shorter wavelengths. An increase in power density will result in a large increase in emissivity (and radiation) at short wavelengths, but only small changes at longer wavelengths where the emissivity is already close to unity.

The transmission properties of the flashlamp envelope will influence the spectral output. Usually the envelope is fused silica, which transmits light between 0.2 and 4 μm. Figure 6.7 shows the uv transmission of doped and undoped fused silica. Germisil and Heliosil, manufactured from fused silica containing a slight admixture of titanium, are sometimes employed as lamp or flow-tube envelope material to prevent the production of ozone in the air or to prevent breakdown of the flashlamp coolant, or to inhibit the formation of color centers in certain solid-state laser materials.

Lamps with cerium-doped quartz envelopes have also been designed. Quartz doped with cerium will absorb radiation below 0.31 μm and will fluoresce at wavelengths between 0.4 and 0.65 μm. Spectral data from these lamps show that a 50-mm × 4-mm bore lamp with 450 torr of pressure operated at 10 J has a 15 % greater output over the spectral range of 0.4 to 0.65 μm compared to a lamp with an undoped envelope.

In the design of pumping geometries for high-power laser systems, it is important to know how opaque the flashtube is to its own radiation. The optical transmission of a xenon flashtube has been measured at wavelengths from 0.25 to 1.0 μm and at currents up to 400 A/cm² [6.3, 4]. From the experimental data

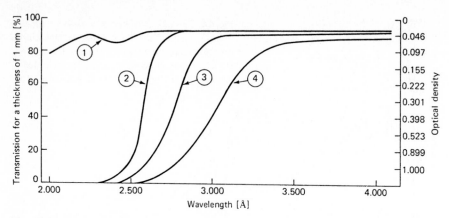

Fig. 6.7. Ultraviolet transmission (for 1-mm thickness) of different types of fused quartz. (*1*) Pursil, (*2*) Germisil, (*3*) Heliosil I, (*4*) Heliosil II

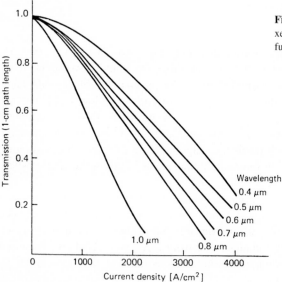

Fig. 6.8. Spectral transmission of the xenon plasma in a flashlamp as a function of current density [6.4]

plotted in Fig. 6.8 it follows that the absorption increases with current and with wavelength. Above about 0.5 μm and a current of 4000 A/cm^2, a discharge tube 1 cm in diameter is nearly opaque. At shorter wavelengths or lower currents, the discharge is fairly transparent. From Fig. 6.8 it also follows that the absorption coefficient varies linearly with the square of the current density [6.5]. Thus in a large-bore flashlamp driven at high current densities, a large percentage of the radiation emanating from the plasma core is absorbed before it reaches the lamp surface. Hence the bulk of the radiation comes from a thin sheath near the lamp surface with a corresponding reduction in lamp efficiency.

Xenon lamps are generally used for high-power pulse applications. The lamp is a relatively efficient device as it converts 40 to 60 % of the electrical input energy into radiation in the 0.2- to 1.0-μm regions [6.1, 6]. Krypton has a radiation efficiency of 25 to 30 %. As noted in Chap. 2, principal pump bands of Nd : YAG are located from 0.73 to 0.76 μm and 0.79 to 0.82 μm. The xenon spectrum has no major line radiation in these bands, so pumping is primarily due to continuum radiation. However, some strong krypton lines fall within the Nd : YAG pump bands. Investigations have shown that krypton flashlamps are more efficient than xenon lamps at low power densities for pumping Nd : YAG lasers [6.6–9]. There are indications that a crossover occurs in the pumping efficiency of krypton and xenon at high peak power densities, corresponding to that power density at which the xenon continuum becomes more efficient than the krypton line structure. This crossover occurs at a power density of approximately 2×10^5 W/cm^3.

In general, the pumping efficiency of Nd : YAG of both xenon and krypton increases as the pressure is increased. In the range 450 to 3000 torr no maxima are found [6.10]. The limitation to further increases appears to be the fact that high-pressure lamps are difficult to trigger. The effect of an increase in gas fill pressure is to reduce the mean free path of the electrons and atoms in the discharge, and thus to increase their collision frequency. This leads to the production of more excited species in the discharge and the emission of more useful line radiation.

Electrical Characteristics

The impedance characteristics of a flashlamp determine the efficiency with which energy is transferred from the capacitor bank to the lamp. The impedance of the arc is a function of time and current density. Most triggering systems initiate the arc as a thin streamer which grows in diameter until it fills the whole tube. The expansion period is fast, of the order of 5 to 50 μs for tubes of bore diameter up to 1.3 cm. During the growth of the arc, lamp resistance is a decreasing function of time (Fig. 6.9). The decreasing resistance arises in part from the increasing ionization of the gas and from the radial expansion of the plasma. After the arc stabilizes, the voltage-current relationship is described by [6.1, 12, 13]

$$V = K_0 i^{1/2} , \tag{6.1}$$

where $K_0 = kl/d$ and describes the impedance characteristics of the particular lamp. The constant k is dependent only on such parameters as gas type and gas pressure, l and d are the length and diameter of the flashlamp bore, respectively. K_0 or k is essentially the only parameter needed to describe the high-current electrical characteristics of a given flashlamp. This parameter is usually supplied by the manufacturers of the flashlamps. K_0 is found simply by flashing the lamp at some reasonable loading, while measuring the voltage and current. The resistance of the flashlamp for the high-current regime with the plasma filling the bore of the tube is obtained from

$$V \cong R_L(i)i = \left(\frac{\varrho(i)l}{A} \right) i , \tag{6.2}$$

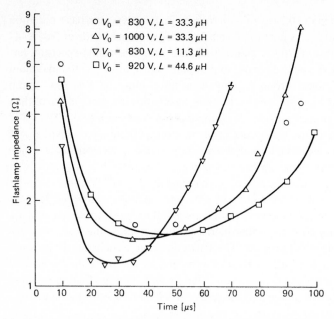

Fig. 6.9. Time dependence of flashlamp impedance for several input energies and series inductance values. The 60-mm-long and 3-mm-diameter flashlamp was filled with 450 torr of xenon. The discharge circuit consisted of a 19-μF capacitor and one inductor which was varied from 11.3 to 44.6 μH [6.10]

where $\varrho(i)$ is the specific resistivity of the xenon or krypton plasma, and is a function of the current density. A is the cross-sectional area of the lamp. In (6.2) the voltage drop at the electrodes, which is of the order of 20 V, has been neglected. In operating regions of interest, the flashlamp resistivity for xenon at 450 torr is related to current density j by [6.12]

$$\varrho(i) = \frac{1.13}{j^{1/2}} \ . \tag{6.3}$$

This relationship is valid for current densities of approximately 400 to 10,000 A/cm^2 [6.14]. The lamp resistance R_L is obtained by introducing (6.3) into (6.2)

$$R_L(i) = 1.27 \left(\frac{l}{d}\right) i^{-1/2} \ . \tag{6.4}$$

Comparing (6.2) with (6.1), we obtain an expression for the flashlamp parameter K_0

$$K_0 = 1.27 \frac{l}{d} \ . \tag{6.5}$$

The value of 1.27 holds for 450-torr xenon lamps. It has been found from many flashlamp experiments that (6.5) can be generalized for other pressures and expressed as

$$K_0 = 1.27 \left(\frac{P}{450} \right)^{0.2} \frac{l}{d} , \qquad (6.6)$$

where P is the flashlamp pressure in torr. The parameter K_0 obtained for krypton has been found to be about the same as that for xenon.

Tube Failure

The flashtube end-of-life can occur in either of two modes, and we can distinguish between catastrophic and nondestructive failures. The factors which contribute to catastrophic failure are explosion of the walls due to the shock wave in the gas when the lamp is fired, or overheating of the lamp envelope or seal and subsequent breakage due to excessive thermal loading of the lamp. The first failure mode is a function of pulse energy and pulse duration, whereas the second catastrophic-failure mode is determined by the average power dissipated in the lamp. When the flashtube is operated well below the rated maximum pulse energy and average power, the lamp usually does not fail abruptly. Rather, the tube will continue to flash with a gradual decrease in light output, which will eventually fall below a level necessary for the particular application. In the latter mode of failure, the reduced light output is caused by the erosion of the flashtube electrodes and of the quartz walls and by the gradual buildup of light-absorbing deposits within the flashtube envelope.

Flashtube Explosion

When a high-voltage trigger pulse is applied between the electrodes of a flashtube, the gas breaks down generally near the axis of the tube and a conducting filament is established. As energy is released into the channel, heating of the ambient gas causes the filament to expand radially, forming cylindrical shock waves. The shock front and its associated plasma travel radially from the tube axis to its wall. The radial velocity of the plasma discharge and the shock amplitude are proportional to the input energy. The velocity was measured to vary from about 90 to 900 m/s for inputs from 60 to 600 J into a 5-cm-long flashlamp with 6-mm bore diameter [6.11]. The cylindrical shock wave and the associated plasma heat cause stress on the inside of the tube wall which is axial in tension toward the electrodes. If the energy discharged exceeds the explosion limit for the lamp, the shock wave will be sufficiently intense to rupture the lamp walls and consequently destroy the lamp.

If the lamp is operated somewhat below the explosion point, catastrophic failure of the lamp is preceded by the formation of hairline cracks and a milky deposit. The microcracks, originating around minute flaws in the quartz envelope, will expand with each flash, eventually resulting in lamp failure. As a result of the high temperature of the expanding plasma, a steep temperature gradient is created near the tube surface when the plasma impinges upon the tube wall. Therefore, in addition to the progressive appearance of cracking and crazing of the inside wall, there is an associated silica deposit from evaporation and deposition [6.15].

Dugdale et al. [6.16] have investigated the effect of pulse heating on glass and ceramic surfaces. If the energy Q_0 per unit area is delivered to a material surface in time t_p, a rise in surface temperature ΔT takes place, given by

$$\Delta T = a Q_0 (\pi K c_p \varrho t_p)^{-1/2} , \tag{6.7}$$

where K is the thermal conductivity, c_p is the specific heat, and ϱ is the density of the material; a is a constant between 1 and 2 depending on the heat pulse waveform ($a = 1.08$ for an exponential waveform). We can relate the pulse energy per unit area to the total flashlamp electrical input energy E_0 by writing

$$Q_0 = \frac{k_1 E_0}{\pi d l} , \tag{6.8}$$

where the parameter k_1 depends on the type of gas and fill pressure used. During the heat pulse, the temperature rise is confined to a depth Δx below the surface, given by

$$\Delta x = 2 \left(\frac{K t_p \varrho}{c_p} \right)^{1/2} . \tag{6.9}$$

This depth lies in the range 10^{-4} to 10^{-2} cm for glasses and ceramics subjected to heat pulses of duration 10^{-6} to 10^{-3} s. The restraint imposed by the unheated substrate may cause stresses σ_T to develop in the hot surface layer of magnitudes up to

$$\sigma_T = \frac{\varepsilon \alpha \Delta T}{1 - \mu} , \tag{6.10}$$

where α is the thermal expansion of the material, μ is Poisson's ratio, and ε is the Young's modulus of elasticity. The temperature gradient ΔT which will cause thermal shock damage of the envelope is obtained by equating σ_T with the rupture stress σ in (6.10). From (6.7, 8, 10) follows the lamp input pulse energy which will generate stresses that exceed those of the wall material. Therefore, the explosion energy E_{ex} of thin-walled flashtubes, defined as the minimum input energy sufficient to crack the lamp catastrophically, is

$$E_{ex} = k_2 L d t_p^{1/2} , \tag{6.11}$$

where k_2 depends on the type of gas and fill pressure as well as on the physical and thermal properties of the lamp envelope. The explosion energy is directly related to the inside-wall surface area of the lamp, and to the square root of the pulse duration. If L and d are measured in centimeters and t_p is in seconds, and if a critically damped single-mesh discharge circuit is assumed, then it has been found empirically that for xenon-filled quartz flashtubes [6.2, 17]

$$E_{ex} = (1.2 \times 10^4) L d t_p^{1/2} . \tag{6.12}$$

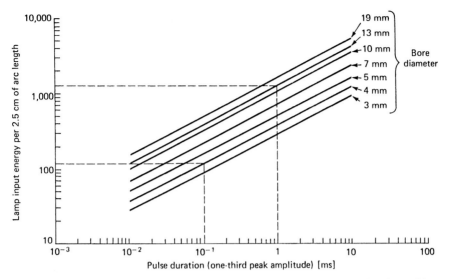

Fig. 6.10. Explosion energy of linear flashlamps (J/2.5 cm) as a function of pulse duration and bore diameter. The wall thickness of the lamps is 1 mm, except for the 19-mm-bore lamp which has a wall thickness of 1.5 mm [6.19]

This relation is useful for lamps with $0.5 \leq d \leq 1.5$ cm and xenon fill pressures of 300 to 450 torr, where t_p is defined as the time between the one-third peak light output points for a damped sinusoidal waveform and the one-half light output points for a quasi-rectangular waveform [6.18]. E_{ex}, as calculated above, is a "free air" value which assumes that no radiation which leaves the lamp is reflected back to be reabsorbed. In an actual laser pumping cavity a certain fraction of the radiated energy – depending on the reflectivity of the cavity, its diameter, and its focal properties – is returned to the flashtube. This markedly increases the erosion and wear accompanying each discharge. In fact, the flashtube often acts as if it were being loaded with as much as 30 % more energy than is actually being dissipated in the tube. Single-pulse explosion energy as a function of pulse duration is shown in Fig. 6.10 for quartz-envelope xenon flashlamps of different bore diameters [6.19]. For a flashlamp with a specific bore size and arc length, the explosion point is a function of lamp input energy and pulse duration.

Having determined the ultimate limit for the tube, the question of the life that the tube will have in a given application has to be addressed. It has been shown that lamp life can be related to the fraction of explosion energy at which the lamp is operated. The life in flashes for a xenon-filled quartz-envelope flashlamp as a function of the single-shot explosion energy is empirically given by [6.2, 20]

$$N \cong \left(\frac{E_{ex}}{E_0}\right)^{8.5}. \tag{6.13}$$

The extremely strong dependence of flashlamp lifetime on the total input energy is a great incentive to underrate the lamp.

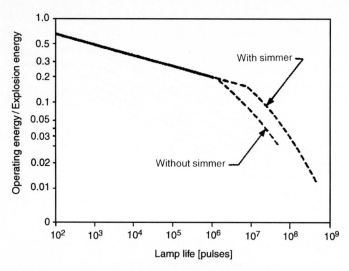

Fig. 6.11. Lamp life as a function of operating energy [6.20]

In Fig. 6.11, life expectancy as a function of lamp input is shown assuming that the interpulse interval is sufficiently long to permit the tube to cool between pulses [6.20]. The lifetime expressed by (6.13) is defined as the number of shots at which the light output is reduced by 50% of initial value. The life expectancy of a flashlamp begins to deteriorate rapidly when peak current exceeds 60% of explosion current. Above 60%, the lamp's life is usually limited to less than 100 discharges, even if all other parameters are optimized. At the 80% level, explosion can occur on any discharge. Most flashlamps, with the exception of lamps installed in single-shot lasers, such as employed in inertial confinement fusion, rangefinders and lasers for research, are operated at only a small fraction of the explosion energy. In this case, cathode degradation is the primary failure mode. The two dotted lines in Fig. 6.11 indicate where cathode end of life begins to occur. As explained below, lamplife is considerably improved by operation in the simmer mode which keeps the lamp continuously ionized at a low current level. Because of the increased lamplife most lamps are operated in the simmer mode today.

Excessive Thermal Wall Loading

Cooling of the end seals and the lamp envelope has to be adequate. The amount of average power a flashlamp can handle is determined by the quartz wall area available to dissipate the heat. The envelope of a flashlamp is under stress which is caused by the temperature and pressure differences between the inner and outer walls, and by stresses caused by the acoustical shock wave. The combined stress permissible at the envelope determines the power dissipation per unit area. The total electrical input power of a gas discharge lamp is divided between the heat dissipated by the electrodes, the quartz walls, and the emitted radiation. Typically 30 to 50 % of the lamp input power is dissipated as heat by the quartz envelope [6.21]. Under high-average-power operating conditions the inner wall is close to the melting temperature of quartz, whereas the liquid-cooled outer wall surface is close to room temperature. As will be discussed in Sect. 6.1.2, the maximum heat which can be dissipated through the liquid-cooled quartz envelope of a lamp is $(300\text{--}400)\,\text{W/cm}^2$ depending on the wall thickness.

In flashlamps a temperature rise from 300 to 12,000 K is experienced during the pulse, which will produce a 36-fold increase in gas pressure. As a result of the additional stresses caused by the pressure increase and the shock wave, flashlamps are operated considerably below the thermal limit. Conservative numbers assuring good life expectancy are 50 to $100\,\text{W/cm}^2$ for water-cooled lamps.

Nondestructive Failure

Operated well below the explosion limit, flashlamp lifetime is limited by electrode erosion with the subsequent formation of opaque deposits on the lamp jacket near the lamp electrodes or by outgassing or tube leakage. Of these, cathode sputtering caused by the bombardment of the xenon ions onto the negative lamp electrode is the most significant. The emitted hot-cathode material deposits on the relatively cool lamp walls and effectively attenuates the amount of radiation released by the plasma. It was observed that the optimum cathode temperature is the most critical parameter in reducing wall deposits. If the temperature is too low, tungsten will sputter onto the walls which will cause darkening. If the temperature is too high, barium will evaporate and deposit onto the envelope.

Tube blackening which occurs in a new tube after a few shots is a sign of electrode impurity. The greatest hazard to a lamp during its final stages of processing is the introduction of water vapor from the atmosphere [6.22]. If the water vapor is not entirely removed from the lamp, tungsten will react readily with water vapor traces to form WO_2 and atomic hydrogen. The tungsten oxide vaporizes from the electrode and condenses on the cooler parts of the envelope. This reaction often accounts for premature blackening of the envelope walls. Besides darkening of the envelope, the useful life of a flashlamp is limited by erratic triggering and the eventual inability to reignite; this indicates that a significant outgassing or tube leakage has occurred.

In all these cases where the input energy is a small fraction of the explosion energy, the above-listed secondary effects dominate and the life of a lamp is largely determined by the lamp process control and the precautions taken during manufacture. For lamps which are operated in this regime, the useful lamp life, rather than being determind by a mechanical failure, is determined by the reduction of lamp radiation below a useful level. The term "lamp life" is used loosely, since no definite standards exist with regard to the determination of the end of lamp usefulness. Lamp manufacturers usually state lamp life as the number of shots after which lamp intensity drops to 50 % of its initial value. The life expectancy is usually stated for a particular set of operating parameters and for a lamp operated in free air. In laser applications, lamp life is usually defined to end when the laser output has degraded a certain amount, for example 10 %, or when the lamp fails to trigger at a specified voltage.

Continued improvements in flashlamp design and fabrication techniques have resulted in ever increasing flashlamp life. Careful attention to the operational aspects of flashlamps have also contributed substantially to longer lifetimes. As illustrated in Fig. 6.11 lamp life is greatly increased by maintaining a low current discharge in the lamp between pulses. This simmer mode minimizes cathode sputtering, and arc formation is more controlled and associated with a smaller shockwave as compared to a high voltage trigger pulse.

About 20 years ago, nominal lifetime for flashlamps employed in industrial lasers was on the order of 10 to 20 million shots. Optimization of the tungsten matrix, improvements in the design of the geometry of the cathode and the sur-rounding quartz envelope has extended flashlamp lifetime into the 10^8 region. Today state of the art flashlamps can operate up to several hundred million shots. Under a special lamp improvement program which included refinement of the cathode tungsten matrix, careful cathode temperature control, ultra pure quartz envelopes and backing the anode in high vacuum and at a high temperature, life-times of up to one billion shots were observed [6.23]. Reliability and performance data of large xenon filled flashtubes are summarized in [6.24].

Examples of Typical Operating Conditions

In selecting a particular lamp the following steps are usually followed: (1) A flashlamp arc length and bore diameter is chosen which matches the rod size. (2) The explosion energy for the lamp chosen is calculated (Fig. 6.10). (3) An input energy is selected which is consistent with the desired lamp life (Fig. 6.11). If the lamp input required to achieve a certain laser output is not consistent with a desired minimum lamp life, the application of a multiple-lamp pump cavity has to be considered. (4) The average power loading of the lamp is calculated and compared with the limit specified by the lamp manufacturer. (5) In designing the pulse-forming network, care must be taken to avoid fast current rise times and/or reverse currents. To obtain the expected life, manufacturers recommend that the rise time of the flashlamp current be greater than $120 \mu s$ for lamps discharging more than a few hundred joules. Both electrode erosion and wall vaporization are accelerated by fast-rising peak currents.

First we will consider a typical small linear flashlamp used in military equip-ment such as target designators or laser illuminators. These systems operate typically at a flashlamp input of 10 J with pulse repetition rates of 20 pps. The lasers typically use 63-mm-long Nd:YAG laser rods of 6.3 mm diameter and have a Q-switched output of 100 to 200 mJ. A 4-mm-diameter and 60-mm-long lamp operated at 10 J input and a pulse width of $300 \mu s$ is operated at only 2 % of the explosion energy. The wall loading of the liquid-cooled lamp operated at 10 pps is 15 W/cm². From the extensive data which exist for this type of lamp operated in a closed coupled laser cavity, it follows that the useful lamp life

(10 % drop in laser performance) is in excess of 10^7 shots. An example of a large lamp employed to pump Nd : glass laser rods is the FX-67B-6.5 made by EG & G. This lamp is 16.5 cm long and has a bore diameter of 13 mm. The lamp has an explosion point of 8450 J for a 1-ms-long pulse. The maximum long-term average power input for the water-cooled lamp is specified by the lamp manufacturer at 10 kW. This corresponds to an effective thermal wall loading of about 60 W/cm². If the lamp is operated at an input energy of 2000 J, a flashlamp life of 150×10^3 shots can be expected before the light output is decreased to 50 % of its initial value.

For a laser operated essentially on a single-shot basis in a research environment, this is not an unreasonable lamplife.

As already mentioned, extremely long lifetimes can be obtained with carefully selected operating parameters and lamp design [6.23]. In this context it is interesting to review the operating parameters of the lamp employed in these life tests. The lamp filled with 760 Torr of xenon had a 6 mm bore and a length of 10 cm. One group of lamps was operated at 15 J input in a 180 μs long pulse. The other lamps were fired at 45 J per pulse at the same pulselength. From Fig. 6.10 follows an explosion energy of 1200 J for this lamp and pulselength. That means the lamps were operated at only 1.3% or 3.8% of the explosion energy. From the lamps tested at the 1.5 J level, most reached half a billion shots and some exceeded one billion. At the 45 J input level most lamps failed between 10^8 and a half billion shots. Additional literature on flashlamp performance can be found in [6.20, 25–27].

6.1.2 Continuous Arc Lamps

For continuous pumping of Nd : YAG lasers, the use of halogen cycle tungsten-filament lamps, noble gas arc lamps, and alkali vapor lamps has been explored. Since the discovery of the Nd : YAG laser in 1964, there has been a continual effort to find more effective means of pumping this material. One of the early devices used for pumping of the Nd : YAG laser was the tungsten-halogen filament lamp. The pump lamps considered next were the inert gases, which are rich in line structure in the near-infrared. Xenon has the highest overall conversion efficiency and is commonly used in arc lamps. However, the infrared line spectra of xenon misses all of the Nd : YAG pump bands, which are located at 0.73 to 0.76, 0.79 to 0.82, 0.86 to 0.89, and 0.57 to 0.60 μm. It has been observed [6.21, 28, 29] that the line spectrum from krypton is a better match to Nd : YAG than the line spectrum of xenon, since two of its strongest emission lines (7600 and 8110 Å) are strongly absorbed by the laser crystal (Fig. 6.12).

The laser output from a krypton-filled lamp is about twice that obtained from a xenon arc lamp operated at the same input power. Krypton arc pumped Nd : YAG lasers are currently the highest continuous power solid-state lasers.

Mechanical Design. Continuous arc lamps are similar in design to linear flashlamps, with the exception that the cathode has a pointed tip for arc stability (see

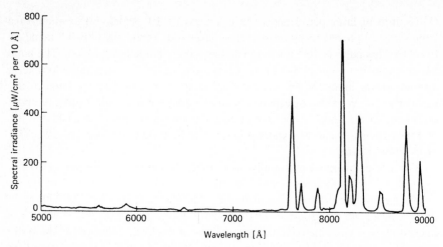

Fig. 6.12. Emission spectrum of a typical cw-pumped krypton arc lamp (6 mm bore, 50 mm arc length, 4 atm fill pressure, 1.3 kW input power). (ILC Bulletin 3533)

Fig. 6.4b,c). In the type of lamp employing a soft solder seal, additional cooling is provided to the tungsten anode cap by means of a water channel that forces some of the cooling flow through the copper cylinder and into close proximity to the tungsten. In the tungsten rod seal lamps the entire heat must be dissipated through the quartz envelope. In order to reduce thermal stresses in the walls, these lamps are often fabricated from 0.5-mm-thick tubing rather than the standard 1-mm thickness.

Cooling of the lamps is accomplished by circulating water in a flow tube surrounding the quartz envelope of the lamps. Often the flow tube will be made out of Germisil to absorb the ultraviolet content of the lamps. The design of the electrode holders and the flow tube must be such that localized boiling of the water is avoided.

Thermal Considerations. The maximum input power of a cw-pumped arc lamp is determined by the permissible stresses in the quartz envelope. The stresses are caused by the temperature gradient between the inner surface of the wall and the outer surface, and by the internal gas pressure during operation of the lamp. Since thermal considerations are of utmost importance in the operation of cw lamps, we will take a closer look at the mechanisms that generate heat at the electrodes and the lamp envelope.

At both electrodes, heat is generated by bombardment from either electrons or ions, conduction of heat from the plasma, and absorption of radiation. The kinetic energy of electrons incident on the anode is very high compared to that of ions impinging on the cathode. Therefore, the anode heats up much more than the cathode. The anode also conducts more heat from the plasma, since the area of the anode that is in contact with the plasma is larger than in the case of the pointed cathode. As far as absorption of radiation is concerned, both electrodes will absorb about the same amount of power. Polished tungsten has a reflectivity of about 60 % in the visible region. Due to emission of electrons from the cathode surface, this electrode is cooled to a certain extent. The heat removed by this process is given by

$$P = \frac{\Delta E I}{e} ,$$ (6.14)

where ΔE is the work function, I is the current, and e is the electron charge. With $\Delta E = 2\,\text{eV}$, $I = 30\,\text{A}$, and $e = 1.6 \times 10^{-19}\,\text{A}$, the heat removed from the cathode is 60 W in a typical lamp.

The total electrical input power of a gas discharge lamp is divided among the heat dissipated by the electrodes, the quartz walls, and the emitted radiation. Generally, it can be assumed that 10 to 20 % of the electrical input power is dissipated as heat through the electrodes and 30 to 50 % is dissipated through the envelope [6.30]. The thermal loading of the tube envelope is caused by the loss of kinetic energy of electrons, ions, and molecules by collision with the wall of the tube. The thermal gradient in the envelope is

$$\Delta T = \frac{Qd}{K} ,$$ (6.15)

where Q is the dissipated power density, d is the wall thickness, and K is the thermal conductivity. Allowing a maximum temperature difference of $\Delta T = 1800°\,\text{C}$ for fused quartz between the inner and outer wall surface, then with $K = 0.017\,\text{W/°C\,cm}$ follows for the maximum power which can be dissipated through the quartz wall a value of $Q = 300\,\text{W/cm}^2$ for a 1 mm thick envelope. In thinner walls thermal loadings as high as $400\,\text{W/cm}^2$ are achieved. Absorption of radiation in a quartz envelope is negligible. However, vaporization of electrode material leads to a gradual darkening of the lamp envelope. The deposits absorb radiation, hence, the wall-temperature gradient increases during the life of the lamps. The pressure of the fill gas when hot increases by up to a factor of 10. The operating pressure depends on the cold fill pressure of the lamp, the temperature of the plasma, and the volume of dead space behind the electrodes. Thermal and pressure stresses occurring in arc lamps can be calculated using equations given by Thouret [6.31]. Calculations carried out on cw arc lamps [6.24] have shown that the inner wall is under compression whereas the outer wall is under tension and the total stress is largest on the outside of the quartz envelope. The rupture stress of quartz is $500\,\text{kg/cm}^2$. Most lamps in commercial applications are operated at about 20 to 30 % of the ultimate tensile strength of quartz.

Electrical Characteristics. Continuous arc lamps are ignited just like flashlamps. For example, a series injection trigger transformer supplying a 30 kV spike will lower the lamp impedance to a level that is within the range of a 300-V dc supply. The operating parameters and impedance depend on current density, arc length, bore size, and fill pressure. Typical krypton arc lamps are operated between 80 and 150 V and 20 to 50 A. Characteristic impedance values are between 2 and

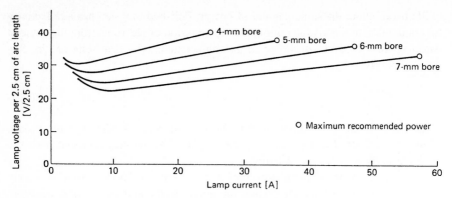

Fig. 6.13. Voltage versus current for different krypton arc lamps. (ILC Bulletin 3533)

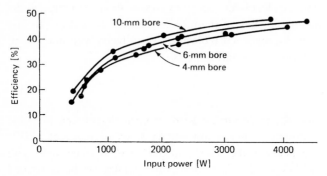

Fig. 6.14. Krypton arc lamp radiative efficiency as a function of input power; 4-, 6-, and 10-mm bore, 75-mm arc length, 4 atm of Kr fill pressure [6.30]

$10 \, \Omega$. Figure 6.13 shows typical current-voltage curves for several arc lamps filled at 4 atm of krypton. The discharge voltage and dynamic resistance increases with the increase in gas pressure.

Optical Characteristics. At current densities obtainable in standard krypton arc lamps the plasma is optically thin. Calorimetric measurements of the total optical output power in the spectral region 0.3 to 1.2 μm reveal a radiation efficiency of about 40 % for most Kr arc lamps. Figure 6.14 shows the dependence of the radiation efficiency on lamp input power and bore diameter. It has been found that the useful light output of krypton arc lamps for pumping Nd:YAG lasers increases with bore size, fill pressure, and input power. For example, the conversion efficiency of a lamp filled with 8 atm of Kr is about 1.2 to 1.5 times as high as that of a lamp filled at 2 atm, depending on the electrical input [6.32]. A summary of the spectral data of a representative Kr arc lamp is given in Table 6.4. Note the large fraction of radiation between 0.7 and 0.9 μm.

Combining the absorption spectra of Nd:YAG with the emission spectra of the krypton lamp, the spectral utilization as a function of sample thickness

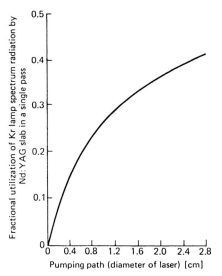

Fractional utilization of Kr lamp spectrum radiation by Nd:YAG slab in a single pass

Pumping path (diameter of laser) [cm]

Fig. 6.15. Fractional utilization of krypton lamp output by Nd : YAG

Table 6.4. Spectral data for cw krypton arc lamps. Data are typical for lamps having a 6–13 mm bore, 7.5–25 cm arc length, 2–3 atm fill pressure, operated at 6–16 kW

Quantity	Definition	Numerical Data
Radiation efficiency	Radiation output/ electrical input	0.45
Spectral output	Fraction of radiation in spectral lines	0.40
	Fraction of radiation in continuum	0.60
Spectral power distribution	Fraction of total radiation below $0.7\,\mu$m	0.10
	Fraction of total radiation between 0.7 and $0.9\,\mu$m	0.60
	Fraction of total radiation between 0.9 and $1.4\,\mu$m	0.30

has been calculated. The result is shown in Fig. 6.15. The curve illustrates rather dramatically the kind of improvement one can achieve by increasing the diameter of a Nd : YAG laser rod in a pumping cavity.

Lamp Life and Operating Characteristics of Typical CW Krypton Arc Lamps. The main degradation mechanism in these lamps is the accumulation of wall deposits. The deposits arise from evaporation and sputtering from the electrodes and from the residual gaseous and high-vapor-pressure impurities left in the lamp after fabrication. The black deposit that builds up eventually completely attenuates the radiative output of the lamp. The underlying quartz becomes overheated, and catastrophic failure of the lamp occurs from excessive thermal stresses developing in the envelope walls. Black anode wall deposits are greatly reduced by the use of an internally water-cooled anode structure.

Krypton arc lamps are available from 5 to 20 cm arc length, with bore diameters from 3 to 10 mm, and are designed to handle electrical input powers of up to 15 kW [6.24]. Design parameters and performance data of typical lamps used in commercial lasers are summarized in Table 6.5. Some of the lamps listed are operated far below the manufacturer's maximum rating in order to obtain good lamp life. Typical overall efficiencies obtainable in krypton-pumped Nd:YAG lasers are between 2 and 3%. For example, at output levels of 100 and 250 W, 2.9% and 2.1% efficiencies were achieved with krypton lamps filled to 4 atm [6.29, 33]. Efficiencies of 3.3% have been attained at input power levels of 3 kW with a 6-mm-bore and 50-mm-long lamp filled at 8 atm of krypton [6.30].

Table 6.5. Typical operating parameters of cw pumped krypton arc lamps

Model number	FK-125-C2.75 EG&G	FK-111-C3 EG&G	5Kr2 ILC
Arc length	70 mm	75 mm	50 mm
Bore diameter	5 mm	7 mm	5 mm
Fill pressure	2 atm	2 atm	4 atm
Typical input power	3 kW	6 kW	2.5 kW
Lifetime	400–600 h	40–60 h	150–200 h
Wall loading (40% of electrical input)	110 W/cm^2	145 W/cm^2	128 W/cm^2
Coolant flow rate	120 cm^3/s	120 cm^3/s	60 cm^3/s
Electrical	100 V	112 V	84 V
characteristics	30 A	56 A	30 A
Current density	150 A/cm^2	140 A/cm^2	150 A/cm^2

6.1.3 Laser Diodes

The most efficient pump source for solid-state lasers is the diode laser. Throughout the last twenty years numerous laboratory devices have been assembled which incorporated single diode lasers, small laser-diode arrays or LED's for pumping of Nd:YAG, Nd:glass and a host of other Nd lasers. The low power output, low packaging density, and extremely high cost of diode lasers prevented any serious applications for laser pumping until the mid 1980's. Overviews and summaries of early work in this area can be found in [6.34, 35].

The major attributes of diode pumping can be summarized as follows:

- Increased System Efficiency
  The high pumping efficiency compared to flashlamps stems from the excellent spectral match between the laser-diode emission and the Nd absorption bands at 808 nm. Actually, flashlamps have a higher radiation output to electrical input efficiency (70%) compared to laser diodes (25–50%), however,

only a small fraction of the radiation is absorbed by the various Nd absorption bands. In contrast, the output wavelength of laser diodes can be chosen to fall completely within an absorption band of a particular solid-state laser.

- Increased component lifetime
  System lifetime and reliability is higher in laser diode pumped solid-state lasers as compared to flashlamp based systems. Laser-diode arrays exhibit lifetimes on the order of 10^4 h in cw operation and 10^9 shots in the pulsed mode. Flashlamp life is on the order of 10^8 shots, and about 500 h for cw operation.

- Improved beam quality
  A concomitant advantage derived from the spectral match between the diode-laser emission and the long-wavelength Nd absorption band is a reduction in the amount of heat which is deposited in the laser material. This reduces thermo-optic effects and therefore leads to better beam quality. In addition, the directionality of diode radiation allows designs with good spatial overlap between pump radiation and low-order modes in the resonator, which in turn, leads to a high-brightness laser output.

- Increased pulse repetition rate
  Besides low repetition rate and cw operation covered also by flashlamps and cw arc lamps, quasi-cw laser diodes permit, in addition, pulsed operation of solid-state lasers in the regime from a few hundred Hz to a few kHz.

- Benign operating features
  The absence of high-voltage pulses, high temperatures and UV radiation encountered with arc lamps lead to much more benign operating features of laser-diode-pumped systems. Furthermore, the high pump flux combined with a substantial UV content in lamp-pumped systems causes material degradation in the pump cavity and in the coolant, which lead to systems degradation and contribute to maintenance requirements. Such problems are virtually eliminated with laser-diode-pump sources.

- Enabling technology for compact and versatile laser systems
  The directionality of the diode output and the small emitting area, as compared to lamp pump sources, made it possible to design whole new classes of solid-state lasers, such as end-pumped systems, microchip lasers and fiber lasers. The flexibility of shaping and transferring the output beam from the pump source to the laser medium provides a great opportunity for the invention of new pump configurations and design architectures. An almost endless variety of optical coupling schemes has been reported.

- Enabling technology for new laser materials
  The most prominent laser materials which are pumped with diode pump sources can also be pumped with flashlamps. However, a number of very useful materials such as $Nd : YVO_4$, $Yb : YAG$ and $Tm : YAG$ have only reached prominence as a result of diode pumps.

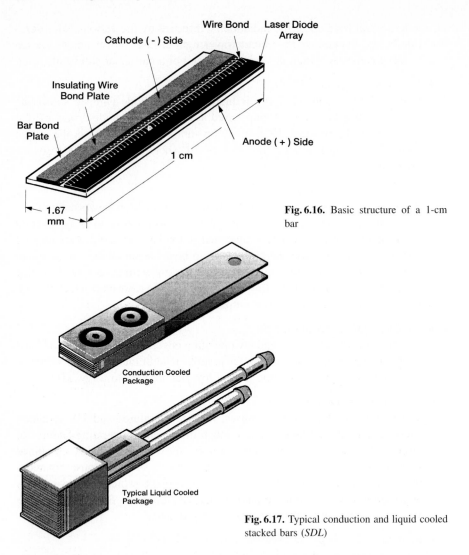

Fig. 6.16. Basic structure of a 1-cm bar

Fig. 6.17. Typical conduction and liquid cooled stacked bars (*SDL*)

The design and performance of laser-diode arrays suitable for solid-state laser pumping will be discussed in this subsection.

Depending on the output-power requirement of the laser and the particular pump configuration chosen, one can select a diode pump source from a number of standard commercial designs. These include small linear arrays with a length of 100 or 200 μm, 1-cm long bars, and stacked diode bars. The most common pump source is a 1-cm diode bar, as illustrated in Fig. 6.16. It consists of a monolithic chip of linear diode arrays which is bonded to an alloy submount. An insulated wire bond plate is soldered to the submount behind the laser-diode array. Electrically the individual diodes in the array are all connected in parallel to a wire bond plate. A number of 1-cm bars can be stacked to form two dimensional arrays, as shown in Fig. 6.17. The stacked arrays can be conduction or liquid cooled. For very small lasers a typical pump source is a 1 W diode array as illustrated in Fig. 6.20b.

Significant progress has been made during the last decade in developing monolithic, linear laser-diode arrays which have become the building blocks for solid-state laser pumps. Output power, slope efficiency, laser threshold and wavelength control have all been dramatically improved due to a combination of new structures and advanced growth techniques. In particular, epitaxial growth based on MOCVD allows close control of material composition, layer thickness and device geometry.

Today essentially all low power solid-state lasers are diode pumped, whereas commercial systems with average powers in excess of 10 W are still, for the most part, flashlamp or cw-arc lamp pumped. In these systems, the diode arrays are the major cost driver which do not make them competitive with lamp-pumped systems. Exceptions are applications where either size and weight, efficiency, beam quality or low maintenance are the overriding considerations. One can expect that a more automated diode fabrication, assembly and testing process combined with higher volume production will eventually decrease the cost per watt to a level where all solid-state lasers will be diode pumped. In order for this to happen, the lifetime of diode bars and 2-D arrays has to increase also another order of magnitude compared to current performance.

In this subsection the following topics will be discussed: Internal structure of laser diodes, packaging configurations, spatial and spectral beam characteristics, input vs. output power, lifetime issues and thermal control.

Internal structure of laser-diodes. The most basic form of a diode laser consists of a semiconductor in which one part is doped with electron donors to form n-type material and the other part is doped with electron acceptors (holes) to produce p-type material. Application of a negative voltage to the n-material and a positive voltage to the p-material drives electrons and holes into the junction region between the n and p doped material. This is called forward biasing the pn-junction. In the junction electron–hole recombination takes place. In this process, electrons from a higher-energy state (conduction band) transfer to a lower-energy state (valence band) and this releases energy in the form of photons. In order to produce stimulated emission in the junction area, population inversion and optical feedback is required. Passing a high current through the junction area provides a population inversion. A Fabry-Perot resonator, formed by cleaving the chip along parallel crystal planes provides optical feedback. Gain in laser diodes is very high, therefore only a small length of active material is needed. Typical edge emitting laser chips have active layers about $500\,\mu$m long. In order to increase efficiency and output power, actual laser diodes contain a number of different layers with the active region sandwiched in between.

Semiconductor-laser technology has produced an amazing variety of new device structures in the past decade. Overviews of this technology made possible by sophisticated growth techniques such as MetallOrganic Chemical-Vapor Deposition (MOCVD) and molecular beam epitaxy (MBE) can be found in [6.36, 37]. Most techniques differ in the way confinement is achieved regarding the width and depth of the gain region and mode volume of the optical radiation. We will illustrate the key design features of diode lasers by describing the Single Quantum Well Separate Confinement Heterostructure (SQW-SCH) because it is the most widely employed design for solid-state laser pumps.

Junctions Width. The output power of a laser diode increases linearly with junction width. Early diode lasers had active regions $100 \, \mu m$ or more in width. Such wide-stripe lasers do emit substantial power, but the current in these devices tends to break up into filaments, rather than spread evenly through the active region. The device susceptibility to uncontrollable filamentary operation leads to localized damage of the faces and substantially reduces the life of the diodes.

In order to achieve more stable laser operation, the broad active stripe is divided into many narrow stripes, each allowing single transverse mode operation. Today, most diode lasers are stripe-geometry types with active regions only a few μm wide. High powers can be generated by laser arrays that contain many parallel laser stripes. Figure 6.18 shows how laser action is divided into narrow stripes. In the gain-guiding concept, illustrated here, the drive current is restricted to small channels of the active layer, so that only in that region recombination of carriers and therefore population inversion takes place. This is achieved by deposition of regions of high-resistivity material beneath the contact area. Therefore, the current is channeled through narrow low-resistance areas of the p-GaAs surface layer.

Double-Heterojunction Design. The active layer shown in Fig. 6.18 is sandwiched between two pairs of layers having a different concentration of Al. The main purpose of the innermost pair of layers is to confine the carriers to the active region, whereas the purpose of the outer layers is to confine the optical beam. In a diode laser, recombination of current carriers takes place in a thin active layer which separates p- and n-doped regions.

Progress in high-power laser diodes for solid-state laser pumping has emphasized the development of quantum-well structures in which laser emission is produced in very thin epitaxial layers (quantum wells) less than $0.02 \, \mu m$ thick.

A quantum well is a thin layer of semiconductor located between two layers with larger bandgap. Electrons in the quantum-well layer lack the energy to escape, and cannot tunnel through the thicker surrounding layers. The quantum well layer in Fig. 6.18 has a composition of Ga and Al indicated by x which defines the emission wavelength. A higher Al concentration increases the bandgap and shifts the output towards shorter wavelength. The quantum-well structure is sandwiched between two thick layers of composition which contains a higher concentration of Al ($y > x$). The higher Al concentration increases the bandgap, thereby defining the quantum well, and the large thickness of the layer prevents tunneling of the carriers out of the quantum well.

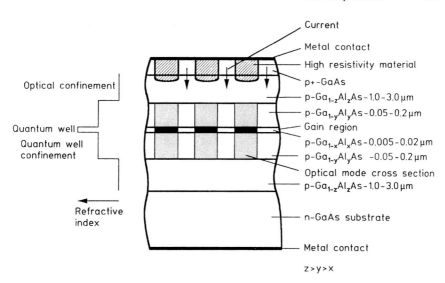

Fig. 6.18. Gain-guided, single quantum well separate confinement heterostructure stripe laser. Note: Thickness of layers is greatly exaggerated. Optical mode cross-section is actually $5\,\mu\text{m}$ wide and $0.5\,\mu\text{m}$ high

Due to the characteristically small dimensions of quantum-well lasers, the injected carriers are subject to quantum effects, that is, the conduction and valence bands, normally considered a continuum, become quantized. Quantum-well structures allow a better utilization of carriers for radiative transitions compared to other designs. Multiple quantum wells can be formed by alternative quantum-well layers with high-bandgap barriers thick enough to prevent quantum tunneling.

Both single- and multiple-quantum-well devices are commercially manufactured. Single-quantum-well devices exhibit slightly lower lasing thresholds and slightly higher differential quantum efficiencies, making them preferable for high-power lasers.

Optical Mode Volume. The thin active region incorporating a quantum-well active layer structure provides low threshold and high electrical-to-optical efficiency. However, such a very small emitting surface poses one problem since the power from a diode laser is limited by the peak flux at the output facet. One can increase the output from these devices by spreading the beam over an area which is larger than the active layer or gain region. The standard approach is to deposit layers next to the active layer, each of which has a slightly lower refractive index than the active layer, thus making a wave guide of the active layer. In the separate carrier and optical confinement heterostructure (SCH) design shown in Fig. 6.18, the refractive index boundary is abrupt, at the layer boundary. Alternatively, the refractive indices of the surrounding layers may be graded, forming a Graded-Index and Separate carrier and optical Confinement Heterostructure (GRINSCH).

In the SQW-SCH structure, the overlap between the optical mode and the gain region is only about 4%, which results in a large effective aperture of the laser on the order of 0.3 to 0.5 μm. This substantially reduces the energy density at the output facet and enhances reliability by minimizing catastrophic facet damage. The cavity length of the stripe lasers is typically between 250 and 350 μm.

Diode Pump-Source Configuration. Commercially available pump sources include small diode arrays used in low-power end-pumped lasers, and single and multiple 1-cm bar stacks for side-pumped as well as end-pumped lasers. Experimental surface emitting arrays for sidepumping of lasers will also be discussed briefly. For high-power lasers, the 1-cm linear bars must be combined into modules at some level to reduce complexity of the electronic drivers, heat removal system and mechanical structure. Stacking of diode arrays is done manually or in a semi-automated mode at the present time. This process allows large-area arrays with dimensions up to about 1 cm^2 to be fabricated. However, the process is labor intensive. The cost of the arrays also increases faster than linearly as the size of the array increases. This is because the selection of wavelength and current threshold of linear bars becomes more critical, as the size of the array increases. All of the bars in the array must be matched for optimum performance. A trade-off exists between increased design complexity using arrays which consist of only a few bars, and reduced efficiency and increased cost with large area arrays comprised of a large number of bars. As a result of this trade-off, a common module for high-power lasers is an array consisting of 5 one-cm bars. Such an array has an aperture of about $10 \times 1 \, \text{mm}^2$.

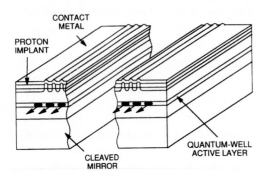

Fig. 6.19. Multistripe laser geometry

Diode Arrays. An array containing a number of stripe lasers on a single chip, is shown in Fig. 6.19. A typical example is a 20-stripe gain-guided, SQW-SCH laser array made by Spectra Diode Laboratories. The diode array contains twenty 5-μm wide stripe lasers on 10-μm centers which produce a total output of 1 W in the cw mode. The output is emitted from an area of 200 μm $\times$ 1.0 μm with a beam divergence of 40° by 10°. The output power vs. input current is shown in Fig. 6.20a. Depending on the composition, the wavelength can be specified

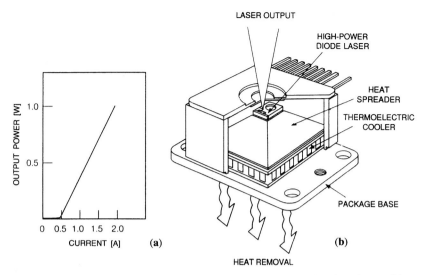

Fig. 6.20. Output power vs. input current of a 1 W diode array (**a**), and packaged array (**b**) mounted on a thermoelectric cooler (SDL-2460)

between 790 and 835 nm. The device is available in different configurations such as an open heat sink, packaged with a window, coupled to a fiber, or mounted on a thermoelectric cooler. In the latter configuration shown in Fig. 6.20b, the array is mounted on a pyramidal heat sink that connects to a thermoelectric cooler. Heat dissipates through the base, while the radiation emerges from the top. The TE cooler permits temperature tuning of the output to match the absorption lines of the solid-state laser. In a GaAlAs diode laser, the wavelength changes with temperature according to 0.3 nm/°C. If instead of the cw mode, pulsed operation is desired, a similar device will produce 2 W peak power from a $200 \times 1.0\,\mu m^2$ aperture.

Linear Bars. In the fabrication of linear bars the ten or twenty stripe sub-arrays discussed before, are serially repeated in a single substrate to form a monolithic bar structure. Practical processing considerations limit the length of such bars to 1 cm. An example of such a 1-cm monolithic linear diode bar is shown in Fig. 6.21. The device, composed of 20 ten-stripe laser sub-arrays spaced on 500 μm centers, produces an output of 10 W in the cw mode (Fig. 6.22). The spacing of the sub-arrays, which amounts to a fill factor of 20%, is determined by thermal limitations. Recent devices are capable of up to 40 W output power in a 1-cm bar.

The use of diode lasers for pumping pulsed Nd lasers requires pulse length on the order of several hundred μs. One limit of the output of diode lasers is the temperature of the junction and associated thermal runaway condition that raises the local temperature of the facet to the melting temperature. Because of the small volume of material involved, this temperature rise occurs in less

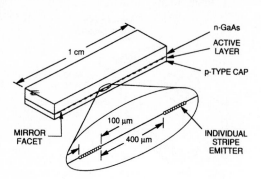

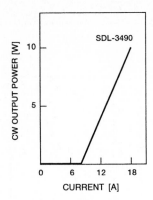

Fig. 6.21 Monolithic 1-cm bar for cw output operation

Fig. 6.22 cw output power vs. input current for a 1-cm bar

than 1 μs. Therefore, the output limit of diodes in the long pulse mode is the same as for cw operation. Thus, long-pulse operation ($\gg$ 1 μs) is called quasi-cw operation. The average temperature of the diodes is governed by the mean power dissipated in the devices, but the instantaneous temperature of the diode junction will undergo excursions above the mean temperature as it is pulsed. Since the average heat dissipation of a pulsed array is considerably lower as compared to a cw bar, the fill factor is increased to 100%. A typical building block for pumping a pulsed Nd:YAG laser is a 1-cm long bar containing up to 1 000 stripe lasers (as compared to 200 stripes in the cw device). The bars are etched periodically to avoid transverse laser emission. Each bar generates 60 W peak power or 12 mJ in a 200 μs pulse. Table 6.6 lists the current status of quasi-cw diode bars.

Table 6.6. Performance of current state-of-the-art quasi cw bars

Electrical efficiency	%	45–50
Power	W/cm	60–100
Linewidth	nm(FWHM)	4–5
Lifetime	shots	10^9
Thermal resistance	°C/Wcm^2	0.3

Bars designed for cw operation are mounted and cooled in individual packages with up to 40 W output power. Pulsed bars typically have 60 W peak power, although up to 100 W and 40 mJ per pulse are available at reduced lifetime. Depending on cooling conditions, a duty cycle up to 20% is possible. Therefore, the maximum average output power for both devices is (20–40) W for a 1 cm bar. Pulsed bars are generally assembled in multiple bar configurations as will be discussed next.

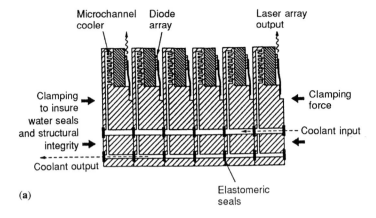

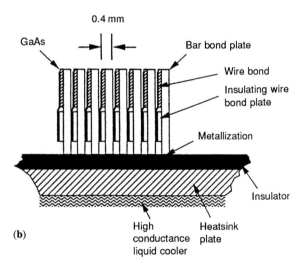

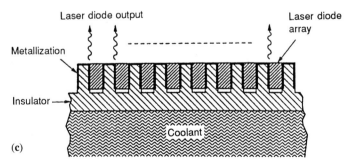

Fig. 6.23a–c. Cooling techniques for multi-bar arrays. (**a**) Microchannel cooling of each bar. (**b**) Bars preassembled and mounted on a common heat sink. (**c**) Bars directly mounted onto a common heat sink

Two-Dimensional Arrays. The bar concept can be extended to a stack of bars to form a two-dimensional array. One of the key issues in the design of a 2D array is the removal of waste heat from highly concentrated thermal sources. The duty cycle of pulsed arrays, i.e. average output power, is entirely dependent on the thermal design. Three concepts, known as "rack and stack", microchannel or back-plane cooled arrays, and "bars in a groove," have evolved with regard to assembly and cooling of 2D arrays. The three approaches of thermal management within an array are illustrated in Fig. 6.23.

In the microchannel cooling approach (Fig. 6.23a), each individual diode bar has its own internal liquid heat exchanger which is produced based on silicon etching technology [6.38]. Two-dimensional arrays are created by stacking individual bar/cooler assemblies. Conductive silicon rubber gaskets between the bars provide seals for the coolant channels.

In the other two approaches, many laser-diode bars are mounted on one liquid-cooled structure.

The most common 2D structure depicted in Fig. 6.23b, uses back-plane cooling of individual bars [6.39]. A number of bars, each containing a diode-array submount for heat conduction and a wirebound plate, are stacked and mechanically bonded via solder to create a 2D array. The whole assembly is then soldered to a common cooled heat sink. Laminar or highly turbulent flow (impingement cooling) can be used to remove heat from the common heat sink.

In the third approach (Fig. 6.23c), a multiplicity of bars are individually mounted in a thermally conductive but electrically insulating structure (BeO) with machined grooves that match the dimensions of the bars [6.40, 41]. Electrical interconnections are achieved by metallizing the side walls of the grooves.

Since the back-plane cooled stacked arrays shown in Fig. 6.23b are the most common devices available commercially, we will describe some typical design features and performance data. In these devices, heat from each diode array is transferred to the common heat sink via a thin copper heat spreader. Higher average powers require thicker heat spreaders, therefore the stacking density depends on the duty cycle of operation.

Bar spacings range from 0.4 to 2.0 mm, depending on the thickness of the copper heat spreader onto which each array is mounted. For low-duty-cycle operation around 3%, the bars can be densely packaged with only 0.4 mm space in between; this allows a power density of 1.5 kW/cm^2. At a 6% duty cycle, the spacing increases to 0.8 mm and 800 W/cm^2 can be achieved. At the highest duty cycle of 20%, bar spacing is 2 mm, which results in a power density from the array of 300 W/cm^2.

In devices intended for quasi-cw operation, the 1-cm bars are stacked up between 4 and 10 bars high to form a small two-dimensional array. Figure 6.24 illustrates the performance characteristics of a 10 bar array. Operation of a 10 stack array, with 200 μs pulses at 40 Hz produces a peak power around 500 W at 35% efficiency. Two different-stacked-array configurations have been employed for pumping high power solid-state lasers depending upon whether the laser medium is in the shape of a rod or rectangular slab. Rod lasers are efficiently

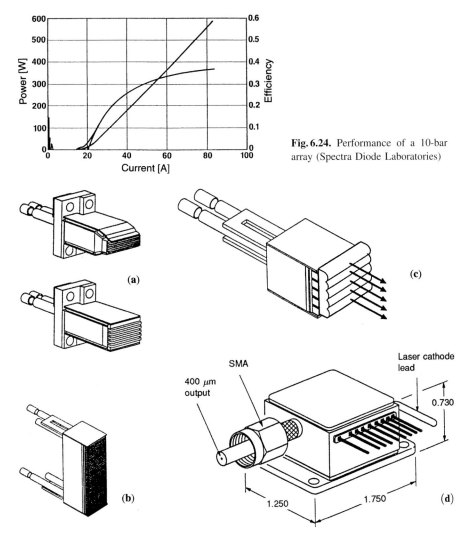

Fig. 6.24. Performance of a 10-bar array (Spectra Diode Laboratories)

Fig. 6.25a–d. Packaging configurations of diode array. High-duty-cycle liquid-cooled arrays for rod pumping (**a**) and slab pumping (**b**); array with microlens (**c**), or single cw array coupled to an optical fiber (**d**)

pumped by small-area arrays consisting of less than ten bars with an output area of a few tenths of a cm². This configuration has become an industry standard for rod pumping. Slab designs generally require diode arrays with several square cm of output area. These arrays couple most efficiently to the large pump face of a rectangular laser slab.

Figure 6.25 illustrates different packaging configurations of diode arrays. The 2D arrays employed in rod pumping (Fig. 6.25a) have typically an emitting area of 1 cm × 0.3 cm or 1 cm × 0.5 cm. The arrays can be symmetrically arranged

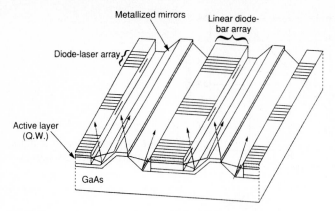

Fig. 6.26. Diode-laser array with integrated 45° mirrors

around the laser rod. Pumping of slabs requires large area pump sources such as the 1 cm × 3 cm depicted in Fig. 6.25b. At a 6% duty cycle, a peak power of 800 W/cm^2 is achievable in these devices. In order to increase the brightness from a laser array, a microlens array optic can be combined with the laser source (Fig. 6.25c). This design is very useful if one wants to concentrate pump radiation into a narrow region of a small-slab crystal. Figure 6.25d displays the package of a single 20 W cw bar coupled to a 400 μm fiber. Typically, 16 W are available at the output end of the fiber. The ability to mount the pump source away from the laser onto a heat dissipating structure is very useful in some applications.

Surface Emitters. The basic method for stacking up linear arrays to create 2-D arrays requires a large number of fabrication processes. An intuitively appealing alternative approach is to fabricate two-dimensional arrays directly at the wafer stage, thereby substantially reducing the manufacturing complexity.

Figure 6.26 exhibits an example of a monolithic structure, in which a mirror at a 45° angle to the active layer deflects laser radiation from the chip surface, rather than the edge. The mirrors are etched into the substrate in these monolithic, surface emitting, quantum-well diode-laser arrays [6.42]. Instead of using integrated mirrors, the diode radiation can be deflected with a distributed-Bragg reflector [6.43], or the resonator can be rotated 90° in the so-called vertical cavity laser [6.44].

Spatial Profile. The output beam emerging from the facet of a diode laser is in the form of a light cone which is oval in shape, as illustrated in Fig. 6.27. In a vertical direction to the emitting stripe, the beam divergence is largest, about 30° to 40° FWHM, and is determined by diffraction from the multilayer structure. If one introduces in (5.6) a wavelength of $\lambda = 0.8\,\mu$m and a spot size for a Gaussian beam of 1 μm the beam divergence at the $1/e^2$ points is 57°. In the horizontal direction, the beam divergence is smaller, about 10°, and depends

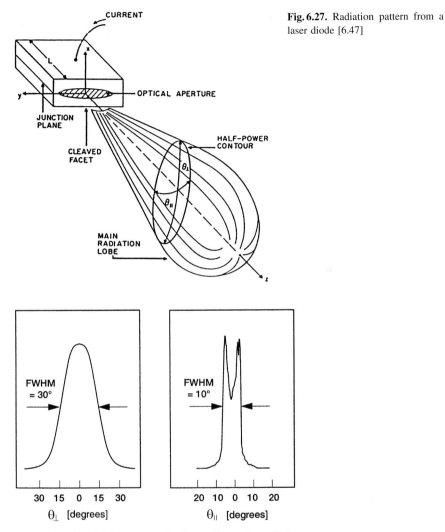

Fig. 6.27. Radiation pattern from a laser diode [6.47]

Fig. 6.28. Typical far field energy distribution from a laser diode

on the stripe width, cavity length and various material parameters (Fig. 6.28). The two lobes in the farfield pattern is characteristic of phase coupling between adjacent stripe lasers in cw arrays. In the so-called leaky mode operation, the optical field width is greater than the width of the gain region and adjacent stripes are phase coupled. The particular gain and mode structure in these lasers causes adjacent stripes to couple 180° out of phase. The farfield pattern, as a result, has two lobes symmetrically positioned about the normal of the facet [6.45, 46].

Reformatting the highly astigmatic pump beam emitted from a line source into a circular beam, which is required for endpumped lasers, is a challenging task.

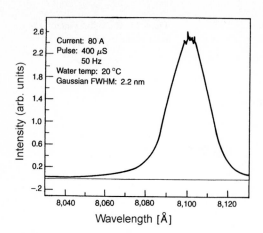

Spectral Properties. The spectral properties of laser-diode arrays which are most critical for the pumping of solid-state lasers are the center wavelength and the spectral width of the emission and the wavelength shift with temperature.

The wavelength of a laser diode is inversely proportional to the energy difference E_g between the conduction and valence band. When this energy gap is expressed in electron volt (eV) the laser wavelength in μm can be calculated according to $\lambda = 1.24/E_g$. The bandgap depends on the crystalline structure and chemical composition of the semiconductor. A binary compound such as GaAs has a fixed wavelength which is 904 nm. Adding a third element, by substituting a fraction of gallium with aluminum in GaAs, changes the bandgap.

In a compound semiconductor, such as GaAlAs, the wavelength of the emitted light can be tailored over a wide range by changing the ratio of Al and Ga. Adding aluminum increases the bandgap, producing lasers with shorter wavelengths. Most commercial GaAlAs lasers have wavelengths of 750 to 850 nm. Nd ions have substantial absorption in the vicinity of $0.807\,\mu$m, which is the emission wavelength of diode lasers with $Ga_{0.91}Al_{0.09}As$ active regions. As far as pumping of Nd lasers is concerned, the output wavelength can be tailored to the peak absorption by adjustments of the Al concentrations. Typically, a change in Al concentration of 1% results in a 1 nm change in wavelength.

Compositional changes and temperature gradients within an array lead to a broader spectral output for the whole array as compared to a single device. Typical spectral width of a single bar or small stacked array is on the order of 4 to 5 nm. State-of-the-art performance is about 2.2 nm for a 10-bar array, as shown in Fig. 6.29. The bandwidth of the Nd : YAG absorption line at 808 nm is 2 nm for an absorption coefficient larger than $3.8\,\mathrm{cm}^{-1}$. Similarly, in Nd : YLF, the absorption line is about 2 nm wide for absorption coefficients larger than $2\,\mathrm{cm}^{-1}$. In diode pumped lasers which have only a short absorption path, it is important to have a narrow spectral emission from the diode array in order to absorb most of the pump radiation. In optically thick materials, such as large diode pumped lasers with rod or slab dimensions on the order of 10–15 mm, spectral width

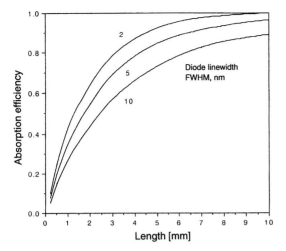

Fig. 6.30. Absorption efficiency of Nd:YAG as a function of absorption length for a pump wavelength of 808 nm [6.48]

becomes less important because eventually all the pump radiation gets absorbed. In Fig. 6.30, the absorption of diode-pumped radiation for Nd:YAG is plotted vs. optical thickness with spectral width as a parameter.

In a GaAlAs structure the peak emission changes 0.3 nm/°C. Therefore, the material composition has to be chosen such that the desired wavelength is achieved at the operating temperature of the junction. Figure 6.31a displays the spectral output from diode array as the repetition rate is increased from 10 to 100 Hz. The output shifts slightly from 807 to 810 nm as a result of the increased junction temperature at the higher repetition rate. In Fig. 6.31b, the wavelength shift of a diode array is plotted vs. change in drive current.

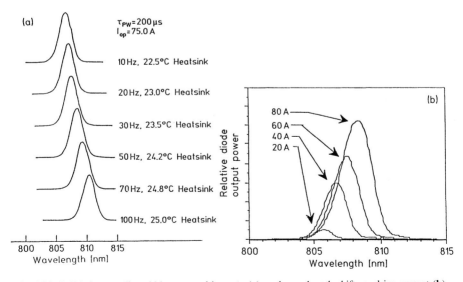

Fig. 6.31a,b. Diode array linewidth vs. repetition rate (**a**), and wavelength shift vs. drive current (**b**)

For laser materials with an absorption band between 750 and 850 nm, pumping with GaAlAs diodes is the obvious choice because of the availability and high state of development of these devices. Lasers which can be pumped with GaAlAs diodes include Nd doped materials and Tm : YAG, as listed in Table 6.7. For pumping Yb : YAG or Yb-sensitized lasers such as the Er : Yb : Glass, a pump source around $0.98\,\mu$m is needed. For this wavelength so-called strained-layer super lattice lasers based upon InGaAs on GaAs substrates have been developed with emission wavelengths between 0.95 and $0.98\,\mu$m [6.49–51]. Traditional diode lasers are restricted to materials which can be lattice-matched to GaAs or InP substrates, because unmatched materials are prone to defects and quickly degrade. However, it was discovered that very thin layers – below a critical thickness of a few tens of nanometers – can accommodate the strain of a small lattice mismatch. This discovery led to the development of a single thin strained layer used as a quantum well which has an emission wavelength not achievable with GaAlAs or InGaAsP laser diodes.

On the short-wavelength end of available diode pump sources are the visible AlGaInP diode lasers operating in the 640–680 nm region. These devices have been used to pump Cr doped lasers, such as Cr : LiCaAlF$_6$ (LiCAF) and Cr : LiSrAlF$_6$ (LiSAF).

Table 6.7. Important lasers and diode pump sources

Laser material	Pump wavelength [nm]	Semiconductor laser material
Nd : YAG	808	GaAlAs
Nd : YLF	798,792	GaAlAs
Nd : YVO$_4$	809	GaAlAs
Tm : YAG	805 (wing)	GaAlAs
	785 (peak)	GaAlAs
Yb : YAG	941	InGaAs
Er : glass	980	InGaAs
Cr : LiSAF	670 (peak)	AlGaInP
	760 (wing)	GaAlAs

Input vs. Output Characteristics. The amount of electric current necessary for the amplification in the diode structure to overcome optical losses defines the threshold current J_{th}. Laser differential quantum efficiency η_d is characterized by the number of photons emitted per injected electron, which can also be expressed by the slope efficiency in watts per ampere. Fig. 6.22 depicts the output power vs. current for a 10 W cw bar. The device has a slope efficiency of 1 W/A and a quantum efficiency of 65%. Overall the total power conversion efficiency is 33%. Forward voltage for diode lasers is typically $V_f = 1.5\,\text{V} + IR_s$, where R_s is the series resistance. For the device characterized in Fig. 6.22 the series resistance is 0.015 ohm.

If quasi-cw bars are stacked into 2-D arrays, the bars are connected electrically in series to avoid high currents and the associated resistive losses. This is illustrated in Fig. 6.24. The drive current for this 10 bar stack is identical to the current of a single bar. The nonlinear behavior of the overall efficiency with increasing current occurs because the ohmic losses in the material increase as the square of the current, whereas the output power is linear in current.

Diode-Laser Lifetime. We will distinguish between catastrophic failures and normal degradation during operation. Laser diodes, like all semiconductors, are very susceptible to damage by electro-static discharges or high-voltage transients. Therefore handling of these components requires electric discharge control such as wearing wrist straps and grounded work surfaces. Also great care has to go into power supply and driver design to protect this very expensive pump source from transients, current surges and high voltage. Operation at excessive currents will lead to catastrophic mirror facet damage. Also, insufficient cooling combined with high currents can melt the solder joints or vaporize the wire bonds within the diode structure.

Under proper operating conditions laser diodes degrade in a fairly predictable manner which results in a decrease in output over time. The higher the operating current and/or the operating temperature, the faster the degradation. For example, a rule of thumb is that for every $10°C$ increase in temperature of the active region the lifetime drops by a factor of two. Also there is a clear correlation between the number of system turn-on and turn-off cycles and lifetime. The longest lifetime is achieved in systems which are operated uninterrupted for long periods of time. This is related to the stresses introduced into the diode structure by thermal cycling.

The degradation of AlGaAs laser diodes, the most common pump source, is usually attributed to oxidation and migration of aluminum under high-power operation. This causes structural defects which spread through the laser diode forming light absorbing clusters. Dark lines or spots in the output beam are a manifestation of this damage.

Lifetimes of laser diodes in the 1 W region are on the order of 10^4 h and selected devices are warranted up to 10^5 h. However, diode bars and particularly stacked bars used in high-power pump sources have not reached the same level of maturity. Lifetime is also a question of economics. Since the cost of the pump source is so high, the devices are usually operated at full current to provide the maximum pump power. As indicated above, operation at reduced current drastically increases the lifetime. However, for a given laser this requires more bars therefore increasing cost of the system.

Degradation of 100 W quasi-cw laser diode bars was evaluated by NASA as part of a complete laser system life test. One bar was operated at 500 Hz at a pump pulse of 200 μs. The diode's output energy decayed by 22.8% after 7 billion shots. Similar tests on 3 other 100 W devices showed a degradation of 20% after 5 billion shots [6.52].

Our own experience shows that, depending on the operating conditions, lifetime in quasi-cw diode stacks can range from 150 million to 15 billion shots. For example, one of our Nd : YAG systems contains four internally liquid cooled arrays, each comprised of 8-bar stacks. The arrays operated at 20% duty cycle are pulsed at a repetition rate of 2 kHz with a pulse width of 100 μs. The peak power from each 1-cm bar is limited to 45 W, and the system is operated for long periods of time without being turned off. Typically, the system is run in 8 hour shifts. The diode pump source, as well as the rest of the system, operated maintenance free for about 2000 hours. During this time the diodes accumulated a total of 15 billion shots. After 10 billion shots, the laser power dropped below the level required for the application and the user increased diode current until units failed catastrophically at 15 billion shots.

The other extreme, in terms of operating condition, is encountered in our systems installed in helicopters. The 1-cm bars are operated at 60 W and the system is subjected to many turn-on and turn-off cycles during each mission. Each system contains 96 five-stack bars operated at a pulse width of 200 μs and a repetition rate of 40 Hz. The low duty cycle arrays are mounted on water-cooled heat sinks. The arrays typically show a 20% drop in output after 1000 hours of operation which represents about 150 million shots. This might not seem impressive in an industrial application, but 1000 hours of maintenance free operation is totally acceptable in a helicopter. In six systems built to date, no catastrophic pump source failures occurred, and the diodes are replaced every 18–24 months during regular maintenance.

Lifetime for 20 W cw arrays can range from 1000 h for devices operated at 75°C in conductively cooled systems, to 10^4 h for slightly derated arrays operated at 25°C. In most commercial systems requiring long life, diode arrays are operated typically at 75% of their rated power levels.

Thermal Management. Diode-array pumping offers dramatic improvements in efficiency of solid-state laser systems. However, the need to maintain the operating temperature within a relatively narrow range requires a more elaborate thermal management system as compared to flashlamp pumped lasers.

In end-pumped systems, the diode wavelength is usually temperature tuned to the peak absorption line of the laser and maintained at that wavelength by controlling the array temperature with a thermo-electric cooler. This approach works well for small lasers, and systems including TE coolers can work over a large range of ambient temperatures.

In large systems, the power consumption of TE coolers is usually prohibitive. For these systems, a liquid cooling loop with a refrigeration stage can be employed which maintains the temperature of the coolant independent of the environment. However, the coefficient of performance for refrigerators is typically no greater than one. Therefore, the electrical input power requirement is equal to the heat load to be controlled.

Recently, the development of diode arrays which can operate at high junction temperatures has eliminated the need for refrigerated cooling loops. Diode arrays can be operated now at junction temperatures as high as 75°C which is higher than ambient in most situations. A simple liquid-to-air cooling system provides the most efficient thermal control system for the laser because only the power consumption of a pump and possibly a fan is added to the total electrical requirements. There is a slight reduction of the diode operating efficiency (about 0.2%/C) at elevated temperatures. However, this reduction in diode efficiency is small compared to the power penalty one would have to pay for a refrigeration cooler.

Thermal resistance, temperature drops and coolant flow rates will be dependent on the exact cooling geometry selected. Heat generated in the most recent generation (backplane-cooled) diode arrays is conducted away from the individual bars by a submount which is typically made from copper or beryllia. This conduction path is characterized by a finite thermal resistance. The resistance depends on the thickness of the mount and therefore on the spacing of the bars in the array. A rough calculation of the resistance is obtained from

$$R_{th} = \frac{\Delta T}{Q} = \frac{l}{kA}$$

where R_{th} is the thermal resistance [C/W], l and A are the length and area of the one cm-wide laser bar submount, and k is the submount conductivity ($\approx$ 3 W/cm-C). For a 1.7 mm long by 0.4 mm thick copper submount, the thermal resistance is 1.4 C/W. Several options are available as coolers for the diode arrays. These include "impingement" and "micro-channel" type coolers. In the impingement cooler, turbulent flow provides a thin boundary layer between the cooler surface and the flowing coolant. In the micro-channel cooler, the flow is forced through small-area channels ($\approx 10^{-4}$–10^{-5} cm^2) to reduce the thickness of the boundary layer to a few micrometers. In either case, heat transfer between the fluid and the metal is improved over conventional coolers.

In a passive cooling system, there is a considerable variation of the cooling fluid temperature depending on the ambient condition. At ambient temperatures much below the operating temperature of the laser diodes, the flow is thermostatically controlled. At higher ambient temperatures, and maximum coolant flow, the diode temperature will vary according to the ambient changes. Provided that diode-array wavelength, spectral width and optical thickness of the solid-state laser material are properly chosen, a fairly large tolerance for temperature fluctuations can be achieved. Actually, temperature insensitive performance closely matching the athermal behavior of lamp-pumped systems can be obtained.

Diode-pumped systems have been designed with constant output power and beam quality over a $-30°$ to $+50°$C temperature range without employing a refrigeration stage. This is only possible in very large diode-pumped systems with absorption lengths of the pump light on the order of 10–20 mm. Because the laser material is optically thick, the tolerances for the emission wavelength and spectral width is much relaxed. In this case, it is actually advantageous to increase the spectral width of the diode in order to provide a more uniform pump distribution throughout the active material. The benefit of a large pathlength in the laser material is revealed dramatically in Fig. 6.32. The data plotted shows that for a diode bandwidth of 15 nm and an absorption depth of 16 mm (8 mm double pass), the absorbed energy (and to a first approximation the laser output energy)

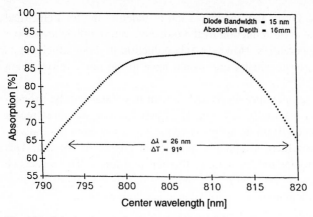

Fig. 6.32. Calculated total absorption in Nd : YAG for a 15 nm pump width and a 16 mm optical path

Table 6.8. Power dissipation from a diode array pumped solid-state laser

System output:	160 W
Diode array:	1300 W
Laser Crystal:	160 W
Electronics/other components:	380 W
Total electrical input:	2000 W

can be held to within ±10% even though the center wavelength of the diode output varies by 26 nm. This 26 nm change in central frequency corresponds to a change in diode temperature of 86°C.

The power dissipation of the various subsystems of a pulsed diode-pumped laser is listed in Table 6.8. The system has a repetition rate of 200 Hz and an output pulse energy of 0.8 J/pulse. By far the greatest heat dissipation occurs in the diode pump arrays, and accounts for approximately 65% of the system input power. The solid-state laser medium itself dissipates heat at a rate approximately equal to the output power from the laser system. The electronics dissipate on the order of 19% of the input power.

The use of liquid cooling is at present the simplest and most cost-effective method for heat removal from diode-pumped laser systems. However, expected material improvements can significantly reduce the heat dissipation and temperature-control requirements on the diodes. These improvements include increases in the efficiency of the diode arrays, and the development of new solid-state laser materials in which the sensitivity to pump wavelength and linewidth is reduced. With continuing improvements, the need for liquid cooling can be eliminated in many applications altogether. For burst mode operation, the possibility exists for the implementation of cooling by latent heat of fusion of low-melting-temperature substances such as paraffin. For space-based operation, radiative cooling can be implemented in many systems using heat pipes for thermal transfer from the dissipative elements to low-temperature radiators.

6.2 Power Supplies

In its basic configuration, a solid-state laser is comprised of a laser head which contains all the optical elements, a power supply to drive the pump source, and a cooling system. However, most lasers contain a number of auxiliary electronic subsystems. Furthermore, in most lasers, the vital system functions are controlled and monitored via digital electronics. Generally, the electronics of solid-state lasers can be divided into the following subsystems:

- *Power conditioning.* This unit converts ac line voltage or power from a battery to the appropriate dc level required for operation of arc lamps or laser-diode arrays.

- *Energy storage.* Capacitors are required for pulsed arc lamps and diode arrays to store the energy transferred to the pump during each pulse.

- *Switching or trigger circuit.* In the case of diode arrays, energy is transferred from a capacitor bank via a semiconductor switch. Flashlamps behave electrically like switches which are turned on by a high-voltage trigger pulse.

- *Auxiliary subsystems.* Besides subsystems which provide energy to the pump source, a number of electronic subsystems are usually required in a laser, such as Q-switch electronics, heaters for second-harmonic generators, mechanical shutters, motor driven translation stages, piezoelectrically driven mirrors, fans, coolers, etc.

- *System microcontroller.* With the exception of laboratory set-ups, most lasers contain a microcontroller which is employed for controlling and monitoring all electrical, electromechanical and thermal subsystems. The output from the microcontroller usually consists of a display indicating the health and status of the laser system. Housekeeping functions which are displayed may contain interlock data, such as temperature and coolant flow, optical data, such as pulse energy and pulse timing, and electrical data, such as voltage and current levels.

- *Timing control board.* This unit is typically comprised of synchronous clocks, logic devices, gate arrays, and drive components to supply the necessary timing signals for control of the pump source and Q-switch driver.

In this section, we are only concerned with the electronics associated with operating the pump source of the laser. This system is often loosely referred to as the power supply of the laser.

6.2.1 Operation of Laser-Diode Arrays

Compared to arc lamps, the electrical operating conditions of laser diodes are much more benign. The operating voltages are low, and a trigger circuit is not required.

CW Diode Arrays

Essentially, only a low-voltage dc power supply of the appropriate voltage and current range is required to operate the cw devices. Despite the very straightforward electronics, there are several challenges for operating laser diodes efficiently and reliably. The series resistance of laser diodes is very low; therefore, the internal resistance of the power supply must be minimized for efficient energy transfer.

Secondly, compared to flashlamps, laser diodes are electrically much more vulnerable. Therefore, overvoltage, reverse bias and current protection are essential features in the design of power sources.

The diode array must be protected from several electrical fault modes both during operation and during installation, repair, and in the "off" mode. The most serious electrical "threat" to diode arrays is the possibility of electrostatic buildup across the diode-array junction. Static charge can build up to a level sufficient to cause electrical breakdown across the diode junction, inducing permanent damage. Thus, the diode array must be shorted or connected across a low resistance at all times. When in use, a 100-Ω resistor is sufficient to prevent static buildup without affecting electrical efficiency of the diodes.

Diodes are high-power devices that are not easily damaged by high currents. Diode drivers must be designed to deliver only slightly more than the design operating current for the arrays. The possibility of a high current being applied to the diode arrays is avoided by limiting the output current of the diode driver power supply. This can be accomplished with a current-limiting power supply.

Arrays have to be protected against reverse voltage bias using a diode across the input to the arrays. This device also protects against reverse voltage transients and static buildup. Excess transient forward bias must be eliminated by careful design of the array driver.

Pulsed-Diode Arrays

The power supply of a diode-pumped solid-state laser is the subsystem where most of the weight and volume savings occur compared to a flashlamp-based laser. High efficiency of a diode-pumped laser, low-voltage operation of the diodes, and the use of electrolytic capacitors for energy storage allow the design of very small and compact power supplies. In a pulsed diode-pumped solid-state laser system, the diode array drivers replace the pulse forming network in a conventional flashlamp-pumped laser system. The diode-array driver is a low-voltage switching network which supplies a constant-current pulse to the diode arrays. The low-voltage requirements for the diode arrays allow the use of power

MOSFET's as switching elements. Electrical energy stored in electrolytic capacitors is transferred to the diode arrays on each laser pulse. Figure 6.33 depicts a schematic diagram of a typical diode-array driver. The inherent simplicity of the device is responsible for the high reliability and efficiency.

A power conditioning unit converts line power or battery power to the appropriate dc voltage which is typically on the order of 20 to 200 V, depending on how many diode bars are connected in series. In order to avoid current surges from the DC power supply during the nominally $200\,\mu$s-long pump pulse, energy is stored in electrolytic capacitors. Connected parallel to the capacitors is a string of laser-diode bars, one or several MOSFETs, and a small series resistor for current sensing. The function of the MOSFET is to generate the desired current waveform by switching the diode current on and off, and by providing current control by means of a current sensing feedback loop. The MOSFET's are switched on and off by input pulses from a timing circuit.

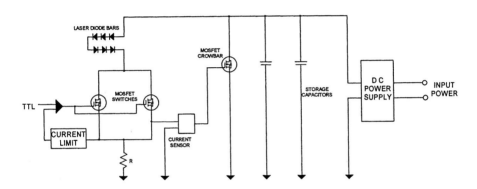

Fig. 6.33. Schematic diagram of a typical laser-diode-array driver

In addition, a diode-array driver includes fault protection to protect the diode arrays against damage caused by catastrophic failure of an array or the driver itself. Fault protection is based on a fast crowbar circuit that senses excess current, duty cycle or temperature. The crowbar circuit can shunt the output of the driver within one pulse in order to remove the power from the diode arrays. This circuitry protects the diodes from several rare but potentially catastrophic system faults. The first mode of failure is for the MOSFET not to switch off. When this happens, all of the energy stored in the capacitor is dumped into the diode arrays. This condition usually results in diode failure, since the stored energy is many times that delivered in a single pulse. Without the crowbar circuit, the heat dissipated when this failure occurs is enough to melt the solder used to attach the bars to the heat sink. A second fault mode arises from accidentally applying an excessively long trigger pulse to the diode drivers. This can occur during set-up and test. Under this condition, the diode driver senses the product of the pulse length and repetition rate. If it exceeds the preset duty-cycle, the driver is shut down, again protecting the diode arrays. The value of fault protection cannot be overemphasized.

Besides protecting the diode arrays electrically from damage due to excessive voltage and currents, another key feature of a well-designed diode driver is a high transfer efficiency of the stored energy to the laser diode arrays. This is not a trivial task due to the very low impedance of laser diode arrays. A high efficiency can be obtained by operating individual diode bars in series, reducing circuit losses to an absolute minimum, and by selecting a large energy storage capacitor.

The impact of the unavoidable circuit losses on efficiency is minimized by connecting as many diode arrays in series as possible. The fraction of diode impedance to the total resistance of the capacitor discharge circuit has to be as close as possible to one. The maximum number of diodes which can be connected in series is set by the limits of the standoff voltage of the diode arrays. Diode drivers have been reliably operated up to 80 bars in series without breakdown.

In addition, MOSFET's and capacitors have to be selected for low internal resistance, and current leads and wires have to be kept as short as possible. Constant current to the laser diodes is maintained by regulating the internal impedance of the MOSFET. A large voltage droop during the discharge cycle is undesirable, because the excessive voltage must be dropped across the power MOSFET which results in a low driver efficiency at the beginning of the pulse. A small voltage droop requires a large capacitance, which stores many times the energy extracted during one pulse. The size of the capacitor is a trade-off between efficiency and physical size constraints of the power supply.

To illustrate some of the criteria in designing diode drivers, we will use, as an example, a diode-pumped Nd : YAG laser which produces 200 mJ in a Q-switched pulse. The laser is pumped by 16 diode arrays, each containing five 1-cm bars. All arrays and bars are electrically connected in series to increase impedance from $0.026\,\Omega$ for each bar to $R_\mathrm{D} = 2.09\,\Omega$. Each bar has a 60-W optical output and requires a 70-A pulse. The voltage drop across each bar is 1. 83 V. With the bars electrically connected in series, a voltage of 146.4 V is required across the diode array.

In series with the diode resistance R_D is the internal resistance R_M of the MOSFET, and a resistor R_L representing circuit losses due to wires, leads and the current sampling resistor. In Fig. 6.33, two MOSFETs are configured in parallel in order to achieve at least a 50 % current derating from manufacturer specifications for improved reliability and lifetime, and also to decrease the internal resistance of this switch.

Typical series resistance of a MOSFET is $0.2\,\Omega$ with two in a parallel configuration, series resistance is reduced to a minimum of $R_\mathrm{M} = 0.1\,\Omega$ for the fully forward biased condition. The wires, connectors, and the series resistance of the energy storage capacitors accounted for a total series resistance of $0.04\,\Omega$. The current sensing resistor R represents an additional impedance of $0.01\,\Omega$. Therefore the discharge side of the energy storage capacitor contains an additional resistive loss of $R_\mathrm{L} = 0.05\,\Omega$.

The combined impedance of the discharge circuit at the end of the pulse is $R = R_D + R_M + R_L = 2.24\,\Omega$ which requires a voltage at the capacitor of $V_2 = 156.8\,\text{V}$. If we design for a 10 % voltage drop during the pulse, then the voltage at the capacitor is $V_1 = 174.2\,\text{V}$ at the beginning of the discharge cycle, and the initial impedance of the MOSFET is $R_i = 0.35\,\Omega$ in order to limit the desired drive current to $I_D = 70\,\text{A}$. Total impedance at the beginning of the discharge cycle is therefore $R_T = 2.49\,\Omega$.

Based on the RC time constant of the circuit and the allowable voltage drop during the pulse, the capacitance of the energy storage capacitor follows from $C = -(t_p/R_T)\ln(1 - \Delta V/V_1)$. With $t_p = 200\,\mu s$, $R_T = 2.49\,\Omega$, $\Delta V/V_1 = 0.1$, one obtains $803\,\mu F$. The energy delivered to the diodes is $E_D = R_D I_D^2 t_p = 2.05\,\text{J}$, and the energy extracted from the capacitor is $E_{EX} = C(V_1^2 - V_2^2)/2 = 2.31\,\text{J}$. The efficiency of the diode driver is therefore $\eta = 0.88$.

Actually, this overall efficiency can be viewed as the average of the efficiency at the beginning and the end of the discharge cycle. In this particular case, efficiency starts at 84 % and increases to 93 % at the end of the current pulse. The energy transfer efficiency can also be expressed by

$$\eta = \frac{R_D}{R_D + R_M + R_L}\left(1 - \frac{\Delta V}{2V_1}\right)$$

The first term on the right is the limit of efficiency determined by circuit losses, whereas the second term is due to the dissipation of excess energy stored in the capacitor by the MOSFET in order to maintain a constant current. In our example, circuit losses limit efficiency to 93 %, and voltage droop limits efficiency to 95 %.

6.2.2 Operation of Arc Lamps

Arc lamps require dc voltage for operation, and they need to be triggered by a high-voltage pulse. Therefore, a power conditioning system which converts line power, or the voltage from an onboard system to the appropriate dc level, together with a trigger unit, are essential parts of the electronics needed for operating arc lamps. In addition, pulsed flashlamps require energy storage with capacitors combined with a pulse shaping device such as pulse forming network or switching electronics.

Power Supply for CW Arc Lamps

A trigger circuit is required to start the lamps, and in some cases when the lamps are filled at high pressure an additional high-voltage boost must be provided. A simplified schematic of a power supply which is employed in a large number of krypton arc pumped Nd : YAG lasers is shown in Fig. 6.34. The input is applied to a three-phase isolation transformer. Each output of the secondary of the 6 kV power transformer is connected to the center of three pairs of series connected SCRs. This is the conventional three-phase bridge rectifier configuration, with the exception that the rectifiers are SCRs. The control unit senses the analog

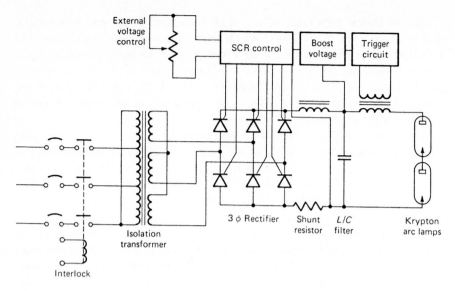

Fig. 6.34. Phase-controlled power supply for use with krypton arc lamps

input from a current-sampling resistor, compares it with a reference voltage, and generates phase-controlled pulses which fire the SCRs. Control of the SCR conductance time provides regulation of the supply. The unit operates as a variable time switch for each leg of the three-phase ac input. The output of a rectifier is applied to a L/C filter network for ripple elimination. The ignitor provides the high-voltage pulse to start the lamp. This is done by discharging a capacitor with a SCR through the primary of a trigger transformer. The transformer steps up this voltage to over $30\,\text{kV}$, which is sufficient to ignite the lamp. For reliable lamp starting, the voltage of the main power supply must be boosted to about $600\,\text{V}–1\,\text{kV}$ during the trigger phase. This is most conveniently done by charging the filter capacitor to this voltage with a small, low-current, high-voltage supply. Power supplies of the type shown in Fig. 6.34 have been built for lamp inputs of up to $20\,\text{kW}$.

Power Supply for Flashlamps

The major components of a power supply employed in a flashlamp-pumped laser are a charging power supply, capacitive energy storage, a pulse shaping device, and a flashlamp trigger unit. A Pulse-Forming Network (PFN) combines the function of energy storage and pulse shaping. A flashlamp pulse can be initiated and controlled essentially in three different ways:

a) The flashlamp serves as a switch. In this case, energy is stored in a pulse forming network which is directly connected to the flashlamp. The flashlamp is ignited by a trigger circuit which then discharges the energy stored in the capacitor of the PFN into the flashlamp. Pulse duration and pulse shape are determined by the PFN. The lamp extinguishes after the capacitor is discharged. Simplicity is the major advantage of this design which requires no high-voltage and current-switching elements. Examples of this technique, which is particularly common in single-shot devices, are illustrated in Figs. 6.37, 38, 40, 41.

b) Discharge controlled by a separate on-switch. In this technique the PFN is connected to the flashlamp via a switch such as an SCR, or a gas or mercury filled tube. These switches can be turned on but not off. The pulse shape is determined by the PFN, as before, and the switch, as well as the flashlamp, is turned off once the capacitor is discharged. There are several reasons why such a switch is advantageous. For example, under high-repetition-rate operation the PFN can be recharged while the flashlamp is still ionized. The switch prevents firing of the lamp in cases where the voltage of the capacitor bank is higher than the ignition voltage of the flashlamp. And foremost, a separate switching device permits operation of the flashlamp in the so-called simmer mode, which is a low-level continuous ionization. This mode of operation increases lamp life, allows monitoring of the lamp status and provides for a more reliable performance. Examples of charging supplies with separate on-switches are shown in Figs. 6.36, 51, 52.

SCR's have also been used as on-switches in situations where the flashlamp energy is directly supplied by a three-phase powerline. This is possible in applications, such as materials processing, which require relatively long pulse width and modest peak currents. An isolation transformer contains a bridge rectifier comprised of phase controlled SCR's. The combination of the rectified ac waveform and the firing point of the SCR's provide control of flashlamp pulse length and energy. The lamp extinguishes when the rectified half-wave voltage reaches zero. The basic power supply is very similar to the unit exhibited in Fig. 6.34 without the L/C filter.

c) Pulse shape controlled by an on-off switch. The availability of high-voltage and current semiconductor switches such as GTO's (gate turn-off SCR), MOSFET's and IGBT (Insulated Gate Bipolar Transistor) make it possible to design very compact and flexible flashlamp charging units if the lamp operating parameters are within the capabilities of these devices. In such a charging unit, a fraction of the energy stored in electrolytic capacitors is transferred to the flashlamp at each pulse. The externally controlled switch determines the pulse width, which in this case can be varied over a wide range. Very large energy storage banks or flashlamps pumped with very short pulses, often require high voltages and currents which are beyond the ratings of these switching devices. An example of a charging unit with a programmable switching device is illustrated in Fig. 6.39.

Charging Unit

The function of the charging unit is to charge the energy storage capacitor to a selected voltage within a specified time, which depends on the desired repetition rate of the laser. The capacitor-charging source usually consists of a transformer followed by a rectifier bridge, a switching element in the primary of the transformer, a current-limiting element, a voltage sensor, and control electronics. The transformer and the rectifier bridge provide the required dc voltage for the energy-storage capacitor. In order to be able to vary this voltage and, therefore, to obtain a variable output energy from the laser, a semiconductor switch is usually included in the primary circuit of the transformer. This control element, which can be either a triac, a solid-state relay, or a pair of inverse parallel SCRs, is turned on at the beginning of the charge cycle and turned off when a preset voltage is reached in the capacitor. Control signals are derived from the capacitor voltage as in conventional dc supply designs. The charging of a capacitor presents a problem insofar as the discharged capacitor constitutes a short circuit. Without a current-limiting device in the power supply, the short-circuit current is limited only by the resistance of the transformer windings. To protect the rectifier diodes and other components in the circuit, current-limiting circuits are required. Ideally, one would like a constant current supply, in which case the capacitor-charging current would be constant over the entire charging cycle. The charging current in this case would be

$$I = CV/t , \tag{6.16}$$

where C is the capacitance of the capacitor, V is the final charging voltage, and t is the charging time. Figure 6.35 depicts a number of circuits which are frequently employed to limit the short-circuit current into a capacitive load. The most straightforward way to charge capacitors is resistance-limited charging from a constant-voltage source, as shown in Fig. 6.35a. The least amount of power dissipated in the current-limiting resistor is equal to that stored and discharged in the capacitor if three or more time constants are used in charging. For shorter charging times, even more power is wasted in the resistor. For low-power, low-repetition-rate systems the heating losses usually can be tolerated. Since capacitor charging to high energy levels, or at high repetition rates, causes appreciable power losses when charging through a resistor, attempts have been made to produce power supplies without the resistor, but retaining the resistor's function of limiting initial capacitor-charging currents to safe values.

Figure 6.35b displays current limitation by an inductor in the primary of the transformer. During initial charging, when the secondary of the transformer is shorted by the discharged capacitor, the peak current is limited by the inductive reactance. The inrush current equals the primary voltage divided by the reactance. As a refinement of this technique, the inductance can be built into the transformer as primary leakage (Fig. 6.35c).

At repetition rates approaching the power-line frequency, it becomes difficult to achieve repeatable output voltage from charge to charge because of the limited

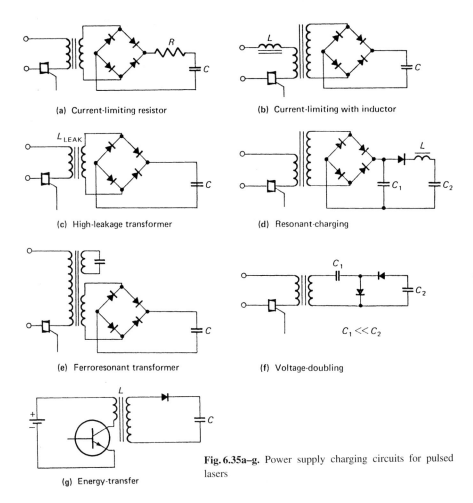

(a) Current-limiting resistor

(b) Current-limiting with inductor

(c) High-leakage transformer

(d) Resonant-charging

(e) Ferroresonant transformer

(f) Voltage-doubling

$C_1 \ll C_2$

(g) Energy-transfer

Fig. 6.35a–g. Power supply charging circuits for pulsed lasers

number of current pulses within the charging period. For these cases it becomes necessary to use resonant charging (Fig. 6.35d). *Glascoe* [6.53] gave a complete analysis of the circuit. Current will flow from the dc source for the first half-cycle of the resonant frequency, charging the capacitor to twice the source voltage. The peak current drawn from the supply depends on the inductor, the resonant frequency of the LC circuit, and the voltage to which the capacitor is charged.

The capacitance C_1 shown in Fig. 6.35d is a filter to aid voltage regulation and is usually ten times the bank capacitance C_2. The constant-voltage and short circuit current characteristics of ferroresonant line regulator transformers have been used to approximate a constant-current source for capacitor charging. Figure 6.35e shows the typical ferroresonant transformer with its associated components. The current-limiting characteristic of the ferroresonant transformer is attained through the use of a magnetic shunt and a resonant tank circuit. In addition to resistive and inductive current limitations, the short-circuit current can be reduced by repetitively transferring a small amount of energy to the energy

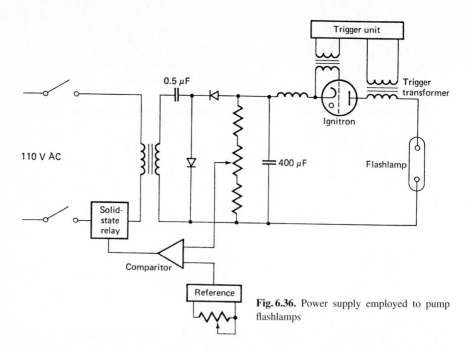

Fig. 6.36. Power supply employed to pump flashlamps

storage capacitor. In the voltage-doubling circuit depicted in Fig. 6.35f, the small capacitor C_1 transfers its charge to the main capacitor C_2 during each cycle of the ac line. A circuit in which magnetically stored energy is transferred to a capacitor is shown in Fig. 6.35g.

In the following, we will describe a few typical power supplies employed in pumping solid-state lasers. Figure 6.36 exhibits the schematic of a power supply used to pump flashlamps of high-energy Nd : glass lasers. The lasers are operated at a repetition rate of a few pulses per minute. The power supply essentially consists of a standard power transformer with a solid-state relay in the primary, a voltage-doubling circuit, an energy storage capacitor, an ignitron, and a trigger circuit. The voltage across the 400 μF energy storage capacitor is sampled by a high-impedance voltage divider, and the voltage is fed to a comparitor. If the input is equal to a reference voltage, an output from the comparitor will turn off the solid-state relay. This disconnects the transformer from the line so the voltage on the capacitor remains at a constant value. By changing the reference voltage the capacitor voltage can be selected. The small 0.5-μF capacitor provides a constant charge rate of the energy-storage capacitor and, as discussed before, provides current limitation. However, many hundred charge cycles are required before the energy from the small capacitor has charged up the energy storage capacitor. Therefore, this charge-transfer technique is useful only for low-repetition-rate systems.

As we will see in the next section, large flashlamps require high voltages of the energy-storage capacitor because of their high impedance. Since these lamps will usually self-trigger at voltages lower than the storage bank voltage, a holdoff

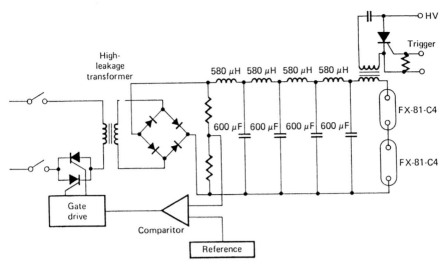

Fig. 6.37. Power supply featuring a high-leakage transformer

device is required. Lamp breakdown during the charge cycle is usually prevented by an ignitron. The ignitron is a mercury-filled tube which is capable of handling very high currents. Triggering of the flashlamp is initiated by a 15- to 20-kV trigger pulse in the series injection trigger transformer and by simultaneously triggering the ignitron to allow bank discharge. In Fig. 6.36 an additional coil for shaping the flashlamp current pulse is shown. For lamps which are operated considerably below their self-triggering voltage, the ignitron may not be required. The holdoff voltage of a 7.5-cm-long linear flashtube is around 3.5 kV, whereas for a lamp with an arc length of 72 cm, holdoff voltage is around 6 kV.

Figure 6.37 shows a power supply which can charge a 1-kJ capacitor bank at a repetition rate of 10 pps. The unit features a high-reactance current-limiting transformer, with two inverse parallel SCRs as a switching element in the primary, and a full-wave rectifier bridge. Since this power supply is for a Nd:YAG laser welder, long pulses of the order of 5 ms are required. As we will see in the next subsection, this requires a multiple-mesh network. The trigger circuit, which we will also discuss in more detail later, consists essentially of a trigger transformer with a capacitor and a SCR in the primary. When the capacitor is discharged, a high voltage is generated at the secondary which breaks down the lamp. The power supply is designed to operate two linear flashlamps (FX81-C4). Each lamp has a bore of 10 mm and a length of 10 cm. This power supply can also be built as a three-phase supply. In this case, three separate transformers and voltage-rectifier bridges are connected in parallel at the dc output end.

For high repetition rates a resonant charging supply is best used. Figure 6.38 shows the simplified schematic of a power supply which is employed to pump a Nd:YAG laser at repetition rates of up to 50 pps. The laser is pumped by a single linear flashlamp with 5-mm bore diameter and 50-mm arc length at a maximum

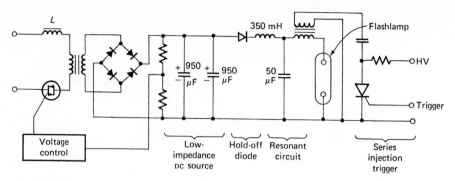

Fig. 6.38. Resonant charging of the energy storage capacitor in a high-repetition-rate system

of 20 J. The dc section of the power supply represents a variable voltage source. In the primary circuit of the transformer is an inductor for current limitation and a triac for voltage control. The stepped-up voltage is rectified by a full-wave bridge and filtered by means of a bank of 950-μF electrolytic capacitors. The dc output voltage is controlled by a circuit which senses the voltage and generates trigger pulses to the gate of the triac. The resonant charging device consists of a hold-off diode, a 50-μF energy storage capacitor, and a 350-mH inductor. The flashtube trigger is generated by switching the energy from a small storage capacitor into the trigger transformer primary. The secondary of this transformer is connected in series with the pulseforming network (series injection triggering). When a flashlamp is operated at high repetition rates, the recovery characteristic of the lamp must be taken into account. After cessation of current flow, the lamp remains ionized for a time on the order of a few milliseconds.

A type of power supply often employed in Nd:YAG and Nd:glass lasers, mainly for materials processing applications, is illustrated in Fig. 6.39. In applications such as spot welding the desired laser pulse is relatively long on the order of (0.5–2) ms. This often reduces the flashlamp voltage to below 2 kV which makes it possible to use controllable on-off switches such as GTO's, MOSFET's and IGBT's. Such a device has the advantage that the flashlamp pulse width is continuously variable. Also, the energy can be stored in electrolytic capacitors which results in a very compact and flexible flashlamp driver unit.

Figure 6.39 depicts a schematic diagram of a flashlamp charging unit which contains an IGBT as switching element. The high-voltage supply on the right-hand side of Fig. 6.39 charges a small capacitor, C_1 to a sufficiently high voltage, about (8–15) kV until the flashlamp is ignited. Once the flashlamp is conducting, the simmer supply takes over through diodes D_1 and D_2, and the trigger unit, no longer needed, is turned off by the automatic controller. Diode D_1 protects the left part of the circuit from the high voltage trigger pulse, and diode D_2 protects the simmer supply from the high voltage when the IGBT is conducting. Since IGBT's have a backward conducting "free wheel" diode built in, diode D_3

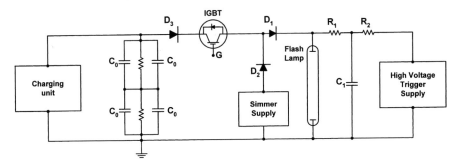

Fig. 6.39. Power supply with programmable pulsewidth control and electrolytic capacitor energy storage

prevents that the simmer supply current is diverted to charge the capacitor bank C_0.

In most cases, strings of diodes, rather than single devices, will be needed for D_1 to D_3 to meet the high-voltage equirements. A constant current power supply charges a capacitor bank comprised of electrolytic capacitors. The capacitor bank stores at least ten times the energy discharged into the flashlamp at each pulse. This is important because electrolytic capacitors are specified for a maximum ripple current. Electrolytic capacitors provide the most compact and economical means of storing electric energy. Firing time and pulse width of each flashlamp pulse is controlled by a trigger input delivered to the gate of the IGBT from a control unit.

We found that in high-current and high-voltage applications IGBT's tend to have a smaller voltage drop during the pulse compared to MOSFET's. The typical saturation voltage across the IGBT's is between 3.5 and 5 V. Currently the most powerful IGBT (Toshiba ST 1000 EX21) has maximum ratings of 2500 V for the collector-emitter voltage and 2000 A forward current under pulsed conditions. The collector-emitter saturation voltage is 5 V.

Power supplies used for military-type Nd : YAG lasers, such as rangefinders, target designators, or illuminators, mostly operate from batteries or on-board dc power lines. Figure 6.40 sketches a simplified schematic of the power supply of an airborne target designator, which operates from 28 V dc. The unit, which consists of a dc-to-ac converter, a step-up transformer, a high-voltage rectifier, a pulse-forming network, and a parallel trigger circuit, is capable of charging the capacitor to an energy of 20 J at a repetition rate of 10 pps. The flashlamp (ILC L-213) has a bore of 4 mm and an arc length of 50 mm. Current is limited by an inductor on the high-voltage dc side of the unit. A feature of this power supply is the fact that the repetition rate of the dc-to-ac converter progressively increases from 1 kHz to about 5 kHz during the charge cycle in order to keep the peak-to-peak rms current drawn from the battery at a low value. This is achieved by a current sensing network which feeds a signal to the logic circuit, which in turn controls the switching frequency of the transistors in the dc-to-ac converter.

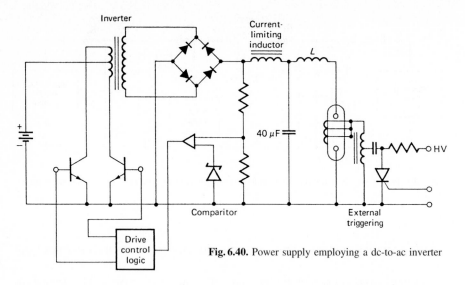

Fig. 6.40. Power supply employing a dc-to-ac inverter

At full capacitor charge, a voltage sensor turns off the driving signals to the transistors. A schematic diagram of another type of power supply used in small military systems is shown in Fig. 6.41. The system operates on the principle of a flyback dc-to-ac converter. The transistor is turned on for a short period of time to allow current flow in the primary of the inductor. When the transistor is turned off, the magnetic energy stored in the inductor is transferred to the energy storage capacitor; at the same time the voltage is stepped up from 28 V to approximately 800 V. The device is short circuit protected since the inductive voltage kickback charges the capacitor after the transistor has turned off. The transistor is operated typically at a repetition rate of 1 to 10 kHz in order to keep the inductor small. For a 10 pps system this allows between 100 and 1000 transfers of energy from the inductor to the capacitor.

Further references on the design of power supplies employed to operate flash-lamps can be found in [6.18, 54–57]. The special problems which arise in power supplies designed to charge capacitor banks with stored energies in the MJ range have been discussed in [6.58].

Pulse-Forming Network

Flashlamps are usually operated from a single- or multiple-mesh LC network. The network stores the discharge energy and delivers it to the lamp in the desired current pulse shape. In most situations, the lamp input energy E_0, the pulse width t_p, and the lamp dimensions have been determined before the pulse-forming network is designed. The above-mentioned parameters completely describe the network's capacitance and inductance as well as the charging voltage and peak current. To familiarize the reader with the design of flashlamp energy-storage networks, we will consider first the elementary model of a pulse discharge circuit (Fig. 6.42). The capacitor C is charged to an initial voltage V_0. At time $t = 0$ the

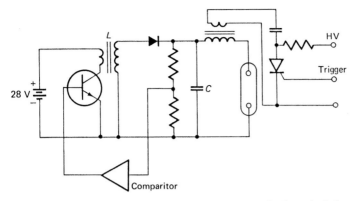

Fig. 6.41. Battery-operated power supply with energy transfer from the inductor L to the PFN capacitor C

switch S is closed and the energy stored in the capacitor is delivered to the load. We assume for the moment that the load is a linear resistor R. In this case the circuit is described by the second-order differential equation

$$L\frac{d^2q}{dt^2} + R\frac{dq}{dt} + \frac{q}{C} = 0 , \qquad (6.17)$$

where q is the charge on the capacitor. With the initial charge on the capacitor $q(0) = CV_0$ and the initial current $dq/dt(0) = 0$, the voltage across R is given by

$$v(t) = V_0(\alpha/\beta)\{\exp[-(\alpha - \beta)t] - \exp[-(\alpha + \beta)t]\} , \qquad (6.18)$$

where $\alpha = R/2L$ and $\beta = (R^2/4L^2 - 1/LC)^{1/2}$.

Depending on the relative values of R, L, and C, the voltage or current waveform across R may be as follows:

Critically Damped. In this case, $\beta = 0$ and

$$R = 2\left(\frac{L}{C}\right)^{1/2} . \qquad (6.19)$$

The capacitor voltage decays exponentially from V_0 to zero. If we define the rise time t_r as the time required for the voltage or current at R to rise from zero to its maximum value, we obtain

$$t_r = (LC)^{1/2} . \qquad (6.20)$$

The rise time measured between the 10 and 90 % points of the voltage is $0.57t_r$. The discharge current as a function of time is

$$\frac{i(t)}{i_p} = e\left(\frac{t}{t_r}\right)\exp\left(-\frac{t}{t_r}\right) , \qquad (6.21)$$

where the peak discharge current is

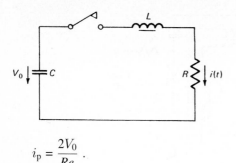

Fig. 6.42. Single-mesh discharge circuit

$$i_p = \frac{2V_0}{Re} \; .$$

Overdamped. The waveform is similar to the critically damped pulse, except that the peak current is reduced and the time required to reach zero is extended. The overdamped case (β real) is defined by

$$R > 2 \left(\frac{L}{C} \right)^{1/2} . \tag{6.22}$$

Underdamped. The resulting pulse shape is characterized by strongly damped oscillations (β imaginary). The frequency of the oscillation depends mostly on L and C, whereas the decaying amplitude is a function of R only. Oscillations and the associated current and voltage reversal in the discharge circuit occur if

$$R < 2 \left(\frac{L}{C} \right)^{1/2} . \tag{6.23}$$

Nonlinear Load. As we have seen in Sect. 6.1, the flashlamp resistance is non-linear and can be described by $R_L(i)$ given in (6.4) or by the lamp impedance parameter K_0 defined in (6.1). The design of single-mesh circuits for driving flashlamps has been considered in [6.13, 59]. In order to apply the results published in the literature, we must make the following substitutions and normalizations:

$$Z_0 = \left(\frac{L}{C} \right)^{1/2} , \quad \tau = \frac{t}{T} , \quad T = (LC)^{1/2} , \quad i = \frac{IV_0}{Z_0} . \tag{6.24}$$

In addition, a damping factor α is defined which determines the pulse shape of the current pulse. It is

$$\alpha = K_0(V_0 Z_0)^{-1/2} . \tag{6.25}$$

Solutions of the nonlinear differential equation of the lamp current for various values of α have been reported in Ref. [6.13]. The current waveform for different values of α is shown in Figs. 6.43 and 44. As can be seen from these curves, $\alpha = 0.8$ corresponds to the critically damped case. The current pulse duration t_p at the 10 % points is approximately $3T$; during this time about 97 % of the energy has been delivered. If we substitute

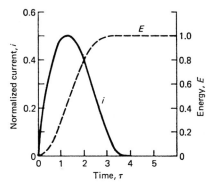

Fig. 6.43. Normalized current and energy of a critically damped flashlamp discharge circuit. Pulse shape factor $\alpha=0.8$

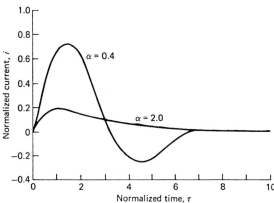

Fig. 6.44. Overdamped and underdamped current waveforms of a lamp discharge circuit

$$\alpha = 0.8 \quad \text{and} \quad t_\mathrm{p} = 3T = 3(LC)^{1/2} \tag{6.26}$$

for a critically damped pulse into (6.24, 25), then we can determine for a given lamp type the relationship between energy input, pulse duration, pulse shape, inductance, capacitance, and operating voltage.

The energy initially stored in the capacitor is

$$E_0 = \tfrac{1}{2}CV_0^2 \; . \tag{6.27}$$

The relation is used to eliminate V_0 from (6.25). With $\alpha = 0.8$ we obtain from (6.25) the value of the capacitor

$$C^3 = 0.09 \frac{E_0 t_\mathrm{p}^2}{K_0^4} \; . \tag{6.28}$$

The inductance can be calculated from (6.26)

$$L = \frac{t_p^2}{9C} .$$
(6.29)

We now have a set of three equations (6.27–29) which, given the specifications of the flashlamp parameter K_0, the desired input energy E_0, and pulse width t_p, provide explicit values of C, L, and V_0. If the lamp parameter K_0 is not specified by the lamp manufacturer, it can be calculated from (6.5). From Fig.6.43 also follows the peak current for a critically damped current pulse

$$i_p = 0.5 \frac{V_0}{Z_0} .$$
(6.30)

The rise time to reach this peak value is

$$t_r = 1.25(LC)^{1/2} \cong 0.4 t_p .$$
(6.31)

Differences between these solutions and those for the linear load LCR circuit are readily apparent. For a given Z_0, V_0, and peak current, the solution of the nonlinear equation is more heavily damped than that of the linear case. The damping parameter α is dependent on V_0, the initial capacitor voltage. This is consistent with the experience that a discharge circuit, critically damped at one voltage, begins to ring as the voltage is increased.

Multiple-Mesh Networks

For pump pulses with a more rectangular pulse shape, multiple-mesh networks are used. These consist of two or more LC networks in series. Switching an open-ended charged transmission line into a resistor equal to its characteristic impedance yields a rectangular pulse across the resistor. The width of the pulse equals twice the propagation time of the line. This same technique can be applied to a lumped parameter line. The principal advantage of the lumped-parameter delay line as an energy-storage pulse-forming network is its nearly constant output over the length of the pulse. This property is especially useful in producing long normal-mode laser pulses of constant power.

Most pulse-forming networks employ the "E"-type circuit of a lumped constant transmission line. In this configuration, all capacitors are of equal value and the inductance values per mesh are nearly identical (Fig. 6.45). Following are the approximate equations of a lumped-type "E" transmission line terminated by its characteristic impedance [6.53]. The characteristic impedance of the network is

$$Z_n = \left(\frac{L_T}{C_T} \right)^{1/2} ,$$
(6.32)

where L_T and C_T are the total inductance and total capacitance. The pulse width at the 70 % point is

$$t_p = 2(L_T C_T)^{1/2} .$$
(6.33)

With the pulse duration known, the total capacitance follows from

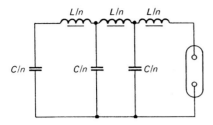

Fig. 6.45. Multiple-mesh network

$$C_T = \frac{t_p}{2Z_n} , \tag{6.34}$$

and the total inductance is

$$L_T = \frac{t_p Z_n}{2} . \tag{6.35}$$

In order to deliver the required energy, the capacitors of the network must be charged to a voltage

$$V = \left(\frac{2E}{C_T}\right)^{1/2} . \tag{6.36}$$

The peak current during the discharge from the pulse-forming network is given by

$$i_p = \frac{V_0}{2Z_n} . \tag{6.37}$$

From (6.34, 35) it follows that the values of L_t and C_t are determined by the desired pulse width and the lamp impedance $R = Z_n$. The desired lamp energy input determines the charging voltage.

Once L_T, C_T, and V_0 are calculated, the desired rise time determines the number of circuit meshes in the network. Experiments have shown that current rise times of 50 μs for lamps pumped up to the 100-J level and rise times of 200 μs for lamps pumped in the kilojoule regime can cause severe electrode sputtering and crazing of the flashtube. The number of sections chosen to make up the pulse-forming network is determined by first considering the rise time and then a convenient number of components. If n is the number of meshes, then the rise time from the 10 % to the 80 % level is

$$t_r = \frac{t_p}{2n} , \tag{6.38}$$

and the fall time (80 % to 10 %) is

$$t_f = 3t_r . \tag{6.39}$$

The capacitance and inductance per mesh are

$$C_M = \frac{C_T}{n} \quad \text{and} \quad L_M = \frac{L_T}{n} . \tag{6.40}$$

Network Components

Inductors for pulse-forming networks are usually air-core coils, except for small systems where an iron core may be used. Besides inductance, other parameters of interest are the operating voltage and the rms current which determine insulation and wire size. Energy storage capacitors ranging from 1 to 10 kV and energy levels from 10 to 5000 J are commonly used. Because space is usually at a premium, energy storage capacitors are customarily designed with a higher dielectric stress than is usually employed for conventional dc capacitors. These low-inductance capacitors are designed for quick charge and discharge. The capacitor's dc life may be entirely different from its pulse discharge lifetime. Large discharge capacitors with storage energies in the kilojoule range are usually castor oil-impregnated paper dielectric capacitors. These capacitors are designed for energy densities of 0.04 to 0.07 J/cm^3 and 30 to 65 J/kg. In pulse discharge capacitors using high-strength dielectric materials, such as polyester or polypropylene film, energy densities of up to 0.5 J/cm^3 and 270 J/kg can be achieved. These capacitors, which are used mostly in military-type systems, have storage energies of up to a few hundred joules. Capacitor lifetime is determined by the charging voltage, the percentage of voltage reversal, and the operating temperature. In most practical situations, where a critically damped or overdamped pulse is generated and operation occurs below the peak rated voltage and at temperatures under 50° C, the capacitor life is not a serious consideration compared to other component failures.

In Figs. 6.46 to 49 typical current waveshapes are shown which are obtained from single- and multiple-mesh networks for various types of lamps. One usually finds that the final selection of the elements of a discharge circuit depends on commercially available components rather than on the exact calculated values. Furthermore, a laser system is in most cases operated over a range of input energies, and theoretically the optimum network would have to be slightly different for each input level.

We will illustrate the procedure for calculating a multiple-mesh network for the pulse shape shown in Fig. 6.46. A 5-ms-long current pulse is desired for two linear lamps connected in series and pumped at a total energy input of 1 kJ. Each of the lamps (EGG-FX81-4) has an arc length of 10 cm and a bore of 1 cm. From (6.4) it follows that with $l = 10$ cm and $D = 1$ cm, there is a resistance of $R = 0.5\,\Omega$ for each lamp if we assume a peak current of $i_p = 650$ A. As we will see, the design of a PFN is an iterative process. First, a peak current is assumed which permits one to calculate the lamp impedance. With this parameter known the network can be designed. However, at the end of the analysis the peak current has to be rechecked. With $Z_0 = 1\,\Omega$ for the total lamp resistance, it follows from (6.34–36) that $C_T = 2500\,\mu$F, $V = 900$ V, and $L_T = 2.5$ mH. Rechecking the peak current, we find from (6.37) a value of 450 A, which means the values of C_T and L_T must be recalculated using another assumed i_p. To obtain a fairly flat 5-ms-long pulse, a four-mesh network, each section containing a 600-μF capacitor and a 580-μH inductor, is chosen. The pulse-forming network is shown in Fig. 6.37

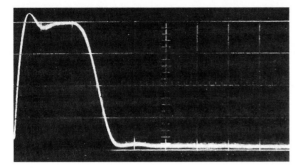

Fig. 6.46. Current waveform obtained from circuit shown in Fig. 6.37 at lamp input energy of 1 kJ and a bank voltage of 900 V. (*Horizontal*) 2 ms/div; (*vertical*) 100 A/div

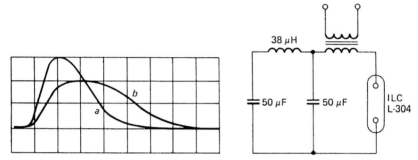

Fig. 6.47. Current pulse obtained from a (*a*) single- and (*b*) double-mesh network for 25-J input (ILC lamp model L-304). (*Horizontal*) 50 μs/div; (*vertical*) 200 A/div

and the current waveform is illustrated in Fig. 6.46. The measured rise time of the current pulse, which is about 0.6 ms, follows also from (6.38).

Figure 6.47 exhibits the current pulse of a single- and a double-mesh network. Each section consists of a 50-μF capacitor and a 38-μH inductor. For the single-mesh network the inductance is actually the saturated inductance of the trigger transformer. The flashlamp is 50 mm long and has a bore of 3 mm (ILC L-304). Total input energy is 25 J. The current pulse into a 75-mm-long, 4-mm-bore flashlamp generated by a three-mesh network is shown in Fig. 6.48. Each section of the network contains a 33-μH inductor and a 120-μF capacitor. The total stored energy is 300 J. Figure 6.49 depicts the current waveform generated by a single-mesh network. The load is a helical flashlamp with a tube inside diameter of 8 mm and a total arc length of 72 cm. The lamp impedance parameter for this lamp was measured to be $K_0 = 150 \, \Omega \, A^{1/2}$. The values for the discharge circuit are obtained if one introduces $t_p = 500 \, \mu s$ and $E_0 = 2700 \, J$ into (6.27–29). The pulses shown in Fig. 6.49 correspond to different charging voltages of 5, 7, 9, and 9.5 kV. As can be seen from this figure, despite the large range in bank voltage, input energy, and peak current, the pulses are all approximately critically damped and have pulse widths of about 500 μs.

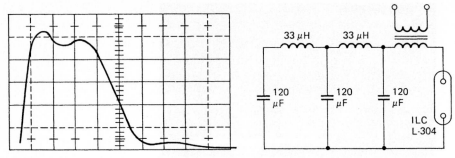

Fig. 6.48. Current pulse obtained from a three-mesh network. Input energy 300 J into ILC lamp model L-305. (*Horizontal*) 100 μs/div; (*vertical*) 200 A/div

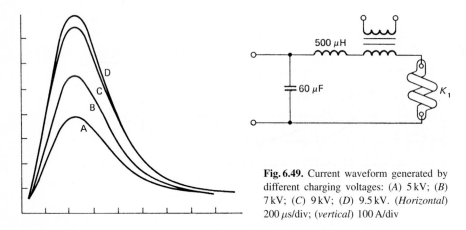

Fig. 6.49. Current waveform generated by different charging voltages: (*A*) 5 kV; (*B*) 7 kV; (*C*) 9 kV; (*D*) 9.5 kV. (*Horizontal*) 200 μs/div; (*vertical*) 100 A/div

Trigger Circuit

The discharge of the stored energy into the flashlamp is generally initiated by a high-voltage trigger pulse. The function of the trigger signal is to create an ionized spark streamer between the two electrodes so that the main discharge can occur. The initial spark streamer is formed by the creation of a voltage gradient of sufficient magnitude to ionize the gas column. The concept of a voltage gradient is important here, since it implies the existence of a stable voltage reference surface in close proximity to the flashlamp. Regardless of the triggering method used, reliable triggering cannot be achieved without this reference. Usually, an equipotential condition on the outside wall of the tube is achieved through the use of a wire wrapped around the flashlamp or by the proximity of the metal parts of the pump cavity. The ignition process can be explained by assuming that areas of the inside glass wall behave as small electrodes capacitively coupled to the reference plane. When the trigger voltage is applied, a discharge takes place between the cathode and the nearby part of the inside wall; this part will then be almost at cathode potential. Next, a discharge takes place between this part of the glass wall and a more remote area still at high potential, and so on until the anode is reached [6.60].

The two most common methods of triggering flashlamps are external and series injection triggering. Other techniques are the simmer and pseudosimmer triggers. In the external trigger device a wire is wrapped around the flashlamp between the electrodes and connected to the secondary of a high-voltage transformer. A high voltage is generated by discharging a capacitor through the primary of the transformer. The switching element can be either a SCR or krytron. The trigger voltages of flashtubes are between 5 and 10 kV. Figure 6.40 shows an example of external triggering. The waveform of a typical flashlamp trigger pulse is given in Fig. 6.50. The pulse shown was generated by discharging a 33-μF capacitor at a voltage of 500 V through the primary of a trigger transformer (EGG model TR132) with a turns ratio of 53 : 1.

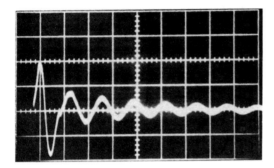

Fig. 6.50. Typical flashlamp trigger pulse. (*Horizontal*) 10 μs/div; (*vertical*) 10 kV/cm

Sometimes it is undesirable to have a wire wrapped around the flashlamp. For example, the possibility of accidental voltage break-down to the surrounding pump cavity is greatly increased by the presence of a trigger wire. In such applications, a triggering scheme which does not require an external trigger wire is highly desirable. In the series injection method, a pulse is generated in a transformer whose secondary winding is in series with the flashlamp. The high-voltage pulse causes initial ionization of the plasma. The large lamp current which follows saturates the transformer so that its inductance is low. Examples of the series trigger technique are shown in Figs. 6.36–38, 41, and 49. Each method has its own particular advantages. External triggering provides greater design flexibility, because the secondary winding is not in the main discharge circuit. External triggering is the simplest, lightest, and most commonly used technique. Series triggering, on the other hand, is the choice when no exposed high-voltage leads are permitted. When a series trigger circuit is used, the secondary of the trigger transformer is part of the lamp discharge circuit, therefore the resistance of the secondary winding should be much less than the flashlamp resistance. This makes the series injection trigger transformer relatively bulky and heavy.

On the other hand, series triggering produces generally better pulse-to-pulse uniformity and reproducibility than external triggering. Also, in some cases longer flashtube life and more efficient pumping have been observed. These observations may possibly be explained by the fact that when a lamp is series-triggered, the core of the trigger transformer is not saturated and initial current flow is limited by the unsaturated inductance of the transformer. It appears that the reduced current growth rate during the arc expansion phase results in a more uniform arc formation and an associated smaller shock wave. An extension of this reasoning has led to the development of the so-called simmer and pseudosimmer triggering schemes. In these schemes a low-current discharge is maintained in the lamp between pulses. This keep-alive discharge is accomplished by employing a low-current dc power supply in parallel with the main supply. A typical circuit is shown in Fig. 6.51.

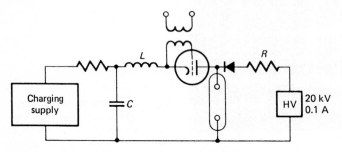

Fig. 6.51. Power supply with keep-alive discharge circuit

The simmer mode of operation requires a switching element between the lamp and the pulse-forming network. The lamp is initially ignited by the open-circuit voltage of 20 kV of the high-voltage power supply. At steady state the voltage drop across the lamp is between 500 V and 1 kV. Using this technique, considerable improvements in system performance can be achieved, including the following:

1) Increased flashlamp life: An order-of-magnitude improvement in lamp life can be achieved in employing a keep-alive discharge [6.20]. Improvement is due mainly to a drastic decrease in cathode sputtering which is caused to a large degree by the high-voltage surge across the lamp in standard trigger schemes.

2) Reduced coolant degradation: The fluorinated hydrocarbons, such as 3M FC-104 and Dupont E-4, possess the temperature characteristics required for military equipment, but they tend to degrade from ultraviolet irradiation. A major source of this radiation is the high-voltage ignition which is eliminated in the simmer mode.

3) Reduced EMI/RFI radiation: A difficult requirement, particularly in military laser systems, is the reduction of broad-band rf radiation. For example, a laser operating at 10 pps presents three major electric-field noise sources, the

flashlamp igniter pulse, flashlamp current pulse, and Pockels-cell switching pulse. The first, and often the largest, source of this type of noise is eliminated in the simmer mode.

4) Increased efficiency: It has been demonstrated that in some cases the simmer mode can provide increases in Kr and Xe flashlamps' pumping efficiency for lamps operated with a pulse length of 50 to 200 μs and pulse energy levels of 5 to 20 J.

5) Lamp status control: By monitoring the keep-alive current, a broken flashlamp or a short circuit can be detected before the energy storage capacitor is discharged. This is particularly important in large systems or in systems employing many flashlamps.

The major disadvantage of this mode of flashlamp operation is the added electronics and the power consumption of the continuous discharge. Small 7.5-cm long linear lamps are operated at about 30 mA dc and consume between 5 and 10 W. To reduce the power consumption of the simmer mode, a pseudosimmer was devised that combines the advantage of the light-weight external trigger transformer with the improvements achieved in a simmer-mode system. Figure 6.52 shows a schematic block diagram of the circuit which has been used in small Nd : YAG systems. The lamp is ignited with an external trigger transformer. Lamp current initially flows through a limiting resistor, which is in parallel with the pulse-forming inductor and the SCR. Current flow through the inductor is prevented by the SCR. After an appropriate time delay, the SCR is turned on, and the high-current pulse is initiated.

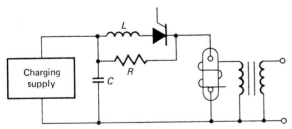

Fig. 6.52. Power supply with pseudosimmer discharge

6.3 Pump Cavities and Coupling Optics

The efficiency in the transfer of radiation from the source to the laser element determines to a large extent the overall efficiency of the laser system. The pump cavity, besides providing good coupling between the source and the absorbing active material, is also responsible for the pump density distribution in the laser element which influences the uniformity, divergence, and optical distortions of the output beam.

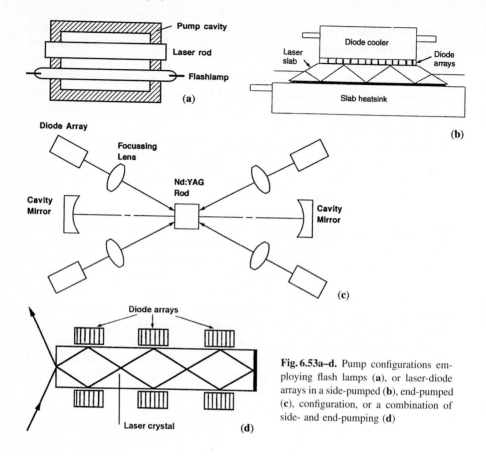

Fig. 6.53a–d. Pump configurations employing flash lamps (**a**), or laser-diode arrays in a side-pumped (**b**), end-pumped (**c**), configuration, or a combination of side- and end-pumping (**d**)

In the development of solid-state lasers, many different optical designs have been employed to transfer the radiation from a light source to the active material. Depending on the shape of the active material and the type of pump source used, one can broadly divide pumping geometries into systems in which the active material is side-pumped, end-pumped, or face-pumped.

In a side-pumped illumination system, the active material, shaped in the form of a cylindrical rod or rectangular slab, is pumped by arc lamps or laser diode arrays. Pump light strikes the active material perpendicular to the laser beam. The most common type of illumination systems fall into this category.

Figure 6.53a exhibits a typical flashlamp-pumped solid-state laser with the pump source and the laser material within a highly reflective housing. In case of diode pumping, this can result in a very simple geometry, as shown in Fig. 6.53b. A laser slab is sandwiched between a diode array and a heatsink. The array and the slab can be air- and liquid-cooled.

With the emergence of laser diodes as important pump sources for solid-state lasers, end-pumping has become a common technique in the design of low power solid-state lasers. In contrast to flashlamps, the directed radiation emitted by a laser-diode array permits shaping and focusing of the pump beam to accommodate a large variety of pump schemes. End-pumping can be accomplished on axis from one or several arrays, on one or both ends, or the beams can be angular multiplexed, as shown in Fig. 6.53c. Also, a combination of end- and side-pumping is possible with laser diodes, as is shown in Fig. 6.53d. By utilizing internal reflections of the laser beam, the pump sources and the mode volume can be arranged for maximum overlap.

6.3.1 Pump Configurations for Arc Lamps and Laser Diodes

Flashlamp pumped laser slabs and rods will be considered first, followed by an overview of the many different designs possible with laser-diode sources.

Pump Cavities for Arc Lamps

In Chap. 7 we will discuss slab geometry designs such as the disk amplifier, the active mirror and the zig-zag laser. The first two schemes are employed in large-aperture Nd : glass lasers, whereas the zig-zag slab laser is of interest in high-average-power laser systems. In all of these designs the flashlamp-reflector housing requires a careful design to provide uniform illumination of the slab as well as having good coupling efficiency. Figure 6.54 shows several reflector geometries which are in use for pumping slab lasers. Illustrated are the cusp-, and "V"-shaped, and semicircular reflectors (Fig. 6.54a–c), surrounding a single lamp, as well as a "V"-shaped reflector in combination with a diffuser to spread out direct radiation and thus provide a more uniform pump intensity profile (Fig. 6.54d). A cusp-shaped reflector for a multiple lamp reflector is illustrated in Fig. 6.54e. A single slab pumped from both sides by "V"-shaped reflectors is shown in Fig. 6.54f, whereas Fig. 6.54g illustrates a reflector design employed for the pumping of two slabs by a single flashlamp. Details of the designs illustrated in Fig. 6.54 can be found in [6.61–63]. In what follows, the more common geometries for side-pumping laser rods will be discussed.

The most widely used pump cavity is a highly reflective elliptical cylinder with the laser rod and pump lamp at each focus. The elliptic configuration is based on the geometrical theorem that rays originating from one focus of an ellipse are reflected into the other focus. Therefore an elliptical cylinder transfers energy from a linear source placed at one focal line to a linear absorber placed along the second focal line. The elliptical cylinder is closed by two plane-parallel and highly reflecting end plates. This makes the cylinder optically infinitely long.

Figure 6.55 illustrates different pump geometries based on the elliptical cylinder. A single elliptical cylinder can have a cross section with a large or small eccentricity (Figs. 6.55a and b). In the former case, the laser rod and lamp are separated by a fairly large distance, in the second case they are close together. If the elliptical cylinder closely surrounds the lamp and rod, then one speaks of a close-coupled elliptical geometry (Fig. 6.55c). As we will see, this geometry usually results in the most efficient cavity. This cavity has the further advantage that it minimizes the weight and size of the laser heads. The semi-elliptical cavity (Fig. 6.55e) is employed in cases where conduction cooling of the laser rod

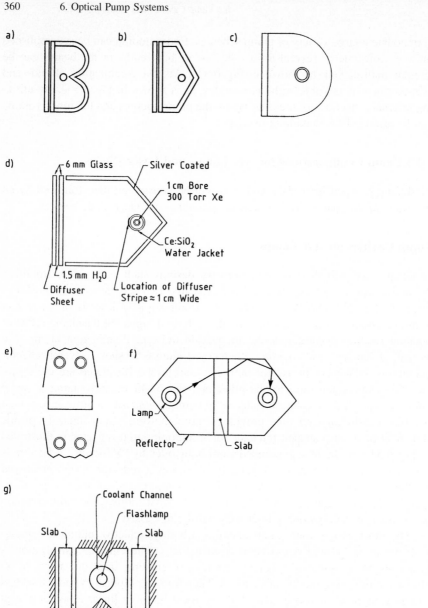

Fig. 6.54. Typical face-pumped Nd:glass disk laser amplifier geometry

is desired. In an elliptical cylinder, radiation passing through the rod is quickly defocused upon further passes within the resonator and no longer reenters the rod. Focusing can be somewhat improved by locating the lamp and rod beyond the foci of the elliptical reflector. With exfocal pumping, however, (Fig. 6.55d),

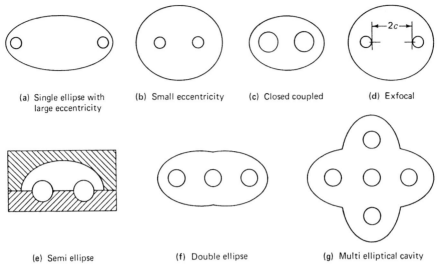

(a) Single ellipse with   (b) Small eccentricity   (c) Closed coupled   (d) Exfocal
large eccentricity

(e) Semi ellipse   (f) Double ellipse   (g) Multi elliptical cavity

Fig. 6.55a–g. Elliptical pump cylinders

the number of reflections from the cavity walls increases considerably, therefore the system efficiency is quite dependent on the wall reflectance. Since the energy delivered to a discharge lamp is limited, schemes to focus the energy from many lamps onto a single crystal are attractive. Figures 6.55f and g show two and four partial elliptical cylinders having one common axis at which the crystal is placed.

Figure 6.56 exhibits examples of close-coupled nonfocusing pump cavities. The lamp and rod are placed as close together as possible, and a reflector closely surrounds them. The reflector can be circular or oval in cross section (Figs. 6.56a and b). The latter type is often used in laboratory setups of low-repetition-rate pulsed lasers. The reflector is simply silver or aluminum foil wrapped around the flash lamp and laser rod. The efficiency of the closely wrapped cavity is found to be about as high as when an elliptical cylinder is used. The advantage of the pump cavity shown in Fig. 6.56b is simplicity; however, this design has a number of disadvantages: nonuniform pumping of the rod cross section and difficulties in providing adequate cooling of the rod, flashlamp, and reflector.

Multilamp close-coupled cavities provide a higher degree of pumping uniformity in the laser rod than single-lamp designs. The reflectors are used for creating the pumping density required when large-diameter rods are employed. The designs shown in Figs. 6.56c, d, and e are typically used to pump long Nd:glass laser rods.

Figure 6.57 displays designs which have diffuse-reflecting surfaces. If a helical lamp is used to pump a laser rod, the lamp is usually surrounded by a ceramic reflector (Fig. 6.57a). Occasionally, rods have been pumped by hollow lamps which were surrounded by a diffusely reflecting surface [6.64–66]. In the helical lamp and coaxial lamp pumping system shown in Figs. 6.57a and b, the efficiency is determined by the ratio of the inner lamp diameter to the rod diame-

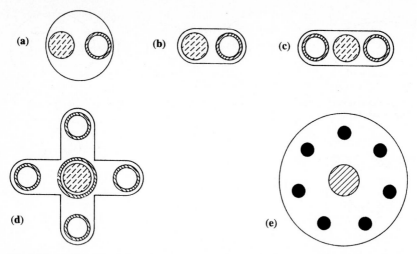

Fig. 6.56a–e. Close-coupled configurations. (**a**) Circular cylinder, (**b**) single-lamp close-wrap, (**c**) double-lamp close-wrap, (**d**) four-lamp close-wrap configuration, (**e**) close-coupled multiple-coaxial design

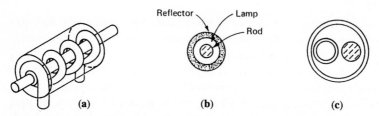

Fig. 6.57a–c. Diffuse reflecting cavities. (**a**) Helical flash tube, (**b**) coaxial flash tube, (**c**) diffuse cylinder

ter. A very simple reflector is a ceramic cylinder with a laser rod and pump lamp in close proximity. In the design depicted in Fig. 6.57c, the reflector sometimes is specular, i.e., aluminum foil or tubing. Ceramic materials have the advantage that they do not corrode or tarnish.

A high-energy transfer efficiency is achieved in pump systems with reflectors in the form of an ellipsoid of revolution [6.67] or a sphere [6.68], as illustrated in Fig. 6.58. In the spherical reflector the lamp and rod are positioned immediately adjacent to each other along a diameter of the reflector; in the ellipsoid of revolution the lamp and rod are positioned along the major axis between the focus and the ellipsoid surface. Somewhat similar to the ellipsoid is a reflecting cone system in which the lamp and laser rod are positioned on the axis of two cones, joined by a cylindrical reflecting surface [6.69].

The elliptical cylinders shown in Fig. 6.55 focus the pumping light in two dimensions, that is, in a plane which is the cross section of the cylinder. In contrast, the configurations illustrated in Fig. 6.58 focus the pumping light in three dimensions, thus a highly uniform rod illumination is achieved. Nevertheless,

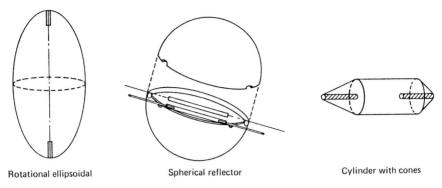

Rotational ellipsoidal   Spherical reflector   Cylinder with cones

Fig. 6.58. Pump cavities with rotational symmetry

ellipsoids, spheres, and axicons have a number of disadvantages. They can be used for short laser rods only, and they are not very attractive from the standpoint of fabrication costs, size, and weight.

Radiation Transfer in Diode-Pumped Systems

In contrast to flashlamps, the emission from laser diodes is highly directional, therefore many arrangements are possible for transferring the pump radiation to the solid-state laser material. Since the output beams of laser diodes can be shaped and focused, a major consideration is the design of a pump geometry which maximizes the overlap of the pumped volume with the volume occupied by a low-order resonator mode. The optimization of the overlap is referred to as mode-matching.

Typically two basic approaches have been followed in diode pumping of solid-state lasers, namely end-pumped and side-pumped configurations. In the end-pumped technique, which is unique to laser diodes, pump radiation is introduced longitudinally into the active material, i.e. co-linear with the resonator axis. In the side-pumped configuration, which is similar in concept to flashlamp-pumped lasers, the radiation from the diode bars enters the active material transverse to the optical axis of the laser radiation. End pumping is the more efficient method for generating diffraction limited performance, as the pump radiation is spatially overlapped with the TEM_{00} lasing mode. Matching the high aspect ratio of the emitting area of diode bars to the intracavity mode size is a very challenging optical task and a number of beam shaping techniques have been developed. Although less efficient, side pumping of laser rods and slabs is easily scaled and is therefore mainly employed for high-power systems. In side pumping, the high aspect ratio of the emitting surface can easily be matched to the laser medium, and heat removal can be accomplished over a larger surface area which reduces thermally related problems.

Figure 6.59 illustrates common techniques of coupling radiation from diode arrays into solid-state laser materials. Figure 6.59a,b,c depicts end-pumped arrangements whereby the output from a small array, single bar or a stack of bars is transferred into the laser medium. This can be accomplished by an optical imaging system, a fiber optic bundle or a non-imaging concentrator. Side pumping of a laser rod or slab, depicted in Fig. 6.59d,e, can be accomplished by mounting the laser bars in close proximity to the surface of the active material.

In large diode-pumped lasers with average output powers of hundreds of watts, the options of pump configurations are usually limited to designs depicted in Fig. 6.59d and e. The objective in these large systems is to pump the full cross-section of the laser material. On the other hand, in low-power lasers where the beam is often smaller than the cross-section of the laser material, there exists a very large diversity of pump schemes. For example, in circumventing the difficulty of reformatting the line source of a diode bar to a small spot, designs evolved where the laser beam inside the resonator is directed to provide maximum overlap with the pump source. These designs, are based on internal reflections of the laser beam inside the active medium (Fig. 6.73). In the following sections, we will illustrate concepts which take advantage of the unique properties of diode lasers as pump sources for solid-state lasers.

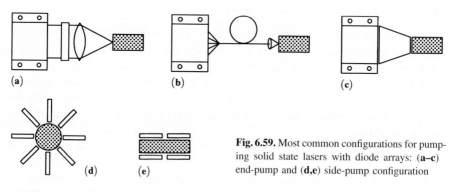

Fig. 6.59. Most common configurations for pumping solid state lasers with diode arrays: (a–c) end-pump and (d,e) side-pump configuration

The excellent spectral match of the diode output and the absorption of the solid-state material, in conjunction with a high spatial overlap between pump radiation and resonator mode volume, are responsible for the high overall efficiencies which can be achieved in diode-pumped lasers. Figure 6.60 illustrates the energy flow in a typical Nd : YAG laser, pumped with a 20 W diode bar at 808 nm. Laser diodes have efficiencies of $\eta_p = (0.3-0.45)$ with quasi-cw diodes at the upper range since they are more efficient pump sources as compared to cw diodes. Transfer of diode radiation and absorption by the lasing medium is expressed by the transfer and absorption efficiencies which in most systems are around $\eta_T \eta_a = (0.7-0.8)$. In end-pumped configurations, diode radiation is, in most cases, not all collected by the optical system and reflection losses further diminish pump radiation. On the other hand, the diode radiation is almost completely absorbed in the longitudinally pumped active medium. In side-pumped configurations, transfer efficiency is close to one, but absorption of the pump

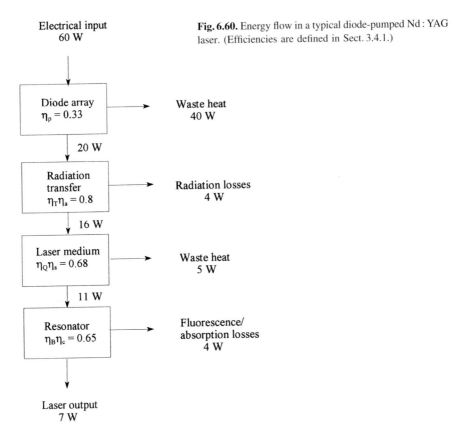

Fig. 6.60. Energy flow in a typical diode-pumped Nd : YAG laser. (Efficiencies are defined in Sect. 3.4.1.)

radiation is often incomplete because a portion of the diode output is transmitted through a small rod or thin slab.

The absorbed pump radiation produces heat in the crystal. The fractional thermal loading, i.e. the ratio of heat generated to absorbed power is 32% in Nd : YAG pumped at 808 nm [6.70]. These losses are accounted for by the Stokes shift and quantum efficiency $\eta_s \eta_Q = 0.68$. Mode matching and optical losses in the resonator are expressed by the beam overlap and coupling efficiency which typically are $\eta_B \eta_c = (0.6–0.8)$. The generic laser postulated in Fig. 6.60 has an overall efficiency of 11.6% and an optical-to-optical efficiency of 35%. In reviewing the energy-flow diagram, it is clear that the laser designer has only control over steps 2 and 4, namely the design of the pump radiation transfer, pump geometry and mode volume. The various pump techniques which have evolved in an effort to optimize the performance of diode pumped solid-state lasers will be considered next.

End-Pumping Configurations. The focused end-pumping configuration, if properly designed to provide matching of the pump light distribution and resonator mode, is the most efficient pump-radiation transfer scheme. Since the pump beam from the diode array is collinear with the optical resonator, the overlap between

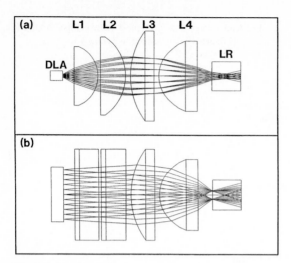

Fig. 6.61. Imaging optics for end-pumping a Nd : YAG crystal with a stack of three 1-cm laser diode bars

the pumped volume and the TEM_{00} mode can be very high. Practical realization of these advantages depends upon the possibility of reshaping a strongly astigmatic diode-laser beam into a beam with a circular symmetry.

Originally, end-pumping was performed with small diode arrays with linear dimensions on the order of 200 μm (Fig. 3.43). The optical design is fairly straight forward in this case. Also in some designs, such as the microchip laser, diode source and active medium are directly coupled or connected by a single fiber, as illustrated in Fig. 5.46. Today, single 1-cm bars and stacks of bars are employed in end-pumping in order to achieve high output power. This requires reformatting a long narrow line source, with an aspect ratio of 10^4, into a circular beam. This is a difficult task because a diode bar has an emitting area of 1 μm $\times$ 1 cm and as a consequence, produces a highly elliptical output beam that is diffraction limited in a plane vertical to the array and more than 1000$\times$ diffraction limited in the array plane.

The most common techniques of transferring and reformatting radiation from a single bar or stack of bars to an end-pumped laser crystal includes imaging optics, fiber optic coupling and nonimaging light concentration.

Imaging optics: Focusing the output from a 1 cm bar or a stack of bars into a laser crystal requires collimation of the emitted radiation, compensation of astigmatism and finally focusing the beam to a circular spot. An example of such an optical design developed in our laboratory is shown in Fig. 6.61 [6.71]. The end-pumping optics consists of two cylindrical lenses, L1 and L2, with 12.7 and 19 mm focal length respectively, followed by a 30 mm focal length plano-convex lens L3 and a 12 mm focal length aspheric L4. The rays traced in the plane perpendicular to the bars span a total angular field of 60° at the source, and those traced in the

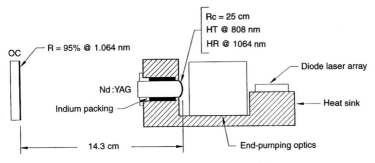

Fig. 6.62. Mechanical design of and end pumped Nd : YAG laser

plane parallel to the bars cover a total field of 10°. The corresponding width at the pump end of the laser rod is ≈ 2 mm.

The design has been used for end-pumping an Nd : YAG crystal with a 3-bar diode stack in a quasi-cw mode. A photo of the system and performance is illustrated in Figs. 3.45 and 3.46. The transmission of the AR-coated optical systems is 89%. It can be noticed that the distribution in the direction parallel to the bars is fairly broad, consistent with the 23° divergence angle at the $1/e^2$ points of the arrays. This translates into a width of 4 mm at the end of the laser rod. In order to achieve a more uniform distribution of pump radiation in the direction parallel to the bars, the lenses were adjusted to move the focus deep into the laser rod.

Figure 6.62 illustrates the mechanical design of the laser pump module. The diode array, end-pumping optics and Nd : YAG rod are mounted on a single aluminum structure. The laser rod was wrapped with one layer of indium wire before mounting. After insertion into the aluminum housing the indium was compressed by a flange having a ring-shaped insert. This mounting technique provides for a very low thermal impedance interface and very low rod-mounting stresses. The whole assembly was conductively connected to the water-cooled heat sink. The laser consists of a nearly hemispherical resonator with a flat output coupler. The $1/e^2$ beam diameter is ≈ 120 μm at the flat mirror, and by adjustment of the cavity length, the resonator mode can be expanded to 1.8 mm at the pump end of the laser rod.

A number of optical designs utilize a microlens for collimation of the highly divergent beam in the fast axis of the diode bar. The design of such a microlens, mounted in close proximity to the emitting surface, can range from a simple 1 mm diameter cylindrical glass rod to a fiber drawn precision lens [6.72–76]. Figure 6.25c depicts a microlens array that collimates the radiation from each of the five pulsed bars of a high duty cycle 1 cm² array.

In one design, which utilizes two cylindrical lenses and an aspheric condenser lens, a highly elliptical pump beam with cross-section 1.1 mm × 0.15 mm is imaged into an Nd : YAG crystal from a 1 μm × 1 cm bar. The optical resonator was designed to create an elliptical TEM$_{00}$ mode to match the pump profile. This was achieved with an anamorphic prism pair and a Brewster angle Nd : YAG crystal [6.72]. Recently a focusing scheme has been reported which permits the output from a 1-cm diode bar to be focused into a circular spot of less than $200\,\mu m$ [6.73]. In this technique the output from the diode bar is first collimated using a fiber lens and subsequently imaged with a combination of lenses into a two-mirror beam-shaping device.

By end-pumping a laser crystal off-axis, as shown in Fig. 6.53c, a number of pump sources can be geometrically multiplexed [6.77].

Fiber coupling. Using a fiber-optic bundle to deliver the diode output to the laser crystal has a number of advantages: the beam at the fiber output has a circular distribution, and the ability to remove heat from the diode remote from the optical components of the laser is an attractive feature. Also the pump source can be replaced relatively easy without disturbing the alignment of the laser. A number of approaches have been proposed to couple the radiation from a line emitter into a fiber. Conceptually, the simplest approach is a line-to-bundle converting fiber-optic coupler. It consists of a large number of small fibers spread out in a linear array which faces the emitting diodes of the laser bar. At the other end, the fibers are arranged into a round fiber bundle. A high numerical aperture of the fibers is required to capture most of the diode output. In one approach [6.78], a hundred $84\,\mu m$ diameter fibers form a linear array at the emitter side and are closely packed into a 0.9 mm diameter bundle at the pump side. With a numerical aperture of NA 0.54, over 95% of the radiation from the diodes was collected. The requirement for a high numerical aperture of the fibers can be reduced if the radiation from the facet in the fast axis is first collimated by a cylindrical microlens which then focuses the radiation into the linear fiber array. In one such approach [6.79] a total of 114 fibers, with a diameter of $80\,\mu m$ each, are combined into a bundle of 1.1 mm. The numerical aperture of the fibers in this case was NA 0.14. Besides a fiber-based beam shape transformation, other techniques include multiple reflections between two plane-parallel mirrors to reshape the emission from a diode bar [6.73] reflections from micro-step mirrors [6.81, 82] and deflection by micro prisms [6.80, 83].

Figure 6.63a depicts an optical schematic of a "z" configuration $Nd : YVO_4$ laser pumped from both ends by two fiber-coupled, 20 W cw diode bars [6.84]. The radiation from each bar is coupled into a fiber bundle by a fiber microlens which reduces the divergence in the fast axis of the diode output. At the pump end, the fibers form a round bundle, and a collimating and focusing lens transfer the pump radiation into the crystal through a dichroic fold mirror. The high efficiency which can be achieved with diode end-pumped solid-state lasers is illustrated in Fig. 6.63b. With a fiber coupling efficiency of 85%, and some reduction in the rated power level of the diode to preserve lifetime, each side of the $Nd : YVO_4$ crystal is pumped with 13.8 W. Pump radiation at the crystal is converted with over 50% conversion efficiency into a polarized TEM_{00} mode.

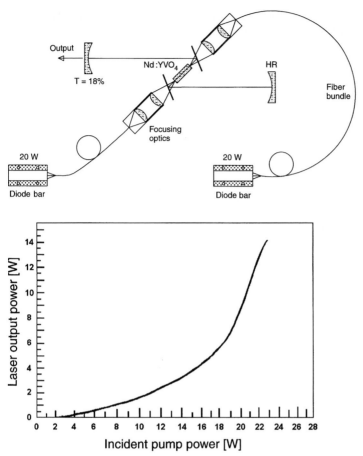

Fig. 6.63. (a) Optical schematic of a Nd : YVO$_4$ laser pumped with fiber-coupled diode bars (Spectra Physics), (b) laser output vs. optical pump power incident on the crystal

Maximum output of the system is 13.8 W. An Nd : YAG laser used in the same configuration produced 11 W of polarized cw output.

Non-imaging light concentrators. Figure 6.64 illustrates end-pumping techniques based on non-imaging devices. In Fig. 6.64a, the output from a stack of bars is first collimated in the fast axis by cylindrical microlenses mounted in front of each bar. The collimated radiation is directed into a lens duct which combines focusing at its curved input face and total internal reflection at the planar faces to channel the diode pump radiation to the laser rod [6.85]. Powerful Nd : YAG, Tm : YAG and Yb : YAG lasers have been pumped with this technique and outputs in excess of 100 W have been reached in these lasers. A slightly different version which has parallel walls in the vertical direction is shown in Fig. 6.64b. The stack width is reduced from 10 mm to 1.5 mm in the horizontal dimension this resulted in an output aperture of the device of $1.5 \times 1.5\,\text{mm}^2$ [6.68]. Concentrators are an attractive solution for large stacks because they are relatively easy to fabricate. The degree of concentration is governed by the radiance theorem which states that the radiance of the light distribution produced by an imaging system cannot be greater than the original source radiance.

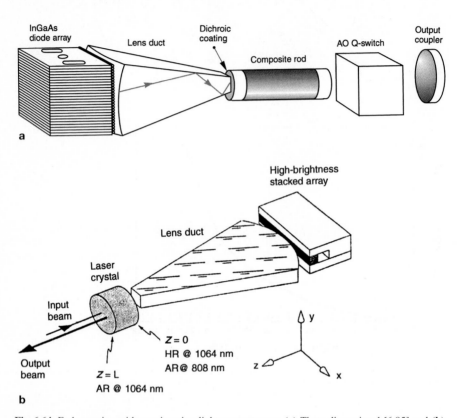

Fig. 6.64. End-pumping with non-imaging light concentrators. (**a**) Three-dimensional [6.85] and (**b**) two-dimensional [6.86] lens duct

Side Pumping of a Laser Rod. In this configuration, the diode arrays are placed along the length of the laser rod and pumped perpendicularly to the direction of propagation of the laser radiation. As more power is required, more diode arrays can be added along and around the laser rod. There are three approaches to couple the radiation emitted by the diode lasers to the rod: a') *direct coupling.* From a design standpoint this is by far the most desirable approach. However, this arrangement does not allow for much flexibility in shaping the pump radiation inside the laser rod. b) *Intervening optics between source and absorber:* In this case, the pump distribution can be peaked at the center of the rod allowing for a better match with the resonator modes. Optical coupling can be achieved by

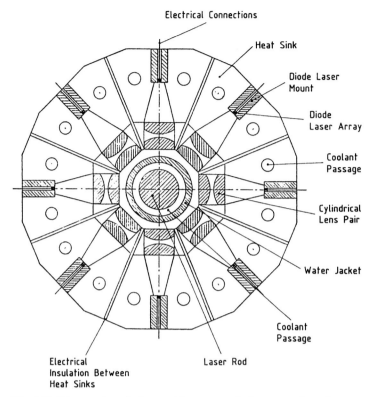

Electrical Connections

Heat Sink

Diode Laser
Mount

Diode
Laser Array

Coolant
Passage

Cylindrical
Lens Pair

Water Jacket

Coolant
Passage

Electrical
Insulation Between
Heat Sinks

Laser Rod

Fig. 6.65. Side pumping of a laser rod in a symmetrical pump configuration

using imaging optics such as lenses or elliptical and parabolic mirrors, or by non-imaging optics such as reflective or refractive flux concentrators. c) *Fiber optic coupling*: Due to the coupling losses combined with the increased manufacturing cost of fiber coupled diode arrays, this technique is not very attractive.

Whether or not to use intervening optics between the arrays and the laser rod depends mainly on the desired beam diameter and optical density of the active medium. Generally speaking, oscillator pump heads usually employ optical elements in order to concentrate the pump radiation in the center for efficient extractions at the TEM_{00} mode, whereas large amplifier rods are direct pumped. Side-pumped lasers have been described in [6.87–95].

Figures 6.65 and 6.66 illustrate typical designs. In Fig. 6.65, the laser rod is surrounded by a quartz jacket to allow for liquid cooling, whereas in Fig. 6.66, the laser rod is conductively cooled by a copper heat sink. The pumping cavity consists of a number of linear arrays symmetrically located around the rod. The arrays are mounted on long heat-sink structures and contain cylindrical optics. The optics allow the arrays to stand-off from the laser crystal and the arrays. The output apertures of the diode bars are imaged near the center of the rod by using a symmetrical doublet of plano-convex cylindrical lenses.

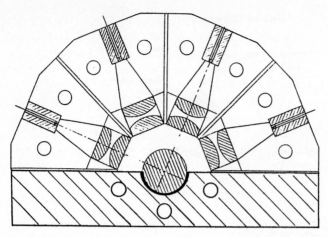

Fig. 6.66. Side pumping of a cylindrical laser rod employing a hemispherical pump geometry

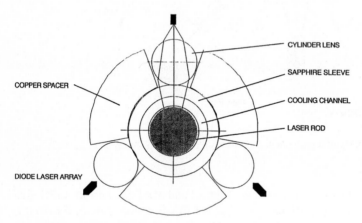

Fig. 6.67. Cross-section of diode-pumped laser head using cylindrical lenses

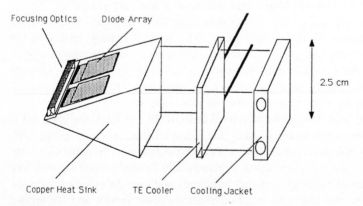

Fig. 6.68. Diode arrays and heat sink mounting and assembly details

A simpler design is shown in Fig. 6.67 where the cylindrical lens doublet is replaced by a glass rod. Figure 6.68 illustrates diode array and heat sink mounting details of an assembly employing focusing optics and a TE cooler for temperature control. Shown are several 1-cm diode bars, forming a small 2-D array, which are mounted and bonded to a common copper tungsten heat sink. Two such arrays are soldered to copper heat-sink blocks which, in turn, are coupled to TE coolers and liquid-cooled heat exchangers. The entire unit consisting of the diode array, coupling optics and heat sink is positioned and mounted around the laser rod.

A pump configuration of a high-power amplifier head which does not use coupling optics is illustrated in Fig. 6.69. The laser rod is surrounded by 8 stacked arrays, each containing 6 bars. Since the bars are 1-cm long, and each bar produces 12 mJ in a 200 μs pump pulse, the rod can be pumped at 576 mJ per cm rod length.

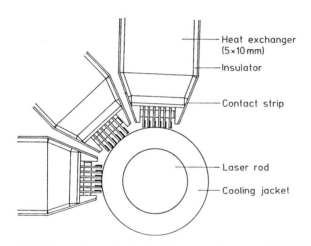

Fig. 6.69. Arrangement of stacked diode array bars around a laser rod

Figure 6.70 displays a photograph of a pump head containing 4 rings of stacked diode arrays. In this design, 4 cm of the 2″ long rod are pumped by the diode arrays.

In order to optimize the uniformity and radial profile of the pump distribution within the gain medium, a number of ray trace programs have been developed. An example of such a programm called CRADLE [6.35] solves the general problem of radiation transport in a side-pumped laser rod using ray tracing with integration over the diode-array pump system. As schematically illustrated in Fig. 6.71, this code calculates the spatial pump distribution $p(r)$ in the laser material as a function of spectral absorption $\alpha(\lambda)$ and the spectral and spatial properties of the radiation source $p(\lambda)$ as modified by the optical surfaces between the source and absorber.

Fig. 6.70. Photograph of diode array pumped laser head (*background*) and stacked diode array bars mounted on copper heat sink (*foreground*)

Figure 6.72b presents the result of such a calculation for a laser rod mounted on a heat sink for conduction cooling and pumped by 4 diode bars. This cw pumped laser head, which contains cylindrical doublets as indicated in Fig. 6.72a, had very good TEM_{00} performance due to the strong concentration of pump radiation at the center of the rod. The analysis was in excellent agreement with the measurements performed with the actual laser head.

Side-pumped Slab Laser. Slab lasers are usually pumped by large densely packed 2-D arrays. In a slab laser, the face of the crystal and the emitting surface of the laser diodes are in close proximity, and no optics is employed (Fig. 6.53b).

The slab is usually pumped from one face only. The opposite face is bonded to a copper heat sink containing a reflective coating to return unused pump radiation back into the slab for a second pass. An antireflection coating on the pump face is used to reduce coupling losses (the diode array is not in contact with the active material). Liquid cooling is employed to remove heat from the crystal and diode heat sink.

The planar pump and cooling geometry achieved with a slab provides the possibility for one-dimensional heat flow, and simplifies heat removal from the laser medium and pump arrays. The design of flashlamp-pumped slab lasers has been extensively studied and the potential advantages and deficiencies of such systems will be discussed in Chap. 7. For flashlamp pumping, the primary advantages of the zig-zag slab are the minimization of stress-induced birefringence and thermally-induced focusing by optical propagation along a zig-zag path. There are additional advantages for diode-pumped Nd:YAG systems. Because of the

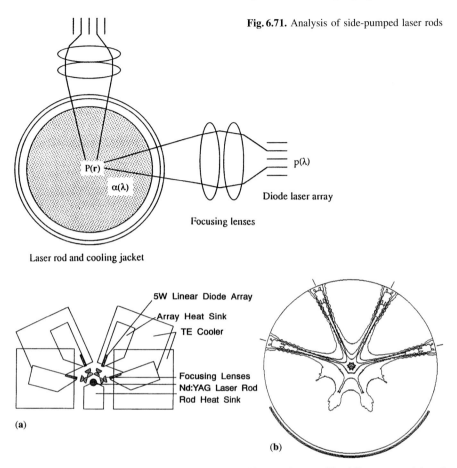

Fig. 6.71. Analysis of side-pumped laser rods

Fig. 6.72. Pump configuration of a conductively cooled laser rod pumped by 4 linear arrays (**a**), and contour plot of pump intensity distribution (**b**)

good spectral overlap of the output of diode lasers with the absorption bands of Nd : YAG, the effective absorption coefficients are large and great care must be taken to achieve the gain uniformity required by most laser sytems. A zig-zag path in the plane of the optical pumping can compensate for non-uniform gain by averaging out the gain inhomogeneities in that plane. Because of reduced sensitivity to thermal effects, relatively simple conduction cooling techniques are possible. The single-sided pumping that is possible with zig-zag slabs adds additional benefits. Heat removal from the slab and diode arrays becomes a simple matter of planar flow to liquid-cooled heat sinks.

Together with these advantages for the use of zig-zag slabs, there are drawbacks as well. The surface and angular tolerances required to achieve typical design goals make slabs more expensive to fabricate than rods. In practice, edge effects and thermally induced stresses degrade their performance from that predicted. The slab geometry also supports a large number of parasitic modes that can limit the useful stored energy. Special coatings to suppress these parasitics, or to provide thermal coupling to the cooling system without spoiling the total internal reflections can add to the cost and design complexity of slabs. The advantages given above and possible approaches to dealing with the disadvantages will be discussed in more detail below.

The first step in a slab-based design is a careful analysis of the trade-offs possible using different slab geometries and an analysis of fabrication issues, including coatings, that will be necessary for optimum performance. A choice between odd-bounce and even-bounce configurations will be the first decision that will have to be made. For even bounce slabs, the entrance and exit faces are parallel. This simplifies fabrication and reduces the cost of the slab compared to a odd-bounce slab with opposing faces. However, for one-sided conduction cooling, the geometry of the odd-bounce slab gives symmetric thermal gradients at the two ends of the slab, which are better compensated by the zig-zag path.

Flatness tolerances on the TIR surfaces of the slabs are an important factor in determining the wavefront distortion of the beam after passing through the slab. Since there will be several bounces per pass, it is important that the flatness of the area intercepted at each bounce be $< \lambda/5$. However, it is not necessary that the overall surface be this flat as long as the two surfaces are sufficiently parallel.

The combination of relatively high localized gain regions, due to strong diode pump absorption coupled with the highly symmetric geometry of the slab can cause Amplified Spontaneous Emission (ASE) and parasitic oscillations which limit the useful stored energy. Coating or roughening the sides of the slab helps to suppress modes that propagate along the relatively high gain region parallel to and near the pump face. In addition, it has been shown that breaking the symmetry of the slab by slightly canting the parallel surfaces can significantly reduce the number of parasitic modes.

A number of slab-design issues are closely linked to single-sided diode pumping. The combination of a large pump absorption cross-section and the susceptibility of slabs to ASE and parasitics make a careful analysis of the effects of varying the Nd concentration and diode pump bandwidth critical. There is a trade-off between overall pump absorption and uniformity of energy deposition.

Pump Configurations Based on Internal Reflections of the Laser Beam. Besides the classical end- and side-pumped configurations discussed in the previous sections, there is an almost unlimited number of pump-configuration variations possible. Most of the more unusual design configurations utilize internal reflections of the laser beam. In Fig. 6.73, designs are illustrated which contain one, two, and many internal reflections of the laser beam. An extension of this concept is the monolithic ring laser which was discussed in Sect. 3.7.

As we have seen in end-pumping schemes, the overlap between pump and mode volume can be optimized. Side-pumping has the advantage that extended pump sources can be used to provide the power necessary for many applications. The key to efficient operation in side-pumped systems is to utilize the pump

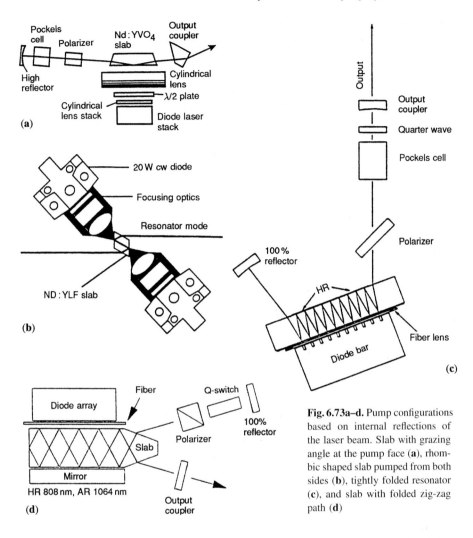

Fig. 6.73a–d. Pump configurations based on internal reflections of the laser beam. Slab with grazing angle at the pump face (**a**), rhombic shaped slab pumped from both sides (**b**), tightly folded resonator (**c**), and slab with folded zig-zag path (**d**)

radiation absorbed near the surface where excitation is highest. Several of the schemes depicted in Fig. 6.73 combine some of the advantages of end-pumping, such as good pump/mode matching with some of those of side-pumping, such as scalability and simplicity.

In the design shown in Fig. 6.73a, side-pumping through a polished face is used to obtain high gain. The resonator mode makes a single grazing-incidence total internal reflection at the pump face, thus remaining in the region of the highest gain throughout its passage through the slab. This design has been used for small Nd:YVO$_4$ and Er:YAG lasers [6.96, 97]. The rhombic-shaped slab employed in the design depicted in Fig. 6.73b can be end-pumped from two sides. The internal reflections of the laser beam simplify the design of separating pump and laser beam. In the tightly folded resonator configuration illustrated in Fig. 6.73c, the zig-zag beam path in the active medium is configured to maximize the overlap with the radiation emitted from each individual diode of the array [6.98]. The design also spreads the heat load over a wide area in the crystal. The rectangular block of Nd : YLF or Nd : YAG has two opposing sides polished. One of the sides is slightly reflective-coated over the full length, whereas the other size is partially covered with a high-reflective coating. The ends are AR-coated for coupling of the laser beam into the slab. The design can be thought of as end-pumped at each bounce of the laser beam. Figure 6.73d exhibits a miniature slab laser with a folded zig-zag path of the laser beam. The entrance and exit faces are cut at Brewster's angle. The beam path in this particular Nd : YAG slab consisted of five bounces [6.99]. The pump radiation is collimated in the plane of the zig-zag path by a fiber to maximize the overlap between pump beam and resonator mode.

Many more design variations have been conceived around these basic schemes and almost every month, a new configuration is reported in the literature.

6.3.2 Energy Transfer Characteristics of Pump Cavities

Theoretical Coupling Efficiencies

The amount of pump energy which is transferred from the source to the laser rod can be approximated by

$$\eta = \eta_{ge} \times \eta_{op} , \tag{6.41}$$

where η_{ge} is the geometrical cavity transfer coefficient. It is the calculated fraction of rays leaving the source which reach the laser material either directly or after reflection from the walls. The parameter η_{op} expresses the optical efficiency of the cavity and essentially includes all the losses in the system. This parameter can be expressed as

$$\eta_{op} = r_w(1 - r_r)(1 - a)(1 - f) , \tag{6.42}$$

where r_w is the reflectivity of the cavity walls at the pump bands, r_r is the reflection losses at the laser-rod surface or at the glass envelopes of the cooling jackets and Fresnel losses of any filters inserted in the cavity, a is the absorption loss in the optical medium between the lamp and the laser rod such as coolant liquid, filters, etc., and f is the ratio of the nonreflecting area of the cavity to the total inside area. This factor accounts for losses due to openings in the reflector which are required, for example, to insert the laser rod and pump lamp. Equations (6.41, 42) are rough approximations based on the assumption that all the lamp radiation undergoes just one reflection. If one were to include direct radiation and multiple reflections from the cavity walls, these equations would have to be developed in series expressions.

The elliptical cavity has been most extensively discussed in the literature [6.100–109]. As mentioned before, in this configuration a linear lamp and a laser rod, possibly with different radii, are placed at the foci of an elliptical

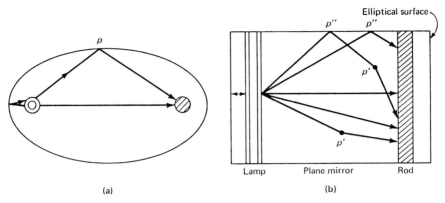

Fig. 6.74a, b. Trajectories of photons emitted from the pump source. **(a)** Cross section. **(b)** Top view of elliptical pump cavity

cylinder. The pump source is usually believed to be a cylindrical radiator having a Lambertian radiation pattern. This implies that it appears as a source having constant brightness across its diameter when viewed from any point.

In Fig. 6.74 trajectories of photons are shown, which originate from a volume element dv of the source. In Fig. 6.74a the rays leave normal to the surface of the source and, therefore, remain in a cross-sectional plane. Figure 6.74b depicts trajectories of photons which leave the source at an angle with respect to a cross-sectional plane of the ellipse. In this case the photons can be reflected off the end-plate reflectors as well as undergo a reflection at the elliptical cylinder. In an elliptical pump cavity, all rays emanating from one point of the source are transformed into parallel lines at the laser rod. Each line corresponds to rays leaving the source inclined at the same angle with respect to the major axis of the ellipse, but at different angles with respect to the cross-sectional plane.

Image formation, in its usual sense, is meaningless for the elliptical cylinder, since rays emanating in different directions from a point on the source converge after reflection from the cavity at altogether different points. In a pump cavity, all that is actually desired is the transfer of radiant energy from the source to the laser rod. From the foregoing considerations, it is obvious that the condition of an infinitely long elliptical pump cavity can be satisfied by enclosing the ends of the cylinder with highly reflective plane mirrors. Therefore, in the analysis of this arrangement the two-dimensional case can be treated by considering the light distribution in a plane perpendicular to the longitudinal axes of the cylinder.

In the theoretical expressions obtained for the efficiency of elliptical configurations, the pump system is usually characterized by the ratio of the rod and lamp radii r_R/r_L, the ratio of the lamp radius to the semimajor axis of the ellipse r_L/a, and the eccentricity $e = c/a$, where $2c$ is the focal separation. Consider any point P on the surface of the cavity with a distance l_R from the crystal, and distance l_L from the lamp as shown in Fig. 6.75. Suppose that the lamp has a radius r_L, then, as a consequence of the preservation of angles, upon reflection

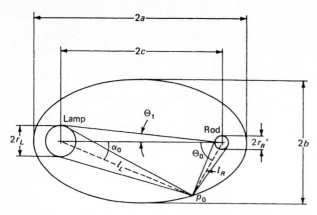

Fig. 6.75. Cross section of elliptical pump cavity. Eccentricity $e = c/a$; focal point separation $c = (a^2 - b^2)^{1/2}$

the image will have the radius $r'_{\text{L}} = r_{\text{L}} l_{\text{R}}/l_{\text{L}}$. This means that the portion of the elliptical reflector nearer the lamp forms a magnified image at the laser rod while that portion nearer the crystal forms a reduced image of the lamp. A point P_0, with corresponding angles α_0 and Θ_0 measured from the lamp and rod axis, respectively, may be defined dividing these two regions. At this point, the ellipse generates an image of the lamp which exactly fills the crystal diameter. We must allow for this effect of magnification and demagnification when determining how much energy is captured by the crystal. From the properties of an ellipse and noting that at P_0, $l_{\text{R}}/l_{\text{L}} = r_{\text{R}}/r_{\text{L}}$, we obtain

$$\cos \alpha_0 = \frac{1}{e}\left[1 - \frac{1-e^2}{2}\left(1 + \frac{r_{\text{R}}}{r_{\text{L}}}\right)\right] , \tag{6.43}$$

and

$$\sin \Theta_0 = \left(\frac{r_{\text{L}}}{r_{\text{R}}}\right)\sin \alpha_0 . \tag{6.44}$$

The geometrical cavity transfer coefficient can be calculated by considering what fraction of the energy radiated by the lamp into an angle $\Delta\alpha$ is trapped by the crystal. Integration over all angles leads to [6.100, 105]

$$\eta_{\text{ge}} = \frac{1}{\pi}\left[\alpha_0 + \left(\frac{r_{\text{R}}}{r_{\text{L}}}\right)\Theta_0\right] . \tag{6.45}$$

This expression is plotted in Fig. 6.76. A certain portion of the reflecting surface behind the lamp is screened from the crystal by the lamp itself. In filament lamps

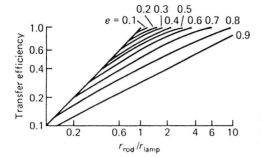

Fig. 6.76. Dependence of transfer efficiency of an elliptical reflector on the quantity r_R/r_L and eccentricity e [6.110]

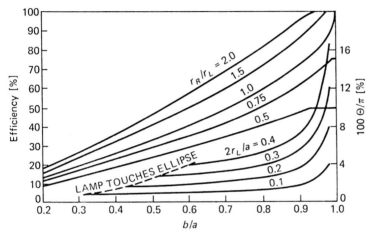

Fig. 6.77. Efficiency of a single-elliptical pumping cavity. Top set of curves is for $2r_L/a = 0$, while bottom set gives the loss due to finite lamp diameter. Loss multiplied by r_R/r_L and substracted from efficiency given by upper curve. Left-hand scale is for top set of curves only. Right-hand scale is for bottom set of curves only [6.105]

and in pulsed flashlamps, which resemble a black-body source, the radiation reflected back into the lamp will be absorbed. On the other hand, the plasma in cw arc lamps is optically thin, and the reflected pump radiation is transmitted through the plasma. If we assume the radiation reflected back to the pump source to be lost, we must reduce the angle Θ_0 in the above formula by Θ_1, where

$$\sin \Theta_1 = \frac{r_L}{4ae} , \tag{6.46}$$

and

$$\eta'_{ge} = \frac{1}{\pi} \left[\alpha_0 + \left(\frac{r_R}{r_L} \right) (\Theta_0 - \Theta_1) \right] . \tag{6.47}$$

This transfer coefficient is obtained from Fig. 6.77 by subtracting the lower set of curves from the upper curves.

It is seen from Figs. 6.76, 77 that the efficiency increases with an increase in the ratio r_R/r_L and with a decrease in the eccentricity e. This result stems from the fact that the magnification of the pump source increases with the eccentricity of the ellipse. A superior efficiency is therefore obtained for an almost circular cavity and as small a pump source as possible.

It should be noted that in these calculations the crystal is assumed to absorb the radiation which falls upon it; no second traverses of light through the crystal are considered. Furthermore, no allowance was made for the reflection loss at the cavity walls or losses due to Fresnel reflection at the surface of the laser rod. Multiple reflections can become significant if the absorption coefficient of the crystal is small compared to the inverse of the crystal diameter [6.101, 104].

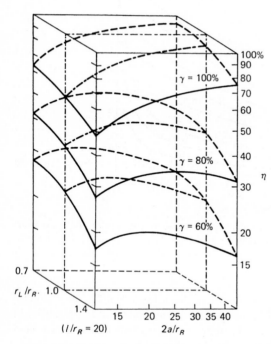

Fig. 6.78. Three-dimensional illustrations of the dependence of total transfer efficiency η upon three parameters that are reflectivity γ of mirrors, r_L/r_R, and $2a/r_R$, where r_L and r_R are the radii of the lamp and the laser rod, respectively, and $2a$ is the major axis of the ellipse $(2c/r_R = 8)$ [6.103]

Figure 6.78 shows a three-dimensional illustration of the dependence of the transfer efficiency in an elliptical cylinder on the reflectivity γ of the end plates, the ratios r_L/r_R, and $2a/r_R$ [6.103]. For the elliptical portion of the cavity a reflectivity of one is assumed. The curves are calculated for a length of the cylinder of $l = 20r_R$ and a distance between the rod and the lamp of $2c = 8r_R$. Therefore, a large value of the variable $2a/r_R$ indicates that the eccentricity of the ellipse $e = c/a$ is small, and vice versa. As may be seen from Fig. 6.78, the efficiency increases if one decreases the radius of the lamp. It should be noted that for $\gamma = 1$ the cavity should have an eccentricity as small as possible in agreement with other theoretical calculations. However, when $\gamma < 1$, the optimum eccentricity increases with decreasing values of the reflectivity. For example, for $r_L/r_R = 1$ and a reflectivity of $\gamma = 0.60$, the optimum ellipse has a major axis of $2a/r_R = 16$ or an eccentricity of 0.5. For a given focal point separation $2c$, an ellipse with a large eccentricity has a small cross-sectional area, which tends to reduce the losses caused by the end mirrors.

The analytical methods developed for the design of pumping configurations imply idealized systems in which reflection and refraction losses are usually ignored and only a single reflection from the walls is considered. In order to analyze pumping systems with fewer idealizations, statistical sampling methods, such as the Monte-Carlo method, have been employed. The Monte-Carlo method consists essentially of tracing, in detail, the progress of large numbers of randomly selected rays over many events (i.e., absorptions, reflections, and refractions) throughout the pumping cavity. The average transfer efficiency of these rays can then be taken as representative of all possible rays in the cavity. Monte-Carlo calculations for a variety of geometrical configurations were reported in [6.111–113]. For the single elliptical pump cavity, the adverse effect of a small value of the quantity r_R/r_L was confirmed. The analysis also showed the efficiency to be strongly dependent on the reflector reflectance. On the other hand, the calculated efficiency values appear to be less dependent on the eccentricity than indicated by the results of earlier work, in which multiple passes were ignored.

Multiple-elliptical laser pump cavities generally consist of a number of partial-elliptical cylinders having a common axis at which the crystal is placed. The results of calculations performed to obtain the efficiency of various multiple cavities with varying sources and crystal diameters show that, for equal diameters of the source and the crystal, the single-elliptical cavity has the highest efficiency. However, a cavity with two or more partial-elliptical cavities will allow higher input power, at some sacrifice in overall efficiency, if a high-power output laser is required. If the diameters of the source and crystal are different, these conclusions must be modified. It should also be pointed out that the multi-ellipse cavity results in more uniform pumping of the laser crystal.

Referring back to the case of a single-elliptical cylinder, (6.45) gives the efficiency for the case when the lamp does not block radiation. For multi-elliptical pump cavities one obtains the expression

$$\eta_{eg} = \frac{1}{\pi}\left[(\alpha_0 - \alpha_2) + \left(\frac{r_R}{r_L}\right)\Theta_0\right] . \tag{6.48}$$

Here, α_0 has been reduced by an angle α_2 which represents that portion of the reflecting wall that was cut away. Light falling in this region is assumed to be lost. For a double-elliptical pumping cavity,

$$\cos \alpha_2 = \frac{2e}{1 + e^2} . \tag{6.49}$$

Equation (6.48) is plotted in Fig. 6.79.

We will briefly discuss pumping systems other than elliptical cylinders. The analysis of a system with a circular reflective cylinder is analogous to that for

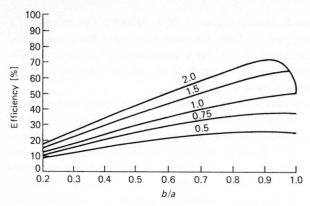

Fig. 6.79. Efficiency of a double-elliptical cavity. Parameters are for r_R/r_L [6.105]

an elliptical system. The lamp and laser rod are symmetrically positioned at the same distance from the cylinder axis. The system efficiency increases with an increase in the ratio r_R/r_L and decreases for increases in the ratios of F/R_0, where F is the distance from the cylinder axis to the axes of the lamp or rod.

The transfer efficiency of spherical and ellipsoidal cavities has been calculated in [6.114], and a review of most common pumping systems has been given by *Kalinin* and *Mak* [6.110].

The transfer efficiency of laser pumping cavities that have diffusely reflecting walls can be estimated if we assume that the light inside the cavity is isotropic. Since all surfaces will be equally illuminated, the light absorption at each surface will be the product of the area and the fractional absorption. Since all radiation must eventually be absorbed somewhere, the proportion of light absorbed at a surface area S_1 having a fractional absorption A_1 will be [6.112]

$$\eta = \frac{S_1 A_1}{\sum_{i=1}^{n} S_i A_i} , \tag{6.50}$$

where the summation covers the entire internal surface of the cavity. If S_1 is the exposed surface area of the laser crystal and A_1 is the effective absorption over this surface, the transfer efficiency of the laser configuration is given by

$$\eta = \frac{S_1 A_1}{S_1 A_1 + S_2 A_2 + S_3 A_3 + S_4} , \tag{6.51}$$

where S_2 and A_2 are the diffusely reflective wall area and absorption coefficient of the cavity, S_3 is the surface area of the lamps, and A_3 is the absorption of pump radiation by the lamp. S_4 is the area covered by holes in the cavity wall; these surfaces are treated with unit absorption, i.e., $A_4 = 1$.

An important parameter in the formula for transfer efficiency is A_1, the fraction of photons which strike the laser-rod surface and are absorbed by the laser material. This parameter, which has been called the capture efficiency, depends on the absorption coefficient, the refractive index, and the radius of the laser rod.

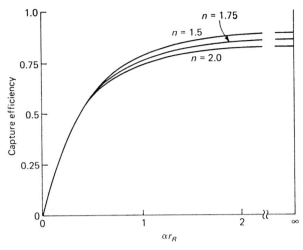

Fig. 6.80. Capture efficiency of a cylindrical laser rod exposed to isotropic light. The parameters n, r, α are the refractive index, radius, and absorption coefficient of the rod [6.112]

Figure 6.80 shows the capture efficiency A_1 of a cylindrical crystal exposed to isotropic light [6.112]. Radiation incident on the rod which is not absorbed is either reflected off the surface or passes through the laser material. For large values of αr_R the capture efficiency is limited only by Fresnel reflection losses on the crystal surface. Using this grossly simplified analysis, no significant discrepancy between it and the Monte-Carlo calculation is usually found.

With the aid of (6.51) and Fig. 6.80, we will calculate the transfer efficiency of the diffusely reflecting cavities shown in Figs. 6.95 and 3.13. Both cavities are 10 cm long and contain a 1-cm-diameter ruby crystal. In the first case, the crystal is pumped by a linear flashlamp and, in the second case, by a helical flashlamp. For ruby with an index of refraction of 1.76 and an average absorption coefficient in the pump bands of $1.0\,\mathrm{cm}^{-1}$, we obtain a capture efficiency of $A_1 = 0.60$ from Fig. 6.80; the surface area is $S_1 = 31.4\,\mathrm{cm}^2$. For both lamps we assume that 10 % of the pump radiation is reflected back to the lamp and is reabsorbed by the plasma ; that is, $A_3 = 0.1$. The linear lamp is of the same size as the laser rod ($S_3 = 31.4\,\mathrm{cm}^2$). The helical lamp is approximated by an annular region of isotropically radiating material with an inner diameter of 1.8 cm. The reflectivity of the diffusely reflecting walls is assumed to be 95 %. The cavity diameter is 4 cm, from which we obtain $S_2 = 150\,\mathrm{cm}^2$, if we include the end walls. The area covered by holes and apertures is about $S_4 = 0.1 S_2$. Introducing these parameters into (6.51), we obtain $\eta = 0.42$ and $\eta' = 0.40$ for the linear and helical lamp-pumped ruby system.

Empirical Data

Theoretical considerations of the efficiency of pump cavities are of utility primarily in their qualitative aspects and in their illustration of the gross features of the energy transfer in the various reflector configurations. Even under very idealized assumptions, calculation of the amount of energy absorbed in the laser rod is very tedious since the spectral properties of the pumping source, the absorption spectrum of the laser material, and all possible path lengths of the elementary rays through the laser rod, determined by the specific reflector configuration, must be taken into consideration. Not even the most thorough computations can take into account the effect of mechanical imperfections in the fabrication of the reflectors, optical obstructions caused by seams and gaps, aberrations by flow tubes, losses due to holes and apertures in the end plates, etc.

Summarizing the experimental work carried out on pump-cavity design, we may draw the following conclusions:

- For maximum efficiency and relatively short rods (up to 10 cm in length) a small single-elliptical cylinder, an ellipsoid of revolution, or a spherical cavity is best. For longer rods, the closed-wrapped geometry is about as efficient.
- Small elliptical cavities with low major axis-to-rod diameter ratios are more efficient than large cavities. In a small elliptical cavity the fraction of direct radiation is high, and most of the pump radiation is incident on the rod after a single reflection on the cavity walls. Therefore, in this geometry, a reflectivity of less than unity, imperfections in the geometry, and obstruction of the light path are less detrimental to the efficiency than is the case in a large elliptical cylinder. Another strong argument for making elliptical cavities as small as possible is the increased probability of the lost radiation being redirected to the laser rod if the ratio of cavity volume to laser rod volume is small.
- In agreement with theory, the highest efficiency in an elliptical cylinder is obtained for a large r_R/r_L ratio. However, lamp-life considerations demand that in most cases a ratio of one is chosen.
- A high reflectivity is extremely important, whereas mechanical tolerances of the elliptical shape are not very critical in small ellipses. This is not surprising because in a close-coupled ellipse the cross sections of lamp and rod are fairly large with respect to the dimensions of the ellipse, which reduces the requirement for accurate imaging.
- The length of elliptical-cylinder cavities should be as large as possible in relation to the diameter, but the laser rod, pump lamp, and cavity should all be the same length.
- Optimum small elliptical-cylinder pump cavities are equally as good as ellipsoid-of-revolution pump cavities.
- Multiple cylindrical ellipse cavities may be utilized to increase the number of pump lamps and to produce symmetrical pumping. If the cavity size is kept small, this can be achieved with a minimum sacrifice in efficiency. Figure 6.81 shows the reflection characteristics of a double-elliptical cavity.

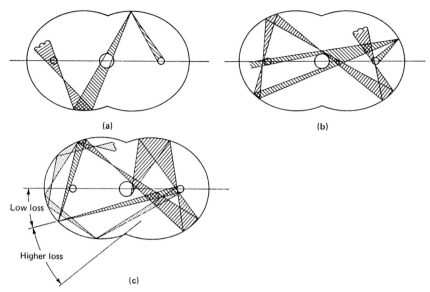

Fig. 6.81a–c. Reflection characteristic of double-elliptical cavities. (a) Radiation focused on the rod but not absorbed is focused back upon the second lamp. (b) Radiation which misses the rod on the first pass is intercepted after a few reflections by one of the lamps or the laser rod. (c) Direct radiation that passes close to the laser rod will eventually be intercepted by one of the pump lamps or the laser rod. Direct radiation which enters the second ellipse far away from the rod is essentially lost

In designing a close-coupled single-elliptical cylinder, the following procedure can be followed: First, the smallest separation of rod and pump source is determined according to

$$2c = s_1 + \frac{d_r + d_L}{2} , \tag{6.52}$$

where d_r and d_L are the clearances required for the rod and lamp assembly, respectively. These two dimensions are normally several times larger than the physical diameter of the rod and lamp in order to allow space for flow tubes, O-rings, rod and lamp supports, rod holders, lamp terminals, cooling channels, etc. s_1 is the separation between the rod and lamp assembly; it is a safety factor to allow for tolerance buildup.

After the separation of the focal points has been determined, we can calculate the major axis of the ellipse:

$$2a = s_1 + 2s_2 + d_r + d_L , \tag{6.53}$$

where s_2 is the clearance of the rod and lamp assembly from the elliptical surfaces. We see from (6.52, 53) that in the extreme, $s_1 = s_2 = 0$, $a = 2c$, and the eccentricity of the ellipse is $e = 0.5$. In practical cases, e will be somewhat lower.

The minor axis of the ellipse follows from

$$b = (a^2 - c^2)^{1/2} . \tag{6.54}$$

Figure 6.89 shows the cross section of a double-elliptical cavity of a cw-pumped Nd : YAG laser designed to yield a close separation between laser rod and lamps. Figure 6.90 depicts the outside configuration of the laser head.

Pump-Light Distribution in the Laser Rod

The pump-light distribution over the cross section of the active material is the result of a combination of three effects: the properties of the pump cavity, refractive focusing occurring in the rod itself, and nonuniform absorption of pump radiation. How dominant these individual effects are depends on the pumping geometry, i.e., focusing or diffuse reflecting, on the treatment of the cylindrical surface of the rod, i.e., rough or polished, and on the optical thickness, i.e., the product of absorption coefficient and radius.

We can superficially distinguish between strongly focused, uniform, and peripheral pump light distributions. A nonuniform distribution of pump energy in the active medium leads to a nonuniform temperature and gain coefficient distribution. Saturation effects, threshold energy, and possible laser rod damage depend on the gain distribution across the rod. The distribution of pump light in a cylindrical laser rod will be considered for two cases of practical interest: a strongly focusing elliptical cylinder and a diffuse reflecting cavity.

Considering first the case of a laser rod in a diffuse reflecting cavity, we find that if a laser rod with a polished surface is exposed to isotropic pump radiation, appreciable focusing results from the refraction of the pump light at the cylindrical surface. Rays which strike the rod periphery tangentially will pass the center of the rod at a radius [6.115]

$$r_i = \frac{r_R}{n} \,. \tag{6.55}$$

Furthermore, all radiation incident on the surface passes through a core bound by the diameter $2r_i$. The pump light intensity in this region is higher than in the boundary region $r_R \geq r \geq r_i$. The focusing action from the refraction of pump light at the surface of a polished rod is modified by absorption [6.110, 116]. The pump intensity as a function of path length x in the rod is given by

$$\frac{I(\nu)}{I_0(\nu)} = \exp[-\alpha(\nu)x] \,. \tag{6.56}$$

Pump radiation propagating to the center of the rod is attenuated by absorption at the periphery. The compensating effects of refractive focusing and pump light absorption can be seen in Fig. 6.82a. Plotted is the relative energy density within a Nd : glass rod situated in an isotropic field as a function of normalized radius for a number of absorption values [6.110]. It is seen that for an optically thin rod, refractive focusing is predominant and the pump density is highest in the central region ($r \ll 0.63r_R$) of the rod. On the other hand, in rods which are optically dense, the pump intensity is highest at the periphery.

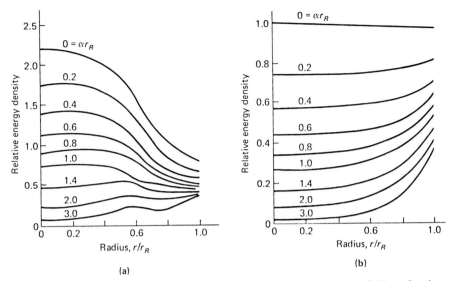

Fig. 6.82a, b. Relative energy density within a Nd : glass rod situated in an isotropic field as a function of normalized radius for different absorption values. **(a)** With polished, **(b)** with frosted lateral surface [6.110]

Several solutions exist to improve the pumping uniformity of laser rods located in diffuse cavities. The internal focusing effect can be reduced by roughening the surface of the rod, immersing the rod in a liquid (ideally an index-matching fluid), and cladding the rod. In the latter case the active core is surrounded by a transparent layer, such as sapphire for ruby rods and glass for Nd : glass rods. This makes the collecting cross section larger than the absorbing cross section.

Figure 6.82b shows the pump-light distribution in a rod with a frosted lateral surface. No focusing effect is observed and the rod exhibits a peripheral pump light distribution which is strictly determined by absorption. At low absorption values αr_R a fairly uniform pump light distribution is obtained in the laser material. Generally, the most uniform pump distribution is obtained in a laser rod which has a roughened side wall, is optically thin, and is pumped by a helical flashlamp.

If the laser rod is in a focusing pump cavity, the light source is imaged on the laser rod. When the region of maximum energy density in free space is smaller than the cross section of the laser rod, a highly illuminated core results, whose diameter is no longer given by the outside diameter of the rod but by the image-forming properties of the pump system and the index of refraction of the active material.

Figure 6.83 shows the reflection of different light cones emerging from a circular light source of an elliptical reflector. Reflection at point P_a gives the beam of greatest diameter, perpendicular to the major axis. Reflection at P_e

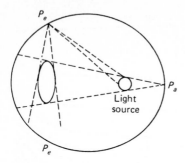

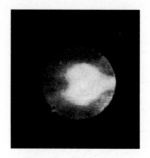

Fig. 6.83. Reflections of different light cones emerging from a circular light source located inside an elliptical reflector

Fig. 6.84. Distribution of fluorescence in a polished ruby rod as a result of illumination in an elliptical reflector [6.118]

gives the beam of least diameter parallel to the major axis. The three cones which are reflected at points P_a and P_e define the region through which the total flux of the light source passes. This oval region has a long axis (ϱ_a) parallel to the minor axis of the ellipse, and a short axis (ϱ_b) parallel to the major axis of the ellipse. We have [6.117]

$$\varrho_a = r_1 \frac{a/e + 1}{a/e - 1} \quad \text{and} \quad \varrho_b = r_1 \frac{(a/e)^2 - 1}{(a/e)^2 + 1} \, , \tag{6.57}$$

where r_1 is the radius of the light source. Note that ϱ_a and ϱ_b are the boundaries in free space, inside a laser rod these dimensions are reduced by the ratio $1/n$.

From these considerations it follows that, while the emission of the light source is axially symmetrical, the energy density distribution at the site of the laser is no longer axially symmetrical in an elliptical cavity. There is a preferred axis perpendicular to the long axis of the ellipse. The region of maximum energy density is of interest for laser operation, since it is here that the material first reaches the threshold inversion. The focusing action is strongest in a laser rod with a polished surface located in an elliptical pumping cavity.

The distribution of the pump energy within the laser rod can be studied by taking near-field pictures of the pattern of fluorescence emission. A typical picture obtained in this way from a ruby rod pumped by a flashlamp in a single elliptical cavity is shown in Fig. 6.84. The pump light is mostly concentrated in a central core approximately elliptical in shape with its major axis orthogonal to the major axis of the elliptical cylinder, which is in agreement with the aforementioned theoretical considerations. Facing the flashlamp, a wedge-shaped bright region, which is caused by direct radiation from the flashlamp, is present in the rod.

Figure 6.85 exhibits the result of a Monte Carlo calculation carried out to determine the pump radiation distribution in a ruby rod pumped by a xenon flashlamp in a single-elliptical cavity. Indicated in Fig. 6.85 are the areas of the ruby crystal which reach threshold as a function of time. The pump nonuniformity in the crystal is again a combination of the focusing action of the ellipse and direct radiation from the pump source.

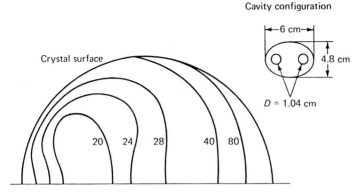

Fig. 6.85. Area of ruby crystal that has reached the threshold condition for lasing as a function of time (arbitrary units) in a single elliptical pump cavity

In order for strong focusing to occur, it is necessary that the rod be fairly transparent to the pump radiation. Standard-doped ruby, Nd : glass, and Nd : YAG rods of up to 1 cm diameter are sufficiently transparent for strong focusing effects. With three-level systems, where a substantial change in the ground-state population will occur during pumping, a rod that is initially opaque may become transparent in the course of the pump pulse. The pump radiation can "bleach" the rod, so to speak, or "burn" its way in. The bleaching of ruby rods can be very substantial when the rods are used as storage devices (amplifiers, Q-switched lasers). Here, no regeneration is present during the pump phase of the laser cycle and the ground state can be almost completely depleted.

One of the consequences of focusing is that the threshold for oscillation is lowered in the focal region. Concentration of pump light at the center of the rod can be desirable in systems which operate at the TEM_{00} mode, since operation in this mode requires maximum gain in the center.

If the lateral surface of the rod is frosted, then the pump radiation entering the rod remains diffuse; in this case radiation from the entire surface arrives at each point within the rod. Frosting of the lateral surface of the rod provides for a considerably more uniform gain distribution, and a larger cross section of the rod is utilized. However, an asymmetric gain distribution remains, since the rod is pumped from one side. To eliminate this asymmetry, a dual flashtube and close-coupled geometry can be utilized. This configuration provides for a relatively uniform gain distribution. If the laser rod is surrounded by a flow tube for water cooling, the glass tube can also be ground to achieve a homogeneous pump light distribution in the laser rod [6.119].

6.3.3 Mechanical Design

In the construction of a pump cavity, several critical design areas can be identi-fied; these are efficient cooling of the laser rod, lamps and reflector; the design of the various O-ring seals; the selection of the reflector base material; polishing and plating procedure; and prevention of arcing. In addition, human engineering aspects have to be considered, such as ease of lamp replacement. In the following we will address these potential problem areas.

The mechanical design of a pump cavity is determined mainly by two consid-erations: the geometry chosen for efficient energy transfer from the pump source to the laser material, and the provisions required for extracting the heat generated by the pump source. Optically pumped lasers have efficiencies of a few percent, therefore, almost all the electrical energy supplied to the lamp will have to be removed as heat from the pump cavity. The causes of this low efficiency can be divided between the poor conversion of electrical energy to energy absorbed by the lasing material (typically 5 to 10 %) and the poor utilization of this absorbed energy in contributing to the output of the laser (approximately 20 to 30 %).

Thermal Load of Cavity Components

Let us consider the radiant energy-transfer processes in the pump cavity of a laser cavity. Figure 6.86 shows in a simplified way the energy balance in a laser system. The electrical input power supplied to the lamp is either dissipated as heat by the lamp envelope and electrodes or emitted as radiation. A portion of the radiation will be absorbed by the metal surfaces of the pump cavity. The radiation reflected from the walls will either be absorbed by the lasing medium or will return to the lamp. The light which is absorbed by the lamp will add energy to the radiation process in the same way as the electrical power does, and the returned light will be radiated with the same efficiency as the power supplied electrically. One consequence of the reabsorption is that a lamp, when enclosed in the pumping cavity, is operated under a higher thermal loading, resulting in shorter lamp life than when operated in the open for the same electrical input power. Since most laser cavities are liquid-cooled, a distinction is made between the radiation actually absorbed by the rod and the radiation absorbed by the surrounding cooling liquid, flow tubes, filters, etc. The pump power absorbed by the laser rod causes stimulated emission and fluorescence at the laser wavelength and other main emission bands. The remainder is dissipated as heat by the laser material. The percentages which appear in Fig. 6.86 are based on the electrical input power to the lamp. The numbers are typical for cw and pulsed pumped Nd : YAG lasers employing arc lamps.

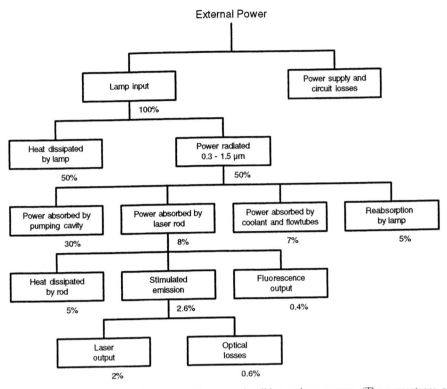

Fig. 6.86. Energy balance in an optically pumped solid-state laser system. (The percentages are fractions of electrical energy supplied to the lamp)

Table 6.9. Energy transfer in a cw krypton arc lamp, pumped Nd:YAG laser

Heat dissipation of lamps	53%
Heat dissipation of laser rod assembly	14%
Heat dissipation of pump reflector	30%
Radiation output	3%
Electrical input to lamps	100%

The specific percentage values of Fig. 6.86 were measured on a Nd:YAG laser pumped by two cw krypton arc lamps contained in a gold-plated double-elliptical pump cavity. The radiation efficiency of the lamps was measured calorimetrically by surrounding a lamp with an opaque liquid calorimeter. The energy balance inside the cavity was also measured calorimetrically by measuring separately the heat removed by the lamps, laser rod, and cavity cooling loops. The transfer efficiency of the pump cavity and the power absorbed by the rod itself was determined by measuring the heat extracted from the laser rod cooling loop if a black-anodized copper rod, a Nd:YAG laser crystal, and a quartz rod were

Table 6.10. Energy transfer in a Nd:glass disk amplifier

Circuit losses	8%
Lamp heat	50%
Heating of pump cavity walls	30%
Ultraviolet absorption	7%
Heating of glass disks	2%
Fluorescence decay	2%
Useful laser energy	1%
Electrical input to lamps	100%

installed in the cavity [6.120]. With a black absorber installed in the cavity in place of the laser crystal, 43 % of the available pump radiation was absorbed. This number represents the cavity transfer efficiency. Table 6.9 summarizes the energy balance in the laser cavity. As a comparison, Table 6.10 lists the results of an energy-transfer analysis performed on a large Nd:glass disk amplifier [6.121]. Useful information for the calculation of heat load and energy balance in pump cavities can be found in [6.122, 123].

Spectral Properties of the Materials Employed in the Design of Pump Cavities

Since the reflectivity of the metal surfaces as well as the transmission of the cooling fluid in the cavity are wavelength dependent, the spectrum of the pump light incident on the laser rod is different from the source emission spectrum. Ideally, in the transmission of the radiation from the source to the laser rod, one would like to have minimum optical losses in the pump bands of the laser material, and total absorption of all pump energy in spectral regions which do not contribute to the laser output. In this way the thermal heat load and the associated optical distortions in the active material would be kept at a minimum. Particularly undesirable is the ultraviolet content of the pump light, because it causes solarization in most materials. Furthermore, ultraviolet radiation leads to rapid deterioration of any organic materials in the pump cavity, such as silicon O-rings. Also undesirable is the formation of ozone by the ultraviolet radiation, because it leads to corrosion of metal parts in the cavity. Pump radiation, which has a longer wavelength than the stimulated emission, does not contribute to the laser output but does heat up the laser crystal and leads to optical distortions. The intensity and spectral content of the pump radiation reaching the laser rod depend on the reflectivity of the cavity walls, spectral filters placed inside the pump cavity, and the cooling medium.

The cavity walls can consist of specular-reflecting metal surfaces, diffuse-reflecting surfaces from ceramics, and compressed powders of inorganic materials or, in special cases, the reflector can be lined with a dielectric thin-film coating. The metals most commonly employed to obtain specular-reflecting surfaces in laser cavities are aluminum, silver, and gold. The reflectance versus wavelength

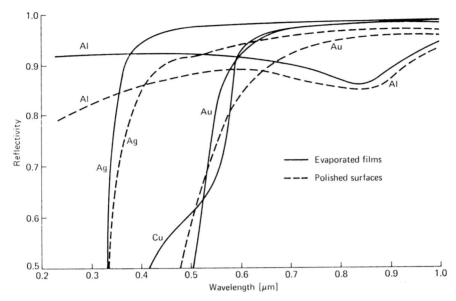

Fig. 6.87. Reflectivity versus wavelength for metals commonly used in the design of laser pump cavities

of these materials is shown in Fig. 6.87. The reflective metal surfaces are usually obtained by evaporation, sputtering, polishing, or electroplating. The reflectance of a good evaporated coating is always higher than that of a polished or electroplated surface.

The cavity walls must have a high reflectivity at the absorption bands of the laser material. Therefore, for pumping ruby with the absorption bands located between 4100 and 5600 Å, only aluminum and silver can be employed, since gold has a low reflectivity at these short wavelengths. Compared to silver, aluminum has a higher reflectivity for wavelengths shorter than about 0.35 μm, whereas for all longer wavelengths the reflectivity of silver is higher. Maintaining the high reflectivity of silver presents a real problem. During the aging of silver in air, a layer of silver sulfide forms on the surface, causing the reflectance to drop. This tarnishing may be prevented by coating the silver with a thin protective layer of a transparent material, such as SiO_2. Alternatively, if the silver reflectors are immersed in an inert cooling fluid or operated in a dry nitrogen atmosphere, they will maintain their high reflectance for long periods of time. Because of the problems associated with silver, the most commonly used metal in pump cavities containing ruby rods is aluminum. For pumping Nd:YAG or Nd:glass, the situation is different. As can be seen from Fig. 6.87, aluminum has a minimum in its reflectance curve in the near-infrared. Therefore, both silver and gold have higher reflectances in the main pump bands of neodymium lasers. For cw-pumped Nd:YAG lasers, where most of the pumping occurs in the wavelength region between 0.7 and 0.9 μm, gold is used exclusively because, in contrast to silver, it does not tarnish. In high-power pumped-pulsed neodymium lasers, a considerable amount of the flashlamp radiation is in the Nd pump bands located around 0.53 and 0.58 μm. In these systems silver-coated reflectors are usually employed.

In focusing geometries the base material for the reflector is either aluminum, copper, or stainless steel. Aluminum is usually employed for light-weight systems. If used in ruby systems, aluminum, when highly polished, does not require plating. However, aluminum reflectors in Nd : YAG lasers require silver plating for pulsed-pumped systems and gold plating for cw-pumped systems. Usually, aluminum is plated with nickel first. Nickel provides a hard surface which polishes very easily. After polishing, a flash of either silver or gold is applied. If weight considerations are not too stringent, a better choice for the reflector base material is copper, since copper has a lower thermal expansion, higher thermal conductivity compared to aluminum, and nickel plating on copper is much more durable than on aluminum. Kanigen-plated copper can easily be polished to very good tolerances. Both aluminum and copper must be nickel-plated before polishing if a good surface is to be obtained. Polishing nickel-plated copper or aluminum reflectors can represent a problem in some cases because of the danger of polishing through the nickel. In this case stainless steel offers an attractive alternative despite its thermal conductivity, which is a factor 10 lower than copper. Stainless steel can be polished to the highest optical finish. A thin layer of gold or silver is electroplated or evaporated onto the surface after the polishing process.

A highly polished and reflective pump cavity is paramount in focusing geometries for the attainment of good efficiency. A poor electroplating quality showing a haze or orange peel on the reflector surface, flaking, pits, and cracks, is of great concern to the designer of laser systems. The improvement in laser output obtained by gold plating a highly polished stainless steel reflector is between a factor of 2 and 3 for a cw-pumped Nd : YAG system. A comparison of the laser output of a pulsed ruby system from a highly polished and an unpolished aluminum reflector revealed an improvement of 50 %.

A diffuse reflector is usually fabricated from ceramics or compressed powder of MgO or $BaSO_4$. The powder is usually contained in the inner space between two concentric quartz tubes. The reflection from these materials is in the order of 90 to 98 % and fairly independent of wavelength for the region of interest. A reflector surrounding a helical lamp can be made from ceramic, aluminum, stainless steel, etc. Diffuse reflecting, molded ceramic inserts do not tarnish or corrode. Some ceramic materials containing impurities, however, show discoloration after prolonged exposure to flashlamp radiation. The difference in output due to different diffuse reflectors in a system containing a helical flashlamp and a ruby crystal is illustrated in Fig. 6.88.

In special cases, where it is important to minimize the heat load of a laser rod, dichroic thin-film coatings can be employed as reflective surfaces. These coatings, which are applied on either glass or metal surfaces, are designed such that they reflect pump radiation but transmit all unwanted radiation. The transmitted radiation is absorbed by the metal walls of the cavity or by an absorber surrounding the glass reflector.

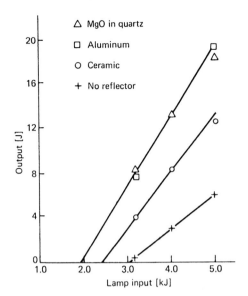

△ MgO in quartz

□ Aluminum

○ Ceramic

+ No reflector

Fig. 6.88. Output from a ruby laser pumped by a helical flashlamp for different reflector materials. Laser was operated in the conventional mode

The purpose of spectral filters placed inside the pump cavity is to provide adequate absorption of the intense ultraviolet radiation from xenon flashlamps. One way to accomplish this is to surround the laser rod or flashlamp with a tube made of Nonex, Pyrex, titanium-doped quartz, samarium-doped glass ED-5 and ED-6, or colored filter glass with a sharp cutoff. Instead of a tube as filter, a flat plate can be inserted, for example, in an elliptical-pump cylinder separating the rod and lamp. Other approaches include the use of ultraviolet-free lamps which are manufactured by employing an ultraviolet-absorbing or reflecting envelope. In glass rods, ultraviolet radiation can be prevented from reaching the active material by cladding the rod with samarium-doped glass.

The spectral properties of the cooling fluid can be utilized to remove some of the unwanted pump radiation. Water, if used as a coolant, is very effective in absorbing radiation at wavelengths longer than $1.3\,\mu$m. Where absorption of ultraviolet radiation by the laser material must be held to a minimum, potassium chromate, potassium dichromate, or sodium nitrite can be added to the cooling water.

The efficiency of Nd:glass and Nd:YAG lasers pumped by xenon flashlamps can be improved by circulating fluorescent dyes around the laser rod [6.124–126]. The dyes absorb in the spectral region in which the Nd ions do not absorb, and become fluorescent at one of the pump bands of Nd^{3+}. Rhodamine 6G dissolved in ethanol was identified as the best dye, giving better than 50 % improvement in the laser output.

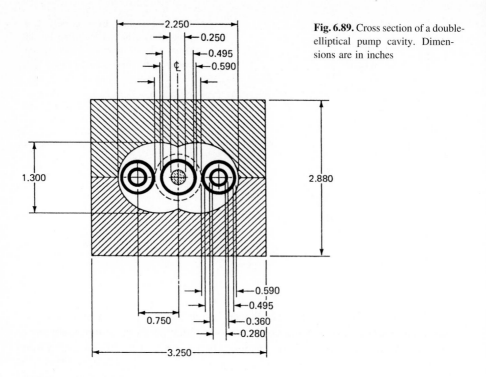

Fig. 6.89. Cross section of a double-elliptical pump cavity. Dimensions are in inches

Cooling Techniques

In elliptical-pump cavities, the laser rod and lamp are often liquid-cooled by circulating the coolant in flow tubes which surround these elements. The inside of the pumping chamber itself is dry. However, in most cases the body of the reflector contains cooling chambers through which the coolant fluid passes.

The pumping cavity usually consists of two parts that separate along the plane of the major axis. The end-plate reflectors, in addition to serving their optical function, can be used to mount and register the elliptical sections and to provide precision mounts for the laser rod assembly, as shown in Fig. 6.90. The rod assembly, consisting of the laser rod mounted in rod holders (Fig. 6.91a) can slide into precision-bored holes in these end plates. O-rings around the rod holder will seal this unit from the cavity. Figure 6.91b shows an example of a large Brewster-angle ruby rod installed in the end plate of an elliptical cavity without the use of rod holders. The drawing also shows the cooling channel and plenum chamber employed to force the cooling water into the annular cooling path defined by the ruby rod and the flow tube.

An O-ring seal, if properly designed, is very reliable and does not present a problem. Seal areas within the pump cavity are considerably more critical because organic materials exposed to the pump radiation will quickly deteriorate. If O-rings made from an organic material are used to seal the laser rod, flow tubes, or lamps, they should be well shielded from radiation. Figure 6.91 displays several examples of O-ring locations which provide maximum protection from

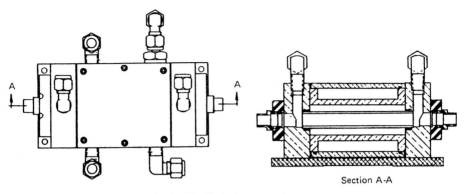

Fig. 6.90. Outside configuration of a double-elliptical pump cavity

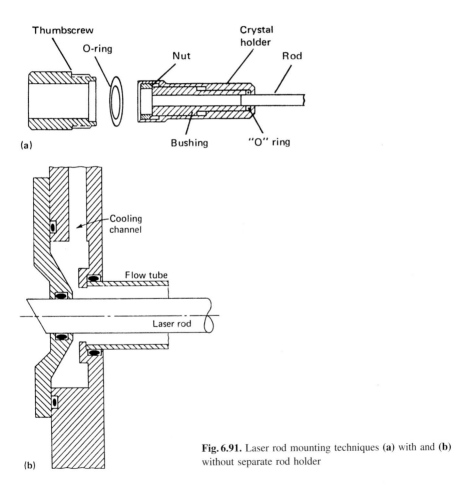

Fig. 6.91. Laser rod mounting techniques (a) with and (b) without separate rod holder

pump light. White silicon O-rings are usually found to be best suited for laser applications.

In the so-called flooded cavity approach, the whole inside of the pump cavity is immersed in cooling fluid. The absence of flow tubes and separate cooling chambers for the reflector makes this type of cavity very compact and simple in design. For example, only one inlet and outlet are required for the cooling loop, whereas in an elliptical cavity with flow tubes, one pair of coolant ports with the associated fittings, tubing, etc., is required for each reflector half, lamp, and laser rod. Also, in the flooded cavity design, lamp and laser rod can be brought very close together, and no reflection losses from additional glass surfaces are encountered. Figure 6.92 presents a photograph of a commercial cw-pumped Nd:YAG laser featuring a single elliptical reflector in a flooded cavity.

Most liquid-cooled military-type Nd:YAG lasers and most commercial cw-pumped Nd:YAG lasers feature this design,because of its compactness and simplicity. Figure 6.93 shows a photograph of a liquid-immersed pump cavity used for a high-repetition-rate military Nd:YAG laser. The cavity is sealed by one large O-ring in the top cover. The reflector inserts are machined from aluminum which is nickel-plated, polished, and silver-plated. The laser head is machined from aluminum which is hard-anodized (Mil. Spec. A-8625) to prevent corrosion.

Figure 6.94 illustrates an exploded view of a single- and double-elliptical pump cylinder using immersion cooling. The laser head is machined from a solid block of acrylic. This material alleviates the problem of arcing and eliminates the need to feed the lamp anode through the laser head. The reflector inserts are machined from copper and are gold-plated.

Great care has to be taken in the design of liquid-immersed cavities of the type displayed in Figs. 6.92–94 to achieve symmetrical high-velocity flow along the lamp and laser rod. Otherwise, the different cross sections and pockets in the cavity result in low-velocity and, in extreme cases, stagnant areas in the cooling loop. Note that in the design depicted in Fig. 6.92 the lamps and laser rod are surrounded by flow tubes to provide symmetrical and highly turbulent flow.

Diffuse-reflecting pump cavities are usually liquid-immersed. Figure 3.13 presents an example of a laser head containing a helical flashlamp. The main elements of such a system are the laser rod, flashlamp, a closely fitted reflector, and a housing containing these elements. In the design shown in Fig. 3.13, the laser rod, flashlamp, and reflector are all immersed in water. In helical-pump lamp systems the laser material is sometimes surrounded by a flow tube to define a cooling channel which forces water first through an annulus between rod and flow tube and then back over the helical lamp.

Figure 6.95 presents a laser head employed to pump laser rods up to 15 cm in length by a single linear flashlamp. Lamp and rod are contained in a diffuse reflector consisting of barium sulfate which is compressed between two quartz tubes. Inside, the pumping chamber is filled with water. The body of the laser head is made from acrylic.

The light-emitting surface of laser diodes cannot come directly in contact with liquid coolants, therefore laser diodes need to be conductively cooled. The design

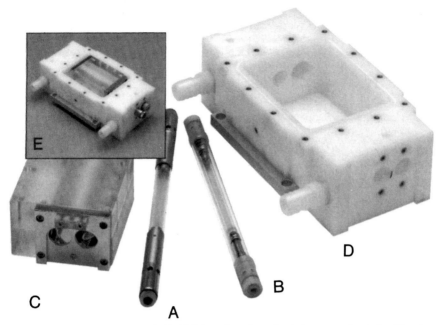

Fig. 6.92. Commercial cw-pumped Nd : YAG laser featuring a flooded pump cavity design. The photo shows the laser rod assembly (*A*), krypton arc lamp assembly (*B*), single elliptical reflector (*C*), and pump housing (*D*). The insert (*E*) shows the assembled pump head with top removed. (Quantronix Corp. Model 114)

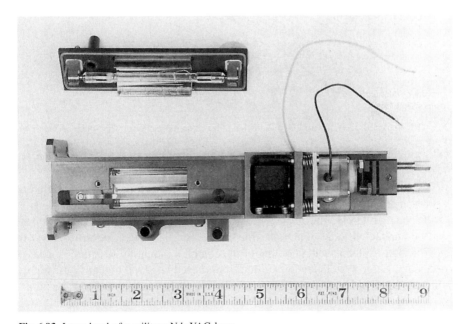

Fig. 6.93. Laser head of a military Nd : YAG laser

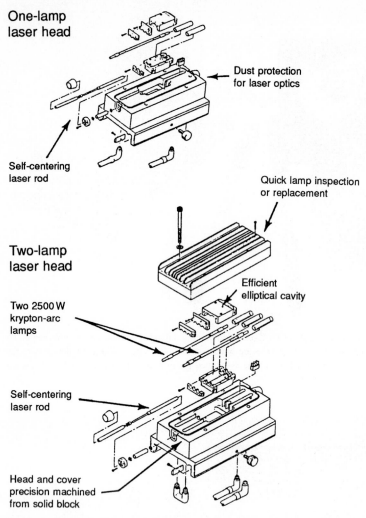

One-lamp laser head

Dust protection
for laser optics

Self-centering
laser rod

Quick lamp inspection
or replacement

Two-lamp laser head

Efficient
elliptical cavity

Two 2500 W
krypton-arc
lamps

Self-centering
laser rod

Head and cover
precision machined
from solid block

Fig. 6.94. Exploded view of a single- and double-elliptical pump cavity of a cw-pumped Nd:YAG laser

of cooled pump heads for diode-pumped lasers requires an entirely different approach compared to flashlamp-based systems.

In smaller lasers, the heat generated by the diode array and laser crystal can be conducted to a mounting structure and carried away by air cooling. In intermediate systems, water cooling of the common mounting structure is usually necessary. Large systems require a liquid-immersed laser rod or slab and a highly efficient liquid coolant flow as close as possible to the junction of the laser diode.

Figure 6.96 exhibits two 15 W cw laser-diode arrays and a small slab Nd:YLF crystal mounted to a copper block. The triangular structure between the diodes and laser crystal contain cylindrical optics for beam collimation and

Fig. 6.95. Example of a diffuse-reflecting pump cavity

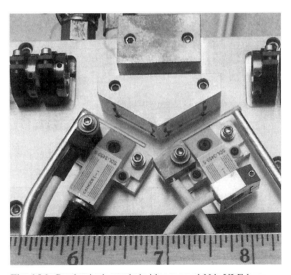

Fig. 6.96. Conductively cooled side-pumped Nd : YLF laser

focusing. The output beams from the two diode arrays arranged in a "V" shaped configuration overlap at the laser crystal. Heat is carried away from the diode array and laser crystal through the water-cooled baseplate onto which all components are mounted.

Large diode-pumped lasers employ liquid-cooled laser rods or slabs similar to flashlamp-pumped systems. The laser-diode arrays are directly mounted on liquid-cooled heat sinks. After many iterations, extremely compact and lightweight structures have emerged which take full advantage of the high packaging density which can be achieved with diode arrays. Figure 6.97a,b displays a design comprised of two end plates, which includes manifolds, and four triangular connecting bars. The laser crystal is inserted into a sapphire sleeve. Cover

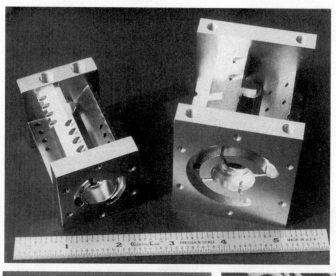

(a)

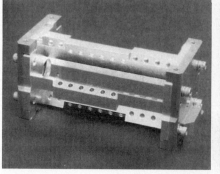

(b)

(c)

Fig. 6.97. Liquid-cooled laser heads for diode-pumped Nd : YAG lasers (**a, b**). Assembled pump head containing 16 diode arrays (**c**)

plates screwed to the end plates, seal and secure the rod. The diode arrays are mounted in a symmetrical pattern around the rod to produce uniform excitation. The flat areas of the triangular bars provide the mounting surface for the diodes and the thermal interface.

To carry off heat from the diodes and laser rod, liquid flows into the pump head through a fitting that screws into a part in the end plate. Directed by the manifolds, the coolant flows back and forth through passages in the triangular structures, and also through the annulus between the rod surface and sapphire sleeve. Eventually, the coolant exits the pump head through a second fitting on the end plate. With the diode arrays mounted on the internally water-cooled bars, the coolant, as it moves through the structure, is close to the heat source, thus facilitating heat transfer. For maximum heat transfer, a thin layer of indium is applied between the mounting surface and the diode arrays.

The monolithic structure comprised of the four bars and two end plates is machined from a single piece of tellurium copper. Because of the monolithic design, only six "O" rings are required. One pair each for sealing the laser rod and flow tube, and one pair for the two end plates which seal the manifolds.

Figure 6.97c presents a photo of an assembled pump head for a side-pumped Nd:YAG rod pumped by 16 diode arrays. Each array is 1 cm long, and the devices are arranged in four pairs symmetrically around the rod. The adjacent 4 pairs are off-set $45°$ in order to achieve an eight-fold symmetry around the laser crystal. The active pump area is 4 cm long and the complete assembly, including the end plates, has a length of 7 cm.

Each diode array contains 5 bars, i.e. the laser crystal is pumped by 80 bars at a pulse energy of 800 mJ in a $200\,\mu\text{s}$-long pulse. Electrical input is 2 J per pulse, which at a repetition-rate of 100 Hz, amounts to about 180 W of heat which has to be carried away by the laser head.

7. Thermo-Optic Effects and Heat Removal

The optical pumping process in a solid-state laser material is associated with the generation of heat for a number of reasons: a) The energy difference of the photons between the pump band and the upper laser level is lost as heat to the host lattice and causes the so-called quantum defect heating; b) similarly, the energy difference between the lower laser level and the ground state is thermalized; c) since the quantum efficiency of the fluorescence processes involved in the laser transition is less than unity, heating due to quenching mechanisms takes place; and d) the broad spectral distribution of arc lamps or flashlamps is such that there is considerable absorption by the host material, mainly in the ultraviolet and infrared bands.

Efficient heat removal, and the reduction of the thermal effects which are caused by the temperature gradients across the active area of the laser medium usually dominate design considerations for high-average-power systems.

One of the advantages of diode-laser pumping is that the waste heat dissipated in the laser medium is greatly reduced by the high efficiency of the pumping process. Quantum-defect heating is reduced because the pump wavelength is closer to the laser-emission wavelength, and heating of the host material by pump radiation located outside the absorption bands of the active ions is completely eliminated. For example, in a diode-pumped Nd : YAG laser the thermal load of the crystal is only about 1/3 that of a flashlamp-pumped system for the same laser output.

Typical examples for the fraction of thermal energy generated in the lasing process have been provided in Figs. 6.60 and 6.86. In the laser-diode-pumped system, an output of 7 W is, at a minimum, associated with 5 W of heating due to the quantum defect and fluorescence quenching mechanisms. In addition, a fraction of the 4 W fluorescence and absorption losses also contributes to the heating of the laser crystal. Let us assume that the 7 W output is produced by a 5 cm long laser Nd : YAG rod with an absorption coefficient of $10^{-3}\,\mathrm{cm}^{-1}$. If the resonator has an output coupler with a reflectivity of 90%, then the intra-cavity power is 133 W which produces about 0.7 W of absorption losses in the Nd : YAG rod. Furthermore, in the high-beam-quality laser postulated in Fig. 6.60 the overlap between the resonator mode and the pumped volume is less than one. As will be explained below, about 10% of the fluorescence power emitted at the outer parts of the rod will appear as heat in the crystal. Therefore, in the example given in Fig. 6.60, the total heat load of the crystal is around 6 W. In general, we can assume a ratio of heat load to laser output of (0.85–1.1) for a diode-pumped

Nd : YAG laser, depending on the crystal quality and the overlap of resonator mode and pump region.

In the example of an arc lamp system illustrated in Fig. 6.86, the measured ratio of heat load to laser output was about 2.5 for a highly multi-mode beam. Any measures to improve the beam quality will invariably result in a reduced overlap between resonator mode and pump region, and therefore, the ratio of heat load to laser output will increase.

In this context it is of interest to point out that the thermal load in an Nd : YAG crystal is different under lasing and non-lasing conditions. This is easily observed in cw or repetitively pumped lasers if the lasing action is interrupted by blocking the beam in the resonator. After lasing is re-established, the beam divergence will show a transient behavior until the thermal load, and therefore thermal lensing, has reached its final value. The reason for this behavior is that the absorbed pump power which is converted into heat depends on whether the upper-state population is extracted by lasing action or depleted by fluorescence. As was mentioned in the context of explaining Fig. 6.60, in a diode-pumped Nd : YAG laser about 43% of the absorbed power ends up as heat under unextracted conditions, whereas under extracted conditions that value drops to about 32%. The reason is that the upper laser level has about a 10% branching ratio into a non-radiative decay, which contributes to the heat load. The other 90% are spontaneous emission. Hence, without extraction 10% of the stored power ends up as heat. But, with laser action, only 10% of the nonextracted power contributes to the heat load.

7.1 Cylindrical Geometry

The combination of volumetric heating of the laser material by the absorbed pump radiation and surface cooling required for heat extraction leads to a nonuniform temperature distribution in the rod. This results in a distortion of the laser beam due to a temperature- and stress-dependent variation of the index of refraction. The thermal effects which occur in the laser material are thermal lensing and thermal stress-induced birefringence.

An additional issue associated with thermal loading is stress fracture of the laser material. Stress fracture occurs when the stress induced by temperature gradients in the laser material exceeds the tensile strength of the material. The stress-fracture limit is given in terms of the maximum power per unit length dissipated as heat in the laser medium.

The particular temperature profile which exists in the laser material depends, to a large degree, on the mode of operation, i.e., cw pumped, single shot, or repetitively pulse pumped. In the case of cw operation, a long cylindrical laser rod with uniform internal heat generation and constant surface temperature assumes a quadratic radial temperature dependence. This leads to a similar dependence in both the index of refraction and the thermal-strain distributions. In a pulse-pumped system, laser action occurs only during the pump pulse or shortly thereafter in the case of Q-switching, therefore the main activity is centered around the time interval of the pumping pulse. Theoretical and experimental investigations have shown that heat transport during the pump pulse, which usually has a duration between 0.2 and 5 ms, can be neglected. Therefore, in single-shot operation, optical distortions are the result of thermal gradients generated by non-uniform pump-light absorption. In repetitively pulse-pumped systems, distortions will occur from the cumulative effects of non-uniform pump processes and thermal gradients due to cooling. Which effects dominate, depends on the ratio of the pulse interval time to the thermal relaxation time constant of the rod. At repetition intervals, which are short compared to the thermal relaxation time of the laser rod, a quasi-thermal steady-state will be reached where the distortions from pumping become secondary to the distortions produced by the removal of heat from the laser material.

7.1.1 CW Operation

Temperature Distribution

We consider the case where the heat generated within the laser rod by pump-light absorption is removed by a coolant flowing along the cylindrical-rod surface. With the assumption of uniform internal heat generation and cooling along the cylindrical surface of an infinitely long rod, the heat flow is strictly radial, and end effects and the small variation of coolant temperature in the axial direction can be neglected. The radial temperature distribution in a cylindrical rod with the thermal conductivity K, in which heat is uniformly generated at a rate Q per unit volume, is obtained from the one-dimensional heat conduction equation [7.1]

$$\frac{d^2T}{dr^2} + \left(\frac{1}{r}\right)\left(\frac{dT}{dr}\right) + \frac{Q}{K} = 0 . \tag{7.1}$$

The solution of this differential equation gives the steady-state temperature at any point along a radius of length r. With the boundary condition $T(r_0)$ for $r = r_0$, where $T(r_0)$ is the temperature at the rod surface and r_0 is the radius of the rod, it follows that

$$T(r) = T(r_0) + \left(\frac{Q}{4K}\right)(r_0^2 - r^2) . \tag{7.2}$$

The temperature profile is parabolic, with the highest temperature at the center of the rod. The temperature gradients inside the rod are not a function of the surface temperature $T(r_0)$ of the rod. The heat generated per unit volume can be expressed as

$$Q = \frac{P_a}{\pi r_0^2 L} , \tag{7.3}$$

where P_a is the total heat dissipated by the rod, and L is the length of the rod. The temperature difference between the rod surface and the center is

$$T(0) - T(r_0) = \frac{P_a}{4\pi KL} . \tag{7.4}$$

The transfer of heat between the rod and the flowing liquid creates a temperature difference between the rod surface and the coolant. A steady state will be reached when the internal dissipation P_a is equal to the heat removed from the surface by the coolant

$$P_a = 2\pi r_0 Lh[T(r_0) - T_F] , \tag{7.5}$$

where h is the surface heat transfer coefficient, and T_F is the coolant temperature. With $F = 2\pi r_0 L$ being the surface area of the rod, it follows that

$$T(r_0) - T_F = \frac{P_a}{Fh} . \tag{7.6}$$

Combining (7.4 and 7.6), one obtains for the temperature at the center of the rod

$$T(0) = T_F + P_a \left(\frac{1}{4\pi KL} + \frac{1}{Fh} \right) . \tag{7.7}$$

Thus, from the geometry, and the appropriate system and material parameters, the thermal profile of the crystal can be determined, except that h must be evaluated. This coefficient is obtained from a rather complex expression involving the thermal properties of the coolant, the mass flow rate of the coolant, the Reynolds, Prandtl, and Grashof numbers, and the geometry [7.2–4]. In Fig. 7.1 the heat transfer coefficient for water is plotted as a function of flow rate for cases of practical interest. The parameters are the rod radius r_0 and the inside radius r_F of a circular cooling channel. The boundary conditions for the heat transfer coefficient are a thermally insulated laser rod ($h = 0$), or unrestricted heat flow from the rod surface to a heat sink ($h = \infty$).

Figure 7.2 shows as an example the radial temperature profile in a Nd : YAG rod calculated from (7.7). The laser, which delivered between 200 to 250 W of output, was pumped at 12 kW of input power. The following parameters have been used in the numerical calculations of the temperature profile in the crystal: rod length $L = 7.5$ cm; rod radius $r_0 = 0.32$ cm; flow-tube inside radius $r_F = 0.7$ cm; power dissipated by the Nd : YAG rod $P_a = 600$ W; mass flow rate of the coolant $m^* = 142$ g/s; fluid temperature entering the cavity $T_F = 20°$ C. As can be seen from this figure, the maximum temperature of the crystal occurring at the center is $114°$ C. The large temperature gradient of $57°$ C between the center of the crystal and the surface is responsible for the high stresses present in the material.

Thermal Stresses

The temperature gradients generate mechanical stresses in the laser rod, since the hotter inside area is constrained from expansion by the cooler outer zone. The stresses in a cylindrical rod, caused by a temperature distribution $T(r)$, can be

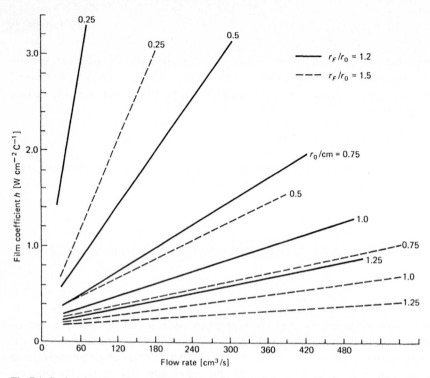

Fig. 7.1. Surface heat transfer coefficient h as a function of flow rate. Parameters are the rod radius r_0 and the flowtube inner radius r_F [7.4]

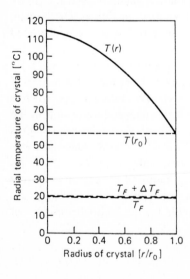

Fig. 7.2. Radial temperature distribution within a Nd:YAG crystal as a function of radius. T_F is the temperature of coolant entering flow-tube assembly; ΔT_F is the axial temperature gradient; and $T(r_0)$ is the rod surface temperature [7.3]

calculated from the equations given by *Timoshenko* and *Goodier* [7.5]. Equations (7.8–10) describe the radial σ_r, tangential σ_ϕ, and axial σ_z stress in an isotropic rod with free ends and a temperature distribution according to (7.2)

$$\sigma_r(r) = QS(r^2 - r_0^2) \,, \tag{7.8}$$

$$\sigma_\phi(r) = QS(3r^2 - r_0^2) \,, \tag{7.9}$$

$$\sigma_z(z) = 2QS(2r^2 - r_0^2) \,, \tag{7.10}$$

where the factor $S = \alpha E[16K(1 - \nu)]^{-1}$ contains the material parameters; E is Young's modulus; ν is Poisson's ratio; and α is the thermal coefficient of expansion. The stress components, σ_r, σ_ϕ, σ_z represent compression of the material when they are negative and tension when they are positive. We notice that the stress distributions also have a parabolic dependence on r. The center of the rod is under compression. The radial component of the stress goes to zero at the rod surface, but the tangential and axial components are in tension on the rod surface by virtue of the larger bulk expansion in the rod's center compared to the circumference.

Figure 7.3 gives the stresses as a function of radius inside the Nd : YAG rod whose temperature profile was shown in Fig. 7.2. From these curves it follows that the highest stresses occur at the center and at the surface. Since the tensile strength of Nd : YAG is 1800 to 2100 kg/cm^2, the rod is stressed about 70% of its ultimate strength. As the power dissipation is increased, the tension on the rod surface increases and may exceed the tensile strength of the rod, thereby causing fracture. It is of interest to determine at what power level this will occur. Using (7.9) we find for the hoop stress

$$\sigma_\phi = \frac{\alpha E}{8\pi K(1 - \nu)} \frac{P_a}{L} \,. \tag{7.11a}$$

The total surface stress $\sigma_{\max}$ is the vector sum of σ_ϕ and σ_z, i.e. $\sigma_{\max} = 2^{1/2}\sigma_\phi$.

We note from (7.11a) that the tension on the surface of a laser rod depends on the physical constants of the laser material and on the power dissipated per unit length of the material, but does not depend on the cross section of the rod. Upon substitution of $\sigma_{\max}$ with the value of the tensile strength of Nd : YAG, we find that with 150 W dissipated as heat per centimeter length of Nd : YAG rod, the tension on the surface of the rod equals the tensile strength of the material. The actual rupture stress of a laser rod is very much a function of the surface finish of the rod, and can vary by almost a factor of three in Nd : YAG [7.6].

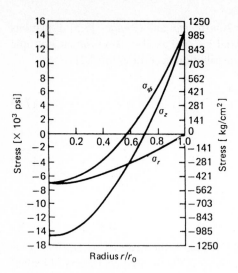

Fig. 7.3. Radial (σ_r), tangential (σ_ϕ), and axial (σ_z) stress components within a Nd:YAG crystal as a function of radius [7.3]

Stress Fracture Limit

The mechanical properties of the laser host material determine the maximum surface stress that can be tolerated prior to fracture. If there were no other constraints, such as stress-induced focusing and birefringence, the thermal loading and thus average output power of a rod laser could be increased until stress fracture occurred. If σ_{max} is the maximum surface stress at which fracture occurs, then we can rewrite (7.11a) as follows

$$\frac{P_a}{L} = 8\pi R$$

where

$$R = \frac{K(1-\nu)}{\alpha E}\sigma_{max} \qquad (7.11b)$$

is a "thermal shock parameter". A larger R indicates a higher permissible thermal loading before fracture occurs. The table below lists typical values for a number of laser materials if we assume a standard surface treatment.

Table 7.1. Thermal shock parameter for different materials

Material	Glass	GSGG	YAG	Al$_2$O$_3$
Thermal shock parameter R [W/cm]	1	6.5	7.9	100

Photoelastic Effects

The stresses calculated in the previous subsection generate thermal strains in the rod, which, in turn, produce refractive index variations via the photoelastic effect. The refractive index of a medium is specified by the indicatrix, which in its most general case is an ellipsoid. A change of refractive index due to strain is given by a small change in shape, size, and orientation of the indicatrix [7.7]. The change is specified by small changes in the coefficients B_{ij} [7.8],

$$B_{ij} = P_{ijkl}\varepsilon_{kl} \quad (i,j,k,l = 1,2,3) \tag{7.12}$$

where P_{ijkl} is a fourth-rank tensor giving the photoelastic effect. The elements of this tensor are the elastooptical coefficients. ε_{kl} is a second-rank strain tensor.

   We will confine our calculation to the case of a Nd:YAG laser rod. The method, however, is applicable to any material if the proper photoelastic matrix is used. Since Nd:YAG is a cubic crystal, the indicatrix is a sphere. Under stress the indicatrix becomes an ellipsoid. Nd:YAG rods are grown with the cylindrical axes along the [111] direction. The light propagates in this direction, and thus the change of the refractive index along the [111] is of interest.

   Since the transverse stresses are in the radial and tangential directions – relative to the coordinate system shown in Fig. 7.4 – the local indicatrix also orients its axis in these directions. In a cylindrical coordinate system the photoelastic changes in the refractive index for the r and ϕ polarizations are given by [7.8]

$$\Delta n_r = -\tfrac{1}{2}n_0^3 \Delta B_r \tag{7.13}$$

and

$$\Delta n_\phi = -\tfrac{1}{2}n_0^3 \Delta B_\phi \ . \tag{7.14}$$

A considerable amount of tensor calculation is required to determine the coefficients ΔB_r and ΔB_ϕ in a plane perpendicular to the [111] direction of the Nd:YAG crystal. The technique used in these calculations consists of introducing the elastooptic coefficients for Nd:YAG into the matrix form P_{mn} of the photoelastic tensor. The published values of these coefficients are given in a [100]-oriented Cartesian coordinate frame [7.9]: $P_{11} = -0.029$; $P_{12} = +0.0091$; and $P_{44} = -0.0615$. The strain tensor is obtained from the stresses calculated in (7.8–10). After a coordinate transformation to bring P_{mn} and ε_{kl} into the same coordinate system, the tensor operation according to (7.12) can be performed. Introducing the expression for ΔB_r and ΔB_ϕ into (7.13, 14), the refractive-index changes are given by [7.10, 11]

$$\Delta n_r = -\frac{1}{2}n_0^3\frac{\alpha Q}{K}C_r r^2 \ , \tag{7.15}$$

$$\Delta n_\phi = -\frac{1}{2}n_0^3\frac{\alpha Q}{K}C_\phi r^2 \ , \tag{7.16}$$

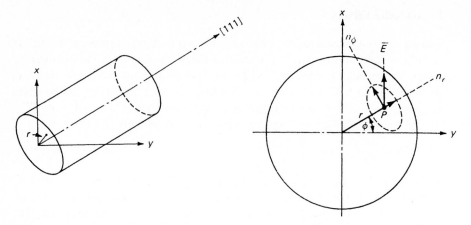

Fig. 7.4. Crystal orientation for a Nd : YAG rod (*left*) and orientation of indicatrix of the thermally stressed Nd : YAG rod in a plane perpendicular to the rod axis (*right*)

where C_r and C_ϕ are functions of the elastooptical coefficients of Nd : YAG.

$$C_r = \frac{(17\nu - 7)P_{11} + (31\nu - 17)P_{12} + 8(\nu + 1)P_{44}}{48(\nu - 1)} \, ,$$

$$C_\phi = \frac{(10\nu - 6)P_{11} + 2(11\nu - 5)P_{12}}{32(\nu - 1)} \, .$$

The induced birefringence is determined from (7.15, 16)

$$\Delta n_r - \Delta n_\phi = n_0^3 \frac{\alpha Q}{K} C_B r^2 \tag{7.17}$$

where

$$C_B = \frac{1 + \nu}{48(1 - \nu)} (P_{11} - P_{12} + 4P_{44}) \, .$$

Inserting the values of the photoelastic coefficients and the material parameters of Nd : YAG, $\alpha = 7.5 \times 10^{-6}/°C$, $K = 0.14 \, \text{W/cm}°\,C$, $\nu = 0.25$, $n_0 = 1.82$, into (7.15–17), one obtains $C_r = 0.017$, $C_\phi = -0.0025$, $C_B = -0.0099$, and

$$\Delta n_r = (-2.8 \times 10^{-6})Qr^2 \, ,$$

$$\Delta n_\phi = (+0.4 \times 10^{-6})Qr^2 \, ,$$

$$\Delta n_r - \Delta n_\phi = (-3.2 \times 10^{-6})Qr^2 \, ,$$

where Q has the dimension of watts per cubic centimeter and r is measured in centimeters.

Thermal Lensing

Having explored the stresses in the laser rod, we now turn to the optical distortions which are a result of both temperature gradients and stresses. The change of the refractive index can be separated into a temperature- and a stress-dependent variation. Hence

$$n(r) = n_0 + \Delta n(r)_T + \Delta n(r)_\varepsilon \ , \tag{7.18}$$

where $n(r)$ is the radial variation of the refractive index; n_0 is the refractive index at the center of the rod; and $\Delta n(r)_T$, $\Delta n(r)_\varepsilon$ are the temperature- and stress-dependent changes of the refractive index, respectively.

The temperature-dependent change of refractive index can be expressed as

$$\Delta n(r)_T = [T(r) - T(0)] \left(\frac{dn}{dT} \right) \ . \tag{7.19}$$

With the aid of (7.2, 4) we obtain

$$\Delta n(r)_T = -\frac{Q}{4K} \frac{dn}{dT} r^2 \ . \tag{7.20}$$

As can be seen from (7.15, 16 and 20), the refractive index in a laser rod shows a quadratic variation with radius r. An optical beam propagating along the rod axis suffers a quadratic spatial phase variation. This perturbation is equivalent to the effect of a spherical lens. The focal length of a lens-like medium where the refractive index is assumed to vary according to

$$n(r) = n_0 \left(1 - \frac{2r^2}{b^2} \right) \tag{7.21}$$

is given by [7.12]

$$f \cong \frac{b^2}{4n_0 L} \ . \tag{7.22}$$

This expression is an approximation where it was assumed that the focal length is very long in comparison to the rod length. The distance f is measured from the end of the rod to the focal point.

The total variation of the refractive index is obtained by introducing (7.15, 16, and 20) into (7.18):

$$n(r) = n_0 \left[1 - \frac{Q}{2K} \left(\frac{1}{2n_0} \frac{dn}{dT} + n_0^2 \alpha C_{r,\phi} \right) r^2 \right] \ . \tag{7.23}$$

As was discussed in the previous subsection, the change of refractive index due to thermal strain is dependent on the polarization of light, therefore the photoelastic coefficient $C_{r,\phi}$ has two values, one for the radial and one for the tangential component of the polarized light. Comparing (7.23) with (7.21) yields

$$f' = \frac{K}{QL} \left(\frac{1}{2} \frac{dn}{dT} + \alpha C_{r,\phi} n_0^3 \right)^{-1} . \tag{7.24}$$

In our final expression for the focal length of a Nd:YAG rod, we will include the contributions caused by end effects. Perturbations of the principal thermal distortion pattern occur in laser rods near the ends, where the free surface alters the stress character. The so-called end effects account for the physical distortion of the flatness of the rod ends. Self-equilibrating stresses causing a distortion of flatness were found to occur within a region of approximately one diameter from the ends of Nd:glass and one radius from the end for Nd:YAG [7.13]. The deviation from flatness of the rod ends is obtained from

$$l(r) = \alpha_0 l_0 [T(r) - T(0)] , \tag{7.25}$$

where l_0 is the length of the end section of the rod over which expansion occurs. With $l_0 = r_0$ and (7.2), we obtain

$$l(r) = -\alpha_0 r_0 \frac{Q r^2}{4K} . \tag{7.26}$$

The focal length of the rod caused by an end-face curvature is obtained from the thick-lens formula of geometric optics [7.7]

$$f'' = \frac{R}{2(n_0 - 1)} , \tag{7.27}$$

where the radius of the end-face curvature is $R = -(d^2 l/dr^2)^{-1}$. From these expressions follows the focal length of the rod caused by a physical distortion of the flat ends:

$$f'' = K[\alpha Q r_0 (n_0 - 1)]^{-1} . \tag{7.28}$$

The combined effects of the temperature- and stress-dependent variation of the refractive index and the distortion of the end-face curvature of the rod lead to the following expression for the focal length:

$$f = \frac{KA}{P_a} \left(\frac{1}{2} \frac{dn}{dT} + \alpha C_{r,\phi} n_0^3 + \frac{\alpha r_0 (n_0 - 1)}{L} \right)^{-1} , \tag{7.29}$$

where A is the rod cross-sectional area, and P_a is the total heat dissipated in the rod. If one introduces the appropriate materials parameters for Nd:YAG into (7.29), then one finds that the temperature-dependent variation of the refractive index constitutes the major contribution of the thermal lensing. The stress-dependent variation of the refractive index modifies the focal length about 20%. The effect of end-face curvature caused by an elongation of the rod is less than 6%.

Ignoring the end effects, we notice that the focal length is proportional to a material constant and the cross section A of the rod and is inversely proportional to the power P_a dissipated as heat in the rod. At first, it may be surprising that the length of the rod does not enter the equations. However, in a longer rod, for example, the reduction in power dissipation per unit length is offset by a longer path length.

We see from (7.29) that we have little flexibility in influencing the focal length. The material constants are determined when we choose the laser material: the dissipated power P_a is determined by the application (even though we may be able to reduce the heat load by avoiding unusable pump radiation); thus the only remaining design parameter is the rod cross section. The focal length can be increased by increasing A, but this is usually not a practical way of solving the problem, since a larger crystal reduces pump power density and therefore leads to lower gain.

According to (7.29), the focal length of a cylindrical laser rod, where heat is generated uniformly within the bulk material, can be written as

$$f = MP_{in}^{-1} , \qquad (7.30)$$

where M contains all the material parameters of the laser rod and an efficiency factor η which relates the electrical input power to the power dissipated as heat in the rod ($P_a = \eta P_{in}$).

We can introduce a laser-rod sensitivity defined as

$$\frac{d(1/f)}{dP_{in}} = M^{-1} . \qquad (7.31)$$

The sensitivity factor describes how much the optical power $1/f$ of a laser rod changes with input power. To understand the importance of this factor, it is necessary to go back to the resonator theory. With the laser operating at a given power level, the designer chooses the resonator optics to provide the desired beam profile. In doing so, he takes into account the lensing of the laser rod. However, no system works at a constant power level. Power-supply fluctuations, lamp aging, general system deterioration, just to mention a few parameters, change the pump power, and thereby, the optical focusing power of the rod. The sensitivity factor tells how sensitive a laser rod is to these changes. The designer must ensure that his resonator design is capable of maintainig the output beam within specifications in spite of these fluctuations. For a Nd : YAG rod 0.63 cm in diameter and assuming that 5 % of the electrical input power to the lamp is dissipated as heat, we obtain a change of focusing power of 0.5×10^{-3} diopters per watt of lamp input variation.

Returning now to (7.29), we can see that the rod acts as a bifocal lens with different focal lengths for radiation with radial and tangential polarization. Since a linear polarized wave or a nonpolarized wave incident on the crystal will always have components in radial and tangential directions, two focal points are obtained. For Nd : YAG one finds a theoretical value of $f_\phi/f_r = 1.2$, whereas measurements show ratios varying from 1.35 to 1.5 [7.14]. A difference in focal length between different polarizations means that a resonator designed to compensate for the rod lensing for radial polarization cannot also compensate for the lensing of tangentially polarized light.

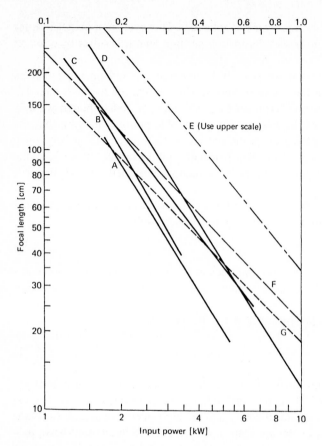

Fig. 7.5. Thermally induced back focal length as a function of lamp input power for a variety of solid-state lasers: *(A–E)* Measurements of the average focal length $(f_r + f_\phi)/2$ of *(A)* B-axis Nd : YAlO$_3$ rod, 7.5 $\times$ 0.62 cm; *(B)* Nd : YAG rod, 10 $\times$ 0.62 cm; *(C,D)* Nd : YAG rods, 7.5 $\times$ 0.62 cm, curves are for different rods and pump cavities; *(E)* Nd : LaSOAP rod, 7.5 $\times$ 0.6 cm (use upper scale for input power). *(F,G)* Theoretical back focal length of a Nd : YAG rod assuming that 5 % of the electrical input power will be dissipated as heat in the 7.5 $\times$ 0.63 cm diameter crystal. Shown is the focal length for the radially polarized *(G)* and tangentially polarized *(F)* beam components. The value for M, see (7.30), is $M_\phi = 22 \times 10^4$ W cm, and $M_r = 18 \times 10^4$ W cm.

In Fig. 7.5, theoretical and measured thermally induced back focal lengths of various laser rods are plotted as a function of lamp input. Experimentally, the focal length is usually determined by projecting a HeNe laser beam through the rod and measuring the position of the beam diameter minimum. Figure 7.5 also depicts the thermally induced focusing in a Nd : LaSOAP rod operated at 20 Hz [7.15]. At this high repetition rate it was shown that the dynamic optical distortions are purely a function of input power, as is the case in cw operation. The data illustrates the very strong thermal lensing which occurs in Nd : LaSOAP as a result of the seven-times-lower conductivity of this material as compared to Nd : YAG.

Comparing the experimental results with (7.29), we find that the focal length does not always vary exactly as the inverse of lamp input power. For example, the curve in Fig. 7.5 for the 10-cm-long rod can be approximated by $f \propto P_{in}^{-1.5}$, whereas for the curve of Nd : LaSOAP we obtain $f \propto P_{in}^{-1.2}$. Applying the sensitivity factor of (7.31) to the measured curves, we find that Nd : LaSOAP is more sensitive to pump fluctuation than Nd : YAG and Nd : YALO. Typical values for Nd : YAG and Nd : YALO are $0.5-1.0 \times 10^{-3}$ diopters/W, and $2.5-4 \times 10^{-3}$ diopters/W for Nd : LaSOAP.

Stress Birefringence

We will now investigate the influence of thermally induced birefringence on the performance of a solid-state laser. Taking Nd : YAG as an example, it was shown in (7.15, 16) that the principal axes of the induced birefringence are radially and tangentially directed at each point in the rod cross section and that the magnitude of the birefringence increases quadratically with radius r. As a consequence, a linearly polarized beam passing through the laser rod will experience a substantial depolarization. We refer to Fig. 7.4, where we have chosen a point $P(r, \phi)$ in a plane perpendicular to the rod axes. At this point we have a radial refractive index component n_r, which is inclined at an angle ϕ with respect to the y axis and a tangential component n_ϕ perpendicular to n_r. Assume that $\overline{E}$ is the polarization vector for incident radiation. Radiation incident at point P must be resolved into two components, one parallel to n_r, and the other parallel to n_ϕ. Since $\Delta n_r \neq \Delta n_\phi$, there will be a phase difference between the two components and the light will emerge elliptically polarized. This will occur for all points of the rod cross section with the exception for points located along the x and y axes in Fig. 7.4. Radiation incident along the y axis will see only one refractive index, n_ϕ, while along the x axis, n_r will be the only refractive index.

Birefringence effects in pumped laser rods can be studied in a polariscopic arrangement in which the expanded and collimated light beam from a HeNe laser serves as an illuminator for the observation of the rod between crossed polarizers [7.16]. Because of thermally induced birefringence, the probe light suffers depolarization and is partially transmitted by the analyzer. The transmitted light forms the so-called isogyres, which display the geometrical loci of constant phase difference. Photographs of conoscopic patterns for various pump powers of a Nd : LaSOAP rod between crossed polarizers are shown in Fig. 7.6.

The isogyre pattern exhibits a cross and ring structure, where the arms of the cross are parallel and orthogonal to the incident polarization. As mentioned before, the crosses correspond to those regions of the crystal where an induced (radial or tangential) axis is along a polarizer axis, so that the induced birefringence results only in a phase delay and not in a polarization rotation. The dark rings correspond to an integral number of full waves of retardation.

A number of operations of solid-state lasers, such as, for example, electro-optical Q-switching, frequency doubling, and external modulation of the beam, require a linearly polarized beam. An optically isotropic material, such as

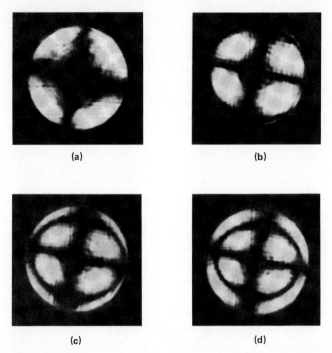

(a)　　　　　　　　(b)

(c)　　　　　　　　(d)

Fig. 7.6a–d. Thermal stresses in a 7.5-cm long and 0.63-cm-diameter Nd : LaSOAP crystal. The rod was pumped at a repetition rate of 40 pps by a single xenon flashlamp in an elliptical pump cavity. Input power (a) 115 W, (b) 450 W, (c) 590 W, (d) 880 W [7.15]

Nd : YAG, must be forced to emit a linearly polarized beam by the introduction of a polarizer in the resonant cavity. In the absence of birefringence, no loss in output power would be expected. However, the thermally induced birefringence causes a significant decrease in output power and a marked change in beam shape.

For a system with an intracavity polarizer, the effect of depolarization involves two phenomena: coupling of power into the orthogonal state of polarization followed by subsequent removal of that component by the polarizer, and modification of the main beam by the depolarization process leading to a distortion of beam shape. When a birefringent crystal is placed between a polarizer and analyzer that are parallel, the transmitted intensity is given by [7.7]

$$\frac{I_{\text{out}}}{I_{\text{in}}} = 1 - \sin^2(2\phi) \sin^2\left(\frac{\delta}{2}\right) , \tag{7.32}$$

where ϕ is the angle between the polarizer and one of the principal birefringence axes, and δ is the polarization phase shift of the light emerging from the crystal. The index difference, $\Delta n_\phi - \Delta n_r$, leads to a phase difference

$$\delta = \frac{2\pi}{\lambda} L(\Delta n_\phi - \Delta n_r) . \tag{7.33}$$

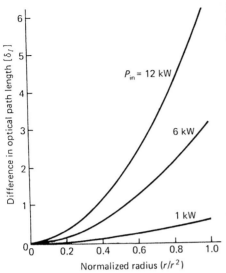

Fig. 7.7. Differences in optical path length as a function of normalized rod radius in a Nd : YAG rod. Parameter is the lamp input power

To illustrate the magnitude of stress birefringence in a Nd : YAG rod, the difference in optical path length

$$\delta_l = \frac{L(\Delta n_\phi - \Delta n_r)}{\lambda} \tag{7.34}$$

is plotted in Fig. 7.7 as a function of pump power. The following constants were used: $P_a = 0.05 P_{in}$, $L = 7.5\,\mathrm{cm}$, $r_0 = 0.31\,\mathrm{cm}$, and $\lambda = 1.06 \times 10^{-4}\,\mathrm{cm}$. As can be seen from this figure, at maximum lamp input power of $12\,\mathrm{kW}$, the path-length difference is approximately six wavelengths. The data plotted in Fig. 7.7 is in excellent agreement with the values obtained from conoscopic interference patterns [7.10, 11, 17]. Assuming a plane wave, we can calculate the total transmitted intensity from (7.32, 33) by integrating over the cross-sectional area of the rod. The following integral must be evaluated

$$\left(\frac{I_{out}}{I_{in}}\right)_T = \left(\frac{1}{\pi r_0^2}\right) \int_{\phi=0}^{2\pi} \int_{r=0}^{r_0} \left[1 - \sin^2(2\phi) \sin^2 \frac{\delta}{2}\right] r\,dr\,d\phi . \tag{7.35}$$

with $\delta = C_T P_a (r/r_0)^2$, where $C_T = 2n_0^3 \alpha C_B / \lambda K$. Integration of (7.35) yields

$$\frac{I_{out}}{I_{in}} = \frac{3}{4} + \frac{\sin(C_T P_a)}{4 C_T P_a} . \tag{7.36}$$

A laser resonator containing a polarizing element is optically equivalent to a laser rod of twice the actual length located between a polarizer and analyzer. Subtracting I_{out}/I_{in} from unity and multiplying the phase difference δ by a factor of two yields the fraction of the intracavity power which is polarized orthogonal to the polarizer. This beam, caused entirely by birefringence, will actually be ejected from the cavity and represents the depolarization loss of the resonator:

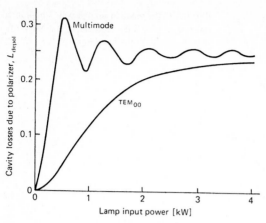

Fig. 7.8. Calculated resonator loss caused by the combination of thermally induced birefringence in a Nd:YAG rod and an intracavity polarizer

$$L_{\text{depol}} = 0.25[1 - \text{sinc}(2C_T P_a)] \; . \tag{7.37}$$

This loss is plotted in Fig. 7.8, where the same constants as for Fig. 7.7 have been used. Note that this curve represents the round-trip loss in the cavity for a plane-parallel beam which was used as an approximation for a highly multimode laser beam.

Also shown is the depolarization loss for a TEM_{00} mode for which it was assumed that the beam radius w is $w = r_0/2$. A similar calculation as the one carried out for a plane wave yields the loss factor [7.18]

$$L_{\text{depol}} = 0.25 \left(1 + \frac{16}{C_T^2 P_a^2} \right)^{-1} \; . \tag{7.38}$$

For the same lamp input power the losses for the TEM_{00} mode are less than for a highly multimode beam. This is expected since the energy in the TEM_{00} mode is concentrated nearer the center of the rod, where the induced birefringence is smaller.

The interaction of a linearly polarized beam with a birefringent laser rod and a polarizer not only leads to a substantial loss in power, but also a severe distortion of the beam shape. Figure 7.9 shows the output beam shape obtained from a cw-pumped Nd:YAG laser with and without an intracavity polarizer. The output is obtained in form of a cross with a bright center. As discussed before, the depolarization losses are smallest in the rod center and in directions parallel and orthogonal to the preferred direction of the polarizer. Areas of high depolarization losses are removed form the cavity by the polarizer and are, therefore, missing in the output beam.

The $\sin(2\phi)$ dependence of birefringence on the azimuth angle ϕ is very nicely illustrated in Fig. 7.9c [7.24]. The plane of polarization is assumed to be in the y-direction, and consistent with observation, the largest depolarization occurs at $45°$ with respect to this direction.

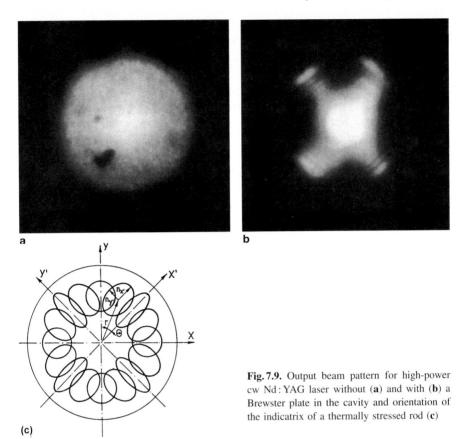

Fig. 7.9. Output beam pattern for high-power cw Nd : YAG laser without (**a**) and with (**b**) a Brewster plate in the cavity and orientation of the indicatrix of a thermally stressed rod (**c**)

Compensation of Thermally Induced Optical Distortions in Cylindrical Laser Rods

Various methods to correct for thermal lensing and birefringence produced in the laser rod have been developed. When in a repetitively pumped laser system, the time between pulses is short as compared to the thermal relaxation time of the laser material, then the thermal profile approaches cw conditions (Sect. 7.1.3). For this reason are thermal compensation techniques, for cw and high repetition rate lasers, treated together here.

Thermal Lensing. Complete compensation of the thermal aberrations produced by a laser rod is difficult because: a) The focal length depends on the operating conditions of the laser and changes with pump power and repetition rate. b) The thermal lens is bifocal due to the stresse-dependent variation of the refractive index. c) Non-uniform pumping leads to non-spherical aberrations. In many pump configurations, with flashlamps as well as diode pump sources, pump radiation is more intense at the center than at the periphery of the rod. The focal length of a given area in the rod is inversely proportional to the intensity of the absorbed pumped radiation. Therefore, in the case of on-axis focusing of pump light, the focal length at the center of the rod is shorter than at the edges. Or expressed differently, the thermally induced refractive index profile contains terms that are higher than quadratic. A negative lens will remove the quadratic term, however higher-order effects cannot be compensated.

A first-order correction of the positive thermal lens of the laser rod can be accomplished by insertion of a negative single-element lens into the resonator, or by designing a resonator configuration which treats the laser rod as a thick positive lens and takes into account its focusing properties. The latter approach has been discussed in detail in Sect. 5.1.10. A first order dynamic compensation, which accommodates variations in pump power can be achieved with a movable optical element. For example, a low-magnification Galilean telescope can be inserted into the resonator, whereby the relative position of the two lenses is adjusted with a motor drive. In another approach, dynamic compensation was achieved with a negative lens in front of the concave rear mirror of the resonator. By adjusting the spacing between these elements, the effective radius of curvature of the rear mirror is altered, thus creating a variable-radius mirror [7.19]. Movement was accomplished by placing the rear mirror on a motorized translation stage. Dynamic compensation of thermal lensing, by means of a microprocessor controlled translation of an optical element, is sometimes required in military systems in order to provide a minimum beam divergence on a first-shot basis, i.e. zero warm-up time.

As already mentioned, insertion of a fixed or movable negative lens corrects only for the quadratic term of the thermally induced refractive-index profile. From the foregoing it is clear that correction of the spherical portion of phase distortion is relatively straight forward. Correction of higher-order thermally induced aberrations are possible, but the implementation is more involved. Essentially two approaches are possible: Dynamic correction of thermally induced wavefront distortions by means of phase conjugation, or static correction with an aspheric correction plate.

In an amplifier, wavefront distortions can be eliminated with an optical phase conjugated mirror as discussed in Sect. 10.4. This approach requires an injection seeded, narrow bandwidth oscillator with a diffraction limited output, and a double pass amplifier. The oscillator typically has a low power output because only aberrations in the amplifier are corrected. Compared to spherical diverging lenses, optical phase conjugation does provide complete correction of wavefront distortions over a large dynamic range.

Thermally induced phase-front distortions can also be corrected with an aspheric optical element in the beam path. Appropriate phase profiles can be generated with currently available techniques such as diamond turning, microlithography and ion etching. Just like optical phase conjugation, such an element can correct for spherical, as well as non-spherical, aberrations. However, a correction can be achieved for only one set of operating conditions. Since static correction requires a thermal aberration profile that is stable over time, this method is particularly well suited to diode pumping. In one of our lasers, intended for a space-based application, a diamond-turned CaF_2 aspheric lens was able to perform a high degree of correction [7.21]. The very compact diode-pumped oscillator-amplifier system generated 46 W of average power in the Q-switched mode within a 1.3× diffraction limited beam. The phase corrector was a single anti-reflection coated CaF$_2$ plano concave plate which was inserted into the amplifier chain. The optical phase profile of the amplifier was measured by monitoring the interference of a probe beam that propagated through the amplifier and a reference path of a Mach–Zehnder interferometer. The phase profile was circularly symmetric, and the predominant aberration was a high-order spherical aberration, which resulted in a stronger focus in the center of the rod compared to the outer region. The interference pattern was quantified with fringe analysis software. This provided the information needed for the fabrication of the optical element. Calcium fluoride was chosen because it can easily be diamond turned and is a very robust material.

Compensation of Thermal Birefringence. The objective of birefringence compensation is to achieve equal phase retardation at each point of the rod's cross-section for radially and tangentially polarized radiation. This can be accomplished by rotating the polarizations either between two identical laser rods or in the same rod on successive passes, such that the radial and tangential components of the polarizations are exchanged.

On passing once through a rod with a radial temperature distribution, a phase difference $[\Delta P_r(r) - \Delta P_\theta(r)]$ is introduced between the radially polarized ray and the tangentially polarized ray at the same radius r. Now, if the two rays pass through a 90° polarization rotator, the radially polarized ray will be converted to tangential polarization, and vice-versa. If the two rays are passed again through the same rod, or through an identical rod, the phase difference between the two rays will be removed.

For example, birefringence compensation in an oscillator containing two identical laser heads can be achieved by inserting a 90° quartz rotator between the laser rods. This rotator can be of crystalline quartz, for example, cut perpendicular to the optic axis. The rotator produces a 90° rotation of every component of the electric field of the laser beam. The part of a mode that is radially polarized in the first rod, is tangentially polarized in the second rod. Since each part of the beam passes through nearly identical regions of the two rods, the retardation induced by one rod is reversed by the other. With this technique, linearly polarized outputs with negligible power loss and TEM$_{00}$ output of 50 to 70 % of the multimode power have been achieved [7.20].

The same approach of birefringence compensation can also be employed between two identical amplifier stages. Figure 7.10 illustrates the improvement one can achieve with this compensation technique. Shown is the measured depolarization of two flashlamp-pumped Nd : YAG amplifiers as a function of average pump power. The Q-switched average output power is also indicated. The two amplifiers are arranged in a double-pass configuration with polarization output coupling similar to the diode-pumped system illustrated in Fig. 4.16. However, instead of an ordinary mirror at the output of the second amplifier, a phase-conjugate mirror based on Stimulated Brillouin Scattering (SBS) is employed to

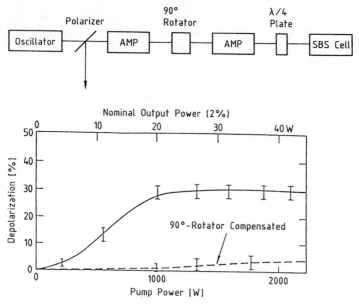

Fig. 7.10. Depolarization losses in a double-pass Nd : YAG amplifier chain with and without birefringence compensation by means of a 90° polarization rotator [7.27]

compensate for optical distortions produced in the amplifier rods. (See Sect. 10.4 for details on SBS). A 90° polarization rotator placed between the two amplifiers compensates for birefringence and greatly reduces the total depolarization.

If the laser system uses only a single amplifier, a Farraday rotator with a 45° rotation, placed between the amplifier and the back mirror or SBS cell, is equivalent to a 90° rotator located between two identical amplifier stages (Fig. 10.51).

A further refinement of the concept of rotating the radial and tangential polarization between two identical and symmetrically loaded rods has been proposed in [7.22]. A necessary condition for complete birefringence compensation, with the technique described above, is the propagation of each ray through exactly the same position in each rod. Because of thermal focusing, the rays do not propagate parallel to the optical axis, and therefore do not retrace identical sections in each rod. This is mainly an issue at high pump powers when thermal focusing is severe. According to theoretical and experimental investigations described in [7.22, 23], a much improved compensation for birefringence and bi-focusing can be achieved if an optical imaging system is added to the polarization rotator. In this approach, the beam in the first rod is imaged into the second rod. Figure 7.11 shows one implementation of the concept. A telescope with magnification one, images the facing principal planes of both rods onto each other and a 90° quartz rotator exhanges the radial and tangential polarization states between the rods.

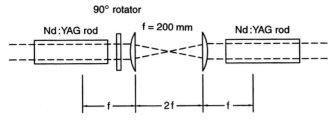

Fig. 7.11. Birefringence compensation with a polarization rotator and a telescope imaging the principal planes of both rods [7.22]

In a laser resonator containing a single rod, the phase difference between the radially and tangentially polarized rays can be compensated by a 45° Faraday rotator placed between the rod and rear mirror. On each round trip, the light passes through the 45° Faraday rotator, thereby undergoing a 90° rotation of the plane of polarization. In the resonator depicted in Fig. 7.12 birefringence compensation, employing a Faraday rotator, is combined with the added feature of four passes through the active medium before the radiation reaches the output coupler [7.19, 26]. In a low-gain laser this can be advantageous in certain situations. In the scheme illustrated in Fig. 7.12, linearly polarized radiation traveling to the right through the polarizer, laser rod and Faraday rotator will be orthogonally polarized on its return path. Therefore, the radiation will be rejected by the polarizer and directed towards the re-entrant mirror. On its second round trip through the laser rod and Faraday rotator, the radiation will pass the polarizer and reach the output coupler.

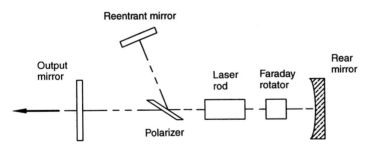

Fig. 7.12. Re-entrant resonator with birefringence compensation [7.19]

Birefringence compensation by double-pass retracing of a ray through the thermally aberrated rod can also be achieved with two optically passive elements, namely a Porro prism and a waveplate. A ray which travels on one side of the laser rod on the first pass, is reflected by the Porro prism about the plane of symmetry, so that it travels down the opposite side of the rod on the second pass at the same radial distance. The phase difference between the radial and tangential polarization state in a thermally stressed rod is only a function of radius and independent of the azimuth angle. Because of this rotational symmetry, the birefringence that a given ray experiences on the two passes through the laser rod has the same magnitude. By modifying the polarization state of the beam between the two passes it is possible to remove any thermally induced depolarizing effects that occur in the first pass with equal and opposite effects in the return pass. With a normal mirror providing reflection after one pass, only a Faraday rotator is able to achieve this. However, if a Porro prism is used, the displacement provided by the Porro places the second pass in a different portion of the laser rod to the first pass, and this allows the use of completely passive components to achieve compensation. The result is a zero-phase retardation upon emergence from the laser rod. The incorporation of a waveplate with the appropriate orientation and phase retardation in front of the Porro prism produces the desired result (see Sect. 8.3 for details).

In Sect. 8.3, oscillators are also described where birefringence is not compensated, but the two orthogonally polarized states of the radiation in the resonator are spatially separated by a polarizer and treated as independent beams.

The aforementioned techniques deal with the issue of minimizing optical losses in the presence of thermally induced stress birefringence. In two other approaches, the need for such a compensation is eliminated. In one approach, naturally birefringent crystals are used as the lasing medium, and the other concept is based on a rectangular rather than a cylindrical geometry, and in particular the zig-zag slab design, which be discussed in Sect. 7.3.

Stress birefringence is avoided in polarized laser materials since the natural birefringence dominates the stress-induced component. In optically isotropic laser crystals, the indicatrix patterns are radially oriented and all linearly polarized waves undergo depolarization. If the host material, however, is an optically anisotropic crystal, such as $YAlO_3$, thermally induced stresses cause a slight modification of the uniform anisotropy. This perturbation is nonuniform but small, since the index difference due to the crystal anisotropy is 10^2 to 10^3 times as large as the index variation due to thermal stresses. Since the indicatrix essentially retains its orientation, linearly polarized waves may propagate through the crystal without suffering depolarization. In some cases, a trade-off between Nd:YAG, Nd:YLF, and Nd:$YAlO_3$ has to be considered. As already mentioned, in the latter two materials thermally-induced birefringence is negligible. However, Nd:YLF is a softer laser material, and the thermal fracture limit is a factor 8 lower than that for Nd:YAG; and large, high-quality crystals of Nd:$YAlO_3$ are difficult to obtain. Furthermore, this material exhibits an astigmatic thermal-lensing coefficient about twice that in Nd:YAG, and therefore requires higher compensation. However, since thermal lensing compensation is much more straightforward as compared to compensation of a large birefringence, Nd:$YAlO_3$ represents a viable alternative to Nd:YAG in some applications.

7.1.2 Single-Shot Operation

If a laser rod is pumped by a single pump pulse, a transient thermal profile will be established which is the result of a fast heating process followed by a slow recovery of the rod to thermal equilibrium. Figure 7.13 sketches the transient temperature distribution in an optically pumped cylindrical laser rod. During the flashlamp pulse, energy is absorbed by the laser crystal. At the end of the pump pulse, the temperature of the rod has increased unformly by ΔT above the coolant temperature if we assume uniform absorption of pump radiation. Since only the surface of the rod is in contact with the coolant, a radial heat flow develops at the end of the pump pulse. The initially uniform temperature distribution changes to a parabolic temperature profile which decays with a certain time constant. Since lasing action occurs during or at the end of the pump cycle, the optical distortion in a single-shot laser arises from the thermal gradients produced during the pump cycle and not from heat flow as a result of cooling.

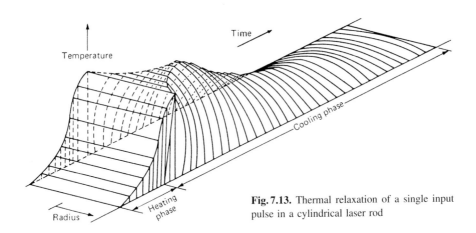

Fig. 7.13. Thermal relaxation of a single input pulse in a cylindrical laser rod

If the temperature of the laser rod increases uniformly over the rod cross section, the bulk temperature rise at the end of the pump pulse is given by

$$\Delta T = \frac{Q}{cV\varrho} , \tag{7.40}$$

where Q is the thermal energy deposited in the rod, c is the specific heat, V is the volume, and ϱ is the density of the laser material. The bulk temperature rise as a function of time depends on the shape of the flashlamp pulse. The energy deposited in the rod is

$$Q = \int_0^\infty P(t)dt , \tag{7.41}$$

where $P(t)$ is the pulse shape of the pump power. The time dependence of the temperature during the pump pulse can be obtained from

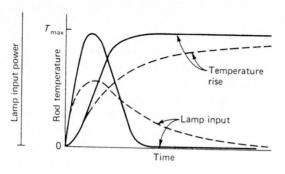

$$T(t) = \frac{1}{cV\varrho} \int_0^t P(t) \, dt \ . \tag{7.42}$$

In Fig. 7.14 the temperature rise for several common pulse shapes is illustrated. Measurements have shown that the temperature increases without any observable delay with respect to the energy deposited in the rod. In air-cooled laser heads, a considerable amount of rod heating can take place after the pump pulse because of conductive and convective heat transfer from the hot flashlamp. In liquid-cooled systems, most of this long IR radiation is absorbed and carried away by the coolant.

If the temperature rises uniformly in the active material, thermal effects in an oscillator are limited to a frequency shift of the emitted radiation and an increase in optical length of the resonator. Nonuniform pumping will also lead to thermal lensing and birefringence. If one assumes a pump-induced thermal profile $T(r)$ at a given time t_1 during the pump cycle, then thermal lensing and birefringence for this particular instant in time can be calculated, as was discussed in Sect. 7.1.1 for the cw case.

The most common technique for observing pump-induced thermal effects is to insert the laser rod in one arm of an interferometer and to observe the time-dependent fringe pattern with a high-speed framing camera [7.25]. Another interferometric technique consists of counting the transient fringes with a fast photodetector mounted behind a pinhole [7.28]. The transient response of a laser rod to the pump pulse has also been studied by photographing the induced stress birefringence by means of a plane polariscope [7.29] or by measuring thermal lensing [7.30] or beam divergence [7.31] during the pump pulse.

The temperature profiles are usually inferred from an interferometric measurement of the optical path-length changes. If we ignore changes in refractive index resulting from thermal stress, then optical path length and temperature changes are related according to

$$\Delta(n_0 l_0) = n_0 l_0 (\alpha_l + \alpha_n)\Delta T \ , \tag{7.43}$$

where $n_0 l_0$ is the undistorted optical length, $\alpha_l = (1/l_0)(dl/dT)$ is the thermal coefficient of linear expansion, and $\alpha_n = (1/n_0)(dn/dT)$ is the thermal coefficient of refractive index.

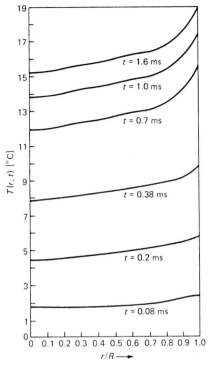

Fig. 7.15. Temperature rise in a 7.5-cm-long and 1.0-cm-diameter Nd : glass rod pumped by a 15-cm-long helical flashlamp operated at 11500 J input [7.25]

Most measurements reveal that the optical path-length change is actually concave and the rod behaves like a negative lens. The implication is that the absorption of pump energy is higher at the outer edge of the rod than in the center. Figure 7.15 shows an example of a concave temperature profile calculated from the fringe pattern of an interferometric measurement [7.25]. The total path-length change in the center of this rod was nine wavelengths.

7.1.3 Repetitively Pulsed Lasers

In the previous subsection we discussed pump-induced thermal distortions which occur for single-shot lasers operated either in normal pulse or Q-switched operation. In this regime, the pump intervals are long with respect to the thermal relaxation of the laser rod, and the active material has returned to ambient temperature before the onset of the next pump pulse. If the laser is operated at a repetition rate at which there exists a residual temperature distribution from the previous pulse, a temperature buildup inside the material will occur. The residual temperature will accumulate for subsequent pulses until a steady-state condition is reached. Figure 7.16 displays the temperature profile of a laser operated at a fairly low repetition rate. The residual temperature has not completely decayed between pulses, and the initial uniform temperature rise caused by the pump source is modified to a slightly parabolic temperature profile.

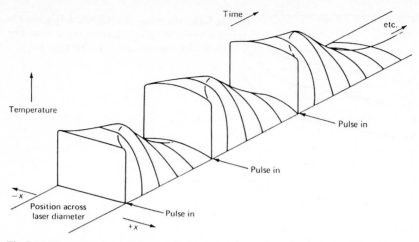

Fig. 7.16. Thermal profile in a laser rod when pump intervals are near the thermal relaxation time

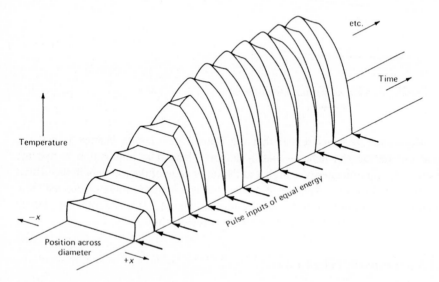

Fig. 7.17. Thermal buildup when the pulse interval time is less than the thermal relaxation time

Figure 7.17 shows a thermal buildup at a pulse interval smaller than the thermal relaxation time. Under these conditions, the laser rod will soon establish a steady-state operating condition where the radial flow of heat out of the rod will be equal to the total heat being put into the rod. During the transition period the cylindrical rod passes through several regions of radial thermal profiles as a steady-state equilibrium is approached. In the extreme case, when the pulse interval time becomes negligibly small as compared to the thermal relaxation time, the thermal profile in the laser rod approaches the cw condition. In this mode of operation the thermal effects, which have been calculated for Nd : glass [7.32–34] and Nd : YAG [7.10, 11, 13], depend only on the average input power.

The transient temperature distribution in a laser rod, as illustrated, for example, in Figs. 7.16 and 17, effects energy output and beam divergence of the laser. Calculations treating the thermal relaxation of a cylindrical laser rod which is suddenly heated by a pump pulse can be found in [7.35–38]. A general theory describing the transient thermal profile in a pulse-pumped rod for a large variety of operating parameters has been presented by *Koechner* [7.4]. The calculations were performed by solving the heat diffusion equation for a cylindrical rod for radial heat flow. The general solution of this equation is given by *Carslaw* and *Jaeger* for an infinite circular cylinder [7.1].

The general expression of the radial temperature profile $T(r, t)$ in a solid rod as a function of time can be expressed in a series of Bessel functions of the first kind and zero-, first-, and second-order (J_0, J_1, J_2)

$$\frac{T(r, t)}{\Delta T} = \sum J_{0,1,2} \left(\frac{t}{\tau}, \frac{r}{r_0}, \frac{\Delta t}{\tau}, M, g, A \right) , \qquad (7.44)$$

where ΔT is the initial temperature rise, t/τ is the normalized time, r/r_0 is the normalized rod radius, $\Delta t/\tau$ is the normalized pulse repetition rate (Δt is the pulse interval time), M is the number of consecutive pulses ($M \to \infty$ at steady state), g specifies the temperature distribution in the rod at the end of the first pump pulse, and A is a cooling parameter which specifies the cooling condition of the rod. Figures 7.18–21 illustrate some of the results of an analysis of (7.44). Shown are thermal distributions in laser rods for a variety of system parameters. Since all quantities are normalized, the results are applicable to any laser host, rod size, cooling flow, pump energy, etc. In order to be able to apply the results to practical cases, we have to discuss briefly the normalization parameters τ, ΔT, and A. The thermal time constant of the rod is

$$\tau = r_0^2/k , \qquad (7.45)$$

where r_0 is the rod radius and k is the thermal diffusivity, which is related to the materials parameters by

$$k = K/c\gamma , \qquad (7.46)$$

where K is the thermal conductivity, γ is the mass density, and c is the specific heat. The initial temperature distribution at the end of the pump cycle is approximated by a parabolic function

$$T(r, 0) = \Delta T \left[1 + g \left(\frac{r}{r_0} \right)^2 \right] , \qquad (7.47)$$

where ΔT is the temperature rise in the center of the rod and g is a measure of the pumping nonuniformity; for example, $g = 0$ indicates a uniform pump distribution across the rod, $g > 0$ represents a pumping nonuniformity leading

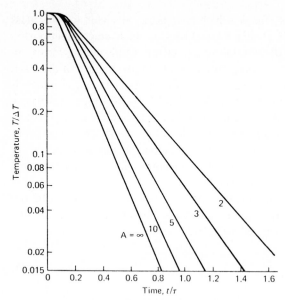

Fig. 7.18. Temperature decay in the center of an optically pumped laser rod versus time. Initial temperature rise is uniform throughout the rod ($g = 0$). Parameter is the cooling factor A [7.4]

to a negative lensing effect, and focusing of pump radiation in the rod can be expressed by $g < 0$. The amount of heat deposited in the laser rod is given by

$$Q = 2\pi l_0 c\gamma \int_0^{r_0} T(r, 0)r \, dr \ . \tag{7.48}$$

From (7.47, 48) it follows that the temperature rise ΔT in the center of the rod at the end of the first pump cycle is

$$\Delta T = \frac{Q}{V c\gamma(1 + g/2)} \ . \tag{7.49}$$

The cooling condition of the rod is specified by a dimensionless parameter

$$A = r_0 h/K \ , \tag{7.50}$$

where h is the surface heat transfer coefficient. Typical values for h were presented in Fig. 7.1.

Figure 7.18 illustrates the thermal relaxation of a cylindrical laser rod pumped by an isolated pulse. The temperature is plotted for the center of the rod as a function of normalized time. The value $A = \infty$ represents the case of infinitely good thermal contact between the laser crystal and the cooling fluid; i.e., the boundary surface of the rod is held at a constant temperature during the heat removal process. With this idealized assumption, the decay of thermal effects depends only on the thermal diffusivity and rod diameter. A thermal relaxation time

$$\tau' = r_0^2/4k \tag{7.51}$$

can be defined at which the temperature in the center of the rod has decayed to $1/e$ of the initial value ΔT.

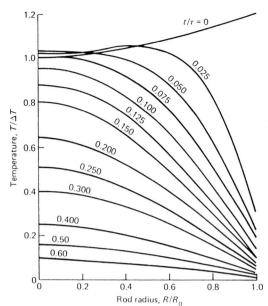

Fig. 7.19. Temperature profile of a solid laser rod for single-shot operation. Pumping coefficient $g = 0.2$, cooling factor $A = 10$ [7.4]

In actual cases, however, it is the temperature of the coolant fluid which is constant and the surface of the rod which varies in temperature. As a result of this assumption, the temperature decay in a rod depends on the material parameters and the conduction of heat across the flow boundary. This condition is expressed by a finite value of A. For most liquid cooled laser rods the value of the rod-surface cooling parameter A is between 2 and 10. Therefore, in practical situations the thermal relaxation processes are considerably slower than in the case where a zero thermal impedance between the rod surface and the cooling fluid is assumed. This explains the discrepancy which existed between the theoretical and measured decay times of thermal effects in laser rods [7.39].

Figure 7.19 shows the temperature profile in a solid rod as a function of time for a cooling factor of $A = 10$. It is assumed that the initial temperature distribution at the end of the pump pulse ($t/\tau = 0$) is parabolic, with the temperature of the cylindrical surface 20 % higher than in the center of the rod ($g = 0.2$). As one can see from this figure, at the water-cooled cylindrical surface the temperature decreases very rapidly, causing large temperature gradients. The temperature in the center changes very little for about 0.075 time constants. Actually, the temperature in the center shows a slight increase due to heat flow toward the center immediately after the flashlamp pulse.

If a laser is repetitively pumped, the thermal buildup in the rod depends on the ratio of the pulse interval time to the thermal time constant. In Fig. 7.20 the temperature buildup at the center of a cylindrical rod is plotted as a function of number of pump pulses. The parameter is the ratio $\Delta t/\tau$. The values for the temperature buildup are obtained at the end of each pump pulse. During the pulse interval time the temperature decreases, of course. The cooling factor was chosen

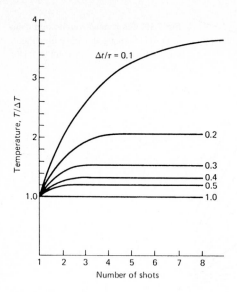

Fig. 7.20. Temperature buildup in the center of a repetitively pumped laser rod versus number of shots. Parameter is the normalized repetition rate. Cooling factor $A = 10$, pumping coefficient $g = 0$ [7.4]

$A = 10$ and the pumping coefficient $g = 0$. In the case of $\Delta t = \tau$, any distortion introduced by the pumping process has a chance to relax out before the onset of the next pump pulse. If the pulse interval time Δt is shorter than the thermal relaxation time τ, a temperature buildup occurs. For example, for $\Delta t/\tau = 0.2$, the incremental temperature rise at the center of the rod is more than twice as high as compared to the single-pulse operation.

Figure 7.21 exhibits temperature profiles at the end of each pump pulse for different repetition rates. As one can see from this figure, the initially concave temperature profile is changed to a convex profile. The absolute temperature as well as the temperature gradient at each point of the rod cross-section increase for the higher repetition rates. Again, $A = 10$ and $g = 0.2$ were chosen for the cooling factor and the pumping coefficient, respectively. After the first pulse, the outside of the rod is at a higher temperature than the center as a result of the assumed nonuniform pump light absorption. In Fig. 7.21a the negative focusing effect created by nonuniform pumping is partially compensated by the interpulse cooling process. In Figs. 7.21b and c the initial diverging condition of the rod changes to a strongly focusing condition after quasi-steady state is reached.

The thermal relaxation in a laser rod can be determined experimentally by measuring the thermally induced birefringence [7.40] or lensing [7.36, 41, 42] or by observing an interference fringe pattern [7.39, 43]. Figure 7.22 shows the temperature decay in a ruby rod, 1 cm in diameter and 12 cm long [7.39]. The temperature was calculated from the interference pattern which was recorded with a movie camera. According to (7.45), the thermal time constant of this rod is about 2 s. The temperature in the center of the rod has decayed to 0.37 of the initial value after 1.1 s. Comparing this data with Fig. 7.18 reveals a cooling parameter of this water-cooled rod of $A = 2$ or a surface heat transfer coefficient of $h = 1.68 \, \text{W/cm}^2 \, \text{K}$.

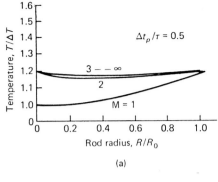

(a)

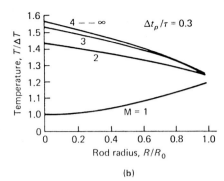

(b)

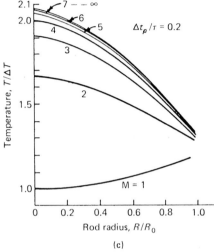

(c)

Fig. 7.21. Thermal profile in a repetitively pumped laser rod at the end of each pump cycle. Parameter is the normalized repetition rate. Δt is the pulse interval time, and τ is the thermal time constant. M is the number of consecutive pulses. Cooling factor is $A = 10$, and pumping coefficient is $g = 0.2$ [7.4]

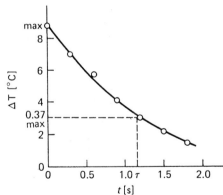

Fig. 7.22. Temperature difference ΔT between center and edge of a ruby rod after the pump pulse [7.39]

The experimental results obtained from birefringence measurements can be divided into three regimes: inherent birefringence present in the laser rod; induced birefringence during the pump cycle; and variations in birefringence after the pump pulse due to thermal conduction within the laser rod.

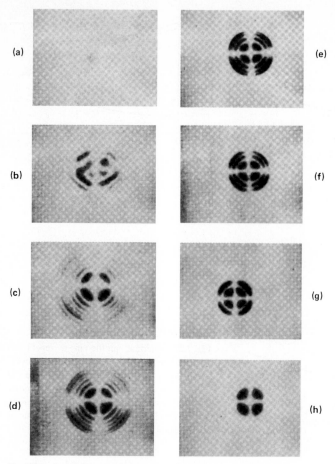

Fig. 7.23a–h. Thermal birefringence in a 15.2–cm-long by 1.5-cm-diameter Nd : glass rod pumped by two flashlamps in a double-elliptical pump cavity. Total pump energy was 7000 J. (**a**) $t < 0$; (**b**) $t = 0.1$ s: (**c**) $t = 2.4$ s; (**d**) $t = 3.3$ s; (**e**) $t = 8.7$ s; (**f**) $t = 9.5$ s; (**g**) $t = 13.1$ s; (**h**) $t = 18.6$ s [7.40]

Figure 7.23 shows the light pattern transmitted through the crossed polarizers for a Nd : glass rod pumped by a single flashlamp pulse [7.40]. A 16-mm movie camera was used to record the pattern on the screen as a function of time. The first photograph is a measure of the static birefringence in the laser rod. The distorted nature of the second pattern indicates that the heat deposition did not possess perfect cylindrical symmetry. However, as the rod thermalizes during cooling, the symmetry does clearly develop. It is of interest to note that even after 18 s, considerable birefringence still remains in the rod. This is not surprising since the thermal time constant for a glass rod of this size is $\tau = 140$ s. From Fig. 7.18 it follows, that for a cooling factor of $A = 10$, it takes at least 80 s before the temperature at the center has decayed to less than 10 % of its initial value.

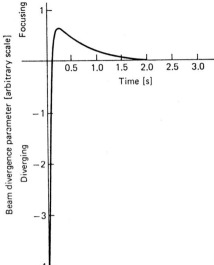

Fig. 7.24. Calculated time-dependent thermal lens effect of a 0.48-cm-diameter ruby rod [7.36]

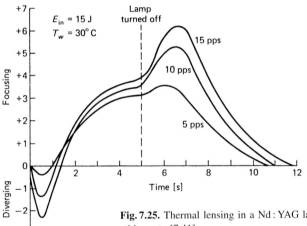

Fig. 7.25. Thermal lensing in a Nd:YAG laser rod as a function of repetition rate [7.41]

The thermally induced lens-like behavior of a water-cooled ruby rod (0.48 cm in diameter) was experimentally measured by passing a collimated HeNe laser probe beam through the rod and observing the behavior of the probe beam during and after the lamp pulse [7.36]. It was determined that during the 400-μs lamp pulse, the ruby rod behaves as a negative lens and about 120 ms later, it behaves as a positive lens that increases in strength and reaches a maximum 250 ms after the lamp pulse, and then completely decays approximately 1.8 s after the lamp pulse. Figure 7.24 represents a calculated curve which best describes the experimentally observed focusing action of the rod. The coefficient of surface heat transfer was determined to be $h = 0.27$ W/cm^2 K. From (7.45, 50) it follows that $\tau = 0.45$ s and $A = 0.15$. Comparing the values with the measured data, one finds that because of the small rod-surface cooling parameter it takes four thermal time constants before the rod has returned to thermal equilibrium.

Figure 7.25 shows thermal lensing in a repetitively pumped Nd : YAG rod. A transient defocusing and a quasi-steady-state equilibrium with positive focusing are identified. At a repetition rate of 5 pps the rod reaches steady state after about 5 s or 25 pulses. For the higher repetition rates steady state had not been reached after 5 s, at which time the system was turned off. From Fig. 7.25 it follows that it takes about 5 to 7 s for the rod to return to equilibrium after the flashlamp has been turned off. The laser rod used in these experiments measured 7.5 cm by 0.63 cm and was pumped by a single linear flashlamp. The surface heat transfer coefficient was caculated to be 0.79 W/cm^2 K. From these parameters we calculate $A = 1.9$, $\tau = 2.1$ s, and $\Delta t_p/\tau = 0.1$. With these values and the graphs presented in [7.4], good agreement between the calculated and measured transient periods is obtained. If this laser rod is operated above 5 pps, the time between pulses is so short compared to the thermal time constant that essentially no temperature decay takes place between pulses. The temperature profile in the rod is then identical to the one obtained at cw operation for the same input power. This is the regime where, for example, thermal lensing is found to be insensitive to pulse repetition rate as long as the average power is constant [7.42].

The transient optical distortions in a large Nd : glass rod for single-shot and repetitive pumping were reported in [7.43]. A 2.5-cm diameter by 30-cm-long rod was pumped by a helical flashlamp in a 0.8-ms-long pulse. The time necessary for the water-cooled Nd : glass rod to return after firing to an initial no-strain condition was on the order of 10 min. Introducing the appropriate material parameters into (7.45), we obtain for the thermal time constant $\tau = 6$ min. Since about 1.7 times this constant is required for the thermal distortions to completely decay, we can estimate a surface cooling parameter of $A = 2$ to 3 from Fig. 7.18. If this rod is pulsed every 25 s (corresponding to $\Delta t/\tau = 0.07$), it takes about 25 shots before the rod has reached steady-state condition.

In a laser-diode-pumped Nd : YAG system, the heat deposited in the crystal per unit of stored energy is 0.7, which is about one-third of the value in flashlamp pumped lasers [7.44a]. If one includes the fluorescence losses during a 200 μs long pump pulse, the ratio of heat deposited per unit of upper state stored energy is 1.1. For flashlamp-pumped Nd : YAG, a value around 3.3 has been reported [7.44b]. The smaller amount of heat deposited by a diode pump manifests itself in a much reduced thermal lensing and birefringence compared to a flashlamp-pumped laser for a comparable output power.

For example, Table 7.2 illustrates the small amount of thermal lensing measured in 8 mm diameter Nd : YAG amplifier rods. The amplifiers were pulsed up to 30 Hz with about 1 J optical input at 808 nm. The thermal focal length scales inversely with the input power to the 1.36 power. This dependence is in agreement with measurements taken with flashlamp pumping where the thermal focal length scales inversely with the pump power to the 1 to 1.5 power. However, the focal length of the diode-pumped laser rods is at least a factor 3 longer compared to flashlamp-pumped systems.

Table 7.2. Thermal lensing in diode-pumped Nd:YAG amplifiers

Repetition rate [Hz]	Total optical diode output [W]	Heat dissipated per rod [W]	Effective focal length [m]	
			One head	Two heads, double pass
10	9.5	1.8	55 ± 10	14 ± 3
30	28.5	5.8	12 ± 2	3 ± 1

7.2 Cooling Techniques

7.2.1 Liquid Cooling

The primary purpose of the liquid is to remove the heat generated in the laser material, pump source, and laser cavity. Sometimes the coolant serves additional functions, such as index matching, thereby reducing internal reflections which could lead to depumping modes, or as a filter to remove undesirable pump radiation. The coolant is forced under pressure to flow over the rod and lamp surfaces. These elements are located either inside flowtubes or in cooling chambers machined out of the main body of the laser head. The temperature difference between the part to be cooled and the liquid is a function of the velocity and the cooling properties of the flowing liquid. At low velocities, the flow is laminar and most of the temperature drop is due to pure conduction across a stationary boundary layer at the liquid interface. For higher velocities, the flow becomes turbulent, leading to a more efficient heat transfer process with a subsequent lower temperature drop. Turbulent flow requires a greater pressure differential for the same volume flow, but the necessary differential usually is still small compared to the total pressure difference associated with the complete cooling system.

The temperature increase of the coolant, as it passes through the laser cavity, is given by

$$\Delta T = \frac{Q}{C_p m} , \tag{7.52}$$

where Q is the extracted heat, C_p is the specific heat of the coolant, and m is the mass flow rate. Water is preferably used as a coolant for solid-state lasers, with the exception of military lasers which have to operate over an extremely wide temperature range. A standard specification for airborne military systems is MIL-E-5400K, class 1, which requires operation of electronic equipment over the range of $-54°$ to $+55°$ C.

The major difficulty in finding coolants for use in systems which have to be stored and operated at freezing temperatures is the strong ultraviolet radiation within the laser cavity. This causes most of the fluids, which would otherwise have suitable optical and physical parameters, to decompose. Extensive tests have indicated that under these conditions, ethylene glycol and water mixtures and fluorinated hydrocarbons are the best choices for laser coolants. Properties of these and other commonly used coolants of solid-state lasers are given in Table 7.3.

From purely heat transfer considerations, water is by far the best fluid. As compared to the other coolants, it has the highest specific heat and thermal conductivity and the lowest viscosity. If we express the mass flow rate by the volumetric flow $f_v = m/\varrho$, where ϱ is the density of the fluid, then after introducing appropriate parameters from Table 7.3 into (7.52), we obtain the temperature rise in water:

$$\Delta T[\mathrm{C}] = \frac{0.24 P[\mathrm{kW}]}{f_v[\mathrm{ltr/s}]} \, , \tag{7.53}$$

where P is the heat carried away by the water. As a comparison, if FC-104 is employed as a coolant, we obtain a temperature rise

$$\Delta T[\mathrm{C}] = \frac{0.56 P[\mathrm{kW}]}{f_v[\mathrm{ltr/s}]} \, . \tag{7.54}$$

For the same heat extraction P and flow rate f_v, the temperature rise in FC-104 is over twice the temperature increase of water. The lower viscosity of water compared to the other coolants results in a smaller pressure drop in the cooling lines.

Water has the additional advantage over all other coolants that it is chemically stable under intensive ultraviolet radiation. Water-alcohol mixtures or fluorinated hydrocarbons have C–H or F–H bonds which break if the energy from the UV radiation exceeds the bond energy. The coolants become acidic, precipitates develop and the optical transmission decreases. The degradation of these coolants can be related to their UV cutoff wavelength, which shifts toward longer wavelengths as the coolant degrades [7.45]. Dissociation followed by complete carbonization of the coolants listed in Table 7.3 have led to the development of UV-free flashlamps. The envelope of these lamps is usually made from ultraviolet-absorbing titanium-doped quartz. Even though most military systems employ flashlamps with UV-absorbing quartz envelopes, the amount of UV remaining may still cause coolant degradation. Once an ethylene glycol and water mixture has been exposed to UV radiation, it continues to build up acidity even when not in use. Stainless steel plumbing has been used with ethylene glycol and water systems to preclude corrosive buildups. When the fluorinated hydrocarbons are exposed to UV radiation, hydrofluoric acid is formed. This acid reacts with silicon. Hence 356 aluminum and other similar alloys cannot be used in fluorinated hydrocarbon coolant systems. Cloudiness that formerly developed in fluorinated hydrocarbon coolants after continued exposure to UV radiation has been eliminated by proper processing and control of the fluid to ensure that all impurities have been removed.

Table 7.3. Room-temperature properties of various coolants

Parameter		Water	Water 60%, methyl alcohol 40%	FC-104[a]	E–4[b]	Ethylene glycol 50%, water 50%
Specific heat	[cal/g °C]	1.0	0.84	0.24	0.24	0.79
Viscosity (poise)	[g/cm s]	1×10^{-2}	0.8×10^{-2}	1.4×10^{-2}	1.4×10^{-2}	3.0×10^{-2}
Thermal conductivity	[cal/cm s K]	1.36×10^{-3}	0.91×10^{-3}	0.33×10^{-3}	0.16×10^{-3}	1.01×10^{-3}
Density	[g/cm^3]	1.0	0.905	1.79	1.76	1.06
Prandtl number N_{Pr}	–	7.4	7.4	10.2	61.5	23.5
Volumetric thermal expansion coefficient	[°C^{-1}]	0.643×10^{-4}	4.14×10^{-4}	9.0×10^{-4}	6.4×10^{-4}	5.7×10^{-4}
Boiling point	[°C]	100	65	104	194	110
Freezing point	[°C]	0	−29	−62	−94	−36

[a] Manufacturer 3 M Company, fluorocarbon "FC" series.
[b] Manufacturer Du Pont, Freon "E" series.

Cooling Equipment

With the exception of lasers cooled directly by tap water, a closed-loop cooling system is usually employed which consists in its most basic form of at least a liquid pump, a heat exchanger, and a reservoir. Commercially available coolers contain, in addition to these components, a particle filter, a demineralizer, gauges, and sensors for monitoring flow, temperature and pressure. If common tap water is used, periodic cleaning is necessary to remove deposition of organic and mineral deposits. In closed-loop systems, if demineralization and filtering are employed, the need to clean surfaces exposed to the cooling fluid is essentially eliminated.

The heat exchanger removes heat from the closed-loop system by thermal coupling to an outside heat sink. This can be accomplished in several ways: In a liquid-to-air heat exchanger the coolant is passed through an array of fins through which air is blown by a fan similar to a car radiator. In a liquid-to-liquid heat exchanger the heat generated within the closed loop is exhausted to external water. Figure 7.26 shows the plumbing diagram of a typical cooler with a liquid-to-liquid heat exchanger. The water flows from the reservoir to a centrifugal pump, through a heat exchanger, into the laser head, and back again into the reservoir. This sequence of components minimizes the static pressure in the laser head. The temperature of the closed-loop water is regulated by a control valve in the external supply line. The valve probe is located in the reservoir. As the temperature at the reservoir increases, the valve is opened, thereby allowing more external cooling water to flow through the heat exchanger. The system contains an in-line honeycomb filter to remove particulate matter and a bypass demineralizer which will maintain low electrical conductivity in the water and minimize corrosion. The return line is monitored by a low-flow interlock and

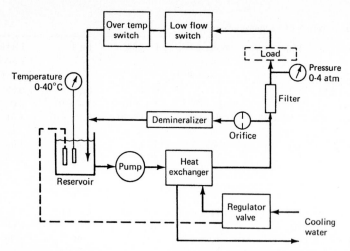

Fig. 7.26. Schematic of a water cooler containing a water-to-water heat exchanger

an over-temperature switch. If flow falls below a preset value or the discharge temperature exceeds a certain limit, the interlock turns off the power supply. The system also contains gauges to display temperature in the return line and the discharge pressure.

In both types of coolers we have discussed so far, the temperature of the closed loop can be regulated only over a relatively narrow temperature range. Furthermore, the temperature of the cooling loop is always above ambient air temperature or the temperature of the external cooling water. Figure 7.27 shows a diagram of a cooler which maintains the cooling water at a precise, reproducible temperature which is independent of the ambient air or water temperature. This cooler contains a thermostatically controlled refrigeration stage between the heat exchanger and the external water supply. Freon is cycled in a close loop between the heat exchanger and a water-cooled condensor by means of a compressor. Any changes in the heat dissipation of the load, or temperature variations in the external lines, are compensated by a hot gas bypass valve which regulates the amount of refrigeration. Smaller units contain aircooled condensors, thus eliminating the need for an external water source.

7.2.2 Air or Gas Cooling

In low average power lasers, especially portable systems, forced air is sometimes used to cool the laser rod and flashlamp. Air flow is generated by employing miniature axial or centrifugal blowers or fans which have been designed for air cooling of electronic equipment. The air flow required for cooling the laser head is calculated from the dissipated heat and the maximum temperature difference along the air stream. For standard air (20° C, 1 atm) we obtain

$$f_v[\text{ltr/min}] = \frac{49P[\text{W}]}{\Delta T[\text{C}]} , \tag{7.55}$$

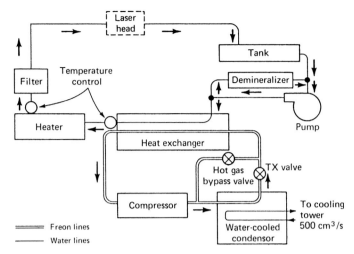

Fig. 7.27. Schematic of a water cooler employing a refrigeration unit

where f_v is the air flow and ΔT is the temperature difference between the inlet air and the exhausted air. The numerical factor contains the thermal properties of air. The air flow which can be obtained from a given fan depends on the static pressure the fan must work against. Small fans which are used for cooling lasers usually provide an air flow of 300 to 600 ltr/min at a static pressure of 30 torr. Allowing a temperature rise of 10° C of the exhausted air, heat removal capacity will be in the 100- to 200-W range. Figure 6.97 shows an example of an air-cooled Nd : YAG laser. A vane axial fan located upstream generates an air flow which passes through the pump cavity and over the rod and flashlamp. The air is exhausted through a nozzle in front of the unit. Besides convection cooling, the laser rod is also cooled by conduction into a copper heat sink to which it is mounted.

In lasers with high power dissipation, difficulties arise if one attempts to force low-pressure air into a small cavity while maintaining a high heat transfer coefficient. As the density of air is increased, for instance by a factor of 20, the pressure drop required to move the same mass flow of air through the cavity is reduced to 1/20th and the pump power required is reduced to 1/400th.

A cooling system which has been employed very successfully in small military laser systems is based on the use of compressed dry nitrogen as the coolant medium to transfer the heat generated in the laser cavity to the ambient air. The compressed nitrogen is circulated through the laser pumping cavity by means of an axial flow blower. Nitrogen exhausted from the cavity is then ducted through fins of a heat exchanger, where it gives up the energy picked up in the laser cavity. The cooled nitrogen is then ducted back through the cavity again to complete the nitrogen cooling loop. A fan provides the required air flow through the heat exchanger. Figure 7.28 shows a typical laser-cooler module which contains a laser crystal, flashtube, trigger transformer, and cooling system. The unit is

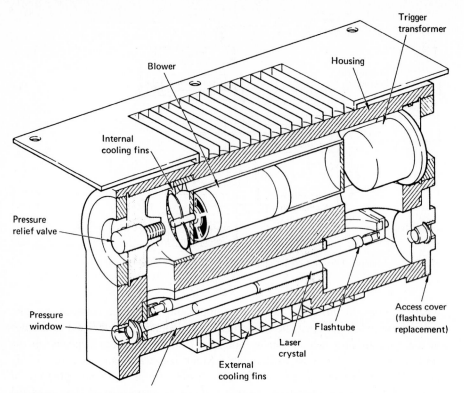

Fig. 7.28. Laser head cooled by pressurized nitrogen. (Courtesy Hughes Aircraft Company)

charged to a pressure of 20 atm; with dry nitrogen at this high pressure the heat transfer capability of compressed nitrogen and FC-104 for example, shows that nitrogen at 20 atm has not only the same specific heat value but also has twice the value of heat transfer coefficient as FC-104 for equal mass flow rates.

7.2.3 Conductive Cooling

In a variety of commercial and military systems the laser rod is mounted directly to a heat sink, as shown schematically in Fig. 7.29. Good conduction cooling of the laser element requires intimate thermal contact between the laser rod and the heat sink. The laser rod can be mechanically clamped, soldered, or bonded to the heat sink. If the laser rod is mechanically clamped to a heat sink, a temperature gradient across the rod-clamp interface will develop. This is

$$\Delta T_1 = Q/hA \,. \tag{7.56}$$

Experimental results have shown that between a ruby rod and an aluminum heat sink the heat transfer coefficient can be as low as $h = 1 \,\mathrm{W/cm^2\,^\circ C}$. Usually, the rod is mounted in a carefully machined saddle of the heat sink with a wetting agent such as indium or gallium at the interface.

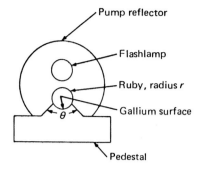

Pump reflector

Flashlamp

Ruby, radius r

Gallium surface

θ

Pedestal

Fig. 7.29. Typical geometry of a conductively cooled laser rod

In a space-born Nd:YAG laser system the rod was soldered to a mounting structure of pure niobium, having a thermal coefficient of expansion which closely matches that of Nd:YAG. The mounting surface for the rod is a groove to provide intimate contact with the rod over 90° of its periphery. The rod was soldered to the heat sink with indium solder after the contact surfaces were gold-plated. The heat was removed from the rod mounting structure by a heat pipe. A fin-type radiator at the condensor end of the heat pipe radiated the heat into space [7.46].

In another technique, a small Nd:YAG rod was bonded to a copper heat sink by means of a silver-filled epoxy adhesive. In this case, ultraviolet-free flash-lamps must be employed to avoid decomposition of the epoxy. The temperature distribution in a conductively cooled rod has been calculated in [7.47] with the assumption that heat is uniformly produced by absorption of radiation and no losses occur except by conduction through the contact surface of the heat sink.

The flat surface of a rectangular slab provides a much better interface for conduction cooling compared to a cylindrical surface. Conduction cooling of slab lasers is a very viable technique and will be discussed in the next section.

7.3 Slab and Disc Geometries

Practical limitations arise in the operation of any solid-rod laser due to the thermal gradients required to dissipate heat from the rod. As was discussed in Sect. 7.1, the thermally loaded cylindrical laser medium exhibits optical distortions which include thermal focusing, stress induced biaxial focusing, and stress induced bire-fringence. These thermally induced effects severely degrade the optical quality of the laser beam and eventually limit the laser output power, either because of an unacceptably poor beam pattern or because of thermal stress induced breakage of the rod. An incremental improvement can be achieved by employing a hollow cylindrical rod with gas or liquid cooling [7.48], or by dividing a large-diameter rod into a bundle of small-diameter parallel rods.

The limitations imposed by the rod geometry have long been recognized and a number of designs have emerged which became known as axial-gradient lasers, slab-geometry lasers, disk and active mirror lasers.

Axial Gradient Laser

In this design, a solid rod is segmented into disks which are perpendicular or at the Brewster angle to the optics axis (similar to stacked coins). The individual disks are face-cooled by forcing a suitable cooling fluid through the spaces between the disks. With this configuration the heat flow paths are essentially parallel to the optical axis, therefore radial distortions should be minimized. If the disks are arranged perpendicular to the optic axis, either an index-matching fluid or antireflective coatings have to be employed. Depending on the geometry and holding structure used, the disks can be circular, elliptical, square, or rectangular. Disk lasers have been built with Nd:glass [7.49], Nd:YAG [7.50], and ruby [7.51,52] as active material. Designs featuring up to 50 disks have been constructed. The disks are typically 10 to 20 mm in diameter, 5 to 10 mm thick, and separated by 0.5 to 1 mm. Although the axial-gradient laser appeared to offer a great potential improvement over solid rods in average power output and beam divergence, because of its better cooling capabilities, the experimental results were very disappointing. Problem areas which are inherent to this type of device are stresses and optical distortions in the disks due to edge cooling effects, optical losses due to surface scatter and attenuation in the coolant, and mechanical problems associated with the holding structure and cooling manifold. Since these engineering problems could not be resolved work stopped on this type of device in the early 1970s.

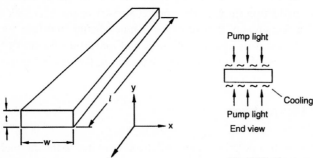

Fig. 7.30. Geometry of a rectangular slab laser

Rectangular-Slab Laser

The rectangular-slab laser provides a larger cooling surface and essentially a one-dimensional temperature gradient across the thickness of the slab. Figure 7.30 shows the geometry of a rectangular slab laser. The z axis coincides with the optical axis of the slab.

   The slab has a thickness t and a width w. The upper and lower surfaces are maintained at a constant temperature by water-cooling, and the sides are uncooled. Provided the slab is uniformally pumped through the top and bottom surfaces the thermal gradients are negligible in the x and z directions and the thermal analysis is reduced to a one-dimensional case, i.e. temperature and stress are a function of y only. This, of course, is only true for an infinitely large plate in x and y, and uniform pumping and cooling. Under these conditions we find that the temperature assumes a parabolic profile (Fig. 7.31a).

The maximum temperature which occurs between the surface and the center of the slab ($y = t/2$) is given by

$$\Delta T = \frac{t^2}{8K} Q \qquad (7.57)$$

where Q is the heat deposition, t is the thickness and K the thermal conductivity of the slab. For example, a heat deposition of $2\,\mathrm{W/cm^3}$, in a $1\,\mathrm{cm}$ thick glass slab ($K = 0.01\,\mathrm{W/cm^3}$) will create a temperature difference between the cooled surfaces and the center of $25°\,\mathrm{C}$.

The temperature rise causes stress in the slab according to

$$\sigma_s = \frac{2\alpha E}{3(1 - \nu)} \Delta T = \frac{\alpha E t^2}{12(1 - \nu)K} Q \qquad (7.58)$$

where σ_s is the surface stress for the slab. The surfaces are in tension and the center is under compression as shown in Fig. 7.31b.

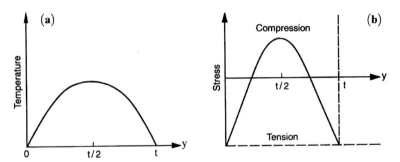

Fig. 7.31a,b. Temperature profile (**a**) and stress (**b**) in a rectangular slab

If we introduce the "thermal-shock parameter" from Sect. 7.1 we can calculate the maximum temperature difference allowed between the surface and the center before thermal fracture occurs; it is

$$\Delta T_{\max} = \frac{3R}{2K} \cdot \qquad (7.59)$$

With $R = 1\,\mathrm{W/cm}$ for Nd:glass, one obtains

$$\Delta T_{\max} = 150°\,\mathrm{C} \ .$$

Stress fracture at the surface limits the total thermal power absorbed by the slab per unit of face area. For slabs of finite width W, the power per unit length at the stress fracture limit is given by

$$\frac{P_a}{L} = 12\sigma_{\max}\frac{(1-\nu)K}{\alpha E}\left(\frac{W}{t}\right) \tag{7.60}$$

where W/t is the aspect ratio of a finite slab. It is interesting to compare the surface stress of a rod and slab for the same thermal power absorbed per unit length. From (7.11, 60) it follows

$$\frac{(P_a/L)_{\text{rod}}}{(P_a/L)_{\text{slab}}} = \frac{2\pi}{3}\left(\frac{t}{W}\right) . \tag{7.61}$$

Thus, for a superior power handling capability relative to a rod, the aspect ratio of the slab must be greater than 2.

The temperature and stress profile leads to a birefringent cylindrical lens. The focal lengths of the birefringent lens are [7.53]

$$f_x = \frac{N-S_1}{2L} ; \quad f_y = \frac{N-S_2}{2L} \tag{7.62}$$

for x and y polarized light, respectively. The parameter N is the contribution from thermal focusing, i.e.,

$$N = \frac{dn}{dT}\left(\frac{Q}{2K}\right) , \tag{7.63}$$

and the parameters S_1 and S_2 are related to stress induced focusing

$$S_1 = \frac{Q\alpha E}{2(1-\nu)K}(B_\perp + B_\parallel) \tag{7.64}$$

$$S_2 = \frac{Q\alpha E}{(1-\nu)K}B_\perp . \tag{7.65}$$

A comparison of the focal length of a rod with that of a rectangular slab for the same heat deposition Q shows, that the slab has approximately twice the optical power of the rod. In a slab, however, for incident radiation polarized along either the x or y directions the stress induced depolarization is zero. In a rectangular rod, the main axes of the resulting index ellipse are oriented parallel to the x and y axes of the rod, and a beam polarized along either one of these axes can propagate in a direction parallel to the z axis without being depolarized. This is, as mentioned earlier, in contrast to the situation in a cylindrical rod, where a plane-polarized beam does suffer depolarization because the direction of the main axes of the index ellipsoid vary from point to point within the cross section of the rod. The advantage of the rectangular geometry is obtained at some cost: The focal lengths are shorter than for the cylindrical rod, and cylindrical rather than spherical compensating optics are needed. The idea of the zig-zag path in a slab which will be discussed next is to eliminate to a first-order focusing in a slab.

Slab Laser with Zig-Zag Optical Path

In the previous section we considered propagation straight through an infinite slab. Due to the rectilinear pumping and cooling geometry of the lasing medium, thermal gradients and thermally induced stresses are present only in the y direction. Therefore, for light polarized in either the x or y directions, stress-induced biaxial focusing and depolarization losses are eliminated. However, the slab still behaves as a thin cylindrical lens with a focal length shorter by a factor of two relative to that of a rod.

The cylindrical focusing in the slab can be eliminated by choosing propagation along a zig-zag optical path. In the zig-zag geometry, the optical beam does not travel parallel to the z axis, as was the case for the straight-through optical path. Instead the beam traverses the slab at an angle with respect to the x-z plane using total internal reflection from the slab y faces. This geometry is depicted in Fig. 7.32.

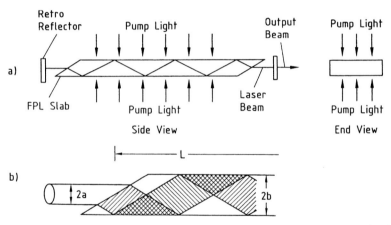

Fig. 7.32a,b. Schematic of a zig-zag laser with Brewster angle faces. **(a)** Schematic, **(b)** Detail of Brewster-angle entrance. The coordinate system is the same, as chosen in Fig. 7.30

The laser beam is introduced into the slab with suitable entrance/exit optics, usually through surfaces at Brewster's angle, as shown in Fig. 7.32b. In order to maintain total internal reflection for the desired path length the two opposing faces have to be highly polished and have to be fabricated with a high degree of parallelism. The same two optical faces are also employed for pumping and cooling of the slab.

In the ideal case, this geometry results in a one-dimensional temperature gradient perpendicular to the faces and a thermal stress parallel to the faces. Since the thermal profile is symmetrical relative to the center plane of the slab the thermal stress averaged from one slab surface to the other is zero. Thus, for a beam traversing from one slab surface to the other, the stress-optic distortion is compensated to a high degree. Also, since all parts of a beam wavefront pass through the same temperature gradients in a surface-to-surface transit, no distortion results from the variation of refractive index with temperature. Thus, in this geometry, thermal distortion effects are fully compensated within the host material in the ideal case.

The advantage of this configuration is the combination of two ideas: the elimination of stress induced birefringence by virtue of the rectangular geometry, and the elimination of thermal and stress induced focusing by optical propagation along a zig-zag path.

The concept of a slab-laser geometry with a zig-zag optical path confined in the slab by total internal reflection was first proposed by *Martin* and *Chernoch* [7.54, 55]. Figure 7.33 shows two different designs of a zig-zag slab laser [7.56]. In Fig. 7.33a a multitude of flashlamps are oriented transverse to the optical axis of the slab, and in Fig. 7.33b a single lamp is placed along the axis of the slab on each side. Appropriate reflectors are employed to distribute the pump light evenly over the slab surface area. Uniform pumping and cooling of the optical surfaces is absolutely crucial for good performance of a zig-zag slab laser, as will be discussed below.

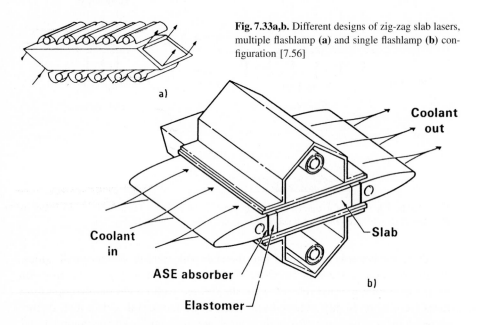

Fig. 7.33a,b. Different designs of zig-zag slab lasers, multiple flashlamp (**a**) and single flashlamp (**b**) configuration [7.56]

Despite the elegance with which the zig-zag slab design addresses the problem of thermal distortions inherent in all solid-state lasers, the performance from these systems has been noticeably poorer than predicted.

In practice, the slab laser approximates only an ideal infinite slab thus far assumed. Pump or cooling induced gradients across the slab width (normal to the plane of reflection), as well as end effects, are often the reason why the slab configuration shows a disappointing optical performance. Distortions can only be eliminated in a slab of infinite extent, which is uniformly pumped and cooled. For a slab of finite width and length, edge and end effects give rise to distortions, and practical solutions of flashlamp pumping and cooling of the slab lead to unavoidable nonuniformities.

For example, gradients in the z-direction are encountered in the transition region between the pumped area and the unpumped ends of the slab. Additional gradients along the z and x axes are caused by nonuniform pumping and cooling, as well as by thermal boundary conditions at the top and bottom of the slab. The major source of gradients in the x direction is a nonuniform illumination of the slab by the flashlamps in conjunction with the reflectors, and cooling of the slab at the edges outside the pumping boundary.

Since there is no compensation in the y and z direction, thermal gradients in these directions must be minimized.

In order to reduce end effects, a smooth temperature profile in the transition region is usually attempted. Figure 7.34 illustrates a particular design whereby a plenum provides for the mixing of flowing with stagnant water to produce a gradually decreasing surface heat transfer coefficient.

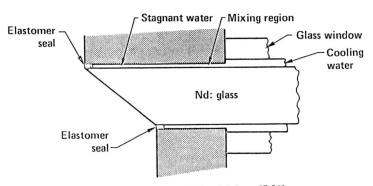

Fig. 7.34. Cooling and sealing of the ends of a slab laser [7.56]

The severeness of surface deformation, and bulging of the entrance and exit faces of a slab under thermal loading is illustrated in Fig. 7.35. These deformations in conjunction with index gradients lead to focusing effects of the slab laser. A detailed model for calculating temperature and stress distributions in an infinitely long slab (plain strain approximation) has been described in [7.53].

Earlier work on slab lasers concentrated on Nd:glass slabs pumped with flashlamps. Besides the designs depicted in Fig. 7.33, systems were developed using large slabs with oscillator and several-pass amplifiers contained in one unit, as shown in Fig. 7.36 [7.57]. Other designs contained two slabs with a bank of flashlamps in between for efficient utilization of the pump radiation [7.58]. In order to increase the average power from a zig-zag Nd:glass slab laser, it was suggested to move the slab between the pumping lamps to distribute the thermal

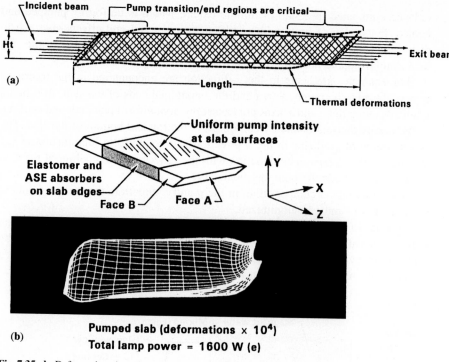

Fig. 7.35a,b. Deformations in a thermally loaded slab laser **(a)** Beam Trajectories, **(b)** 3-D deformations computed in pumped glass slabs [7.56]

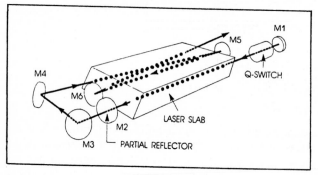

Fig. 7.36. Example of a multipass zig-zag slab laser. Shown is an oscillator and a 3 pass amplifier

loading over the area of the glass while maintaining high gain in the pumped volume [7.59].

Despite the inherent advantages, at least on a conceptual level, of a zig-zag slab to overcome the thermal limitations of rod lasers, the acceptance of these systems has been slow. The reason for this is cost considerations in combination with a number of practical engineering issues which have to be solved before the potential advantages of the zig-zag slab laser over the rod geometry can be realized. a) *Low Efficiency.* The awkward rectangular beam geometry which requires conversion to a circular beam in most applications, combined with the fact that the full width of the slab can usually not be used because of distortions at the edges, results in a poor utilization of the pumped volume. Therefore all slab lasers built to date exhibit rather modest efficiencies. b) *Residual Distortions.* Deformation of the ends and pump faces of the slab due to thermal strain have resulted in beam qualities considerably below expectations. c) *High Fabrication Cost.* The high optical fabrication costs of the slab, combined with the very demanding and therefore expensive laser head design caused by the complicated mechanical mounting and sealing geometry, and stringent pumping and cooling requirements, have also prevented wider application of this technology.

For the last 20 years, continued engineering improvements have been made to overcome some of the problems encountered in the earlier systems. Efforts have been directed to improve sealing and stress free mounting of slabs, and to reduce edge effects. In order to avoid perturbation of the total internal reflections by "O" ring seals due to coupling of evanescent waves into the seals, side faces of slabs are coated over the whole length with a thin layer of SiO_2 [7.60, 61] or Teflon [7.62]. A uniform thermal loading profile is achieved by thermally insulating the top and bottom faces by hollow chambers with thin insulating and compliant material at the contact areas. To avoid thermal distortions due to the long unpumped Brewster tips, sealing and mounting has been changed to allow pumping and cooling up to the very end of the slab.

In addition to these structural changes, major performance improvements have been realized as a result of a thermally superior slab material, improved resonator designs and the use of more efficient pump sources. Emphasis shifted from Nd:glass to Nd:YAG for most modern slab lasers. Improvements in the growth technology of Nd:YAG boules has made it possible to fabricate Nd:YAG slabs up to 20 cm in length and 2.5 cm wide. The superior thermal properties of this crystal, compared to glass has led to dramatic performance improvements, and also opened the door for conductively cooled slabs in combination with diode pumping. Beam divergence of slab lasers has been reduced substantially by utilizing unstable resonators which employ graded-index reflectors. In all but the largest slab lasers, flashlamps have been replaced by diode arrays which results in improved overall system efficiency and reduced heat load in the slab. The combination of component and design changes have led to a new generation of slab lasers.

In recent years, design efforts have concentrated mainly on systems utilizing Nd:YAG slabs. Flashlamp systems have been built in the multi kilowatt regime, and diode-pumped slab lasers range from small conductively-cooled systems to lasers with kW output. Examples of state-of-the-art slab systems will be described next.

The performance of a system which yielded the best beam quality obtained so far from a flashlamp-pumped high power Nd:YAG slab laser is depicted in Fig. 7.37 (lower two curves). Xenon flashlamps pump a 6 mm × 25 mm × 200 mm

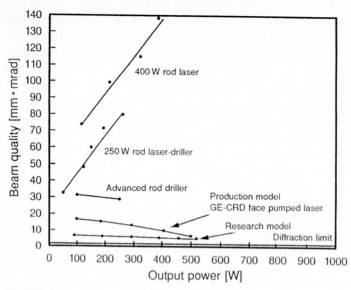

Fig. 7.37. Beam quality vs output power for rod lasers and advanced slab lasers [7.69]

Nd : YAG slab. The research model is designed for four passes through the slab, whereas in the production model two passes are used. The laser beam makes 20 bounces in the Nd : YAG slab. With a resonator of 4 meters in length, an output power of 500 W was achieved in a 3× diffraction limited beam [7.69]. A comparison with typical rod lasers is also shown in Fig. 7.37. Advanced lamp-pumped rod lasers have a beam quality on the order of 30 mm-mrad, whereas standard systems have a much larger beam divergence.

Figure 7.38 exhibits an example of a recent zig-zag slab design which contains a 0.7 cm × 2.6 cm × 19.1 cm Nd : YAG slab pumped symmetrically by four krypton arc lamps in a gold plated reflector [7.61]. The sides are cooled transversely with water at a flow rate of 23 ltr/min each. The faces of the slab are protected with a 2.2 μm thick quartz layer to avoid coupling of the laser radiation into the O-ring seals. Top and bottom are thermally insulated by a 200 μm thick silicone layer. The laser was operated in the long-pulse mode with pump pulses of 4 ms duration and a repetition rate of 22 Hz. A system comprised of three slab-pump heads in an oscillator-amplifier configuration produced a maximum of 2.3 kW output. Overall efficiency of the system was 4.3%. The laser had a beam divergence about an order of magnitude lower than a comparable rod system, which was mainly the result of a negative-branch unstable resonator design, which also contained gradient reflective mirrors. At an output of 1.5 kW, the laser had a beam diameter times divergence product of 16 mm-mrad (FWHM).

A single oscillator containing a Nd : YAG slab of 1 cm × 2.5 cm × 15.2 cm pumped by four krypton flashlamps produced an output of 700 W with a beam divergence of 17 × 8 mrad [7.63]. The overall efficiency for this system was 2.4% when operated at a 5–7 Hz repetition rate and at a pulsewidth of 1–10 ms.

100 mm

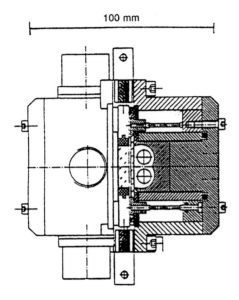

Fig. 7.38. Cross-sectional view of a flash-lamp pumped Nd : YAG slab laser head [7.61]

Diode-array-pumped, and conductively cooled Nd : YAG slab lasers for low and medium power output are very attractive for a number of applications and have received considerable attention in recent years. In these lasers, one face of the slab is thermally bonded to a metal heat sink. Pump radiation enters the slab through the opposite face. At the slab-heat sink interface, a SiO_2 coating preserves total internal reflections, and a gold coating reflects the pump light twice through the slab.

Conductively cooled slabs permit great flexibility in thermal management. Heat from the laser diodes and heat sink can be dissipated into the air via a fin structure and a small fan, or the laser can be attached to a larger structure which provides a thermal sink. An obvious option is liquid cooling. In this case, the liquid is passed through channels in the heat sink and laser-diode mounting structure, and does not come into contact with optical surfaces. This greatly simplifies the design. Waste heat can also be absorbed by a phase change material in cases where the laser is operated only intermittently.

A photograph of a compact zig-zag slab Nd : YAG laser which is air-cooled and battery-powered is depicted in Fig. 7.39a [7.64]. At a repetition rate of 10 pps, the system produces 270 mJ per pulse at the frequency doubled wavelength of 532 nm. The slotted compartment shown in the photograph houses the pump module which consists of the Nd : YAG slab, diode arrays, heat sink and cooling fans. The zig-zag slab is pumped by twelve 18-bar arrays with a total optical output of 2.6 J at a pulsewidth of 200 μs. The slab is designed for 13 bounces and near normal incidence AR coated input faces. This type of slab geometry allows both S and P polarizations to be transmitted without loss. The slab dimensions are 138 mm $\times$ 9.5 mm $\times$ 8 mm. The narrow side of the slab, mounted to the heat sink, was coated with a high reflection coating at 808 nm to allow two passes of the pump light through the slab.

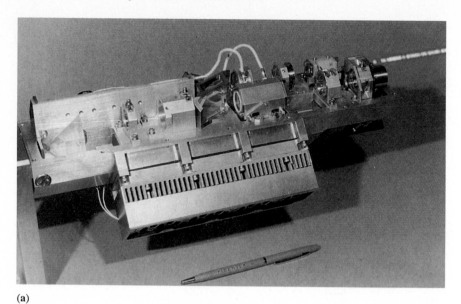

(a)

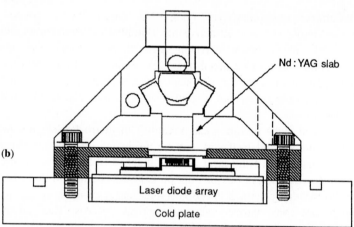

Fig. 7.39. Photograph of an air-cooled and battery-operated zig-zag slab Nd:YAG laser (**a**), and cross-section of pump module (**b**)

One of the engineering difficulties of conduction-cooled laser crystals is stress-free mounting of the slab onto the heat sink. If the slab or the heat sink have even a very slight bow, stress is introduced after the two components are contacted. The application of slight uniform pressure along the length of the slab is not a trivial task. A cross section of the pump module is shown in Fig. 7.39b. The slab was bonded to a thermal expansion-matched heat sink which is clamped to the bridge assembly. This clamping method was employed to reduce mounting and thermal cycling stresses in the laser crystal. A samarium doped glass window isolates the laser diodes from the Nd:YAG slab. The window protects the slab from possible laser-diode bar failure, minimizes contamination near the laser-diode facets, and helps suppress parasitic modes in the zig-zag slab. The bridge assembly and laser-diode arrays are mounted to a common cold plate, where heat is removed by forced convection. Four small fans are used to remove excess heat and regulate the laser-diode temperature.

Typical applications for this kind of system are for space applications, range finders, target designators, and airborne lasers for altimetry, as well as medical lasers and small industrial systems.

Other examples of small conductively-cooled diode-pumped slab lasers can be found in [7.65], with only a single diode array, 2 mJ pulses with a duration of 4 ns, at repetition rates up to 90 Hz have been obtained.

A remarkable achievement is the generation of 40 W cw output in a TEM_{00} mode from a Nd:YAG slab laser which was pumped with only 212 W of diode pump power. The fiber-coupled diode arrays pumped a 1.7 mm × 1.8 mm × 58.9 mm slab [7.62].

Zig-zag slab lasers can be scaled to the kW level employing diode arrays. The large face area available in a slab geometry allows stacking of a sufficient number of diode arrays to achieve a high output power. The large flat surfaces eliminate the need for intervening optics between the pump source and the laser crystal. This facilitates scaling to large systems. For example, one of the first diode-pumped Nd:YAG lasers to reach an average output power at the kW level was reported in [7.66, 67]. In this system a 9 cm × 1.6 cm × 0.4 cm slab was located between two diode modules, each face pumping one side of the Nd:YAG slab. Each pump module consisted of a large number of individual 1 cm long diode bars arranged in a rack and stack fashion, and produced 8 kW of optical pump power in a 150 μs pulse at 2.2 kHz.

The arrangement of the diode arrays and cross-section of the pump module of a similar high-power diode laser is shown in Fig. 7.40 [7.68]. A 20 cm × 2 cm × 0.7 cm Nd:YAG slab is pumped by 14 diode arrays on each side. Each array is comprised of 16 bars with a peak output of 50 W at a 30% duty cycle. The total optical pump power to the slab is 6.72 kW at an electrical input of 15 kW. The system was designed for a pulse repetition rate of 2 kHz and a pump pulse length of 150 μs. With a 2 m long resonator, the pump head illustrated in Fig. 7.40 generated 740 W of average Q-switched power in a 7× diffraction limited beam. Peak power was 1 MW. The oscillator required four acousto-optic Q-switches optically connected in series to prevent prelasing.

Active Mirror Amplifier

The active-mirror concept was invented at General Electric [7.70] and extensively developed at the Laboratory for Laser Energetics, University of Rochester [7.71–73]. In this approach, a single circular or rectangular disk is pumped and cooled from the backside, and laser radiation to be amplified enters from the front, as shown in Fig. 7.41. The front face of the disk has an anti-reflection coating for the laser radiation, whereas the backside has a thin-film coating which is highly reflective for the laser radiation, and transparent for the flashlamp pump radia-

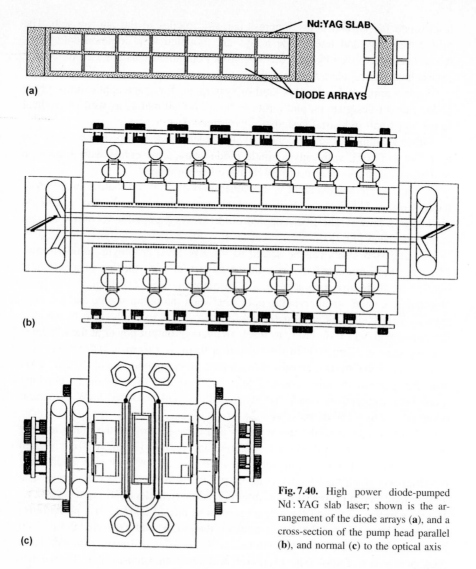

Fig. 7.40. High power diode-pumped Nd:YAG slab laser; shown is the arrangement of the diode arrays (**a**), and a cross-section of the pump head parallel (**b**), and normal (**c**) to the optical axis

tion. The laser radiation passes twice through the disk thus improving extraction efficiency. The advantage of this type of amplifier is that it can be pumped uniformly and efficiently; the disadvantage is that the slab can be pumped only through one face, and this unsymmetrical pumping leads to some thermal distortion. Originally designed for Nd:glass systems requiring large apertures such as needed for fusion research, the concept of an active-mirror design has also been applied to crystalline hosts such as Nd:Cr:GSGG [7.74] and Yb:YAG [7.75].

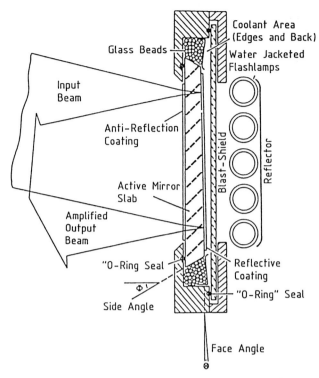

Input
Beam

Glass Beads

Anti-Reflection
Coating

Active Mirror
Slab

Amplified
Output
Beam

"O-Ring Seal

Side Angle

Coolant Area
(Edges and Back)

Water Jacketed
Flashlamps

Blast-Shield

Reflector

Reflective
Coating

"O-Ring" Seal

Face Angle

Fig. 7.41. Schematic of active mirror amplifier design [7.73, 76]

Disk Amplifiers

Solid-state glass-disk amplifiers were considered very early in the development
of high-brightness pulsed laser systems in order to solve problems of cooling,
aperture size, gain uniformity, and pumping efficiency. In a disk laser amplifier
the surfaces are set at Brewsters angle, as shown in Fig. 7.42. The slab faces have
a high-quality optical finish to minimize scattering loss. The beam is linearly polarized in the p-plane to avoid reflection loss. Flashlamps are used for pumping,
with lamp radiation incident on the disk faces and transverse to the beam direction. Thus, the disks are said to be face pumped. Nd : glass amplifiers containing
glass disks were first built in the late 1960s at General Electric, University of
Rochester, Los Alamos Laboratory and at the Naval Research Laboratory. Over
the years this type of amplifier has been highly developed at laboratories engaged
in laser-fusion studies, and particularly at the Lawrence Livermore National Laboratory [7.77–79].

A thorough treatment of all optical, mechanical and electrical aspects of disk
laser amplifiers can be found in the Lawrence Livermore National Laboratory
Laser Program Annual Reports. Also the book by *Brown* [7.73] provides a detailed discussion of the various engineering issues related to disk lasers. In the
earlier designs, the disks were elliptical in shape and mounted in a cylindrical

a)

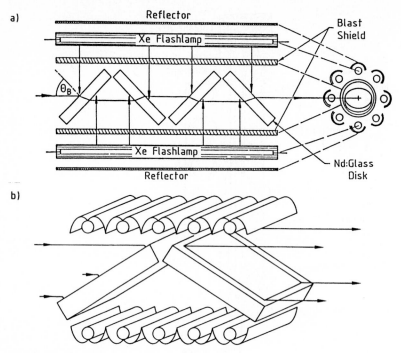

b)

Fig. 7.42a,b. Schematic diagram of disk amplifiers **(a)** cylindrical geometry, **(b)** rectangular geometry

pump geometry as indicated in Fig. 7.42a. The amplifiers for the more recent fusion systems such as NOVA have a rectangular pump-geometry because of their greater pump efficiency (Fig. 7.42b). Disk amplifiers with apertures as large as 74 cm have been built and operated.

Generally, an even number of disks is used in an amplifier since beam translation occurs due to refraction in a single disk. Nonlinear beam steering in disks can be minimized if the disks are placed in an alternate rather than parallel arrangement. The entire periphery of each disk is generally surrounded by a metal frame. A blast shield is normally placed between the flashlamp array and the disks, this makes it possible to completely encase the disk assembly. It also protects the disks in case of a flashlamp explosion.

The concept of face pumping, as opposed to pumping through the edges of a disk or slab, is important in that the pump radiation can be distributed, in an easily controlled manner, uniformly over the faces of the slab. Pumping in this way yields uniform gain over the aperture for a laser beam, but more importantly, the slab heating is uniform laterally, thus allowing transverse temperature gradients to be minimized. Such temperature gradients give rise to distortion in the slab just as in the rod geometry. In contrast to the rod geometry, however, the transverse gradients in the slab geometry arise only from spurious effects in pumping and cooling, whereas in the rod geometry, transverse temperature gradients arise from the principal cooling mechanism.

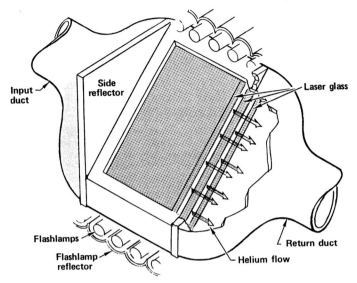

Fig. 7.43. Concept of a gas cooled disk amplifier module designed for high average power operation

The Brewster's-angle disk lasers have been primarily developed for very high peak-power and relatively low average-power operation, with the slabs and coolant all approaching thermal equilibrium in the interval between pulses. To operate the Brewster's-angle slab laser at high average power requires adequate cooling, as, for example, with forced convection over the faces. In addition, since the laser beam must pass through the coolant, the problem of thermal distortion in the coolant must be dealt with. This operating mode requires that the slabs be sufficiently thin so that the cooling time is several times less than the laser repetition period. Figure 7.43 shows a configuration of a gas cooled disk module which could potentially be developed for high average power operation, particularly if the Nd : glass slabs are replaced with crystalline laser host materials [7.76].

7.4 End-Pumped Configurations

So far, we have discussed cylindrical rods and rectangular slab configurations which are pumped from the side. As we have seen, analytical solutions generally assume uniform heating in an infinitely long rod or slab structure. End effects are treated separately as a distortion of the two-dimensional temperature distribution.

With the availability of high-power diode lasers, end-pumping, or longitudinal-pumping, of laser crystals has become a very important technology. End-pumped lasers have been described in Chaps. 3 and 9. Here, we are concerned with the thermal aspects of end pumping. The thermal profile in an end-pumped laser is quite different from the systems discussed so far, due to very localized heat deposition. This leads to highly nonuniform and complex temperature and stress profiles. This is illustrated in Fig. 7.44 which shows the calculated temperature distribution in an end-pumped laser rod [7.81].

An end-pumped laser rod has a temperature profile across the pumped region which is a function of the distribution of pump radiation. From the edge of the pumped region, the temperature decays logarithmically to the cooled cylindrical surface of the rod. Along the axis of the rod, the temperature profile will decay exponentially due to the exponential absorption of pump radiation. Temperature and stress profiles and thermal lensing in end-pumped lasers, employing different pump and rod geometries, and cooling techniques, have been described in the literature. The fraction of pump power which is converted to heat inside the laser material acts as heat source.

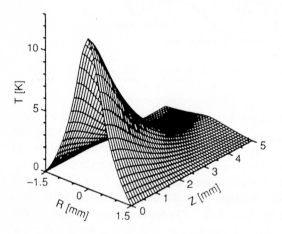

Fig. 7.44. Temperature distribution in an end-pumped laser rod [7.81]

Different cooling arrangements and rod shapes lead to different boundary conditions. The major distinctions are convective or conductive cooling at the rod surface; and face-cooled thin disks or edge-cooled long cylinders.

With the temperature at the radial surface of a cylindrical rod clamped, which is typical of a crystal mounted in a temperature controlled copper heat sink, thermal profiles have been calculated in [7.80–82]. For the boundary condition of a convective heat transfer to the surrounding medium which is typical of a water-cooled crystal, models have been developed and are represented in [7.83, 84]. End-pumped thin disks, either edge- or face-cooled, were treated in [7.85]. Most models have been developed to treat Nd : YAG crystals. A finite-element analysis of temperature, stress and strain distributions which includes, besides Nd : YAG, also Nd : GSGG and Nd : YLF can be found in [7.86]. Another important boundary condition is the radial profile of the pump beam. In the above-mentioned models, top hat profiles were analyzed in [7.82, 85] and Gaussian pump beams have been assumed in [7.80, 81, 86]. A specific investigation of pump-beam profiles and their effect on thermal distortion is found in [7.84].

For most crystalline laser materials, the stress contributions to the refractive index are small. For example, in Sect. 7.1.1, we calculated that in Nd : YAG, the temperature-dependent change of the refractive index represents 74% of the optical distortion. Therefore, most theoretical models consider only the temperature effect on the index of refraction. However, stress components of the refractive index are included in the treatment reported in [7.82, 86].

The large thermal gradients and associated stresses in an end-pumped laser are illustrated in Figs. 7.45a and 7.46a. Shown are the temperature and stress

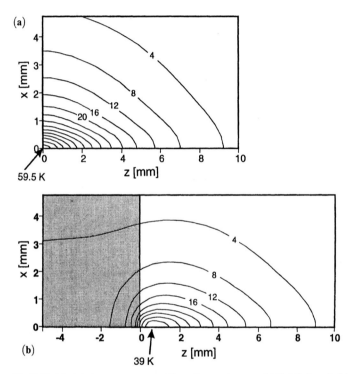

Fig. 7.45. Modeled temperature profiles of an end-pumped Nd : YAG rod, (**a**) with uniform doping and (**b**) with a 5 mm long undoped end cap. The maximum values and their locations are indicated by arrows. The temperature difference beween two isotherms is 4 K [7.87]

distributions calculated by finite-element analysis for a Nd : YAG crystal [7.87]. In this analysis, a 15 W pump beam from a diode array was assumed to be focused onto a Nd : YAG rod of 4.75 mm radius. The pump beam, which enters the laser crystal from the left along the z-axis in Figs. 7.45, 46, has a Gaussian intensity distribution and a spot-size radius of 0.5 mm in the x-direction. It was assumed that 32% of the incident pump radiation is converted to heat. In the case considered here, the absorption length is much larger than the pump spot size.

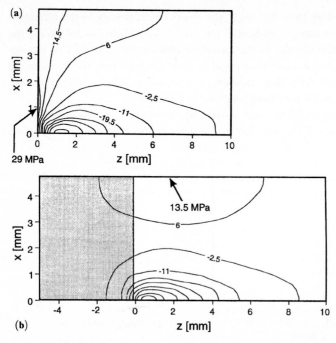

Fig. 7.46. Distribution of the stresses (megapascals) generated by the temperature gradients illustrated in Fig. 7.45 for the uniformly doped crystal (**a**) and the rod with an undoped end cap (**b**) [7.87]

For the 1.3 at.% doped rod an absorption coefficent of 4.5 cm^{-1} was assumed. Therefore, 90% of the pump radiation is absorbed in a 5 mm long path length.

Figure 7.45a illustrates the temperature distribution in one-half of the cross-section of the rod. The isotherms represent the temperature difference from the heat sink surrounding the cylindrical surface of the rod. The temperature reaches a maximum of 59.5 K at the center of the pumped surface. The temperature gradients lead to thermal stress. The corresponding stresses in the Nd:YAG rod are indicated in Fig. 7.46a. Positive values of the isobars indicate tensile stresses, and negative values show areas under compression. The core of the crystal is in compression while the outer regions are in tension. The highest stress occurs on the face of the end-pumped rod, but in contrast to the temperature profile, not at the center, but at the edge of the pump beam (about $2w_p$ from the center). The heat generated by the pump radiation causes an axial expansion of the rod, i.e., the front face begins to bulge as the crystal heats up. In the example above, the bending of the pumped surface resulted in a dioptic power of 1.4 m^{-1}. The axial expansion of the rod leads to the high tensile stress at the pump face, as indicated in Fig. 7.46a.

The above-mentioned example illustrates a number of characteristic features of end-pumped systems. In order to maximize gain, and to match the TEM$_{00}$ resonator mode, the pump beam is tightly focused, which leads to high-pump-power irradiance incident on the end of the laser rod. As a result, the input face is under a high thermal load. The thermal stress, which leads to strong thermal lensing, is often high enough to cause fracture of the end face of the laser rod.

The thermal management of end-pumped lasers can be greatly improved by the use of composite rods. Laser crystals, such as Nd : YAG, Nd : YVO$_4$ and Yb : YAG, have become available with sections of undoped host material on one or both ends. These end caps are diffusion bonded to the doped laser crystal. Composite rods are proven to provide a very effective way to reduce temperature and stresses at the face of end pumped lasers. This is illustrated in Figs. 7.45b and 7.46b, which show the result of an analysis carried out for a Nd : YAG rod with a 5 mm long undoped YAG section at the pumped front end. The other parameters are the same as for the uniformly doped crystal. Because the undoped region is transparent to the impinging pump radiation, there is no thermal load generated at the rod pump face.

The diffusion bond provides uninhibited heat flow from the doped to the undoped region. A large temperature reduction of about 35% is achieved for the composite rod, where a significant part of heat flow occurs into the undoped end cap. The undoped end cap prevents the doped pump face from expanding along its axis, therefore no surface deformation takes place at this interface. This leads to strong compressive stresses (as shown in Fig. 7.46b) instead of tensile stresses, as was the case for the unrestrained surface (Fig. 7.46a). Since compression is not dangerous with respect to rod damage, this change from tensile to compressive stress is really the key to the much higher thermal loading that can be tolerated with the use of the undoped end caps.

Besides increasing the average power capability of the laser, as a result of better thermal management, a composite rod has a number of other advantages. In end-pumped lasers, a dichroic coating is generally applied to the pump face, i.e., the coating is highly reflecting for the laser wavelength at 1.06 μm and highly transmitting for the pump wavelength at 808 nm. The durability of this coating can be affected by the considerable mechanical and thermal stress present at this surface. In a composite rod, the coating temperature is spatially uniform and surface deformation and associated stresses are absent.

Another advantage of undoped end caps is associated with cooling the ends of the laser rod. Typically, the last few millimeters of a rod cannot be effectively cooled by circulating water because of the space required for o-ring seals. The non-absorbing end sections of a composite laser rod solve this thermal management problem.

An important parameter for the design of the optical resonator is the amount of thermal lensing caused by the temperature distribution in an end-pumped laser crystal. In a first approximation, the thermally-induced lens can be described by considering only the temperature-dependent part of the refractive index. Also assuming only radial heat flow in a rod which is in contact with a thermal heat sink of fixed temperature, an analytical solution for the focal length of the thermal lens was derived in [7.80]. A Gaussian pump beam incident on the crystal has been assumed

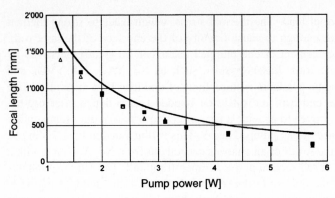

Fig. 7.47. Thermal lensing in an endpumped Nd:YAG laser measured values from [7.88]

$$I(r, z) = I_0 \exp(-2r^2/w_p^2) \exp(-\alpha z) \tag{7.66}$$

where α is the absorption coefficient, and w_p is the $(1/e^2)$ Gaussian radius of the pump beam. With P_H the fraction of the pump power that results in heating, the effective focal length for the entire rod can be expressed by

$$f = \frac{\pi K_c w_p^2}{P_H(dn/dT)} \left(\frac{1}{1 - \exp(-\alpha l)} \right) \tag{7.67}$$

where K_c is the thermal conductivity of the laser material, and dn/dT is the change of refractive index with temperature. From (7.67) follows that the effective focal length depends on the square of the pump-beam radius w_p. Therefore, it is desirable to use the largest pump-beam radius consistent with resonator-mode match.

The dependence of thermal lensing on pump power is illustrated in Fig. 7.47. An end-pumped Nd:YAG rod with a length of 20 mm and a radius of 4.8 mm was pumped with a fiber-coupled laser-diode array. The output from the fiber bundle was imaged onto the crystal surface into a pump spot with radius $w_p = 340\,\mu$m. Figure 7.47 displays the measured thermal lens and the calculated values according to (7.67). In using this equation, we assume that 32% of the pump power results in heating. The material parameters required for Nd:YAG are $dn/dT = 7.3 \times 10^{-6} K^{-1}$, $\alpha = 4.1\,\mathrm{cm}^{-1}$, $K_c = 0.13\,\mathrm{W/cm\,K}$.

8. Q-Switching

A mode of laser operation extensively employed for the generation of high pulse power is known as Q-switching. It has been so designated because the optical Q of the resonant cavity is altered when this technique is used [8.1, 2]. As was discussed in Chap. 3, the quality factor Q is defined as the ratio of the energy stored in the cavity to the energy loss per cycle. Consequently, the higher the quality factor, the lower the losses.

In the technique of Q-switching, energy is stored in the amplifying medium by optical pumping while the cavity Q is lowered to prevent the onset of laser emission. Although the energy stored and the gain in the active medium are high, the cavity losses are also high, lasing action is prohibited, and the population inversion reaches a level far above the threshold for normal lasing action. The time for which the energy may be stored is on the order of τ_f, the lifetime of the upper level of the laser transition. When a high cavity Q is restored, the stored energy is suddenly released in the form of a very short pulse of light. Because of the high gain created by the stored energy in the active material, the excess excitation is discharged in an extremely short time. The peak power of the resulting pulse exceeds that obtainable from an ordinary long pulse by several orders of magnitude.

Figure 8.1 shows a typical time sequence of the generation of a Q-switched pulse. Lasing action is disabled in the cavity by a low Q of the cavity. Toward the end of the flashlamp pulse, when the inversion has reached its peak value, the Q of the resonator is switched to some high value. At this point a photon flux starts to build up in the cavity, and a Q-switch pulse is emitted. As illustrated in Fig. 8.1, the emission of the Q-switched laser pulse does not occur until after an appreciable delay, during which time the radiation density builds up exponentially from noise.

8.1 Q-Switch Theory

A number of important features of a Q-switched pulse, such as energy content, peak power, pulsewidth, rise and fall times, and pulse formation time, can be obtained from the rate equations discussed in Chap. 1. In all cases of interest the Q-switched pulse duration is so short that we can neglect both spontaneous emission and optical pumping in writing the rate equations.

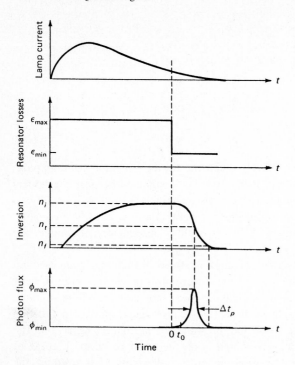

Fig. 8.1. Development of a Q-switched laser pulse. Shown is the flashlamp output, resonator loss, population inversion, and photon flux as a function of time

From (1.61, 58), it follows that

$$\frac{\partial \phi}{\partial t} = \phi \left(c\sigma n \frac{l}{l'} - \frac{\varepsilon}{t_R} \right) \tag{8.1}$$

and

$$\frac{\partial n}{\partial t} = -\gamma n \phi \sigma c . \tag{8.2}$$

In (8.1) we expressed the photon lifetime τ_c by the round-trip time $t_R = 2l'/c$, and the fractional loss ε per round trip according to (3.7). Also, a distinction is made between the length of the active material l and the length of the resonator l'. Q-switching is accomplished by making ε an explicit function of time (e.g., rotating mirror or Pockels cell Q-switches) or a function of the photon density (e.g., saturable absorber Q-switching). The losses in a cavity can be represented by

$$\varepsilon = -\ln R + L + \zeta(t) , \tag{8.3}$$

where the first term represents the coupling losses, the second term contains all the incidental losses such as scattering, diffraction, and absorption, and $\zeta(t)$ represents the cavity loss introduced by the Q-switch. For a particular explicit form of $\zeta(t, \phi)$, the coupled rate equations can be solved numerically with the boundary condition $\zeta(t < 0) = \zeta_{max}$; $\zeta(t \geq 0) = 0$. In many instances Q-switches are so fast that no significant change of population inversion takes place during the switching process; in these cases ζ can be approximated by a step function.

In the ideal case, where the transition from low Q to high Q is made instantaneously, the solution to the rate equations is particularly simple [8.3, 4]. In this case we assume that at $t = 0$ the laser has an initial population inversion n_i, and the radiation in the cavity has some small but finite photon density ϕ_i. Initially, the photon density is low while the laser is being pumped and the cavity losses are $\varepsilon_{max} = -\ln R + L + \zeta_{max}$ as illustrated in Fig. 8.1. The losses are suddenly reduced to $\varepsilon_{min} = -\ln R + L$. The photon density rises from ϕ_i, reaches a peak ϕ_{max} many orders of magnitude higher than ϕ_i, and then declines to zero. The population inversion is a monotone decreasing function of time starting at the initial inversion n_i and ending at the final inversion n_f. We note that the value for n_f is below the threshold inversion n_t for normal lasing operation. At n_t the photon flux is maximum and the rate of change of the inversion dn/dt is still large and negative, so that n falls below the threshold value n_t and finally reaches the value n_f. If n_i is not too far above n_t, that is, the initial gain is close to threshold, then the final inversion n_f is about the same amount below threshold as n_i is above and the output pulse is symmetric. On the other hand, if the active material is pumped considerably above threshold, the gain drops quickly in a few cavity transit times τ_R to where it equalizes the losses. After the maximum peak power is reached at n_t, there are enough photons left inside the laser cavity to erase the remaining population excess and drive it quickly to zero. In this case the major portion of the decay proceeds with a characteristic time constant τ_c, which is the cavity time constant.

The equations describing the operation of rapidly Q-switched lasers involves the simultaneous solution of two coupled differential equations for the time rate of change of the internal photon density in the resonator, (8.1), and the population inversion density in the active medium, (8.2). From the work of *Wagner* et al. [8.3], who first derived solutions to the rate equations, we can express the output energy of the Q-switched laser as follows:

$$E = \frac{h\nu A}{2\sigma\gamma} \ln\left(\frac{1}{R}\right) \ln\left(\frac{n_i}{n_f}\right) \tag{8.4}$$

where $h\nu$ is the laser photon energy, and A is the effective beam cross-sectional area. The initial and final population inversion densities, n_i and n_f, are related by the transcendental equation

$$n_i - n_f = n_t \ln\left(\frac{n_i}{n_f}\right) \tag{8.5}$$

where n_t is the population inversion density at threshold, i.e.,

$$n_t = \frac{1}{2\sigma\ell}\left(\ln\frac{1}{R} + L\right) . \tag{8.6}$$

The pulse width of the Q-switch pulse can also be expressed as a function of the inversion levels, n_i, n_f, n_t

$$\Delta t_p = \tau_c \frac{n_i - n_f}{n_i - n_t \left[1 + \ln\left(n_i/n_t\right)\right]} \,. \tag{8.7}$$

The equations for pulse energy, pulse width, and therefore peak power, are expressed in terms of the initial and final population inversion densities which depend not only on the particular choice of output coupler, but which are also related via a cumbersome transcendental equation. Thus, in order to optimize a given laser for maximum efficiency, it is generally necessary to obtain numerical solutions of these equations.

Degnan [8.5] managed to derive analytical solutions to the optimized operation of Q-switched lasers. He showed that key parameters such as optimum reflectivity, output energy, extraction efficiency, pulse width, peak power, etc., can all be expressed as a function of a single dimensionless variable $z = 2g_0\ell/L$, where $2g_0\ell$ is the logarithmic small-signal gain, and L is the round-trip loss. We will summarize the results of this analysis because the expressions and graphs presented in this work are particularly useful to the laser designer. The following expression for the optimum reflectivity was derived

$$R_{opt} = \exp\left[-L\left(\frac{z - 1 - \ln z}{\ln z}\right)\right] \,. \tag{8.8}$$

This function is plotted in Fig. 8.2a for a number of typical loss parameters L. The energy output for an optimized system is

$$E_{out} = E_{sc}(z - 1 - \ln z) \tag{8.9}$$

which is plotted in Fig. 8.2b. E_{sc} is a scale factor with the dimension of energy which contains a number of constants

$$E_{sc} = Ah\nu L/2\sigma\gamma$$

where A is the beam cross-section, $h\nu$ is the photon energy, σ is the stimulated emission cross-section, L is the round-trip loss, and γ is the degeneracy factor. Figure 8.2b contains also a plot of the FWHM pulse width vs. z, which is obtained from

$$t_p = \frac{t_R}{L}\left(\frac{\ln z}{z[1 - a(1 - \ln a)]}\right) \tag{8.10}$$

where t_R is the cavity round-trip time, and $a = (z - 1)/(z \ln z)$.

In the limit of large z, the output energy approaches the total useful stored energy in the gain medium

$$E_u = \frac{Ah\nu L}{2\sigma\gamma}z = \frac{Vh\nu n_i}{\gamma} \,. \tag{8.11}$$

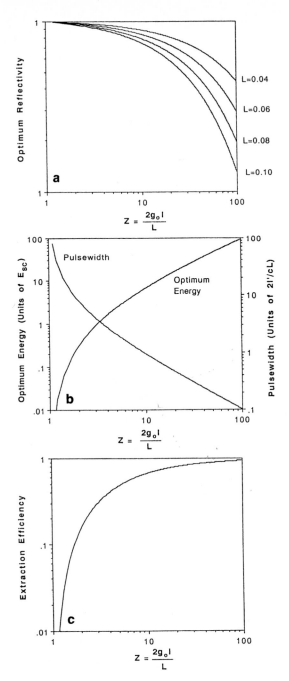

Fig. 8.2. (a) Optimum reflectivity of the output coupler of a Q-switched laser as a function of z. Parameter is the round-trip loss L. **(b)** Normalized energy output and pulsewidth of a Q-switched laser as a function of z. **(c)** Q-switch extraction efficiency as a function of z

With (8.9, 11) one can define an energy extraction efficieny

$$\eta_{eq} = 1 - \left(\frac{1 + \ln z}{z} \right) \tag{8.12}$$

which is plotted in Fig. 8.2c.

As one would expect, a high gain-to-loss ratio leads to a high Q-switch extraction efficiency. For a ratio of logarithmic gain to loss of about 10, an extraction efficiency of 70 % is achieved. For higher factors of z, the extraction efficiency increases only very slowly.

The reader is reminded that the shape of the curve for η_{eq} is similar to the results obtained for the free-running laser discussed in Chap. 3, see (3.73) and Figs. 3.6, 7, which also depends only on the ratio of $2g_0\ell/L$. It is also important to remember that besides the Q-switch extraction efficiency expressed by (8.12), the total energy extraction from a Q-switched laser depends also on the fluorescence losses and ASE depopulation losses prior to opening of the Q-switch. The overall efficiency of the Q-switch process was defined in Sect. 3.4.1 as the product of the Q-switch extraction efficiency, storage efficiency and depopulation efficiency.

Laser-design trade-offs and performance projections, and system optimization can be accomplished quickly with the help of the graphs presented in Fig. 8.2. For example, we consider the design parameters for a Q-switched Nd : YAG laser with a desired multimode output of 100 mJ. The laser crystal has a diameter of 5mm and the laser resonator is 30 cm long. Assuming a 5 % round-trip cavity loss ($L = 0.05$), we calculate $E_{sc} = 3 \times 10^{-3} J$. This requires a ratio of $E_{out}/E_{sc} = 33$ in order to achieve the desired output energy. From Fig. 8.2b, one obtains therefore a value $2g_0\ell/L = 38$, or a single-pass power gain of the rod of $G = \exp(g_0\ell) = 2.6$. Extraction efficiency follows from Fig. 8.2c to be around 88 % for $z = 38$, and the optimum output coupler has a reflectivity of $R = 0.63$ for $z = 38$ and $L = 0.5$ according to the graph in Fig. 8.2A. Since the cavity transit round-trip time for the given resonator length is about 2 ns, the expected pulse width from the laser is $t_p = 12$ ns according to Fig. 8.2b. The peak power of the Q-switch pulse follows from the parameters already calculated and is $P = E_{out}/t_p = 8$ MW.

Slow Q-Switching

Instead of a step function, we will now consider the case of a resonator loss that varies in time [8.6, 7].

A form of Q-switch used in the past is a rotating prism driven by a high-speed motor. A typical spinning rate is 24,000 rpm. At this rate of rotation it takes about 40 ns to sweep over an angle of 0.1 mrad. The loss rate variation for this case can be assumed to be of the form

$$\varepsilon = -\ln R_1 + L + B \cos(\omega t). \tag{8.13}$$

In a slow Q-switch the development of a Q-switch pulse depends on the ratio of pulse build-up time t_D to the switching time. This is illustrated in Fig. 8.3

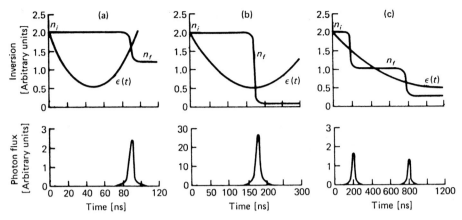

Fig. 8.3. Inversion, cavity losses, and pulse development as a function of time in a slow Q-switched system [8.6]

for three extreme cases. In Fig. 8.3a, the pulse build-up time is longer than the switching time of the Q-switch. The photon flux starts to increase exponentially at $t = 0$, and at the time ϕ_{max} is reached and a pulse is emitted the Q-switch has already passed the point of ε_{min}. The pulse energy is relatively small because the cavity losses are not at a minimum at the time of maximum photon density. In Fig. 8.3b, Q-switch operation is optimized. The Q-switch is emitted when the cavity losses are minimum, i.e., pulse build-up time and switching times are equal. In Fig. 8.3c, the opening time of the Q-switch is much slower than the pulse build-up time. This leads to the emission of several Q-switched pulses. We note that the first pulse is larger than the second and occurs at a larger value of ΔN; the second pulse occurs near the minimum loss level and is much lower in energy. We can explain the occurrence of multiple pulsing as follows: At $t = 0$ the resonator losses are low enough for the photon density to grow exponentially. After 200 ns the photon density has built up to its maximum level and a pulse is emitted. After the pulse is emitted the inversion reaches the level n_f. Because n_f is below threshold, the photon density ϕ is very small and a steady state is reached until the slowly changing loss rate term ε decreases enough so that the condition for photon density buildup occurs. The cycle then starts over, but because it now begins with a lower initial value of the inversion a small pulse is produced.

It should be noted that for this multiple pulsing to occur, the first pulse must be emitted at the point of a relatively high loss rate so that another pulse can build up and be emitted when the output loss rate factor is lower and probably near its minimum.

Multiple pulsing can be avoided either by shortening the switching time of the Q-switch (higher rotational speed) or by increasing the pulse delay time, for example, by increasing the cavity length. From the foregoing consideration it follows that in a laser with fixed pumping level and mirror separation, maximum output is obtained only at one particular speed.

Continuously Pumped, Repetitively Q-Switched Systems

A very important class of laser systems, employed extensively in micromachining applications, is the cw-pumped, repetitively Q-switched Nd : YAG laser. In these laser systems the population inversion undergoes a cyclic variation, as shown in Fig. 8.4. Between Q-switches the population inversion rises from a value n_f to a value n_i. The buildup of the inversion under the influence of a continuous pumping rate and spontaneous decay is described as a function of time by

$$n(t) = n_\infty - (n_\infty - n_f) \exp\left(-\frac{t}{\tau_f}\right) , \qquad (8.14)$$

where n_∞ is the asymptotic value of the population inversion which is approached as t becomes large compared to the spontaneous decay time τ_f. The value n_∞, which depends on the pump input power, is reached only at repetition rates small compared to $1/\tau_f$. For repetition rates larger than $1/\tau_f$, the curves representing the buildup of the population are shorter segments of the same exponential curve followed for the lower repetition rates.

During the emission of a Q-switch pulse, the inversion changes from n_i to n_f. Figure 8.5 shows the development of the Q-switched pulse on an expanded time scale [8.8]. At $t = 0$ the cavity Q factor starts to increase until it reaches its maximum value Q_{max} at $t = t_1$. Pulse formation ensues until the full pulse output is achieved at $t = t_2$. Stimulated emission ceases at $t = t_3$; at this time continued pumping causes the inversion to start to increase. At the point where the inversion begins to increase, $t = t_3$, the cavity Q begins to decrease, reaching its minimum value at $t = t_4$. During the time period t_3–t_5, the inversion is allowed to build up to its inital value n_i.

The theory of *Wagner* and *Lengyel* [8.3], summarized at the beginning of this chapter for the case of single Q-switched pulses, can also be applied to the case of repetitive Q-switches, with some modifications that take into account the effects of continuous pumping. Equations (8.1, 2) are applicable, however, we will set $\gamma = 1$ and $n = n_2$ because all repetitively Q-switched lasers of practical use are four-level systems. The inversion levels n_f and n_i are connected by (8.7). During the low-Q portion of the cycle, the inversion n_2 is described by the differential equation

$$\frac{dn_2}{dt} = w_p(n_{tot} - n_2) - \frac{n_2}{\tau_f} . \qquad (8.15)$$

With the assumption that $n_2 \ll n_{tot}$ and with

$$n_\infty = w_p \tau_f n_{tot} \qquad (8.16)$$

we obtain (8.14) as a solution. For repetitive Q-switching at a repetition rate f, the maximum time available for the inversion to build up between pulses is $t = 1/f$. Therefore,

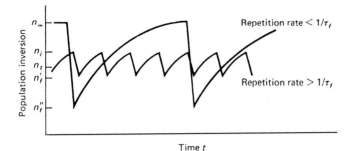

Fig. 8.4. Population inversion versus time in a continuously pumped Q-switched laser. Shown is the inversion for two different repetition rates. At repetition rates less than $1/\tau_f$, the inversion approaches the asymptotic value n_∞

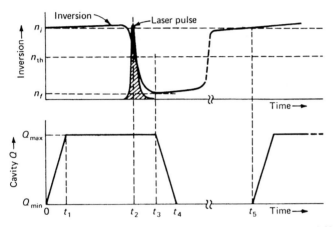

Fig. 8.5. Development of a Q-switched pulse in a cw-pumped system [8.8]

$$n_i = n_\infty - (n_\infty - n_f) \exp\left(-\frac{1}{\tau_f f}\right) \tag{8.17}$$

in order for the inversion to return to its original value after each Q-switch cycle.

During each cycle a total energy $(n_i - n_f)$ enters the coherent electromagnetic field. Of this a fraction, $T/(T + L)$, appears as laser output. Therefore, the Q-switched average power P_{av} at a repetition rate f is given by

$$P_{av} = \frac{Tf}{T + L}(n_i - n_f)h\nu V \ . \tag{8.18}$$

The peak power P_p of the Q-switched pulses is obtained from (8.6). The effective pulse width Δt_p can be calculated from the peak and average powers according to $\Delta t_p = P_{av}/P_p f$. It is convenient to calculate the ratios P_p/P_{cw} and P_{av}/P_{cw}, where P_p is the Q-switched peak power, and P_{cw} is the cw power from the laser at the same pumping level. For cw operation the time derivatives in (8.1 and 2) are zero, and we have

$$P_{\text{cw}} = \frac{T}{T+L}\left(\frac{n_\infty - n_t}{\tau_f}\right) h\nu V \ . \tag{8.19}$$

Because of the transcendental functions expressed by (8.6, 7, and 17), the ratios P_p/P_{cw} and $P_{\text{av}}/P_{\text{cw}}$ cannot be expressed in closed form. The result of numerical calculations is shown in Figs. 8.6 and 7 [8.9].

In Fig. 8.6 the ratio of Q-switched peak power output to the maximum cw output is plotted versus repetition rate for a Nd:YAG laser. For repetition rates below approximately $800\,\text{Hz}$ ($\tau_f f \approx 0.20$) the peak power is independent of the repetition rate. At these low repetition rates there is sufficient time between pulses for the inversion to reach the maximum value n_∞. In the transition region between 0.8 and 3 kHz, peak power starts to decrease as the repetition rate is increased. Above 3 kHz, the peak power decreases very rapidly for higher repetition rates.

Figure 8.7 shows the ratio of Q-switched average power to cw power as a function of repetition rate. Above a repetition rate of approximately 10 kHz, the Q-switch average power approaches the cw power. At low repetition rates the average power is proportional to the repetition rate. In Fig. 8.8 the experimentally determined peak power, average power pulse width, and pulse buildup time is plotted as a function of repetition rate for a Nd:YAG laser. In accordance with theory, for higher repetition rates the pulse width and the pulse buildup time increase as a result of the reduction of gain.

In the following sections we will describe and compare different Q-switch techniques.

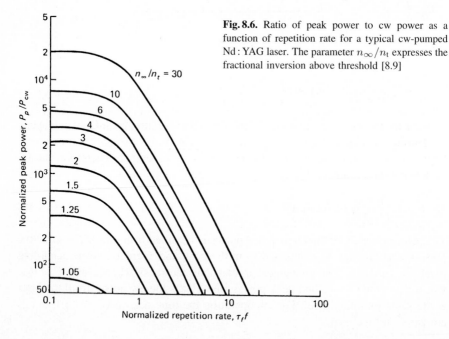

Fig. 8.6. Ratio of peak power to cw power as a function of repetition rate for a typical cw-pumped Nd:YAG laser. The parameter n_∞/n_t expresses the fractional inversion above threshold [8.9]

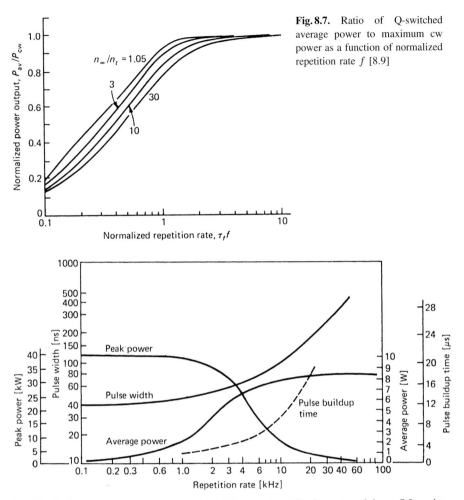

Fig. 8.7. Ratio of Q-switched average power to maximum cw power as a function of normalized repetition rate f [8.9]

Fig. 8.8. Performance of a cw-pumped Nd:YAG laser system. The laser, containing a 7.5-cm by 0.5-cm rod doped with 0.8% Nd, delivered 0.5 W at the TEM_{00} mode. Lamp input power was 5.5 kW. Plotted is the peak power, average power, pulse buildup time, and pulse width as a function of repetition rate

8.2 Mechanical Devices

Q-switches have been designed based upon rotational, oscillatory, or translational motion of optical components. These techniques have in common that they inhibit laser action during the pump cycle by either blocking the light path, causing a mirror misalignment, or reducing the reflectivity of one of the resonator mirrors. Near the end of the flashlamp pulse, when maximum energy has been stored in the laser rod, a high Q-condition is established and a Q-switch pulse is emitted from the laser.

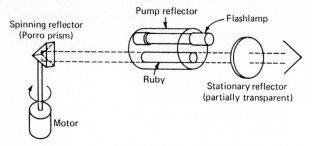

Fig. 8.9. Diagram of a ruby laser employing a spinning prism Q-switch

The first mechanical Q-switch consisted of nothing more than a rotating disc containing an aperture [8.10]. This method was soon abandoned in favor of rotating mirrors or prisms which allow much faster switching times [8.11]. In some cases one of the resonator mirrors was attached to one end of a torsional rod driven at its mechanical resonance frequency [8.12]. Minute translations of one mirror have been utilized to change the Q of the cavity. One technique consists of replacing the output mirror of the laser oscillator by a two-plate Fabry-Perot resonator. By modulating the spacing between the two plates, for example, with a piezoelectric transducer, the device can be shifted from a transmission peak to a reflection peak [8.13]. Another technique makes use of frustrated total internal reflection [8.14]. In this case, a piezoelectric transducer changes the spacing between a roof prism and a second prism. If the two prisms are brought to within a fraction of a wavelength to the surface of the internally reflecting roof prism, the total internal reflection is destroyed. Consequently, the radiation incident on the roof prism is transmitted rather than totally reflected.

The spinning reflector technique for the generation of Q-switched pulses, as shown in Fig. 8.9, involves simply rotating one of the two resonant cavity reflectors so that parallelism of the reflectors occurs for only a brief instant in time. If a plane mirror is employed as the rotating element, the axis of rotation must be aligned to within a fraction of a milliradian parallel to the face of the opposing reflector. This difficulty can be overcome by using a roof prism as the rotating element. If the roof of the prism is perpendicular to the axis of rotation, then the retroreflecting nature of the prism assures alignment in one direction, while the rotation of the prism brings it into alignment in the other direction. Experiments have shown that an angular accuracy of $\pm 3°$ is sufficient to assure optimum performance.

Synchronization of the flashlamp with the mirror position is usually obtained by the use of a magnetic pickup which senses the position of the rotating shaft and generates an electric signal to trigger the flashlamp at the appropriate time. The time delay is adjusted so that at maximum population inversion the rotating reflector is parallel to the fixed mirror. For example, for a prism rotating at 24,000 rpm, the flashlamp has to be triggered 115° prior to optical alignment if a delay of 800 μs is desired. For a typical laser resonator, 30 or 40 cm in length, a rotational speed on the order of 20,000 rpm is needed to ensure that the laser output contains just a single pulse. For example, the angular speed of a prism rotating at 24,000 rpm is $\omega = d\alpha/dt = 2.5$ mrad/μs. Experiments have shown that laser action is inhibited in common laser geometries for a mirror misalignment on the order of 1 mrad. This angle, in conjunction with the rotational speed mentioned above, results in a switching time of 400 ns.

For mechanically Q-switching, cw-pumped Nd : YAG lasers, multisided rotating prisms rather than single prisms or mirrors have been employed for the attainment of high repetition rates. As we have seen from Fig. 8.7, in a Nd : YAG laser the average power increases with repetition rate up to about 8 kHz. A repetition rate of 5 kHz was obtained from a Nd : YAG laser which was Q-switched by a 12-sided prism rotating at 25,000 rpm [8.15].

Rotating-mirror devices are simple and inexpensive. They are insensitive to polarization and, therefore, birefringence effects. Hence, more energy from the laser can be extracted under certain conditions as compared to electrooptic Q-switches. However, the mechanical Q-switches suffer from the tendency to emit multiple pulses. The devices are also very noisy, and they require frequent maintenance because of the relatively short lifetime of the bearings. Due to these disadvantages the rotating prism Q-switch has been replaced by the acousto-optic Q-switch in cw-pumped Nd : YAG lasers.

Rotating prisms attracted renewed interest for Q-switching Er : glass and IR solid-state lasers in the 1.5 to 3 μm region. The flashlamp pumped Er : glass laser with an output wavelength of 1.54 μm is particularly important for applications in compact hand-held eye-safe rangefinders [8.16–19]. Plastic dye Q-switches normally used in this type of equipment for Nd : YAG lasers are not available at 1.54 μm. Only recently have passive Q-switches been developed for this wavelength regime as discussed in Sect. 8.5.

8.3 Electrooptical Q-Switches

Very fast electronically controlled optical shutters can be designed by exploiting the electrooptic effect in crystals or liquids. The key element in such a shutter is an electrooptic element which becomes birefringent under the influence of an external field. Birefringence in a medium is characterized by two orthogonal directions, called the "fast" and "slow" axes, which have different indices of refraction. An optical beam, initially plane-polarized at 45° to these axes and directed normal to their plane, will split into two orthogonal components, traveling along the same path but at different velocities. Hence, the electrooptic effect causes a phase difference between the two beams. After traversing the medium, the combination of the two components results, depending on the voltage applied, in either an elliptical, circular, or linearly polarized beam [8.20]. For Q-switch operation only two particular voltages leading to a quarter-wave and half-wave retardation are of interest. In the first case, the incident linearly polarized light is

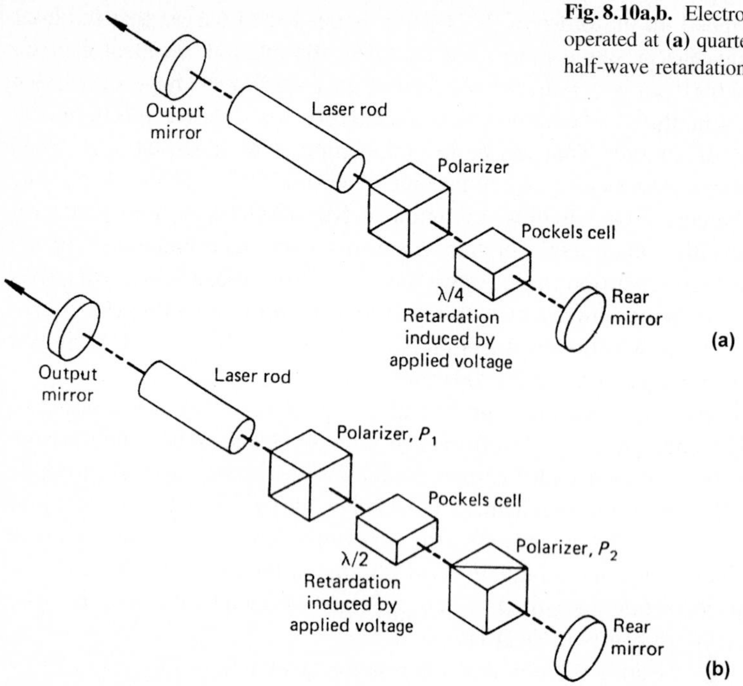

Fig. 8.10a,b. Electrooptic Q-switch operated at **(a)** quarter-wave and **(b)** half-wave retardation voltage

circular polarized after passing the cell, and in the second case the output beam is linearly polarized; however, the plane of polarization has been rotated 90°.

The two most common arrangements for Q-switching are shown in Fig. 8.10. In Fig. 8.10a the electrooptic cell is located between a polarizer and the rear mirror. The inclusion of the polarizer is not essential if the laser radiation is polarized, such as, for example, in Nd:YLF. The sequence of operation is as follows: During the flashlamp pulse, a voltage $V_{1/4}$ is applied to the electrooptic cell such that the linearly polarized light passed through the polarizer is circularly polarized. After being reflected at the mirror, the radiation again passes through the electrooptic cell and undergoes another $\lambda/4$ retardation, becoming linearly polarized but at 90° to its original direction. This radiation is ejected from the laser cavity by the polarizer, thus preventing optical feedback. Towards the end of the flashlamp pulse the voltage on the cell is switched off, permitting the polarizer-cell combination to pass a linearly polarized beam without loss. Oscillation within the cavity will build up, and after a short delay a Q-switch pulse will be emitted from the cavity.

In the arrangement of Fig. 8.10b, an electric voltage must first be applied to the cell to transmit the beam. In this so-called pulse-on Q-switch, the cell is located between two crossed polarizers. As before, polarizer P_1, located between the laser rod and the cell, is not required if the active medium emits a polarized beam. During the flashlamp pulse, with no voltage applied to the cell, the cavity Q is at a minimum due to the crossed polarizers. After a period of time a voltage

$V_{1/2}$ is applied to the cell which causes a 90° rotation of the incoming beam. The light is therefore transmitted by the second polarizer P_2. Upon reflection at the mirror the light passes again through polarizer P_2 and the cell, where it experiences another 90° rotation. Light traveling toward the polarizer P_1 has experienced a 180° rotation and is therefore transmitted through P_1.

Two types of electrooptic effects have been utilized in laser Q-switches: the Pockels effect, which occurs in crystals which lack a center of point symmetry, and the Kerr effect, which occurs in certain liquids. Pockels cells, which require a factor of 5 to 10 lower voltage than Kerr cells, are the most widely used active devices for Q-switching pulsed lasers.

Pockels Cell Q-Switch

Q-switches are classified into two types: cells in which the electric field is applied along the direction of the optical beam, and cells in which the electric field is perpendicular to the direction of the optical beam [8.21, 22].

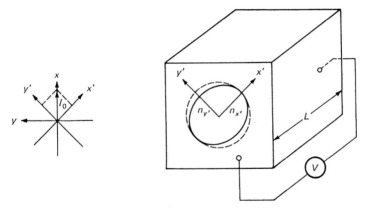

Fig. 8.11. Change of the index ellipsoid in a KDP crystal when an electric field is applied parallel to the z axis, I_0 is an incident wave polarized in the x direction; x and y are the crystallographic axes; x' and y' are the electrically induced axes

Pockels Cells with Longitudinal Field. In this class of Q-switches the electric field is applied parallel to the crystal optical axis and in the same direction as the incident light. The dependence of the index of refraction on the electric field can be described in terms of a change in orientation and dimensions of the index ellipsoid. The crystals used for longitudinal electrooptic Q-switches are uniaxial in the absence of an electric field; that is, there is only one value of refractive index in the direction of light propagation [8.23–27]. The index ellipsoid is an ellipse of revolution about the optic (z) axis. As indicated in Fig. 8.11, the index ellipsoid projects as a circle on a plane perpendicular to the optic axis. The circle indicates that the crystal is not birefringent in the direction of the optic axis. When an electric field is applied parallel to the crystal optic axis; the cross section of the ellipsoid becomes an ellipse with axis x' and y', making a 45° angle with the x and y crystallographic axes. This angle is independent of the magnitude of the electric field. The length of the ellipse axes in the x' and y' directions are proportional to the reciprocals of the indices of refraction in these two directions.

We will now express the phase shift between the orthogonal components corresponding to a wave polarized in the x' and y' directions. Changes of the refractive index Δn are related by the electrooptic tensor r_{ij} of rank 3 to the applied field.

$$\Delta\left(\frac{1}{n_i^2}\right) = \sum_{j=1}^{3} r_{ij}E_j \ , \tag{8.20}$$

where $i = 1, \ldots, 6$ and $j = 1, \ldots, 3$.

Generally there exist 18 linear electrooptic coefficients r_{ij}. However, in crystals of high symmetry, many of these vanish. For phosphates of the KDP family, r_{63} is the only independent electrooptic coefficient which describes the changes in the ellipsoid when a longitudinal field is applied to the crystal. The change of refractive index in the x' and y' directions is [8.25]

$$n_{x'} = n_0 + \tfrac{1}{2}n_0^3 r_{63}E_z \ , \quad n_{y'} = n_0 - \tfrac{1}{2}n_0^3 r_{63}E_z \ , \tag{8.21}$$

where n_0 is the ordinary index of refraction and E_z is the electric field in the z direction. The difference in the index of refraction for the two orthogonal components is then

$$\Delta n = n_0^3 r_{63}E_z \ , \tag{8.22}$$

For a crystal of length l, this leads to a path-length difference Δnl and a phase difference of $\delta = (2\pi/\lambda)\Delta nl$.

The phase difference δ in a crystal of length l is related to the voltage $V_z = E_z l$ applied across the faces by

$$\delta = \frac{2\pi}{\lambda}n_0^3 r_{63}V_z \ . \tag{8.23}$$

It should be noted that δ is a linear function of voltage and is independent of the crystal dimensions. If linearly polarized light is propagated through the crystal with the direction of polarization parallel to the x or y axis, as shown in Fig. 8.11, the components of this vector parallel to the electrically induced axes x' and y' will suffer a relative phase shift δ. In general, orthogonal components undergoing a relative phase shift produce elliptically polarized waves. Thus, the application of voltage in this configuration changes linearly polarized light to elliptically polarized light. If the light then passes through a polarizer, the resulting light intensity will be a function of the ellipticity and therefore the voltage applied to the crystal. A simple derivation shows that, with the analyzer axis oriented at right angles to the input polarization direction, the voltage and the transmitted light intensity I are related by [8.20]

$$I = I_0 \sin^2 \frac{\delta}{2} , \qquad (8.24)$$

where I_0 is the input light intensity.

For Q-switch operation, two particular values of phase shift are of interest; these are the $\lambda/4$ and $\lambda/2$ wave retardations which correspond to a phase shift of $\frac{\pi}{2}$ and π. With linearly polarized light being applied, for example, in the x direction, as shown in Fig. 8.11, the output from the crystal is circularly polarized if $\delta = \frac{\pi}{2}$. For $\delta = \pi$ the output beam is linearly polarized, but the plane of polarization has been rotated $90°$.

From (8.23) it follows that the voltage required to produce a retardation of π is

$$V_{1/2} = \frac{\lambda}{2n_0^3 r_{63}} . \qquad (8.25)$$

Deviations from this voltage will change the transmission $T = I/I_0$ of the Pockels cell according to

$$T = \sin^2 \left(\frac{\pi}{2} \frac{V}{V_{1/2}} \right) . \qquad (8.26)$$

This equation is obtained by combining (8.23–25).

A number of crystals which have been used in Q-switches and electro-optic modulators are listed in Table 8.1. The most widely used crystal is the deuterated form of KH_2PO_4 (KDP) which is KD_2PO_4 (KD*P) because it possesses a low half-wave voltage. For this reason almost all commercially available Pockels cell Q-switches with longitudinal field employ KD*P crystals. The crystal CD*A has actually a lower value, but this advantage is offset by a high loss tangent. KD*P and its isomorphs have been characterized in great detail, their properties are reviewed in [8.21, 27–33]. They are grown at room temperature from a water solution and are free of the strains often found in crystals grown at high temperature. Excellent crystals as large as 5 cm in any dimension can be obtained commercially at nominal cost. Although the crystals are water-soluble and fragile, they can be handled, cut, and polished without difficulty. The crystals have a high optical transmission in the range 0.22 to 1.6 μm, and they possess relatively large electrooptic coefficients. In KDP the half-wave voltage is of the order of 7.5 kV. By using KD*P this figure is reduced by a factor of more than 2. The half-wave voltage can be further reduced by a factor of 2 by utilizing two crystals optically in series and electrically in parallel.

All crystals listed in Table 8.1 are hygroscopic and must be protected from atmospheric water. This protection is typically provided by enclosing the crystal in a cell which is hermetically sealed or filled with index-matching fluid. As a

Table 8.1. Electrooptic parameters of $\bar{4}2m$-type crystals

Material	Index of refraction at 0.55 μm	Electrooptic constant, τ_{63} [μm/V$\times 10^{-6}$]	Typical half-wave voltage [kV] at 0.55 μm
Ammonium dihydrogen phosphate (ADP)	1.53	8.5	9.2
Potassium dihydrogen phosphate (KDP)	1.51	10.5	7.5
Deuterated KDP	1.51	26.4	3.0
Rubidium dihydrogen phosphate (RDP)	1.51	15.5	5.1
Ammonium dihydrogen arsenate (ADA)	1.58	9.2	7.2
Potassium dihydrogen arsenate (KDA)	1.57	10.9	6.5
Rubidium dihydrogen arsenate (RDA)	1.56	14.8	4.9
Cesium dihydrogen arsenate (CDA)	1.57	18.6	3.8
Deuterated CDA	1.57	36.6	2.0

result, six surfaces are encountered by a laser beam on a single transit through the cell. Transmission losses can be reduced by the use of antireflection coatings on the cell windows. While hard, damage-resistant AR coatings cannot be applied to KDP and its isomorphs, losses at the crystal itself can be very much reduced by the use of an index-matching liquid such as 3M's FC77. Other techniques employed to minimize insertion losses of Pockels cells include a Brewster-angle design [8.34] and uncoated crystals with plane-parallel faces accurately aligned to the cavity mirrors [8.35].

Transmission of typical electrooptic shutters with liquid-immersed crystals and AR-coated windows is about 90 %. The electric field is usually applied to the crystal by means of a pair of metal electrodes containing apertures which are either bonded or evaporated onto the square ends of the crystal or simply held in place by compression. The main drawback of end-plate electrodes is their geometry, which gives rise to a nonuniform electric field across the clear aperture. The field strength in the aperture varies from a maximum around the inner edge of the rings to a minimum at the geometric center. Fringing necessitates operation at considerably higher voltages than if the field were uniformly applied. Partial compensation is attained by making the crystal length roughly 30 % greater than the clear aperture diameter.

Typical commercially available Pockels cells employing KD*P have quarter-wave voltages between 3.5 and 4 kV at 0.69 μm and 5 to 6 kV at 1.06 μm. Apertures range from 9 to 25 mm, and the maximum safe operating level for most KD*P Pockels cell Q-switches is around 200 MW/cm^2.

A considerable improvement in fringe uniformity can be achieved by using cylindrical band electrodes applied to the end of the barrel of cylindrical-shaped

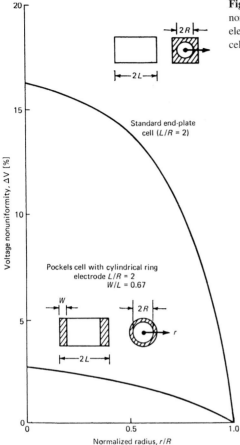

Fig. 8.12. Voltage nonuniformity ΔV versus normalized radius r/R for a cylindrical ring electrode (*lower curve*) and a standard end-plate cell [8.37]

crystals, as shown in Fig. 8.12. Band electrodes allow an optical transmission uniformity to within a few percent across the clear aperture. This represents an order-of-magnitude improvement compared to standard cells containing rect-angular crystals and end-plate electrodes. The Fig. 8.12 also shows calculated curves of the voltage nonuniformity ΔV as a function of the normalized radius r/R for a conventional endplate cell and a cylindrical ring electrode cell. The value of ΔV is the amount by which the voltage on the electrodes must be raised to obtain full half-wave retardation at a point r/R as compared to $r = R$. The more uniform electric field distribution within the crystal having cylindrical band electrodes results in higher extinction ratios and lower half-wave voltages. By use of low-reactance connecting leads, rise times in Pockels cells on the order of 500 ps and shorter have been achieved. In designing fast switching circuits, the impedance of a Pockels cell is a critical parameter. Since electro-optical Q-switches are comprised of insulating dielectrics, the impedance of a Pockels cell is determined primarily by its capacitance. Calculations of the capacitance for a variety of electrode configurations can be found in [8.36, 37].

Pockels Cells with Transverse Fields. In this configuration, the electric field is perpendicular to the direction of the beam. For this geometry, the half-wave voltage depends on the ratio of thickness to length of the crystal, which has the advantage that by proper choice of the crystal geometry a considerably lower voltage is required as compared to longitudinally applied fields. In Fig. 8.13a a KDP crystal is shown where the optical propagation direction is at $45°$ to the x and y axes while the field is applied along the z axis. The half-wave voltage $V_{1/2}$ for this case is given by

$$V_{1/2} = \frac{\lambda d}{n_o^3 r_{63} l} , \qquad (8.27)$$

where l and d are the crystal length and thickness, respectively. The light is linearly polarized at $45°$ to the z axis. In another arrangement, the optical propagation direction in the KDP crystal is at an angle of $45°$ to the y and z axes while the electric field is applied along the x axis, as shown in Fig. 8.13b. If the light is polarized at $45°$ to the x axis, then the half-wave voltage is given by [8.20]

$$V_{1/2} = \left(\frac{1}{n_o^2} + \frac{1}{n_e^2} \right)^{3/2} \frac{\lambda}{r_{41}} \frac{1}{2(2)^{1/2}} \frac{d}{l} , \qquad (8.28)$$

where n_o and n_e are the ordinary and extraordinary indices of refraction, respectively.

Since propagation in the $45°$ z cut or $45°$ y cut KDP and its isomorph is at an angle other than parallel or normal to the optic axis, the extraordinary and ordinary ray propagate in different directions through the crystal. To compensate for this walk-off as well as for temperature effects, two crystals of suitable orientation are usually employed [8.38–43]. Because of the need to compensate for the angular and temperature dependence of birefringence in KDP-type crystals, transverse field devices have not become popular as Q-switches in this class of materials.

However, the material LiNbO$_3$ has found widespread applications as a Q-switch [8.44–48]. This crystal, unlike KD*P, is not hygroscopic and can be operated in a transverse electrode orientation with the light propagation along the optic axis. It does not require the sealed-cell assemblies necessary for KD*P crystals, therefore transmission of an AR-coated LiNbO$_3$ crystal can be as high as 98 %. In an arrangement as shown in Fig. 8.13c, light propagates along the c axis of the crystal. If the light is polarized parallel to the a axis and an electric field is applied parallel to the a axis, the half-wave retardation is

$$V_{1/2} = \frac{\lambda d}{2 r_{22} n_o^3 l} , \qquad (8.29)$$

where l is the length of the crystal in the c direction, d is the distance between the electrodes along the a axis, and r_{22} is the electrooptic coefficient.

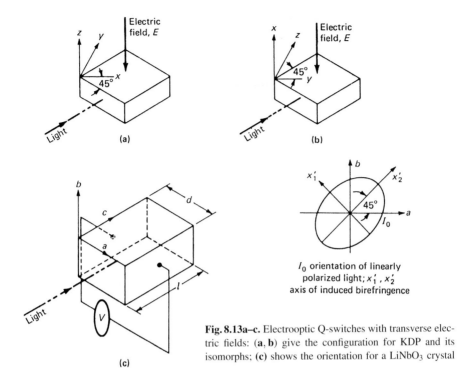

I_0 orientation of linearly
polarized light; x_1', x_2'
axis of induced birefringence

Fig. 8.13a–c. Electrooptic Q-switches with transverse electric fields: (**a, b**) give the configuration for KDP and its isomorphs; (**c**) shows the orientation for a LiNbO₃ crystal

The half-wave voltage is directly proportional to the distance between the electrodes and inversely proportional to the path length. At a wavelength of $\lambda = 1.064\,\mu m$, the linear electrooptic coefficient is $r_{22} = 5.61 \times 10^{-6}\,\mu m/V$ and the refractive index of the ordinary ray is $n_o = 2.237$ in LiNbO₃. The theoretical half-wave voltage obtained from (8.29) for a typical crystal size of $9\,mm \times 9\,mm \times 25\,mm$ is 3025 V. This is the value for a dc bias voltage; for pulsed Q-switch operation a 30 to 40% higher voltage is required [8.45]. In pulsed operations, the value of r_{22} is smaller than for dc operation, which is explained by the fact that for static fields the electrooptic effect is equal to the sum of the intrinsic electrooptic effect and that induced by the piezooptic effect [8.49, 50]. $r_{22} = r_{22}' + p_{2k}d_{2k}$, where p_{2k} and d_{2k} are the elastooptic and piezoelectric constants of the crystal. When the bias voltage is turned on or off rapidly, the piezooptic effect is absent, since deformation of the crystal cannot occur during the fast switching times typical for Q-switch operation.

Since most military applications involve operation over a large temperature range, lithium niobate is also often chosen for its smaller variation in electrooptic coefficients. This feature minimizes the requirements for temperature and optical feedback stabilization controls. At the present time many military rangefinders and target designators utilize lithium niobate Q-switches in a high repetition rate (10 to 60 pps), low power density (5 to 10 MW/cm²) mode. In these applications crystal temperatures may reach 70° C. In contrast, while KD*P Q-switches can be operated at that temperature, they should not be used above 55° C when the electric field is applied continuously, because this leads to a fogging effect on the crystal surfaces. The major drawback of lithium niobate is that it is not suitable for high-power-density applications and users place a limit of between 10 to $50\,MW/cm^2$ on its peak optical power-handling capabilities. KD*P Q-switches are usually specified where large apertures, as large as 40 mm, and high peak power densities are involved. Q-switches employing $LiNbO_3$ crystals have apertures up to 9 mm and require half-wave voltages between 2.5 and 6 kV at $1.06\,\mu m$ depending on the crystal geometry.

Prelasing and Postlasing Occurring with Electrooptical Q-Switches. Theoretically, an infinite loss is introduced into the cavity by a Pockels cell operated at a proper $\lambda/4$ or $\lambda/2$ voltage. In practice, field strength nonuniformities, inhomogeneities in the crystal, birefringence introduced by mechanically clamping the crystal, and nonperfect polarizers limit the extinction ratio of an electrooptic Q-switch to about a hundred. The extinction ratio is defined as the ratio of the maximum to the minimum transmitted intensity obtained between crossed polarizers as the applied voltage is varied through a half-wave voltage.

If the gain of the laser exceeds the loss produced by the Q-switch, normal lasing will occur before the instant of Q-switching. This phenomenon, which is termed "prelasing", is due to the Pockels cell and polarizer combination not acting as a perfect shutter, so that there is still some feedback from the resonator mirror. Prelasing is most likely to occur just prior to the time of Q-switching since the population inversion, the stored energy, and the gain are largest at that time.

Because of the high gain of Nd : YAG, depopulation of the stored energy by prelasing is a particular problem with this material. Tests have also shown that prelasing may be a major cause of damage to the Pockels cell crystal or polarizers. Prelasing, if it occurs, allows a pulse buildup from a "seed" pulse in a small region of the laser rod. In the area of the laser rod where prelasing occurs, the Q-switched pulse develops more rapidly compared to the rest of the pulse, which must develop from spontaneous emission. As a result, a very high peak power density will occur in this small part of the rod.

Appropriate design precautions must be undertaken to ensure that prelasing does not occur. The first major requirement is to operate the Q-switch at its optimum extinction ratio. This requires that the correct dc bias voltage is applied and that the Pockels cell crystal and polarizers are aligned properly. Typically the optic c axis of the Pockels cell crystal must be parallel to the laser beam direction to within 10 arc min or less. Alignment of the c axis is best performed by centering the optic-axis figure (the Maltese-cross pattern) on the resonator optical axis. This is done by illuminating the crystal with a diffuse light source and observing the crystal between crossed polarizers. A pattern of a cross surrounded by a series of circles appears. The line connecting the center of the cross to the point at which the observation is made is exactly parallel to the c axis. Furthermore, to establish the proper condition for Q-switching, the crystal must be aligned so

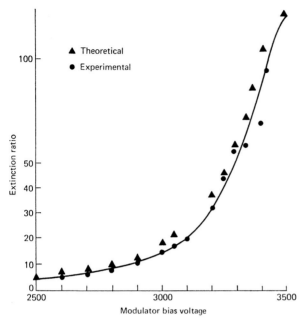

Fig. 8.14. Extinction ratio versus applied voltage of a LiNbO$_3$ Pockels cell-calcite prism assembly [8.50]

that either the a or b axis is parallel to the polarization direction of the laser (LiNbO$_3$ crystals and longitudinal field KD*P). Figure 8.14 shows the extinction ratio versus bias voltage for a 9 mm × 9 mm × 25 mm LiNbO$_3$ crystal [8.51].

The main Q-switched pulse may be followed by one or more pulses of lower amplitude (postlasing). The second pulse may occur from several hundred nanoseconds to several tens of microseconds after the main pulse. Postlasing results from piezooptic effects in the electrooptic crystal. The piezoelectric action of the applied voltage compresses the crystal, and when that voltage is removed the crystal remains compressed for some time. This compression generates a retardation of the optical wave by means of a strain-birefringence effect, thus creating a loss in the cavity which becomes smaller with time as the compression relaxes. Figure 8.15a shows a waveform picture of the voltage conventionally applied to a Pockels cell, and Fig. 8.15b shows the loss versus time for an actual Pockels cell switch. It was found [8.46] that the loss drops to about 25 % and then decays to zero in approximately 400 ns. If the resulting decay time of the loss is longer than the output pulse buildup time, the main laser pulse will be emitted before the loss reaches its minimum. Thus some energy will remain in the rod after the first output pulse. This residual energy may produce a second pulse when the cavity loss reaches its minimum. By creating this time-dependent loss, the elastooptic effect seriously affects the efficiency of a Q-switched laser.

At low input energy levels there is a long time delay between the switching time and the time the output pulse actually appears, as shown in Fig. 8.15c,

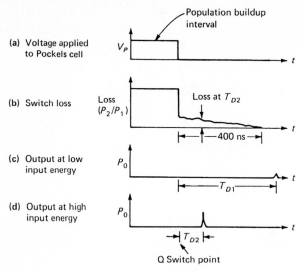

Fig. 8.15. Piezooptic effect in LiNbO$_3$ [8.46]

and thus there is negligible loss in efficiency. However, at higher input energy levels the time delay t_D becomes shorter and thus the laser suffers a considerable output loss due to the switch loss, as shown in Fig. 8.15d. This, besides prelasing, accounts for the roll-off in output efficiency in the Q-switched laser at higher input energy levels.

The elastooptic effect, which is quite pronounced in LiNbO$_3$ but is also observed in KD*P, can be minimized by switching the bias voltage to a negative value rather than to zero.

KTP, a crystal which is used extensively as an efficient nonlinear-optical crystal, has also been explored as a Q-switch because piezoelectric effects have not been observed in this material [8.52].

Examples of Pockels Cell Q-Switched Lasers. Output energies between 100 and 250 mJ are obtained from Nd:YAG oscillators. The energy extraction is limited by the high gain of this material, which leads to prelasing and subsequently to a depopulation of the inversion.

Ruby and Nd:glass laser oscillators produce, depending on size, between 0.1 and 10 J in the Q-switched mode. For example, a ruby rod 75 mm long and 6 mm in diameter pumped by a linear flashlamp at about 150 J will produce around 100 mJ of Q-switched output. Ruby and Nd:glass rods, 100 mm long and 10 mm in diameter pumped at 2000 J in a 500-μs flashlamp pulse, will typically generate pulses with an energy content of 2 J. Large rods, for example 150 mm by 20 mm, pumped around 5 kJ, will produce Q-switched pulses up to 10 J in energy. The width of the Q-switched pulses is usually between 10 and 25 ns, and the pulse buildup time t_D is between 50 and 100 ns in typical systems. In the

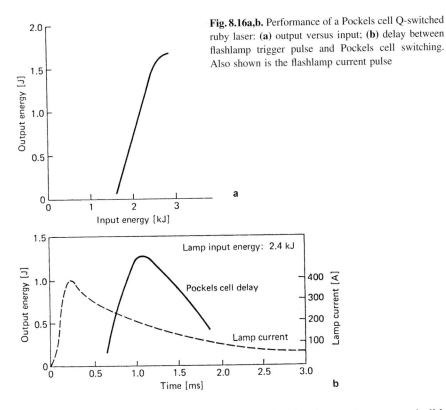

Fig. 8.16a,b. Performance of a Pockels cell Q-switched ruby laser: (**a**) output versus input; (**b**) delay between flashlamp trigger pulse and Pockels cell switching. Also shown is the flashlamp current pulse

larger systems, the energy output is actually limited by the maximum permissible power density on the optical components. In order to avoid optical damage, most commercial solid-state laser oscillators are operated at power densities around $150\,MW/cm^2$. Therefore, from an oscillator containing a 1-cm-diameter Nd:glass or ruby rod a peak power $\approx 100\,MW$ can be expected and $\approx 500\,MW$ from a large, 2-cm-diameter rod.

Figure 8.16a shows a plot of output versus input energy of a typical Q-switched ruby laser containing a 10-cm by 1-cm rod pumped by a helical flashlamp. In Fig. 8.16b the output from the laser at fixed lamp input is plotted as a function of the delay between the flashlamp trigger pulse and the opening of the Pockels cell. The maximum energy extraction is obtained for a delay of 1.1 ms. This time corresponds to a point near the end of the flashlamp pulse at which the lamp current has dropped to 50 % of its initial value. Prior to this time the stored energy in the rod increases because of the pumping action of the lamp, and after the optimum time of 1.1 ms, population inversion decreases because of the diminishing pump power and the increase of spontaneous emission losses.

Figure 8.17 shows the dependence of the pulse width of a Q-switched ruby laser on the energy output, with cavity length and output reflectivity as a parameter. According to (8.9), one would expect that the pulse width Δt_p increases for a system with a long cavity (large τ_c), a low gain (small n_i) and a low output

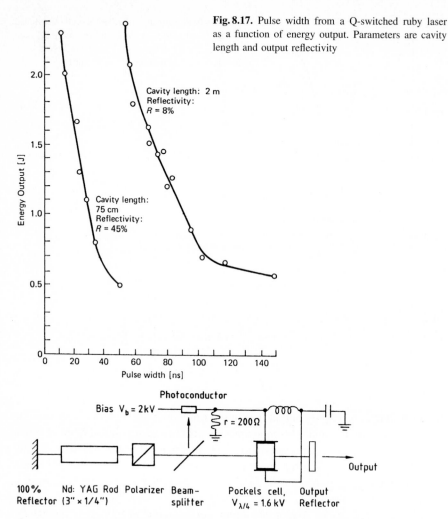

Fig. 8.17. Pulse width from a Q-switched ruby laser as a function of energy output. Parameters are cavity length and output reflectivity

Cavity length: 2 m
Reflectivity:
$R = 8\%$

Cavity length:
75 cm
Reflectivity:
$R = 45\%$

Energy Output [J]

Pulse width [ns]

Photoconductor

Bias $V_b = 2\,kV$

$r = 200\,\Omega$

Output

100%   Nd: YAG Rod   Polarizer   Beam–
Reflector   (3″ × 1/4″)   splitter

Pockels cell,   Output
$V_{\lambda/4} = 1.6\,kV$ Reflector

Fig. 8.18. Electrooptic feedback control for stretching of Q-switched pulses [8.53]

reflectivity (high n_t). The experimental results illustrate this dependence of the pulse width on the system parameters. Decreasing the output energy, achieved by reducing the input power, will result in a low gain and low initial inversion n_i. By combining the effects of low gain, high threshold, and long cavity, pulses up to 150 ns can be produced from Q-switched lasers.

Employing a feedback loop to control the switching of a Pockels cell, pulse durations of up to 1.4 μs have been obtained from Q-switched Nd : YAG lasers [8.53, 54]. The principle of electrooptical feedback control is depicted in Fig. 8.18. The resonator incorporates a Pockels cell as a Q-switching element. Negative feedback to the circulating power is applied by means of the photo-detector-derived voltage on the Pockels cell. In this way the stored energy in the laser rod is released at a controlled rate.

Optimization of Pockels Cell Q-Switched Lasers in the Presence of Thermally Induced Birefringence In low-repetition-rate systems, or at low input powers depolarization losses can be ignored, but in higher-average-power systems depolarization losses become significant.

The polarizer, required for Q-switch operation employing a Pockels cell, rejects any radiation not polarized in the proper plane of polarization. This can lead to large depolarization losses in the presence of thermally-induced birefringence, as may occur in high-repetition-rate systems.

Basically the same techniques, as dicussed in Sect. 7.1.1 can be employed in order to reduce depolarization losses in electro-optic Q-switched resonators. In large Q-switched oscillators, two laser heads with a 90° phase rotator inserted between the laser rods are sometimes used. In an oscillator containing a single laser head, a Faraday rotator inserted between the laser rod and rear mirror can be employed. Smaller Q-switched oscillators, in particular military lasers, employ mostly a waveplate-Porro prism combination for birefringence compensation. This approach will be discussed in more detail at the end of this section.

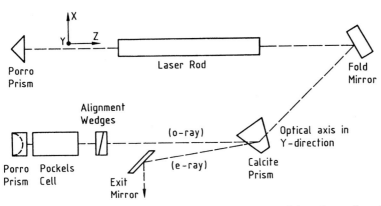

Fig. 8.19. Variable reflectivity resonator which allows extraction of the orthogonally polarized beam

Also several techniques have been developed in conjunction with Pockels-cell Q-switches which act on both polarizations. In general, the techniques are based on the use of a calcite polarizer in the resonator which separates the two orthogonally polarized beams. The methods differ in the ways these two beams are treated in the resonator. In the design depicted in Fig. 8.19, the depolarized component of the intracavity radiation, rather than representing a loss, becomes part of the output from the laser [8.55]. The resonator is comprised of a pair

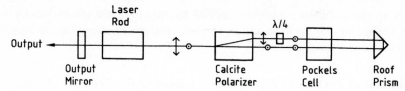

Fig. 8.20. Polarization insensitive Q-switch based on spatial separation of orthogonally polarized beams

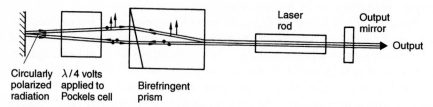

Fig. 8.21. Polarization insensitive Q-switch based on angular separation of orthogonally polarized beams

of Porro-prisms and a calcite prism, which acts as polarizer and also allows extraction of the depolarized component of the laser radiation. Rotation of one of the Porro prisms in azimuth will change the ratio of the power in the ordinary and extraordinary beam and therefore represents a means for optimizing the output coupling.

Rankin et al. [8.56] used a calcite polarizer which separates the two orthogonal polarization components into two parallel beams. A quarter-wave plate is inserted into one of the beams, as shown in Fig. 8.20, before both beams are passed through a large aperture Pockels cell. A roof prism returns the beams back to the Pockels cell/polarizer assembly.

A design which exploits directional differences between the two orthogonal beam components after passing through a birefringent prism has been proposed in [8.57]. A schematic diagram of the laser is shown in Fig. 8.21. The resonator contains a birefringent prism that replaces the linear polarizer commonly used in Q-switched lasers. After passing through the birefringent prism, radiation is split up into two orthogonally polarized components, the extraordinary (e) and the ordinary (o) rays, that propagate in two slightly different directions. When these two rays are reflected by the back mirror, a further separation of the two rays will occur. In this case the losses in the cavity will be very high and laser action will be suppressed. However, if a quarter-wave voltage is applied to the Pockels cell, the e and o rays returning to the prism will be interchanged. In this case, the walk off occurring in the first pass will be canceled by an equal and opposite walk off during the return pass, hence the beam returning to the laser rod will have low losses, allowing a Q-switched pulse to develop.

A very common resonator in military systems is a polarization coupled design such as depicted in Fig. 8.22. The resonator typically contains one or two Porro prisms to take advantage of the alignment insensitivity of these components. If

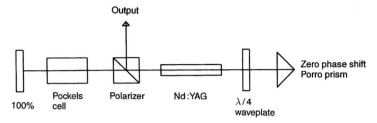

Fig. 8.22. Polarization output coupled resonator with birefringence compensation

such a laser is operated at high average power, the depolarization caused by thermally induced birefringence will drastically distort the output beam profile. An output beam resembling a Maltese cross will be obtained from a thermally stressed laser rod. However, in such a resonator, a birefringence compensation technique is possible by adding to the Porro prism a waveplate with the proper phase retardation and orientation.

Like the compensation with a Faraday rotator between the rear mirror and laser rod, or the placement of a 90° rotator between two laser rods, this technique relies on the fact that two passes occur in the thermally stressed laser rod before the beam is passed through the linear polarizer. However, there is a difference with regard to the above mentioned techniques. The phase difference between radial and tangential rays caused by birefringence is not cancelled by two passes of each ray through the same region of the laser rod, but by passing through two different, but symmetrically located points in each half of the rod. Each half being defined by the Porro prism bisecting plane. If the prism is oriented with its roof edge bisecting the rod axis, then each ray passing through one part in the rod is returning at the opposite side of the bisecting plane through a section which is symmetrically opposite and at the same radial distance from the rod center. In a uniformly pumped laser rod, birefringence is radially symmetric and the amount of phase distortion depends only on the radius.

By changing the polarization state of the beam between the two passes, it is possible to cancel the relative phase difference between the tangential and radial components of polarization that occur in the first pass with equal and opposite phase changes in the return pass. The appropriate change of polarization between the two passes can be accomplished with a waveplate or optical rotator in front of the Porro prism [8.59, 60]. For example, the combination of a zero-phase-shift Porro prism and a quarter-waveplate with an orientation of 45° between its fast axis and the prism apex can accomplish this task. Any thermally induced depolarization effects that occur in the first pass are removed in the return path [8.59].

A Porro prism with zero phase shift (Fig. 8.22) is not a common optical component. As was explained in Chap. 5.1, Porro prisms introduce a phase shift to a linearly polarized beam which depends on the refractive index and the angle between the plane of polarization and the roof edge. A Porro prism which does not introduce a phase shift requires dielectric coatings on the totally reflecting prism surfaces. Another combination, which results in a cancellation of the thermally induced depolarization is a half-waveplate and a π-shift Porro prism at a relative orientation of 22.5°. Again the Porro prism requires dielectric coatings on the internally reflecting surfaces. The combination of a quarter- or half-waveplate and a Porro prism are only special cases of many possible combinations. They are of interest because these waveplates are readily available, however the use of either of these waveplates requires a specially dielectrically coated Porro prism. Using the Jones matrix method, a waveplate with a suitable phase change and orientation can be found for any uncoated Porro prism in order to achieve birefringence compensation. In the references [8.59, 60], curves are provided for waveplate-phase angle and Porro prism refractive index combinations which permit the compensation of thermally induced depolarization in a laser rod.

The output coupling of the resonator, depicted in Fig. 8.22, can be optimized by a rotation of the Porro prism-waveplate combination while maintaining the relative angular position between these two components. As already mentioned, a change of the angle between the plane of polarization and the roof edge introduces a phase change in the return beam. However, in most situations it is found that it is more convenient to add a second waveplate to the resonator for control of the output coupling. Also, the 100% mirror shown in Fig. 8.22 can be replaced by a second Porro prism to add alignment stability in two directions. In addition, space constraints drive the design very often to a folded configuration, shown for example in Fig. 5.31.

In high-average power Q-switched Nd:YAG lasers thermally induced bire-fringence in the Q-switch crystal can be a problem. Similar to the cancellation of thermally induced depolarization in a laser rod, the problem can be solved with two Pockels cells and a polarization rotator inserted between the two electro-optic crystals [8.58].

Kerr Cell Q-Switch

Like the Pockels cell, the Kerr cell is a device which can produce a controllable birefringence by the application of a voltage to a cell. In this case, the cell contains a liquid, usually nitrobenzene, instead of a crystalline solid. In the Kerr cell the birefringence is proportional to the square of the applied voltage. The difference between the index of refraction for light polarized parallel and orthogonal to the direction of the inducing field is given by

$$\Delta n = \lambda B E^2 , \tag{8.30}$$

where B is the Kerr constant and E is the transverse field strength. From (8.30) it follows that the phase difference is

$$\delta = 2\pi l B E^2 , \tag{8.31}$$

where l is the length of the region in which the transverse field exists. Since $E = V/d$ for parallel plates, where d is the electrode separation and $l = 2L$ because the light in a resonator travels twice through the cell of length L, we obtain for the half-wave retardation voltage $(\delta = \pi)$

$$V_{1/2} = \frac{d}{2(BL)^{1/2}} \; . \tag{8.32}$$

For nitrobenzene, $B = 26 \times 10^{-6}$ cgs at 6943 Å and the half-wave voltage becomes

$$V_{1/2} = \frac{30d}{(L)^{1/2}} \; , \tag{8.33}$$

where d and L are in centimeters and V is in kilovolts. For typical values of d and L, V must be 10 to 40 kV. The axes of induced birefringence in a Kerr cell are at 45° with respect to the applied field. Because of the much higher voltage requirement of Kerr cells as compared to Pockels cells, problems associated with high leakage currents due to impurities in the nitrobenzene, and difficulties in constructing leakproof cells, this type of Q-switch has only been used in some of the very early lasers.

Drivers for Electrooptic Q-Switches

In the operation of an electrooptically Q-switched laser, it is necessary to switch voltages electrically in the range between 1.5 kV and 15 kV depending on the crystal material, geometry, and optical wavelength. The driver for the Pockels cell must be a high-speed, high-voltage switch which also must deliver a sizeable current. The cell has a few tens of picofarads capacitance which is charged (or discharged) to several kilovolts in a few nanoseconds. The resulting current is of the order of 10 to 20 A. Common switching techniques include the use of vacuum tubes, cold cathode tubes, thyratrons, SCRs, and avalanche transistors.

Vacuum tubes are usually applied if a rapid sequence of Q-switch pulses must be generated for holographic applications, for example; or if a particular waveform of the electronic signal is desired, for example, to achieve longitudinal mode selection [8.61]. A cold cathode gas tube, such as the EG&G KN-6 krytron, is an attractive device, being of reasonable size and having a very fast switching time. Nevertheless, it is a gas tube with a limited lifetime, and the operating voltage is restricted to values below 8 kV. A typical circuit diagram for a Pockels cell driver using a krytron is shown in Fig. 8.23a. The LiNbO$_3$ crystal operated at a half-wave retardation voltage of 3.3 kV is switched to a negative bias voltage of 800 V to compensate for the piezoelectric effect. The bias voltage is of the same polarity as that of the holdoff voltage but applied to the opposite electrode. The net result is that when the holdoff voltage is removed (at the instant of Q-switching), the bias across the modulator becomes negative. In systems which must operate over a large temperature range, such as military systems, a circuit

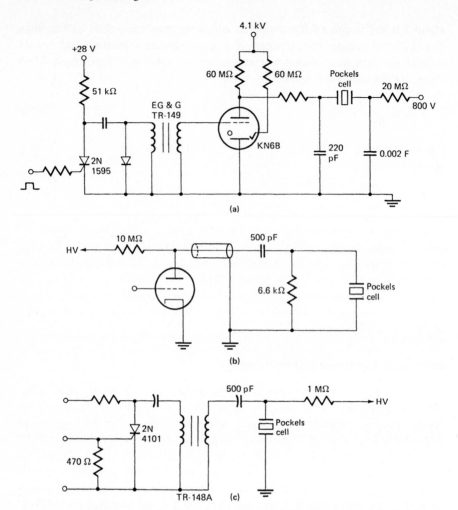

Fig. 8.23. Circuit diagram of Pockels cell drive electronics using **(a)** a krytron; **(b)** a hydrogen thyratron; **(c)** an SCR

is usually incorporated into the system which controls the quarter-wave voltage level at the krytron high-voltage supply.

Hydrogen thyratrons can be operated at higher voltages than krytrons. A circuit for operation of a cell at the pulse-on mode is shown in Fig. 8.23b. When the thyratron conducts, it short-circuits the HV cable between the thyratron and Pockels cell, inducing a negative high-voltage equivalent to the half-wave voltage across the very-high-impedance Pockels cell. The voltage remains nearly constant for a time equal to twice the cable transit time. The voltage across the cell then decays exponentially.

The most important electro-optic Q-switch material is $LiNbO_3$ employed in Nd : YAG or Nd : YLF lasers. Operated at the 1/4-wave voltage, the switch requires 1500 V. Semiconductor devices such as SCR'S, avalanche transistors, and MOSFET's, have been successfully employed to drive this Q-switch.

One approach is to use a voltage of several hundred volts which is stepped up to the 1/4 -wave voltage by a high-frequency step-up transformer. A transformer-driven Pockels cell switched by a SCR is presented in Fig. 8.23c [8.48].

Another approach of switching a $LiNbO_3$ Pockels cell is to use a Marx bank which can have as many as 12 stages of high-voltage transistors and capacitors in series. This design charges the capacitors in parallel and discharges them in series. The advantage of the Marx-bank driver, similar to the step-up transformer approach, is the fact that only a power supply of a few hundred volts is required instead of the full 1/4-wave voltage. On the other hand, the multiple stages contain a large number of components which raises a reliability issue.

A design based on avalanche transistors, uses high-voltage transistors and relies heavily on the transistor electrical breakdown voltage. For a $LiNbO_3$ Q-switch, the design also requires as many as 12 transistor stages in series. Again, all these stages lead to reliability concerns.

Advanced and very reliable Q-switch drivers are based on MOSFET derived systems. These devices have operating voltages up to 2 kV. Therefore, two MOS-FET's in series can handle most Q-switch applications. We have found at our company, that MOSFET based drivers use far fewer components, compared to other designs, and operate very reliably without failure to over 10^9 shots.

8.4 Acoustooptic Q-Switches

In acoustooptic Q-switches, an ultrasonic wave is launched into a block of transparent optical material, usually fused silica. A transparent material acts like an optical phase grating when an ultrasonic wave passes through it. This is due to the photoelastic effect, which couples the modulating strain field of the ultrasonic wave to the optical index of refraction [8.62–67]. The resultant grating has a period equal to the acoustic wavelength and an amplitude proportional to the sound amplitude.

If a light beam is incident upon this grating, a portion of the intensity will be diffracted out of the beam into one or more discrete directions (Fig. 8.24). By properly choosing the parameters, the diffracted beam can be deflected out of the laser resonant cavity, thereby providing an energy loss which is sufficient to Q-spoil the cavity.

The ultrasonic wave is typically launched into the Q-switch block by a piezo-electric transducer which converts electrical energy into ultrasonic energy. The laser is returned to the high Q-state by switching off the driving voltage to the transducer. With no ultrasonic wave propagating through it, the fused silica block returns to its usual state of high optical transmission, and a Q-switch pulse is emitted [8.9, 68–70]. Two different types of diffraction effects are observed depending on the optical and acoustic wavelengths λ and Δ, and the distance l over which the light interacts with the acoustic beam.

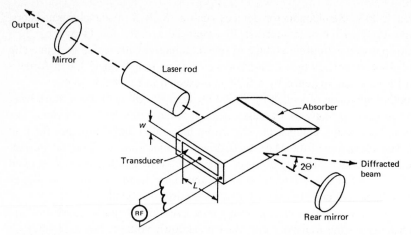

Fig. 8.24. Acoustooptic Q-switch employed in a cw-pumped Nd:YAG laser. The major components of the Q-switch are a very thin quartz crystal transducer having a thickness of a half acoustic wavelength, a fused silica block to which the transducer is epoxy-bonded, an inductive impedance-matching network, an acoustical absorber at the side opposite to the transducer, a water-cooling jacket system to conduct away generated heat, a thermal interlock which automatically turns off the rf power if the temperature in the modulator rises beyond allowable limits, a Bragg angle adjustment, a 50 MHz rf driver containing a pulse generator, and logic which allows adjustment of the repetition rate from 0 to 50 kHz

Raman-Nath Scattering

Raman-Nath scattering occurs when either the interaction path is very short or when the ultrasonic frequency is very low, i.e.,

$$l\lambda \ll \Delta^2 . \tag{8.34}$$

Maximum light scattering is observed if the light beam and ultrasonic wave are perpendicular with respect to each other. The light beam is scattered symmetrically in many higher diffraction orders, as shown in Fig. 8.25a. The intensity of the individual orders is [8.66]

$$\frac{I_n}{I_0} = J_n^2(\Delta\phi) , \tag{8.35}$$

where I_n is the intensity of the nth order, I_0 is the intensity of the incident light, J_n is the Bessel function of nth order, and $\Delta\phi = 2\pi\Delta n l/\lambda$ is the amplitude of the phase grating. The amplitude $\Delta\phi$ has been shown to be [8.63, 71]

$$\Delta\phi = \pi \left(\frac{2}{\lambda_o^2} \frac{l}{w} M_2 P_{ac} \right)^{1/2} , \tag{8.36}$$

where λ_o is the optical wavelength, P_{ac} is the acoustic power, l and w are the dimensions of the flat rectangular transducer, and M_2 includes a group of material parameters known as the acoustooptic figure of merit,

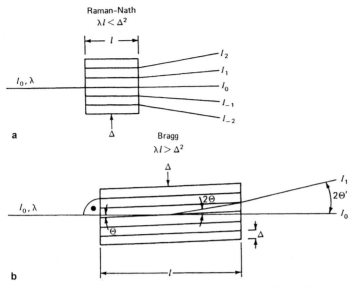

Fig. 8.25. (a) Raman-Nath and (b) Bragg angle acoustooptic Q-switches

$$M_2 = \frac{n^6 p^2}{\varrho v^3} \, , \tag{8.37}$$

where n is the index of refraction, p is the appropriate photoelastic coefficient, ϱ is the density, and v is the acoustic velocity.

Bragg Scattering

When the frequency of the ultrasonic wave is raised and the interaction path is lengthened, higher-order diffraction is eliminated and only two light beams of zero and first order become predominant. The condition for Bragg scattering to occur is

$$l\lambda \gg \Delta^2 \, . \tag{8.38}$$

In the Bragg regime the optical and ultrasonic beams are offset slightly from normal incidence to interact at the Bragg angle

$$\sin \Theta = \frac{\lambda}{2\Delta} \, . \tag{8.39}$$

It should be noted that λ, Δ, and Θ are measured inside the medium (that is, $\lambda = \lambda_0/n$). As shown in Fig. 8.25b, if measured externally the angle between the incident light and the acoustic wave is $\Theta' = n\Theta$ and the scattering angle is twice that, i.e.,

$$2\Theta' = 2n\Theta \approx \frac{\lambda_0}{\Delta} \, . \tag{8.40}$$

The intensity I_1 of the scattered beam is [8.71]

$$\frac{I_1}{I_0} = \sin^2\left(\frac{\Delta\phi}{2}\right) , \tag{8.41}$$

where the phase amplitude is the same as defined in (8.36).

Depth of Modulation

It is apparent from (8.35–41) that the amount of diffracted power depends on the material parameters expressed by M_2, the ratio of length to width of the interaction path, and the acoustical power P_{ac}. In a given material such as, for example, fused silica, the value of the photoelastic coefficient p in (8.37) depends on the plane of polarization of the light beam with respect to the ultrasonic propagation direction and on the type of ultrasonic wave, i.e., longitudinal or shear wave. With shear wave generation the particle motion is transverse to the direction of the acoustic wave propagation direction. In this case the dynamic optical loss is independent of polarization in isotropic materials such as fused quartz [8.65, 72]. Table 8.2 lists the pertinent material parameters for an acoustooptic Q-switch fabricated from fused silica.

Since at a small depth of modulation the sine function in (8.41) can be taken equal to its argument, the diffracted power is proportional to the figure of merit M_2. From Table 8.2 it follows, that in a Bragg angle device employing longitudinal-mode ultrasound, light polarized perpendicular to the acoustic wave vector is deflected five times stronger than light polarized parallel to this direction. The shear wave device, which operates independent of the light polarization, has a higher diffraction efficiency than parallel-polarized light in a longitudinal device, but a substantially lower efficiency compared to the perpendicular-polarized light in a longitudinal modulator.

Table 8.2. Material parameters of acoustooptic Q-switches employing fused silica (Bragg and Raman-Nath devices)

Acoustic wave	p coefficient	Polarization of optical beam with respect to acoustic wave vector	Velocity of sound $\times 10^5$ [cm/s]	Figure of merit $\times 10^{-18}$ [s^3/g]	Acoustical power P_{ac} [W] for 1% deflection ($l/w = 10$)
Shear wave	$p_{44} = 0.075$	Independent	3.76	0.47	0.42
Longitudinal	$p_{11} = 0.121$	Parallel	5.95	0.30	0.67
Longitudinal	$p_{12} = 0.270$	Perpendicular	5.95	1.51	0.13

The fraction of incident light which is scattered by the acoustooptic modulator determines whether the Q-switch can hold off laser action. In order to Q-switch an unpolarized laser system, such as Nd : YAG, the lower of the efficiency factors of a longitudinal device determines the extinction ratio of the device. Therefore, unpolarized lasers are usually Q-switched with shear-wave devices, whereas for polarized laser radiation longitudinal modulators are employed. Because a longitudinal Q-switch in which the large p_{12} coefficient is utilized requires substantially lower rf powers compared to a shear wave device, techniques were devised to Q-switch also unpolarized lasers with this modulator. One commercially available Q-switch, for example, contains two longitudinal modulators orientated at $90°$ with respect to each other [8.73]. In another device the modulator is sandwiched between two quarter-wave plates which provide a $90°$ rotation of the plane of polarization of the light beam after each reflection from a resonator mirror [8.74].

The amount of acoustical power required to achieve a certain diffraction efficiency can be calculated from (8.35–41). For the practical case of a longitudinal Bragg angle device having a length of $l = 50\,mm$ and a transducer width of $w = 3\,mm$, we find from the values given in Table 8.2 that a scattered fraction of 2.1 %/W of acoustic power is theoretically obtained for light polarized in the direction of the acoustic wave propagation. Typically, one measures a scattered fraction of 0.8 to 1 % per watt of electrical power, which indicates a conversion efficiency of the order of 40 to 50 %. For light polarized perpendicular to the acoustic beam vector, a value of 4 to 5 % per watt is typically obtained. For example, at an application of about 20 W to the transducer, a 20 % single-pass loss is achievable for parallel-polarized light. At this level the dynamic optical loss rises up to 70 % for perpendicularly polarized light. At 70 % the deflection process is noticeably saturated, and further increase in the loss is obtained only at a much higher expenditure of driving power. High-powered cw-pumped Nd : YAG lasers usually require a single-pass dynamic loss in excess of 20 %. For example, in a Bragg angle shear wave device, experimentally achievable diffraction efficiencies are of the order of 1 % per watt of electrical power. Commercially available units of this type are operated up to 60 W of rf power to the transducer, which results in a 40 % diffraction efficiency. The dynamic resonator loss introduced by the Q-switch is $L = \eta(2 - \eta)$, where η is the deflection efficiency. In our example, where $\eta = 0.4$, we obtain a double-pass insertion loss of 64 %, which is sufficient to Q-switch most high-power Nd : YAG systems.

Design Features of Acoustooptic Q-Switches

We will calculate the parameters of a typical shear mode Bragg angle device employed for Q-switching a cw-pumped Nd : YAG laser. Essentially all commercially available Q-switches consist of a fused silica block to which a crystalline quartz or a LiNbO$_3$ transducer is bonded. Both the transducer and the fused silica contain vacuum-deposited electrodes to allow for electrical connections. An inductive impedance-matching network usually couples the signal of the rf generator to the quartz transducer. Virtually all acoustooptic Q-switches are single-pass devices; i.e., the acoustic wave generated by the transducer is absorbed after traveling across the interaction medium. The absorber, consisting of a piece of lead attached to the tapered end of the quartz block, prevents reflected acoustical waves from interfering with the incident light beam.

However, Q-switches have been built in which an ultrasonic standing wave was allowed to build up resonantly by reflection from the parallel face opposite to the transducer. This has the advantage that intensities at least ten times higher can be achieved than would exist in single-pass devices [8.75]. The disadvantage of this technique is that the laser can be Q-switched at only one fixed repetition rate.

Typical acoustooptic Q-switches can be operated in five different modes: internally driven Q-switch operation, externally gated pulse operation, an externally gated cw operation, a single-shot mode, and normal cw operation (with no rf modulation on the Q-switch).

Although the figure of merit M_2 of fused quartz is quite low, its optical high quality, low optical absorption, and high damage threshold make it superior to other, more efficient acoustooptic materials, such as lithium niobate ($LiNbO_3$), lead molybdate ($PbMoO_4$), tellurium dioxide (TeO_2), and dense flint glass. These materials are usually employed in low-power light modulators and optical scanners [8.66, 71, 76]. For an optical wavelength of $1.064\,\mu m$, a transducer drive frequency of $50\,MHz$, an index of refraction $n = 1.45$, and an acoustical velocity of $v = 3.76 \times 10^5$ cm/s for a shear wave in quartz, one obtains an acoustical wavelength of $\Lambda = 75\,\mu m$, a Bragg angle of $\Theta \approx 17$ arc min, and a scatter angle of $2\Theta' \approx 49$ arc min. For the light and acoustic waves to intersect at the Bragg angle, a parallelogram angle of the fused quartz block of $89° 43'$ must be chosen (Fig. 8.25b). For a cell $50\,mm$ long, the Bragg condition (8.38) is satisfied. The width of the transducer perpendicular to the acoustic and optical propagation direction is typically $3\,mm$, which is about twice the beam diameter for TEM_{00}-mode Nd:YAG lasers.

To be able to deflect the beam out of the cavity, the frequency of the rf signal driving the transducer is in the 20- to 50-MHz range. For these frequencies we obtain from (8.40) scattering angles in silica between 0.3 and 0.8° for an optical wavelength of $1.06\,\mu m$. The Q-switch must be able to switch from the high-loss to the low-loss state in less than the time required for the laser pulse to build up if maximum output energy is to be achieved. If this condition is met there will be no appreciable loss of laser output energy due to scattering by the switch. The overall turnoff time of an acoustooptical Q-switch is dominated not by electronic switching time but by the transit time of the sound wave across the beam diameter. Because the ultrasound is traveling at an acoustic velocity v in the block, which is typically in the range of $5\,mm/\mu s$, the transit time of the sound wave across the beam diameter D is D/v and will be of the order of $200\,ns/mm$ of optical beam diameter. This time is short compared to the Q-switch pulse evolution time in many laser systems, but may be too long for some high-gain lasers. The low-gain characteristics of cw-pumped solid-state lasers do not require very high contrast for Q-switching but do demand an exceptionally low insertion loss. Since the best optical-quality fused silica with antireflection coatings can be used as the active medium in the acoustooptical Q-switch, the overall insertion loss of the inactive Q-switch can be reduced to less than 0.5 % per pass. The low insertion loss of the acoustooptic Q-switch offers the convenience of converting from Q-switched to cw operation simply by removing the rf drive power. Performance data of an acoustooptic Q-switched Nd : YAG laser are shown in Fig. 8.8.

8.5 Passive Q-Switch

A passive Q-switch consists of an optical element, such as a cell filled with organic dye or a doped crystal, which has a transmission characteristic as shown in Fig. 8.26. The material becomes more transparent as the fluence increases, and at high fluence levels the material "saturates" or "bleaches" which results in a high transmission. The bleaching process in a saturable absorber is based on saturation of a spectral transition. If such a material with high absorption at the laser wavelength is placed inside the laser resonator, it will initially prevent laser oscillation. As the gain increases during a pump pulse and exceeds the round-trip losses, the intracavity flux increases dramatically causing the passive Q-switch to saturate. Under this condition the losses are low and a Q-switched pulse builds up.

Since the passive Q-switch is switched by the laser radiation itself it requires no high voltage, fast electro-optic driver or rf modulator. As an alternative to active methods, the passive Q-switch offers the advantage of an exceptional simple design, which leads to very small, robust and low-cost systems. The major drawbacks of a passive Q-switch are the lack of a precision external trigger capability and the lower output compared to electro-optic or acousto-optic Q-switched lasers. The latter is due to the residual absorption of the saturated passive Q-switch which represents a rather high insertion loss.

Originally, saturable absorbers were based on different organic dyes, either dissolved in an organic solution, or impregnated in thin films of cellulose acetate. The poor durability of dye-cell Q-switches, caused by the degradation of the light sensitive organic dye, and the low thermal limits of plastic materials severely restricted the applications of passive Q-switches in the past. The plastic Q-switch, however, did find applications in single-shot Nd : YAG rangefinders which resulted in very compact and simple designs. The plastic Q-switch consisted of a Kodak polyester sheet glued between two glass or sapphire windows in order to better dissipate the absorbed energy and to reduce optical distortions. Recently, Kodak has discontinued the production of this plastic Q-switch.

The emergence of crystals doped with absorbing ions or containing color centers have greatly improved the durability and reliability of passive Q-switches.

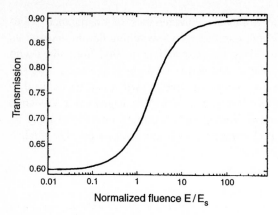

Fig. 8.26. Nonlinear transmission of a saturable absorber vs. fluence normalized to the saturation fluence E_s of the absorber

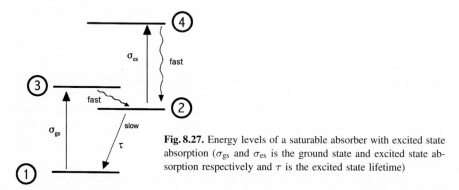

Fig. 8.27. Energy levels of a saturable absorber with excited state absorption (σ_{gs} and σ_{es} is the ground state and excited state absorption respectively and τ is the excited state lifetime)

The first new material to appear was the $F_2^- : LiF$ color center crystal. The color centers are induced in the crystal by irradiation with gamma, electron or neutron sources. Passive Q-switched Nd : YAG laser action with $F_2^- : LiF$ crystals has been reported in [8.77, 78, 85]. Today, the most common material employed as passive Q-switch is $Cr^{4+} : YAG$. The Cr^{4+} ions provide the high absorption cross-section of the laser wavelength and the YAG crystal provides the desirable chemical, thermal and mechanical properties required for long life.

A material exhibiting saturable absorption can be represented by a simple energy-level scheme such as that shown in Fig. 8.27. For the moment we will only consider levels 1–3. Absorption at the wavelength of interest occurs at the 1–3 transitions. We assume that the 3–2 transition is fast. For a material to be suitable as a passive Q-switch, the ground state absorption cross-section has to be large, and simultaneously the upper state lifetime (level 2) has to be long enough to enable considerable depletion of the ground state by the laser radiation. When the absorber is inserted into the laser cavity it will look opaque to the laser radiation until the photon flux is large enough to depopulate the ground level. If the upper state is sufficiently populated the absorber becomes transparent to the laser radiation, a situation which is similar to a three-level laser material pumped to a zero inversion level.

Solutions of the rate equation first carried out by Hercher [8.79], lead to an absorption coefficient which is intensity dependent

$$\alpha(E) = \frac{\alpha_0}{1 + E/E_s} \tag{8.42}$$

where α_0 is the small-signal absorption coefficient, and E_s is a saturation fluence

$$E_s = h\nu/\sigma_{gs} \tag{8.43}$$

where σ_{gs} is the absorption cross-section for the 1–3 transition.

Important characteristics of a saturable absorber are the initial transmission T_0, the fluence E_s at which saturation becomes appreciable, and the residual absorption which results in a T_{max} of the fully bleached absorber.

The small signal transmission of the absorber is

$$T_0 = \exp(-\alpha_0 \ell_s) = \exp(-n_0 \sigma_{gs} \ell_s) \tag{8.44}$$

where ℓ_s is the thickness of the bleachable crystal and n_0 is the ground state density. In order to calculate the transmission as a function of fluence, the photon flux and population density must be considered as a function of position within the absorbing medium.

Identical to the situation which occurs in pulse amplifiers (Sect. 4.1) differential equations for the population density and photon flux have to be solved. The solution is the Frantz-Nodvik equation which is identical to (4.11) except that gain G, G_0 is replaced by transmission T_i, T_0. Therefore the energy transmission T_i of an ideal saturable absorber is given by

$$T_i = \frac{E_s}{E} \ln[1 + (e^{E/E_s} - 1)T_0] \tag{8.45}$$

Equation (8.45) reduces to $T_i = T_0$ for $E < E_s$ and $T_i = 1$ for $E > E_s$.

In practical saturable absorbers, the transmission never reaches 100%. The reason for that is photon absorption by the excited atoms. A passive Q-switch requires a material which exhibit saturation of the ground state absorption. However, most materials also exhibit absorption from an excited state. This is illustrated in Fig. 8.27 by the transition from the excited state (level 2) to some higher level 4 which has an energy level corresponding to the laser transition. As the ground state is depleted, absorption takes place increasingly between levels 2 and 4. Excited State Absorption (ESA) results in a residual loss in the resonator when the ground state absorption has been saturated. The 2–4 transition does not saturate because of the fast relaxation of level 4. A saturable absorber is useful for Q-switching only as long as $\sigma_{gs} > \sigma_{es}$.

A saturable absorber with ESA can be described by a four level model [8.79, 80]. In this case, maximum transmission T_{max} is given by

$$T_{max} = \exp(-n_0\sigma_{es}\ell_s) \tag{8.46}$$

where σ_{es} is the cross-section for excited state absorption.

In [8.80] an approximation in closed form has been derived which gives the shape of the transmission vs. fluence curve in the presence of ESA. For a non-ideal absorber the transmission T_n can be approximated by

$$T_n = T_0 + \frac{T_i - T_0}{1 - T_0}(T_{max} - T_0) \tag{8.47}$$

where T_i is the transmission of an ideal absorber given by (8.45), and T_0 and T_{max} are the lower and upper limits of the transmission.

In Fig. 8.26, T_n is plotted for the case of $T_0 = 60\%$ and $T_{max} = 90\%$. For a given saturable absorber one needs to know the cross-section σ_{gs}, σ_{es}, the ground state concentration n_0 and the thickness l_s of the material. With these quantities known one can calculate T_0, T_{max} and E_s, and plot a transmission vs. energy density curve as determined by (8.47).

Because of the importance of Cr^{4+} : YAG as a passive Q-switch material, we will briefly review the properties of this material. In order to produce Cr^{4+} : YAG, a small fraction of chromium ions in YAG are induced to change valence from the normal Cr^{3+} to Cr^{4+} with the addition of charge compensating impurities such as Mg^{2+} or Ca^{2+}. Cr^{4+} : YAG has broad absorption bands centered at 410, 480, 640 and 1050 nm. Published values for the cross-section of the ground state vary greatly. The most recent measurements indicate $\sigma_{gs} = 7 \times 10^{-18}\,cm^2$ and $\sigma_{es} = 2 \times 10^{-18}\,cm^2$ for the excited-state absorption at the Nd : YAG wavelength [8.80]. The excited state lifetime (level 2 in Fig. 8.27) is 4.1 μs and the lifetime of the higher excited state (level 4 in Fig. 8.27) is 0.5 ns. With $h\nu = 1.87 \times 10^{-19}$ J at 1.06 μm and the above value for σ_{gs} one obtains a saturation fluence of $E_s = 27$ mJ/cm^2 for Cr^{4+} : YAG.

Figure 8.28 illustrates the transmission of a 2.65 mm thick sample of Cr^{4+} : YAG. The transmission starts at about $T_0 = 53.5\%$ for low power, increases linearly, and then saturates at about $T_{max} = 84\%$ for high fluence [8.81]. From (8.44 or 46) follows a Cr^{4+} ion density of $n_0 = 3.4 \times 10^{17}\,cm^{-3}$ for this sample [8.80]. Commercially available Cr^{4+} : YAG passive Q-switches are specified by the low power transmission at the laser wavelength. Typical transmission values range from 30% to 50%, and the crystal thickness is usually between 1–5 mm. Values of the small signal absorption coefficient α_0 vary from 3–6 cm^{-1}. For example, for $\alpha_0 = 4$ cm^{-1} and $\ell_s = 2$ mm the low power transmission is $T_0 = 45\%$.

Passive Q-switching of pulsed and cw pumped Nd : lasers employing Cr^{4+} : YAG are described in [8.77, 81–84]. In a passively Q-switched, cw pumped laser, once a certain threshold is reached, peak power and pulse width do not change further, but the repetition rate increases approximately linearly with pump input power [8.85].

The compactness which can be achieved with a passive Q-switched laser is illustrated in the following example. For a space-borne high-resolution lidar system an extremely small, conduction-cooled laser transmitter was required having pulse width of less than 2 ns. The optical layout is sketched in Fig. 8.29a. The laser uses a single 100 W quasi-cw diode bar to pump a Nd : YAG slab of 13 mm length and 1×0.75 mm^2 cross-section. The dimensions of the slab were chosen to provide a ten bounce zigzag path through the crystal. The ends of the slab were uncoated parallel Brewster faces. The resonator comprised of a flat and curved mirror has an optical length of 6.75 cm. The passive Q-switch is a 2 mm thick Cr^{4+} : YAG crystal. The pump radiation from the diode array is collimated with a cylindrical lens to confine the light into a 500 μm wide region within the slab. This arrangement permits a small pump mode volume which well overlaps the resonator fundamental mode.

Pumping for 200 μs produces 20 mJ of pump power, of that approximately 4.5 mJ is stored for Q-switching. The single pass, small signal gain is $G_0 = 8$. The laser crystal and laser diode are conductively cooled and maintained at a constant temperature with a thermoelectric cooler. The AR coated Cr^{4+} : YAG crystal has a thickness of 2 mm and an optical density of 0.5 or $T_0 = 0.32$ in the unsaturated state. With the above values for σ_{gs} and σ_{es} follows from (8.44 and 46) a saturated transmission of $T_{max} = 0.72$. This results in a round trip loss of about $L = 0.6$ if one includes 4% losses in the crystal slab. Typical of passive Q-switched lasers, the insertion loss is dominated by the residual absorption in the Q-switch.

Although not important in this application, a passive Q-switch allows the laser to preferentially operate in a single longitudinal mode, as discussed in Sect. 5.2.3. Also the output can be linearly polarized without introducing additional optical components by orientating the Cr^{4+} : YAG crystal at the Brewster angle [8.81]. Because of the Brewster ends of the slab, this laser is already polarized.

The laser produces a TEM$_{00}$ beam with an output energy of 0.85 mJ and a pulse length of 1.4 ns(Fig. 8.29b). The system can operate at any repetition rate between 1 and 100 Hz limited only by the duty cycle of the laser-diode array. An amplifier employing an identical slab but pumped by 3 diode arrays increases the output to 5.2 mJ.

From (8.10) follows that the generation of a short pulse width requires a short resonator length and a high gain in the active medium. The passive Q-switch permits the design of a resonator which is only a few cm in physical length having a roundtrip time of $t_r = 0.45$ ns. Generally good agreement is obtained between the calculated pulse width from (8.10) and the measured values. In Fig. 8.2b, pulse width is plotted vs. normalized gain $z = \ell n G_0^2/L$. With the above values for G_0 and L, one obtains $z = 6.9$. From the graph, Fig. 8.2b follows for the normalized pulse width $t_p L/t_r = 1.5$. With L and t_r given above, the predicted pulse width is 1.1 ns.

The output beam has a diameter of about 0.5 mm which gives a fluence of 0.45 J/cm^2. Since the output coupler has a reflectivity of $R = 20\%$, the fluence within the saturable absorber is about 0.56 J/cm^2. This is 21 times above the

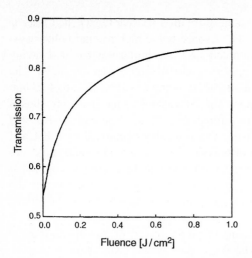

Fig. 8.28. Transmission vs fluence for a 2.65 mm thick Cr^{4+}:YAG absorber [8.81]

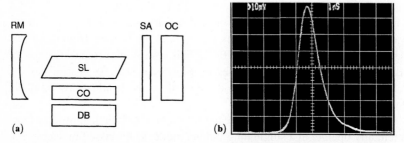

(a)   **(b)**

Fig. 8.29. (a) Optical schematic of a passively Q-switched Nd:YAG oscillator. SL is the Nd:YAG slab, DB diode bar, CO coupling optics, OC output coupler, RM rear mirror, SA saturable absorber, (b) oscilloscope trace of the Q-switched output pulse (Horizontal: 1 ns/div)

saturation level for Cr^{4+}:YAG and according to the curves Fig. 8.26 and 28 it is close to the maximum transmission for this material.

The output mirror has a reflectivity which is lower than the optimum value. Energy output vs. mirror reflectivity is not a strongly peaked function, and it is sometimes desirable to select a lower reflectivity for the output mirror rather than the optimum value in order to make the laser more robust from the point of view of optical damage and multiple pulsing. A lower reflectivity of the output coupler reduces the fluence in the cavity which reduces the likelihood of optical coating damage. Multiple Q-switched output pulses occur if the unsaturated transmission of the absorber and/or the output mirror reflectivity is too high. In this case, Q-switching occurs early in the pump pulse. With the proper values and ratio of output coupling to absorption in the Q-switch, a single pulse can be generated which should occur near the end of the pump pulse.

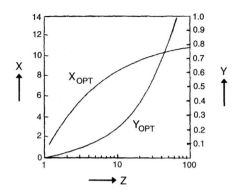

Fig. 8.30. Dependence of the optimum mirror reflectivty x and absorber transmission y on laser gain z. All parameters are normalized according to (8.48)

For a given pump power, i.e. gain in the laser medium, there is an optimal choice of output coupler reflectivity and unsaturated absorber transmission. Design procedures which permit optimization of passively Q-switched lasers have been reported in [8.86, 83]. Adapted from *Degnan*'s paper [8.86] is the graph Fig. 8.30 which permits the optimum choice of the unsaturated Q-switch transmission T_0 and the output mirror reflectivity R for a given gain G_0. The above quantities are normalized as follows:

$$x = \frac{1}{L} \ln \left(\frac{1}{R} \right) \; ; \quad y = \frac{-\ln T_0}{\ln G_0} \; ; \quad z = \frac{\ln G_0^2}{L} \tag{8.48}$$

where

$$L = L_R + 2\sigma_{es} n_0 \ell_s \; . \tag{8.49}$$

The terms in (8.49) are the combined roundtrip loss in the resonator caused by dissipative losses L_R and the residual absorption in the fully saturated Q-switch.

The optimization of a passively Q-switched laser can proceed along the following path: First one determines the single pass, small signal gain G_0 and optical losses L_R according to the procedure outlined in Sect. 3.4.2. From the manufacturer data which usually provides the optical density of T_0 of the crystal, one can calculate n_0 and therefore L. With G_0 and L known, the parameter z can be calculated and with it x_{opt} and y_{opt} follow from Fig. 8.30.

Rearranging (8.48) one obtains

$$R_{\text{opt}} = \exp(-x_{\text{opt}} L) \; , \quad T_{\text{opt}} = \exp(-y_{\text{opt}} \ln G_0) \; . \tag{8.50}$$

For the laser described above, we obtained a value of $z = 6.9$ for the normalized gain. From Fig. 8.30 follows then for $x_{\text{opt}} = 2.0$ and $y_{\text{opt}} = 0.56$. The optimum values for unsaturated transmission and output coupler reflectivity are therefore $R_{\text{opt}} = 30\%$ and $T_{\text{opt}} = 31\%$.

The most-important application for passive Q-switches is for military rangefinders. Modern systems operate at the wavelength of $1.5\,\mu$m to reduce eyesafety hazards. In this case, the wavelength of a passive Q-switched Nd : YAG laser has to be converted by means of a parametric oscillator or Raman cell (Chap. 10). An alternative approach for eyesafe rangefinders is based on flashlamp or diode pumped Er: glass lasers which directly emit at the desired wavelength. The most popular method of Q-switching such lasers is the rotating Porro prism. However, recently, passive Q-switches based on uranium doped CaF_2 have been developed for this laser [8.87, 88].

8.6 Cavity Dumping

A means for generating extremely short Q-switched laser pulses involves Q-switching the laser with 100 % mirrors on both ends of the cavity and then, at the peak of the circulating power, rapidly switching the output mirror from 100 to 0 % reflection. This leads to a rapid dumping of the entire optical energy from within the cavity. One of the advantages of this technique is the production of Q-switched pulses whose width is primarily a function of the oscillator cavity length, rather than the gain characteristics of the laser medium. Specifically, the laser pulse width at the half-power points will be equivalent to the round-trip transit time in the cavity, with the condition that the Q-switch employed be switched within this same time period. Thus, based on allowable cavity dimensions, pulse widths in the range of 2 to 5 ns are feasible for oscillators whose pulse widths are of the order of 10 to 20 ns in the normal Q-switch mode.

Figure 8.31 shows the optical layout of a ruby oscillator employed to generate short pulses by the cavity dumping mode. We will explain the operation of the system by assuming that the ruby c plane is perpendicular to the plane of the paper. When the flashlamp is fired, the horizontally polarized ruby fluorescence is transmitted by the thin-film or calcite polarizer, thereby preventing regeneration. Upon reaching peak-energy storage in the ruby, the Pockels cell is biased to its half-wave retardation voltage. The resulting vertically polarized light is reflected by the polarizer to the off-axis mirror, and regeneration occurs in the cavity. When the power in the cavity reaches its peak value, the bias is removed from the Pockels cell in a time period of about 2 ns. The cavity energy then literally drains out of the cavity in the time required for the radiation to travel one round trip in the optical cavity. The combination of the polarizer, Pockels cell, and 100 % mirror comprises what amounts to a high-speed voltage-variable mirror whose reflectivity is changed from 0 during the pumping cycle to 100 % during the pulse buildup, and back to 0 during the cavity dumping phase.

To illustrate the practical realization of this technique, we will consider a typical ruby oscillator consisting of a 10-cm by 1-cm ruby rod, a Pockels cell, a thin-film polarizer, and two 99 % mirrors. If we assume a 75-cm-long cavity, we obtain a round-trip transit time of 5 ns. The ruby rod is pumped by a 1-ms-long flashlamp pulse, and the Pockels cell is switched the first time after about 0.8 ms. The time delay between switching the Pockels cell and the occurrence of peak power in the cavity is typically 60 ns. In order to extract the stored cavity energy,

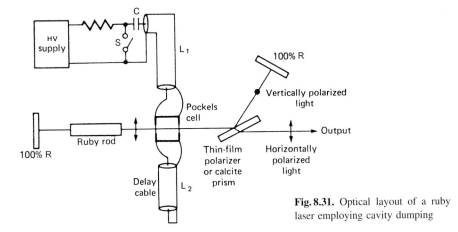

Fig. 8.31. Optical layout of a ruby laser employing cavity dumping

the bias on the Pockels cell is reduced to zero after this time delay. This can be accomplished by means of the circuit shown in Fig. 8.31. In this arrangement, the Pockels cell is connected in-line between coaxial cables L_1 and L_2. Closing the switch S will discharge the capacitor C into the transmission line L_1. When the voltage pulse reaches the Pockels cell, the optical beam will experience a 90° polarization rotation and the Q-switch pulse will start to build up from noise. Assuming a perfect 50 Ω impedance of the Pockels cell, no reflection will occur at the cell and the voltage pulse will travel to the end of the shorted transmission line L_2. At that point the pulse will be reflected with a 180° phase shift. When the reflection reaches the cell, the voltage on the crystal will be zero. Therefore the length of cable L_2 determines for how long the voltage is applied to the Pockels cell.

The performance of a ruby oscillator having the above-mentioned system parameters is illustrated in Fig. 8.32. Shown is the power inside the resonator if the energy is not dumped (Fig. 8.32a). The measurement was made by monitoring the leakage radiation through one of the 99 % mirrors with a fast detector and oscilloscope. The peak power and energy inside the resonator were determined to be 190 MW and 1 J, respectively. Figure 8.32b depicts the circulating power monitored through the same mirror if the energy is dumped and the system is operated as a PTM oscillator. The internal cavity power reaches almost the same value as in Fig. 8.32a, then falls in about 5 ns to almost zero. This revals that all but a small percentage of the available energy has been dumped from the cavity. Figure 8.32c reveals that the dumped pulse is triangular, with a 10 to 90 % rise time of 3.0 ns and a pulse width of 5.3 ns. This width compares exactly with the cavity round-trip transit time within experimental error. The total energy of this pulse was measured to be 0.75 J. The peak power of this pulse is thus 1.4×10^8 W. The rise time of the output pulse is determined by the switching speed of the hydrogen thyratron which was used to discharge capacitor C. The experimental data reveals that 75 % of the stored energy was extracted from the resonator.

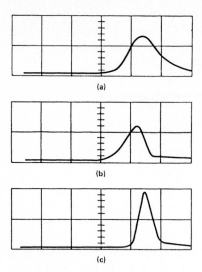

Fig. 8.32a–c. Performance of a ruby oscillator with cavity dumping. Photo scales 10 ns/div. **(a)** Internal cavity power at 99 % mirror without dumping. Pulse rise time 7 ns, pulse width 12 ns, peak power 190 MW. **(b)** Internal power, with cavity dumping, drops from 180 MW to almost zero in 5 ns. **(c)** Power dumped from cavity; rise time 3.0 ns, pulse width 5.3 ns, energy 0.74 J, and peak power 140 MW

In practical situations the design of cavity dumped lasers is completely dominated by the requirement of keeping the power density within the cavity below the damage level. The usual calcite polarizer, being the component with the lowest damage threshold, has been replaced in contemporary oscillators by highly damage-resistant thin-film polarizers. These components permit oscillators to operate at power densities up to 300 MW/cm².

It is not necessary in a cavity dumped system to use the same Pockels cell for both the Q-switch initiation and cavity dumping. Earlier systems employed two Pockels cells for these functions [8.89–91]. Also, instead of a fixed-delay transmission line a more precise synchronization between the peak power in the cavity and the switching of the Pockels cell can be achieved if the cavity radiation is monitored by a detector mounted behind one of the 99 % mirrors [8.92]. Other variations of cavity dumping are described in [8.93–95], this technique can be employed in any solid-state laser; for example, Nd : glass oscillators have produced 3-ns pulses with energies up to 180 mJ [8.95]. Further examples of cavity dumped systems are given in Sects. 9.3.2 and 9.4.2 in connection with diode-pumped actively mode locked oscillators or with regenerative amplifiers. Cavity dumping is also possible with cw-pumped lasers; this will be discussed next.

Cavity Dumping of CW-Pumped Lasers

Cavity dumping can be compared with the Q-switching of a continuously pumped laser. In both cases energy is assumed to be discharged from the laser in the form of a repetitive train of light pulses. However, energy accumulation and storage between output pulses are primarily in the optical field for cavity dumping, and primarily in the population inversion for Q-switching.

The finite buildup time of the field inside the laser cavity and the time required to repump the inversion set an upper limit to the repetition rate available from Q-switched lasers. This maximum value of repetition rate for Q-switched Nd : YAG lasers is of the order of 50 to 100 kHz. Cavity dumping of continuously pumped lasers is a way to obtain pulsed output at higher repetition rates than are available by Q-switching. Repetition rates from 125 kHz up to several megahertz for cavity dumping were achieved with cw-pumped Nd : YAG lasers [8.96, 97].

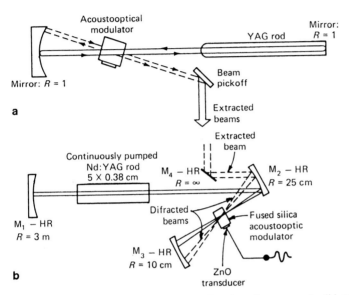

Fig. 8.33. Common arrangements for cavity dumping of cw-pumped solid-state lasers. The broken lines indicate the beams which are diffracted by the modulator

Figure 8.33 exhibits two common arrangements employed for cavity dumping of cw-pumped solid-state lasers [8.96, 97]. Essentially all systems of this type employ acoustooptic modulators as the switching element. In order to obtain fast switching action, the incident beam must be focused to a narrow waist inside the modulator. The two oscillator designs differ in the way the optical beam is focused into the modulator.

In Fig. 8.33a the modulator is located at a beam waist created by a concave mirror and by the thermal lens properties of the laser rod. The acoustic wave in the fused silica causes Bragg scattering of the forward- and backward-traveling light beam in the resonator. The two diffracted beams which are obtained from the cavity-dumped oscillator are initially traveling in opposite directions, therefore their frequencies are shifted to a value of $\omega + \Omega$ and $\omega - \Omega$, where ω is the frequency of the incident beam and Ω is the frequency of the acoustic wave [8.98, 99]. The two diffracted beams are extracted from the cavity as a single beam and deflected out of the system by a mirror. In Fig. 8.33b the cavity is formed by three high-reflectivity mirrors M_1, M_2, and M_3. The mirror curvature and the distance between M_2 and M_3 are chosen such that the light beam between M_2 and M_3 is focused to a small diameter at the center of curvature of M_3. The modulator is inserted at the waist of the optical beam.

Acoustooptic modulators employed for cavity dumping differ from their counterparts used in Q-switch applications in several respects:

1) Compared to Q-switching, the cavity-dump mode requires much faster switching speeds. The rise time in an acoustooptic modulator is approximately given by the beam diameter divided by the velocity of the acoustic wave. In order to obtain rise times around 5 ns, a value which is required for efficient cavity dumping, the incident beam must be focused to a diameter of approximately 50 μm.

2) For efficient operation in the cavity-dump mode, it is important that esentially all the circulating power be diffracted into the first diffraction order. In a Bragg device the diffraction efficiency increases with the rf carrier frequency, therefore modulators employed in cavity dumpers operate at considerably higher frequencies, i.e., 200 to 500 MHz as compared to acoustooptic Q-switches.

3) In order to generate an output pulse in the cavity-dump mode, a short rf pulse is applied to the modulator, whereas in an acoustooptic Q-switch the rf carrier is turned off for the generation of an output pulse.

4) The cavity is never kept below threshold condition as in the Q-switched mode of operation. If the cavity is dumped of all its energy, the field has to build itself from the noise level. Repetitive cavity dumping was observed to become unstable in this case. If the repetition rate is lowered, the laser material is pumped higher above threshold between pulses, therefore, a larger fraction of the stored energy is extracted from the system (see Sect. 8.1). The lower limit of the dumping repetition rate is reached when the internal laser energy decreases to one photon immediately after dumping. The upper limit of the cavity dumping repetition rate is set by the switching speed of the modulator. Repetition rates as high as 10 MHz have been reported. From a cw-pumped Nd:YAG laser capable of 10 W of cw power, peak powers of 570 W with a pulse duration of 25 ns have been obtained at a 2 MHz repetition rate [8.100]. For high-data-rate communications systems, cavity dumping of cw-pumped lasers is sometimes combined with mode locking [8.101, 102].

The technique of cavity dumping is also employed in regenerative systems. In a regenerative laser a pulse is injected into a laser resonator containing an amplifier. The pulse passes several times through the same amplifier medium and is then switched out. Figure 8.34a shows an optical schematic of a laser which employs this technique [8.103]. Illustrated is a diode-pumped Nd:YAG crystal located between two highly reflective mirrors, a Pockels cell Q-switch and a polarizer. The Nd:YAG crystal is cw-pumped and repetitively cavity dumped at a 10 kHz repetition rate. During the 100 μs pump time, the Pockels cell is

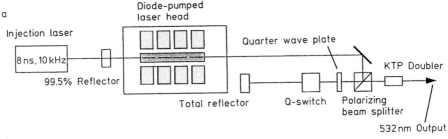

a

Injection laser

8 ns, 10 kHz

99.5% Reflector

Diode-pumped
laser head

Quarter wave plate

Total reflector   Q-switch

Polarizing
beam splitter

KTP Doubler

532 nm Output

b

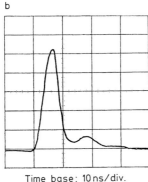

Time base: 10 ns/div.

Fig. 8.34a,b. Cavity dumping of an injection seeded laser. Optical schematic of laser system (**a**) and output pulse shape (**b**)

operated at zero wave retardation. This is the low Q-condition because radiation is transmitted out of the resonator through the polarizer. At the end of the pump pulse, the injection laser seeds a 8 ns pulse into the resonator through the rear reflector. At the same time, the Pockels cell is switched to 1/4 wave retardation which establishes the high Q-condition. Radiation is building up between the two highly reflective mirrors. The injected pulse is regeneratively amplified in the cavity for about 360 ns or 120 passes. The Q-switch is then returned to zero wave retardation which dumps the amplified pulse through the polarizer. The KTP crystal converts the wavelength to 532 nm and shortens the pulse to about 5 ns. The sequence is repeated at a 10 kHz pulse repetition frequency. Figure 8.34b displays the output pulse at 532 nm.

An analysis of the extraction efficiency of cavity-dumped regenerative lasers can be found in [8.104]. The results are quite similar to the Q-switched case, and the extraction efficiency depends only on the amplifier gain and resonator losses.

9. Mode Locking

As we have seen in the previous chapter, the minimum pulse width obtainable from a Q-switched laser is on the order of 10 ns because of the required pulse buildup time. With the cavity dumping technique, the pulse width can be reduced to a minimum of 1 to 2 ns. The limitation here is the length of the cavity, which determines the pulse length. Ultrashort pulses with pulse widths in the picosecond or femtosecond regime are obtained from solid-state lasers by mode locking. Employing this technique, which phase-locks the longitudinal modes of the laser, the pulse width is inversely related to the bandwidth of the laser emission.

The output from laser oscillators is subject to strong fluctuations which originate from the interference of longitudinal resonator modes with random phase relations. These random fluctuations can be transformed into a powerful well-defined single pulse circulating in the laser resonator by the introduction of a suitable nonlinearity, or by an externally driven optical modulator. In the first case, the laser is referred to as passively mode-locked because the radiation itself, in combination with the passive nonlinear element, generates a periodic modulation which leads to a fixed phase relationship of the axial modes. In the second case, we speak of active mode locking because a rf signal applied to a modulator provides a phase or frequency modulation which leads to mode locking.

Utilizing organic dyes as saturable absorbers, mode locking was first observed in solid-state lasers in the mid 1960's. For about ten years, the systems of choice for picosecond pulses were flashlamp-pumped ruby, Nd : glass and Nd : YAG lasers employing saturable absorbers. Employing a dye cell as the mode locking element has the major drawback of poor shot-to-shot reproducibility. In pulsed solid-state lasers, the presence of a saturable dye absorber will result not only in mode locked, but also in Q-switched operation. For each flashlamp pump pulse, a short burst of mode-locked pulses is generated with a duration of a typical Q-switched pulse, i.e. a few tens of nanoseconds. For each flashlamp pulse, the mode- locked pulses build up from noise. Due to the statistical randomness of this process, large variations in the shot-to-shot output from the laser are observed. The problem of poor output reproducibility is exacerbated by the instability of the dye solutions which degrade with time and decompose when exposed to light.

As a result of the difficulty in obtaining reliable and consistent mode-locked output pulses, or pulses shorter than about 10 ps, emphasis shifted away from pulsed mode-locked solid-state lasers to mode-locked organic dye lasers. The large gain-bandwidth product of dye lasers in combination with novel mode-locking techniques made the generation of pulses as short as tens of femtoseconds possible. Therefore, despite the inherent disadvantages of handling and maintaining dye solutions, the dye laser became the workhorse for ultra-fast studies during the 1980's.

At about the same time, tunable solid-state laser materials were developed. Tunable lasers, notably Ti:sapphire, have a gain-bandwidth product equal or greater than organic dyes and are therefore ideally suited for the generation of femtosecond pulses. Motivated by the availability of broadband lasers, novel mode-locking techniques, such as additive pulse mode locking and Kerr-lens mode locking, have been developed. We can describe lasers in which Q-switching and mode locking take place simultaneously as pulsed mode-locked lasers. In cw passive mode locking, a constant train of mode-locked pulses is emitted from the laser. Today, cw passive mode-locked solid-state lasers are capable of producing reliable pulses on the order of tens of femtoseconds. In particular, the argon-pumped Ti:sapphire laser, cw passively mode-locked via Kerr-lens modulation, has become the standard for femtosecond research.

The progress of actively mode-locked solid-state lasers was equally dramatic. Initially developed to mode-lock krypton arc or tungsten lamp cw pumped Nd:YAG lasers by inserting either electro-optic or acousto-optic modulators into the resonator. In contrast to passive mode locking, the method of phase or amplitude modulation has not changed much over the years, but the bulky and inefficient lamp-pumped solid-state lasers have been replaced by extremely compact and efficient diode-pumped lasers. Very stable and reliable trains of mode-locked pulses can be generated with these lasers. Individual pulses from these continuously mode-locked lasers can be selected to provide a seed pulse for further amplification.

Today, laser-diode end-pumped Nd:YLF or Nd:YAG lasers, actively mode-locked with an acousto-optic modulator provide output pulses with pulsewidths on the order of 10–20 ps. Such systems having a high degree of reliability and long-term reproducibility of their performance are offered commercially from a number of companies.

9.1 Pulse Formation

In a free-running laser, both longitudinal and transverse modes oscillate simultaneously without fixed mode-to-mode amplitude and phase relationships. The resulting laser output is a time-averaged statistical mean value. Restricting oscillation of the laser to the TEM_{00} mode for the moment, we recall from Chap. 5 that in a typical laser cavity there are perhaps a few hundred axial modes which fall within the frequency region where the gain of the laser medium exceeds the losses of the resonator.

As an example for illustrating the formation of a mode-locked pulse, we consider a passively mode-locked laser containing as nonlinear element a bleachable

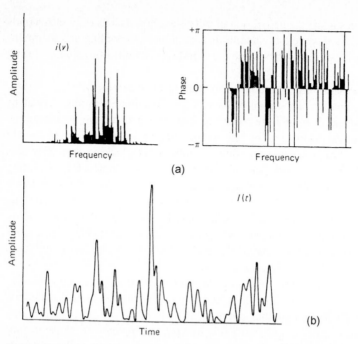

Fig. 9.1a,b. Signal of a non-mode-locked laser. In the frequency domain (**a**) the intensities $i(\nu)$ of the modes have a Rayleigh distribution about the Gaussian mean and the phases are randomly distributed. In the time domain (**b**) the intensity has the characteristic of thermal noise

dye absorber. In order for the passive mode-locking process to start spontaneously from the mode beating fluctuations of a free-running laser, the nonlinear element must create an amplitude instability so that an intensive fluctuation experiences lower losses compared to less intensive parts of the radiation. A further requirement is that the reaction time of the nonlinear element be as short as the fluctuation itself in order to lock all the modes oscillating in the resonator.

In Fig. 9.1 the spectral and temporal structure of the radiation inside a laser cavity are shown for a non-mode-locked laser. In the frequency domain, the radiation consists of a large number of discrete spectral lines spaced by the axial mode interval $c/2L$. Each mode oscillates independent of the others, and the phases are randomly distributed in the range $-\pi$ to $+\pi$. In the time domain, the field consists of an intensity distribution which has the characteristic of thermal noise.

If the oscillating modes are forced to maintain a fixed phase relationship to each other, the output as a function of time will vary in a well-defined manner. The laser is then said to be "mode-locked" or "phase-locked". Figure 9.2 shows the output signal of an ideally mode-locked laser. The spectral intensities have a Gaussian distribution, while the spectral phases are identically zero. In the time domain the signal is a single Gaussian pulse. As can be seen from this figure, mode locking corresponds to correlating the spectral amplitudes and phases.

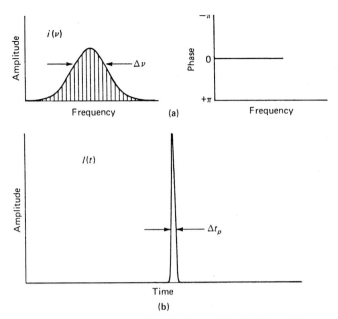

Fig. 9.2a,b. Signal structure of an ideally mode-locked laser. The spectral intensities **(a)** have a Gaussian distribution, while the spectral phases are identically zero. In the time domain **(b)** the signal is a transform-limited Gaussian pulse

When all the initial randomness has been removed, the correlation of the modes is complete and the radiation is localized in space in the form of a single pulse.

Because the intensity profiles $I(t)$ shown in Figs. 9.1b and 2b circulate around inside the cavity with a repetition rate determined by the round-trip transit time, these signals will repeat themselves and appear in the laser output at a rate of $c/2L$. Therefore, mode locking results in a train of pulses whose repetition period is twice the cavity transit time, i.e.

$$\text{PRF} = 1/\varDelta T = c/2L \tag{9.1}$$

where PRF is the pulse repetition frequency, $\varDelta T$ is the separation of individual pulses, and L is the resonator length.

If we make some simplifying assumptions, we can obtain a general idea about the pulse width and peak power of the mode-locked output pulses. The structure of an optical pulse is completely defined by a phase and an intensity. Whether these refer to time or to frequency is immaterial since, if the description in one domain is complete, the profiles in the other are obtained from the Fourier transform. However, there is no one-to-one correspondence between the two intensity profiles $I(t)$ and $i(\nu)$, since each depends not only on the other but also on the associated phase function. The only general relationship between the two is

$$\varDelta\nu\varDelta t \geq K \ , \tag{9.2}$$

where Δt and $\Delta \nu$ are the full width half maximum of $I(t)$ and $i(\nu)$, respectively, and K is a constant of the order of unity. In particular, the shortest pulse obtainable for a given spectral bandwidth is said to be transform-limited; its duration is

$$t_p = \frac{K}{\Delta \nu} \ . \tag{9.3}$$

For mode-locking purposes $\Delta \nu$ corresponds to the gain bandwidth $\Delta \nu_L$ of the laser. The number of axial modes which are contained within the oscillating bandwidth is $N = \Delta \nu_L t_R$, where $t_R = 2L/c$ is the round-trip time in the optical resonator. The width of the individual mode-locked pulses is therefore

$$t_p \approx \frac{1}{\Delta \nu_L} \approx \frac{t_R}{N} \ . \tag{9.4}$$

Equation (9.4) expresses the well-known result from Fourier's theorem that the narrower the pulse width t_p, the larger the bandwidth required to generate the pulse. From (9.4) follows also the interesting fact that the pulse width of the mode-locked pulses roughly equals the cavity round-trip time divided by the number of phase-locked modes.

For simplicity, the spectral intensity in Fig. 9.2a was chosen to be a Gaussian function and hence the temporal profile has also a Gaussian distribution. The most general Gaussian optical pulse is given by

$$E(t) = \left(\frac{E_0}{2} \right) \exp\left(-\alpha t^2\right) \exp\left[\mathrm{j}(\omega_p t + \beta t^2)\right] \ . \tag{9.5}$$

The term α determines the Gaussian envelope of the pulse, and the term $\mathrm{j}\beta t$ is a linear frequency shift during the pulse (chirp). From (9.5) follows for the pulse width at the half-intensity points

$$t_p = \left(\frac{2 \ln 2}{\alpha} \right)^{1/2} , \tag{9.6}$$

and a bandwidth again taken at the half-power points of the pulse spectrum [9.1]

$$\Delta \nu_p = \frac{1}{\pi} \left[2 \ln \left(2 \frac{\alpha^2 + \beta^2}{\alpha} \right) \right]^{1/2} \ . \tag{9.7}$$

Note how the frequency chirp contributes to the total bandwidth. The pulsewidth-bandwidth product is a parameter often used to characterize pulses. For Gaussian pulses, the pulsewidth-bandwidth product is given by

$$t_p \Delta \nu_p = \left(\frac{2 \ln 2}{\pi} \right) \left[1 + \left(\frac{\beta}{\alpha} \right)^2 \right]^{1/2} \ . \tag{9.8}$$

For the important special case $\beta = 0$ (i.e., no frequency chirp), one obtains $K = t_p \Delta \nu_p \approx 0.441$.

Before leaving this subject, we have to mention one other important pulse shape, namely the hyperbolic secant function. The steady-state solution of the differential equation describing the pulse envelope $I(t)$ of a cw mode-locked pulse is a function of the form [9.2]:

$$I(t) = I_O \operatorname{sech}^2(t/\tau_p) \tag{9.9}$$

The FWHM pulsewidth of a sech^2 pulse is τ_p where $\Delta t_p = 1.76\tau_p$, and a transform-limited sech^2 pulse has a pulsewidth-bandwidth product of

$$\Delta t_p \Delta \nu = 0.315 . \tag{9.10}$$

Mode-locked lasers can have high peak powers, because the power contained in the entire output of the uncoupled laser is now contained within the more intense ultrafast pulses. From (9.4) follows for the ratio of pulse-on-to-pulse-off time, i.e., the duty cycle, a value of $\Delta t_p/t_R = 1/N$, so that the peak power of the pulse is N times the average power resulting from incoherent phasing of the axial modes.

$$P_{\text{peak}} = N P_{av} . \tag{9.11}$$

Typical output-pulse formats from pulsed and cw mode-locked lasers are illustrated in Fig. 9.3. The output of a flashlamp-pumped, mode-locked solid-state laser consists of a burst of pulses with amplitudes which fit underneath the envelope of a Q-switched pulse. A cw mode-locked laser produces a train of pulses with equal amplitude. In both cases are consecutive pulses separated by the cavity round-trip time.

A modern diode end-pumped Nd:YAG laser, cw mode-locked with an acousto-optic modulator, produces pulses of about 20 ps in duration and has an average output power of about 200 mW. If we assume a resonator length of $L = 1.2$ m, we obtain from (9.1), a pulse separation of $\Delta T = 8$ ns or a PRF of 125 MHz. The pulsewidth of 20 ps in combination with (9.4) indicates that about 400 longitudinal modes are phase locked together ($N = 400$). From (9.11) follows a peak power of 80 W.

The homogeneously broadened line in Nd:YAG is $\Delta \nu = 150$ GHz. Therefore, the ultimate limit in short-pulse generation in Nd:YAG is according to (9.10) about two picoseconds. Passively cw mode-locked Nd:YAG have approached this limit. The modulation index one can achieve in active mode locking is not large enough to support phase locking over a very large bandwidth. Therefore, actively mode-locked lasers emit longer pulses compared to cw passively mode-locked systems.

The generation of mode-locked pulses from a laser requires that the longitudinal modes be coupled together. This is achieved experimentally by placing inside the laser cavity either an externally driven loss or phase modulator, or a passive device which exhibits saturable absorption. Details of the generation of mode-locked pulses with active and passive devices will be discussed in the

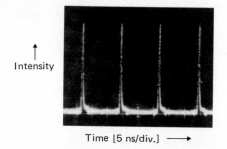

Intensity

Time [5 ns/div.] ⟶

(a)

Fig. 9.3a,b. Output pulses from a cw mode locked (**a**) and a pulsed mode-locked (**b**) laser

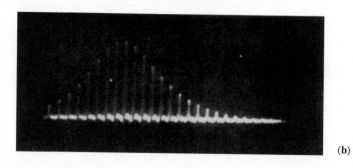

(b)

following sections. Excellent tutorial discussions of the work on mode-locked lasers with extensive references can be found in [9.2–6].

9.2 Passive Mode Locking

The nonlinear absorption of saturable absorbers was first successfully employed for simultaneously Q-switching and mode-locking solid-state lasers in 1965 [9.7, 8]. The saturable absorbers consisted of organic dyes that absorb at the laser wavelength. At sufficient intense laser radiation, the ground state of the dye becomes depleted, which decreases the losses in the resonator for increasing pulse intensity.

In pulsed mode-locked solid-state lasers, pulse shortening down to the limit set by the gain-bandwidth is prevented because of the early saturation of the absorber which is a result of the simultaneously occurring Q-switching process. Shorter pulses and a much more reproducible performance is obtained if the transient behavior due to Q-switching is eliminated. In steady-state or cw mode locking, components or effects are utilized which exhibit a saturable absorber-like behavior, i.e. a loss that decreases as the laser intensity increases. In this section, we will review pulsed and cw passive mode locking.

9.2.1 Pulsed Passive Mode Locking

As mentioned in Chap. 8, a saturable absorber has a decreasing loss for increasing pulse intensities. The distinction between an organic dye suitable for simultaneous mode-locking and Q-switching, as opposed to only Q-switching the laser, is the recovery time of the absorber. If the relaxation time of the excited-state population of the dye is on the order of the cavity round trip, i.e. a few nanoseconds, passive Q-switching will occur, as described in Chap. 8.

With a dye having a recovery time comparable to the duration of mode-locked pulses, i.e. a few picoseconds, simultaneous mode-locking and Q-switching can be achieved. Fast recovery times usually arise from non-radiative decay in the dyes, whereas the slower relaxation times are due to spontaneous emission.

We will now discuss the pulse formation in a passively mode-locked laser. A computer simulation of the evolution of a mode-locked pulse train from noise is shown in Fig. 9.4 [9.9]. This figure shows the transformation of irregular pulses into a single mode-locked pulse. In Figs. 9.4a–c the noise-like fluctuations are linearly amplified, however, a smoothening and broadening of the pulse structure can be seen. In Figs. 9.4d–f the peak-to-peak excursions of the fluctuations have increased and, in particular, the amplitude of the strongest pulse has been selectively emphasized. In Fig. 9.4f the background pulses have been completely suppressed.

The simultaneous Q-switched and mode-locked pulse evolution from noise can be divided into several stages.

Linear Amplification Stage. At the beginning of the flashlamp pulse, the laser gain is not high enough to overcome the loss of the saturable absorber. Population inversion of the laser transition and the transmission of the bleachable dye are not affected by the radiation. As a result, both the amplification and absorption processes can be considered to be linear. The intensity pattern is that of spontaneous emission with a spectral content about equal to the fluorescence bandwidth. Interference of the laser modes with random phase relations leads to fluctuations of the light intensity. The total number of fluctuations is large, being of the order of the number of cavity modes, but there is a small number of intensity peaks exceeding the average intensity significantly. The chaotic sequence of fluctuations shown in Fig. 9.4a represents the laser radiation in the early stage of pulse generation.

As the pump process continues, the gain increases above threshold and the noise-like signal is amplified. During the linear amplification a natural mode selection takes place because the frequency-dependent gain favors cavity modes in the center of the fluorescence line. As a result of the spectral narrowing caused by the amplification process, a smoothening and broadening of the amplitude fluctuations occurs, as shown in Figs. 9.4b and c.

The following numbers are typical for the linear amplification process in Nd:glass. A large number of longitudinal modes is initially excited. For a typical cavity length of ~ 1 m, one calculates $\sim 4 \times 10^4$ cavity modes in Nd:glass with

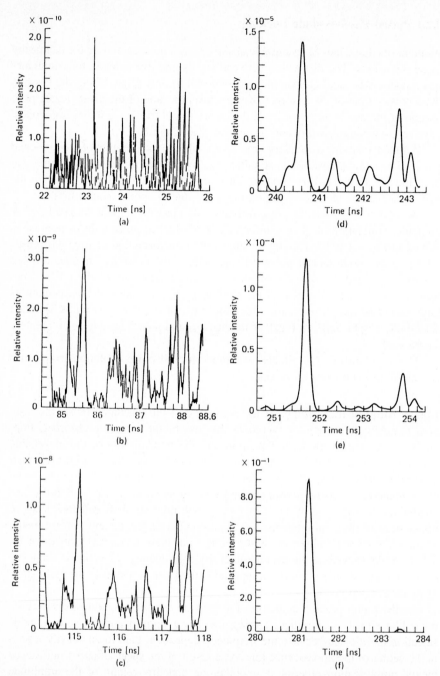

Fig. 9.4a–f. Computer simulation of the evaluation of a mode-locked pulse from noise. **(a–c)** regime of linear amplification and linear dye absorption, **(d–e)** nonlinear absorption in the dye cell, **(f)** regime of nonlinear amplification, dye completely bleached [9.9]

$\delta \nu_F \approx 7500$ GHz. Assuming a typical pulse build-up time of $10\,\mu$s, the linear stage comprises ≈ 1500 round trips. The light intensity rises by many orders of magnitude to approximately 10^7 W/cm^2.

Nonlinear Absorption. In this second phase of pulse evolution, the gain is still linear but the absorption of the dye cell becomes nonlinear because the intensity peaks in the laser cavity approach values of the saturation intensity I_s of the dye (Fig. 8.26). In the nonlinear regime of the mode-locked laser we note two significant processes acting together:

First, there is a selection of one peak fluctuation or at least a small number. The most intense fluctuations at the end of the linear amplification phase preferentially bleach the dye and grow quickly in intensity. The large number of smaller fluctuations, on the contrary, encounter larger absorption in the dye cell and are effectively suppressed.

The second effect is a narrowing of the existing pulses in time, which broadens the frequency spectrum. The shapes of the pulses are affected by the nonlinearity of the dye because the wings of the pulse are more strongly absorbed than the peak. The second phase ends when the absorbing transition in the dye cell is completely saturated. Under favorable conditions the final transmission is close to one; i.e., the dye is transparent. The nonlinear action of the absorber at the intermediate power regime was illustrated in Figs. 9.4d and e.

Nonlinear Amplification. The final phase of the pulse evolution occurs when the intensity is sufficiently high for complete saturation of the absorber transition to take place and for the amplification to be nonlinear. This is the regime of high peak power. During the nonlinear stage the pulse intensity quickly rises within ≈ 50 cavity round trips to a value of several gigawatts per square centimeter. As was shown in Fig. 9.4f, at this point the background pulses have been almost completely suppressed. Successive passages of the high-intensity radiation pulse through the resonator result in a pulse train appearing at the laser output. Finally, the population inversion is depleted and the pulse decays.

The design elements of interest in pulsed passive mode-locked lasers are the resonator configuration, dye cell, and the gain media.

Resonator. One major requirement in the resonator design of a mode-locked system is the complete elimination of reflections which can occur from components located between the two cavity mirrors. This is accomplished by employing laser rods with Brewster's angle at the ends, placing the dye cell at Brewster's angle in the resonator, and by using cavity mirrors which are wedged. Reflection from an optical surface which is parallel to the cavity mirrors will create a secondary resonator. The mode-locked pulse will be split into several pulses which will circulate inside the resonators with different round-trip times. The result is a very erratic output usually consisting of several superpositioned pulse trains or containing subsidiary pulses in the train. With all optical surfaces inside the

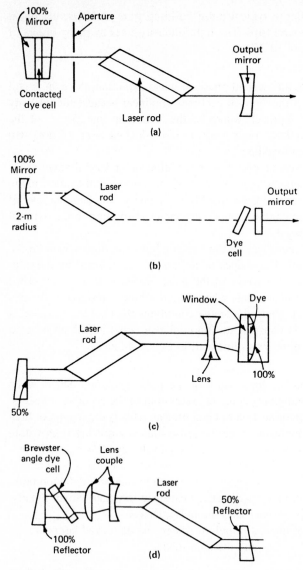

Fig. 9.5a–d. Experimental arrangements for passively mode-locked lasers. Oscillator cavity with optically contacted dye cell **(a)** and with dye cell tilted at Brewster angle **(b)** arrangements, including telescopes in the resonator, are shown in **(c)** and **(d)**

resonator, either at Brewster angle or antireflection-coated and tilted away from the resonator axis, coupled resonator structures can be avoided and the occurrence of satellite pulses is minimized. Similar attention must be paid to avoid back-reflection into the cavity from external components.

Figure 9.5 shows various configurations of mode-locked oscillators. In order to reduce the number of reflective surfaces in the laser cavity and, therefore, minimize the possibility of secondary reflections, the dye cell and rear mirror are often combined. As shown in Fig. 9.5a, the 100 % mirror takes the place of the rear dye-cell window. A saturable absorber, where the dye is in contact with the cavity mirror, not only provides the most reliable mode-locking operation but also yields the shortest pulses. Typically, dye cells vary in length from 1 cm to 0.1 cm. However, it has been found that the width of the individual pulses in a mode-locked train have a direct relationship to the optical path length of the dye cell. The most reliable mode locking and the shortest pulses are obtained when the saturable absorber is placed in contact with one of the dielectric mirrors and the dye thickness l satisfies the condition $l \leq c\tau/2n$, where τ and n are the relaxation time and refractive index of the dye, respectively, and c is the speed of light. For Eastman dyes A9740 and A9860, $l \approx 1$ mm. The relationship between dye thickness and pulse duration has been investigated in [9.10] and the shortest pulses have been measured for $l \approx 30\,\mu$m.

The resonator length is usually chosen between 1 and 1.5 m; in this case the pulse separation is of the order of 10 ns, which makes the selection of a single pulse with an external gate relatively simple. The reflectivity of the front mirror is typically between 50 and 60 %. The output reproducibility from a mode-locked laser is drastically improved by using at least one curved mirror instead of two plane mirrors. Very often a mode-selecting aperture is inserted into the resonator of a mode-locked oscillator because in multitransverse-mode systems the power density is so large in localized areas that component damage frequently occurs. Sometimes a Galilean telescope or a single lens is included in the laser cavity to increase the beam diameter, which reduces the optical power density inside the dye and at the rear mirror.

If the pulse from a mode-locked laser is too short, and one desires to stretch it, a tilted etalon can be inserted into the resonator. This will reduce the number of longitudinal modes and according to (9.4) leads to a longer pulse duration.

Liquid Dye Saturable Absorber. A saturable absorber employed for mode locking must have an absorption line at the laser wavelength, a linewidth equal to or greater than the laser line width, and a recovery time on the order of the width of the mode-locked pulses. Table 9.1 lists the recovery time and saturation flux of four dyes commonly employed in mode-locked lasers [9.11]. For a discussion of the structure and properties of organic dyes, the reader is referred to [9.12].

Nd : glass and Nd : YAG lasers are mode-locked with Eastman 9740, 9860 or 14 015 dye suitably diluted with 1,2-dichloroethane or chlorobenzene. The dye concentration is usually chosen to produce a linear transmission between 50 and 80 % at 1.06 μm through the cell. Experimental data concerning the properties of these dyes are published in [9.13, 14].

Practically all dyes useful in the generation of mode-locked pulses decompose when exposed to ultraviolet light. In particular, Eastman 9740 and 9860 are extremely sensitive to light in the uv region. To eliminate dye breakdown due to uv radiation from flashlamp and ambient light, the dye cell should be well light-shielded with only a small aperture exposed for the laser beam. UV-absorbing

Table 9.1. Saturation density I_s and recovery time τ_s of various
dyes employed for mode-locking

Dye	Eastman No. 9740	Eastman No. 9860	DDI	Cryptocyanine
I_s [W cm^{-2}]	4×10^7	5.6×10^7	$\approx 2 \times 10^7$	5×10^6
τ_s [ps]	8.3	9.3	14	22
Laser	Nd	Nd	Ruby	Ruby

glass and quartz can be used for the dye cell windows, and special uv-free fluorescent lighting around the laser area is usually installed. A suitable fluorescent
lamp is the General Electric Series F96 T12/Gold. Otherwise, while handling
these dyes, fluorescent lights should be turned off and a tungsten-filament lamp
should be used for illumination.

Ruby lasers are usually mode locked with either cryptocyanine or 1, 1'-
diethyl-2, 2'-dicarbocyanine iodide (DDI). The former dye can be diluted with
nitrobenzene, acetone, ethanol, or methanol. The most consistent mode-locked
operation is achieved with acetone, mainly because the absorption peak for cryptocyanine in acetone exactly coincides with the ruby line, while in the other
solvents it is displaced by more than 100 Å compared to the ruby wavelength.
Shorter mode-locked pulses and a more reliable operation is achieved with DDI
diluted in methanol or ethanol.

Saturable absorbers should be replenished with fresh solution periodically.
By far the most reliable performance from mode-locked dye systems is obtained
if the dye is circulated through the cell from a large reservoir. The pumping
action assures uniform mixing of the dye; because of the large volume, the
dye's concentration remains constant over a long period of time and fresh dye
is exposed to each laser pulse.

It has also been found that for maximum reliability of operation, it is necessary to temperature-control the dye solution. A change in absorption of 1%
per degree was found in the diluted Eastman dyes 9740 and 9860. Stable and
reproducible performance was achieved with a flowing dye solution temperature-
controlled to $\pm 0.1°$C. Such a system usually consists of a pump and temperature-
controlled dye reservoir, micropore filter, and a laminar-flow dye cell. Materials
which come into contact with the dye solution must be limited to stainless steel,
teflon, and glass.

Gain Medium. In the past, flashlamp-pumped Nd : glass lasers have been of
particular interest for mode-locking since these lasers have a broader gain bandwidth as compared to Nd : YAG and ruby, and were therefore expected to produce
shorter pulses.

Typical flashlamp-pumped mode-locked Nd : glass lasers produce pulse trains
about 50 to 200 ns wide, containing pulses of 5 to 20 ps in duration. The Brewster-
ended rods have dimensions ranging from 8 to 20 cm in length and up to 1.5 cm
in diameter. Mode locking is achieved with dye cells having a path length of

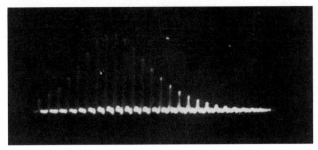

Fig. 9.6. Oscilloscope trace of a typical passively mode-locked Nd : glass laser. Horizontal 20 ns/div. Rise and fall times and pulse width are all detector-limited in this photograph

several millimeters. Mirrors or windows are wedged at least 30 arc/min so that they do not act as mode selectors. The total energy output can vary from 25 mJ to several hundred millijoules, and the energy content of the pulses in the middle of the train is between 1 to 10 mJ.

Figure 9.6 shows an oscilloscope trace of a typical transient passively mode-locked Nd : glass laser.

From ruby lasers of approximately similar geometry as the glass lasers mentioned above, pulse durations from 5 ps to 30 ps have been achieved. Thicknesses of dye cells range typically from several millimeters to a fraction of a millimeter. Optimum absorber cell transmission for most systems is between 0.6 and 0.8. For a typical ruby rod, 10 cm long and 1 cm in diameter, the total energy in the mode-locked pulse train is around 50 mJ and the energy per single pulse is about 2 to 5 mJ in the center of the train.

Mode-locked Nd : YAG lasers typically employ laser rods ranging from 3×5 mm to 6×75 mm in size. With 1-mm-thick dye cells pulses between 20 and 40 ps in duration have been produced. Depending on dye concentration and flashlamp energy, the pulse trains are usually 10 to 80 ns long. Figure 9.7a shows a typical mode-locked train generated by a Nd : YAG laser. A portion of the pulse in the center was switched out in order to measure the pulse width. Figures 9.7b and c show an oscillogram and a two-photon absorption measurement of this pulse.

Despite the relatively simple construction of a passively mode-locked laser oscillator, the output will be very unpredictable unless dye concentration, optical pumping intensity, and resonator alignment are carefully adjusted. It is not uncommon to find that the average pulse duration from one train of picosecond pulses to the next changes significantly and that the pulse train envelopes are not reproducible. For instance, one may find that, in ≈ 10 shots, optimum mode locking occurred only once. With the proper saturable absorber and with a judiciously chosen optical system, the probability of obtaining clean mode-locked pulse trains which are free of subsidiary pulses is typically 0.6 to 0.7 in pulsed passively mode-locked solid-state lasers. Furthermore, mixing and handling the dye solution and maintaining proper dye concentration proved cumbersome. As

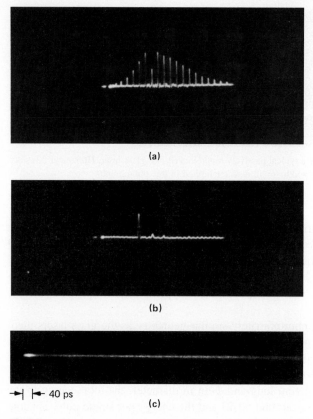

(a)

(b)

→| |← 40 ps

(c)

Fig. 9.7a–c. Passively mode-locked Nd : YAG laser. Oscillograms of **(a)** the pulse train, **(b)** a single pulse, and **(c)** a two-photon absorption measurement

a result of the inherent shortcomings of pulsed passive mode-locking, this technology has been superseded by cw mode locking.

9.2.2 Continuous Wave Passive Mode Locking

In recent years, several passive mode-locking techniques have been developed for solid-state lasers whereby fast saturable absorber-like action is achieved in solids. Most of these novel optical modulators utilize the nonresonant Kerr effect. The Kerr effect produces intensity-dependent changes of the refractive index. It is generally an undesirable effect because it can lead to self focusing and filament formation in intense beams, as explained in Sects. 4.3.1 and 11.3.

In contrast to the absorption in bleachable dyes, the nonresonant Kerr effect is extremely fast, wavelength-independent, and allows the generation of a continuous train of mode-locked pulses from a cw-pumped laser. We can broadly distinguish between two categories of passive mode-locking techniques based on the optical Kerr effect: coupled-cavity or Additive Pulse Mode locking (APM), and Kerr-Lens Mode locking (KLM). In addition to these two technologies, we shall also briefly review cw mode locking by means of passive semiconductor reflectors. This promising approach, based on molecular beam epitaxy technology, can be applied either for cw mode locking of solid-state lasers or for providing the start-up mechanism for KLM.

We can describe here only the most basic aspects of ultra-fast pulse generation via passive cw mode locking. The promise of compact, reliable and low-maintenance laser sources with output pulses in the femtosecond regime has created a large interest in this field. As a result, over a hundred articles are published every year on this topic and special sessions at conferences are devoted to ultra-fast pulse generation. For an in-depth treatment of the subject, the reader is referred to [9.2–4, 15].

Additive Pulse Mode Locking

In this technique, mode locking is induced by feeding back into the laser part of its output after it has been nonlinearly modulated in an external cavity. This technique has been termed Additive Pulse Mode locking (APM) because the pulse-shaping mechanism depends on the coherent addition of pulses that are fed back from the external cavity to pulses in the main laser cavity.

In most cases, the external cavity contains an optical fiber in which a nonlinearity is produced via the intensity-dependent index of refraction. Mode locking employing a coupled cavity with a fiber was first demonstrated in the soliton laser [9.16]. Later, the technique was mainly used to enhance the performance of actively mode-locked lasers. *Goodberlet* et al. [9.17] extended the technique to achieve self-starting, passive APM in a tunable Ti : Al_2O_3 laser. Figure 9.8a illustrates the basic set-up which we will use to illustrate the principle of APM.

The main laser cavity consists of a four-mirror arrangement and an argon laser pumped Ti : Al_2O_3 crystal cut at Brewster angle. The wavelength can be adjusted with a birefringent plate. The output mirror M_o had a 15% transmission. A beam splitter (BS) directed half of the output energy via a GRIN lens into a 5 μm core diameter fiber. The optical beam is reflected back to the main resonator with a mirror that is attached to the end of the fiber. For a given fiber length, the time of flight of a pulse in the external cavity has to be adjusted to be an integral multiple of the main laser cavity. This ensures that pulses that are reinjected from the external cavity are synchronized with the pulses in the main cavity. When a pulse from the main laser is coupled into the external optical fiber, it experiences an intensity-dependent phase shift and will interferometrically recombine with the pulse in the main laser cavity at the output mirror M_o.

When the relative phases are set appropriately, the peak of the pulse constructively interfereres, while the wings of the pulse destructively interfere. This situation is illustrated in Fig. 9.8.b. The nonlinearly coupled cavity can be replaced by an intensity-dependent mirror reflection $R(I)$ which is shown as a function of the optical center frequency ν. An intensity-dependent phase shift

(a)

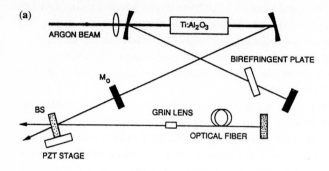

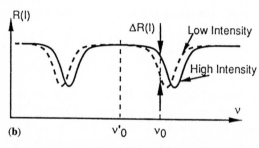

(b)

Fig. 9.8a,b. Additive pulse mode locking using an optical fiber in the external cavity (**a**) [9.17], and intensity dependent phase shift in the coupled cavity (**b**)

causes the reflectivity curve as a function of ν to shift sideways. In order to achieve mode locking, an intensity-dependent reflectivity, which is high at high intensity and low at low intensity, is required. Therefore, the cavity length has to be actively held at the lasing center frequency ν_0. Since feedback from the nonlinear external cavity is interferometric, the relative lengths of the main cavity and the external cavity must be stabilized to within a fraction of a wavelength.

While the interferometric stability is required for the relative phases, the absolute cavity mismatch is less critical and can be on the order of $10\,\mu$m. Once mode locking is achieved, the length of the external cavity is controlled by the use of locking electronics. The system depicted in Fig. 9.8a produced a stable train of 1.4 ps pulses at an average power of 300 mW if pumped at 5 W from an argon laser.

Additive-pulse mode locking has been applied to a number of other solid-state lasers, such as Nd : YAG [9.18–20] Nd : YLF [9.21] and Nd : glass [9.22, 23].

Figure 9.9 shows a schematic diagram of a diode-pumped Nd : YAG laser [9.20]. The end-pumped plane/Brewster design yielded a cw output of 250 mW TEM_{00} mode for 0.9 W diode-pump power. The nonlinear external cavity is formed by a beam splitter, a one-meter long single-mode fiber and a highly reflecting mirror at the end of the fiber. Coupling into the fiber was accomplished with GRIN lenses. To achieve self-starting of the APM process, it is desirable to have a large nonlinearity in the coupled cavity. The phase shift of this Kerr nonlinearity can simply be enhanced by increasing the length of the fiber. The nonlinear phase shift scales as the product, $n_2 L_f I$, n_2 being the nonlinear index of refraction of the fiber, I the intensity in the fiber, and L_f the fiber length. The

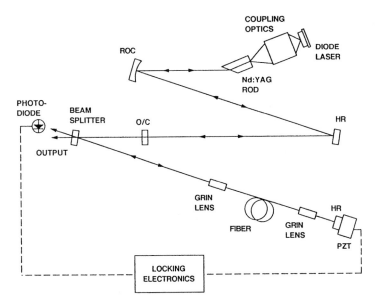

Fig. 9.9. Schematic diagram of an additive pulse mode locked Nd : YAG laser [9.20]

laser depicted in Fig. 9.9 generated 2 ps pulses at 110 mW average output and peak powers of 410 W. With the absolute length of the external cavity adjusted for good pulse overlap, and the relative cavity length stabilized to $\lambda/2$, amplitude fluctuations were reported to be about 1%.

Like with all APM systems, an electronic servo system is required to maintain the relative phase difference of the two cavities. Typically, one of the mirrors of the external cavity is mounted on a PZT. Matching of the cavity lengths is accomplished by micrometer adjustment, with fine tuning via dc bias to the PZT. Once stable operation is achieved, an electronic feedback loop is required to maintain mode locking. Most APM lasers employ a stabilization scheme originally developed for soliton lasers [9.24]. In this design, the correct relative phase is maintained by utilizing the fact that not only the pulsewidth, but also the average power, is a sensitive function of the relative phases. The laser output is monitored and compared to a reference signal, an integrated error signal is then applied to the PZT. By changing the reference voltage, the phase difference between the two cavities can be adjusted for optimum performance.

The disadvantage of coupled cavity mode locking employing optical fibers is the requirement for interferometric stabilization of the cavities. Electronic feedback loops are capable of maintaining such precision over certain time periods. However, the lasers are very susceptible to environmental changes, because any perturbation which exceeds the dynamic range of the feedback loop will cause the lasers to break into relaxation oscillators or self Q-switching. Clearly the need for interferometric control and precision focusing of the beam into a single-mode fiber restricts the use of these systems to the research laboratory.

For this reason, interest has focused on cw passive mode-locking concepts which do not require this amount of precision and complexity.

Kerr-Lens Mode Locking (KLM)

In the previous section, we have seen that an intensity-dependent phase shift in an optical fiber can be exploited for passive mode locking. However, the Kerr effect leads also to an intensity-dependent variation of the beam profile. For a Gaussian beam, the Kerr effect focuses the radiation towards the center, essentially an intensity-dependent graded-index lens is formed. The action of a fast saturable absorber can be achieved if an aperture is introduced in the resonator at a position where the mode size decreases for increased intensity.

The transformation of the power-dependent change in the spatial profile of the beam into an amplitude modulation is illustrated in Fig. 9.10. Transmission through the aperture is higher for an intense beam, as compared to a low-power beam. This technique which provides an extremely simple means for ultra-short pulse generation in tunable lasers has been termed Kerr-Lens Mode locking (KLM).

The laser crystal acts as a lens with a focal length which changes with the intracavity intensity. With the assumption of a parabolic index variation and a focal length much longer than the Kerr medium, we obtain to a first approximation [9.34]

$$f = \frac{w^2}{4n_2 I_0 L} \tag{9.12}$$

where w is the beam waist, n_2 is the nonlinear index, I_0 is the peak intensity and L is the length of the Kerr medium.

Assuming a waist size of 50 μm in a 4 mm long Ti : Al_2O_3 crystal, and a peak power of 150 kW, we obtain with $n_2 = 3.45 \times 10^{-16}$ cm^2/W a focal length of $f = 24$ cm.

As will be discussed in Sect. 11.3, beam focusing due to the Kerr effect can lead to catastrophic beam collapse with the beam breaking up into filaments. The critical power level at which beam collapse will set in is usually defined by

$$P_{cr} = \frac{a\lambda^2}{8\pi n_0 n_2} \tag{9.13}$$

where the factor a can take on values between 3.77 and 6.4 depending on the severity of nonlinear phase distortion [9.25]. For the materials and wavelengths of interest, one finds $P_{cr} = (2-3)$ MW. For example, $P_{cr} = 2.6$ MW for Al_2O_3 and BK7 glass at 0.8 μm. In Kerr-lens mode-locked lasers, the peak power has to be large enough to produce a strong nonlinearity, but needs to be well below the critical power for beam collapse.

Returning to Fig. 9.10, for a hard aperture of radius R, the rate of change in transmission T with respect to the beam waist w for a Gaussian beam is given by

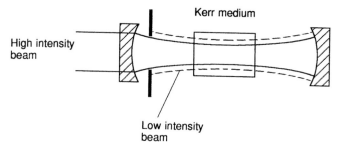

Fig. 9.10. Intensity dependent amplitude modulation of the resonator mode

$$dT/dw = (-4R^2/w^3)\,\exp(-2R^2/w^2)\,. \tag{9.14}$$

If one assumes that the aperture size is close to the mode size $R \approx w$, one obtains $\delta T \approx -(1/2w)\delta w$. The limiting aperture introduces a loss discrimination for low power (cw) and high peak power (mode-locked) operation which is proportional to mode size changes.

Since the mode size w is power dependent due to the Kerr effect, the resonator losses are expressed as

$$L = L_0 - KP \tag{9.15}$$

where L_0 are the fixed losses, P is the intracavity power and K is the nonlinear loss coefficient. For the design of KLM systems, it is more convenient to use instead of K, the small-signal relative spot size variation δ, where $\delta \sim -K$. The parameter δ, also called the Kerr-lens sensitivity is an important factor for analyzing and designing KLM resonators. It is defined as

$$\delta = \left(\frac{1}{w}\frac{dw}{d(P/P_{\mathrm{cr}})} \right) \tag{9.16}$$

where w is the spot size at a given place inside the resonator. The second term is the slope of the spot size vs. normalized power, taken at $P = 0$.

In Fig. 9.11, the beam waist at the center of a Ti:Al$_2$O$_3$ crystal is plotted [9.26]. The calculations were performed for a 100 fs pulsewidth, 3 W intracavity average power, and a 80 MHz pulse repetition rate. The beam radius decreases with a slope of approximately $-1.5 \times 10^{-5}\,\mu$m/W. Introducing this value into 9.16 and normalizing to the spot size w and critical power P_{cr} gives $\delta = -1.36$.

A typical resonator used for Kerr-lens mode locking of a Ti:Al$_2$O$_3$ crystal pumped by an argon laser, consists of two plane mirrors and two focusing mirrors. The crystal is at the beam waist as shown in Fig. 9.12. An aperture is usually located close to one of the flat mirrors. As will be explained later, the spot size variation δ at the aperture as a function of the circulating power, critically depends on the position x of the Kerr medium around the focus of the beam, and on the separation z of the two focusing mirrors.

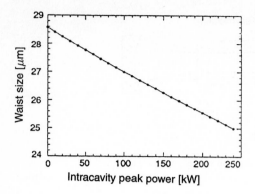

Fig. 9.11. Beam-waist radius as a function of peak power [9.26]

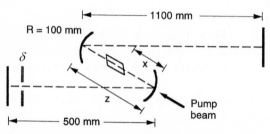

Fig. 9.12. Typical resonator used for Kerr lens mode locking

We will first consider the properties of the empty resonator illustrated in Fig. 9.12. The equivalent resonator consists of two flat mirrors with two internal lenses essentially forming a telescope with magnification of one, as shown in Fig. 9.13. For a slightly misadjusted telescope, the lens spacing is $z = 2f - \Delta z$ where Δz measures the misadjustment. The focal length of the lens assembly is then $f_{ef} = f^2/\Delta z$, where f is the focal length of the individual lenses. The distances from the resonator mirrors and the principal planes of the lenses are l_1 and l_2. The g-parameters for a resonator comprised of two flat mirrors and an internal telescope are [9.27]

$$g_1 = 1 - (l_2/f_{ef}) \quad \text{and} \quad g_2 = 1 - (l_1/f_{ef})$$

or

$$g_1 = 1 - (l_2\Delta z/f^2) \quad \text{and} \quad g_2 = 1 - (l_1\Delta z/f^2)$$

(9.17)

The operating region of the passive resonator is depicted in the stability diagram (Fig. 9.14). For $\Delta z = 0$, the lenses are in focus and the resonator configuration becomes plane-parallel. As the lenses are moved closer together, a stability limit is reached for $g_2 = 0$. As both g-parameters become negative, a second zone of stable operation is reached. As Δz becomes larger, a point is reached for $g_1 g_2 = 1$, where the lens assembly focuses the beam onto one of the flat mirrors, i.e. the spot size becomes zero at the mirror. If the two legs of the resonator are of unequal length, $l_1 \neq l_2$, two stability regions exist, as indicated by curve A in

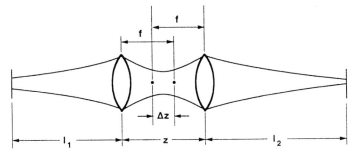

Fig. 9.13. Equivalent resonator of the arrangement shown in Fig. 9.12 (f = 50 mm, l_1 = 500 mm, l_2 = 1100 mm)

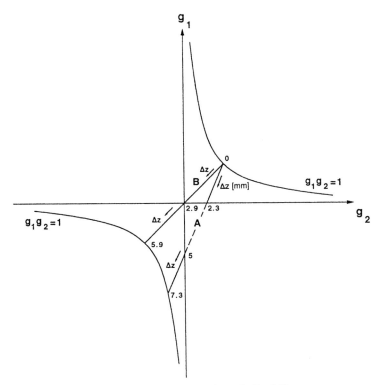

Fig. 9.14. Stability diagram for the resonator shown in Fig. 9.13

Fig. 9.14. Curve B gives the stability range of a resonator of the same total length as before, but with $l_1 = l_2$ = 850 mm. The two stability regions are joined and the resonator changes from plane-parallel, confocal, to concentric for increasing values of Δz (or decreasing mirror separation z).

An analysis of the actual resonator depicted in Fig. 9.12 has to include the optical length of the crystal and the change of the refractive index with beam intensity. A particularly useful and practical design procedure for KLM lasers

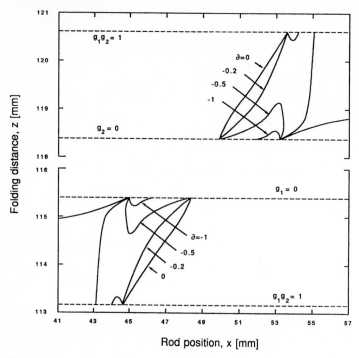

Fig. 9.15. Contour plots of Kerr lens sensitivity δ as a function of mirror separation z and rod position x (see also Fig. 9.12) [9.30]

has been developed by *Magni* et al. [9.28, 29]. They introduced a nonlinear ABCD ray matrix to treat the Gaussian beam propagation in a Kerr medium. The dependence of the spot size w on the power P is given in closed form by a set of equations. Based on the formalism provided in [9.28, 29] contour plots can be generated of the spot size variation δ as a function of x and z for a given spot size and power of the input pump beam.

The dimensions of x and z determine the Kerr-lens sensitivity, resonator stability and the beam waist inside the crystal. Figure 9.15 exhibits such a contour plot for the resonator depicted in Fig. 9.12. The operating region is within the two stability zones, as explained earlier. For KLM operation, the additional constraint is for a δ which is negative and large. As is evident from Fig. 9.15, this reduces the operating range of the resonator to a tiny area within the allowed stability region. It was found experimentally [9.30], that the region for the large mirror separation z (upper half of Fig. 9.15) was more stable as compared to the smaller separation (lower half of Fig. 9.15). In the stability diagram Fig. 9.14, the more stable operation corresponds to operation in the first quadrant.

The maximum Kerr-lens sensitivity is achieved for [9.29]

$$|\delta_1|_{\max} = \frac{1}{\sqrt[4]{g_1 g_2 (1 - g_1 g_2)}} \, . \tag{9.18}$$

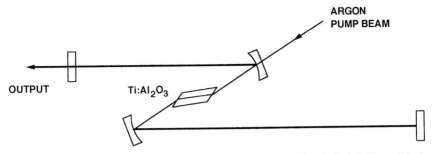

Fig. 9.16. Schematic diagram of the resonator configuration for a self-mode-locked Ti : sapphire laser

Therefore, KLM resonators operate very close to the stability limit, i.e. $g_1 g_2 = 1$, $g_1 = 0$ or $g_2 = 0$. For example, we notice that in Fig. 9.15, the largest δ is achieved if one of the g-parameters is close to zero. Of course, this means also that the resonator is very susceptible to environmental effects, therefore the operating point has to be a compromise between a large δ and stable and reliable performance.

The Kerr-lens medium can be the laser crystal itself or a separate element such as a piece of glass. Also, the resonator can include a hard aperture, as discussed so far, or a soft aperture provided by the spatial profile of the gain in the laser crystal. Particularly in longitudinally pumped lasers, the gain has a Gaussian profile with the peak at the center where the pump intensity is highest. Therefore, a smaller beam will experience higher gain. Gain aperturing is mainly observed in Ti : Al_2O_3 lasers because this material has a much higher saturation intensity, as compared to Nd : YAG or Nd : YLF. The higher gain, which a smaller mode experiences, is limited by gain saturation. The latter effect causes defocusing, which increases the beam cross-section, thereby canceling nonlinear effects.

The simplest embodiment of Kerr-lens mode locking utilizes the nonlinear index of the laser crystal as the Kerr medium and the soft aperture formed by the gain profile within the lasing material. Kerr-lens mode locking was actually first observed by *Spence* et al. [9.31], in an argon pumped Ti : Al_2O_3 laser which did not contain any additional passive or active element in the resonator. Therefore, the laser was called *self-mode locking*. Figure 9.16 illustrates the simplicity of the set-up in which self-mode locking was observed.

Because KLM mechanism simulates intracavity saturable absorbers, no cavity length control is required for passive mode locking. KLM uses the nonlinearity of the whole laser crystal (bulk effect). It is well-suited for mode locking solid-state lasers which have high intracavity intensities. In contrast, APM can be employed in low power lasers since nonlinear phase shift can be integrated over the length of the fiber. Subsequent to its demonstration using Ti : sapphire, Kerr-lens mode locking has been demonstrated in a number of different materials: Cr : LiSAF, Cr : LiCAF, Nd : Glass, Nd : YAG, Nd : YLF, and Cr : Forsterite.

Kerr-lens mode locking is the preferred method for the generation of femtosecond lasers. Pulses on the order of a few tens of femtoseconds have been obtained from solid-state tunable lasers. The ultimate limit of pulse duration for Ti : sapphire is somewhere in the range of $\sim$ 3 fs, or about 1 cycle of light at 800 nm. Some of the engineering aspects of femtosecond lasers such as self-starting of the KLM process, dispersion compensation of the resonator and elimination of spurious reflections, will be discussed in Sect. 9.5. The reader interested in analytical models describing the KLM effect is referred to [9.25, 28, 29, 32, 33, 34].

Semiconductor Saturable Absorber Mirrors

In the early 1990's a new mode-locking device was explored which is based on absorption bleaching in a semiconductor. Molecular-Beam-Epitaxy (MBE) technology allows the design of multi-layer quantum-well structures which have a resonant nonlinearity at the laser wavelength. In these devices, reflectivity increases with increasing beam intensity. The devices have absorbing quantum well layers on top of a highly reflective coating. During the pulse duration increasing numbers of carriers are created causing absorption bleaching. The resonant nonlinearity can be custom-designed by band-gap engineering.

Nonlinear mirrors based on absorption bleaching of a multiple quantum well have been employed for mode-locking of solid-state lasers or to provide a start and stabilization mechanism for KLM modelocking [9.35–39]. The advantage of a semiconductor nonlinear mirror is the inherent device simplicity; essentially such a mirror simply replaces one of the flat cavity mirrors. A number of problems had to be solved before a semiconductor component could be placed inside a laser resonator. In general, semiconductors introduce a high insertion loss, and they have a small saturation intensity and a low damage threshold. Because of the high loss, quantum-well reflectors have been initially used in nonlinear coupled cavities. However, if the absorber layer is integrated into a device structure, key parameters can be adjusted to be compatible with the laser operating parameters. Therefore, subsequent improvements lead to the design of intracavity semiconductor saturable mirrors.

One solution is to integrate the absorbing nonlinear layer inside an anti-resonant Fabry-Perot resonator. Typically, a InGaAs/GaAs semiconductor saturable absorber is monolithically embedded between two reflecting mirrors. The top reflector is a dielectric mirror and the bottom reflector a Bragg reflector or a layer of silver coating. The thickness of the saturable absorber is adjusted for anti-resonance. In an anti-resonant Fabry-Perot, the intensity inside the device is smaller than the incident intensity, which decreases the device loss and increases the saturation intensity. The high damage threshold top reflector determines the amount of radiation that can penetrate and bleach the absorber. Therefore, the top reflectivity determines also the saturation fluence and the insertion loss.

The MBE growth temperature is a variable parameter which adjusts the nonlinear reflectivity. Typical saturation fluence is around $60\,\mu\mathrm{J/cm}^2$ for a InGaAs/GaAs device. Depending on the reflectivity of the top mirror, the effective saturation fluence can be increased up to two orders of magnitude. Depending on the reflectivity and structure of the top and bottom reflectors and the design of the quantum-well saturable absorber, many different configurations have been explored. Comprehensive overviews of this technology can be found in [9.40, 41]. As mentioned, these semiconductor devices can be employed to mode-lock lasers, but they can also be used to start and sustain stable Kerr-lens mode locking. The latter issue will be addressed in Sect. 9.5. Since these quantum-well structures are resonant devices, the tunability of the laser is somewhat restricted. However, with bandgap engineering the wavelength can be adjusted to provide absorption for most solid-state lasers. For example, in a Ti:sapphire laser, pulses as short as 6.5 fs have been generated with an intracavity saturable absorber which assisted Kerr-lens mode-locking [9.40], and 60 fs pulses from a diode pumped Nd:glass laser have been demonstrated which was mode-locked by a low finesse anti-resonant Fabry-Perot saturable absorber [9.42].

9.3 Active Mode Locking

By placing inside a laser cavity either a phase modulator (FM) or an amplitude modulator (AM) driven at exactly the frequency separation of the axial modes, one can cause the laser to generate a train of mode-locked pulses with a pulse repetition rate of $f_m = c/2L$. Active mode locking, performed on cw-pumped lasers such as the Nd:YAG system, is achieved by inserting into the resonator an electrooptic or acoustooptic modulator, as will be discussed in Sect. 9.3.1.

A cw actively mode-locked laser generates a train of equal pulses at a repetition rate typically in the range of (80–250) MHz with pulse energies in the nJ range. If more energy per pulse is required, one pulse is selected from the train and directed to a regenerative amplifier, as described in Sects. 9.4.2. However, one can also obtain more powerful mode-locked pulses from a single oscillator by combining mode locking with Q-switching and cavity dumping. This approach will be discussed in Sect. 9.3.2.

9.3.1 CW Mode Locking

Figure 9.17 presents an example of an end-pumped Nd:YAG laser in a folded resonator containing a LiNbO$_3$ phase modulator for active mode locking of the laser [9.43]. Although the system architecture looks identical, we have to distinguish between AM and FM modulation.

AM Modulation. From a frequency-domain viewpoint, introducing a time-varying transmission $T(t)$ through an amplitude modulator inside the laser resonator creates sidebands on each oscillating axial mode which overlap with adjoining axial modes. The operation can be described by assuming that the mode with the frequency ν_0, nearest the peak of the laser gain profile, will begin to oscillate first. If a loss modulator operating at a frequency f_m is inserted into

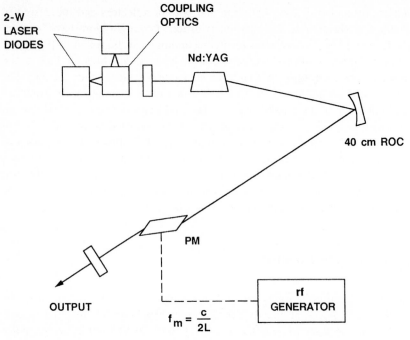

Fig. 9.17. Schematic of a FM mode locked Nd:YAG laser

the resonator, the carrier frequency ν_0 will develop sidebands at $\pm f_m$. If the modulating frequency is chosen to be commensurate with the axial mode frequency separation $f_m = c/2L$, the coincidence of the upper ($\nu_0 + f_m$) and the lower ($\nu_0 - f_m$) sidebands with the adjacent axial mode resonances will couple the $\nu_0 - f_m$, ν_0, and $\nu_0 + f_m$ modes with a well-defined amplitude and phase. As the $\nu_0 + f_m$ and $\nu_0 - f_m$ oscillations pass through the modulator, they will also become modulated and their sidebands will couple the $\nu_0 \pm 2 f_m$ modes to the previous three modes. This process will continue until all axial modes falling within the laser linewidth are coupled.

Viewed in the time domain, the same intracavity modulating element, with its modulation period equal to the round-trip transit time $2L/c$, can reshape the internal circulating field distribution repeatedly on each successive round trip inside the cavity. For example, light incident at the modulator during a certain part of the modulation cycle will be again incident at the same point of the next cycle after one round trip in the laser resonator. Light suffering a loss at one time will again suffer a loss on the next round trip. Thus, all the light in the resonator will experience loss except that light which passes through the modulator when the modulator loss is zero (Fig. 9.18a). Light will tend to build up in narrow pulses in these low-loss time positions. In a general way we can see that these pulses will have a width given by the reciprocal of the gain bandwidth, since wider pulses will experience more loss in the modulator, and narrower pulses

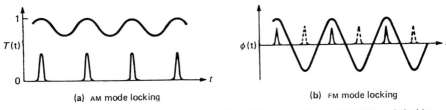

(a) AM mode locking   **(b)** FM mode locking

Fig. 9.18a, b. Mode-coupling behavior in the case of **(a)** AM mode locking and **(b)** FM mode locking

will experience less gain because their frequency spectrum will be wider than the gain bandwidth.

Kuizenga [9.44, 45] developed an elementary analysis of mode locking in homogeneous lasers by following a single mode-locked pulse through one round trip around the laser cavity. For steady-state mode locking the pulse shape should be unchanged after a complete round-trip. The self-consistent solution carried out for a Gaussian pulse leads to a simple expression which shows the dependence of the mode-locked pulse width on linewidth, modulation frequency, depth of modulation, and saturated gain.

For an acoustooptic AM modulator operating in the Bragg regime, as well as for electrooptic AM modulators, the round trip amplitude transmission is

$$T(t) \approx \cos^2 (\delta_{AM} \sin \omega_m t) , \tag{9.19}$$

where δ_{AM} is the modulation depth and $\omega_m = 2\pi f_m$ is the angular frequency of the modulation. In the ideal mode-locking case, the pulse passes through the modulator at the instant of maximum transmission. This occurs twice in every period of the modulation signal ω_m, and hence one drives these modulators at a modulation frequency equal to half the axial mode spacing of the laser. Expanding (9.19) at the transmission maximum gives for the round-trip modulation function

$$T(t) = \exp (-\delta_{AM} \omega_m^2 t^2) , \tag{9.20}$$

which results in a pulse width for the AM mode-locked laser [9.46] of

$$t_p(AM) = \gamma \frac{(gl)^{1/4}}{(\delta_{AM} f_m \Delta \nu)^{1/2}} \tag{9.21}$$

where $\gamma = 0.53$ for Bragg deflection and $\gamma = 0.45$ for Raman-Nath modulation, g is the saturated gain coefficient at the line center, $\Delta \nu$ is the gain bandwidth of the laser, and l is the length of the active medium. The AM mode-locked pulses have no frequency "chirp"; i.e., $\beta_{AM} = 0$ in (9.5).

The pulsewidth-bandwidth product for AM modulation is

$$t_p(AM) \times \Delta \nu(AM) = 0.440 . \tag{9.22}$$

From (9.21) follows that the pulse duration of cw AM mode-locked lasers is inversely related to the product of the modulation depth and modulation frequency, $t_p(AM) \sim (\delta_{AM} f_m)^{-1/2}$. It follows that pulse duration can be shortened by increasing the modulation depth or the frequency of the mode locker. The pulsewidth is also inversely proportional to the gain bandwidth $\Delta\nu$. Therefore Nd:Glass and Nd:YLF produce shorter pulses as compared to Nd:YAG.

As is the case with passive mode-locked systems, etalon effects due to intracavity elements will reduce the bandwidth of the system and broaden the mode-locked pulses. In a cw-pumped Nd:YAG laser, even a weak etalon effect due to the modulator or rod surfaces can decrease the effective value of $\Delta\nu$ by a large amount. Therefore, one of the most important considerations in a practical mode-locking system is the elimination of residual reflections and optical interference effects in the laser cavity.

On the other hand, it is possible to use a tilted etalon inside a mode-locked laser to deliberately lengthen the pulse width. With uncoated quartz etalons of thickness between 1 and 10 mm, good control of the pulse width can be achieved.

FM Modulation. Light passing through an electrooptic phase modulator will be up- or down-shifted in frequency unless it passes through at the time when the intracavity phase modulation $\delta(t)$ is stationary at either of its extrema. The recirculating energy passing through the FM modulator at any other time receives a Doppler shift proportional to $d\delta/dt$, and the repeated Doppler shifts on successive passes through the modulator eventually push this energy outside the frequency band over which gain is available from the laser medium. The interaction of the spectrally widened circulating power with the narrow laser linewidth leads to a reduction in gain for most frequency components. Thus, the effect of the phase modulator is similar to the loss modulator, and the previous discussion of loss modulation also applies here. As shown in Fig. 9.18b, the existence of two phase extrema per period creates a phase uncertainty in the mode-locked pulse position, since the pulse can occur at either of two equally probable phases relative to the modulating signal. The quadratic variation of $\delta(t)$ about the pulse arrival time also produces a frequency "chirp" within the short mode-locked pulses.

In the FM case, the internal phase modulator introduces a sinusoidally varying phase perturbation $\delta(t)$ such that the round-trip transmission through the modulator is given approximately by

$$T(t) \approx \exp\left(\pm j\delta_{FM}\omega_m^2 t^2\right),$$ (9.23)

where δ_{FM} is the peak phase retardation through the modulator. The $\pm$ sign corresponds to the two possible phase positions at which the pulse can pass through the modulator, as mentioned earlier. With these parameters the pulse width of phase mode-locked pulses is given by [9.45]

$$t_p(FM) = 0.54 \left(\frac{gl}{\delta_{FM}}\right)^{1/4} \frac{1}{(f_m\Delta\nu)^{1/2}}.$$ (9.24)

The time-bandwidth product is given by

$$t_p(FM) \times \Delta\nu(FM) = 0.626.$$ (9.25)

In an electro-optic phase modulator the phase retardation is proportional to the modulating voltage, hence $\delta_{FM} \propto P_m^{1/2}$, where P_m is the drive power into the modulator. Therefore, we obtain from (9.24) for the pulse width $t_p(FM) \propto P_m^{-1/8}$, which indicates that the pulses shorten very slowly with increased modulator drive. More effective in shortening the pulses is an increase of the modulation frequency. Since $f_m = c/2L$, the pulse width will be proportional to the square root of the cavity length.

In order to calculate the pulse width from (9.24), we can calculate the saturated gain coefficient g by equating the loop gain with the loss in the resonator

$$2gl \approx \ln\left(\frac{1}{R}\right) , \tag{9.26}$$

where R is the effective reflection of the output mirror and includes all losses. For a typical Nd:YAG laser with 10% round-trip loss, that is, $R = 0.9$, a resonator length of 60 cm, and a linewidth of 120 GHz, the pulse length is given by $t_p(FM) = 39(1/\delta_{FM})^{1/4}$. For $\delta_{FM} = 1$ rad, which is easily obtainable, pulses of 39 ps can be generated. The mode-locked pulses obtained through FM modulation show a linear frequency shift

$$\beta_{FM} = \pi^2 \left(\frac{\delta_{FM}}{4gl}\right)^{1/2} \Delta\nu f_m , \tag{9.27}$$

where β_{FM} was defined in (9.5).

It has been observed that actively mode-locked, diode-pumped solid-state lasers generate pulses with durations considerably shorter than predicted by (9.21 and 24). The shorter pulsewidth is attributed to spatial hole burning. Diode-pumped lasers are typically end-pumped with the gain medium at one end of the resonator. This causes pulse self-overlap near the end mirror which creates a population grating that inhomogeneously broadens the gain. Inhomogeneous broadening, caused by spatial hole burning, results in shorter pulses and larger time-bandwidth products, as discussed in [9.43].

9.3.2 Transient Active Mode Locking

Techniques such as simultaneous mode locking and Q-switching or cavity dumping, can substantially increase the pulse energy from a mode-locked laser. In the case of a Q-switched and mode-locked laser, a burst of mode-locked pulses is emitted which are contained within the envelope of a 100–200 ns Q-switch pulse. If the laser is cavity-dumped, a single mode-locked pulse is emitted from the oscillator.

Schematic diagrams of such systems are shown in Figs. 9.19, 20. As compared to cw mode locking, the resonator contains now a Q-switch as an additional element. In order to understand the operation of the two systems depicted in Figs. 9.19, 20, it is important to consider first the build-up time required for the development of transform limited pulses.

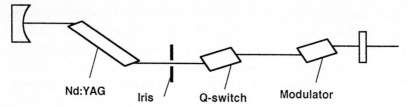

Fig. 9.19. Mode-locked and Q-switched laser

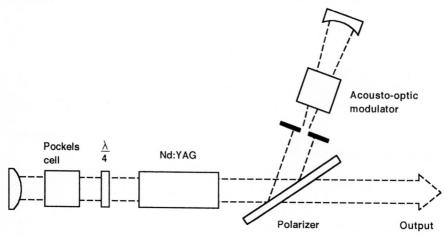

Fig. 9.20. Optical schematic of a mode-locked, Q-switched and cavity-dumped Nd : YAG laser

In a typical Q-switched laser, radiation builds up from noise within a few round trips. For mode-locked pulses to reach their transform-limited pulsewidth, several thousand round trips are required. The time needed to approach the steady-state pulse length is dependent on the modulator depth δ_m and frequency f_m of the mode locker, the saturated gain gl, and the linewidth $\Delta\nu$ of the gain medium. *Kuizenga* [9.47, 48] has shown that after M round trips, the pulse width is given by

$$t_{tr} = \frac{t_p}{[\tanh(M/M_0)]^{1/2}} \tag{9.28}$$

where

$$M_0 = \frac{1}{\sqrt[4]{gl}\,\delta_m}\frac{\Delta\nu}{f_m}. \tag{9.29}$$

The transient pulsewidth t_{tr} is within 5% of the steady-state value t_p after $M = 1.52M_0$ round trips or

$$M = \frac{0.38\Delta\nu}{(gl)^{1/2}\delta_m f_m}. \tag{9.30}$$

Since $\Delta\nu/f_\mathrm{m}$ is a very large number, and the other parameters are less than one, it takes many round trips for the mode-locked pulses to develop.

For a typical diode-pumped Nd:YAG laser, one obtains the following parameters: $\Delta\nu = 150\,\mathrm{GHz}$, $gl = 0.4$, $\delta_\mathrm{m} = 0.5$ and $f_\mathrm{m} = 80\,\mathrm{MHz}$. Therefore, according to (9.30), a minimum of 2250 round trips are required for the pulses to reach steady state. With a round-trip time of $12.5\,\mathrm{ns}$, the time required for the pulses to develop is on the order of $28\,\mu\mathrm{s}$. However, the limiting time constant for repetitive Q-switching is not the time needed to reach a steady state for the mode-locked pulse, but the time required for relaxation oscillations to die out. Each pump cycle produces transient dynamics, which strongly perturb the steady-state level in the laser, therefore the coupling between the population inversion and the intra-cavity radiation results in oscillations of both quantities until a steady-state is established. This usually takes 0.5–$2\,\mathrm{ms}$ before spike-free cw mode locking is established. Therefore, the following timing sequence is followed in the system depicted in Fig. 9.19 for obtaining simultaneous mode locking and Q-switching.

First, the Nd:crystal is quasi-cw pumped at full pump level for about 0.5–$1\,\mathrm{ms}$. The rf signal to the acousto-optic Q-switch is adjusted such that the laser is just above threshold. The rf signal to the mode locker is also switched on together with the pump. Mode-locked pulses develop during this phase in competition with strong fluctuations from relaxation oscillations. After the relaxation oscillations have died out, the Q-switch is rapidly turned-off. Due to the high gain, the mode-locked pulse is amplified by many orders of magnitude. During the Q-switch process, mode-locked pulses are emitted within a bell-shaped envelope which has a width of about 100–$200\,\mathrm{ns}$. The time separation between pulses corresponds to the resonator length, i.e. $2L/c$. If one assumes a pulse separation of typically $10\,\mathrm{ns}$, then a burst of 10–20 mode-locked pulses is obtained, with the strongest pulse at the center. The pump cycle is repeated at a rate of tens of Hz.

The system depicted in Fig. 9.20 emits only one mode-locked pulse during each pump cycle. The resonator is formed by two highly reflective mirrors, and a polarizing beam splitter. The active elements in the resonator are the laser crystal, a Pockels cell and an acousto-optic modulator. Without a voltage applied to the Pockels cell, the resonator has a low Q because radiation leaks out through the beam splitter. Likewise, with a $\lambda/4$ voltage applied to the Pockels cell, radiation is reflected between the two resonator mirrors and the Q of the cavity is high.

Mode locking is established in the resonator similar to the technique described previously. Together, with the pump pulse, rf power is applied to the mode locker, and a bias voltage is applied to the Pockels cell to increase the Q of the resonator just enough for the laser to exceed threshold. The bias voltage to the Pockels cell is usually generated by applying a $\lambda/4$ voltage to electrode 1 and a voltage close to $\lambda/4$ to electrode 2. Operation of the laser slightly above threshold establishes mode locking. After the appropriate build-up time and waiting period for relaxation oscillations to dampen out, the voltage on electrode 2 of the Pockels cell is switched to zero. The mode-locked pulse circulates between the mirrors

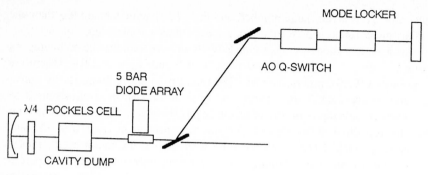

Fig. 9.21. Q-switched, mode locked and cavity dumped diode-pumped Nd : YAG laser

and the power quickly increases by several orders of magnitude. A photo diode is used to monitor the amplitude of the pulse circulating in the resonator. Once the pulse has reached its maximum amplitude, the voltage on electrode 1 of the Pockels cell is quickly switched to zero, and the pulse is coupled out through the polarizer.

Cavity dumping of a mode-locked pulse is an elegant method for producing a single short pulse on the order of 100 ps. However, the requirement of switching high voltages within nanoseconds is not a trivial task. Furthermore, fast and clean switching of the Pockels cell is complicated by the piezoelectric effect in materials such as $LiNbO_3$ which causes ringing, i.e. partial opening and closing of the switch. To minimize the switching problem, the function of the Pockels cell in Fig. 9.20 is sometimes divided into two switches, as shown in Fig. 9.21. Here the acousto-optic mode locker and Q-switch develop a strong mode-locked pulse in a two-step process, as described before. Once the mode-locked pulse has reached maximum intensity, the Pockels cell rotates the polarization and the pulse is coupled out through the thin-film polarizer. This system, developed in the author's laboratory, produced single output pulses at 20 Hz with a pulse length of 95 ps and an energy of 1 mJ [9.49]. The laser rod was a 3.5 mm diameter by 25 mm long Nd : YAG crystal side-pumped diode arrays. The combined output of the two arrays was 300 mJ in a 600 μs long pulse. The laser resonator consisted of a 3 m concave reflector and a flat reflector spaced 86.6 cm apart. Two thin-film polarizers were used, one acting as the output coupler for the cavity-dumped pulse. The rf drive frequency to the acousto-optic mode locker is 80 MHz and the cavity round-trip time is 6.25 ns. Cavity dumping was accomplished using a KD*P Pockels cell and a quarter-wave plate. The 600 μs long pump pulse was required to allow the mode-locked pulse to approach steady state and to allow the relaxation oscillations to dampen. After about 500 μs stable mode locking was established, the acousto-optic Q-switch was turned off, allowing the circulating pulse to build up in approximately 80 ns. The Pockels cell was then switched from $\lambda/4$ voltage to zero, thus dumping the pulse through a thin-film polarizer.

Figure 9.22 depicts an oscilloscope trace of the pulse build-up after the acousto-optic Q-switch is turned off. Also shown is the single output pulse of the laser.

Although Q-switching and cavity dumping in combination with mode locking appears to be a simple method for obtaining a single, powerful mode-locked pulse from a single oscillator, the technique is not easy to implement. The major problem is the presence of relaxation oscillations caused by the transient nature of the pulse evolution. An alternative approach of generating single, powerful mode-locked pulses is by means of a cw mode-locked oscillator, followed by a pulse slicer and a regenerative amplifier, as will be described in Sect. 9.4.2. Shorter pulses, and generally a more reliable performance, can be achieved with the latter approach. Further examples of transient mode locking can be found in [9.50, 51, 86].

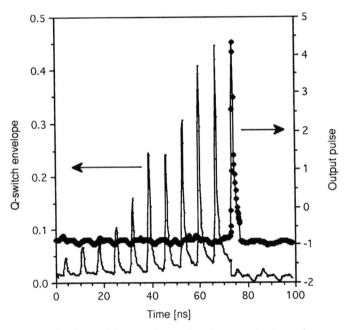

Fig. 9.22. Q-switch build-up envelope and cavity dumped output pulse

9.4 Picosecond Lasers

Today, diode-pumped cw Nd:YAG or Nd:YLF lasers mode-locked with an acousto-optic modulator, are the standard sources for the production of picosecond pulses. In this section, we will review the design and performance of actively mode-locked, diode-pumped Nd oscillators. Since the energy of individual pulses from these oscillators is quite low, amplification by means of a regenerative amplifier is frequently necessary. The design and operation of regenerative amplifiers for the amplification of mode-locked pulses is somewhat unique and will therefore also be discussed in this section.

9.4.1 Oscillators

Traditionally, actively mode-locked solid-state lasers have been pumped with cw arc lamps. Considerably higher efficiency, shorter pulses, superior stability and greater compactness of diode-pumping has resulted in a replacement of flash-lamps in favor of all solid-state lasers. As an example, actively mode-locked Nd : YAG lasers pumped with cw krypton arc lamps produce pulses of about 100 ps in duration and the overall efficiency of the system is around 0.1%. In contrast, an end-pumped Nd : YAG laser, actively mode-locked, has an overall efficiency of typically 3% and produces 20 ps long pulses. Other problems of flashlamp-pumped systems are the vibrations caused by the water pump, which are transmitted to the laser, as well as instabilities of the plasma arc in the flash-lamp. These effects tend to interrupt the mode-locking process and cause the laser to break into relaxation oscillations. End-pumping with diode lasers eliminates liquid cooling and the associated vibration problems, and the beam pattern output from diode lasers is very stable over a long period of time.

A typically end-pumped and actively mode-locked Nd laser is shown in Fig. 9.23 [9.52]. The output from a GaAlAs diode array is focused onto the end of the laser crystal. Spot sizes of the pump beam range from 50 μm to a few hundred μm. A common resonator configuration for diode-pumped actively mode-locked lasers is a three mirror arrangement. A folded cavity geometry provides a beam waist at both the laser crystal and the modulator, as well as astigmatic compensation. In commercial mode-locked lasers, the resonator is usually folded a few times to decrease the overall length of the system. The length of the cavity is usually a compromise between the need for short pulses, which requires a short resonator and a high modulation frequency, and the ability to slice out a single pulse with a Pockels cell.

Typical of end-pumped lasers is the highly reflective coating applied to the back face of the crystal (Fig. 9.23) to form one of the resonator mirrors. Small thermal or vibrational changes in the pump radiation readily induce relaxation oscillations in mode-locked systems. These relaxation oscillations cause fluctuations in amplitude, accompanied by damped oscillations at frequencies in the 30 to 150 kHz range. Therefore, careful alignment and isolation is required for clean mode locking. The modulator is usually placed close to the front mirror at the beam waist created by the folding mirror. The resonator length is typically on the order of one meter or longer to provide a sufficiently large separation between pulses for pulse selection by an external switching device. Actively mode-locked lasers are very sensitive to cavity length detuning. In practice, cavity length changes on the order of one micrometer can cause serious degradation in pulse

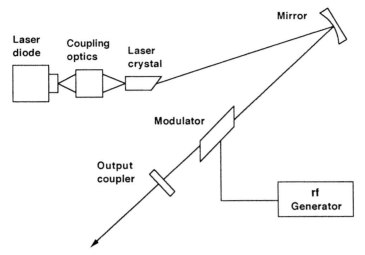

Fig. 9.23. Typical end-pumped and actively mode-locked Nd laser

quality. Therefore, active length stabilization of the resonator by a feedback loop is needed to maintain long-term stability of mode locking.

Active mode locking of a laser can be achieved by using a tunable rf oscillator and adjusting the modulator frequency to agree with the cavity length; or, alternatively, selecting a fixed frequency and adjusting the mirror spacing.

AM Mode Locking. Amplitude modulation for mode-locking cw lasers is usually performed with an acousto-optic modulator. These devices are different from the modulators employed for Q-switching lasers because they are operated at resonance. The modulator material, such as a quartz block, has parallel faces and the sound wave is reflected back and forth in the material. The length of the quartz cell is cut such that the length is equal to an integer number of half wavelength of the sound wave. In such a standing-wave pattern, the diffraction loss of the optical beam will be modulated at twice the frequency of the sound wave since the diffraction loss reaches a maximum wherever the standing-wave pattern has a maximum. The transducer of the acousto-optic modulator is therefore driven at an rf frequency of $\nu_{rf} = c/4L$. Operation of the modulator at resonance requires considerably less rf power compared to a traveling-wave device. For Q-switching of lasers, this approach is not feasible because the high Q of the acoustic resonance prevents fast switching.

Acousto-optic mode locking requires a precise match of the drive rf with one of the acoustic resonances of the modulator, and also with the frequency spacing of the cavity modes. Since an acousto-optic mode locker operates at a fixed frequency, exact synchronism between the modulation frequency and the cavity length is achieved by mounting one of the resonator mirrors on a piezoelectric translator for cavity-length adjustment. A second feedback loop is usually required to maintain synchronism between the rf drive frequency and the resonance frequency of the high-Q modulator. This is achieved by sampling the rf reflected back from the modulator by means of a directional coupler. This reflected rf power provides an error signal for slight frequency adjustments caused by a shift of the modulator's resonance due to thermal effects.

Usually, fused quartz is used as the Bragg deflector in mode lockers because of its excellent optical quality, which ensures a low insertion loss. A transducer, such as $LiNbO_3$, is bonded to the quartz block and launches an acoustic wave into the substrate. Such mode lockers are driven by an rf generator anywhere from 40 to 120 MHz and at power ranges from 1 to 5 W. For example, two commercially-available units at the low and high end of the frequency range have the performance characteristics, as listed in Table 9.2.

The modulation depth δ_m is usually determined experimentally by measuring the one-way diffraction efficiency of the device according to the relationship

$$E = 0.5[1 - J_0(2\delta_m)] \tag{9.31}$$

where $J_0(x)$ is the zero-order Bessel function of x.

In cases where a modulation frequency considerably higher than about 100 MHz is desired, or in cases where a very large modulation depth is required, a material other than quartz has to be chosen. As was discussed in Sect. 9.3.1, a higher modulation frequency or a larger depth of modulation decreases the duration of a mode-locked pulse. Therefore, the generation of mode-locked pulses with a duration of 20 ps or less requires a departure from the standard transducer-fused quartz Bragg cell. Since the acoustic losses rapidly increase in fused quartz for acoustic frequencies above 100 MHz, sapphire has been chosen for Bragg cells operating around 500 MHz. Pulses with less than 10 ps in duration have been obtained with a ZnO-sapphire acousto-optic modulator at this frequency [9.53].

A technique which will substantially increase the modulation depth is the use of the transducer itself as the optical loss modulator. In conventional acousto-optic modulators, a piezoelectric transducer is used to launch an acoustic wave into the Bragg deflector. In a so-called *piezoelectrically-induced strain optics modulator*, the piezoelectric material itself is selected as the deflection medium. Operated at resonance, a very efficient refractive-index grating can be created in the bulk of the piezoelectric material. The shortest actively mode-locked pulses are generated with this type of mode locker which employs usually GaP or $LiNbO_3$ as Bragg material [9.87, 88]. Actually, the shortest pulses reported from an actively mode-locked diode-pumped Nd : YLF laser, with 6.2 ps duration, has been achieved with a GaP acousto-optics mode locker [9.54]. Since both gallium phosphide (GaP) and lithium niobate ($LiNbO_3$) have large strain-optic coefficients, a very efficient refractive index grating can be created in the bulk piezoelectric transducer itself. This technology has, of course, its own difficulties. The optical quality of the material is inferior to fused quartz and impedance matching the rf generator to a thick transducer creates considerable difficulties.

Table 9.2. Modulation depth and rf power requirements for typical mode lockers

Frequency [MHz]	Modulation (δ_m)	Drive power [W]	Corresponding resonator length
38	0.8	5	197.4 cm
120	0.63	1	62.5 cm

Figure 9.24 depicts the equivalent rf circuit for a piezoelectric transducer. The two conductively coated surfaces of the transducer with the piezoelectric crystal in between represents a capacitor C_0. Stray capacitance associated with the package and electrical leads is accounted for by C_s, whereas R_0 accounts for the dielectric losses. Connecting an inductance L_0 parallel to the transducer creates a resonance circuit with a resonance frequency $1/\nu_R = 2\pi\sqrt{L_0(C_0 + C_s)}$ and a quality factor $Q = 2\pi\nu_r L_0/R_0$. The electrical resonance frequency ν_R has to correspond to the acoustic standing-wave resonance of the transducer which is controlled by the thickness of the transducer crystal. As we recall from the previous discussion, the laser light is loss modulated at twice the rf drive frequency ν_R owing to the acoustic standing wave in the mode-locker substrate. The inductance L_0 is actually the secondary winding of a transformer required to match the impedance of the transducer circuit to a $50\,\Omega$ coaxial transmission line. The transducer can be made very thin if it is used to launch an acoustic wave into a Bragg cell, such as a quartz block. In this case, C_0 is large compared to the stray capacitance C_s.

If a transducer is employed in a dual role of launching the acoustic wave and modulating the optical beam, it has to be fairly thick for the laser beam to pass through. This reduces C_0, and in order to achieve resonance at ν_R, a very large inductance has to be matched to the transducer. A small C_0 makes it very difficult to efficiently couple the rf signal into the transducer because the stray capacitance provides a parallel loss path to the transducer.

Optimum mode locking is achieved by adjusting the cavity length such that its resonant frequency matches the frequency of the mode locker. This is usually accomplished by fixing the frequency of the mode locker and tuning the resonator length to match the mode locker's frequency. The cavity length is tuned by translating one mirror mounted on a piezoelectric translator.

Figure 9.25 shows feedback control of a laser mode-locked with an acousto-optic Bragg cell. The mode locker is driven by an rf generator whose output frequency is divided by 2. Manually, the cavity length is adjusted until a region of stable mode locking is achieved. This is accomplished by adjusting the reference voltage of a differential operational amplifier which then provides a dc voltage to the piezoelectric transducer. At that point, the feedback loop is closed. The train of mode-locked pulses is detected by a photo diode. Appropriate filtering of the photo-diode output provides a sine wave which corresponds to the resonator

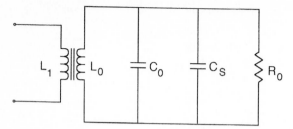

Fig. 9.24. Equivalent rf circuit for a piezoelectric transducer with impedance matching transformer

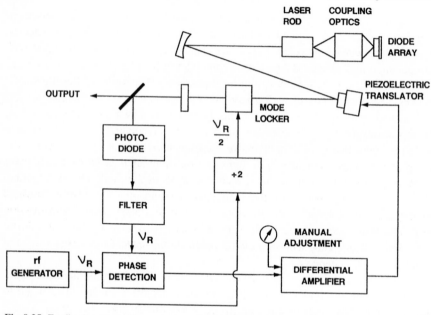

Fig. 9.25. Feedback control of an acousto-optic mode-locked laser

round-trip repetition frequency. The phase of this waveform is compared with the waveform of the rf generator in a phase detector. A deviation or drift between these two frequencies produces an error signal which is directed to the differential amplifier which, in turn, provides a dc voltage to the piezoelectric translation stage.

FM Mode Locking. To achieve FM mode locking, an intra-cavity electro-optic phase modulator is driven at the cavity resonance frequency. Electrically, the modulator consists of a pair of electrodes between which the crystal is placed. An inductive loop and a tuning capacitor are shunted across these electrodes and are used to achieve the desired resonant frequency.

Lithium niobate is usually the crystal employed in electro-optic modulators. With the electric field applied transversely to the optical beam, the r_{33} electrooptic coefficient can be utilized and one obtains

$$\Delta n_z = \frac{-1}{2} r_{33} n_e^3 E_z \qquad (9.32)$$

where n_e is the extraordinary index of refraction. The light propagates in the x-direction, and the beam is polarized in the z direction, in the same direction as the applied electric field.

The total phase change in the crystal is $\delta = 2\pi \Delta na/\lambda_0$, where a is the length of the crystal in the x direction. If a voltage $V = V_0 \cos \omega_m t$ is applied across the crystal in the z direction, the peak single-pass phase retardation of the modulator is

$$\delta'_{FM} = \frac{\pi r_{33} n_e^3 V_0 a}{\lambda_0 d} , \qquad (9.33)$$

where d is the dimension of the crystal in the z direction.

For LiNbO$_3$, the materials parameters are: $n_e = 2.16$ at $1.06\,\mu m$, $r_{33} = 30.8 \times 10^{-10}$ m/V. Typically, a phase retardation of about $\delta_{FM} = 1$ rad is achieved with 300 V across a 5 mm $\times$ 5 mm $\times$ 20 mm crystal. At a frequency of a few hundred MHz, this requires an rf power of a few watts.

We will illustrate the performance of an FM mode-locked Nd:YLF laser which has an optical design, as shown in Fig. 9.23 [9.52]. The Brewster cut LiNbO$_3$ phase modulator had dimensions of 15 mm $\times$ 6 mm $\times$ 2 mm and was driven by 1 W of rf power at a frequency of 200 MHz. The Nd:YLF rod was pumped by a 3 W cw laser diode. The output from the 500 μm $\times$ 1 μm diode emitting aperture was focused to a 300 μm spot at the laser crystal. The mode-locked oscillator generated 14 ps pulses at an output power of 830 mW at an output wavelength of 1.047 μm.

In the case of an FM mode-locked laser, a simple mode-locking system can be built by allowing the laser to determine its own drive frequency, as shown in Fig. 9.26. This can be accomplished by using a high-speed photodetector to sense the first beat frequency $c/2L$ of the oscillator. This signal is amplified, phase-shifted a variable amount, and then applied to the intracavity modulator. The adjustable phase shifter compensates for the delay in the feedback loop. When the phase of the electric signal on the modulator and the signal derived from the laser equal an integer times π, the loop goes into regeneration and the laser is mode-locked. The resultant mode-locking system is a closed-loop oscillator using the laser cavity as the basic reference. Such a system will automatically track changes in cavity length.

In a phase-modulation system there is an ambiguity in the phase relationship between the laser pulse train and the modulator driving signal, as the pulses may pass through the modulator at either of the two extreme voltage points. Because of this phase ambiguity, two possible pulse trains can be obtained – one at 0° phase with respect to the modulator drive and one at 180°. The laser will always operate with only one of these pulse trains running at a time, because to do otherwise would result in a larger net loss in the laser cavity. However, small perturbations to the laser can cause random shifting from one phase to

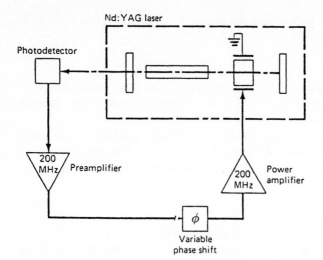

Fig. 9.26. Direct-drive mode-locking system

the other, the phase change requiring several microseconds. An automatic phase-modulating mode-locked system, which operates by synchronizing the modulator drive signal to the first-order mode beat of the laser, suffers from sporadic loss of lock as the laser hops from one stable phase to the other.

In a phase-modulated laser one can avoid this failure mode which occurs due to a spontaneous switch of 180° in the phase of the output pulse train by utilizing the second-order beat frequency. A mode-locking control system that has proven to be quite stable is shown in schematic form in Fig. 9.27 [9.55]. The oscillator is phase-modulated at the fundamental beat frequency of 200 MHz by a modulator driven from a voltage-controlled oscillator (VCO). A photodetector mounted behind the rear mirror of the oscillator samples a portion of the beam. A bandpass filter selects the second harmonic of the cavity beat frequency. The resultant amplified 400-MHz signal is compared in a phase detector with a 400-MHz comparison signal derived by multiplying the output of the VCO by 2. The phase-detected output is amplified and used to control the VCO. Initial mode locking is performed manually by adjusting the bias of the VCO.

Information about diode-pumped, actively mode-locked lasers employing AM and FM modulation in a number of Nd materials such as Nd:YAG, Nd:YLF, Nd:glass, and Nd:YVO$_4$ can be found in [9.52, 56–60].

9.4.2 Regenerative Amplifiers

Regenerative amplifiers produce energetic picosecond pulses at repetition rates up to several kHz from a train of low-energy pulses emitted, by a mode-locked oscillator. The design and operation of these amplifiers is distinct from single- or double-pass amplifiers described in Sect. 4.1 or injection-seeded oscillators discussed in Sect. 5.2.3. A regenerative amplifier selects individual pulses from a train of mode-locked pulses and allows many passes through the gain medium. The energy of an individual pulse may increase by as much as 10^6–10^7 before

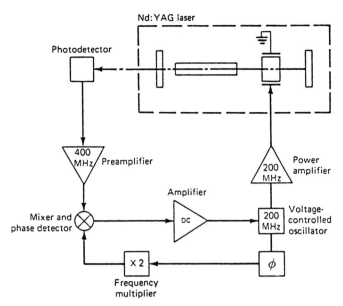

Fig. 9.27. Automatic closed-loop mode-locking system

being switched out from the amplifier. Originally, regenerative amplifiers have been pumped by flashlamps or by cw arc lamps [9.61–66]. More recently, the amplifiers employ diode arrays for pumping Nd : YAG or Nd : YLF crystals [9.67–70] or Nd : Glass [9.71].

Figure 9.28 shows a schematic of a diode-pumped Nd : YLF regenerative amplifier. The end-pumped Nd : YLF crystal is at one end of an astigmatically compensated three-mirror resonator. The other side of the resonator contains a thin-film polarizer, quarter-wave plate and a Pockels cell. These three components are characteristic for a regenerative amplifier. Another feature of this type of amplifier is the Faraday rotator, $\lambda/2$ plate and thin-film polarizer which separate the input and output beams. The operation of a regenerative amplifier can be explained by distinguishing three operating phases:

Pump Phase with Pockels Cell Voltage at Zero. The laser crystal can be either cw or pulsed pumped. During the pump phase, laser action is prevented by the $\lambda/4$ plate and rear-mirror combination, which causes a 90° rotation of the horizontally emitted radiation by the Brewster-cut laser rod.

A horizontally polarized pulse from a mode-locked oscillator will pass through the thin-film polarizer P2, experience a 45° rotation at the $\lambda/2$ plate and a further 45° rotation at the Faraday rotator. The now vertically polarized pulse is reflected by the thin-film polarizer P1 into the resonator. After reflection from the rear mirror M1 and passing through the $\lambda/4$ plate, the pulse will be horizontally polarized. Therefore, the pulse passes through the polarizer P1 and completes one round trip. On its second round trip, the pulse is rotated again

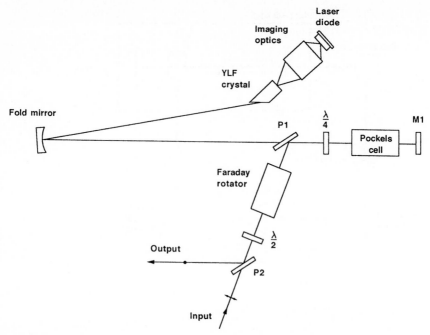

Fig. 9.28. Schematic of a diode-pumped Nd: YLF regenerative amplifier [9.68]

90°, and being now vertically polarized, it is reflected off the polarizer P1. The 45° rotation from the Faraday rotator is compensated by the $\lambda/2$ wave plate and therefore the vertically polarized pulse is directed to the output port of the amplifier by polarizer P2.

The ratio of average power at the output with and without the pump source turned on is a measure of the small-signal double pass gain.

Amplification Phase, with Pockels Cell Switched to $\lambda/4$ Retardation. In order to trap a pulse in the resonator, a $\lambda/4$ voltage is applied to the Pockels cells as soon as a pulse has left the Pockels cell in the direction of the fold mirror. It takes about 5–8 ns to switch the voltage from zero to the level required for $\lambda/4$ rotation. Therefore, switching at this particular point provides the longest time period for the pulse to return or for a new pulse to arrive. With the $\lambda/4$ voltage applied to the Pockels cell, the pulse returning from the gain medium no longer experiences a polarization rotation and therefore stays in the cavity.

Once the Pockels cell is switched to $\lambda/4$ voltage, pulses arriving from the oscillator also no longer experience a polarization rotation. Therefore, the pulses are ejected from the amplifier after being reflected by mirror M1.

The build-up time for the circulating pulse to reach maximum energy depends on the gain in the system and is characteristic of the Q-switch envelope of a cavity dumped pulse. The higher the gain, the shorter the build-up time. Typical time periods are from less than 100 ns for a very high gain system to a few hundred nanoseconds for smaller systems.

Cavity Dump Phase, Pockels Cell Voltage Switched to $\lambda/2$ Retardation or Zero Voltage. After many passes in the resonator, the energy of the pulse reaches a maximum because of gain saturation. At that point, the Pockels cell is switched to a $\lambda/2$ retardation voltage. Again as before, switching is started when the pulse is just leaving the Pockels cell. On its return, the pulse experiences now a 90° polarization rotation and the vertically polarized pulse is reflected by polarizers P1 and P2 to the output port of the amplifier.

Beside the single amplified pulse emitted by the amplifier, there are smaller pulses preceding and following the main pulse. As the circulating pulse inside the resonator increases in energy, a small fraction is reflected from the amplifier each time the pulse passes through P1. This is due to the finite extinction ratio of the polarizer and the limited contrast ratio of the Pockels cell. A small misalignment, or thermal effects leading to birefringence, or small voltage changes in the Pockels cell, will cause the pulse to have a small vertical polarization component. This leakage gives rise to a sequence of premature pulses separated by the round-trip time of the regenerative amplifier.

Another source of background noise is a second pulse trapped in the resonator. For equal resonator length of oscillator and regenerative amplifier, a new pulse arrives as one pulse is leaving the resonator. If the two resonators are not equal in length, there are always two pulses present inside the resonator separated by the difference in round-trip time. If this pulse is approaching the Pockels cell as it is switching, a small fraction of this pulse will also stay in the resonator and be amplified. The contrast ratio between the energy of the main pulse and the integrated background is on the order of 20:1 for most systems.

Performance of Regenerative Amplifiers. The energy which can be extracted from a regenerative amplifier is the most important consideration. *Lowdermilk* et al. [9.61, 62] derived analytical expressions which permit the calculation of the maximum fluence level in the resonator in terms of physical parameters of the gain medium and the resonator. Figure 9.29 summarizes the parameters which are needed to describe the output of a regenerative amplifier. The gain medium is characterized by the small-signal gain $G_o = \exp(g_o l)$, the saturation fluence I_s, and a gain recovery coefficient p. The amplifier performance depends on the gain recovery from lower-level relaxation between successive passes, therefore the parameter p is a function of the time interval between passes and the lower-level relaxation time. For complete gain recovery, p corresponds to $p = 0.5$ and for no gain recovery between pulses $p = 1$. The parameter I_{in} is the input fluence of the seed pulse, and I_o is the fluence of the output pulse which is the same as the peak fluence in the resonator since the pulse is switched out as soon as it reaches maximum intensity. The total losses L are represented by a single-pass transmission T through the resonator $T = \exp(-L)$.

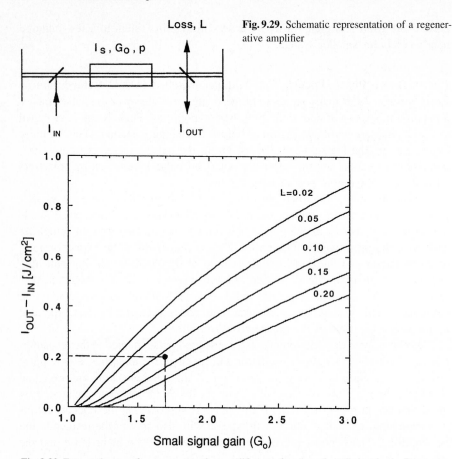

Fig. 9.29. Schematic representation of a regenerative amplifier

Fig. 9.30. Extracted energy from a regenerative amplifier as a function of small signal gain. Parameters are the resonator losses

With the parameters defined in Fig. 9.29, it can be shown that the fluence increases with the number of passes according to [9.61].

$$I_{K+1} = T I_s \ln\{G_K[\exp(I_K/I_S) - 1] + 1\} \tag{9.34}$$

and the gain decreases after each pass according to

$$g_{K+1} = g_K - (p/I_S)[(I_{K+1}/T) - I_K] . \tag{9.35}$$

In these equations, uniform gain across the laser medium and complete spatial overlap of input and output pulse is assumed.

Equations (9.34, 35) can be iterated numerically to model multi-pass amplifiers. An analytical solution for the case in which changes of the gain are relatively small during each pass has also been derived in [9.61]. Gain saturation is reached when gain and losses are equal and the maximum fluence is

$$I_{\text{OUT}} = (I_s/p) \ln\left[TG_o\left(\frac{1-T}{T(G_o-1)}\right)^{1-T}\right] + I_{\text{IN}} . \tag{9.36}$$

The function $(I_{\text{OUT}} - I_{\text{IN}})$ vs. G_o has been plotted in Fig. 9.30 with the resonator losses L as parameter. The curves are for Nd : YAG for which $I_s = 0.66\,\text{J/cm}^2$ and $p = 0.72$ was chosen. The value of the saturation fluence $I_s = h\nu/\sigma$ is obtained from the values for the photon energy $h\nu$ and stimulated emission cross-section σ given in Table 2.4.

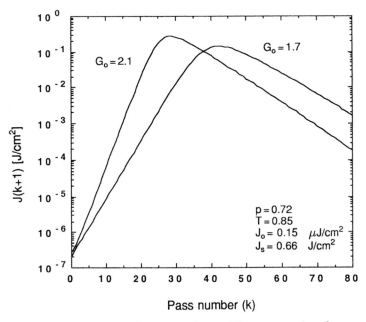

Fig. 9.31. Increase of energy in a regenerative amplifier versus number of passes

Figure 9.31 illustrates the increase of pulse energy during each pass in a regenerative Nd : YAG amplifier calculated from (9.34, 35). Input fluence for $K = 0$ was assumed to be $1.5 \times 10^{-7}\,\text{J/cm}^2$. With a small-signal gain of $G_o = 1.7$, pulse energy reaches a maximum at about 40 passes through the gain medium. The linear portion of the curve indicates exponential gain up to about 30 passes, after which gain saturation starts to have an effect. Fewer passes are required to reach gain saturation, and the maximum fluence is higher also, for the case in which a higher gain has been assumed.

Examples. An end-pumped Nd : YLF laser in a resonator configuration, as shown in Fig. 9.28, has produced $88\,\mu\text{J}$, 11 ps pulses at a 1 kHz repetition rate [9.68]. The crystal was pumped by a 2 W cw laser diode and was operated at $1.047\,\mu\text{m}$. Output energies of up to 0.5 mJ have been obtained in another end-pumped configuration utilizing a 15 W cw diode array pump [9.69]. Side-pumping of Nd : YAG or Nd : YLF can produce even higher energy levels [9.70, 67].

As an example of a powerful all solid-state picosecond laser source, we will describe a regenerative amplifier of a system built in the author's laboratory [9.72]. The system produces 28 ps pulses at a repetition rate of 2 kHz and a pulse energy of 2.5 mJ. The schematic of the amplifier is sketched in Fig. 9.32. The resonator consists of a concave-convex mirror combination and two turning mirrors to reduce the physical length of the system. The resonator is designed according to the procedure described in Sect. 5.1.10. With an overall length of 1.5 m, mirror curvatures of $R_1 = 0.5$ m and $R_2 = -2.0$ m, and taking the thermal lensing of the laser rod into account, the resonator operates in a high stability zone. The resonator mode is largest close to the concave mirror, therefore all components are located as close as possible to this mirror. Side-pumping of laser rods creates a relatively large pump cross-section which needs to be matched by a large TEM_{00} resonator mode for good energy extraction efficiency. Furthermore, a large beam waist reduces the possibility of damage in the $LiNbO_3$ crystal.

Input seed pulses were derived from an end-pumped acousto-optic mode-locked oscillator which generated 16 ps pulses with pulse energies of 1.5 nJ at a repetition rate of 100 MHz. The pump head of the regenerative amplifier contains a Nd : YAG rod, 4 mm in diameter and 68 mm long, which is pumped by 32 diode arrays arranged in a four-fold symmetry around the crystal. The arrays are pulsed for 100 μs at a repetition rate of 2 kHz. Peak power from each array is 50 W or 1.6 kW total. At this 20% duty cycle, the average optical pump power is 320 W. The pump head, tested in a very short cavity as a long-pulse oscillator, produced 80 W of average multi-mode power. At a pump power of 1.6 kW, a single-pass small-signal gain of $G_o = 1.7$ was measured for the regenerative amplifier configuration. Operation and switching sequence are identical to the device described previously. An acoustically damped $LiNbO_3$ Q-switch, with a switching time of 5 ns was employed in the amplifier. In this Q-switch, the acoustic modes which are excited by the piezoelectric effect are transmitted into a strongly attenuating material that is acoustically impedance matched to $LiNbO_3$. The Pockels cell is switched to $\lambda/4$ retardation at the end of the 100 μs-long pump cycle. After about 100 ns, the circulating pulse has reached maximum energy, and the Pockels cell is switched to a $\lambda/2$ retardation.

Figure 9.33a,b shows oscilloscope traces of the circulating pulse inside the resonator up to the point of cavity dumping, as well as a pulse train for the case where the Pockels cell is not switched a second time to eject the pulse from the cavity. Both traces were obtained by monitoring the leakage through the rear mirror. The output pulse emitted from the regenerative amplifier is displayed in Fig. 9.33c.

Figures 9.30, 31 have been generated to model this particular regenerative amplifier. We will compare the predictions from the model with the experimental results. The model assumes uniform gain and a top-hat beam profile. The actual system had a fairly uniform gain profile in the center of the laser rod, but a Gaussian beam profile. The TEM_{00} mode size was 2 mm at the $(1/e)^2$ points. A

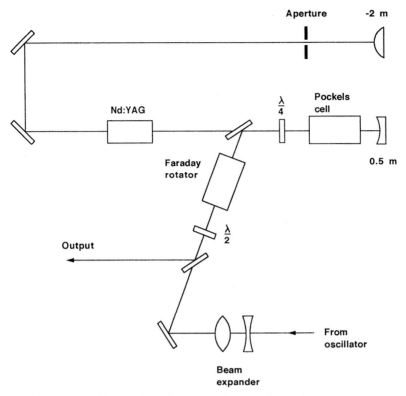

Fig. 9.32. High repetition rate laser diode pumped regenerative amplifier

uniform beam with the same energy content as the Gaussian beam would have a beam width of 1.25 mm. This beam diameter was used in the comparison of the data. The measured small-signal gain and output fluence are indicated in Fig. 9.30. From this data follows a single-pass loss of 11% in the system. The high losses in the amplifier are mostly dynamic losses associated with heating of the Pockels cell, and incomplete opening and closing. The former is a result of the high average power, and the second loss factor is due to piezoelectric ringing in the crystal. The result in Fig. 9.31 indicates that 40 single passes or 200 ns are required in the 1.5 m long resonator before the fluence reaches its maximum value. The actual measurements showed a time delay of 240 ns. Considering the simplifying assumptions made in the model, the experimental and theoretical data agree fairly well. More important is the ability of the model to predict the performance of the amplifier if changes are made during the optimization process.

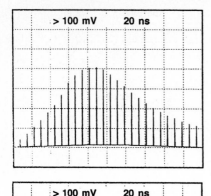

Fig. 9.33a. Circulating pulses of mode-locked laser if system is not cavity dumped

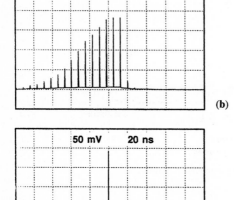

(b)

(c)

Fig. 9.33b,c. Pulse train of mode-locked laser up to cavity dumping (**b**), and output pulse (**c**)

9.5 Femtosecond Lasers

The goal of all solid-state laser sources for ultra-fast pulse generation has motivated research efforts world wide. Kerr-lens mode locking has revolutionized femtosecond pulse generation in solid-state materials because the KLM technique produces the shortest pulses, and it is relatively simple to implement. If the peak power of the pulses obtained from a Kerr-lens mode-locked oscillator is insufficient for a particular application, the pulse can be amplified by means of a chirped pulsed amplifier. In this section, we will describe some of the design considerations for the development of femtosecond oscillators and amplifiers.

9.5.1 Oscillators

We will briefly review the available laser materials suitable for femtosecond production and then address the issues of resonator design, dispersion compensation and start-up of the KLM process. Finally, a few representative examples of femtosecond lasers will be discussed.

Laser Materials. The laser crystal most actively explored for ultra-short pulse generation is $Ti:Al_2O_3$. This laser has the largest gain bandwidth and is therefore capable of producing the shortest pulses; it also provides the widest wavelength tunability. Typically, the $Ti:Al_2O_3$ crystal is pumped by a cw argon laser. The future trend will be to replace the argon laser with a diode-pumped, frequency-doubled $Nd:YAG$ laser in order to create an all-solid-state source.

With regard to providing a compact, reliable femtosecond source, $Cr:LiSAF$ ($Cr:LiSrAlF_6$) is an intriguing material because it can directly be pumped by laser diodes. Compared to $Ti:Al_2O_3$ systems, $Cr:LiSAF$ does not have as wide a tuning range, but the much longer upper-state lifetime ($67\,\mu s$) and a broad absorption band centered at $640\,nm$ permit pumping with red AlGaInP diodes. Diode-pumped $Cr:LiSAF$ lasers are currently the most compact femtosecond pulse sources.

A third material which has a broad gain bandwidth is $Cr:Forsterite$. This laser can be pumped with the output from a diode-pumped $Nd:YAG$ laser to produce femtosecond pulses at around $1.3\,\mu m$. Table 9.3 lists key parameters of these three important materials for femtosecond pulse generation in solid-state lasers.

Resonator Design. A resonator commonly employed for KLM is an astigmatically compensated arrangement consisting of two focusing mirrors and two flat mirrors. In order to obtain a high nonlinearity, the Kerr medium is inserted into the tightly focused section of the resonator, as shown in Fig. 9.34a. Several researchers have analyzed such resonators based on nonlinear ABCD matrix representations [9.2, 28–30, 75, 76]. The results of these calculations provide guidelines for the design of KLM resonators. Taking into account the astigmatism of the Brewster cut crystal and the tilted mirrors, the resonator shown in Fig. 9.34a has to be evaluated in the tangential and sagittal plane. The curved mirrors correspond to lenses with focal lengths f, which are different for the two planes, the same is true for the equivalent lengths l of the laser crystal. Figure 9.34b depicts the equivalent resonator. The function of the two prisms is to provide dispersion compensation in the resonator, as will be explained in the next subsection. For the purposes of this discussion, we will omit these elements.

As was explained in Sect. 9.2.2, the operation of a KLM laser is a trade-off between output power, stability and tolerance to the exact position of the components. An analytical treatment of nonlinear resonators has shown that for a given pump power and pump spot size, the most critical parameters are: (a) the

Table 9.3. Spectroscopic properties of broadband solid-state laser materials [9.73, 74, 112]

	Ti : sapphire	Cr : LiSAF	Cr : Forsterite
Peak emission cross section [$\times 10^{-20}$ cm^2]	30	4.8	14.4
Emission peak [nm]	790	850	1240
Gain bandwidth [nm]	230	180	170
Upper state lifetime [μs]	3.2	67	2.7
Nonlinear index n_2 [$\times 10^{-16}$ cm^2/W]	3.2	1.5	2.0

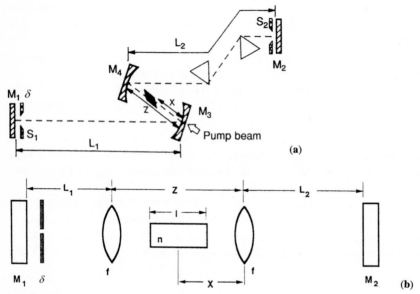

Fig. 9.34. Resonator for Kerr-lens mode locking ($M_3 = M_4 = 100$ mm **(a)** and equivalent resonator **(b)**

distance z of the two focusing mirrors; (b) the location x of the Kerr medium with respect to the mirrors, and (c) the spot size variation δ at the aperture.

As we have seen in Sect. 9.2.2, the Kerr-lens sensitivity δ is highest near the limit of the stability range of the resonator, therefore loss-modulation efficiency has to be traded off against stable laser performance. From the analysis presented in [9.30], contour curves for δ are plotted in Fig. 9.35 for the resonator depicted in Fig. 9.34a. The plots are for different lengths L_1 and L_2. As we have seen in Sect. 9.2.2, for unequal lengths L_1 and L_2, two stable regimes exist for the resonator. As already mentioned, the region in the first quadrant of the stability diagram (Fig. 9.14) was found to be less sensitive to alignment changes. Therefore, only the contour plots are shown for which more stable and reproducible operation was achieved.

The plots are for the tangential plane of the resonator, because δ is larger in this plane as compared to the saggital plane. The curves for δ are bound by the stability limits of the resonator. Experiments performed on an actual system

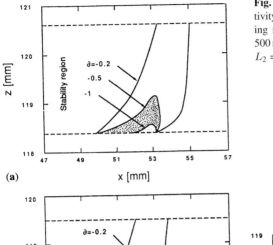

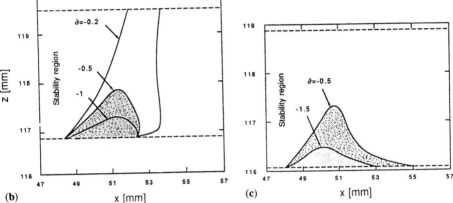

revealed that for reliable mode locking, a value of $|\delta| \geq 0.5$ was necessary. This defines the operating regime of the laser system indicated by the shaded areas. Comparing the operating regime for the three resonator configurations, it is clear that the parameter space for x and z becomes larger as the resonator approaches a symmetric configuration. Experiments verified that the best compromise between a large δ and a reasonably stable performance is achieved for a symmetric resonator $L_1 = L_2$ and operation close to $g_1 g_2 = 0$ [9.77]. For this configuration, which is equivalent to a confocal resonator, the tolerance for stable operation is about ± 0.6 mm in the z-direction, with the resonator operated about 0.5–1 mm from the lower stability limit. From the results illustrated in Fig. 9.35, it is clear that the alignment of a KLM resonator is very critical, and the tolerance in length adjustments are fractions of a millimeter.

Because of the Brewster-angle design, these resonators behave differently in the tangential and sagittal planes. This is illustrated in Fig. 9.36 which shows the change in beam size in two orthogonal directions at the flat mirror for a similar resonator, as shown in Fig. 9.34 [9.78]. Since the spot size changes essentially in only one direction, a slit rather than an aperture is usually employed for the adjustment of the proper loss modulation.

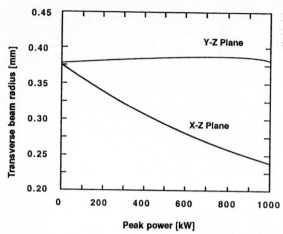

Fig. 9.36. Spot sizes as a function of intracavity peak power for two orthogonal directions [9.78]

The purpose of this discussion was to acquaint the reader with the basic issues related to KLM resonator design and to provide an idea about the dimensional tolerances which have to be observed in aligning these systems. Specific resonator designs can be analyzed by following the formulas developed in the referenced papers. The design usually starts with the assumption of a pump beam which has a certain spot size and power in the Kerr medium (defined by n_2 and P_{cr}). From this a nonlinear matrix is developed, which combined with the linear matrix of the passive elements of the resonator, describes the behavior of the system.

Another issue crucial for reliable mode locking performance is the absence of spurious reflections, as demonstrated experimentally [9.79] and studied theoretically [9.80]. These studies revealed that even a very small fraction on the order of 10^{-5} of the output power fed back into the resonator can seriously impair mode locking. It was reported that even the back face of a high reflectivity mirror may adversely affect mode locking.

Dispersion Compensation

For ultra-short pulse generation, the round-trip time t_R in the resonator for all frequency components of the mode-locked pulse must be frequency independent, i.e. $t_R(\nu) = d\phi/d\nu = $ constant, where ϕ is the phase change after one round trip. Otherwise, frequency components which experience a cumulative phase shift no longer add constructively and are attenuated. This limits the bandwidth of the pulse and leads to pulsewidth broadening. The frequency-dependent phase shift of the pulse during one round trip can be expressed in a Taylor series about the center frequency ν_0.

$$\frac{d\phi}{d\nu} = \phi'(\nu_0) + \phi''(\nu_0)\Delta\nu + \frac{1}{2}\phi'''(\nu_0)\Delta\nu^2 \tag{9.37}$$

where ϕ', ϕ'', ϕ''', are the derivatives of the phase with respect to frequency. When ϕ'' is non zero, the pulse will have a linear frequency chirp, while a non-zero third-order dispersion will induce a quadratic chirp on the pulse.

The two major sources of dispersion in a mode-locked laser are self-phase modulation which is part of the Kerr effect, and normal dispersion in the laser crystal or any other optical component in the resonator.

Self Phase Modulation. Besides modifying the spatial profile of the beam leading to a self-induced quadratic index gradient, the Kerr effect also causes a phase shift among the frequency components as the pulse propagates through the crystal. The time varying phase shift $\phi(t)$ or phase modulation produced by the pulse itself can be expressed by

$$\Delta\phi(t) = (2\pi/\lambda)n_2 I(t)l , \qquad (9.38)$$

where $I(t)$ is the intensity and l is the length of the Kerr medium. Since the Kerr effect always leads to an increase of n for increasing intensity, the rising edge of the pulse experiences a medium which is getting optically denser, i.e. $dn/dt > 0$. The increasing index of refraction at the rising edge of the pulse delays the individual oscillations of the electric field, which is equivalent of red-shifting of the leading edge. The opposite occurs at the trailing edge of the pulse, the trailing edge is blue-shifted. Self phase modulation will thus cause a frequency chirp of the pulse and prevent the pulsewidth from becoming transform-limited.

Normal Dispersions. The change of group velocity with frequency is usually expressed by

$$\frac{dv_g}{d\nu} = -v_g^2 \beta'' \qquad (9.39)$$

where β'' is the group dispersion of the medium. Materials in the visible region of the spectrum have positive or normal dispersion, i.e. $\beta'' > 0$. Therefore in a laser crystal v_g decreases with increasing frequency ν, i.e. longer wavelengths travel faster than short ones, causing a red shift of the pulse.

Negative Dispersion. From the foregoing discussion follows that a high-intensity mode-locked pulse is red-shifted due to self phase modulation and normal dispersion. Positive self phase modulation and positive group-velocity dispersion in the Kerr medium can be compensated by a dispersive delay line based on a prism pair which intentionally introduce negative dispersion into the resonator. Although the glass of the prisms have normal dispersion, the geometry of the ray path can be arranged such that the blue components of the pulse traverses the two prisms in a shorter time than do the red components. Although a number of prism arrangements can be devised, usually two prisms are used at minimum deviation and Brewster's angle incidence at each surface. In Fig. 9.37, the entrance face of prism II is parallel to the exit face of prism I, and the exit face of prism II is parallel to the entrance of prism I. The prisms are cut so that the angle of minimum deviation is also Brewster angle. The plane MM' normal to the rays is a plane of symmetry.

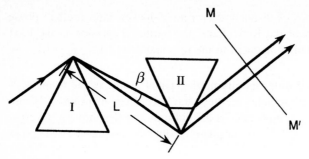

Fig. 9.37. Dispersive delay line employing a prism pair

A detailed description of the design of dispersive delay lines based on prism pairs can be found in [9.81, 82]. Referring to Fig. 9.37, it can be shown that the optical path that contributes to dispersion is $l = 2L \cos \beta$. Group-velocity dispersion is the second derivative of the path length with respect to wavelength. From [9.81] we obtain

$$\frac{d^2l}{d\lambda^2} = 4L \left\{ \left[\frac{d^2n}{d\lambda^2} + \left(2n - \frac{1}{n^3} \right) \left(\frac{dn}{d\lambda} \right)^2 \right] \sin \beta - 2 \left(\frac{dn}{d\lambda} \right)^2 \cos \beta \right\} . \quad (9.40)$$

The second part of (9.40) is responsible for negative dispersion; therefore the first part has to be made as small as possible. The term $L \sin \beta$ expresses the distance of the beam form the apex of the first prism. This term is minimized by placing the beam as close to the apex as possible. In actual systems, the incident beam is adjusted to pass at least one beam diameter inside the apex of the first prism. We introduce $L \sin \beta = 4w$, where w is the beam radius and with $\cos \beta \approx 1$, and $2n \gg 1/n^3$, obtain

$$\frac{d^2l}{d\lambda^2} = 16w \left[\frac{d^2n}{d\lambda^2} + 2n \left(\frac{dn}{d\lambda} \right)^2 \right] - 8L \left(\frac{dn}{d\lambda} \right)^2 . \quad (9.41)$$

For sufficiently large prism separation L, the right-hand side of (9.41) can be made negative, as illustrated by the following example. We assume a Ti : sapphire laser, with two SF10 glass prisms for dispersion compensation. With $n = 1.711$, $dn/d\lambda = -0.0496\,\mu\text{m}^{-1}$, $d^2n/d\lambda^2 = 0.1755\,\mu\text{m}^{-2}$ at 800 nm and a beam radius of $w = 1$ mm

$$\frac{d^2l}{d\lambda^2} = 0.294 - 0.0197L(\text{cm}) \quad [1/\text{cm}] . \quad (9.42)$$

Therefore, for a prism separation larger than about 15 cm, negative group-velocity dispersion is obtained.

Generally speaking, in the design of the femtosecond lasers, dispersion effects have to be minimized. A small positive dispersion in the resonator requires a short laser crystal. As the pulsewidth approaches a few tens of femtoseconds, third-order dispersion from the prism pair and the laser crystal become critical. Minimizing the optical path length in the prisms will reduce third-order dispersion.

Experimental results, supported by theoretical analysis, has actually shown that, the shortest pulses are not obtained for an overall zero dispersion in the cavity, but for the case of a slight negative group-velocity dispersion [9.83].

Self Starting of KLM. The Kerr nonlinearity is usually not strong enough for the cw mode-locking process to self start. In order to initiate KLM, usually a strong fluctuation must be induced by either perturbing the cavity or by adding another nonlinearity to the system. The simplest method to start KLM in a laboratory set-up is to slightly tap one of the resonator mirrors. Disturbing the cavity mirrors will sweep the frequencies of competing longitudinal modes, and strong amplitude modulation due to mode beating will occur. The most intense mode beating pulse will be strong enough to initiate mode locking. The condition for self-starting has been analyzed numerically in [9.84].

Several methods for self-starting of KLM have recently been developed which are practical enough to be implemented on commercial lasers. A simple approach for starting KLM is to mount one of the resonator mirrors on a PZT and introduce a vibration on the order of a few μm at a frequency of tens of Hz [9.85, 30].

Semiconductor saturable absorbers of the type discussed in Sect. 9.2.2 can be employed in a coupled cavity or directly in the resonator to provide a strong nonlinearity which will start KLM [9.35, 41].

In synchronous pumping, the Kerr material is pumped from a source which produces already mode-locked pulses. For example, a Ti : Al$_2$O$_3$ crystal can be pumped with a frequency-doubled Nd : YAG laser which is actively mode-locked [9.89].

Another approach for starting passively mode-locked lasers is by means of regenerative feedback [9.90]. In this case, an acousto-optic modulator receives a drive signal from a frequency component of the output from the laser oscillator. Therefore, the drive frequency to the acousto-optic device is automatically matched to the resonator frequency. This technique, originally applied to actively mode-lock lasers [9.91], has been applied to initiate Kerr-lens mode locking. Any initial fluctuations at the cavity frequency owing to mode beating are detected, amplified and fed back to the acousto-optic modulator. Once the pulses have sufficient short duration, the optical nonlinearities in the Kerr medium will dominate and lead to femtosecond pulse generation.

Examples of Femtosecond Lasers

Passive mode locking produces shorter pulses compared to active mode locking because the nonlinearity continues to increase as the peak intensity of the pulse increases and the pulse is shortened. In active mode locking, the modulator response is independent of the pulse duration.

Today, passive mode locking via the Kerr-lens effect is almost always the choice for femtosecond-pulse generation. The laser crystals most frequently employed are $Ti:Al_2O_3$, $Cr:LiSAF$ and $Cr:Forsterite$. Usually a cw argon laser is used to pump $Ti:Al_2O_3$. However, all three crystals can be pumped by solid-state laser sources. A diode-pumped $Nd:YAG$ laser can be employed to pump $Cr:Forsterite$, a doubled $Nd:YAG$ can pump $Ti:Al_2O_3$, and $Cr:LiSAF$ can be pumped directly with laser diodes. The following examples illustrate femtosecond-pulse generation in all solid-state laser sources which have the potential for compactness, reliability and fieldability.

Ti : Sapphire Laser. Figure 9.38 exhibits the schematic of a KLM Ti : sapphire laser which was pumped by a frequency-doubled Nd : YLF laser [9.3, 92]. The pump laser was actively mode-locked to provide picosecond seed pulses in order to get passive mode locking started via the KLM process. The external initiation of mode locking resulted in femtosecond-pulse generation in a continuously self starting manner. The $Ti:Al_2O_3$ crystal provided the Kerr-medium and provided soft aperturing due to the spatial gain profile. In addition, a hard aperture was sometimes incorporated close to the end mirror M4. Dispersion compensation was achieved by translating one of the prisms along its symmetry axis. The resonator consisted of two 10 cm radius focusing mirrors and two flat end mirrors. With a 15 mm long $Ti:Al_2O_3$ crystal and SF10 prisms, the laser produced pulses of 60–70 fs in duration.

In order to reduce third-order dispersion, the crystal length was shortened to 4 mm, the prisms were replaced with fused silica prisms. In ultra-short pulse generation, third-order dispersion in the resonator becomes an important factor. The dominant contribution to third-order dispersion was from the refraction in the prisms used to correct for second-order dispersion in the resonator. A figure of merit for the optimum prism material is the ratio of second- to third-order dispersion. For the wavelength of Ti : sapphire, SiO_2 has the largest ratio. After the changes in crystal length and prism material, the laser produced 15 fs pulses over a tuning range of 790–850 nm. For an absorbed pump power of 3 W, and with a 4.5% output coupler, the output was about 3 nJ per pulse and 200 kW peak power. Average output power was 200 mW. At this pulsewidth, the pulses contain only 5 optical cycles within FWHM and are shorter than 5 μm in space.

Design information such as the parametric dependence of the mode locking on mode radius, reduction of amplitude fluctuations and optimization of intracavity dispersions can be found in [9.93].

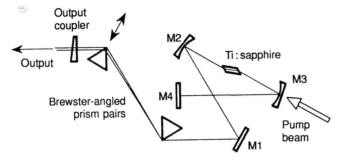

Fig. 9.38. Femtosecond Ti : sapphire laser pumped by a frequency-doubled mode-locked Nd : YLF laser

Cr : LiSAF Laser. This broad-band laser material is particularly interesting for ultra-fast pulse generation because it can directly be pumped by laser diodes at an absorption band around 670 nm. Compared to diode-pumped actively mode-locked Nd : YAG lasers discussed in Sect. 9.3, diode pumping of Cr : LiSAF is a difficult task because this material has a lower gain than Nd : YAG, and the laser diodes available at this shorter wavelength have a lower brightness compared to their counterparts at 800 nm. Since the KLM process is nonlinear, it is strongly dependent on power. Despite these drawbacks, sufficiently high gain has been achieved in a number of mode-locked Cr : LiSAF oscillators end-pumped by laser diodes [9.94–96]. Figure 9.39 presents the schematic of a KLM diode-pumped Cr : LiSAF laser [9.97]. The laser crystal is pumped with a single 250 mW laser diode. The resonator is a standard X-fold design with the 6 mm long Cr : LiSAF crystal in the focal region of the two folding mirrors M2 and M3. Mirrors M1 and M4 are flat, whereas mirrors M2 and M3 have a 100 mm radius of curvature. In accordance with the theoretical results [9.98] discussed earlier, the lowest threshold was found for equal legs of the resonator ($l_1 = l_2 = 885$ mm). The focusing-mirror separation z was found to be very critical. The optimum performance was achieved with $l = 103$ mm with a tolerance on the order of ± 0.1 mm. The crystal was off-set from the beam waist between mirrors M2 and M3. Best results were obtained for $x = 51.5$ mm (see Fig. 9.34a for the definition of coordinates). The spot sizes at both crystal surfaces was 20 and 60 μm, respectively. The large spot size matched the waist of the pump beam.

Kerr-lens mode locking was initiated with an acousto-optic modulator that was regeneratively driven. As explained in the previous section, pulse formation in this case proceeds through regenerative mode locking and evolves to a KLM steady state. Two Brewster-cut prisms provide dispersion compensation. Mode locking was achieved by adjusting the slit S1, the second slit S2 is used to tune the wavelength. The laser produced a train of self mode-locked pulses at a repetition rate of about 80 MHz and with an average power of 2.7 mW. The tuning range was between 845 and 890 nm. The pulses had a duration of 97 fs. The measured spectral bandwidth was 8.8 nm which gives a time-bandwidth product of 0.35.

In a subsequent improvement of the system, the laser crystal was symmetrically pumped from both sides with two 400 mW laser diodes. In this laser con-

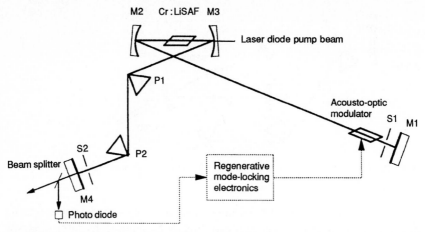

Fig. 9.39. KLM diode-pumped Cr : LiSAF laser [9.97]

figuration, the average output increased to 48 mW and pulse duration decreased to 27 fs.

Cr : Forsterite. This crystal which can be pumped with the output of a Nd : YAG laser at 1.06 μm has a wavelength tunability near 1250 nm. Self mode locking can be achieved in this material with initiation via active mode locking, regeneratively mode locking, or just by tapping one end mirror or moving a prism [9.99–102].

Figure 9.40 exhibits the schematic of a synchronously pumped KLM Cr : Forsterite laser [9.103]. Pulses from an actively mode-locked Nd : YAG laser provide the starting mechanism for self mode-locked operation. The 12 mm long Cr : Forsterite crystal was placed between two 100 mm radius mirrors, in a standard X-fold resonator. A prism pair provided dispersion compensations. The laser produced a train of pulses at 82 MHz with a maximum power of 45 mW for 3.9 W of absorbed pump power. The transform limited pulses had a duration of 50 fs.

9.5.2 Chirped Pulse Amplifiers

The femtosecond oscillators described in the previous subsection generate a train of pulses at a repetition rate of around 100 MHz and with an average power of a few hundred milliwatts. Therefore, the energy of each mode-locked pulse is on the order of a few nanojoules. These pulses have been amplified by as much as a factor of 10^{10}, or up to the joule level, by regenerative amplification followed by a chain of power amplifiers. The chirped-pulse amplification technique was developed to reduce the enormous peak powers in these amplifiers, which would otherwise limit the output energy because of intensity-dependent pulse distortions and damage of the amplifier components.

The technique uses the dispersion of gratings to expand the pulsewidth at the input of the amplifier chain by many orders of magnitude and thereby reduce

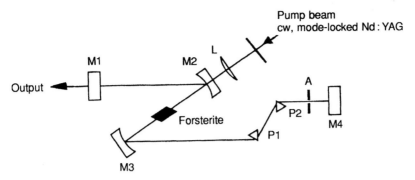

Fig. 9.40. Synchronously pumped KLM Cr : Forsterite laser

peak power. The chirp process is then reversed after the final amplifier, dispersion is now employed to compress the pulse to near its original duration.

A pair of plane ruled gratings arranged in tandem and with their faces and rulings parallel, has the property of producing a time delay that is an increasing function of wavelength. A grating pair can thus be used to compress optical chirped pulses or to generate chirped pulses. The variation of group-velocity delay with wavelength can be understood by reference to Fig. 9.41 [9.104]. The path length PABQ for the wavelength λ is less than the similar path PACR for the longer wavelength λ'. Thus, the grating provides large negative group-velocity dispersion. If a telescope is added between the gratings, the sign of the dispersion can be inverted and positive group-velocity dispersion can be obtained [9.105].

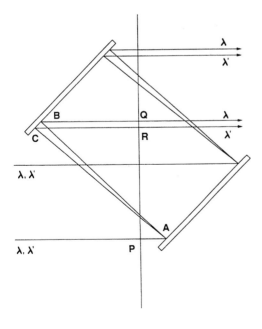

Fig. 9.41. Diffraction grating used for pulse compression

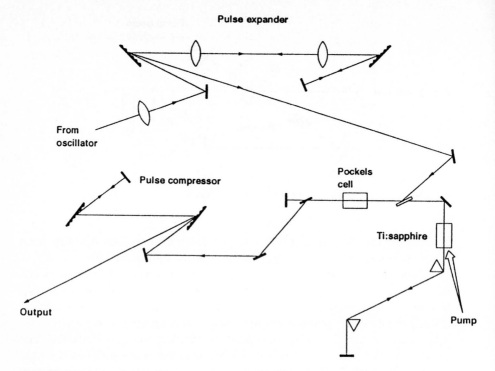

Fig. 9.42. Femtosecond laser with pulse expander, regenerative amplifier and pulse compressor

Typically, pulses are stretched with a dispersive delay line consisting of a pair of gratings arranged to provide positive group-velocity dispersion and compressed with a grating pair which provides negative group-velocity dispersion.

Figure 9.42 illustrates the architecture of a typical chirped pulse amplifier system. Pulses arriving from a femtosecond oscillator are stretched before injection into the regenerative amplifier by two gratings arranged in an anti-parallel configuration and separated by a unit-magnification telescope. The amplified pulses are then recompressed by using a pair of parallel diffraction gratings. The gratings, typically separated by 1–2 meters to provide a large dispersion, can stretch and compress pulses by over 3 orders of magnitude. For example, a 200 fs pulse from a Ti : sapphire laser was stretched to 600 ps, or 6000 times, using the type of stretcher illustrated in Fig. 9.42. The theory and design of diffraction gratings employed for expanding and compressing optical pulses was described in [9.104, 105].

Chirped-pulse amplification was first developed for radar and later adapted to the optical regime [9.106]. It is now a powerful technique by which the energy of femtosecond pulses can be increased to the joule level which results in peak powers in the terawatt regime. The chirped-pulse amplification technique has been implemented in Ti : sapphire [9.107-111], Cr : LiSAF [9.112, 113], and Ti : sapphire/Nd : Glass [9.114], Ti : sapphire/Alexandrite [9.115] amplifier chains.

For example, a Cr:LiSAF flashlamp-pumped laser system consisting of a regenerative amplifier and three additional double-pass amplifiers produced 90 fs pulses at a peak power of 8 TW. The 1 nJ input pulse is stretched 2000 times to 170 ps before injection into the regenerative amplifier [9.112]. Similarly, a multi-stage Ti:sapphire laser chain produced 95 fs pulses at an energy of 0.45 J per pulse. In this case, the pulses are stretched to 500 ps before they enter the amplifier chain [9.108].

10. Nonlinear Devices

Nonlinear optical devices, such as harmonic generators and parametric oscillators, provide a means of extending the frequency range of available laser sources. In 1961, *Franken* and coworkers detected ultraviolet light at twice the frequency of a ruby laser beam when this beam was propagated through a quartz crystal [10.1].This experiment marked the beginning of an intense investigation into the realm of the nonlinear optical properties of matter.

Frequency conversion is a useful technique for extending the utility of high-power lasers. It utilizes the nonlinear optical response of an optical medium in intense radiation fields to generate new frequencies. It includes both elastic (optical-energy-conserving) processes, such as harmonic generation, and inelastic processes (which deposit some energy in the medium), such as stimulated Raman or Brillouin scattering.

There are several commonly used elastic processes. Frequency doubling, tripling, and quadrupling generate a single harmonic from a given fundamental high-power source. The closely related processes of sum- and difference-frequency generation also produce a single new wavelength, but require two high-power sources. These processes have been used to generate high-power radiation in all spectral regions, from the ultraviolet to the far infrared. Optical parametric oscillators and amplifiers generate two waves of lower frequency from a pump source. They are capable of generating a range of wavelengths from a single frequency source, in some cases spanning the entire visible and near-infrared regions.

As far as inelastic processes are concerned, the Raman process can be utilized in solid-state lasers for the generation of additional spectral output lines. The strongest interaction is for the output shifted towards a longer wavelength (first Stokes shift), but at sufficiently high pump intensities additional lines at longer as well as shorter wavelengths with respect to the pump wavelength will appear. (Stokes and anti-Stokes lines.)

Although it produces a small wavelength shift, stimulated Brillouin scattering is mainly of interest for the realization of phase-conjugating mirrors. The application of phase conjugation, or wavefront reversal, via stimulated Brillouin scattering offers the possibility of minimizing thermally-induced optical distortions [10.2] which occur in solid-state laser amplifiers.

Nonlinear optical effects are analyzed by considering the response of the dielectric material at the atomic level to the electric fields of an intense light beam. The propagation of a wave through a material produces changes in the spatial and temporal distribution of electrical charges as the electrons and atoms react to the electromagnetic fields of the wave. The main effect of the forces exerted by the fields on the charged particles is a displacement of the valence electrons from their normal orbits. This perturbation creates electric dipoles whose macroscopic manifestation is the polarization. For small field strengths this polarization is proportional to the electric field. In the nonlinear case, the reradiation comes from dipoles whose amplitudes do not faithfully reproduce the sinusoidal electric field that generates them. As a result, the distorted reradiated wave contains different frequencies from that of the original wave.

In a given material, the magnitude of the induced polarization per unit volume P will depend on the magnitude of the applied electric field E. We can therefore expand P in a series of powers of E and write:

$$\begin{aligned} P_l(\omega_j) = \; & X_{lm}^{(1)} E_m(\omega_j) + \ldots + X_{lmn}^{(3)} E_m(\omega_r) E_n(\omega_s) \\ & + X_{lmnp}^{(4)} E_m(\omega_r) E_n(\omega_s) E_p(\omega_t) + \ldots \\ & + X_{lmn}^{(6)}(-i\omega_r) E_m(\omega_r) B_n(\omega_s) + \ldots \end{aligned} \tag{10.1}$$

where P and E are vectors linked by tensors of second ($X^{(1)}$), third ($X^{(3)}$, $X^{(6)}$), and fourth ($X^{(4)}$) rank. The values of the tensor coefficients are functions of frequency and temperature. The subscripts j, r, s, and t denote different frequency components, and l, m, n, and p are Cartesian indices that run from 1 to 3. In (10.1) only those terms are listed which give rise to optical phenomena treated in this book.

For small field strengths the polarization is proportional to the electric field E and is accounted for by the polarizability tensor $X_{lm}^{(1)}$. Linear optics encompass all the interaction of light and dielectrics where the first term of (10.1) is a valid approximation. In linear optics the index of refraction is given by

$$n = (1 + 4\pi X^{(1)})^{1/2} = \varepsilon^{1/2} \tag{10.2}$$

where ε is the dielectric constant of the material.

The $X^{(3)}$ term is responsible for second-harmonic generation, optical mixing, and the Pockels effect. The nonlinear polarization tensor $X^{(3)}$ vanishes in crystals that have a center of symmetry. In these crystals second-harmonic generation is not possible. The third-rank tensor $X^{(3)}$ in general has 27 components. As a result of crystal symmetry, many of the components of $X^{(3)}$ will be zero or equal to other components of the tensor. Furthermore, for those crystals of main interest to us, there is usually one predominant coefficient associated with a single light propagation direction which yields maximum harmonic power. If $\omega_r = \omega_s$ we obtain the relationship of second-harmonic generation

$$P_l(2\omega) = X_{lmn}^{(3)} E_m(\omega) E_n(\omega) . \tag{10.3}$$

The generation of harmonics is a special case of optical mixing in nonlinear materials. The simultaneous application of two fields with frequencies ω_r and ω_s

produces a polarization at the sum and difference frequencies. The polarizations produced are of the form

$$P_l \begin{pmatrix} \omega_r + \omega_s \\ \omega_r - \omega_s \end{pmatrix} = X^{(3)}_{lmn} E_m(\omega_r) E_n(\omega_s) . \tag{10.4}$$

In general, if three waves ω_r, ω_s, and $\omega_i = \omega_r + \omega_s$ are superimposed in a nonlinear medium, each wave is coupled to the other two through polarization waves. This is called the parametric interaction of three waves. Therefore, it is possible to convert energy into radiation at the sum and difference frequencies. From the point of view of parametric amplification, the second-harmonic generation is a special case of interaction between two waves with a common frequency.

The Pockels effect is obtained if one of the electric fields is taken to be a dc field applied across a suitable crystal. For $\omega_s = 0$ and $E_n = E_{dc}$ one obtains $\omega_j = \omega_r$, and the index of refraction becomes a function of E_{dc}, i.e., $P_l(\omega_j) = [X^{(3)}_{lmn} E_{dc}] E_m(\omega_j)$.

The $X^{(4)}$ term of (10.1) couples the nonlinear polarization to three electric field vectors. If $\omega_r = \omega_s = \omega_t$ is the fundamental frequency, then $\omega_j = 3\omega$ and $P_l(3\omega)$ will generate the third harmonic.

For $E_n = E_p$ and $\omega_s = -\omega_t$ the $X^{(4)}$ term transforms to

$$P_l(\omega_j) = [X^{(4)}_{lmnn} E^2_{av}] E_m(\omega_j) \tag{10.5}$$

where E_{av} is the average electric field strength. In this case the index of refraction becomes a function of the light intensity, which leads to nonlinear processes, such as the Kerr effect, and electrostriction.

With the application of a dc magnetic field ($\omega_s = 0$) the $X^{(6)}$ term describes the Faraday effect.

10.1 Harmonic Generation

In this section, we will review the basic theory and discuss system parameters and material properties which affect harmonic generation.

10.1.1 Basic Equations of Second-Harmonic Generation

The process of harmonic generation by an incident wave of frequency ω_1 must be viewed as a two-step process: First, a polarization wave at the second harmonic $2\omega_1$ is produced which has a phase velocity and wavelength in the medium which are determined by n_1, the index of refraction for the fundamental wave, that is, $\lambda_p = c/2\nu_1 n_1$. The second step is the transfer of energy from the polarization wave to an electromagnetic (em) wave at frequency $2\nu_1$. The phase velocity and the wavelength of this em wave are determined by n_2, the index of refraction for the doubled frequency, that is, $\lambda_2 = c/2\nu_1 n_2$. For efficient energy transfer it is necessary that the two waves remain in phase, which implies that $n_1 = n_2$. Since almost all materials have normal dispersion in the optical region, the radiation will generally lag behind the polarization wave. The phase mismatch between the polarization wave and the em wave for collinear beams is usually expressed as the difference in wave number

$$\Delta k = \frac{4\pi}{\lambda_1}(n_1 - n_2) \, . \tag{10.6}$$

If Maxwell's equations are solved for a coupled fundamental and second-harmonic wave propagating in a nonlinear medium, then the ratio of the power generated at the second-harmonic frequency to that incident at the fundamental is given by [10.3]

$$\frac{P_{2\omega}}{P_\omega} = \tanh^2 \left[lK^{1/2} \left(\frac{P_\omega}{A}\right)^{1/2} \frac{\sin \Delta kl/2}{\Delta kl/2} \right] \, , \tag{10.7}$$

where

$$K = 2\eta^3 \omega_1^2 d_{\text{eff}}^2 \, , \tag{10.8}$$

l is the length of the nonlinear crystal, A is the area of the fundamental beam, η is the plane-wave impedance $\eta = \sqrt{\mu_0/\varepsilon_0\varepsilon} = 377/n_0 \, [V/A]$, ω_1 is the frequency of the fundamental beam, and d_{eff} is the effective nonlinear coefficient of the nonlinear polarizability tensor $\mathbf{X}^{(3)}$ in (10.1). The dimension of d_{eff} in (10.8) is given in the MKS system and includes ε_0, the permittivity of free space, thus $d_{\text{eff}} \, [\text{As/V}^2]$. Some authors exclude ε_0 from the d coefficient, in this case $d \, [\text{As/V}^2] = 8.855 \times 10^{-12} d \, [\text{m/V}]$. The conversion from the cgs system to MKS units becomes $d \, [\text{As/V}^2] = 3.68 \times 10^{-15} d \, [\text{esu}]$.

For low conversion efficiencies, (10.7) may be approximated by

$$\frac{P_{2\omega}}{P_\omega} = l^2 K \frac{P_\omega}{A} \frac{\sin^2 (\Delta kl/2)}{(\Delta kl/2)^2} \, . \tag{10.9}$$

For a given wavelength and a given nonlinear material, K is a constant. The conversion efficiency, therefore, depends on the length of the crystal, the power density, and the phase mismatch. For a crystal of fixed length, the second-harmonic power generation is strongly dependent on the phase mismatch expressed by the $\sin^2$ function, as is illustrated in Fig. 10.1. In this case, a variation of Δk was obtained by changing the crystal temperature. The harmonic power is at maximum when $\Delta k = 0$, that is, at the exact phase-matching temperature. For a fixed Δk, the second-harmonic power as a function of distance l along the crystal grows and decays with a period of $\Delta kl/2 = \pi$. Half of this distance has been termed the coherence length l_c. It is the distance from the entrance face of the crystal to the point at which the second-harmonic power will be at its maximum value. The reader is warned not to confuse this parameter with the coherence length l_c of the laser beam, which was defined in Chap. 5.

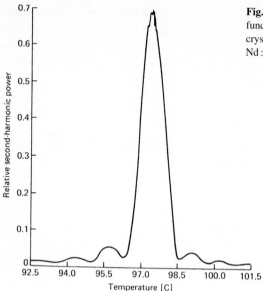

Fig. 10.1. Second-harmonic generation as a function of temperature in a $Ba_2NaNb_5O_{15}$ crystal employed to frequency-double a Nd:YAG laser

For normal incidence the coherence length is given by

$$l_c = \frac{\lambda_1}{4(n_2 - n_1)} \;.$$

(10.10)

Expressing the phase mismatch Δk in terms of coherence length in (10.9), one obtains

$$\frac{P_{2\omega}}{P_\omega} = l_c^2 K \frac{4}{\pi^2} \frac{P_\omega}{A} \sin^2\left(\frac{\pi l}{2l_c}\right) \;.$$

(10.11)

The oscillatory behavior of (10.11) is shown in Fig. 10.2 for several values of l_c. For the ideal case $l_c = \infty$, the second-harmonic conversion efficiency is proportional to the square of the crystal length, at least in the small-signal approximation

$$\frac{P_{2\omega}}{P_\omega} = l^2 K \frac{P_\omega}{A} \;.$$

(10.12)

Clearly, if the crystal is not perfectly phase-matched ($l_c = \infty$), the highest second-harmonic power we can expect to generate will be the signal obtained after the beam propagates one coherence length, no matter how long the crystal is. The decrease in harmonic power, for example, between l_c and $2l_c$, is explained by a reversal of the power flow. Instead of power being coupled from the polarization wave into the em wave, it is coupled from the em wave into the polarization wave: i.e., the power is coupled back into the input beam. Thus we find that the power oscillates back and forth between the harmonic and the fundamental wave.

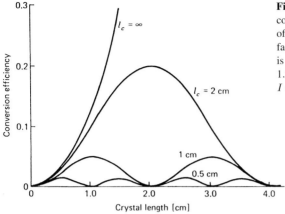

Fig. 10.2. Second-harmonic power conversion efficiency as a function of distance l from the entrance surface of a CDA crystal. Parameter is the coherence length l_c ($K = 1.3 \times 10^{-9}\,\text{W}^{-1}$; $\lambda = 1.06\,\mu\text{m}$; $I = 100\,\text{MW/cm}^2$)

In almost all practical cases the coherence length is limited by the beam divergence and the bandwidth of the laser beam and by angular and thermal deviations of the crystal from the phase-matching angle and temperature.

Index Matching

With typical dispersion values in the visible region, the coherence length in most crystals is limited to about $10\,\mu\text{m}$. For this reason the intensity of second-harmonic power is small. Only if n_1 can be made substantially equal to n_2 will relatively high efficiencies of frequency doubled power be obtained.

An effective method of providing equal-phase velocities for the fundamental and second-harmonic waves in the nonlinear medium utilizes the fact that dispersion can be offset by using the natural birefringence of uniaxial or biaxial crystals [10.4, 5]. These crystals have two refractive indices for a given direction of propagation, corresponding to two orthogonally polarized beams; by an appropriate choice of polarization and direction of propagation it is often possible to obtain $\Delta k = 0$. This is termed phase matching or index matching.

We shall restrict our discussion to the most frequently encountered case of birefringence, that of uniaxial crystals. These crystals, to which the very important nonlinear crystals KDP and its isomorphs and LiNbO₃ belong, have an indicatrix which is an ellipsoid of revolution with the optic axis being the axis of rotation, as shown in Fig. 10.3. The two directions of polarization and the indices for these directions are found as follows: We draw a line through the center of the ellipsoid in the direction of beam propagation (line $0P$ in Fig. 10.3). Then we draw a plane perpendicular to the direction of propagation. The intersection of this plane with the ellipsoid is an ellipse. The two axes of this ellipse are parallel to the two directions of polarization and the length of each semi-axis is equal to the refractive index in that direction.

We now examine how the indices of refraction vary when the direction of propagation is changed. We notice that for the direction of polarization perpendicular to the optic axis, known as the ordinary direction, the refractive index is

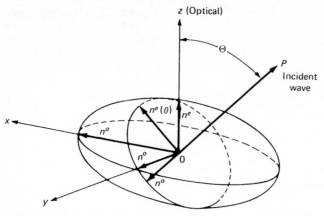

Fig. 10.3. Indicatrix ellipsoid of a uniaxial crystal. Shown is also a cross section perpendicular to the light propagation direction P

independent of the direction of propagation. For the other direction of polarization, known as the extraordinary direction, the index changes between the value of the ordinary index n_0 when $0P$ is parallel to z, and the extraordinary index n_e when $0P$ is perpendicular to z. When the wave propagation is in a direction Θ to the optic axis, the refractive index for the extraordinary wave is given by [10.6]

$$n^e(\Theta) = \frac{n^o n^e}{[(n^o)^2 \sin^2 \Theta + (n^e)^2 \cos^2 \Theta]^{1/2}} , \tag{10.13}$$

where the superscripts "o" and "e" refer to the ordinary and the extraordinary rays.

Changing the point of view, we look now at the shape of the wavefronts for these two rays instead of their direction and polarization. Assuming that the input is a monochromatic point source at 0, the expanding wavefront for the o ray is spherical, whereas the spreading wavefront for the e ray is an ellipsoid. This property of crystals is described by the index surface, which has the property that the distance of the surface from the origin along the direction of the wave vector is the refractive index. For an uniaxial crystal this surface has two sheets – a sphere for ordinary waves polarized perpendicular to the optic axis with index n^o, and an ellipsoid for extraordinary waves with index $n^e(\Theta)$. By definition the optic axis is that direction at which the o and e rays propagate with the same velocity. If the value of $n^e - n^o$ is larger than zero, the birefringence is said to be positive and for $n^e - n^o$ smaller than zero the birefringence is negative; the corresponding crystals are called positive or negative uniaxial. Figure 10.4 shows a cross section of the index surface of a negative uniaxial crystal (for the moment we consider only the solid lines n_1^o and n_1^e). The complete surfaces are generated by rotating the given sections about the z axis. The wavefront velocity

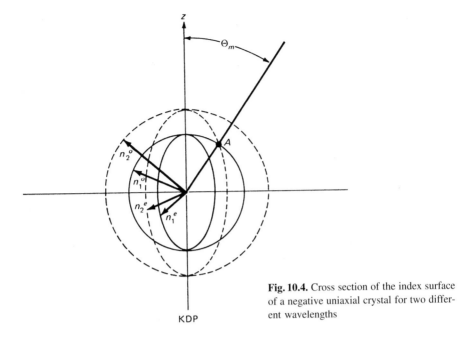

Fig. 10.4. Cross section of the index surface of a negative uniaxial crystal for two different wavelengths

v and the refractive index n are related by $v = c/n$, where c is the velocity of light.

Both refractive indices n^o and n^e are a function of wavelength. Figures 10.4 and 10.5 illustrate how the dependence of the refractive index on beam direction, wavelength, and polarization can be utilized to achieve angle-tuned phase matching. The dashed lines in these figures show the cross section of the index surfaces n_2 at the harmonic frequency. As can be seen, the negative uniaxial crystal has sufficient birefringence to offset dispersion, and the matching condition can be satisfied for a beam deviating from the z axis by the angle Θ_m.

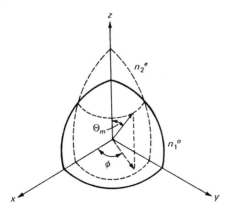

Fig. 10.5. Direction for phase-matched second-harmonic generation (type-I) in a uniaxial crystal, where Θ_m is the phase-matching angle measured from z, and ϕ is the azimuth angle measured from x

The directions for phase-matched second-harmonic generation are obtained by considering the intersection of the index surfaces at the fundamental and harmonic frequencies. As was mentioned earlier, frequency doubling may be considered as a special case, where two incident waves with electric fields E_m and E_n are identical wave forms. There are two types of processes in harmonic generation, depending on the two possible orientations for the linear polarization vectors of the incident beams. In the type-I process both polarization vectors are parallel: in the type-II process the polarization vectors are orthogonal.

In a negative uniaxial crystal there are two loci where the index surfaces intersect and $\Delta k = 0$ [10.10]

$$n_2^e(\Theta_m) = n_1^o \quad \text{type-I}$$
$$n_2^e(\Theta_m) = \tfrac{1}{2}[n_1^e(\Theta_m) + n_1^o] \quad \text{type-II} ; \tag{10.14}$$

first, in a symmetrical cone at Θ_m (type-I) about the optic axis, where two o-rays at ω are matched to an e-ray at 2ω; second, in a cone at Θ_m (type-II), where an o-ray and an e-ray at ω are matched to an e-ray at 2ω.

The harmonic power is not independent of the azimuthal angle of the phase matched direction (Fig. 10.5). In general, d_{eff} is a combination of one or several coefficients of $X^{(3)}$, and the angles Θ and ϕ which define the direction of the wave propagation vector. For example, for KDP and its isomorphs and type-I index matching, one obtains [10.7, 8]

$$d_{\text{eff}} = d_{14} \sin 2\phi \sin \Theta_m .$$

The phase matching angle Θ_m is obtained by combining (10.14 and 13)

$$\sin^2 \Theta_m = \frac{(n_1^o)^{-2} - (n_2^o)^{-2}}{(n_2^e)^{-2} - (n_2^o)^{-2}} .$$

Maximum interaction is achieved when a single beam incident on the crystal is equally divided into two fundamental beams propagating along the crystal's x and y axis. Therefore, one obtaines $\phi = 45°$ and $E_m = E_n$ in (10.3). In type-I phase matching the fundamental beams are polarized perpendicular to the optical axis and the harmonic beam is polarized parallel to the optical axis of the crystal.

For type-II phase matching one obtains

$$d_{\text{eff}} = d_{14} \cos 2\phi \sin 2\Theta_m .$$

In this case the beam propagation vector is orientated at $90°$ with respect to the x axis ($\phi = 90°$) for maximum interaction. A single linearly polarized fundamental beam incident on the crystal may be equally divided into o and e rays by orienting the polarization vector at $45°$ with respect to the x axis. The polarization vector of the harmonic beam is rotated $45°$ from the fundamental beam with the polarization parallel to the optical axis.

Comparing type-I and -II phase matching, we find that type I is more favorable when Θ_m is near $90°$, whereas type-II leads to a higher d_{eff} when Θ_m lies near $45°$. For LiNbO$_3$ and type-I phase matching we obtain

$$d_{eff} = d_{31} \sin \Theta_m + d_{22} \cos \Theta_m (4 \sin^3 \phi - 3 \sin \phi) .$$

Critical Phase Matching

If phase matching is accomplished at an angle Θ_m other than $90°$ with respect to the optic axis of a uniaxial crystal, there will be double refraction. Therefore, the direction of power flow (Poynting vector) of the fundamental and second harmonics will not be completely collinear but occur at a small angle.

For a negative uniaxial crystal and type-I phase matching, this angle is given by [10.9]

$$\tan \varrho = \frac{(n_1^o)^2}{2} \left(\frac{1}{(n_2^e)^2} - \frac{1}{(n_2^o)^2} \right) \sin 2\Theta . \tag{10.15}$$

The angle ϱ has the effect of limiting the effective crystal length over which harmonic generation can take place. The beams completely separate at a distance of order

$$l_a = a/\varrho \tag{10.16}$$

called the aperture length, where a is the beam diameter. Of course, at only a fraction of this distance the reduction of conversion efficiency due to walk-off becomes noticeable and has to be taken into account.

For weakly focused Gaussian beams, the aperture length can be expressed as [10.9]

$$l_a = w_0 \sqrt{\pi}/\varrho \tag{10.17}$$

where w_0 is the fundamental beam radius.

For a given crystal, the effect of beam separation due to double refraction can be accounted for, to first order, by a reduction of the material constant K in (10.8). Walk-off over the crystal length reduces K according to

$$K_w = K/(1 + l/l_a) . \tag{10.18}$$

Another limitation of angular phase matching is due to the divergence of the interacting beams. For second-harmonic generation in a negative crystal we find from (10.6) that $\Delta k = 0$ if $n_1^o = n_2^e(\Theta_m)$, which is exactly true only at $\Theta = \Theta_m$. In Sect. 10.1.2 it will be shown that there is a linear relationship between small deviations $\delta\theta$ from the phase matching angle and Δk. The change of Δk as a function of $\delta\theta$ can be large enough to limit the conversion efficiency in real devices. Consider as an example second-harmonic generation in KDP at $1.064\,\mu m$ with $\Theta_m = 42°$. The linear change in Δk with Θ is sufficiently great to restrict the divergence from the phase-matched direction to approximately 1 mrad if the coherence length is to be greater than 1 cm. Phase matching under these unfavorable conditions is termed "critical phase matching". Critical-phase matching can be made to approximate noncritical phase matching by dividing the crystal into a number of short segments arranged so that alternate segments have opposite birefringence-induced walk-off directions [10.24].

Noncritical Phase Matching

If the refractive indices can be adjusted so that $\Theta_m = 90°$, by variation of a parameter such as the temperature or chemical composition of the crystal, the dependence of Δk on angular misalignment $\delta\theta$ is due to a much smaller quadratic term, instead of a linear relationship as is the case in critical phase matching.

In addition, at $\Theta_m = 90°$ there are no walk-off effects due to double refraction. For example, second-harmonic generation in $LiNbO_3$ occurs at 90° for 1.064-μm radiation provided the crystal is at a temperature of $\approx 47°C$. The allowable divergence from the phase-matched direction is greater than 10 mrad if the coherence length is to be 1 cm. For these reasons, 90° phase matching is often called noncritical phase matching. Provided that n_1^o and n_2^e are nearly equal and $d(n_1^o - n_2^e)/dT \neq 0$, noncritical phase matching can be achieved by temperature tuning the crystal.

Because of their importance, we have emphasized phase matching in negative uniaxial crystals. The reader interested in a systematic review of phase matching in uniaxial and biaxial crystals is referred to the paper by *Hobden* [10.10]. Frequency doubling with focused Gaussian beams has been discussed in [10.12, 13]. The subject of nonlinear optics has been treated in several books [10.3, 11, 14–16] and tutorial review articles [10.17, 18]. For a general introduction to crystal optics the reader is referred to standard texts [10.6, 19, 20].

10.1.2 Parameters Affecting the Doubling Efficiency

High-efficiency second-harmonic conversion depends on parameters which are related to the laser source, such as power density, beam divergence, and spectral linewidth, and parameters associated with the harmonic generator, such as the value of the nonlinear coefficient, crystal length, angular and thermal deviation from the optimum operating point, absorption, and inhomogeneities in the crystal. The dependence of the second-harmonic power on these parameters will be examined in this subsection.

The Dependence of Harmonic Generation on Laser Parameters

Power Density. From (10.9) it follows that the conversion efficiency is proportional to the power density of the fundamental beam, whereas the harmonic power itself is quadratically proportional to the fundamental power. At conversion efficiencies above 20 %, the second-harmonic generation starts to deviate markedly

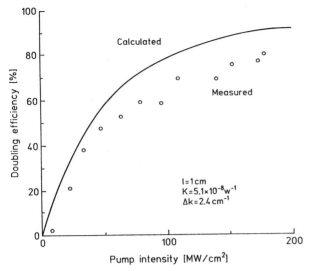

Fig. 10.6. Second-harmonic conversion efficiency versus power density for a KTP crystal pumped by a Q-switched Nd : YAG laser at $1.06 \, \mu m$

from the linear relationship of (10.9) because of depletion of the fundamental beam power. At these high efficiencies, (10.7) should be used.

Figure 10.6 shows experimental data and predicted performance for a harmonic generator comprised of a 10 mm long KTP crystal. The fundamental beam is derived from a Q-switched diode-pumped Nd : YAG laser at $1.064 \, \mu m$. The laser had a maximum pulse energy of 40 mJ in a 12 ns pulse. At the peak intensity of 175 MW/cm² for the pump beam, the conversion efficiency approached 80%. The solid curve is obtained from (10.7) if one introduces the following system parameters: crystal length l = 10 mm, effective nonlinear coefficient d_{eff} = 3.2 pm/V, fundamental wavelength λ_1 = 1.064 μm. The fundamental beam was a weakly focused Gaussian beam with a spot radius of w_0 = 0.75 mm. The beam area is therefore $A = \pi w_0^2 = 1.8 \times 10^{-2}$cm².

As will be demonstrated in the next section, phase-mismatch can be expressed as $\Delta k = \beta_\Theta \delta\Theta$, where β_Θ is the angular sensitivity and $\delta\Theta$ is the beam divergence of the fundamental beam. From (10.24) and Table 10.2 follows β_Θ = 0.22 (mrad cm)⁻¹ for KTP at $1.06 \, \mu m$ and type-II phase matching. The beam divergence of the weakly focused beam was about 12 mrad or $\Delta k = 2.4 \, \text{cm}^{-1}$.

Walk-off reduces the region of overlap between the beams, thus reducing the effective gain length of the crystal. From (10.17) we obtain for the aperture length l_a = 290 mm if we use a beam walk-off angle of $\varrho = 4.5$ mrad (Table 10.2). Since the aperture length is large compared to the crystal length, the correction factor defined by (10.18) is close to 1.

The theoretical doubling efficiency calculated using (10.7) gives quite acceptable agreement with the experimental data. As expected, the theory predicts higher efficiency than can be achieved in practice due to the imperfect nature of the pump beam and crystal quality.

Beam Divergence. When collinear phase-matched second-harmonic generation is used, the light waves will have a small but finite divergence. It is necessary to consider the mismatch Δk of the wave vector for small deviations $\delta\Theta$ from the phase-matched direction. An expansion for $n_1^o - n_2^e(\Theta)$ taken for a direction close to the perfect phase-matching direction Θ_m
yields

$$n_1^o - n_2^e(\Theta) = \frac{\partial[n_1^o - n_2^e(\Theta)]}{\partial\Theta}(\Theta - \Theta_m) \ . \tag{10.19}$$

The expression given in (10.13) for the dependence of n_2^e on the angular direction in the crystal can be very well approximated by [10.21]

$$n_2^e(\Theta) = n_2^o - (n_2^o - n_2^e) \sin^2 \Theta \ . \tag{10.20}$$

Introducing (10.20) into (10.19) gives

$$n_1^o - n_2^e(\Theta) = \delta\Theta(n_2^o - n_2^e) \sin 2\Theta_m \ , \tag{10.21}$$

where we note that $\partial n_1^o/\partial\Theta = 0$, and $\partial n_2^o/\partial\Theta = 0$.
    Introducing (10.21) into (10.6) yields

$$\Delta k = \beta_\Theta \delta\Theta' \tag{10.22}$$

where

$$\beta_\Theta = \frac{4\pi}{\lambda_1 n_1^0} (n_2^o - n_2^e) \sin 2\Theta_m$$

is a material constant and expresses the angular sensitivity of the crystal. The expression for β_Θ has been divided by n_1^0, because $\delta\Theta'$ is now the angular misalignment measured external to the crystal.
    From (10.9) follows that the conversion efficiency will be reduced to one-half of its peak value for

$$\frac{\Delta k l}{2} = 1.39 \ . \tag{10.23}$$

If we combine (10.22) and (10.23), we obtain an expression for the angular tolerance. The full angle $\Delta\Theta = 2\delta\Theta'$ measured external to the crystal which determines the full width half maximum (FWHM) of the conversion process is given by

$$\Delta\Theta = \frac{5.56}{l\beta_\Theta} \ . \tag{10.24}$$

For example, KTP has an angular bandwidth of $\Delta\Theta = 25$ mrad-cm or an angular sensitivity of $\beta_\Theta = 0.22 \, (\text{mrad-cm})^{-1}$.

Under noncritical phase-matching conditions ($\Theta_m = 90°$), we can make the approximation $\sin 2(90° + \delta\Theta) \approx 2\delta\Theta$, and instead of (10.24) we obtain

$$\Delta\Theta = 0.66 \left(\frac{\lambda_1 n_1^0}{l(n_2^0 - n_2^e)}\right)^{1/2} \tag{10.25}$$

Spectral Linewidth. Expanding $n_1^0 - n_2^e(\Theta)$ for small wavelength changes around the central wavelength λ_0 at which phase matching occurs results in [10.25]

$$n_1^0 - n_2^e(\Theta) = \left(\frac{\partial n_1^0}{\partial\lambda_1} - \frac{1}{2}\frac{\partial n_2^e(\Theta)}{\partial\lambda_2}\right)(\lambda - \lambda_0) . \tag{10.26}$$

From (10.6, 23, 26) we obtain an expression for $\Delta\lambda = 2\delta\lambda$, where $\delta\lambda = \lambda - \lambda_0$ is the deviation from the phase-matching wavelength at which the doubling efficiency drops to one-half

$$\Delta\lambda = 0.44\lambda_1/l \left(\frac{\partial n_1^0}{\partial\lambda_1} - \frac{1}{2}\frac{\partial n_2^e(\Theta)}{\partial\lambda_2}\right) . \tag{10.27}$$

Spectral Brightness. From the foregoing considerations it becomes clear that in order to achieve maximum second-harmonic power, the laser source should have a high power density, small beam divergence, and narrow linewidth. These properties of the laser can also be expressed by a single parameter, namely the spectral brightness [W/cm^2 sterad Å]. For high-efficiency second-harmonic generation the laser must exhibit a high spectral brightness, which can be achieved by transverse and longitudinal mode selection. Experimental data showing the dependence of second-harmonic generation on the mode structure of the laser can be found in [10.26, 27]. In general, one finds that Nd : YAG lasers have sufficiently narrow linewidths for efficient harmonic generation. In these lasers attention is focused mainly on obtaining a diffraction limited beam, i.e., TEM$_{00}$-mode operation. In Nd : glass, however, the broad linewidth of the laser without axial mode selection can present a severe limitation in obtaining high conversion efficiencies.

Very high conversion efficiency for the entire pulse requires that all the radiation incident on the nonlinear crystal be converted efficiently, regardless of its intensity, bandwidth, polarization, or amplitude and phase variations. The range of intensity over which the efficiency is high is especially important. The intensity is necessarily nonuniform because it drops to zero at the spatial and temporal edges of the beam, and there may also be amplitude nonuniformities. The standard technique of frequency conversion uses a single crystal of a nonlinear material for frequency doubling. With a single nonlinear crystal, the dynamic range is necessarily limited to moderate values. *Eimerl* [10.28, 29] found that using two crystals for each conversion step, the dynamic range can be much larger. The two crystals are arranged so that the output wave generated in the first one is polarized orthogonally to that generated in the second. The two harmonics therefore add in quadrature, which gives the technique its name: quadrature frequency doubling. The conversion efficiency of these quadrature arrangements is much less-sensitive to laser pulse nonuniformities than that in single-crystal methods. Consequently, very high conversion efficiency is possible for typically nonuniform laser pulses.

Parameters of the Harmonic Generator which Affect Doubling Efficiency

Temperature. The indices of refraction of the crystal at the phase-matching temperature T_0 can be expanded in a Taylor series for a nearby temperature [10.30]

$$n_1 - n_2 = (T - T_0)\frac{d(n_2 - n_1)}{dT} .$$ (10.28)

With the aid of (10.6, 23) we obtain an expression of the temperature sensitivity of the doubling crystal with respect to second-harmonic generation,

$$\Delta T = \frac{0.44\lambda_1}{l \, d(n_2^e - n_1^o)/dT} ,$$ (10.29)

where ΔT is the full width at half-maximum of the temperature range over which second-harmonic generation is possible in a particular crystal (Fig. 10.1). Temperature changes of the doubling crystal may be the result of ambient temperature variations, or they may be caused by absorption losses in the crystal. Experimentally determined thermal tuning ranges for several crystals are shown in Fig. 10.7.

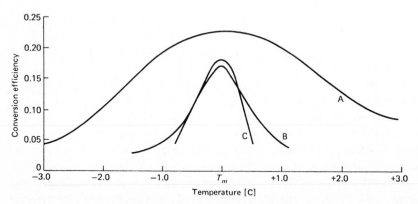

Fig. 10.7. Thermal tuning curves of several nonlinear crystals. A: CDA, $1.06\,\mu$m, $I = 100$ MW/cm^2, 1.8-cm-long crystal; B: ADP, $0.53\,\mu$m, $I = 10$ MW/cm^2, 4-cm-long crystal; C: RDA, $0.69\,\mu$m, $I = 100$ MW/cm^2, 1.5-cm-long crystal

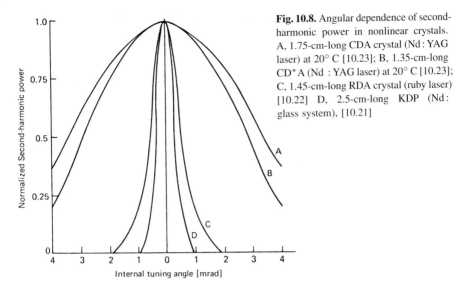

Fig. 10.8. Angular dependence of second-harmonic power in nonlinear crystals. A, 1.75-cm-long CDA crystal (Nd : YAG laser) at 20° C [10.23]; B, 1.35-cm-long CD*A (Nd : YAG laser) at 20° C [10.23]; C, 1.45-cm-long RDA crystal (ruby laser) [10.22] D, 2.5-cm-long KDP (Nd : glass system), [10.21]

Phase-Matching Angle. The sensitivity required to angular-adjust a doubling crystal and maintain its alignment is given by (10.24). The angle $\Delta\Theta$ now becomes the full acceptance angle over which the crystal can be tilted before the second-harmonic power decreases to less than half. Figure 10.8 illustrates the angular tuning range for several crystals.

Absorption. Absorption in the crystal will lead to thermal gradients as well as thermally induced stresses. The associated refractive index nonuniformities severely restrict the crystal volume over which phase matching can be achieved [10.31, 32]. Heating of the nonlinear crystal due to absorption is particularly a problem in doubling experiments involving high-power cw beams. A self-induced thermal distribution in the nonlinear material can be caused by absorption of the fundamental or harmonic beams or by multiphoton absorption processes [10.33].

Optical Homogeneity. Index inhomogeneities in nonlinear optical crystals are an important limiting factor in second-harmonic generation, since the phase-matching condition cannot be satisfied over the whole interaction path of the beams. Effects of crystal inhomogeneities, such as striae, have been discussed by several authors [10.30, 34].

Nonlinear Coefficient. From purely theoretical considerations it would seem that the crystal with the highest nonlinear coefficient would be the most desirable material to use in a doubling experiment. However, in practical situations it turns out that damage threshold, optical quality, angular and thermal tuning range, acceptance angle, etc., are equally important parameters. From (10.11) it follows that it is possible to trade off the nonlinear coefficient d for the interaction length l_c. A material with a low nonlinear coefficient but with properties which allow a long interaction length can be as efficient as a short crystal with a high nonlinear coefficient.

Figure of Merit. The parameters which affect the doubling efficiency can be combined in a figure of merit which characterizes either the laser source, the crystal, or both. For example, *Eimerl* [10.35] has shown that the performance of a nonlinear frequency doubler is uniquely determined by the output brightness of the pump laser and the figure of merit of the nonlinear material. Furthermore, the analysis revealed that the maximum conversion efficiency is only dependent on the optical properties of the material, and the brightness of the pump laser, but is independent of the physical size of the nonlinear crystal. The size is adjusted to avoid optical damage in the material. The constraint is that the material under consideration must be available in large enough sizes to fit the requirements of the laser system. The nonlinear frequency doubling crystal is characterized by the nonlinear index and the parameters defining the tolerance to angular, wavelength, and temperature detuning from phase-matched conditions.

In the analysis mentioned above, second-harmonic conversion is characterized by two parameters relating the conversion efficiency, the pump intensity and the nonlinear properties of the material. These are the nonlinear "drive" which is the source term for generation of the electric field at the second harmonic, and the detuning which is the phase mismatch between second-harmonic waves at the exit and entrance planes of the crystal. The drive is given by

$$\eta_0 = C^2 I l^2 , \tag{10.30}$$

where C is proportional to the nonlinear coefficient of the material, I is the pump intensity, and l is the crystal length. If d_{eff} is expressed in pm/V and λ in μm, one obtains

$$C = \frac{2.75 d_{\text{eff}}}{\lambda_2 n^{3/2}} .$$

The detuning is given by

$$\delta = 0.5 \, \Delta k l , \tag{10.31}$$

where Δk is the wavevector mismatch given by (10.22).

The drive and detuning parameters determine the conversion efficiency

$$\eta = \eta_o (\sin \delta / \delta)^2 . \tag{10.32}$$

The use of these equations is illustrated in the following example. The theoretical conversion efficiency of a Q-switched Nd:YAG laser with an output energy of 1 Joule per pulse in a 15 ns pulse is calculated. The laser beam has a divergence of 0.3 mrad (FWHM) and a cross-sectional area of 0.5 cm^2 at the doubling crystal. The harmonic generator is a 2.5 cm long KD*P crystal oriented for type-II doubling. Introducing d_{eff} and n for KD*P, one obtains $C = 0.97\,\text{GW}^{-1/2}$ and from the other parameters follows a drive of $\eta_0 = 0.78$. The detuning parameter $\delta = 0.47$ is obtained from $\beta_0 = 2.5\,(\text{mr-cm})^{-1}$ for KD*P, $2\delta\theta' = 0.3\,\text{mrad}$ and $l = 2.5\,\text{cm}$. Introducing these two parameters into (10.32) yields a theoretical conversion efficiency of $\eta = 0.72$.

10.1.3 Properties of Nonlinear Crystals

Survey of Materials

Tables 10.1 and 10.2 list properties and phasematching parameters for several important nonlinear materials. For the design of a frequency converter the following properties of the nonlinear crystal are of key importance: value of the nonlinear coefficient, damage threshold, phase-matching and transparency range, available crystal size and optical homogeneity, and chemical and mechanical stability. The selection of a particular nonlinear material for use in a solid-state laser is predicated upon high damage threshold and good optical quality, with secondary emphasis placed on the magnitude of the nonlinear coefficient.

Table 10.1. Properties of important nonlinear materials

Material	Phase-matching type	Effective nonlinear coefficient* $[10^{-12}\,\text{m/V}]$	Refractive index $n_0(w)$	Damage threshold $[\text{GW/cm}^2]$	Absorption $[\text{cm}^{-1}]$
KD*P	II	0.37	1.49	0.5	0.005
KTP	II	3.18	1.74	0.5	0.010
LBO	I	1.16	1.56	2.5	0.005
BBO	I	1.94	1.65	1.5	0.005
LiNbO$_3$ (5% MgO)	I	4.7	2.23	0.10	0.002

* For 1064 nm to 532 nm second-harmonic generation

For example, despite the lower nonlinear coefficient of KD*P compared to the other materials discussed in this section, conversion efficiencies as high as 80% have been obtained from these crystals. Since a high conversion efficiency is actually the result of material properties as well as pump source characteristics, the high peak power, narrow spectral bandwidth, small beam divergence and clean spatial and temporal beam profiles obtained from Nd:glass lasers employed in fusion research, make possible such high harmonic conversions even in crystals with modest nonlinearity.

At the present time, the leading candidate material for nonlinear experiments with solid-state lasers is KTP, since the crystal has a large nonlinear coefficient, high damage threshold and large angular and temperature acceptance range. Its major drawbacks are the limited size (about 1–2 cm^3) and the high cost associated with the difficult growth process. KDP is superior to any other nonlinear crystal with regard to availability in large sizes combined with an excellent optical

Table 10.2. Phasematching parameters for 1064 to 532 nm conversion

Material	Phase matching angle	Walk-off angle	Tolerance parameters (FWHW)		
			Angular [mr-cm]	Thermal [°C-cm]	Spectral [nm-cm]
KD*P	53.7°	1.45°	2.2	6.7	0.66
KTP	24.3°	0.26°	25	25	0.56
LBO	–	–	4.3	3.6	0.75
BBO	22.8°	3.19°	0.5	55	0.66
LiNbO$_3$ (5% MgO)	90° (1)	0	47	0.6	0.23

(1) at $T = 107°C$

quality. Crystals with diameters as large as 27 cm have been fabricated. Crystals grown from the melt such as LiNbO$_3$ or KTP are generally hard, chemically stable and can easily be polished and coated. Solution grown crystals such as KDP and its isomorphs are soft and hygroscopic. These crystals need to be protected from the atmosphere.

The reader is reminded that considerable differences exist in the literature regarding the values of the nonlinear coefficient and damage threshold. The latter depends on the beam quality and spot size within the crystal and whether surface or bulk damage is considered. The nonlinear coefficients listed in Table 10.1 are from [10.36] with the exception of LBO which is taken from [10.37].

The most important application for harmonic generators is the frequency conversion of Nd lasers. In Table 10.2, the phase-matching conditions for important nonlinear crystals are listed for 1064 nm to 532 nm harmonic generation.

The dependence of phase-matching conditions on temperature, angle, and wavelength variations from the ideal condition is expressed by ΔT, $\Delta\Theta$, and $\Delta\lambda$. These parameters, given in (10.24, 25, 27, 29, 55) define the interval between the half-power points of the curve of second-harmonic power against temperature, angle, and wavelength, respectively.

If birefringence is a slow function of T, then the crystal has a wide central peak, making it easier to maintain peak output in the presence of small temperature fluctuations. In this regard, KTP and BBO outperform other linear materials.

Crystals which have a birefringence which is not strongly dependent on wavelength allow efficient doubling of laser beams having a broad spectral bandwidth. As was explained earlier, in a 90° phase-matching condition the effect of beam divergence on the interaction length is minimized. This explains the large acceptance angle of LiNbO$_3$.

KDP and Its Isomorphs. The crystals of this family have proven to be an important group of useful second-harmonic generators. The most prominent members of this group of nonlinear crystals are potassium dihydrogen phosphate, KH$_2$PO$_4$ (Symbol KDP); potassium dideuterium phosphate, KD$_2$PO$_4$ (Symbol KD*P); cesium dideuterium arsenate, CsD$_2$ASO$_4$ (Symbol CD*A); and Ammonium dihydrogen phosphate, $NH_4H_2PO_4$ (Symbol ADP). The crystals which are all negative uniaxial, belong to point group $\bar{4}2\,m$ and, thus, have a tetragonal symmetry. The crystals are grown at room temperature from a water solution which makes large, distortion-free, single crystals available. Transparency exists from 0.22 to 1.6 μm for the phosphates, and from about 0.26 to 1.6 μm for the arsenates. Deuteration increases the infrared limit to about 1.9 μm. The greatest attributes of this family of crystals are their resistance to laser damage and their high optical quality. Opposing these advantages, there are several disadvantages. The materials have fairly low refractive indices, typically 1.50 to 1.55, and therefore they have small nonlinear coefficients. All of the KDP isomorphs are water-soluble and have a maximum safe operating temperature of about 100° C. The crystals are sensitive to thermal shock, and should be heated slowly at a rate of less than about 5° C/min.

ADP and KDP were the first crystals used for the demonstration of phase-matched second-harmonic generation. Isomorphs of these materials have similarly been used in nonlinear optics, the most widely known isomorphs being deuterated KDP, which is normally designated as KD*P. Some of the other isomorphs have been used because the temperature dependence of their refractive index allows 90° phase matching for particular interactions. CDA and CD*A 90° phase-match the important 1.06-μm transition of Nd:YAG and Nd:glass [10.23, 38].

Calculations of various basic properties of KDP related to the generation of the second, third, and fourth harmonics of 1 μm laser radiation for five different conversion processes can be found in [10.39]. The paper contains a detailed tabulation of phase matching angles, sensitivity to angular mismatch, wavelength, and temperature. Figure 10.9 shows the crystal and electric vector orientation for harmonic generation in KDP and its isomorphs. KDP and KD*P crystals are commercially available for either type I or type II angle-tuned phase matching. Type II is more efficient than type I, but places more restrictions on input beam quality. The crystals are usually supplied in cells sealed in dry inert atmosphere or index matching fluid.

Potassium Titanyl Phosphate (KTP)

The crystal $KTiOPO_4$ (KTP) is a unique nonlinear optical material that is being widely used for second-harmonic generation of Nd lasers emitting around 1 μm. KTP is also very attractive for various sum- and difference- frequency and optical parametric applications over its entire transparency range from 0.35 to 4.0 μm. Although a few specific characteristics of other materials are better, KTP has a combination of properties that make it unique for second-order nonlinear optical applications, and second-harmonic generation of Nd lasers, in particular. The crystal has large nonlinear coefficients; the adequate birefringence in the y-z and the x-z planes allows phase matching for the more effective type-II process over a large wavelength range. It has wide acceptance angles, an unusually large temperature bandwidth, relatively good thermal properties and a high damage

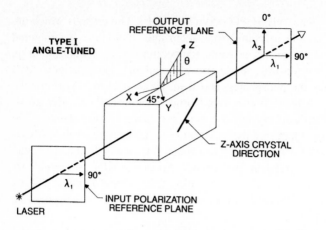

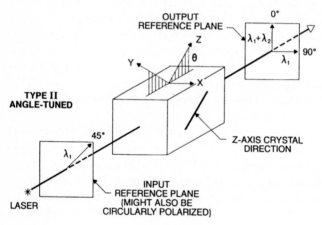

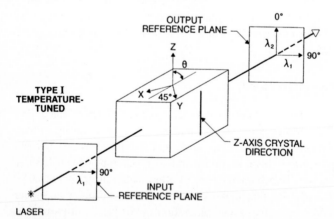

Fig. 10.9. Crystal and electric vector orientation for harmonic generation in KDP and its isomorphs [10.40]

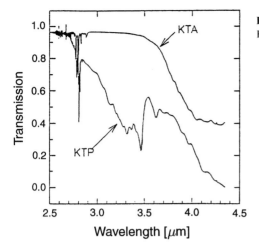

Fig. 10.10. Optical transmission of KTP and KTA [10.42]

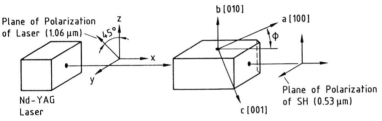

Fig. 10.11. Orientation of KTP for type-II interaction at $1.06\,\mu$m. $\phi = 26°$ for hydrothermally and $\phi = 21°$ for flux-grown material [10.41]

threshold. KTP is the best nonlinear material for Nd:lasers to emerge in recent years. The major drawback is the difficult growth process required to produce these crystals, which leads to high cost and small-size crystals.

KTP decomposes on melting ($\approx 1150°$C) and hence normal melt processes cannot be used to grow this material. However, single crystals of KTP can be grown by both hydrothermal and flux techniques. Currently crystals up to 20 mm in length are commercially available. KTP crystallizes in the orthorhombic point group mm2. The crystal structure, refractive indices, and phase match parameters have been reported in [10.36, 41–54].

Figure 10.10 shows the transmission curve for KTP. The material is transparent from $0.35\,\mu$m to about $4.0\,\mu$m. The optical spectrum is structure-free except for traces of OH$^-$ absorption bands observed at 2.8 and $3.5\,\mu$m.

Figure 10.11 shows the crystal orientation for the phase-match condition of the type-II interaction at $1.06\,\mu$m. It should be noted that the phase-match angle ϕ measured from the x-axis in the xy plane is different for flux and hydrothermally grown crystals. The phase match direction in KTP for second-harmonic generation of $1.06\,\mu$m radiation results in a walk off angle of 4.5 mrad between the fundamental and second-harmonic beam.

Frequency-doubling efficiencies in excess of 65% are now routinely obtained from KTP pumped by high-quality laser beams. However, KTP suffers from gradual photochemical degradation (gray tracking) which is cumulative with exposure to a combination of second-harmonic and fundamental radiation. This degradation leads to increased absorption in the crystal which can eventually cause crystal failure. The photochemical effect is reversed by operating the KTP at an elevated temperature. Crystals operating at flux levels of $150\,MW/cm^2$ and temperatures of $80°C$ have shown lifetimes in excess of 20 million pulses with $> 60\%$ conversion efficiency. However, there have been failures due to bulk damage in crystals operated at $65°C$ and lower.

Potassium Titanyl Arsenate (KTA)

KTA is isomorphic with KTP, but has lower absorption losses in the 3–$5\,\mu m$ region (see Fig.10.10). It has about the same nonlinear coefficients and high damage threshold of KTP. Since KTA does not phase match a $1.06\,\mu m$ laser for second-harmonic generation, the main attraction of this material is for applications in optical parametric oscillators (see Sect.10.2.2). Properties of this material can be found in [10.55, 56].

Lithium Triborate (LBO)

LBO is a nonlinear optical crystal characterized by good UV transparency, a relatively high optical-damage threshold, and a moderate nonlinear optical coefficient [10.37, 57, 58]. These properties, along with its mechanical hardness, chemical stability, and nonhygroscopicity, make LBO an attractive material for certain nonlinear optical processes. Because the birefringence in LBO is smaller than that in BBO, it tends to limit the phase-matching spectral range. However, it also leads to the possibility of noncritical phase-matching and larger acceptance angle for frequency-conversion applications in the visible and near-IR. Commercial availability of this crystal is still very limited.

Beta-Barium Borate. The material β-BaB_2O_4 (BBO) is a nonlinear optical crystal which possesses excellent properties for nonlinear frequency conversion in a spectral range that extends from the ultraviolet to the mid-infrared. This material has a moderately large nonlinear coefficient, a large temperature tolerance, low absorption and a very high damage threshold. The principal shortcoming with BBO is the low angular tolerance of 0.5 mr-cm, which requires a diffraction limited-beam for efficient frequency doubling.

BBO is of particular interest for frequency doubling into the blue region. The transmission band for BBO extends to 200 nm in the UV so that single or multiphoton absorption is not a problem with this material. BBO has been used to double the output of a Ti:Sapphire laser with efficiencies up to 60%. Relevant data can be found in [10.36, 58–63].

Lithium Niobate (LiNbO₃). This material is nonhygroscopic and hard, taking a good polish readily. The uniaxial crystal, belonging to the trigonal point group 3 m, has a large nonlinear coefficient relative to KDP. The crystals of lithium niobate are transparent in the region 0.42 to 4.2 μm. Temperature sensitivity of birefringence is such that, by varying the temperature, phase matching can be achieved at 90° to the optical axis.

Unfortunately, LiNbO₃ is particularly susceptible to photorefractive damage from visible or UV radiation [10.64, 65]. This optical-index damage is reversible if the ambient temperature of the crystal is increased to about 170°C which is above the annealing temperature for the photorefractive damage. Doping of LiNbO₃ with 5% of MgO reduces photo-refractive damage and permits noncritical phasematching at a somewhat lower temperature, namely 107°C [10.66].

Despite the high nonlinear coefficient, and the ease of handling and polishing, the need for operation at elevated temperatures combined with a relatively low damage threshold have severely limited practical applications for this material.

The emergence of Periodically Poled LiNbO₃ (PPLN), which will be discussed in Sect. 10.2.2, has attracted renewed interest in this material. In PPLN, the largest nonlinear index of LiNbO₃ can be utilized. This allows efficient nonlinear conversion at modest power levels which are well below the damage threshold.

10.1.4 Intracavity Frequency Doubling

In the previous section we discussed frequency doubling by placing a nonlinear crystal in the output beam of the laser system. Frequency doubling a cw-pumped laser system in this manner results in an unacceptably low harmonic power because large conversion efficiencies require power densities which are not available from a cw-pumped laser. One obvious solution to this problem is to place the nonlinear crystal inside the laser resonator, where the intracavity power is approximately a factor $(1 + R)/(1 - R)$ larger than the output power. The power is coupled out of the resonator at the second-harmonic wavelength by replacing the output mirror with transmission T by one which is 100% reflective at the fundamental and totally transmitting at the second harmonic. Functionally, the second-harmonic crystal acts as an output coupler in a manner analogous to the transmitting mirror of a normal laser. Normally the transmitting mirror couples out power at the laser frequency, whereas the nonlinear crystal inside the laser couples out power at twice the laser frequency. Because advantage is taken of the high power density inside the laser cavity, it is only necessary to achieve a conversion efficiency equal to the optimum mirror transmission to convert completely the available output at the fundamental to the harmonic. For example, for a cw-pumped Nd:YAG laser with an optimum output coupling of $T = 0.1$, an intracavity conversion efficiency of 10% will produce an external conversion of 100%, in the sense that the total 0.53-μm power generated in both directions by the nonlinear crystal is equal to the maximum 1.06-μm power which could be extracted from the cavity without the nonlinear crystal.

There are several disadvantages associated with intracavity doubling. A nonlinear crystal of poor optical quality will drastically degrade the performance of the laser. Amplitude fluctuations are strongly magnified by the combination of the nonlinear process and the gain of the active material. The harmonic power is generated in two directions, which requires an additional dichroic mirror for combining the two beams. The other alternative to intracavity doubling of a cw laser is external doubling with a strongly focused beam. However crystal acceptance angles are frequently too narrow to permit tight focusing of the beam onto the crystal. For pulsed-pumped operation, on the other hand, the fraction of intracavity power that is coupled out is so high that there is little advantage in placing the crystal inside the laser.

The choice of nonlinear materials for intracavity frequency doubling of the Nd:YAG, Nd:YLF and Nd:YVO$_4$ lasers are KTP, BBO, LiNbO$_3$ and PPLN. Compared to pulsed-pumped systems, in internally frequency-doubled cw lasers the average power is usually high and the peak power low. Therefore, a high nonlinear coefficient, small absorption losses, and good optical quality are the determining factors for the selection of a particular crystal. In order to increase the conversion efficiency of cw pumped systems, the power density can be increased by employing an acousto-optic Q-switch or modelocker in the resonator.

Simplification of the basic intracavity doubling design is achieved by using a gain medium that also functions as the nonlinear material. Work on diode-pumped self-doubling lasers is still in the early phases of development. The most attractive nonlinear gain medium is Nd:YAB, which is a dilute form of the stoichiometric neodymium compound neodymium aluminum borate (NAB). Diode-pumped Nd:YAB lasers with output powers in the milliwatt range have been demonstrated [10.67].

Cavity Configuration

There are three primary factors which affect the choice of a laser cavity configuration. In order to achieve efficient harmonic generation it is important to obtain a high power density inside the nonlinear crystal. Concurrently, the intracavity beam must be large enough inside the laser medium to utilize the maximum volume which can contribute to TEM$_{00}$-mode oscillation. This generally requires that the beam cross-sectional area be at least an order of magnitude larger inside the laser medium than inside the nonlinear crystal. Finally, since intracavity harmonic generation produces a beam of harmonic power in each of two directions, it is desirable to select a cavity configuration which permits recovery of both beams. A resonator configuration which allows for a large TEM$_{00}$-mode volume in the laser rod and provides a high power density in the nonlinear crystal is shown in Fig. 10.12. The cavity, formed with one 10-m and one 0.5 m radius-of-curvature mirror, represents a near-hemispherical configuration. Both mirrors have high-reflection coatings for the 1.06-μm radiation and the front mirror is transparent for the 0.53-μm radiation. Only the green light, emitted toward the flat mirror, leaves the cavity; the green light traveling in the opposite direction is absorbed in the Nd : YAG crystal. The nonlinear crystal is located at the position of minimum spot size.

Barium sodium niobate ($Ba_2NaNb_5O_{15}$) was a very popular nonlinear material in the early 1970's, since it has a very large nonlinear coefficient (about 38 times larger than that of KDP). However, material issues associated with poor optical quality and low damage threshold could not be solved and the crystal is no longer available commercially.

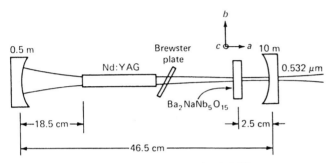

Fig. 10.12. Internal second-harmonic generation [10.68]

The Brewster-angle plate in the cavity serves to polarize the laser in the proper plane. Also, the position of the laser rod in the cavity was chosen so that self-aperturing by the laser rod permitted the laser to operate only in the TEM_{00} mode. With this arrangement a total cw power of 1.1 W at 0.532 μm has been produced, which equals the available 1.064-μm TEM_{00} output of the basic Nd : YAG laser [10.68].

If it is desired to obtain all the harmonic power in a single output beam, it is necessary to employ a dichroic mirror to reflect one of the beams back in the same direction as the other. Figure 10.13 shows a technique which combines into a single output beam the second-harmonic which is generated in both directions by the frequency-doubling crystal [10.69]. The frequency-doubled beam has a polarization which is rotated 90° with respect to the polarization of the funda-mental beam. The dichroic mirror M_1 is designed to reflect the 1.06-μm beam completely and transmit virtually all the orthogonally polarized 0.53-μm beam. Mirror M_2 is a 100% reflector for both 1.06-μm and 0.53-μm mirrors. In this way the forward and reverse green beams are combined into one.

To prevent any destructive interference effects, the phase shifts due to disper-sion between the 1.06- and 0.53-μm beams in the air path between the doubling crystal and the dielectric coating of mirror M_2 must be adjusted so that they are in phase as they reenter the doubling crystal after reflection from M_2. This is accomplished by providing a translation adjustment for the mirror M_2 to move it toward or away from the doubling crystal, such that the net phase shift between the two waves is a multiple of 2π. The computed phase shift, based on the mea-

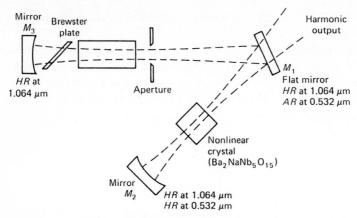

Fig. 10.13. Three-mirror, folded cavity configuration allows harmonic power generated in two directions to be combined into a single output beam [10.69]

sured $1.06-0.53\,\mu$m dispersion of air of 3.6×10^{-6}, is approximately 29.0°/cm. For doubling of the 1.06-μm line of Nd : YAG, this requires a mirror movement of no more than 3.28 cm [10.70], and the accuracy required to correct for the phase shift is of the order of millimeters.

The beam diameter at various locations inside a laser cavity of the type shown in Fig. 10.13 is a function of mirror curvature, laser rod size and face curvature, nonlinear crystal orientation, and the placement and spacing of all these and any other intracavity elements (such as Brewster plates, intracavity lenses, or apertures), as well as of the thermal focusing present in the rod. A computer program which takes into account all of the factors mentioned above in determining the beam diameter at various intracavity locations is presented in [10.71].

Intracavity frequency doubling of a laser operating in many longitudinal modes will generate fluctuations in the instantaneous as well as the average harmonic power. The nonlinear process in the resonator actually magnifies the fluctuations occurring due to mode beating. Single-longitudinal-mode operation or mode locking avoids this problem. Figure 10.14 shows an intracavity frequency-doubled Nd : YAG laser containing two quartz etalons in the resonator to obtain single-frequency operation [10.72]. The cavity is designed to obtain an adequate TEM_{00}-mode volume in the laser rod and a small waist at the nonlinear crystal. The beam waist is generated by a short-radius mirror and the Nd : YAG rod with one end curved to act as an internal focusing element. Frequency-doubled output in a single direction is achieved with a three-mirror cavity similar to Fig. 10.13. The multiple-dielectric-layer 45° mirror favors s-polarized laser oscillation at $1.06\,\mu$m and very effectively transmits the second-harmonic p-polarized radiation. The system shown in Fig. 10.14 produces unidirectional power outputs of 0.3 to 0.5 W at $0.53\,\mu$m.

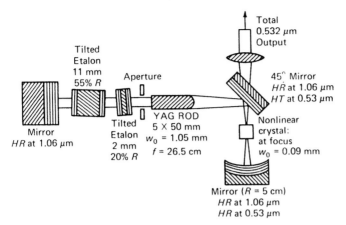

Fig. 10.14. Single-ended output by use of three-mirror, folded cavity configuration (similar to Fig. 10.13). The intracavity 45° mirror is highly reflecting at 1.06 μm but transparent at 0.53 μm. The short-radius curved mirror is highly reflecting at both 1.06 and 0.53 μm [10.72]

An analysis of optical second-harmonic generation internal to the laser cavity has been presented in [10.68, 73]. The steady-state condition for intracavity doubling can be determined if we equate the round-trip saturated gain of the laser to the sum of the linear and nonlinear losses

$$\frac{2g_0 l^*}{1 + I/I_s} = L + K'I \,, \tag{10.33}$$

where g_0 is the unsaturated gain coefficient, l^* is the length of the laser medium, I is the power density in the laser rod, and I_s is the saturation power density of the active material. All linear losses occurring at the fundamental frequency are lumped together into the parameter L; the quantity $K'I$ is the nonlinear loss. The nonlinear coupling factor K', defined by

$$I(2\omega) = K'I^2(\omega)$$

is related to K in (10.12); it is $K' = \kappa l^2 K$, where κ accounts for different power densities in the laser rod and nonlinear crystal in the case of focused beams; that is, $\kappa = I_{\text{crystal}}/I_{\text{rod}}$ and l is the crystal length.

From the theoretical treatment of intracavity doubling, it follows that a maximum value of second-harmonic power is found when

$$K'_{\text{max}} = \frac{L}{I_s} \,. \tag{10.34}$$

The magnitude of the nonlinearity required for optimum second-harmonic production is proportional to the loss, inversely proportional to the saturation density, and independent of the gain. Thus, for a given loss, optimum coupling is achieved for all values of gain and, hence, all power levels. As we recall from Chap. 3, in a laser operating at the fundamental wavelength the optimum output coupling of the front mirror is gain-dependent. The usefulness of (10.34) stems from the fact that by introducing the material parameters for a particular crystal into (10.8) and combining with (10.34), the optimum length of the nonlinear crystal is obtained. For example, from (10.8) we obtain $K' = 1.5 \times 10^{-6}l^2\kappa$ for $Ba_2NaNb_5O_{15}$. With (10.34) and $I_s \approx 10^3$ W/cm^2 for Nd:YAG, the optimum crystal length is $l^2 = 660\,L/\kappa$ [cm^2]. Assuming a round-trip loss of $L = 0.02$ and a power-density enhancement of $\kappa = 50$, a crystal of $l = 0.5$ cm is sufficient to optimally couple the second harmonic from the laser.

From the theory presented in [10.73], it follows that if the loss due to the inserted second-harmonic crystal is small compared to the total internal loss, the value of K' is the same for fundamental and second-harmonic output coupling. Also, the maximum second-harmonic power equals the fundamental power obtainable from the same laser.

In most systems, the second-harmonic power is considerably below the output power which can be achieved at the fundamental wavelength. Insertion losses of the nonlinear crystal, or any other additional intracavity element, are often the reason for the poor performance, or in the case of Nd:YAG, thermally induced depolarization losses can have a major effect. In low-gain lasers employed for intracavity harmonic generation, even small insertion losses are very detrimental to the system performance. The effect of intracavity losses on the laser performance has been analyzed in detail by Smith [10.73]. For example, in a laser with a round-trip gain coefficient of $2g_0\ell = 0.2$, the output drops 20% if the internal loss is increased from 2% to 3%. Furthermore, availability of crystals and damage considerations lead often to non-optimum conditions as far as crystal length and pump-power density are concerned.

As an example, the performance of a typical intracavity doubled Nd:YAG laser is illustrated in Fig. 10.15. The output at the harmonic wavelength is compared to the TEM$_{00}$ output available at 1.064 μm from the same laser. The Nd:YAG laser employed in these experiments was cw pumped by diode arrays and intracavity doubled with KTP. After measuring the output at 1.064 μm, a KTP crystal was inserted in the folded resonator (similar to the arrangement shown in Fig. 10.14) and the output coupler was replaced with a mirror of high reflectivity at 1.064 and 0.53 μm. For the short crystals, the nonlinear coupling factor K' is too small and far from optimum. A higher output at 532 nm would require an increase of κ by thighter focusing of the beam. However, this makes alignment very critical and sacrifices mechnical and thermal stability of the laser. A longer nonlinear crystal is therefore a better solution. However, a 15 mm long crystal is about the longest crystal available for KTP.

Continuously pumped, intra-cavity doubled lasers usually exhibit strong amplitude fluctuations in the green output. These instabilities, which are caused by nonlinear interactions of the longitudinal modes, have hindered the development of these systems in the past. One solution of eliminating these output fluctuations, referred to as the green problem [10.74], is to operate the laser in a single longitudinal mode. Also a very large number of longitudinal modes, obtained in a very long resonator, tend to stabilize the output. The largest output fluctuations have

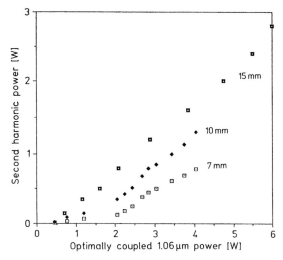

Fig. 10.15. Comparison of fundamental and second harmonic output from a cw pumped intracavity frequency doubled Nd:YAG laser for different KTP crystal lengths

been observed in relatively short resonators, typical of dioded-pumped lasers, in which only relatively few longitudinal modes are present.

Solutions of the nonlinear rate equations, which describe each longitudinal mode and the gain of each mode, show instability and chaotic behavior [10.74, 78]. A steady-state solution derived in [10.78] predicts that the stability of the laser increases with round-trip time and the number of longitudinal modes. It was indeed demonstrated experimentally [10.76] that output fluctuations and random behavior in cw frequency-doubled systems can be greatly reduced by a long resonator. The measurements were performed with a lamp-pumped Nd:YLF laser, intra cavity frequency doubled with LBO, which provided as much as 13.5 W in a 1.9 m long resonator. In smaller systems, instabilities can be eliminated by operating the laser in a single mode. This can be accomplished by insertion of an etalon, as shown in Fig. 10.14, or by means of a birefringent filter [10.75, 79]. Also the design of a laser utilizing a unidirectional ring resonator is an effective means of eliminating longitudinal mode coupling and the associated output fluctuations [10.77]. All these approaches require the insertion of multiple elements in the cavity, and in the case of birefringent filters a very accurate temperature stabilization is necessary. Therefore, in a number of commercial lasers, a long resonator is the preferred approach.

In applications requiring high amplitude stability, one can combine intra cavity frequency doubling with repetitive Q-switching or mode-locking.

In a typical system of the first type the laser contains, in addition to the laser crystal and harmonic generator, an acoustooptic Q-switch in the resonator. In the case of an internally doubled Q-switched laser, the harmonic power generated cannot be simply related to the harmonic coupling coefficient, since the effects of pump depletion and pulse stretching must be considered. However, the harmonic power extracted from the laser can be related to the pumping rate above threshold, and to the normalized harmonic coupling coefficient [10.80]. Due to the higher intensities in a Q-switched laser, as compared to a cw system, the optimum second-harmonic output occurs at a lower value K'. The lower value of K' permits the use of a mirror configuration which yields a fairly large beam waist. The configuration is thus relatively insensitive to slight misalignment and motion of the resonator elements. In the case of the cw system, the optimum second-harmonic output is obtained for a resonator operating with an extremely small beam waist (and thus larger K') which is sensitive to both transverse and axial motions of the resonator elements.

Information on repetitively Q-switched and internally frequency doubled lasers can be found in [10.81–87].

The inherent instability of cw systems can also be overcome by mode-locking which locks the longitudinal modes in phase and thus eliminates mode beating.

Since the harmonic conversion efficiency is highest at the peak of the mode-locked pulses, the harmonic conversion process tends to flatten and broaden the pulses, opposing the modulator's efforts to sharpen the pulses [10.88].

The pulse-lengthening mechanism can be understood by considering the losses introduced to the resonator by the mode-locking modulator and the harmonic generator. The loss due to second-harmonic generation is caused by the transmitted output power; this loss decreases with longer pulses because peak power and, hence, harmonic conversion efficiency decreases with longer pulses. The loss introduced by the mode-locking modulator, on the other hand, increases with longer pulses, since this loss results from the tails of the pulse passing through the modulator under less than optimum conditions. The pulse length of the mode-locked laser is affected by both these losses. To a first approximation, the pulse length will be that which minimizes the sum of these two losses, as is illustrated in Fig. 10.16 for a typical Nd : YAG laser [10.89]. The pulse-lengthening effect decreases the peak power as well as the average power of the laser. Therefore, the enhancement of second-harmonic power due to mode locking is not as large as one would expect. Theoretical expressions for the enhancement of second-harmonic generation to be expected from mode locking have been given in [10.90, 91].

Internally frequency-doubled and mode-locked Nd : YAG lasers have produced 0.5 to 1.5 W of output power at $0.53\,\mu\text{m}$ [10.89, 92]. Pulse widths are typically 250 to 330 ps. With proper temperature control of the nonlinear crystal, it is even possible to perform both the electrooptic modulation and optical harmonic generation in a single intra-cavity nonlinear crystal [10.89, 92–95].

10.1.5 Third-Harmonic Generation

Frequency up-conversion to the third harmonic has attracted considerable attention for Nd : glass laser beams for fusion applications. Third-harmonic generation is a two-step process involving two harmonic generators. In the first crystal ("doubler") some fraction of the fundamental radiation is converted to the sec-

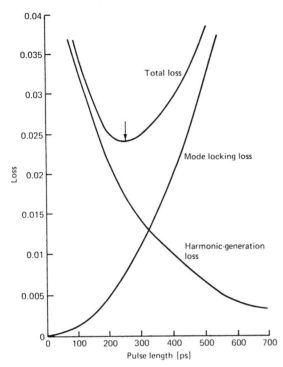

Fig. 10.16. Loss introduced by the mode-locking modulator and loss introduced by intracavity second-harmonic generation, calculated as a function of pulse length for a Gaussian pulse in a typical Nd : YAG laser [10.89]

ond harmonic, followed by a second crystal ("tripler"), in which unconverted fundamental radiation is mixed with the second harmonic to produce the third harmonic. The equations governing frequency mixing in nonlinear crystals have been given in [10.96]:

$$dE_1/dz = -jK_1E_3E_2^* \exp(-j\Delta k \cdot z) - \tfrac{1}{2}\gamma_1 E_1 \quad,$$

$$dE_2/dz = -jK_2E_3E_1^* \exp(-j\Delta k \cdot z) - \tfrac{1}{2}\gamma_2 E_2 \quad, \tag{10.35}$$

$$dE_3/dz = -jK_3E_1E_2 \exp(j\Delta k \cdot z) - \tfrac{1}{2}\gamma_3 E_3 \quad.$$

Here the E_j's are the complex electric vectors of waves propagating in the z direction with frequencies ω_j, where $\omega_3 = \omega_1 + \omega_2$. The electric field of wave j is the real part of $E_j \exp(i\omega_j t - ik_j z)$, and the phase mismatch $\Delta k = k_3 - (k_1 + k_2)$ is proportional to the deviation $\Delta\theta$ of the beam path from the phase-matching direction. The γ_j's are absorption coefficients. For tripling, $\omega_2 = 2\omega_1$, $\omega_3 = 3\omega_1$, $K_2 \simeq 2K_1$, and $K_3 \simeq 3K_1$.

Equations (10.35) describe single rays as well as complete beams [10.96–98], and solutions are presented in Fig. 10.17 for the phase-matched case ($\Delta k = 0$). The efficiency refers to the combined fundamental and second-harmonic input power to the tripler.

The parameter M is the ratio of second-harmonic power to total power in the tripler:

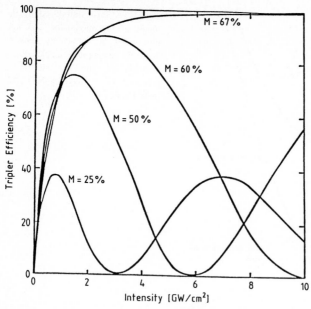

Fig. 10.17. Tripling efficiency of a 9 mm thick phase-matched KDP type-II crystal as a function of total input intensity, for various percentages M of the second harmonic in the input. A small absorption of $0.04\,\mathrm{cm}^{-1}$ is included for the fundamental [10.98]

$$M = P_{2\omega}/(P_\omega + P_{2\omega}). \tag{10.36}$$

If the input photons at ω and 2ω are matched $1:1$, then $P_{2\omega}+2P_\omega$ and $M = 0.67$. In principle, complete conversion of the input beams to the third harmonic can be achieved. For other values of M, after a ray has propagated an optimum distance Z_{opt} into the crystal, the number of photons in one of the components depletes to zero, and as z increases beyond Z_{opt} the mixing process reverses and the third-harmonic radiation reconverts.

Successful tripling is made difficult because the distance Z_{opt} is a sensitive function of both M and intensity; for example, Fig. 10.17 reveals that input at $4\,\mathrm{GW/cm}^2$ will convert with an efficiency of 25% if $M = 50\%$ but 80% if $M = 60\%$.

Efficient tripling depends on the fundamental and second-harmonic photons emerging from the first crystal in a ratio of $1:1$ over a broad intensity range. *Craxton* [10.98] has demonstrated that this requirement can be achieved by means of an appropriate choice of polarization angle in the doubler. Figure 10.18 shows three tripling schemes which can lead to high efficiency of the third-harmonic generation. In each case, two out of three photons of the ordinary beam propagating in the doubling crystal at the fundamental wavelength ω are converted to an extraordinary beam at 2ω. Depending on crystal orientation and thickness, the unconverted photons of the fundamental beam can emerge from the doubling crystal either elliptically polarized at $45°$ to the o- and e-axes, or plane polarized parallel to the o-axis, or plane polarized parallel to the e-axis.

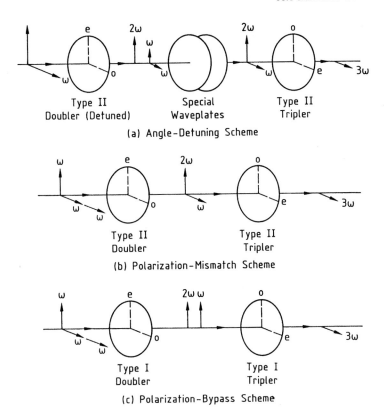

Fig. 10.18a–c. Different tripling schemes. Fundamental wavelength emerges from the doubling crystal (**a**) elliptically polarized at 45° to the o- and e-axes, (**b**) plane polarized parallel to the o-axis, and (**c**) plane polarized parallel to the e-axis [10.98]

A detailed discussion of the relative merits of the different configurations illustrated in Fig. 10.18 can be found in the above-cited references.

10.1.6 Examples of Harmonic Generation

Examples of several different types of laser systems which incorporate harmonic generators will illustrate the current state-of-the-art of this technology. Employing a KTP-crystal, harmonic generation with an output power of 3.9 W at 532 nm by intra-cavity frequency doubling in a Nd:YAG laser oscillator acousto-optically Q-switched at 10 KHz has been demonstrated [10.99]. The laser head contained a 25 mm by 3 mm Nd:YAG crystal side pumped by 6 one-centimeter long diode arrays which were arranged symmetrically around the rod in 2 sets of 3 arrays each. The optical pump power at 807 nm was 60 W. At the fundamental output, the laser oscillator produced 12 W multimode and 5.5 W TEM$_{00}$ Q-switched average power. The optimal fundamental power was obtained with a 96% reflecting mirror.

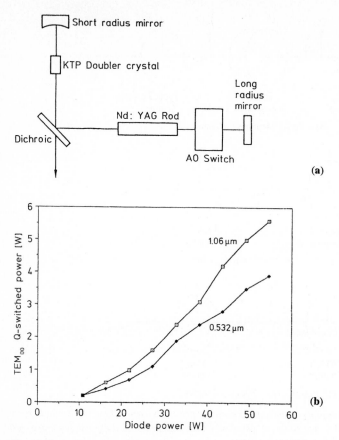

(a)

(b)

Fig. 10.19. Optical schematic (**a**) and fundamental and second harmonic average output power from a repetitively Q-switched cw pumped Nd:YAG laser (**b**)

A hemispherical laser cavity was employed to obtain TEM_{00} operation with reasonable extraction efficiency, while at the same time giving a tight beam waist for intra-cavity doubling. Using a type-II doubling crystal, the intra-cavity flux incident upon the crystal was maximized by placing it at the waist of the cavity mode. The length of the KTP crystal was 15 mm. To improve the doubling efficiency, and to obtain a uni-directional output, the cavity was folded with a dichroic mirror, and the 1.06 μm output coupler used to obtain optimum fundamental output was replaced with a high reflector at 1.06 and 0.532 μm.

Figure 10.19a illustrates the optical schematic of the folded resonator-configuration. The performance of the acousto-optically Q-switched laser is depicted in Fig. 10.19b, which exhibits the average power produced at 1.06 μm (for optimum coupling), and at 0.532 μm when the output coupler is replaced by a total reflector at 1.06 and 0.532 μm.

Figure 10.20 shows the schematic of an end-pumped intra cavity frequency doubled and Q-switched Nd:YAG laser [10.100]. The system uses the same

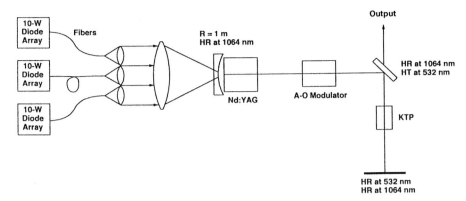

Fig. 10.20. Schematic of an end-pumped Q-switched and internally frequency-doubled cw-pumped Nd:YAG laser [10.100]

type of three-mirror 'L' shaped resonator, as shown in the previous figure. Also, the average output in the green is about the same as in the previous example. However, due to the more efficient pump arrangement, only 30 W of optical pump power is required, compared to 60 W for the side-pumped laser. In the design depicted in Fig. 10.20, the Nd:YAG crystal is pumped with the combined output of three fiber-coupled diode laser arrays. The fiber output of each pump was first collimated and then focused with a single lens onto one end of the Nd:YAG rod. The laser produced 3.5 W average power in the green at a repetition rate of 50 kHz in a nearly diffraction limited beam.

At the high end of intra-cavity second-harmonic generation, one finds krypton arc lamp and diode pumped Nd:YAG lasers, AO Q-switched and intracavity doubled with KTP or LBO, which have generated in excess of 100 W at the harmonic [10.82, 83, 85].

An illustration of such a high average power second-harmonic generation system is given in Fig. 10.21 [10.85]. This design features a "Z" shaped resonator, containing two laser heads, a Q-switch and a KTP crystal. The laser was repetitively Q-switched with an acousto-optic device at repetition rates ranging from 4 to 25 kHz. The laser operated in high-order multi transverse mode. An interesting feature of the design is an optical relay system formed by the two curved mirrors between the Nd:YAG rod and the KTP crystal. One plane of this optical relay is at the end face of the YAG rod closest to the KTP crystal and the other plane is in the center of the KTP crystal. The purpose of the relay optics is to keep the spot size in the KTP crystal as constant as possible, despite large changes of thermal lensing in the rods. The optical relay images the hard aperture of the laser rod into the KTP crystal, and therefore the multi-mode spot size in the KTP is fairly constant over a large range of pump power. The hard aperture at the rod was 4.9 mm, the relay optics was adjusted for a magnification of 2.5, therefore a multi-mode spot size in the KTP crystal of about 2 mm was created. Particularly at these high power levels, a well controlled and fairly large

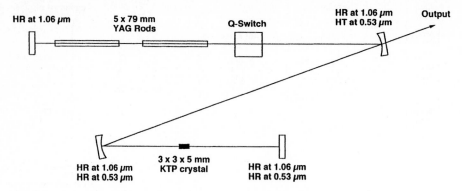

Fig. 10.21. High power second-harmonic generation in an internally frequency doubled Nd : YAG laser featuring a "Z" resonator with relay optics

spot size in the nonlinear crystal is very important in order to avoid material damage.

The system produced 106 W of average power at 0.532 μm when Q-switched at 17 kHz. The laser was also operated at 1.32 μm with a different set of optics. In this case, 23 W at 0.659 μm has been generated. The nonlinear crystal for doubling both wavelengths was KTP. For 0.532 μm generation, the crystal was cut such that critical phase matching was obtained at the usual x-y plane, at $\theta = 90°$ and $\phi = 23.5°$. The cut used for 0.694 μm generation was such that phase matching in the x-z plane was achieved at $\phi = 0°$ and $\theta = 60°$.

Output of up to 40 W average power from an externally frequency-doubled, diode-pumped Nd : YAG oscillator amplifier system has been obtained. The system architecture is shown schematically in Fig. 10.22a [10.101]. The oscillator was Q-switched at a repetition rate of 100 Hz. The output from the amplifiers was 0.7 J per pulse at 1.064 μm in a 2× diffraction limited beam. At 0.532 μm, the energy per pulse was 400 mJ in a 17 ns pulse. Each laser head contains a Nd : YAG rod side-pumped with 5-bar laser diode arrays operated at 60 W/bar. In the oscillator, 80 bars pump a 5-mm diameter rod, whereas in the amplifiers 160 bars are used each, to pump the 9 mm rods.

Thermal lensing and birefringence of the amplifier is compensated with a lens and a 90° phase rotator. The output is doubled to the green with a 1-cm long KTP crystal heated to 100°C. As shown in Fig. 10.22b, near 60% conversion efficiency was achieved.

The requirement for generation of short wavelengths from inertial-confinement-fusion lasers has prompted the development of large-aperture harmonic converters for doubling, tripling and quadrupling of the Nd : glass laser output. As an example, we will consider conversion experiments originally carried out on the now dismantled Argus laser at the Lawrence Livermore National Laboratory [10.102]. Argus was an image-relayed, spatially filtered laser system capable of delivering up to 1 kJ output energy in a 600 ps pulse. With a 10 cm diame-

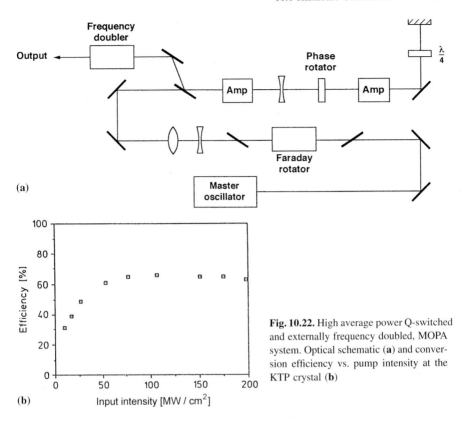

Fig. 10.22. High average power Q-switched and externally frequency doubled, MOPA system. Optical schematic (a) and conversion efficiency vs. pump intensity at the KTP crystal (b)

ter, and 2.29 cm thick KDP crystal up to 346 J of laser energy at 532 nm was generated. This corresponds to a 83 % conversion efficiency. Tripling and quadrupling yielded efficiencies of 76 % and 70 %, respectively. A summary of the performance of the conversion experiments is shown in Table 10.3.

The variation of the doubling efficiency vs. input at the fundamental intensity for the type-I KDP crystal is shown in Fig. 10.23. Since the crystals were uncoated, the internal conversion efficiency refers to the measured overall efficiency corrected for Fresnel reflection losses.

Following the early experiments on the Argus system, KDP crystal arrays with apertures of 74 cm have been built for conversion of the Nova laser output. These large aperture converters are constructed from 15 cm or 27 cm KDP segments, held together by an eggcrate stainless steel support structure [10.103].

Table 10.3. Frequency conversion efficiencies achieved on the Argus laser

Wavelength [nm]	Conversion efficiency	Maximum energy generated [J]	Output power [GW]
532	83%	346	495
355	55%	41	68
266	51%	50	83

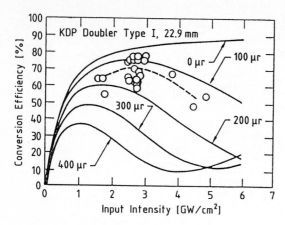

Fig. 10.23. Frequency-doubling efficiency as a function of input power

10.2 Parametric Oscillators

As was mentioned at the beginning of this chapter, two beams with different frequencies incident on a nonlinear crystal will generate a traveling polarization wave at the difference frequency. Provided the polarization wave travels at the same velocity as a freely propagating electromagnetic wave, cumulative growth will result. For reasons which will soon become clear, the two incident beams are termed "pump" and "signal" waves, having a frequency of ν_p and ν_s, and the resulting third wave is termed "idler" wave with frequency ν_i. Under proper conditions, the idler wave can mix with the pump beam to produce a traveling polarization wave at the signal frequency, phased such that growth of the signal wave results. The process continues with the signal and idler waves both growing, and the pump wave decaying as a function of distance in the crystal.

Since each pump photon with energy $h\nu_p$ is generating a photon at the signal ($h\nu_s$) and idler frequency ($h\nu_i$), energy conservation requires that

$$\frac{1}{\lambda_p} = \frac{1}{\lambda_s} + \frac{1}{\lambda_i} \tag{10.37}$$

In order to achieve significant parametric amplification, it is required that at each of the three frequencies the generated polarization waves travel at the same velocity as a freely propagating electromagnetic wave. This will be the case if the refractive indices of the material are such that the k vectors satisfy the momentum-matching condition $k_p = k_s + k_i$. For collinearly propagating waves this may be written

$$\frac{n_p}{\lambda_p} - \frac{n_s}{\lambda_s} - \frac{n_i}{\lambda_i} = 0 \tag{10.38}$$

where n_s, n_i, and n_p are the refractive indices at the signal, idler, and pump frequency, respectively.

Since the three indices of refraction depend on the wavelength, the direction of propagation in the crystal and on the polarization of the waves, it is generally possible by using birefringence and dispersion to find conditions under which (10.38) is satisfied.

Tunability is a fundamental characteristic of all parametric devices. With the pump providing input at the fixed wavelength λ_p, small changes of the refractive index around the phase matching condition, will change the signal and idler wavelengths such that a new phase-matching condition is achieved. Tuning is possible by making use of the angular dependence of the birefringence of anisotropic crystals, and also by temperature variation. Rapid tuning over a limited range is possible by electro-optic variation of the refractive indices.

Figure 10.24 illustrates different configurations which make use of the parametric-interaction process of three waves. The simplest device is a non-resonant configuration, namely an optical parametric amplifier (OPA) exhibited in Fig. 10.24a. In this case, a pump beam and a signal beam are present at the input. If the output of a Q-switched laser is focused into the crystal and if the intensity of the pump is sufficiently high, and phase matching conditions are met, gain is obtained for the signal wave and at the same time an idler wave is generated. Typically, an OPA is used if the signal obtained from an optical parametric oscillator is too weak and further amplification is desired.

The most common optical parametric device is the singly resonant oscillator depicted in Fig. 10.24b and c. In this device, the crystal is located inside a resonator that provides feedback at either the signal or idler frequency. In the example illustrated, the pump beam enters through a mirror which is highly transmitting at the pump wavelength and highly reflective for the signal wavelength. In Fig. 10.24b, the opposite mirror which is the output coupler, has typically 80–95% reflectivity for the signal wavelength, and high transmission for the idler and pump beam. Only the signal wavelength is resonant in the cavity and a small fraction is coupled out through the front mirror. In the configuration in Fig. 10.24c, the pump beam is reflected at the output mirror and makes one return pass through the crystal. Since the input mirror through which the pump enters is highly transmissive for this wavelength, no resonance condition is set up for the pump wavelength. However, threshold is lowered, because on the return path the pump beam provides gain for the resonant signal wave.

Figure 10.24d depicts a doubly resonant oscillator (DRO), which provides feedback at both the signal and idler wavelengths. The double-resonance condition, in which both the signal and the idler waves are simultaneously resonant within the optical cavity, lowers the threshold significantly. However, this advantage of a DRO is off-set by a reduction in stability and tunability. Maintaining the doubly resonant condition in a common resonator requires that the pump be single frequency and stable, and that the OPO cavity length be actively controlled to maintain the resonance condition. The considerably lower threshold of a DRO makes it possible to obtain parametric gain at the low power densities achievable under cw conditions, therefore the DRO configuration is employed mostly in cw OPO's.

Since the gain of a parametric device depends on peak power density, the design of a cw OPO is a particular challenge. Doubly resonant, or even triply res-

a) Optical Parametric Amplifier (OPA)

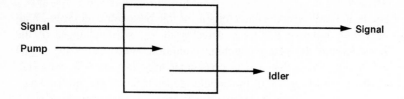

b) Singly Resonant Optical Parametric Oscillator (SRO)

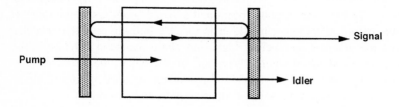

c) Singly Resonant Oscillator with Pump Beam Reflected

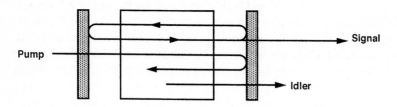

d) Doubly Resonant OPO (DRO)

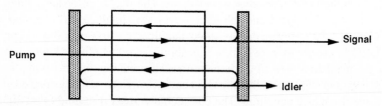

Fig. 10.24a–d. Configurations for parametric interactions

onant devices with tightly focused beams can compensate for low pump powers. However, a practical implementation of a cw pumped DRO which produces even modest output power over an acceptable tuning range has not yet been realized. For this reason, DRO's have attracted only modest attention to date. The reader interested in DRO technology is referred to [10.104].

The most widely used OPO's are pumped with Q-switched outputs from Nd:YAG lasers. However, parametric conversion of the output from mode-locked lasers has also received a great deal of interest. Pump sources generating pulses in the pico- and femtosecond regime pose a difficult problem. The short pulse width precludes the use of the standard OPO cavity in which the oscillating fields build up during the duration of the pump pulse. Such a cavity would have to be on the order of a few millimeters long for a picosecond system, and far less for femtosecond pulses, and therefore is not feasible. Two basic approaches have been used for parametric frequency conversion of pico- and femtosecond pulses, namely synchronous pumping of an OPO [10.105–107] or a traveling-wave optical parametric generator (OPG) [10.108–110]. An overview of this particular class of devices, which will not be discussed further, can be found in [10.104, 111].

The first successful operation of an optical parametric oscillator was achieved by *Giordmaine* and *Miller* [10.112] in 1965. Their oscillator employed lithium niobate as the nonlinear material and used a 530 nm pump signal derived by frequency-doubling the 1.06-μm output of a Nd:CaWO$_4$ laser.

Since then considerable effort has been expended in understanding and improving the device performance, and numerous nonlinear materials were evaluated for possible use in practical devices. While the process of optical parametric conversion is conceptually simple and elegant, many difficulties have been encountered in realizing practical systems. Among the most serious has been optical damage to the nonlinear crystal, caused by the high electric fields necessary for nonlinear conversion. Typical damage thresholds for nonlinear materials are in the same range as the intensities required for efficient conversion. Another limitation for achieving high parametric conversion has been the poor beam quality of the pump sources themselves.

In recent times, however, these two problems have been overcome: first, because new nonlinear crystals with high damage thresholds have been developed (for example, KTP, BBO, LBO) and second, because diode pumping has provided a new generation of efficient high-power solid-state lasers with single-transverse-mode outputs and a high degree of pulse-to-pulse stability. With the introduction of these new nonlinear materials in conjunction with diode-pumped solid-state lasers, renewed interest has developed in optical parametric oscillators. These advances in lasers and materials technology have considerably improved the prospects for efficiently generating tunable radiation using optical parametric oscillators (OPO).

As an example, Figure 10.25 depicts the wavelength coverage of Nd:YAG or Nd:YLF laser pumped OPO's employing the above-mentioned nonlinear crystals. The optical parametric oscillator has been the subject of many papers and

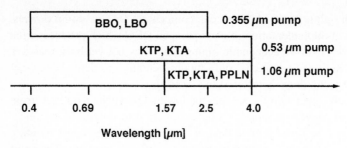

Fig. 10.25. Wavelength coverage of Nd : YAG or Nd : YLF pumped OPO's

review articles; detailed discussions of the theory and summaries of earlier work on OPO's can be found in [10.113–116].

10.2.1 Performance Modeling

Of greatest interest for the design of parametric devices are simple models which describe gain, threshold, phase matching and conversion efficiency as a function of device and input parameters.

For the amplifier depicted in Fig. 10.24a, a parametric gain coefficient for the amplification of the signal wave can be defined [10.116]

$$g = \sqrt{\kappa I_p} \tag{10.39}$$

where I_p is the pump flux and κ is a coupling constant

$$\kappa = \frac{8\pi^2 d_{\text{eff}}^2}{\lambda_s \lambda_i n_s n_i n_p \varepsilon_0 c} . \tag{10.40}$$

The effective nonlinear coefficient d_{eff} connects the pump, signal, and idler fields. The parameters n_p, n_s, n_i and λ_s and λ_i are the refractive indices of the three waves, and the wavelengths of the signal and idler wave, respectively.

The single-pass power gain of the parametric amplifier in the high gain limit can be approximated by

$$G = \frac{1}{4} \exp(2gl) \tag{10.41}$$

where l is the length of the crystal.

A phase mismatch between the waves can be expressed by

$$\Delta k = k_p - k_s - k_i \tag{10.42}$$

where $k_j = 2\pi n_j / \lambda_j$, with $j = p, s, i$, are the propagation constants of the three waves.

In the presence of phase mismatch, the gain coefficient is reduced according to

$$g_{\text{eff}} = \left[g^2 - \left(\frac{1}{2} \Delta k \right)^2 \right]^{1/2} .$$

(10.43)

The reduction of gain resulting from momentum or phase mismatch is clearly evident. Maximum gain is achieved for $\Delta k l = 0$. Typical values for the coupling constant are on the order of $\kappa = 10^{-8}$/W, as will be discussed later. Therefore, a power density of at least $100\,\text{MW/cm}^2$ is required to obtain a gain coefficient of $g = 1\,\text{cm}^{-1}$ and a modest power gain of $G = 1.8$ in a 1 cm long crystal. At a 1 μm wavelength, the propagation constant is $k \approx 10^5$/cm in a material with a $n = 1.7$ refractive index. In order to minimize the effect of phase mismatch, we require from (10.43), that $\Delta k / 2 < g$. This means that the propagation constants have to be phase matched to better than 10^{-5}/cm.

From these very basic considerations, one can already draw several conclusions which govern the design of parametric devices. The optimum configuration of a parametric converter depends critically on the pump intensity in the nonlinear crystal. There is a strong incentive to operate at the highest attainable levels for a given pump source. However, the practical and acceptable pump intensity depends strongly on the optical damage threshold of the crystal and its coatings. The importance of a high nonlinear coefficient d_{eff} is also clearly evident from these equations, as is the detrimental effect of phase mismatch.

Singly-Resonant OPO: The singly-resonant OPO is the most common configuration due to the ease of mirror and resonator design, good conversion efficiency, frequency and power output stability. These advantages more than offset the disadvantage of increased threshold relative to the DRO. The pump wave makes a single pass through the nonlinear crystal. The generated signal and idle waves increase during the single pass in the pump-wave direction. Following reflection and backward trip in the resonator, the signal wave travels again with the pump wave and is amplified.

Threshold condition for steady-state operation for a singly-resonant OPO is given by

$$g^2 l^2 = 2 \sqrt{1 - R_s} (1 - L)$$

(10.44)

where R_s, the reflectivity of the output coupler for the signal wave, and L are combined reflection and absorption losses.

Threshold Condition: Singly-resonant OPO's are usually pumped with the output from a Q-switched Nd:YAG laser at either the fundamental or a harmonic of 1.06 μm, depending on the desired output wavelength and nonlinear material. The signal wave is amplified from an initial noise power, as it makes m cavity transits. A model which calculates the threshold pump intensity of a pulsed singly-resonant oscillator has been described by Brosnan et al. [10.115]. The model assumes that only the signal wave is resonated, whereas the idler wave is free to accept the profile of its driving polarization. The pump flux required to reach threshold is given by

$$I_{th} = \frac{1.12}{\kappa g_s l_{eff}^2} \left(\frac{L}{t_p c} \ln \frac{P_s}{P_n} + 2\alpha l + \ln \frac{1}{\sqrt{R}} + \ln 2 \right)^2 . \tag{10.45}$$

The pump flux I_{th} on the left-hand side of this equation relates to gain, see (10.39), and the right-hand side represents loss terms.

The coupling constant κ has been previously defined, g_s is the signal spatial mode coupling coefficient defined as

$$g_s = \frac{1}{1 + (w_s/w_p)^2} \tag{10.46}$$

The signal spot size w_s for a Gaussian beam is always smaller than the pump beam w_p because of the nonlinear conversion process. The effective parametric gain length l_{eff} is determined by the walkoff length

$$l_{eff} = \frac{\sqrt{\pi} \, w_p}{2\rho} \tag{10.47}$$

where ρ is the double refraction walkoff angle. The walkoff length is closely related to the aperture length introduced in (10.17). In short crystals, or under noncritically phase-matched conditions, the effective length can equal the crystal length l, i.e., $l_{eff} = l$. From (10.47) follows that walk-off has a greater influence on a small beam. The pump spot size dependence of parametric gain for critically phase-matched interactions is well known [10.117]. Actually for maximum efficiency, the beam diameter must be increased until the effective walk-off length is equal to the crystal length.

The first term in the bracket of (10.45) represents the loss of the pump beam due to the build-up time required to increase signal power from the noise level P_n to a signal level P_s defined as threshold. It was found that best agreement between the model and experimental data was obtained if threshold signal power P_s was defined as an increase of 10^{14} over the noise floor, or $\ln(P_s/P_n) = 33$. The logarithmic term is divided by the number of round trips it takes to reach P_s. Since L is the cavity length, t_p is the $(1/e^2)$ intensity full width of the pump pulse and c is the velocity of light. If more round trips $(t_p c/L)$ are allowed for the noise to build up to the signal level, then the pump-power density to reach threshold is lowered. In practice, this means a long pump pulse and a short OPO cavity lower the threshold.

The second term in (10.45) describes the absorption losses in the crystal where α is the absorption coefficient and l is the crystal length. The next term is the cavity output coupling loss, determined by the reflectivity R of the output mirror. The final term is due to SRO operation.

Inspection of (10.45) reveals that a high d_{eff} and long l_{eff} are the key to a low threshold for a parametric oscillator. The threshold input power is inversely proportional to the square of these terms. A long pump pulse and short cavity also tend to decrease the threshold somewhat as was already mentioned.

The model assumes a collimated Gaussian beam with a uniform wavefront. A transverse phase variation in the pump beam, caused by the laser or by intervening optics, acts as an effective phase mismatch in the presence of beam walk-off. If the beam has a transverse phase variation, and propagates at the walk-off angle with respect to the resonator axis, the generated signal wave sees a pump beam with a changing phase. As a result, threshold can be noticeably increased if the pump beam optical quality is non-uniform.

Singly-Resonant OPO with Return Pump Beam: The threshold in a singly-resonant OPO can be reduced by reflecting the pump radiation for a second pass through the crystal. This creates signal gain on both the forward and backward passes through the crystal. The changes in the OPO design are minimal; it only requires that the coating of the output mirror has to reflect both the resonated signal wave and the pump wave. If we let γ be the ratio of backward to forward pump intensity inside the crystal, then the threshold condition is as follows:

$$I_{th} = \frac{1.12}{\kappa g_s l_{eff}^2 (1 + \gamma)^2} \left(\frac{L}{t_p c} \ln \frac{P_s}{P_n} + 2\alpha l + \ln \frac{1}{\sqrt{R}} + \ln 4 \right)^2 . \tag{10.48}$$

The reduction in threshold can be clearly seen from this equation. It should also be noted, that in this scheme, the fluence of the forward and backward pump beams are superimposed in the crystal, which leads to higher power densities compared to the previous case, and therefore damage considerations are important.

We found the model describing singly-resonant OPO's to be of adequate accuracy for design purposes, particularly in view of the fact that different qualities of nonlinear crystals can easily change threshold by a factor of two.

We use (10.48) to calculate the threshold of an OPO build in our laboratory. The OPO contains a 15 mm long KTP crystal as the nonlinear material which is pumped by a Q-switched Nd:YAG laser at 1.064 μm. The maximum output energy of the pump laser was 10 mJ, produced in a 50 ns pulse. This OPO is not used for generating a tunable output, but for providing an eye-safe wavelength at 1.6 μm [10.118–120]. The crystal was positioned to achieve type-II noncritical phase matching for a pump wavelength of 1.06 μm (Nd:YAG) and a signal wavelength near 1.6 μm; the idler wavelength was therefore near 3.2 μm. Noncritical phase matching maximizes the effective nonlinear coefficient and essentially eliminates walk-off. The crystal was positioned inside an optical cavity formed by a pair of plane-parallel mirrors, as shown in Fig. 10.26. The input mirror, through which the pump enters, was anti-reflection coated at the pump wavelength (1.06 μm) and highly reflecting at the signal wavelength (1.6 μm); while the output mirror was highly reflecting at the pump wavelength and 10% transmitting at the signal wavelength. The laser output was focused by a 100 cm focal length lens, carefully positioned so as to mode match the waist of the pump to the cavity mode of the OPO resonator.

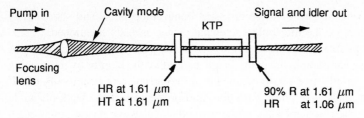

Fig. 10.26. Plane-parallel resonator for optical parametric oscillator, showing mode matching of focused pump beam to OPO cavity mode

In this noncritically phase-matched configuration, the wavelengths are λ_s = 1.61 μm, λ_i = 3.2 μm, λ_p = 1.06 μm and the respective indices of refraction are n_s = 1.7348, n_i = 1.7793 and n_p = 1.7474. One can approximate the effective nonlinear coefficient for both type II doubling and this OPO by use of [10.50]

$$d_{\mathrm{eff}} \approx (d_{24} - d_{15}) \sin 2\phi \sin 2\theta - (d_{15} \sin^2 \phi + d_{24} \cos^2 \phi) \sin \theta \qquad (10.49)$$

which reduces to $|d_{\mathrm{eff}}|$ = $0.15 d_{15} + 0.84 d_{24}$ for θ = $90°$ and ϕ = $23°$. With the values of d_{24} = 3.64 pm/V and d_{15} = 1.91 pm/V [10.49], one obtains $d_{\mathrm{eff}} \approx 3.3$ pm/V. Using these parameters, we calculate from (10.40): $\kappa = 1.2 \times 10^{-8}$ W^{-1}.

In order to calculate the threshold of a SRO with pump reflection according to (10.48), we need the following system parameters: because of the noncritical phase-matching condition, l_{eff} is as long as the physical length of the crystal, i.e., l_{eff} = 1.5 cm, the round-trip loss is about $2\alpha l$ = 0.01, the optical length of the OPO resonator is L = 5 cm, output reflectivity is R = 0.9, pump length is t_p = 50 ns. At threshold, we can make the following approximation: the ratio of backward to forward pump intensity is $\gamma \approx 1$, and the mode sizes are about equal at this low power level with $w_s \approx w_p$, one obtains for the mode-coupling coefficient $g_s = 1/2$. Introducing these parameters into (10.48) yields a value for the threshold power density of I_{th} = 105 MW/cm^2.

The performance of this OPO is illustrated in Figs. 10.27 and 28. In the experiment, a confocal configuration was also evaluated and its performance is displayed. In the confocal design, the plane mirrors were replaced with a pair of 5 cm concave mirrors with identical coatings to those used in the plane parallel cavity. The use of this cavity allows the pump to be more tightly focused while still maintaining good matching of the pump, to reduce the pump energy threshold for the device. This matching proved critical to successful operation of the OPO. Any mismatch between the pump mode and the TEM$_{00}$ mode of the OPO cavity will cause a reduction in gain for optical parametric oscillation and a subsequent increase in threshold. For this reason, a single-transverse-mode pump is essential for obtaining high OPO efficiency. The OPO employing a plane-parallel resonator reached threshold at a pump energy of 1.5 mJ, with a maximum output energy of 2.5 mJ obtained at 10 mJ pump energy. This corresponds to an energy conversion

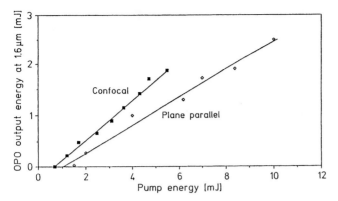

Fig. 10.27. Signal (1.6 μm) energy vs. pump energy for confocal and plane-parallel OPO resonator configurations

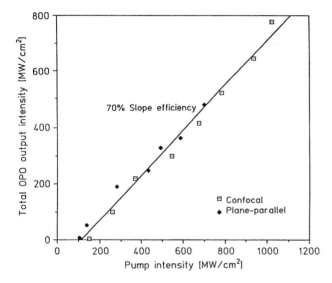

Fig. 10.28. Total OPO output intensity vs. pump intensity

efficiency of 25%. Since the 25 ns duration signal pulse is only half as long as the pump pulse, the power conversion efficiency of the OPO is 50%.

Higher conversion efficiencies were observed with the confocal resonator than with the plane-parallel resonator, due to the lower threshold energy required in the former configuration. Threshold energy is reduced because the pump is focused more tightly into the KTP crystal to obtain optimum matching for the confocal resonator. The threshold pump energy of this device was only 0.8 mJ, considerably lower than that of the plane-parallel configuration. A maximum output energy at 1.6 μm of 1.8 mJ was obtained for 5.5 mJ input pump energy. This corresponds to an energy-conversion efficiency of 35%. In the confocal configuration, the power conversion efficiency of 1.06 μm pump to 1.61 μm signal exceeds 50%, and the overall power conversion efficiency of the OPO, to both signal and idler outputs, exceeds 70%.

Optical parametric oscillation is intensity dependent, so the pump intensity is a more meaningful parameter for gauging OPO performance than is pump energy. In Fig. 10.28, total output intensity is plotted vs. pump intensity. From the data presented in this figure follows that both resonator configurations produce similar results. The calculated threshold intensity agrees reasonably well with the measured results. Earlier published results of this device used a nonlinear coefficient which was much higher, therefore agreement was not very good.

Saturation and Pump Beam Depletion: When the Q-switched pump pulse is incident on the nonlinear crystal, the signal and ideal waves are amplified from the initial noise level. The number of round trips in the optical cavity necessary to amplify the signal and idler waves, multiplied by the cavity round-trip time leads to the rise time necessary to achieve threshold. Above threshold, after a short transition period during which a steady-state condition is established, the pump power is limited at the threshold value. Any pump input power above threshold is divided into power at the signal and idler beams. Since $\nu_3 = \nu_1 + \nu_2$, it follows that for each input pump photon above threshold, one photon at the signal and idler wavelength is generated. In other words, the energy of the pump beam is depleted, and in a lossless system, the depleted energy goes into the signal and idler beams. The oscilloscope trace in Fig. 10.29, taken from the OPO described on the previous pages, shows the dynamics of signal generation very nicely. The dashed curve is the input pump pulse. The two solid curves display the signal pulse and the depleted pump pulse at the output of the OPO. During the early part of the pump pulse, there is a transient period during which the oscillation builds up from noise. After the transient period, the pump beam is clamped at its threshold value until the pump power falls below threshold.

The onset of depletion to full limiting occurs quite rapidly. The fast-falling part of the pump beam has a time constant characteristic of the rise time of the signal wave. The build-up time required to achieve parametric oscillation causes a temporal compression of the OPO output with respect to the pump pulse. It is also clear from these considerations that the pump power must remain on long enough to allow the fields to build up to the threshold value. This leads to the consideration of a minimum pump fluence as well as flux.

Conversion efficiency: In the plane-wave approximation, the conversion efficiency of a SRO for the ideal case of perfect phase matching and zero losses is given by [10.121]

$$\eta = \sin^2 gl \tag{10.50}$$

where g and l have been defined in (10.39, 41).

Using this expression, total conversion of the pump can be achieved in theory. As the point of total conversion is exceeded, power starts to couple back into the pump field at the expense of the signal and idler fields, and the conversion efficiency decreases.

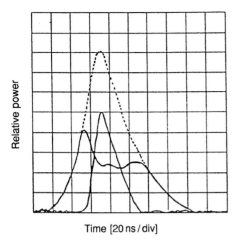

Fig. 10.29. Oscilloscope trace showing the depleted pump and the generated OPO output at $1.61\,\mu\mathrm{m}$ (solid curves) and the input pump (*dashed curve*)

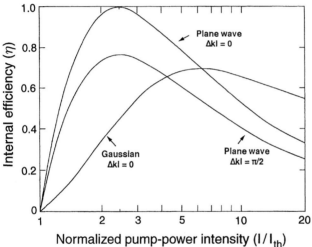

Fig. 10.30. Theoretical conversion efficiencies for a phase-matched plane wave and Gaussian pump beam and a plane-wave beam with a phase mismatch of $\pi/2$

In practice, Gaussian beams are employed rather than plane waves. In this case, as in the case of frequency doubling, the maximum theoretical conversion efficiency is reduced. Although the center of the beam may be achieving total conversion, the intensity in the wings is considerably lower than the optimum. As one increases the pump power to convert the wings, the center exceeds the optimum and some back conversion occurs, resulting in an overall decrease in the conversion efficiency.

Theoretical conversion efficiencies for a plane wave and a Gaussian pump beam are plotted in Fig. 10.30 as a function of pump intensity above threshold. For a phase-matched uniform plane wave, the conversion can be 100% and occurs for $I_p/I_{th} = (\pi/2)^2$, where I_p, and I_{th} are the input and threshold pump intensities,

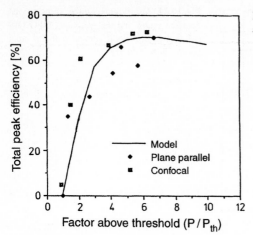

Fig. 10.31. Efficiency of an OPO vs. factor
above threshold

respectively. For a phase mismatch of $\Delta kl = \pi/2$, efficiency is around 75% for the plane wave pump beam. Using a Gaussian pump beam, maximum efficiency is about 71% and occurs for $I_p/I_{th} \approx 6.5$.

The dependence of the conversion efficiency on pump power is illustrated in Fig. 10.31. Plotted is experimental data from several OPO experiments carried out with KTP as the nonlinear crystal and a Nd : YAG laser pump at 1.064 μm. The solid curve was obtained by integrating the equations for plane wave OPO's over the spatial profile of the Gaussian beam, as described in [10.121].

One can derive a generic relationship between the factor by which pump power exceeds threshold power, and the OPO conversion efficiency. While there is some spread in the data, the general agreement is surprisingly good. The general "rule of thumb" for design of an efficient OPO, is to pump at least 4 times above threshold for maximum efficiency. Thus one can calculate the threshold for a given OPO, then multiply this value by a factor 4 to determine the required pump intensity.

It is important to remember that the conversion efficiency discussed so far is for the conversion of the pump power into both the signal and idler beams. The energy or power between the signal and idler beams is divided according to the ratio of the photon energies, i.e.

$$\frac{E_s}{E_i} = \frac{\lambda_i}{\lambda_s} \; . \tag{10.51}$$

From this follows the energy of the signal compared to the total energy E_t converted by the OPO

$$E_s = \frac{E_t}{1 + \lambda_s/\lambda_i} \; . \tag{10.52}$$

In the so-called degenerate mode, for which $\lambda_s = \lambda_i = 2\lambda_p$, each pump photon generates two photons at twice the pump wavelength. In this case, two orthogonally polarized beams at the same wavelength are emitted from the OPO.

Phase Matching: In order to achieve significant gain in the parametric device, the pump, signal and idler waves have to be phase matched according to (10.38). In a medium without dispersion, all waves propagate with the same velocity, hence $\Delta k = 0$. In reality, Δk is not zero for different wavelengths, and just like in harmonic generation, dispersion is compensated by use of birefringence. The index of refraction at a given frequency is a function of the direction of propagation in the crystal as well as orientation of polarization.

When the signal and idler waves are both ordinary rays one has type-I phase matching. When either one is an ordinary ray, it is referred to as type-II phase-matching.

In a uniaxial crystal, if the signal and idler are ordinary rays (type I) with indices of refraction n_s and n_i then the index of refraction for the pump wave necessary to achieve phase matching is

$$n_p = (\lambda_p/\lambda_s)n_s = (\lambda_p/\lambda_i)n_i . \tag{10.53}$$

In uniaxial crystals, the index of refraction n for an extraordinary wave propagating at an angle θ to the optic axis is given by

$$n^{-2} = n_o^{-2}\cos^2\theta + n_e^{-2}\sin^2\theta , \tag{10.54}$$

where n_o and n_e are the ordinary and extraordinary indices of refraction, respectively. The index n is thus limited by n_o and n_e. If n_p (10.53) falls between n_o and n_e, then an angle θ_m exists for which $n = n_p$. Propagation with $\Delta k = 0$ results, and all three waves travel at the phase-matching angle Θ_m with respect to the optical axis.

Phase matching with a propagation direction normal to the optic axis (90° phase matching) is preferred if possible. In that direction, the double refraction angle is zero and the nonlinear interaction is not limited by the effective gain length but the crystal's physical length. Non-90° phase-matching does, however, allow angle tuning.

Linewidth Control and Wavelength Tuning: The gain line width of an OPO is set by crystal dispersion; therefore the device has a rather broad linewidth, limited only by the phase-matching bandwidth of the crystal. Small dispersion and birefringence leads to relatively large bandwidths and tuning ranges, and vice versa. In [10.116], a simple relationship is given for the bandwidth $\delta\nu$ (in wave numbers) and birefringence Δn,

$$\delta\nu[\text{cm}^{-1}] \approx \frac{1}{\Delta n l[\text{cm}]} \tag{10.55}$$

where l is the crystal length. Crystals with small birefringence have larger bandwidth than crystals with large birefringence.

The linewidths of OPO's are generally too broad for spectroscopic applications. Line-width control can be achieved by using highly dispersive resonators which contain etalons, diffraction gratings, or resonant reflectors [10.115, 122].

A more recent development is line width control by means of injection seeding of the OPO. Similar to the technique described for linewidth control of a laser oscillator, a cw beam is injected into the pulsed SRO cavity. The length of the oscillator cavity is adjusted so that it is resonant at the injected frequency. When the pump is then applied, the desired mode starts from the level of the injected signal, whereas the competing modes begin from spontaneous emission. Provided the injected power level is sufficiently high, the injected signal will be the first to build up to a high power level and will deplete the pump.

Tuning curves for parametric oscillators can be determined by solving the phase-matching equations (10.37, 38) for signal and idler frequencies at a given pump frequency as a function of the tuning variable. The most common tuning methods are varying the direction of propagation with respect to the crystal axes or by changing the crystal temperature. To carry out the calculations, the indices of refraction must be known over the entire tuning range. For most nonlinear crystals of interest, the indices can be obtained from the Sellmeier equation.

We will consider KTP pumped by a Nd : YAG laser at 1.06 μm, as an illustration of the type of calculation which has to be performed to determine the tuning range and appropriate phase-matching angle for an optical parametric oscillator [10.123].

Potassium Titanyl Phosphate (KTP) belongs to the orthorombic crystal system, and therefore is optically biaxial (having two optical axes). The mutually orthogonal principal axes of the index ellipsoid are defined such that $n_x < n_y < n_z$, and the optical axes lie in the x-z plane (at 18° to the z axis in KTP [10.124] (Fig. 10.32)).

The indices of refraction for any propagation direction are given by the index ellipsoid defined by

$$k_x^2/(n_{\omega j}^{-2}-n_{x,\omega j}^{-2}) + k_y^2/(n_{\omega j}^{-2}-n_{y,\omega j}^{-2}) + k_z^2/(n_{\omega j}^{-2}-n_{z,\omega j}^{-2}) = 0, \quad j = \mathrm{p,s,i} \quad (10.56)$$

where $k_x = \sin\theta \,\cos\phi$, $k_y = \sin\theta\,\sin\phi$, $k_z = \cos\theta$; θ is the angle to the z axis and ϕ is the angle to the x axis in the x-y plane. The subscript "j" refers to either the pump, signal, or idler frequency. Equation (10.56) must be solved to determine the refractive indices ($n_{\omega j1}$, $n_{\omega j2}$) for the two eigen-polarizations perpendicular to the propagation direction for each wavelength. The equations can be solved numerically using the technique described in [10.47, 124]. In our work, this has been done using a spreadsheet (Excel) program.

The indices of refraction are obtained from the appropriate Sellmeier equation [10.48, 49].

$$n_x^2 = 3.0065 + \frac{0.03901}{\lambda^2 - 0.04251} - 0.01327\,\lambda^2 \,,$$

$$n_y^2 = 3.0333 + \frac{0.04154}{\lambda^2 - 0.04547} - 0.01408\,\lambda^2 \,, \quad\quad (10.57)$$

$$n_z^2 = 3.3134 + \frac{0.05694}{\lambda^2 - 0.05658} - 0.01682\,\lambda^2 \,,$$

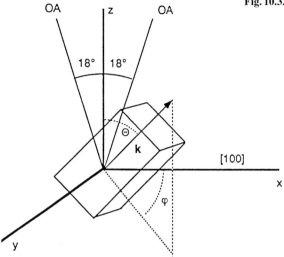

Fig. 10.32. KTP phase-matching angles

where λ is in micrometers.

The phase matching condition for optical parametric conversion in KTP is given by [10.47, 124]:

$$\omega_p n_{\omega p1} = \omega_s n_{\omega s2} + \omega_i n_{\omega i2} \quad \text{(type I)}, \tag{10.58}$$

or

$$\omega_p n_{\omega p1} = \omega_s n_{\omega s1} + \omega_i n_{\omega i2} \quad \text{(type IIa)}, \tag{10.59}$$

or

$$\omega_p n_{\omega p1} = \omega_s n_{\omega s2} + \omega_i n_{\omega i1} \quad \text{(type IIb)}. \tag{10.60}$$

In KTP, type-I interactions have very low nonlinear coefficients, and therefore are not useful. For propagation in the x-z plane ($\phi = 0$) in KTP, type-II interactions correspond to type-II phase matching in a positive uniaxial crystal (e.g., $\omega_p n_p^o = \omega_s n_s^o + \omega_i n_i^e$), where "o" and "e" represent the ordinary and extraordinary rays, respectively.

Figure 10.33 exhibits the calculated phase-matching angles for a KTP optical parametric converter pumped at $1.064\,\mu$m. The two curves correspond to the signal and idler wavelengths for type-II phase matching with $\phi = 0$. The polarization of the pump wave is along the y axis (o-wave) of the crystal as is that of the signal wave. The idler wave is polarized in the x-z plane (e-wave). The degenerate point at $2.12\,\mu$m corresponds to a phase matching angle of $\theta = 54°$.

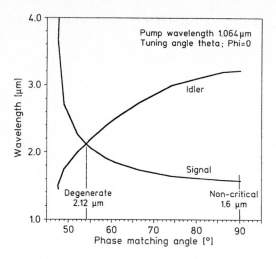

Fig. 10.33. Tuning range of an OPO pumped by a 1.06 μm Nd laser and employing a Type-II KTP crystal

10.2.2 Materials

The nonlinear crystals used for parametric oscillators must be transparent and have low loss at the pump as well as the signal and idler wavelengths. Very important for reliable OPO operation is a high damage threshold, and a large ratio of optical damage threshold to pump threshold. Table 10.4 lists a number of nonlinear materials for use in OPO's. The critical parameters for OPO crystals are slightly different to those for doubling where one requires broad angular acceptance, low-temperature sensitivity, and broad frequency acceptance. In OPOs, absence of these qualities merely leads to broad-linewidth output from the OPO due to the inherent tunability of the OPO. For example, exceeding the acceptance angle prevents phase matching at one pair of signal and idler wavelengths, however, the OPO will tune these wavelengths until phase matching is recovered, producing an output of broader linewidth.

Listed in Table 10.4 is also the maximum crystal size currently available from commercial sources. A minimum crystal length of at least 1 cm is usually required for the achievement of a high conversion efficiency, and a sufficient crystal cross-section is important to keep the pump intensity substantially below the damage threshold. For example, when pumping at the joule level, one typically needs to expand the pump beam to lower the pump intensity. In this case, operation at 100 MW/cm^2, with a typical pulse length of 20 ns, requires a beam diameter of 0.8 cm. Thus, one requires a crystal size on the order of 1 cm^3. The use of a larger pump-beam diameter also reduces the effect of walk-off; for example, the walk-off angle for a 200 μm pump beam diameter in a mid-IR KTP OPO is 2.6°, which reduces the effective interaction length to 4 mm. However, if the pump-beam diameter is increased to 8 mm, the interaction length is dramatically increased to 160 mm, so that walk-off is negligible over the typical 15 mm long crystals employed in practice.

Table 10.4. Common crystals employed in OPO's

	LBO	BBO	KTP	KTA	AgGaS$_2$	LiNbO$_3$	PPLN
d_{eff} [pm/V]	1.16	1.94	3.64	4.47	13.5	5.1	17.2
Transparency [μm]	0.16–2.6	0.19–2.5	0.4–3.5	0.4–4.0	0.5–12	0.5–5	0.5–5
Damage threshold [GW/cm^2]	2.5	1.5	0.5	0.5	0.03	0.2	0.2
Typical length [cm]	0.5–1	0.5–1	1–2	1–2	2–4	2–5	2–6

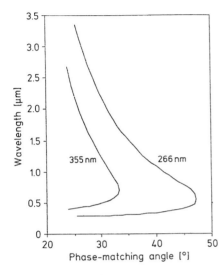

Fig. 10.34. Tuning curves for BBO in a type-I configuration pumped at the wavelength of 266 nm and 355 nm

Most of the materials listed in Table 10.4 have already described in Sect. 10.1.3. For OPO's operating in the UV, visible or near-infrared BBO and LBO are the preferred materials because of the transparency range which extends to very short wavelengths. As an example, Fig. 10.34 displays tuning curves for OPOs using BBO in a type-I configuration pumped at the tripled or quadrupled output from a Nd laser [10.125, 126].

KTP is the material of choice for the design of OPO's pumped with Nd lasers at 1.064 μm [10.118–120, 127–129]. This material has already been described in the previous sections. It has been observed that the damage threshold of KTP when used in an OPO can be considerably lower compared to second-harmonic generation [10.42].

Potassium Titanyl Arsenate (KTA) is a new nonlinear material which is isomorphic with KTP. Substitution of arsenic for phosphorous extends the transparency range further into the infrared. However, the major difference beween the two materials is the considerably lower absorption losses in the 3–5 μm region of KTA, as illustrated in Fig. 10.10. Usually, the wavelength beyond 3 μm is the idler wavelength when a KTA or a KTP OPO is pumped by a Nd:YAG or a Nd:YLF laser. For example, a pump wavelength of 1.06 μm and signal wavelength of 1.54 μm produce an idler wavelength of 3.44 μm. Since KTA does not have any appreciable absorption at the idler wavelength, this material has fewer problems associated with thermal lensing and heating of the crystal in high-average-power OPO's. In terms of damage threshold and nonlinear coefficient, the materials are quite similar although d_{eff} is somewhat higher in KTA. With a $1.064 \, \mu$m pump beam from a Nd : YAG laser noncritical phase matching produces a signal at $1.538 \, \mu$m in KTA, whereas in KTP the signal output is at $1.574 \, \mu$m. OPO's featuring KTA crystals are described in [10.42, 130, 131].

Although silver gallium selenide (AgGaSe$_2$) has a low damage threshold, it is of interest for the design of low-power OPO's because its transparency reaches far into the infrared. Optical parametric oscillators using AgGaSe$_2$ have been pumped by Nd : YAG lasers to produce output from 6 to $14 \, \mu$m. This tuning range was achieved in a two stage OPO system, with a Nd : YAG laser pumping a noncritically phase matched KTP OPO at fixed wavelength of $1.57 \, \mu$m, which was used to pump a AgGaSe$_2$ OPO [10.132]. Also continuously tunable output from 6.7 to $9.8 \, \mu$m was achieved by directly pumping a AgGaSe$_2$ OPO with a Q-switched pulse from a Nd : YAG laser [10.133].

LiNbO$_3$ is a nonlinear crystal which has been employed widely in the past. Unfortunately, it has a much lower damage threshold compared to newer materials such as KTP and KTA which cover the same wavelength regime. However, very recently a different version of this material, namely periodically poled LiNbO$_3$ (PPLN) has become avaible. PPLN has a very large nonlinear coefficient, therfore it is well suited for low-peak-power OPO's in the near- and mid-infrared. Because of the unique concept underlying PPLN, it will be described in more detail below.

Periodically Poled Lithium Niobate (PPLN)

In this subsection we will briefly review the concept of quasi phase matching (QPM),and summarize the properties of periodically poled LiNbO$_3$. Although QPM is applicable to any nonlinear process, the subject is covered here because of the importance of PPLN and its application in OPO's for the generation of mid-infrared radiation.

PPLN is the first commercially available crystal in which efficient nonlinear conversion processes, such as harmonic generation or parametric interactions, are not based on birefringence phase matching but on a periodic structure engineered into the crystal. We will briefly explain this technique, termed quasi-phase matching, by referring to Sect. 10.1.1. An in-depth treatment of this subject can be found in [10.134–137].

As illustrated in Fig. 10.2 and expressed by (10.9), in a phase-matched condition $\Delta k = 0$ the harmonic power increases with the square of the interaction length ℓ. In the situation of a fixed phase mismatch Δk, energy flows back and forth sinusoidally between the fundamental and harmonic beams with a period of $\Delta k \ell / 2 = \pi$ as the waves propagate through the crystal. Half of this period is the coherence length ℓ_c given by (10.10) which is the maximum distance over which the second harmonic can grow. In a non-phase matched condition, this distance is on the order of (2–20) μm for conversion processes of interest. The objective of birefringence phase matching is to make this distance as long as possible by taking advantage of the wavelength and polarization dependence of the refractive indices in a crystal. By arranging proper balance between dispersion and birefringence the technique of phase matching increases ℓ_c by a factor of 10^3.

Even before the concept of birefringence phase matching was invented, it was proposed that a periodic structure in the crystal which corrects the phase of the propagating beams each time it reaches π would enable continued energy flow from the fundamental to the harmonic beam [10.96]. An implementation of this concept is a crystal where the sign of the nonlinear coefficient is reversed after each distance ℓ_c. In this case the relative phase between the waves is inverted after the conversion has reached its maximum. Therefore, on the average the proper phase relationship between the beams is maintained and the second-harmonic power increases with the square of crystal length similar to the birefringence phase-matched case. The nonlinear coefficient d_Q for quasi phase matching is, however, reduced as compared to the phase matched interaction according to [10.134].

$$d_Q = \frac{2}{\pi} d_{\text{eff}} .\tag{10.61}$$

A phase reversal is equivalent of slicing a crystal in thin wafers and stacking the wafers by rotating alternate wafers by $180°$. The periodicity $\wedge$ of this structure is twice that of ℓ_c, i.e.

$$\wedge = 2\ell_c .\tag{10.62}$$

Because of the micrometer size thickness of alternating layers, the practical realization of quasi phase matching had to wait until it was possible to engineer a periodic phase reversal into a monolithic crystal. Only recently has the technique of periodically reversed polarization domains in ferro-electric crystals, combined with advances in lithography, made it possible to realize this concept. A reversal of the ferroelectric domains corresponds to a sign reversal of the nonlinear coefficient. In this process, standard lithography produces a patterned electrode with a period between 5 and 30 μm on the surface of a ferroelectric crystal such as $LiNbO_3$. A high-voltage pulse is applied to the crystal which is sandwiched between the patterned electrode and a uniform electrode. The high electric field strength of the voltage pulse permanently reverses the sign of the nonlinear coefficient in a pattern determined by the electrode structure.

This electric-field-poling technique is employed to reproducibly manufacture periodically poled lithium niobate (PPLN) suitable for applications in infrared parametric oscillators and second-harmonic generation.

We will now derive an expression for the quasi-phase matching condition in OPO's for periodically poled crystals. Parametric conversion requires energy conservation and momentum matching conditions, as given by (10.37 and 38). A momentum mismatch expressed as a phase mismatch between the pump, signal and idler waves has been defined in (10.42). In the frequency domain, a periodic structure can be represented by a grating wave vector

$$k_g = 2\pi / \wedge \tag{10.63}$$

where $\wedge$ is the period of the grating.

In the presence of a grating structure in the crystal, the equation for phase mismatch (10.42) includes as an additional term the grating wave vector k_g [10.135]

$$\Delta k = k_p - k_s - k_i - k_g , \tag{10.64}$$

the first three terms being the conventional phase matching condition. It is assumed that all wave vectors are collinear with the grating vector.

The objective of high parametric conversion is to eliminate the phase mismatch caused by dispersion by selecting the appropriate crystal orientation, temperature and polarization such that $k = 0$ is achieved. The grating vector in (10.64) provides an additional adjustable parameter that is independent of inherent material properties.

Differences of the three wave vectors k_p, k_s, k_i, can be compensated by an appropriate choice of the grating vector k_g such that $\Delta k = 0$ can be achieved. Introducing (10.63) into (10.64) and with $\Delta k = 0$ the grating period under quasi phase matching conditions is given by

$$\wedge_g = \frac{2\pi}{k_p - k_s - k_i} . \tag{10.65}$$

Equation (10.65) can also be expressed by

$$\frac{1}{\wedge_g} = \frac{n_p}{\lambda_p} - \frac{n_s}{\lambda_s} - \frac{n_i}{\lambda_i} . \tag{10.66}$$

This equation is the equivalent of (10.38) which was derived for the birefringence phase matching case. If one substitutes Δk in (10.6) with k_g from (10.63) one obtains the condition for harmonic generation.

$$\wedge_g = \frac{\lambda_1}{2(n_2 - n_1)} . \tag{10.67}$$

The fact that the domain thickness is a free parameter which can be customized for a particular nonlinear process offers significant advantages over birefringent phase matching. For example, QPM permits wavelength selection over the entire transmission window of the crystal, it allows utilization of the largest nonlinear coefficient, and it eliminates problems associated with walk-off since all interactions are non-critical.

For example, in LiNbO$_3$ interactions with all waves polarized parallel to the crystal optic axis utilizes the largest nonlinear coefficient $d_{33} = 27$ pm/V. PPLN permits noncritical phase matching with this coefficient. Birefringent phase matching requiring orthogonally polarized beams can only be accomplished with the smaller coefficient d_{31} = 4.3 pm/V. Therefore, PPLN has a parametric gain enhancement over single-domain material of $(2d_{33}/\pi d_{31})^2 \approx 20$.

The grating period of PPLN for a particular OPO can be calculated from (10.66) with the indices of refraction obtained from the Sellmeier equation. For example, if it is desired to shift the wavelength of a Nd:YAG laser with output at λ_p = 1.064 μm to the eye-safe region of λ_s = 1.540 μm then one obtains from (10.37) an idler wavelength of λ_i = 3.442 μm. The next step is the determination of the indices of refraction at these wavelengths. The most recent coefficients of the Sellmeier equation for the extraordinary index of refraction have been published in [10.138]. At an operating temperature of 120°C we obtain n_p = 2.160, n_s = 2.142 and n_i = 2.085. Introducing these values into (10.66) yields Λ_g = 29.8 μm.

PPLN is commercially available as 1 mm-thick crystals, with lengths ranging from 20 to 60 mm, and a width depending on the number of different gratings desired. For example, a width of 11 mm may contain up to 8 separate grating periods. Translation of the crystal normal to the beam moves a different grating into the beam, which shifts the signal and idler waves to a new wavelength band. Fine tuning can be accomplished by changing the temperature.

LiNbO$_3$ has a relatively low damage threshold in comparison to, for example, KTP. Also, PPLN, like single-domain LiNbO$_3$, has to be heated to eliminate photo-refractive effects. Although photo-refractive damage is mainly caused at shorter wavelengths it can be a problem also in infrared OPO's because a certain amount of visible radiation is generated in these devices by second-harmonic and sum-frequency processes.

Pulsed and cw optical parametric oscillators using PPLN, and pumped by Nd:YAG lasers, have been operated over the wavelength range from 1.4 μm to 4 μm and with outputs from cw [10.136, 139] to Q-switched pulses in the nanosecond regime [10.140–142] up to femtosecond pulses [10.139, 143, 144]. Tuning has been accomplished by translation of the PPLN crystal or by temperature changes or both.

Peak powers and pulse energies are limited by the small aperture of available crystals and the relatively low damage threshold of this material. Therefore, PPLN has become the crystal of choice for OPO's pumped by cw or high-repetition-rate Nd:YAG lasers producing up to several watts of mid-infrared radiation. These lasers have pulse energies and peak powers that are relatively low and provide a natural match for PPLN. In addition, the high gain and noncritical phase matching permit the design of highly-efficient OPO's.

Although PPLN is used mainly in OPO's for infrared interactions, the five times higher nonlinear coefficient compared to KTP allows efficient single-pass cw frequency doubling external to the resonator [10.145]. The efficiency obtained is comparable to the performance of intracavity doubling but without the complications associated with output fluctuations discussed in Sect. 10.1.4. Since dispersion is larger at shorter wavelengths, the period of the domain structure must be reduced according to (10.67). This poses a fabrication problem since the domain period for frequency doubling a Nd : YAG laser is only 6.5 μm for a PPLN crystal operated at 200°C. As mentioned earlier, suppression of photo-refractive damage requires operation of a PPLN at a fairly high temperature. However, it seems that PPLN is more resistant to photo-refractive damage compared to ordinary LiNbO₃ [10.146].

Although not available commercially, other ferroelectric crystals such as KTP, RTA (rubidium titanyl arsenate) and LiTaO₃ (lithium tantalate) have been periodically poled, and promising results have been achieved with these materials in harmonic generators and OPO's [10.147–150]. Since these crystals have high damage thresholds they hold promise for efficient nonlinear interactions at higher pulse energies compared to PPLN.

10.2.3 Design and Performance of Optical Parametric Oscillators

Generation of tunable radiation in the UV, visible and IR region of the spectrum is of great interest in many applications. The Optical Parametric Oscillator (OPO) has long been regarded as a convenient means of achieving tunable output from solid-state lasers. However, progress in the development of OPOs has been slow owing to a lack of suitable nonlinear materials and pump sources of desirable beam quality. As mentioned at the beginning of this section, the advent of new nonlinear materials such as KTP, KTA, BBO, LBO and PPLN in combination with the high beam quality obtained from diode-pumped lasers has revived interest in OPOs for the generation of tunable laser radiation. In this subsection, we will illustrate the design and performance of state-of-the-art OPOs which are pumped either by the fundamental or by one of the harmonic wavelengths of a Nd laser.

The first example illustrates the combination of the mature Nd laser technology with a PPLN OPO to produce mid-infrared radiation at the one-watt level [10.141]. As illustrated in Fig. 10.35, wavelength conversion was achieved in a 2 step process. A diode-pumped and repetitively Q-switched Nd : YLF laser at 15 kHz contained an intracavity noncritically phase-matched KTP OPO which generated about 1.8 W at the 1.54 μm pump wavelength for the second OPO.

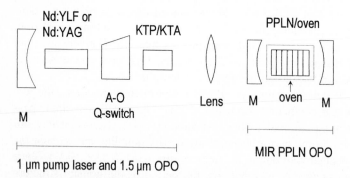

Fig. 10.35. Optical schematic of MIR laser system

Just like in intracavity frequency doubling, placing an OPO inside the resonator increases the flux in the nonlinear element by a factor $(1 + R)/(1 - R)$, R being the reflectivity of the output coupler for normal operation. In cw or high-repetition-rate Q-switched lasers the power density is often too low for efficient OPO operation. Instead of very alignment sensitive focusing optics, an attractive option to overcome this difficulty is an intracavity OPO configuration.

The OPO depicted in Fig. 10.35 is a monolithic, singly resonant device with one side coated AR for $1.05\,\mu$m and highly reflective for $1.54\,\mu$m. The output coupler was coated highly reflective for $1.05\,\mu$m and transmissive for $1.54\,\mu$m. The idler wave is mostly absorbed in the KTP crystal.

A lens focused the pump beam to approximately a $100\,\mu$m diameter spot at the center of the 0.5 mm thick PPLN crystal. The input mirror had a high transmission at the pump wavelength of $1.54\,\mu$m and a high reflectivity from 2.5 to $4\,\mu$m. The output coupler had a high reflectivity at the pump and a high transmission at $4\,\mu$m and about 10% transmission at $2.5\,\mu$m. The PPLN OPO resonator is therefore singly resonant at the signal wave near $2.5\,\mu$m. The pump beam made a complete round trip inside the resonator.

The 19 mm long PPLN crystal had five grating periods across a width of 11 mm. Tuning with temperature and different grating periods is illustrated in Fig. 10.36. The theoretical curves were calculated with the nonlinear-optics program SNLO and using the Sellmeier equation for PPLN [10.151]. The combined output of signal and idler was close to 1 W, which yields a conversion efficiency from the $1.54\,\mu$m pump of 55%.

In a subsequent system, Nd : YLF was replaced by Nd : YAG, and KTP was substituted with KTA. The new combination yields about the same output wavelength and power output, however Nd : YAG is a stronger material and can be more readily fabricated into small slabs required in this conduction-cooled system, and KTA does not have heating related problems caused by the absorption of the idler wavelength.

In a further evolution of the work described above, the PPLN OPO was placed inside the Nd : YAG resonator [10.142]. The Nd : YAG laser was Q-switched from 30 to 100 kHz, and the $1.06\,\mu$m pump beam was converted in one step to a signal wave near $1.45\,\mu$ and an idler wave with one-watt output near $4\,\mu$m. A further improvement is the replacement of the 0.5 mm thick PPLN with a 1 mm-thick crystal which shows promise for scaling up the $4\,\mu$m infrared output power to several watts.

Optical parametric oscillation in the UV, visible and near-infrared region has been demonstrated in BBO and LBO which have been pumped with Q-switched, frequency tripled [10.152–155] and quadrupled [10.156] Nd : YAG lasers. An example of such a device is the singly resonant OPO, shown in Fig. 10.37 [10.156]. A 20.5 mm long BBO crystal pumped by a Q-switched and quadrupled Nd : YAG laser with output at 266 nm provided tunable output from 0.33 to $1.37\,\mu$m with different sets of mirrors. The BBO crystal was cut for type-I phase matching at $39.1°$. The optical cavity length was 7.5 cm, which allowed about 20 round-

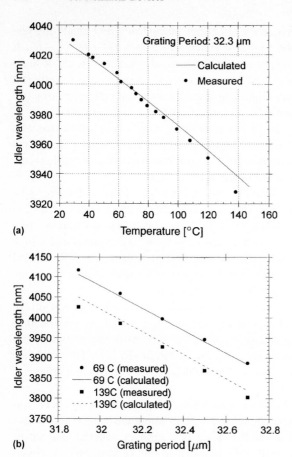

Fig. 10.36. Idler wavelength as a function of temperature (**a**) and grating period (**b**)

trip passes through the crystal during the 9 ns-long pump pulse. Threshold was obtained at 4.5 mJ or 23 MW/cm^2.

Optical damage of the mirrors is a major consideration in the design of OPOs. In a singly-resonant OPO, trichroic dielectric mirrors are required which are highly transmissive at the pump and idler wavelengths, and highly reflective at the signal wavelength. In addition, the mirrors must have a high damage threshold. With a pump wavelength in the ultraviolet, demands on mirror coatings are particularly severe. An interesting feature of this OPO is the introduction of the pump beam into the OPO resonator by means of a separate set of mirrors oriented at Brewster's angle. The advantage of this design is the fact that the resonator mirrors M1 and M2 of the OPO do not have to transmit the intense UV pump beam.

Figure 10.38 illustrates a configuration of an intra-cavity doubled OPO designed to produce output in the UV/blue wavelength region [10.157]. The device generates radiation between 760 and 1040 nm, which is internally doubled to produce tunable output from 380 to 520 nm. The OPO employs a 15 mm long KTP crystal, antireflection coated at 911 nm, cut for normal incidence at $\phi = 0°$ and

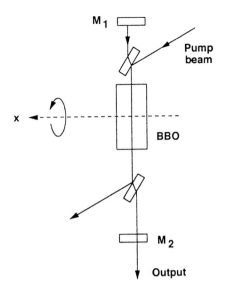

Fig. 10.37. Optical parametric oscillator featuring a BBO crystal pumped by the fourth harmonic of a Nd:YAG laser

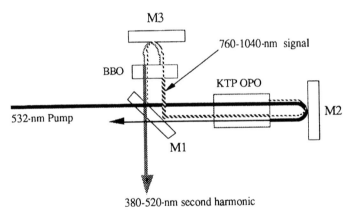

Fig. 10.38. Layout of intracavity OPO, showing the different pump wavelengths involved: the 532-nm pump, the 760–1040-nm signal, and the 380–520-nm intracavity doubled signal [10.157]

$\theta = 69°$. The angles correspond to type-II phase matching for a signal wavelength of 911 nm.

The pump source is a diode-pumped Nd:YAG laser which is frequency doubled by a 1 cm long KTP crystal. The waist of the pump beam is positioned at the center of the 5 cm long OPO cavity.

The 532-nm pulse enters the OPO cavity through a flat dichroic turning mirror (M1; highly reflecting at 800–950 nm and highly transmitting at 440–540 nm). The 532-nm pulse passes through the OPO crystal and is reflected off the flat rear cavity mirror (M2; highly reflecting at 532 nm and 880–950 nm), to leave the cavity through mirror M1 after making a second pass through the OPO crystal.

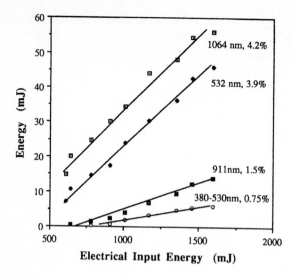

Fig. 10.39. Output at various wavelengths of the OPO depicted in Fig. 10.38 as a function of electrical input to the Nd:YAG pump laser

All mirrors have minimal reflectivity at the idler wavelength, making the OPO singly resonant. The intracavity 911-nm flux is reflected by the turning mirror (M1) to pass through a 3-mm-long BBO type-I doubling crystal. The final cavity mirror (M3) is highly reflective at both 880–950 nm and 410–480 nm, which allows the doubled output to be extracted through the turning mirror (M1) after making another pass through the doubling crystal. The use of two-pass nonlinear generation in both the doubler and the OPO significantly improves the conversion efficiency of these processes. The advantage of this configuration is that the KTP crystal is not exposed to the blue-UV flux, which could cause damage problems.

The output energies obtained at the various wavelengths are shown in Fig. 10.39, plotted as a function of the electrical input energy to the diode arrays of the Nd:YAG pump laser.

Figure 10.40 illustrates the technique of injection seeding an OPO in order to generate narrow-bandwidth output [10.122]. A three-mirror L-shaped cavity configuration, containing a 50 mm long LiNbO$_3$ crystal, was used to couple the pump and seed beam into the crystal. The OPO was pumped at 1.064 μm by a Nd:YAG laser with a pulse length of 8 ns and a pump energy of 75 mJ. The injection source was a grating tuned, external cavity InGaAsP diode laser tunable from 1.49 to 1.58 μm with a line width of 150 kHz.

The rear mirror (M1) had a reflectivity of 90% at the signal wavelength and was mounted upon a piezoelectric stack that was used to adjust the length of the cavity. Mirrors M2 and M3 had signal reflectivities of 74% and 95%, respectively. All three mirrors had high transmission at 1.06 μm. A 50-cm lens was used to focus the seed radiation into the cavity.

The nonlinear medium used for the OPO was a 5 cm long lithium niobate crystal cut at 47° to the optic axis and antireflection coated at the signal wavelength. The crystal was kept at a constant temperature of 38°C and was angle tuned by type-I phase matching. In the unseeded operation, a pump energy of

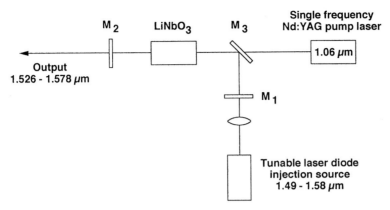

Fig. 10.40. Schematic diagram of an injection-seeded OPO

75 mJ generated an output of 22 mJ at 1.535 µm with a bandwidth of 50 GHz. With injection seeding, the output was 7.6 mJ and the bandwidth was reduced to 180 MHz. Injection seeding was achieved over a signal wavelength ranging from 1.526 to 1.578 µm, with a corresponding idler of 3.26 to 3.51 µm.

10.3 Raman Laser

The Raman laser which is based on Stimulated Raman Scattering (SRS) has shown to present a practical way to access wavelengths not directly available from solid-state lasers. Stimulated Raman scattering was first observed by *Woodbury* and *Ng* [10.158] in experiments with ruby lasers in 1962. In its basic form, the Raman laser consists of a high-pressure gas cell and resonator optics. If this completely passive device is pumped by a high-power laser, a fraction of the laser wavelength is shifted to a longer wavelength. The particular wavelength shift depends on the gas in which SRS takes place. Also, if the power level of the laser is increased, additional spectral lines will appear at longer as well as shorter wavelengths with respect to the pump wavelength.

10.3.1 Theory

The basic Raman effect is an inelastic light scattering process. The energy levels of interest for Raman scattering are shown in Fig. 10.41. An incident quantum $h\nu_p$ is scattered into a quantum $h\nu_s$ while the difference in energy $h(\nu_p - \nu_s) = h\nu_R$ is absorbed by the material. In Fig. 10.41 u is the upper state of the molecule, and i and f are the initial and final states. In principle, the excitation of the material may be a pure electronic excitation, or a vibrational or rotational excitation of a molecule. Solid-state-laser-pumped Raman lasers typically employ gases such as hydrogen or methane, therefore level i and f are the vibrational levels of the ground state of the molecule. The upper level u can be a real state or a "virtual" upper state. The frequency ν_s is called a Stokes frequency and is lower than the incident light frequency ν_p. The difference between ν_p and ν_s

$$\nu_p - \nu_s = \nu_R \tag{10.68}$$

is the Raman shift which is characteristic of the material in which the Raman process is observed.

If the system is in an excited state to begin with, it may make a transition downward while the light is scattered. In that case the scattered light contains anti-Stokes frequencies which are higher than the incident frequency.

In the stimulated Raman effect, the pump laser at the frequency ν_p excites molecules to the level u, and if a population inversion exists between levels u and f, it can produce lasing action. In this case the radiation ν_s becomes amplified, while the pump radiation ν_p loses energy. The process has typical laser characteristics, such as a pump energy threshold, exponential gain and narrow linewidth. The emission in Fig. 10.41a is called the first Stokes line, usually written S_1. If a high-power laser is focused into a Raman medium, additional lines will appear at the output. Actually, a single laser frequency interacting with molecules will produce a "comb" of frequencies, each separated from its neighbor by the frequency spacing ν_R. These additional lines will be to the left and right of the wavelength scale with regard to the laser pump wavelength.

The additional lines are produced by parametric four-wave mixing of the various waves propagating in the Raman medium. As an example Fig. 10.41b illustrates the generation of one such line, having a wavelength shorter than the pump wavelength. This so-called anti-Stokes line is the result of the interaction of ν_p and ν_{s1} both propagating in the same direction. The parametric four-wave mixing process does not require a population inversion between w and i, therefore there is no well defined threshold. The simplest way of looking at this interaction is that the two frequencies beat together to produce polarization (induced dipole moments in the molecules) at the difference frequency. This polarization then modulates the laser-molecule interaction and produces light beams at the side frequencies.

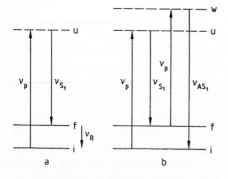

Fig. 10.41a,b. Raman process. Generation of first Stokes light **(a)** and first anti-Stokes light **(b)**

Stokes lines have lower frequency (longer wavelength) and anti-Stokes lines have higher frequency (shorter wavelength). In each case the line is labeled first, second, etc., by counting the number of frequency shifts from the pump laser.

The gain equations given below, which are taken from [10.159], provide the basic design parameters of the Raman lasers. A rigorous mathematical treatment of stimulated Raman scattering can be found in [10.160–162]. These references also provide comprehensive introductions to all areas of coherent Raman spectroscopy.

SRS can be described as a nonlinear interaction involving the third-order nonlinear susceptibility χ^3. At a medium's Raman resonance, the third-order susceptibility reduces to the peak Raman susceptibility χ_R'', where the double prime indicates the imaginary part of the total susceptibility.

The growth of the electric field at the Stokes wavelength and depletion of the pump field is governed by the equations:

$$\frac{\partial E_p}{\partial z} = -\frac{\omega_p}{2cn_p} \chi_R'' |E_s|^2 E_p \tag{10.69a}$$

$$\frac{\partial E_s}{\partial z} = \frac{\omega_s}{2cn_s} \chi_R'' |E_p|^2 E_s \tag{10.69b}$$

where $\omega_p - \omega_s = \omega_R$ are the frequencies, n_s and n_p are the indices of refraction, and c is the velocity of light. For a constant pump field, the Stokes field grows exponentially with a power gain given by

$$P_s(l) = P_s(0)e^{g_s l} \tag{10.70}$$

where g_s is the gain coefficient and l the interaction length in the Raman medium.

$$g_s = \frac{\omega_s \chi_R'' |E_p|^2}{n_s c} = \frac{4\pi \chi_R'' I_p}{\lambda_s n_s n_p \varepsilon_0 c} . \tag{10.71}$$

If one expresses the third-order Raman susceptibility, χ_R'' in terms of spontaneous Raman scattering cross-section, $d\sigma/d\Omega$, one obtains

$$g_s = \frac{\lambda_p \lambda_s^2 N(d\sigma/d\Omega)I_p}{n_s^2 hc\pi \Delta\nu_R} \tag{10.72}$$

where λ_s is the Stokes wavelength, I_p is the pump intensity, N is the number density of molecules, h is Planck's constant, and $\Delta\nu_R$ is the full-width, half-maximum Raman linewidth.

From these equations follows that the gain for a single-pass Raman medium is proportional to the incident intensity, the active media cross-section (which includes pressure and linewidth dependencies), and the length of the Raman cell.

The threshold of a Raman laser is usually defined as the gain required to achieve an output power at the Stokes wavelength that is of the same order as the pump radiation. For example, to achieve a 1 MW Stokes shifted power output, one requires a gain-length product of $g_s l = 36$ in the Raman medium in order for the radiation to build up from the initial spontaneous noise level which is $P_s = h\nu_s \Delta\nu_s \approx 10^{-19}$ W in the visible. Quantitative information such as gain coefficient and Stokes wavelength for many gases, liquids and solids are listed in [10.163]. Table 10.5 summarizes the data for the most important Raman media.

Table 10.5. Stokes shift and Raman scattering cross-section for several gases

Medium	ν_R [cm^{-1}]	$\dfrac{d\sigma}{d\Omega}\left[\dfrac{\text{cm}^2}{\text{Ster}}\right]$
H$_2$	4155	8.1×10^{-31}
CH$_4$	2914	3.0×10^{-30}
N$_2$	2330	3.7×10^{-31}
HF	3962	4.8×10^{-31}

The maximum theoretical conversion efficiency of a Raman laser is

$$\eta = \frac{\nu_p - \nu_R}{\nu_p} . \tag{10.73}$$

If one introduces into (10.73) the numbers given for ν_R in Table 10.5 it is obvious that the conversion efficiency can be very high. For example, a frequency-doubled Nd:YAG laser, Raman shifted with CH$_4$ provides an output at 630 nm. With $\nu_R = 2914$ cm^{-1} and $\nu_p = 18,797$ cm^{-1} one obtains $\eta = 84\%$.

10.3.2 Device Implementation

Raman lasers employed to shift the wavelength of solid-state lasers are restricted to gases as the nonlinear medium. In liquids, stimulated Raman scattering is complicated by the onset of Brillouin scattering and by self-focusing which leads to filament formation. Although Raman shifting is simple, in principle, there are numerous design parameters to consider in engineering a practical device. SRS in gases compete with optical breakdown and with stimulated Brillouin scattering. In order to avoid these competing nonlinear effects, and to produce significant energy conversion, pump pulse length, input beam diameter, focusing and interaction length – which are all interrelated – must be carefully optimized. Raman lasers have been designed in a number of configurations such as the single-pass cell, Raman resonator, oscillator-amplifier and waveguide system. A short description of the salient features of these different designs will be given below.

The simplest Raman laser is based on the single-pass emission in a gas-filled cell, as shown in Fig. 10.42a. The output beam quality is similar to that of the input pump. While the optics may be simply designed to prevent optical damage to the windows at focal intensities high enough to produce significant energy conversion, many nonlinear processes may occur to limit the conversion efficiency at high- energies. Copious second Stokes and anti-Stokes production

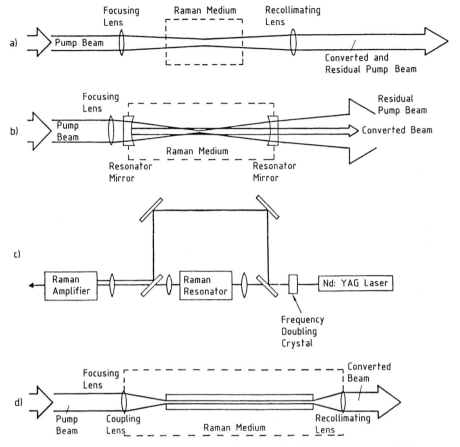

Fig. 10.42a–d. Raman laser configuration: **(a)** single pass cell, **(b)** Raman resonator, **(c)** Raman oscillator-amplifier, **(d)** waveguide

may occur, as well as stimulated Brillouin scattering and optical breakdown. The single-pass cell does not provide discrimination against these other nonlinear processes at high-energy inputs.

By using mirrors at each end, feedback can be selectively enhanced at only the first Stokes wavelength.

A gas Raman laser utilizing a single resonator is shown schematically in Fig. 10.42b. The mirror coatings selected allow all the pump light to pass into the cavity, while inducing resonator action at the Raman shifted frequency. The resonator transforms a multi-mode pump laser beam into a nearly diffraction-limited output beam with a slightly narrower pulse width. The high-quality beam is a result of keeping the pump intensity below single-pass threshold so that only multiple reflections of the lowest-order mode will achieve sufficient intensity to reach the stimulated scattering threshold.

The resonator length must be balanced between two conflicting requirements. Long cells will allow a large beam size on the cell windows, which is necessary to prevent optical damage. The number of round-trip passes in the cell, however, varies inversely with the length of the cell, so the best resonator action is produced with cells whose length is much shorter than the equivalent pump laser pulse width. Alignment difficulties will also become troublesome at the longest lengths. For these reasons resonators should be limited to lengths less than about 50 cm. The type of resonator is also an important parameter.

In order to have a large spot size on the windows and a small waist at the center of the cell, a nearly concentric arrangement is required. However, as the resonator is approaching the true concentric condition, the beam waist becomes very small and the laser intensity increases dramatically. Higher-order nonlinear effects are then produced which limit the conversion efficiency, particularly at higher energies.

For high-energy applications an oscillator-amplifier arrangement, as shown in Fig. 10.42c, may be considered. A portion of the pump beam bypasses the resonator, and, along with the Raman shifted resonator output, enters the amplifier cell. The two input beams entering the amplifier must be propagating precisely parallel or antiparallel. The low-energy oscillator can be designed to produce only the first Stokes wavelength, and discriminate against all others. The amplifier has to be designed for a gain low enough that self-oscillation is prevented, yet it must saturate at a sufficiently strong Stokes signal.

The use of a glass capillary to confine the pump beam to a long interaction length, as sketched in Fig. 10.42d, has been exploited in the waveguide Raman laser. Almost complete conversion may be obtained due to the long interaction length, even at relatively low pump intensities. Because of the physical nature of the waveguide, the intensity at the inner walls of the capillary is near zero so there is little danger of damage. Similarly the input and output coupling optics can be in low intensity regions. One has to realize, however, that the long interaction length possible in a waveguide makes it a high-gain device not just for the first Stokes component. At input intensities large enough to reach threshold easily in the pump pulse, generated intensities may be large enough to produce second Stokes components, as well as other nonlinear effects.

10.3.3 Examples of Raman-Shifted Lasers

Table 10.6 summarizes the wavelengths and energies available from a Raman laser which consists of a gas cell filled with hydrogen and receives an input of 85 mJ in a 5 ns pulse at 560 nm, generated from a Nd:YAG pumped dye laser [10.164]. The maximum energy is obtained at the first Stokes wavelength which occurs at 730 nm and then falls off rapidly for the additional Stokes and anti-Stokes lines.

Figure 10.43 displays the conversion efficiency of a Raman laser which is pumped by a frequency doubled Nd:YAG laser [10.159]. Plotted is the conversion efficiency to the first Stokes wavelength for three gases versus the normalized

Table 10.6. Wavelengths, energies and optimum gas pressures of a commercial Raman shifter. S_i is the ith Stokes wavelength; AS_i denotes ith anti-Stokes wavelength. The pump laser had 85 mJ of energy in a 5 ns pulse at 560 nm. The Raman medium was hydrogen

Wavelength [nm]	Energy [mJ]	Pressure [psi]
195 (AS_8)	0.0031	125
213 (AS_7)	0.0091	125
234 (AS_6)	0.024	110
259 (AS_5)	0.054	115
290 (AS_4)	0.10	145
330 (AS_3)	0.26	160
382 (AS_2)	0.78	190
454 (AS_1)	2.10	200
730 (S_1)	17.00	90
1048 (S_2)	6.20	300
1855 (S_3)	0.60	275

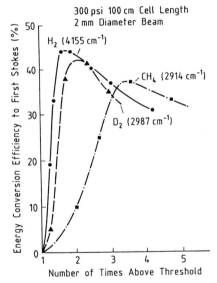

Fig. 10.43. Stimulated Raman conversion efficiency versus number of times above threshold for 0.532 μm pumping

pump intensity expressed in times above threshold. The experiments were carried out with a 2 mm diameter beam in a 1 meter long cell. The collimated pump beam at 532 nm was emitted from a frequency doubled Q-switched Nd:YAG laser. Hydrogen has the lowest threshold and also the highest conversion efficiency. Threshold was achieved at an input of 8 mJ with an 8 ns long pulse.

The design and performance of a state-of-the-art hydrogen gas Raman laser which converts the frequency doubled output of a Nd:YAG laser from 532 to 683 nm with an efficiency of up to 40% will be illustrated below [10.165]. The pump laser consisted of a Q-switched Nd:YAG oscillator, a single-pass amplifier

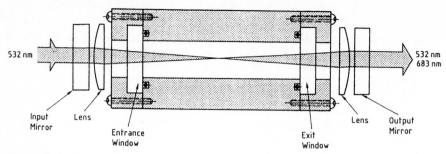

Fig. 10.44. Schematic diagram of a high-energy Raman cell

followed by a frequency doubler. The 532 nm pump beam emerging from the CD*A doubler had an output energy up to 210 mJ, a pulsewidth of 24 ns, a beam diameter of 4.9 mm and a beam divergence of 7 mrad. The repetition rate of the system was 2 Hz. The Raman laser consisted of the gas-pressure cell and the resonator optics, as shown in Fig. 10.44. The Raman cell is made from a block of aluminum bored with a 16 mm diameter hole, fitted with O-ring grooves and windows held together by end caps. The 25 mm diameter by 12.5 mm thick fused silica windows are antireflection coated at both the pump and the first Stokes wavelengths. The output of the Raman laser was measured as a function of input power with gas pressure, resonator configuration and length as parameters. Figure 10.44 depicts the concentric resonator configuration comprised of two flat mirrors and a pair of plano-convex lenses. The resonator length was 20 cm. The entrance mirror has high transmission at 532 nm and high reflection at 683 nm. The exit mirror has high transmission for the pump wavelength at 532 nm and 50% reflectance at the first Stokes wavelength. The reflectance for the second Stokes (954 nm) and first anti-Stokes (436 nm) wavelength was only a few percent at both mirrors.

Figure 10.45 shows the output for different gas pressures and input energies. Optical breakdown leads to a drop in energy at the highest energies and pressures. The misalignment sensitivity of the resonator axis to the pump axis is shown in Fig. 10.46. At low energies, the tilt is important because the gain is low and a good overlap of the beams is necessary. At the higher energies, however, tilt is not as important. It should be noted that the resonator itself was aligned and the misalignment refers only to the tilt with respect to the pump beam axis.

Figure 10.47 exhibits the output vs. input for the resonator at the exact length for the concentric geometry (solid line) and for slight variations in length. As is apparent from the data, the resonator length becomes more critical for the higher input energies, and a change of (1–2)% in resonator length has a noticeable effect. The output energy of 65 mJ at 683 nm for 185 mJ input at 532 nm was the highest energy achieved with the concentric resonator. Optical breakdown, probably caused by small particles and impurities in the gas set this upper limit.

By removing the lenses shown in Fig. 10.44, experiments were also performed with a 35 cm long flat/flat resonator. Hydrogen has enough gain to allow high

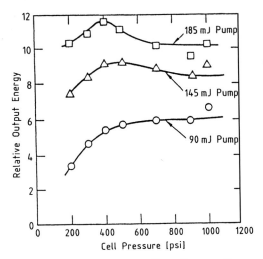

Fig. 10.45. Output energy for different Raman cell pressures and input energies

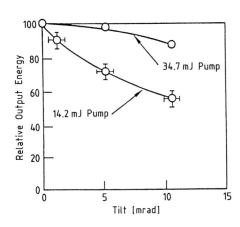

Fig. 10.46. Output energy from a tilted concentric resonator. Hydrogen pressure 1000 psi, output reflector 50 % at 683 nm, 20 cm resonator

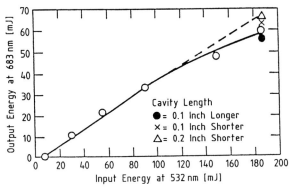

Fig. 10.47. Concentric resonator output energy for different pump input energies and cavity lengths. The solid line corresponds to the concentric position. (Hydrogen pressure: 1000 psi, output reflector 50 % at 683 nm, 20 cm resonator)

conversion efficiency without an internal focus in the Raman cell. Since there is no focus in the cell, the problem of optical breakdown is reduced.

For the flat/flat resonator configuration the Raman-laser output increased monotonically with gas pressure. The output vs. input energy at 1100 psi, the highest pressure applied, is plotted in Fig. 10.48. The average input intensity was $100\,MW/cm^2$ and the output reflectance at 683 nm was 50%. The threshold increased from 8 to 80 mJ compared to the concentric resonator, however, the total output at the highest input energy was the same, i.e. 65 mJ. The highest output energy of 78 mJ was achieved at 1700 psi gas pressure and an input energy of 195 mJ. It required also a change in output mirror, which for this case had a reflectivity of 27% at 683 nm and 86% at 532 nm.

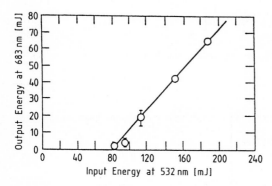

Fig. 10.48. Flat/flat resonator output energy for different input energies. 35 cm flat/flat resonator, output reflector 50% at 683 nm, hydrogen pressure 1100 psi

10.4 Optical Phase Conjugation

Optical phase conjugation, also referred to as wave-front reversal, has been demonstrated in solids, liquids and gaseous media using a number of nonlinear optical interactions, such as three-wave and four-wave mixing, stimulated Brillouin scattering (SBS), stimulated Raman scattering (SRS), and photon echoes.

From the standpoint of solid-state laser engineering, phase conjugation via stimulated Brillouin scattering is particularly important because it provides the simplest and most efficient interaction, and initial work in this area carried out by Zel'dovich et al. [10.2] revealed that optical aberrations produced by a laser amplifier stage could be corrected. The concept of reflecting an amplifier beam off an SBS cell and passing it back a second time through the amplifier was first used to compensate phase distortions in a ruby amplifier [10.166]. Following that initial work, SBS mirrors have been investigated in a number of solid-state lasers, using different nonlinear materials and optical arrangements [10.167–169]. Comprehensive descriptions of the principle of phase conjugation are contained in [10.170–173].

10.4.1 Basic Considerations

From a mathematical point of view, phase conjugation can be explained by considering an optical wave of frequency ω moving in the $(+z)$ direction,

$$E(x, y, z, t) = A(x, y) \exp\{j[kz + \phi(x, y)] - j\omega t\} \tag{10.74}$$

where E is the electric field of the wave with wavelength $\lambda = 2\pi/k$. The transverse beam profile is given by the function $A(x, y)$ and the phase factor $\phi(x, y)$, indicating how the wave deviates from a uniform, ideal plane wave. In particular, the phase factor carries all the information about how the wave is aberrated.

If the beam given by (10.74) is incident upon an ordinary mirror, it is reflected upon itself and the sign of z changes to $(-z)$, all other terms of the equation remain unchanged. However, if the beam is incident upon a phase conjugate mirror it will be reflected as a conjugated wave E_c given by

$$E_c(x, y, z, t) = A(x, y) \exp\{j[-kz - \phi(x, y)] - j\omega t\} \tag{10.75}$$

In addition to the sign change of z, the phase term has changed sign, too. The conjugate beam corresponds to a wave moving in the $(-z)$ direction, with the phase $\phi(x, y)$ reversed relative to the incident wave. We can think of the process as a reflection combined with phase or wave-front reversal. The phase reversal expressed by (10.75), for example, means that a diverging beam emitted from a point source, after reflection at a phase-conjugate mirror will be converging and be focused back to the point of origin.

A practical application of phase reversal in a laser system is depicted in Fig. 10.49. Shown is an oscillator which produces an output with a uniform flat wavefront that is distorted in the amplifier medium. An ordinary mirror merely inverts the distortion as it reflects the beam, thereby keeping the distortion fixed with respect to the propagation direction. With a second pass through the amplifier, the distortion is essentially doubled. A phase-conjugate mirror, on the other hand, reverses the wavefronts relative to the wave propagation direction; hence, the same region of the amplifier that originally created the distortion compensates for it during the second pass through the amplifier.

However, it should be noted that although the wavefront is conjugated, the polarization state of the backward-going field is not conjugated. As far as the polarization vector is concerned, the SBS mirror behaves like an ordinary mirror. As we shall see, this has important implications with regard to the compensation of optical distortions caused by thermally induced birefringence in the amplifier rods.

As already mentioned, stimulated Brillouin scattering of laser radiation in liquids or gases is the preferred approach for designing phase-conjugate mirrors. Stimulated Brillouin scattering involves the scattering of light by sound waves, i.e., acoustic phonons, or pressure/density waves. The incoming laser radiation generates an ultra-sound wave by electrostriction corresponding to a spatial density modulation in the material. This modulation changes the refractive index and therefore a refractive index grating is created in the material.

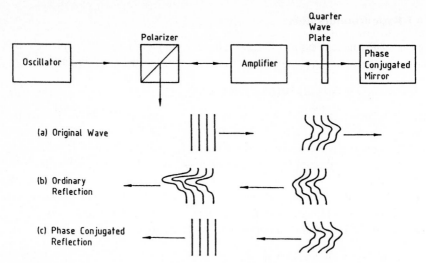

Fig. 10.49. Compensation for optical phase distortions caused by an amplifying medium using optical phase conjugation

   The resultant Stokes scattered light is down-shifted in frequency by a relatively small amount ($\approx 10^9$ to 10^{10} Hz). The gain of the Stokes-shifted wave is generally the highest in a direction opposite to that of the incident beam. The efficiency of the SBS process (defined as the ratio of the Stokes-shifted, backward-going energy or power, to the incident optical energy or power) can be as high as 90%. SBS has been studied extensively in liquids, solids, and gases. The steady-state gain factor of SBS is generally large, making SBS the dominant simulated scattering process in many substances. Phase-conjugate mirrors can be employed to compensate thermal lensing in double-pass amplifiers if the beam intensity is high enough to provide sufficient reflectivity from the SBS process.

   Reflectivity is an important parameter in designing a SBS mirror since it greatly influences the overall laser-system efficiency. The calculation of the reflectivity of SBS mirrors requires the solution of coupled differential equations for the incident and reflected light waves and the sound wave grating. The calculations of the scattering processes require a great numerical effort, assuming steady-state, or plane wave approximations solutions have been derived in [10.174–176].

   The reflectivities of SBS mirrors depend on the excitation condition and the material properties of the nonlinear medium. Reflectivity vs. pump power or energy can be described by a threshold condition and a nonlinear power relationship. Threshold depends mainly on materials properties and cell geometry, whereas the increase of reflectivity is a function of the pump-beam power. Both threshold and the slope of reflectivity depend on the coherence properties of the beam. As will be explained later, there is a minimum requirement on the coherence or linewidth of the incident beam such that a coherent spatial and temporal interaction with the sound-wave grating is possible.

Design parameters which affect the performance of a SBS mirror will be examined in the following section.

10.4.2 Material Properties

Pressurized gases such as CH_4 and SF_6, or liquids such as CCl_4, acetone and CS_2 are usually employed to provide efficient SBS for a $1.06 \mu m$ pump. SBS materials can be characterized by a gain coefficient and an acoustic-decay time. The gain coefficient g determines threshold and slope of the reflectivity curve. Materials with a lower gain show a higher threshold and a slower increase of the reflectivity with increased pumping.

The acoustic-decay time τ establishes a criteria for the coherence or linewidth of the pump beam. For efficient interaction with the sound-wave grating, a coherent interaction on the order of the acoustic decay time is required. An inverse bandwidth of the pump pulse which is larger than the acoustic-phonon damping time describes the steady-state condition, i.e., $(\Delta \nu)^{-1} > \tau_B$. Maximum gain is achieved if this condition is satisfied. Most calculations, as well as the simple relationships for gain and threshold, given below, are for this case.

Table 10.7 lists steady-state SBS gain coefficient g and acoustic-decay time τ of representative gases and fluids.

Quite noticeable is the longer phonon lifetime in gases compared to liquids. Single longitudinal mode Q-switched lasers have a pulse length of 10 to 20 ns, therefore the coherence requirement for steady-state conditions, stated above, will not be met in gases. SBS in gases will be highly transient, while in liquids, it will reach a steady state. This has the important implication that the effective gain for gases is actually considerably lower than the values in Table 10.7. The effective gain in a transient SBS process can be as much as 5 to 10 times less than the value given in the table. The lower SBS threshold and higher gain, is the reason, while most current systems employ liquids as the nonlinear medium. Also, liquid cells are not as long as gas filled cells. Liquid cells are typically 10–15 cm long, whereas gas cells are between 30–100 cm in length.

Gases employed as nonlinear elements have the advantage that they eliminate problems which can arise in liquids due to optical breakdown caused by bubbles or suspended particles.

Table 10.7. Steady-state SBS gain coefficients and acoustic decay time of representative gases and fluids, taken from [10.167, 169, 177]

Medium	g [cm/GW]	τ [ns]
CH_4 (30 atm)	8	6
CH_4 (100 atm)	65	17
SF_6 (20 bar)	14	17
SF_6 (22 atm)	35	24
CCL_4	6	0.6
Acetone	20	2.1
CS_2	130	5.2

The gain of a stimulated Brillouin scattering process is a function of the frequency ν of the pump radiation, the phonon lifetime τ_B, the electrostrictive coefficient γ, the velocity of light and sound c and v, and the refractive index n and density ρ_0 of the material [10.176].

$$g_B = \frac{\nu^2 \gamma^2 \tau_B}{c^2 v n \rho_0} \; . \tag{10.76}$$

The threshold pump power for SBS is a parameter of practical importance. If one measures reflectivity vs pump power input, one observes that at a particular input power, reflectivity changes over many orders of magnitude for just a few percent increase in power. In this regime of rapid signal growth, a threshold is usually defined if reflectivity exceeds 2% [10.175] or if the gain has reached exp(30) [10.169, 178].

The gain of the backward reflected Stokes wave over a length l of the nonlinear medium is $\exp(g_B I_P l)$ for a plane wave of pump intensity I_P. If the threshold P_{th} is defined as the condition when the gain reaches exp(30), then one obtains [10.169]

$$P_{th} = 30 A_p / g_B l \tag{10.77}$$

where A_p is the area of the pump beam.

For a Gaussian pump beam, one obtains [10.179]

$$P_{p,th} = \frac{\lambda_p}{4 g_B} \left\{ 1 + \left[1 + 30 / \arctan \left(L / b_p \right) \right]^{1/2} \right\}^2 \tag{10.78}$$

where b_p is the confocal parameter defined in (5.8), and L is the length of the SBS cell. If the cell is long compared to the confocal parameter $L \gg b_b$, the lowest threshold is obtained, since $\arctan(L/b) = \pi/2$ and (10.78) reduces to

$$P_{th} = 7.5 \frac{\lambda_p}{g_B} \; . \tag{10.79}$$

10.4.3 Focusing Geometry

A SBS mirror in its most common form consists of a lens which focuses the incident beam into the bulk nonlinear material. The first buildup of the sound-wave amplitude is concentrated in the focal region of the incident beam. However, the sound wave can have a wide distribution over large parts of the cell.

As confirmed in experiments [10.176], in a first approximation, a change in focal length should not affect threshold or reflectivity, since a shorter focal length, which leads to a shorter interaction length l in (10.77) is compensated by stronger focusing leading to a smaller area A_p of the pump beam. Choosing short focal lengths will lead to a very compact sound-wave distribution with the advantage of a lower demand on the coherence of the laser radiation. The main disadvantage of very short focal length is the high intensity and therefore the occurrence of optical breakdown.

Depending on the focal length of the lens, the sound wave may be sharply concentrated in the cell or distributed over the whole cell length.

There is also a difference whether the lens is the front window of the cell, and the nonlinear medium fills the space between the lens and the beam waist, or if the lens is positioned outside the cell and only the area around the beam waist is covered by the medium. The choice of a focusing lens, for an incident laser beam of given pulse energy and duration, has to be such that at the focal region the intensity is well above threshold but below optical breakdown.

10.4.4 Pump-Beam Properties

The pump power has to be well above the threshold of the nonlinear reflectivity curve. Otherwise, the fidelity of the reflected signal becomes small, because the parts of the beam away from the central peak, with energies below this threshold value are not reflected. The useful power range of the incident radiation is limited by the SBS threshold at the lower limit and by optical breakdown at the upper end.

A typical curve for SBS reflectivity vs input energy is shown in Fig. 10.50. These experimental data were obtained by focusing a Q-switched Nd:YAG laser into a CCl_4 cell with an 80 mm long focal length lens [10.177]. The general shape of the increase in reflectivity with intensity can be approximated by a simple expression [10.174]

$$R = \frac{C(E_{in}/E_{th})}{1 + C(E_{in}/E_{th})} \quad \text{for} \quad E_{in} \gg E_{th} \tag{10.80}$$

where E_{in}, E_{th} is the input energy of the incident beam and threshold value, respectively, and C is a constant which depends on the beam and nonlinear material properties.

In order to obtain a low SBS threshold and high reflectivity, an efficient coherent wave interaction is necessary between the pump beam and the sound-wave grating. There is a spatial and temporal requirement on the coherence of the incident beam. First, the coherence length of the beam should exceed the longitudinal sound-wave extension. If the pump beam has a Gaussian spatial profile, the sound wave has a longitudinal extent approximately given by the confocal parameter. Second, the coherence time, (i.e., the inverse of the linewidth, $1/\Delta\nu$) of the pump beam should exceed the response time τ_B of the acoustic phonons. Typically, τ_B is on the order of nanoseconds for liquids, and tens of nanoseconds for pressurized gases. Therefore, high-efficiency stimulated Brillouin scattering requires linewidth narrowing or preferably single-longitudinal-mode operation. If the laser operates in a single longitudinal mode, this means that the pulse length should be longer than τ_B.

If these conditions are not met, the steady-state SBS process changes to transient stimulated Brillouin scattering with a correspondingly lower gain parameter. The effect of multimode lasers on the SBS threshold has been discussed

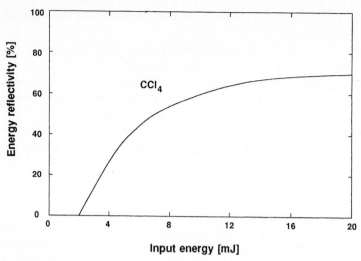

Fig. 10.50. SBS energy reflectivity of CCL_4 as a function of pump energy for a Q-switched Nd:YAG laser [10.177]. Curve represents a best fit of experimental data

in [10.180]. Experiments which address the issue of multimode operation will be discussed next.

Carr and *Hanna* [10.169] investigated SBS with a Q-switched single transverse and longitudinal mode Nd-YAG laser which had a pulse duration of 30 ns and a coherence length of 6 meters. The one-meter long cell filled with CH_4 at 30 atm results in a 6 ns acoustic decay time and a gain coefficient of 8 cm/GW. The beam was focused to a waist of $w_T = 150\,\mu$m, from which follows with $\lambda = 1.06\,\mu$m, a confocal parameter of $b = 13$ cm, see (5.8). Introducing these parameters into (10.77) yields a threshold for SBS of $P_{th} = 100$ kW. Even for a ratio of pulse width to phonon decay time of a factor of five ($t_p/\tau_p = 5$), a significant degradation from steady-state conditions can be expected. The theory predicts a threshold 2.5 times greater than for steady-state conditions or 250 kW in this particular case [10.181]. The actual observed threshold was 400 kW. Since the coherence length l_{coh} was 6 meters and the confocal parameter b was 13 cm, the condition $l_{coh} \gg b$ is clearly satisfied.

The coherence length is reduced for multimode operation according to $l_{coh} = 2L/n$, where L is the length of the resonator, and n the number of longitudinal modes. For the 1-meter long resonator, the condition $l_{coh} \gg b$ is satisfied even in the presence of several modes. The experimental results confirmed that the SBS threshold was little affected if the laser was operated in several modes. However, when the oscillator was operated on 15 modes ($l_{coh} = 15$ cm or a bandwidth of ≈ 2 GHz) threshold increased by 75%. Phase conjugation still occurred, but was less reliable and resulted in poor beam reconstruction.

The dependence of SBS reflectivity on coherence length and input energy is illustrated in Fig. 10.51. In this experiment, radiation from a Nd:YAG laser

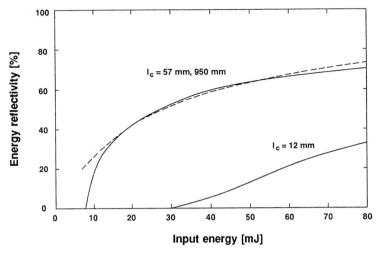

Fig. 10.51. SBS energy reflectivity in SF_6 for a Q-switched Nd : YAG laser with different coherence lengths l_c. Dashed line is an approximation according to (10.80) with $C = 0.25$

was focused with a lens of 100 mm focal length into a cell containing SF_6 at a pressure of 20 bar [10.177].

The lens was positioned about 40 mm in front of the cell entrance window. In Fig. 10.51, measured energy reflectivities with different coherent laser pulses are shown. As can be seen from this figure, the energy reflectivity is different for the shortest coherence length of 12 mm compared to the two others. No difference was detectable between the pulses with 57 mm and 950 mm coherence length. In both cases, the coherence length was longer than the whole interaction length between the entrance window of the cell and the focus of the beams inside. The measured data agreed quite well with predicted values calculated from the theory described in [10.176].

10.4.5 System Design

In Fig. 7.10 we have already exhibited the most common configuration of compensating thermally induced lensing in amplifiers with a SBS mirror. The output from the oscillator passes through the amplifiers, is reflected by the SBS mirror, and propagates a second time through the amplifier chain. Output coupling is realized by a polarizer and a $90°$ polarization rotation carried out by a $\lambda/4$ wave plate which is passed twice. Conjugate reflectivities in this type of configuration are typically 50% to 80%, so that the efficiency of the amplifier stage is reduced by a fraction which depends on the ratio of energy extraction between the first and second pass in the amplifiers.

As already pointed out, phase conjugation will invert the wave front and correct thermal lensing introduced by the amplifier chain; however, it will not correct for depolarization caused by birefringence. Depolarization causes distortion of

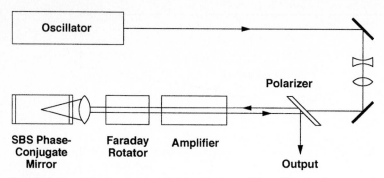

Fig. 10.52. Depolarization compensation in a single amplifier with a Faraday rotator in a double pass configuration using a SBS mirror [10.169]

the output beam profile, leads to polarization losses and the wrongly polarized light is back reflected towards the oscillator.

Several approaches have been explored to deal with this problem. One approach is to minimize the effect of birefringence as one would in any system independent of the use of a SBS mirror. This includes rotation of the birefringence axis in single- and double-stage amplifiers and the use of a naturally birefringent laser crystal. The second approach is unique to SBS and involves separating the depolarized beam into its two components of orthogonal polarization and individual conjugation of each polarization state. These approaches will be illustrated in more detail in the following section.

One approach of minimizing birefringence is utilized in the design exhibited in Fig. 7.10 in which a 90° rotator is placed between two identical amplifier stages. A ray passing through the amplifier has its electric field component along the fast birefringence axis rotated by 90° so that it lies along the slow axis in the second amplifier, and vice versa. Thus, regardless of the orientation of the fast and slow axes, the birefringence in one amplifier is canceled in the second. A commercially-available Nd : YAG oscillator-amplifier system with a SBS mirror uses this architecture.

If the laser system employs only a single amplifier, a Faraday rotator with 45° rotation placed between the amplifier and mirror is equivalent to a 90° rotation between two identical amplifier stages. With the arrangement shown in Fig. 10.52, and using methane at 30 atm, nearly 300 mJ of diffraction limited output was obtained in a Nd : YAG amplifier operating at 15 Hz [10.169].

Thermally-induced birefringence effects can be greatly reduced if an anisotropic crystal such as Nd : YAlO or Nd : YLF is used as laser rod. These crystals have a poorer thermal conductivity compared to Nd : YAG, and therefore exhibit higher thermal lensing. But the latter can be compensated with a phase conjugate mirror. Using this rationale, *Eichler* et al. [10.182] constructed a Nd : YAlO oscillator-amplifier system which produces 100 W average power in a 1.2 diffraction limited beam. The amplifier had negligible depolarization losses but a strong thermal lens of less than 15 cm at the maximum lamp input of 8 kW. The ex-

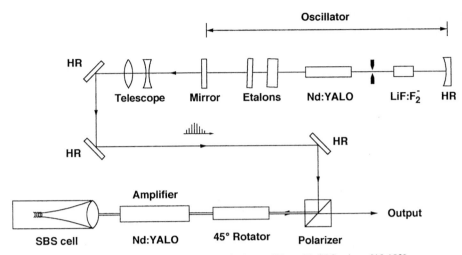

Fig. 10.53. Nd: YA1O oscillator and double pass single amplifier with SBS mirror [10.182]

perimental set-up is depicted in Fig. 10.53. The oscillator, which was passively Q-switched with a LiF:F_2 color center crystal, contained two intra-cavity etalons with thicknesses of 2 and 20 mm, which resulted in a coherence length of 2.5 meters. The output from the amplifier was focused with a 200 mm lens into the SBS cell, which contained CS_2 as the nonlinear material.

In the approaches described so far, birefringence effects are minimized before the beam enters the SBS cell. There is a limit to the compensation capabilities of these schemes. Compensation will be acceptable if bifocusing effects are not too strong, otherwise the second-pass beam will not return exactly along the path of the first beam and the fidelity of beam restoration will suffer [10.183]. The best compensation can be achieved if the polarization is uniform and linear at the SBS cell.

An approach investigated by *Basov* et al. [10.184] is to split the beam into two orthogonal polarizations after passing through the amplifier. One of the beams is then rotated 90° so it has the same polarization as the other. The two beams will cooperatively reflect from the SBS phase-conjugate mirror. The back-reflected phase-conjugated beams then pass back through their respective optical paths and are recombined at the beam splitter before propagating through the amplifier for a second time. The beam emerging from the amplifier has the same beam quality and polarization as the original incident beam. A more recent implementation of this general approach was described in [10.185].

From a design point, the question that needs to be answered is whether a phase-conjugation system is necessary, or whether insertion of internal elements such as lenses, phase plates or 90° rotators, is sufficient to remove troublesome distortions associated with the amplifier stages. Other techniques which reduce beam aberrations, are diode pumping and a zig-zag slab design. The former reduces the heat load on the laser crystal, and the latter compensates to a first-order beam distortion by virtue of its geometry. Of course, these techniques are not exclusive of each other and can be combined in one system. It should be mentioned that besides compensating for thermal lensing, a phase-conjugating mirror has some additional interesting features. Misalignment of a conventional rear mirror caused by mechanical, thermal or vibration effects which will lead to beam wander and power loss is eliminated because the beam reflected from the phase-conjugated mirror is self-aligned. It greatly depends on the particular application and average-power requirements of the system, when it is justified to remove the troublesome aberrations associated with the amplifier stages by means of a phase-conjugate mirror. Also, a simple SBS mirror arrangement does not remove aberrations and losses due to birefringence, and the schemes for correction of stress birefringence by resolving the depolarized beam into components for separate conjugation are too complicated and require a large number of additional optical components.

Conceptually, the replacement of a conventional mirror at the end of an amplifier chain with a SBS cell and a lens appears straight forward. However, a number of issues have to be resolved in a practical implementation of this technique. In order to achieve an acceptable overall system efficiency, the oscillator requires longitudinal mode selection, and the cell has to be operated at high beam intensities without causing optical breakdown. The latter is always a concern because impurities in the liquid could cause breakdown at the GW/cm^2 power levels, which are required to achieve a high SBS reflectivity.

Oscillators with SBS Q-Switching Mirrors: By far the most common application of SBS mirrors is in double-pass amplifier configurations. However, SBS mirrors have been also employed in oscillators. Since the amplified spontaneous emission is too small to start the SBS process, a conventional mirror is placed behind the SBS cell to start lasing action. As soon as the reflectivity of the SBS mirror is high enough, it will take over and form the main resonator. Since the losses decrease (i.e., reflectivity increases) as the flux in the resonator builds up, Q-switching occurs. Oscillators employing SBS Q-switching mirrors have been described in [10.186, 187].

11. Damage of Optical Elements

Optically induced damage to the laser medium and to components of the laser system generally determine the limit of useful performance of solid-state lasers. The performance of a laser system in terms of pump efficiency, gain, nonlinear conversion, brightness, and beam quality increases with power density of the beam. Power density is fairly independent of the size of the laser, output power scales mostly with beam diameter rather than beam intensity. As a consequence, it is desirable to operate all lasers at the highest beam intensity consistent with reliable operation which, in turn, is determined by the optical-damage threshold of the various components. Therefore, an understanding of the mechanisms which cause radiation damage to optical components, and a knowledge of the damage threshold of the materials employed in a laser, are of great importance to the engineer who is designing a laser system.

Damage may occur either internally, or at the surface of an optical component due to a number of intrinsic and extrinsic factors. Intrinsic processes that limit the optical strength of materials include linear absorption, color-center formation and a variety of nonlinear processes such as self-focusing, multiphoton absorption and electron avalanche breakdown. Extrinsic factors include impurities, and material defects (voids, dislocations, etc.), and in particular surface scratches and digs, and surface contamination. In general, damage threshold is determined by surface finish and optical coatings, rather than by the bulk material. The problem of surface damage is compounded because most laser systems contain many surfaces.

There are different ways of defining damage threshold. One might define damage as the physical appearance of a defect in the material or by a degradation in the output performance of the laser system. From the standpoint of the user of the laser system, the performance deterioration is of more importance than the physical appearance of a defect in the material. Often the occurrence of a small blemish or defect does not alter system performance. In this case it is important to determine whether the defect remains constant or increases with time.

Quantitative measurements of the damage threshold of an optical component should be performed following a procedure specified by the International Standards Organization [11.1–3]. At test sample is irradiated with a well characterized laser at different sites. The fluence levels are chosen so that at the higher fluence a high probability of damage exists, whereas at the lower levels the probability of damage is low. The percentage of failures is plotted vs. fluence. A least square linear fit to this data is calculated.

The linear data extrapolation to 0% damage frequency is defined as the damage threshold. The testing sites are inspected for damage with a Nomarski microscope at a magnification of 150. Figure 11.1 shows a damage frequency plot according to ISO 11254.

The results of experimental and theoretical investigations of laser-induced damage in dielectric and sol-gel coatings, optical glasses, metal films, laser crystals and glasses, crystals used in polarizers, Q-switches, harmonic generators and parametric oscillators, are presented annually at the Boulder Damage Symposium held since 1969. The earlier proceedings of these symposia have been published by the National Bureau of Standards [11.4] and since 1990 by the International Society for Optical Engineering [11.5]. Selected papers from the earlier symposia have also been published in a book [11.6].

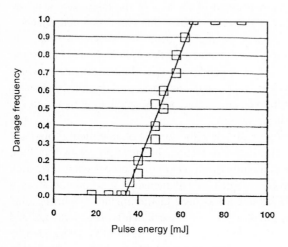

Fig. 11.1. Damage frequency plot for defining 0% damage probability according to ISO 11254

11.1 Surface Damage

Undoubtedly the weakest parts in a well-designed laser are the surfaces of the optical components. It has been found that surfaces, even those that are scrupulously cleaned, have a lower damage threshold than the bulk.

One reason for the lower surface-damage threshold compared to bulk damage stems from the fact that even after the polishing process used in the finishing of optical elements, there are residual scratches, defects, and imperfections on the surface. At these points, the electric field of the light wave is greatly intensified so that the effective field value just inside the surface is much greater than the average field. When breakdown or failure of the surface occurs, it occurs near one of these imperfections [11.7]. Furthermore, as a consequence of the polishing process, minute inclusions of polishing material or other impurities are embedded in the surface. These impurities may be strongly absorbing and, if so, can provide a nucleus around which damage may occur. Other factors tending to lower the damage threshold at the surface are contamination of the surface due to airborne particles, fingerprints, and residue from outgasing of nearby components.

From the foregoing it follows that surface damage is caused either by absorption of submicrometer inclusions or by the formation of a plasma at the surface because of electron avalanche breakdown in the dielectric. In the former case, microexplosions of isolated inclusions occur just under the surface [11.8]. In the latter case, the electric field produced by an intense laser beam is sufficiently high that the dielectric material will break down in the optical field [11.9].

This avalanche ionization represents the ultimate failure mode of dielectric materials. In this process, a few free charge carriers from easily ionized impurities, or as a result of multi-photon ionization, absorb energy from the oscillating electric field of the light wave. Hence, they gain kinetic energy and multiply by impact ionization of the solid-state lattice. As a result of avalanching the concentration of electrons increases significantly, which leads eventually to a complete absorption of radiation. The avalanche breakdown is usually associated with a luminous plasma formed in the material. The avalanche process is the high-frequency analogue of dc breakdown. Avalanche ionization is the primary cause of surface damage, once absorbing impurities are removed. Due to field intensification effects, avalanche ionization is more likely to occur in the vicinity of pits and scratches at the surface.

The reduction of damage threshold at imperfect surfaces due to multiple reflections has been discussed theoretically by Bloembergen [11.7]. The presence of surface pits, grooves, cracks, nodules or voids can increase the local field intensity by a factor between n and n^4 depending on the defect geometry, where n is the refractive index. This explains why in some cases a dielectric coating applied to a surface raises the damage threshold. Micropits and scratches are filled in by the coating material which then reduces the reflectivity and standing wave ratio in association with smoothing of the surface. This effect has been observed in such cases as MgF_2 on glass, ThF_4 on $LiNbO_3$, and sol-gel AR coatings on fused silica.

Super-polishing techniques, such as ion polishing, and chemical etching, lead to higher damage thresholds than those for conventionally polished optical surfaces. Superpolish produces surfaces with a rms roughness of less than 1 Å whereas a normal high quality polish has a surface roughness of (10–20) Å.

Surface damage typically takes the form of a pit, and is observed more frequently at the exit surface rather than the entrance surface of an optical component. Several models have been proposed to explain these observations.

In one model, pit formation and the difference of damage between front and back surface has been explained to be the result of reflections occurring at the various boundaries [11.10, 11]. The formation of pits in glass is assumed to be the result of a standing wave which is formed inside the glass near the surface. The standing wave is caused by reflections from the plasma-glass boundary. After plasma initiation there is a layer of plasma several wavelengths thick contiguous to the surface. This plasma is densest at its source, that is, next to the surface. The electric field at the antinodes of this standing wave can become twice as large as the electric field in the incident beam. The first antinode is within one-quarter wavelength of the surface. It is this large electric field at the antinode that initiates pitting at, or just inside, the surface.

The fact that a sample of transparent dielectric material exposed to a collimated beam of light will usually damage at the exit surface at a lower power level than at the entrance surface can be explained by the effect of Fresnel reflection. When a light pulse enters an isotropic optical material at near-normal incidence, there is a reflection at the air-to-sample boundary. Since the index of refraction n of the material is higher than that of air, the reflected light wave suffers a phase shift of 180° with respect to the incident light wave. This phase shift results in a partially destructive interference of the two light waves within a distance of $\lambda/4$ of the entrance surface. As a result the light intensity I_{en} at the entrance surface is related to the incident intensity I_0 by

$$I_{en} = \frac{4}{(n+1)^2} I_0 . \tag{11.1}$$

At the exit surface, the reflected wave suffers no phase shift, so that the intensity at the exit surface I_{ex} is related to the intensity incident on the sample by

$$I_{ex} = \left(\frac{4n}{(n+1)^2} \right)^2 I_0 . \tag{11.2}$$

Thus the ratio of the intensity inside the medium at the exit surface to the intensity inside the medium at the entrance surface is

$$\frac{I_{ex}}{I_{en}} = \left(\frac{2n}{n+1} \right)^2 . \tag{11.3}$$

For $n = 1.55$, this ratio is 1.48.

A model which explains another difference between contaminated entrance and exit surfaces was put forward in [11.12]. However, the same conclusion is reached, namely, that particles on the front surface cause less damage than on the back surface. On the exit surface, a plasma is formed and confined between two solid materials i.e. the substrate and the particle. The high density of the plasma and associated pressure causes the particle to be ejected rapidly on heating. The reaction force leads to a small pit on the surface which then leads to damage. On the front surface, the plasma can expand and only a weak shockwave is launched into the substrate, and the surface is not pitted or damaged. Small particles may be completely vaporized in the process.

Consistent with the particle ejection model is the observation that internal reflective surfaces of Porro prisms and corner cubes have generally a higher damage threshold if the reflective surfaces are coated with SiO_2 coatings of about $2\,\mu$m thickness. The coating prevents the field of the evanescent wave from reaching the particles and therefore prevents the initiation of reaction pits.

11.2 Inclusion Damage

Optical components tend to exhibit minute inhomogeneities, which generally form in the process of fabricating the material. The inhomogeneities can consist of small bubbles, dielectric inclusions, or platinum particles in the case of laser glass. In crystalline materials, although some microinhomogeneities do occur during the growth process, there is less evidence that particulate damage is the limiting factor. In glass, however, during the course of melting, crucible material, finds its way into the melt and upon cooling condenses out in the form of particulate inclusions. This has been a particularly difficult problem in the fabrication of laser glass, which is manufactured in platinum containers. The problem of platinum inclusions has been solved by glass manufacturers either by using crucibles of ceramic and clay materials or by carefully controlling the oxygen content in the melt. If the material contains minute inhomogeneities such as particulate inclusions, especially metallic inclusions, they can absorb the laser radiation, giving rise to local heating and melting of the surrounding laser host material. This in turn causes stress concentrations sufficient to rupture the material. Elaborate theoretical treatments of the problem of stress formation in the vicinity of an absorbing particle have been published [11.13].

To summarize the considerations that apply, let us consider an absorbing particle embedded in laser material, illuminated by a brief and intense pulse of light. If the particle is a sphere of diameter d, it presents an absorbing area to the incident light of $\pi d^2/4$. The total energy absorbed in the particle is approximately

$$\Delta E = It\left(\frac{\pi d^2}{4}\right)(1 - e^{-\alpha d}) \tag{11.4}$$

where I is the power density, t is the pulse duration, and α is the material absorption coefficient. The maximum temperature rise experienced by the particle will be given by

$$\Delta T = \frac{\Delta E}{C_p(\pi d^3/6)\gamma}, \tag{11.5}$$

neglecting heat conduction during the pulse. In the above equation C_p is the specific heat, and γ is the density of the inclusion.

Equation (11.5) indicates that the temperature of metal particles subject to a 20-J/cm^2, 30-ns laser pulse can exceed 10 000 K for typical particle sizes. These high temperatures produce stress in the glass adjacent to the particles which can exceed the theoretical strength of glass and result in failure.

11.3 Self-focusing

Self-focusing is a consequence of the dependence of the refractive index in dielectric materials on the light intensity.

As we discussed in Sect. 4.3.1, an intense laser beam propagating in a transparent medium induces an increase in the index of refraction by an amount proportional to the beam intensity. If we express the nonlinear part of the refractive index given in (4.50) by the beam intensity I we can write

$$\Delta n = n_2 I \ . \tag{11.6}$$

For optical glass typical values are $n_2 = 4 \times 10^{-7} \, \mathrm{cm^2/GW}$. Thus at a power density of 2.5 GW/cm^2 the fractional index change is 1 ppm. Although a change of this magnitude seems small, it can dramatically affect beam quality and laser performance.

For numerical calculations the following conversion factors between esu and mks units are helpful:

$$n_2 \, [\mathrm{cm^2/W}] = 4.19 \times 10^{-3} (n_2/n_0) \quad [\mathrm{esu}]$$

Since laser beams tend to be more intense at the center than at the edge, the beam is slowed at the center with respect to the edge and consequently converges. If the path through the medium is sufficiently long, the beam will be focused to a small filament, and the medium will usually break down via avalanche ionization.

This convergence is limited only by the dielectric break-down of the medium. Due to this intensity-induced index change, small beams (5 mm or so) tend to collapse as a whole (whole-beam self-focusing), forming a damage track in the material when the intensity exceeds the damage level. Larger beams tend to break up into a large number of filaments, each of which causes damage (small-scale self-focusing) [11.14–16].

A variety of physical mechanisms result in intensity-dependent contributions to the refractive index, such as electronic polarization, electrostriction and thermal effects. The latter two mechanisms are responsible for stimulated Brillouin and Rayleigh scattering which form the basis for optical phase conjugation treated in Chap. 10. For short-pulse operation in solids, electronic polarization, also called the optical Kerr effect, which is caused by the nonlinear distortion of the electron orbits around the average position of the nuclei, usually dominates. The optical Kerr effect is utilized in Kerr lens mode-locking discussed in Chap. 9. Here we treat the negative aspects of this physical process as far as the design of lasers is concerned.

11.3.1 Whole-Beam Self-focusing

Self-focusing of laser radiation occurs when the focusing effect due to the intensity-dependent refractive index exceeds the spreading of the beam by diffraction. For a given value of n_2, two parameters characterize the tendency of a medium to exhibit self-focusing. The first, the so-called critical power of self-focusing, is the power level that will lead to self-focusing that just compensates for diffraction spreading. The second is the focusing length which represents the distance at which an initially collimated beam will be brought to a catastrophic self-focus within the medium for power levels in excess of the critical power.

A first-order description of the phenomenon can be obtained by considering a circular beam of constant intensity entering a medium having an index nonlinearity Δn proportional to power density. We define the critical power P_c to be that for which the angle for total internal reflection at the boundary equals the far-field diffraction angle. The critical angle for total internal reflection is

$$\cos \Theta_c = \frac{1}{1 + (\Delta n / n_0)} . \tag{11.7}$$

If we expand the cosine for small angles and the right-hand side of (11.7) for small $\Delta n / n$, we obtain

$$\Theta_c^2 = 2\Delta n / n_0 . \tag{11.8}$$

The beam expands due to diffraction with a half cone angle according to

$$\Theta_D \approx 1.22 \frac{\lambda_0}{n_0 D} , \tag{11.9}$$

where D is the beam diameter and λ_0 is the wavelength in vacuum. For $\Theta_c = \Theta_D$, and with

$$P = I(\pi/4)D^2 \tag{11.10}$$

one obtains from (11.6, 8, 9), the critical power

$$P_{cr}' = \frac{\pi (1.22)^2 \lambda_0^2}{8 n_0 n_2} . \tag{11.11}$$

Theoretical analysis of self-focusing based either upon numerical solutions of the nonlinear wave equation, ray tracing, or quasi-optic approximation [11.14–19] lead to formulas describing the behavior of the self-focusing Gaussian beam. For Gaussian beams the equation for P_{cr} has the same functional form as (11.11) and differs by just a numerical factor [11.20]

$$P_{cr} = a \frac{\lambda_0^2}{8\pi n_0 n_2} \tag{11.12}$$

where n_2 is in cm^2/W, and the value for a represents a correction factor which takes into account the fact the assumption of an initial Gaussian beam profile that remains Gaussin during the self-focusing process tends to overestimate the nonlinear effect. In most analyses, a value of $a = 4$ is assumed [11.2, 22].

With the assumption that the shape of the amplitude profile is unchanged under self-focusing, one finds the following relation between power and self-focusing length: For a Gaussian profile whose spot size $w(z)$ changes by a scale factor $f(z) = w(z)/w_0$, the width of the beam, varies with z according to [11.14]

$$f^2(z) = 1 - \left(\frac{P}{P_{cr}} - 1\right)\left(\frac{\lambda z}{\pi w_0^2}\right)^2 \tag{11.13}$$

where w_0 is the beam waist radius at the entrance of the nonlinear material. For $P \ll P_{cr}$ this expression is identical to (5.5) and describes the divergence of the beam determined by diffraction. For a beam with $P = P_{cr}$ diffraction spreading is compensated by self-focusing and $f(z) = 1$ for all z; this is referred to as self-trapping of the beam. For larger powers $P > P_{cr}$ self-focusing overcomes diffraction and the beam is focused at a distance

$$z_f = \frac{\pi w_0^2}{\lambda}\left(\frac{P}{P_{cr}} - 1\right)^{-1/2} \approx w_0(2n_0n_2I_0)^{-1/2} \tag{11.14}$$

As can be seen, the self-focusing length is inversely dependent on the square root of the total laser power. Equation (11.14) may be used even when the constant phase surface passing through $z = 0$ is curved, corresponding to a converging or diverging incident beam. One need only replace z_f by

$$z_f(R) = (1/z_f - 1/R)^{-1}, \tag{11.15}$$

where R is the radius of curvature of the incident phase surface ($R < 0$ for converging beams) and $z_f(R)$ is the new position of the self-focus.

Since the power in a laser pulse changes in time, a related change occurs with the instantaneous self-focusing length z_f in (11.14). The temporal amplitude change of the incident pulse, causing moving foci, will lead to a string of heavily damaged spots along the track. For example, when a laser beam is weakly focused into a sample of dielectric material, the self-focus first appears at, or near, the geometrical focal point, then splits into two foci, one of which travels down stream and one up-stream. The up-stream focus reaches its minimum z_f value at the peak power of the pulse. It will dwell at this position for a while and this presumably allows massive damage to occur at the head of the track. As the beam intensity decreases, the focus will tend to move back down-stream. Therefore, the damage track of microscopic bubbles observed in a broad range of materials is the fossil record of the moving focus generated by dynamic self-focusing of the incident light. This picture of bulk damage has been given by several authors [11.14–16, 19]; it is in agreement with the observation of very small diameter damage tracks in transparent dielectrics when irradiated with short duration, intense laser pulses [11.19, 23, 24].

Of course, (11.14) is only valid when the length of the medium is greater than z_f so that catastrophic self-focusing can occur. If the medium is shorter, it acts as a nonlinear lens. If the medium is so short that the beam profile does not change significantly, the equivalent focal length of the lens is

$$f = \frac{w^2}{4n_2 I_0 L} \tag{11.16}$$

where I_0 is the peak intensity, and L is the length of the medium. This equation is the same as (9.12) which has been used in the section on Kerr lens mode locking.

11.3.2 Small-Scale Self-focusing

Actual laser beams do not have perfectly smooth envelopes as was assumed in treating whole-beam self-focusing. Experimentally it is found that beams containing amplitude irregularities on propagating in a nonlinear medium eventually break up into separate filaments [11.23, 25–29]. The physical description of the process is that interference between a strong beam and weak beams produces index variations due to the intensity-dependent refractive index and that the diffraction of the strong beam by means of the resulting phase grating spills energy into the weak beams.

A theory first suggested by *Bespalov* [11.30] treats self-focusing as an instability phenomenon wherein an initial perturbation (e.g., dust, refractive index inhomogeneities), no matter how small, grows to what are ultimately catastrophic proportions. Linear stability theory predicts that a small, one-dimensional intensity modulation of spatial wave number K, superimposed upon a uniform intensity background, will undergo approximately exponential growth with a gain coefficient g given by [11.31]

$$g = K \left(\frac{n_2 I}{n_0} - \frac{K^2 \lambda^2}{16\pi^2} \right)^{1/2} , \tag{11.17}$$

where λ is the wavelength in the medium, and n_0 is the linear refractive index. For a range of K from 0 to $(4\pi/\lambda)(n_2 I/n_0)^{1/2}$, g is real and the ripples grow exponentially. The maximum growth rate occurs at

$$K_{\mathrm{m}} = \left(\frac{2\pi}{\lambda} \right) \left(\frac{2n_2 I}{n_0} \right)^{1/2} \tag{11.18}$$

and has the value

$$g_{\mathrm{m}} = \left(\frac{2\pi}{\lambda} \right) \left(\frac{n_2 I}{n_0} \right) . \tag{11.19}$$

An initial perturbation $\delta = |\delta E_0|/E_0$ of the fastest growing instability mode leads to self-focusing at a distance given by [11.25]

$$z_{\mathrm{f}} = (g_{\mathrm{m}})^{-1} \ln \left(\frac{3}{\delta} \right) . \tag{11.20}$$

Since the self-focusing length of small-scale fluctuations is generally much shorter than that of large-scale intensity variations, the self-focusing of fluctuations determines the effective damage threshold of beams propagating in nonlinear dielectrics.

The validity of the linearized theory of small-scale self-focusing was verified by *Bliss* et al. [11.26] who used a shear plate interferometer to generate a perturbation of known amplitude and spatial frequency on an intense beam. It was found that the theory provided an accurate description of the modulation growth (Fig. 11.2).

The laser designer is primarily interested in the threshold for self-focusing, since it is an effect he seeks to avoid rather than to study. Therefore, particular interest is attached to the values of n_2 for subnanosecond pulses in laser media. Several workers have measured the effective value of n_2 in various materials for pulses from 20 ps to 250 ps duration. It was reported that the results obtained were not very sensitive to pulse duration. Table 11.1 contains a summary of published values of n_2 for several important optical materials [11.32, 33].

BK-7 and fused silica (SiO_2) are widely used for lenses and mirrors, the next four materials are Nd doped silicate and phosphate laser glasses, the host materials YAG and Al_2O_3, are used in a number of doped laser crystals, and the last group in Table 11.1 represent crystals employed in polarizers (calcite), harmonic generators and Q-switches.

Self-focusing is of primary concern in Nd : glass lasers. Compared to other solid-state lasers, in glass systems, beams of extremely high peak powers pass through long sections of dielectric materal. For a typical laser glass with $n_2 = 1.5 \times 10^{-13}$ esu and $\lambda_0 = 1.06$ μm one obtains from (11.12) a value of $P_{cr} = 711$ kW. Typical Q-switched Nd : glass oscillator-amplifier systems are capable

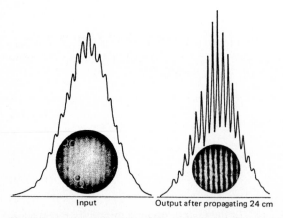

Input   Output after propagating 24 cm

Fig. 11.2. Growth of interference fringes in a 24-cm-long unpumped ED-2 glass. Fringes are spaced 2 mm apart on a pulse with a peak intensity of 5 GW/cm^2 [11.26]

Table 11.1. Linear and nonlinear refractive index for various optical materials

Material	Refractive index n	Nonlinear refractive index $n_2 \ [10^{-14} \ \mathrm{esu}]$	$n_2 \ [10^{-16} \ \mathrm{cm}^2/\mathrm{w}]$
BK-7	1.52	14.6	4.0
SiO_2	1.48	7.4	2.1
LG-680	1.56	16.0	4.3
LSG-91H	1.55	15.8	4.3
LG760	1.51	10.2	2.8
LHG-5	1.53	12.8	3.5
YAG	1.83	30	6.9
Al_2O_3	1.76	13	3.2
$CaCO_3$	1.59	11	2.9
KTP/KTA	1.77	100	23.7
BBO	1.6	11	2.9
$LiNbO_3$	2.2	48	9.1

of generating peak output powers around 1 GW at beam diameters of 20 mm. Assuming a Gaussian beam radius of $w_0 = 5$ mm and $\lambda = 1.06 \, \mu$m, we obtain $z_f = 420$ cm from (11.14) for the self-focusing length in the final amplifier rod. For a mode-locked system producing 1 J in a 20 ps pulse and $w_0 = 5$ mm, one obtains $z_f = 60$ cm. Large-aperture mode-locked Nd : glass laser systems employed for fusion experiments operate at up to 10^5–10^6 times the critical power in the beam. Clearly, in these lasers break-up of the laser beam into filamentary regions with consequent damage to laser components represents the major limitation in achieving higher powers.

For a $1.06 \, \mu$m laser beam at an intensity of 10 GW/cm² and with $n_0 = 1.5$ and $n_2 = 1.5 \times 10^{-13}$ esu, we obtain from (11.18, 19) a gain coefficient of $g_m = 0.16 \, \mathrm{cm}^{-1}$ for a disturbance which has a spatial wavelength of $\lambda_{rms} = 0.43$ mm. Assuming that the disturbance has a depth of modulation of 1 % ($\delta = 0.01$), then according to (11.20) the perturbation will focus at a distance of $z_f = 35$ cm.

From these considerations, it follows that spatial beam quality is of paramount importance in high-power systems. Small perturbations are always present in the beam profile, arising from diffraction around dirt and defects on surfaces, or from bubbles or inclusions in the media, surface irregularities, or the random fluctuations of refractive index within the optical media.

In large Nd : glass lasers built for fusion research several approaches are taken to reduce the growth of small-scale instabilities. First, by maintaining a high degree of cleanliness and uniformity on optical surfaces and within the bulk media, the basic level of the perturbation is kept to a minimum. Secondly, by inserting spatial filters at suitable intervals in the amplifier chain, an exponential increase of the perturbation amplitude is prevented. Finally, by choice of suitable materials for both active and passive components in the laser system, the coefficient n_2 is kept to a minimum value. Even with these precautions, system design must be carried out carefully to reduce the effect of the index nonlinearity and the ensuing phase distortion.

11.3.3 Examples of Self-focusing in Nd : YAG lasers

In the past, optical damage due to self-focusing was mainly an issue in large Nd : glass laser systems designed for inertial confinement fusion. Improvements of the damage threshold of optical coatings, and of the optical quality of crystals, in combination with advanced resonator designs and diode pumping, have considerably increased the beam brightness obtainable from solid-state lasers. As a result, modern Nd : YAG oscillators and amplifiers are operated at peak powers and fluence levels which can cause self-focusing. In diode-pumped systems, catastrophic self-focusing can actually be observed under certain conditions, because optical breakdown manifests itself by generation of bright flashes in the laser crystal. In flashlamp pumped lasers, these flashes are obscured by the pump light. Even in diode pumped lasers it is not always possible to see these flashes, particularly if the ends of the laser rod are located deep inside the laser head.

In our lasers we have seen these bright light flashes to occur in both oscillator and amplifier stages. Close examination of the laser crystal reveals tiny damage tracks which are very difficult to see and do not seem to affect laser performance. After many hours of operation the Nd : YAG crystal will show gray tracks which eventually lead to a performance degradation. In glass, catastrophic self-focusing leads to a highly visible streak of bubbles. In Nd : YAG, due to the high melting temperature and higher material strength, the bubbles tend to be very small. Also, we believe, because Nd : YAG is a single crystal, a certain amount of re-crystallization and annealing may take place after each optical breakdown. The gray tracks are most likely caused by color centers which are created by absorption of UV radiation emitted during the electron avalanche process.

Arisholm [11.34] reported the occurrence of light flashes and associated damage tracks in a Nd : YAG oscillator rod and explained it as the result of catastrophic self-focusing. He also provided a simulation model which reproduces self-focusing of short spikes, and computes the probability of an event consistent with the observed rate of flashes in his laser. Multi-longitudinal mode operation can lead to intensity peaks due to mode-beating. Such a spike can be powerful enough to produce intensity induced self-focusing which then builds-up on successive passes through the resonator. Eventually the reduction in beam cross-section during each round trip increases the power density to the point of optical breakdown.

The laser in which damage tracks and flashes were observed had a pulse energy of 45 mJ and a pulse width of 30 ns. Temporal pulse shapes and the occurrence of flashes were correlated, and it was observed that smooth pulses never caused beam collapse, whereas pulse shapes with superimposed spikes did frequently lead to catastrophic self focusing. The detection system was not fast enough to measure the width of individual spikes, but based on the bandwidth of the laser, spikes were assumed to have a width on the order of 40 ps. The peak intensity of the beam incident on the output mirror was calculated to be 90 MW/cm^2 and the ($1/e^2$) beam radius was 1.5 mm. The average intensity in the resonator is not high enough to start self-focusing in the laser crystal. With a 10 cm long Nd : YAG rod and a 2.5 cm long LiNbO$_3$ crystal, and using the values for n_2 for both materials from Table 11.1, one obtains from (11.16) a combined focal length due to self-focusing of about $f = 25$ m. For the intensity, a value of $I = 270$ MW/cm^2 has been introduced into (11.16) because in counter propagating beams, the return beam contributes twice as much to the refractive index change as the forward wave [11.35].

Because of multi-longitudinal mode operation, there can be random spikes of much higher intensity. Some of these spikes can be high enough to cause a collapse of the beam due to self-focusing. The simulation was run with different spike intensities and it revealed that a spike with an initial intensity about 12 times the mean intensity will cause catastrophic self-focusing. Assuming such a spike, the new focal length of the nonlinear lens, comprised of the Nd : YAG rod and Pockels cell, is on the order of $f = 6$ m. This value is obtained by introducing a value of $I_0 = 1.1$ GW/cm^2 into (11.16) because for a spike circulating in the resonator, the effect of counter-propagating fields can be ignored. The short spike is not likely to meet an equivalent spike circulating in the opposite direction through the nonlinear medium.

This new focal length is short enough to have a significant effect on a spike that propagates for many round trips. If the spike gets focused, the effect of the nonlinearity will grow for each round trip because the peak intensity increases. Because of the reduction in beam diameter, the spike reached a peak intensity around 40 GW/cm^2 at which point catastrophic self focusing occurred. The probability that the field in the resonator has a spike exceeding a certain value has been addressed also in [11.34]. It was assumed that the laser was running on more than 100 modes based on the spectral width. Interference between these modes can give rise to short spikes with high intensity. The intensity of a spike relative to the mean intensity is proportional to the number of modes interfering in phase. The analysis assumes lasing on multiple longitudinal modes with independent random phases and uniform probability distribution. If the number of modes is large, and the output is not dominated by a few modes, the probability that the average field $\langle E \rangle = \sigma$ has a spike exceeding a value $E = \beta\sigma$ during a round trip time t_r, of the resonator is given by [11.36]

$$P_t(\beta) = \frac{t_r}{\tau_c}\sqrt{\frac{4\ln 2}{\pi}}\beta e^{-\beta^2}$$

where τ_c is a correlation time of the signal. The laser has a spectral width of 20 GHz indicating a correlation time of about 50 ps. With $t_r/\tau_c = 150$ for the laser described above, the probability for a spike to occur with an intensity 12 times the average ($\beta = 3.5$) is $P_t(3.5) = 2.4 \times 10^{-3}$. Since the laser is operated at 20 Hz, one would expect to observe 2.9 spikes per minute. A number which was in agreement with observation.

In one of our diode-pumped MOPA high-brightness military lasers we observed initially a light flash about every second in the last amplifier stage. The system was comprised of an oscillator and 2 amplifier stages with Faraday rotators between stages for isolation. The system was folded by three corner cubes into a very small compact system. The complete laser system contained 47 individual pieces of optics such as laser rods, Porro prisms, waveplates, lenses, alignment wedges, polarizer, Faraday rotators, Q-switch, windows etc. The complete optical train from the rear mirror of the oscillator to the exit window contained a total of 46 cm of Nd : YAG crystals, 42 cm of glass, and 6 cm of other crystalline material such as $LiNbO_3$, calcite and BBO. The laser produced 1 J of output in a 16 ns long pulse from a 9 mm diameter aperture. The beam quality was $5\times$ diffraction limited.

The bright light flashes emerging from the laser crystal in the last amplifier stage could be clearly seen, because the rod ends protruded from the laser head. The laser was run for many hours during initial alignment, system tests, and in performing nonlinear conversion experiments before the issue of self-focusing could be addressed. No measurable degradation in performance was observed during this time. Clearly the large amount of glass and crystalline material in this laser provides an ideal medium for self-focusing to occur. We believe catastrophic self-focusing was caused by a combination of temporal spikes forming in the oscillator as a result of longitudinal mode interference, and small ripples on the spatial beam profile due to diffraction effects.

Elimination of self-focusing, aimed at reducing temporal and spatial beam fluctuations can be achieved in a number of ways. The most effective way clearly is to injection seed the oscillator to achieve single-mode operation, and to incorporate a spatial filter to eliminate spatial disturbances in the beam profile. Space limitations and operational considerations precluded the incorporation of these solutions into our system. Self focusing was eventually eliminated by a number of incremental changes such as: insertion of an etalon in the resonator to reduce the bandwidth of the laser, which increases the pulse width and lowers the peak intensity of the spikes; insertion of a thin nonlinear crystal in the resonator which converts a few percent of the average intensity into the second harmonic but represents a large nonlinear loss for high peak intensities; alignment changes of the optical path to avoid overfilling of the amplifier rods which produces diffraction rings; and stretching of the pulse width to reduce peak power.

Self-focusing is difficult to characterize because it depends not only on peak power but also on intensity ripples in the beam. In oscillators, the dominant process is believed to be whole-beam self-focusing originating from temporal spikes. Whereas in amplifiers the dominant process seems to be small scale self-focusing originating from intensity or phase ripples in the beam.

11.4 Damage Threshold of Optical Materials

The data presented here on bulk, surface and dielectric thin-film damage levels should only be used as a guide, since damage thresholds depend on a great number of laser and materials parameters, as well as sample cleanliness and prior history. Laser parameters, such as wavelength, energy, pulse duration, transverse and longitudinal mode structure, beam size, location of beam waist, all play an important part in the definition of damage threshold. For example, multimode beams usually contain "hot spots" at which the peak power is many times the average value estimated from the pulse energy, pulse duration, and beam cross section.

Experience also shows that the damage threshold can vary greatly even within one particular sample due to compositional variations, localized imperfections, absorption centers, etc. It has also been shown that the prior history of a sample can have a large influence on the damage level.

11.4.1 Scaling Laws

Although the specific threshold at which laser damage is observed depends on a great many factors, empirical scaling laws offer a guide to the parametric variation of the thresholds for a given material.

In this subsection, the dependence of damage threshold on pulse width, wavelength, beam diameter and laser conditioning is discussed.

Pulse Width . We can distinguish two different behavioral regimes. For pulses longer than about 100 ps, damage is caused by conventional heat deposition resulting in melting and boiling [11.37, 38]. Energy is absorbed by conduction band electrons from the external field and transferred to the lattice. In this regime, the source of the initial seed electrons can be local defects or impurities. For damage to occur, the temperature near the local absorption centers must be sufficiently high to cause fracture or melting.

Because the limiting process for temperature growth is thermal conduction and diffusion into the lattice, the fluence required to raise the heated region to some critical temperature is proportional to $t_p^{0.5}$ where t_p is the pulse duration [11.39]. This is in reasonably good agreement with numerous experiments which have observed a t_p^{α} scaling with α between 0.3 and 0.6 in a variety of dielectric materials from 20 ps to over 100 ns including samples with defects.

Figure 11.3 illustrates the dependence of damage threshold on pulse width for a number of materials. The data presented are the results of numerous measurements performed at the Lawrence Livermore National Laboratory over a period of 10 years [11.40]. The materials include silicate glasses, such as borosilicate BK-7, low expansion TiO_2-SiO_2 (ULE), fused silica, a commercial phosphate glass (LG 750), fluorophosphate glass, a sol-gel AR coating and a high reflectivity dielectric coating HfO_2 on a fused silica substrate.

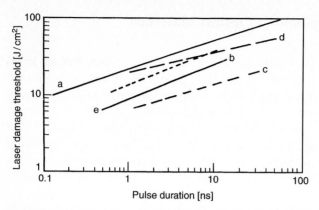

Fig. 11.3. Damage threshold vs. pulse duration for fused silica, BK-7, ULE, LG750 and flurophosphate glass at 1064 nm (**a**); fused silica at 355 nm (**b**), HfO₂/SiO₂ HR coating at 1064 nm, unconditoned (**c**) and laser conditioned (**d**); sol-gel AR coating on fused silica at 1064 nm and 355 nm (**e**) [11.40]

The pulse-width dependence ranges from $t_p^{0.5}$ for fused silica at 355 nm and the sol-gel AR coating, to $t_p^{0.4}$ for fused silica, BK-7, ULE, LG 750 and fluorophosphate at 1064 nm, and to $t_p^{0.3}$ for the HfO₂ coating.

As the pulse width decreases below 100 ps, a gradual transition takes place from the long pulse, thermally dominated regime, to an ablative regime dominated by collisional and multiphoton ionization and plasma formation [11.41–43]. For pulses shorter than about 20 ps, electrons have insufficient time to couple to the lattice during the laser pulse. The damage threshold continues to decrease with decreasing pulse width, but at a rate slower than $t_p^{0.5}$. This departure is accompanied by a qualitative change in the damage morphology indicative of rapid plasma formation and surface ablation. The damage site is limited to only a small region where the laser intensity is sufficient to produce a plasma with essentially no collateral damage.

Optical breakdown occurs due to an electron avalanche in which conduction-band electrons, oscillating in response to the laser field, achieve energy equal to the bandgap, and subsequent impact ionization promotes another valence electron into the conduction band. Because the pulses are so short, collisional heating of the electrons occurs before there is significant transfer of energy from the electrons to the lattice. Impact ionization eventually results in an electron avalanche [11.9, 44].

Wavelength Dependence. The wavelength dependence of the damage threshold does not lend itself to a simple scaling law. Typically, the damage threshold decreases for the shorter wavelength and drops off rather sharply at the shortest wavelengths. A typical behavior is shown in Fig. 11.4, which summarizes the results of over 1000 damage measurements on KDP crystals [11.45].

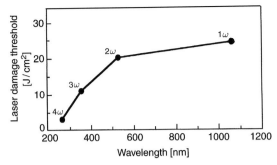

Fig. 11.4. Bulk laser damage threshold of KDP crystals measured with 3-ns pulses at the fundamental and harmonics of a Nd:YAG laser [11.45]

Other experimental data show that for bulk and sol-gel AR-coated fused silica, the damage threshold is reduced by a factor (1.8–2.4) in changing the wavelength from $1.06\,\mu$m to $0.35\,\mu$m [11.40, 46].

Spot-Size Dependence. Surface damage is usually initiated by defects, such as pits, inclusions, or scratches. Probing the surface with a beam smaller than the separation of isolated defects will measure the damage threshold of the basic substrate or coating material. A larger spot size will increase the probability that surface defects will be within the beam area. Therefore, there is a dimensional dependence of the damage threshold. Depending on the density and distribution of these imperfections at a certain spot size, the dimensional dependence will disappear.

The dimensional dependence indicates that the failure mechanism is due to defect-driven localized failures rather than the basic coating or substrate material. In testing for damage, it is important that the beam spot size is large enough to provide meaningful results. Figure 11.5 illustrates a typical result of a dielectric coating tested at different spot diameters [11.47]. The dimensional dependence of the damage threshold disappears for spot sizes larger than $100\,\mu$m.

Laser Conditioning. It has been consistently observed in a number of optical materials, both bulk and thin film, that the damage threshold can be substantially and permanently improved by a process which consists of exposing a component to incrementally higher and higher fluence levels beginning well below the single shot damage threshold [11.48–51].

Multi-layer HfO_2 coatings on fused silica substrates show the most consistent improvement with laser conditioning. In general, conditioning produces about a two-to-three fold improvement in the damage threshold. The curves (*c* and *d*) in Fig. 11.3 illustrate this improvement. The conditioning effect has been associated with the gentle removal of nodular-shaped defects, which are known to limit the damage threshold of these multilayer coatings. Other dielectric coatings which have been reported to respond to laser conditioning are MgF_2, LaF_3, ZrO_2

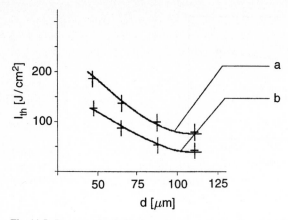

Fig. 11.5. Damage threshold dependence on beam diameter of a ZrO$_2$/SiO$_2$ (*curve a*) TiO$_2$/SiO$_2$ (*curve b*) multilayer coating [11.47]

[11.52, 53]. The bulk damage of KDP crystals can also be increased by a factor of 2 or more over the unconditioned threshold [11.45].

Actually, laser conditioning may represent the greatest single practical improvement in thin film damage thresholds in the past ten years. As a general rule of thumb at least five shots are needed to condition the optic to $\approx 85\%$ of the maximum conditioned threshold. Furthermore, the first shot in the conditioning sequence should be at a fluence of about one-half the unconditioned damage threshold.

11.4.2 Laser Host Materials

The damage threshold of the laser medium is usually determined by the coatings rather than the bulk material. Internal damage occurs relatively seldom, unless an internal focus is created due to self-focusing, or as a result of back reflections from an optical element.

Bulk damage using 47 ps pulses from a mode-locked Nd : YAG laser at 1.06 μm has been measured for a number of materials by focusing the beam inside the sample [11.54]. The samples tested included LiSAF, LiCAF and LiS-CAF, with different chromium dopings, along with Nd : YAG, fused silica and BK-7. The threshold value of the Cr-doped laser crystals was about 20 J/cm^2 or 400 GW/cm^2, which are comparable with those of fused silica and BK-7 and about a factor of two larger than Nd : YAG.

Bulk damage threshold measurements with a Q-switched Nd : YAG laser which had a pulse width of 10 ns are reported in [11.55]. The lowest power density for which damage was observed in the Nd : YAG sample was 3.6 GW/cm^2, and 2.6 GW/cm^2 for Nd : YALO. As a comparison, the intrinsic damage thresholds for LiNbO$_3$ and calcite were at least a factor 6 lower compared to the laser crystals. Bulk damage thresholds for a number of Nd : glasses have already been summarized in the previous section.

Although this chapter is concerned with laser-induced damage to optical materials, it might be of interest to some readers to briefly mention damage to laser crystals induced by gamma and proton radiation. Solid-state lasers in space are exposed to charged particles, such as electrons, protons and high-energy cosmic rays. The effect of gamma rays and high-energy protons on Nd:YAG and Nd:YLF crystals has been described in [11.56]. It was found that both forms of radiation create color centers in the crystals which reduced optical output primarily by absorbing the laser radiation.

For Nd:YAG the induced absorption loss reached a limiting value of about $0.015\,\text{cm}^{-1}$ at 100 krad exposure levels; the loss in Nd:YLF was higher and did not reach a limiting value. Codoping Nd with Cr^{3+} significantly reduced radiation susceptibility of Nd:YAG. The result of the experiments suggested that diode-pumped Nd:YAG lasers can be effectively radiation hardened to withstand exposure to charged particles in space. Design features which improve radiation hardness include: codoping of the Nd:YAG crystal with Cr^{3+}, a short laser crystal, and pulsed Q-switched rather than cw operation, because the higher gain can more readily overcome the induced losses created by color centers.

11.4.3 Optical Glass

Data on bulk damage thresholds for a number of glasses have been provided in Sect. 11.4.1. In inclusion-free silica, damage threshold is limited by the surface. Extensive measurements at different wavelength and pulse width gives a simple empirical relationship governing the safe operating limit for fused silica [11.46]

$$D = (22\text{--}25)t_p^{0.4} \quad \text{at} \quad 1064\,\text{nm},$$
$$D = (9\text{--}14)t_p^{0.5} \quad \text{at} \quad 355\,\text{nm}.$$

The lower values are for bare surface fused silica, and the higher values are measured after a sol-gel AR coating is applied to the surface. D is the damage threshold (J/cm^2) and t_p is the pulse length (ns). In agreement with the common trend in optical materials, the measurements also reveal a substantial decrease of the damage threshold for the shorter wavelength.

For pulses shorter than 20 ps, the damage fluence no longer follows the $t_p^{0.5}$ dependence. As explained earlier, for such short pulses there is insufficient time for lattice coupling. Figure 11.6 summarizes measurements of the damage fluence for fused silica at very short pulse widths [11.43]. Summaries of the damage threshold of a large number of different optical glasses including absorption filters can be found in [11.57].

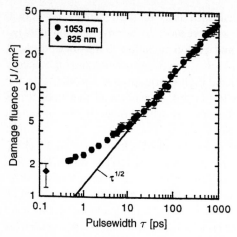

Fig. 11.6. Threshold damage fluence for fused silica as a function of pulse width [11.43]

11.4.4 Damage Levels for Nonlinear Materials

From all the optical components and materials employed in solid-state laser systems the largest spread in damage thresholds occurs with nonlinear crystals. Table 11.2 summarizes the nominal damage levels of important crystals employed in nonlinear processes of solid-state lasers. It shows a spread of over 2 orders of magnitude from KDP/KD*P to the $AsGaSe_2/AsGaS_2$ crystals.

In addition to exhibiting damage mechanisms operative in other dielectric materials, such as thermal heating, and creation of an electron avalanche by impact ionization, nonlinear materials are subject to damage by virtue of their nonlinear properties. It is frequently observed that the damage levels observed in harmonic generation are far lower in the presence of the harmonic than those observed with just the fundamental present. This effect is often caused by multiphoton absorption which plays a major role in laser-induced damage in nonlinear crystals. In this process, 2-photons or 3-photons can generate radiation at a wavelength which is below the absorption edge of the crystal. Absorption of UV radiation typically leads to color center formation.

Laser-induced damage in KDP and KD*P has been studied in great detail because these crystals are employed in inertial confinement fusion lasers for the generation of the second and third harmonics from Nd : glass lasers. Laser damage threshold for KDP as a function of wavelength is depicted in Fig. 11.4. The damage threshold for KD*P is about a factor 2 lower compared to KDP. As already mentioned the threshold for bulk damage can be increased by subjecting the crystals to a series of gradually increasing fluence levels beginning below unconditioned threshold levels [11.58, 59].

The borate crystals LBO and BBO also have very high damage thresholds. LBO is emerging as an alternative to KTP for harmonic generation of Nd : lasers. This crystal does not suffer from nonlinear absorption or photo-refractive effects;

Table 11.2. Typical damage thresholds for nonlinear crystals

Nonlinear crystal	KDP	KD*P	LBO	BBO	KTP KTA	LiNbO₃ PPLN	AsGaS₂ AsGaSe
Damage threshold [GW/cm²]	8	4	2.5	1.5	0.5	0.2	0.03

however, the nonlinear coefficient and thermal acceptance parameter of LBO are about one third those of KTP. Lifetimes in excess of 1000 hours have been reported in LBO at a second-harmonic output close to 150 W [11.60]. BBO has a transmission band which extends to 200 nm; therefore multiphoton absorption is not a problem. This material is of particular interest for harmonic generation in the blue region of the spectrum.

KTP crystals are primarily employed for frequency doubling of Nd: lasers, and in OPO's for the generation of the eye-safe wavelength of $1.5\,\mu$m. Although the bulk and surface damage thresholds for KTP are quite high at $1.06\,\mu$m, employed in a second harmonic generator or as an OPO, a number of failure mechanisms set in which are related to the conversion process.

At around $1.06\,\mu$m, laser-induced damage in KTP is associated with surface coatings and macroscopic flaws in crystals. The damage threshold of KTP at $1.06\,\mu$m, measured with a Q-switched Nd: YAG laser with a pulse width of 20 ns, is depicted in Fig. 11.7.

Nonlinear effects were minimized in these tests by appropriate orientation of the crystal. Surface failure for uncoated crystals is typically 15 to 20 J/cm² for pulse width of 10 to 20 ns. Bulk failure thresholds have been measured as high as 30 GW/cm² [11.61, 62].

When KTP is used in a non-critical phase-matched OPO, it is subject to thermal heating due to $3.3\,\mu$m idler absorption. In high-average-power systems this absorption results in a substantial lowering of the damage threshold. With a 20 Hz Nd: YAG laser, massive bulk and surface damage was observed at fluence levels as low as 5 J/cm² [11.61, 62].

KTA, a crystal very similar to KTP, has much lower absorption at the idler wavelength (Fig. 10.10). KTA avoids the problems associated with thermal heating due to absorption in the $(2.8–3.5)\,\mu$m region. Consequently KTA, rather than KTP, is usually employed in high average power OPO's designed for the generation of infrared radiation from a Nd: laser.

Damage of KTP in a second harmonic generator results from a photochromic process which produces darkening, or gray tracks, in the material. Absorption of both the fundamental and harmonic radiation leads eventually to catastrophic failure of the crystal. The gray-tracking behavior of KTP is complex. The usual causes of damage are related to chemical impurities (Ti^{3+} or Fe^{3+}), physical defects (such as precipitates), and nonstoichiometric effects (such as oxygen vacancies).

The initial step leading to damage involves the introduction of charges in the lattice. The charges are believed to be generated through 2-photon or 3-photon

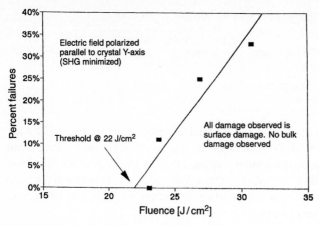

Fig. 11.7. Damage frequency plot of KTP at 1064 nm [11.61]

absorption processes. KTP is found to be susceptible to gray-track damage as a result of radiation generated by multiphoton processes, which is shorter than the absorption edge of 355 nm. In typical doubling applications, a combined 1064 and 532 nm process is likely. However, gray track damage has also been observed with irradiation of only 532 nm light and could be due to a 2-photon process as well. A 3-photon process could result from the sum of three 1064 nm photons. Gray-tracking in KTP has been extensively investigated [11.63–71].

Gray track damage can be removed by heating the crystal during operation to about 70 to 100°C. At this elevated temperature darkening due to the photochromic process is annealed out fast enough to achieve long-term stable operation. Higher temperatures will accelerate the annealing process even more; and temperatures as high as 170°C have been applied to the crystal. However, a higher temperature has to be weighted against potential contamination problems from outgassing of the hot components of the crystal oven (Sect. 11.5).

In a test conducted at 75°C, exposure to 12 J/cm² and 10,000 shots produced no bulk darkening; the same experiment performed at 30°C caused a transmission loss ranging from 4 to 8% [11.65].

Intracavity-doubled Nd : YAG lasers have produced over 100 W of second-harmonic power using KTP. In one of these high-power systems a degradation rate was measured of 0.04% per hour during a 300 hour test run at 150 W of green output and a repetition rate of 13 kHz. Most of the degradation was attributed to gray tracking in the KTP crystal which was operated at 30° [11.60].

In our laboratory, we have built a number of diode-pumped Nd : YAG lasers which produce between 250 and 500 mJ per pulse of green output at a pulse width of 10 to 20 ns and repetition rates up to 100 Hz. At a fluence level of about (1–2) J/cm² at 1.06 μm, degradation in KTP is minimal for a pulse count up to $(10^8–10^9)$. The KTP crystals are operated typically at 90°C. One investigation revealed that the dependence of the induced absorption of 532 nm radiation is nonlinear and has a threshold of 80 MW/cm² [11.70].

Experiments have also shown that the peak power threshold for gray tracking damage strongly depends on the repetition rate of the laser. An exponential decrease of damage threshold as a function of repetition rate was observed, ranging from 125 MW/cm^2 at 1 kHz, to 18 MW/cm^2 at 6.3 kHz, to a few MW/cm^2 for frequencies above 10 kHz [11.68].

Lithium niobate has a relatively low damage threshold in comparison with other common nonlinear materials. For example, at 1.06 μm, a surface damage threshold of 180 MW/cm^2 has been measured for a 10 ns pulse with a Gaussian intensity profile [11.72]. At a wavelength of 1.54 μm, a damage level of 20 J/cm^2 with a 50 ns long pulse has been reported [11.47].

LiNbO$_3$ is plagued by photo-refractive effects resulting from propagation of visible or UV radiation. Photo-refractive effects cause refractive index inhomogeneities which seriously degrade beam quality and distort the index-match condition necessary for efficient second-harmonic generation [11.73, 74]. The occurance of refractive-index gradients is reversible if the temperature of the crystal is increased during operation to about 160°C. The index inhomogeneities are associated with imperfections in the crystal. Photons at the shorter wavelength have enough energy to break the binding energy which traps electrons at the vacant lattice site. The electrons are then free to move away, leaving behind a local electric field distribution in the crystal. Because of the large electro-optic effect of LiNbO$_3$, this local electric field causes refractive index gradients. Annealing the crystal at an elevated temperature increases the mobility of electrons in the crystal enough to permit neutralization of the charged vacancy, thereby removing the refractive index gradients. Also, doping with MgO has been used to reduce photorefractive damage; but doped crystals do not exhibit the same high quality that is available in undoped crystals.

There have been several reports that periodically poled lithium niobate is more resistant to photo-refractive damage than are homogenously poled crystals [11.75–77].

The large transparency range of 0.8 to 18 μm make silver gallium selenide (AgGaSe$_2$) an excellent candidate for OPO's in the infrared region. Unfortunately, the crystal has a very low damage threshold which must be taken into account when engineering specific OPO devices. Measurements with a pump laser at 2.09 μm revealed an average surface damage threshold for coated and uncoated crystals of around 1 J/cm^2 at a pulse width of 50 ns [11.78].

Silver gallium sulfide (AgGaS$_2$) with a transmission range of (0.5–12) μm is another crystal for OPO devices in the mid- and far infrared region. A surface damage threshold of (25–50) MW/cm^2 for 10 ns pulses at 1.064 μm has been measured [11.79]. Since the bulk damage threshold of this crystal is an order of magnitude higher, further improvements of the crystal might lead to a higher surface damage threshold.

11.4.5 Dielectric Thin Films

Multilayer dielectric films are generally the weakest elements of any laser system. Because of the relative complexity and variety of deposition factors, thin films vary greatly in purity, morphology, uniformity and composition, depending, to a great extent, on the specifics of the fabrication process. In addition, residual stress, substrate cleanliness, and other extrinsic factors tend to reduce the thin film damage level well below that expected for the same material in bulk form.

The usefulness of a dielectric coating for high-power applications depends on the choice of coating material, thin-film design, and deposition technique. Today's knowledge of coating failures distinguishes between homogeneous background absorption in the film and absorption at localized defects. Many investigations have shown that microscopic absorbing defects in the film are the initial sites of failure [11.80–83].

During evaporation, ejected particles of coating material from the source, or particles flaking from the chamber walls, can become embedded in the film and can initiate the growth of so-called nodular defects. Inclusions form a discontinuity in the coating stack that disrupts the electric field distribution, resulting in localized electric field enhancement in the vicinity of the nodule. This leads eventually to significant heating of the localized region. The effects of laser conditioning are now commonly seen to be a result of the heating and subsequent ejection of such inclusion.

Numerous measurements have conclusively shown that the defect density and film roughness, caused by overcoated particles, are key indicators in determining the threshold of a dielectric coating [11.81, 82, 84]. In AR coatings, the field intensity is near maximum at the substrate surface, where the coating is most vulnerable due to surface imperfection, embedded polishing materials and poor film structure.

Therefore, the ultimate limit for AR coatings is the damage threshold of the uncoated substrate material. With HR coatings the substrate surface is shielded from the high field strength, and the ultimate limit of a perfect coating is determined by bulk values of the coating materials.

Thin films and multi-layer stacks for optical components of solid-state lasers are usually produced by electron-beam evaporation. Advanced deposition techniques include ion-assisted deposition, molecular-beam deposition, and HF magnetron sputtering. Very large pieces of optics, such as employed in inertial confinement fusion drivers, are AR coated by a sol-gel process [11.85].

Sol-gel coatings are a very cost effective method to produce coatings on large optical substrates since these coatings can be applied in a dip or spin process at room temperature. The procedure involves the preparation of a solution of metal alkoxides which is converted by hydrolysis or heat to an oxide in a gelation stage. For example, from a solution of $Si(OCH_3)_4$ a SiO_2 coating can be produced. Very high damage thresholds can be achieved with sol-gel coatings. Drawbacks include their fragile nature and difficulty in cleaning the optical components. For this reason, sol-gel coatings are not often used in standard solid-state lasers. The AR coating of choice is usually vacuum deposited MgF_2.

We will briefly review the laser-induced damage thresholds and major issues for thin-film coatings from the UV to the mid-infrared regime, which covers the range from frequency-quadrupled Nd : YAG lasers to OPO's operating in the (2–4) μm region.

The most detailed damage studies have been carried out for high-reflectivity HfO_2/SiO_2 coatings and for anti-reflection sol-gel coatings at the wavelengths of 1054 nm and 355 nm, because these coatings are used for the optical components of the large Nd : glass lasers at the Lawrence Livermore National Laboratory.

UV Spectral Region. The damage threshold of dielectric thin films is lower in the UV region, as compared to the longer wavelengths. One reason is the increased absorption, the other is the higher photon energy. Ionization and multi-photon absorption is more likely to occur at shorter wavelengths. The substrate materials transparent in the ultraviolet region are fused silica, MgF_2 and CaF_2. The choice of coating materials at the shortest wavelengths around 266 nm include MgF_2, LaF_3, SiO_2 and Al_2O_3. At the wavelength around 355 nm, ZrO_2 and HfO_2 can be added to the above-listed materials.

Table 11.3. Damage threshold at 248 nm of various substrates and coatings [11.86]

		J/cm^2
Substrate	MgF_2 crystal	20
	Quartz crystal	10
	Suprasil 300	8
	CaF_2	3
AR coatings	MgF_2	9
	LaF_3/MgF_2	10
	Al_2O_3/SiO_2	3
	Al_2O_3/MgF_2	3
HR coatings	Al_2O_3/SiO_2	25
	LaF_3/MgF_2	20

Table 11.3 lists laser-induced damage thresholds for common substrates and thin films employed in the UV region [11.86]. At 355 nm, the damage threshold of a HfO_2/SiO_2 high-reflectivity coating was measured between (5–8) J/cm^2 for a 3 ns pulse. The damage threshold for ZrO_2/SiO_2 coatings was slightly lower, whereas Al_2O_3/SiO_2 coatings had damage thresholds around 10 J/cm^2 [11.84].

Visible and Near-Infrared Region. Multi-layer thin films for this wavelength region usually consist of alternating layers of a metal oxide which provides the high-index material and SiO_2. Common combinations are HfO_2/SiO_2, ZrO_2/SiO_2,

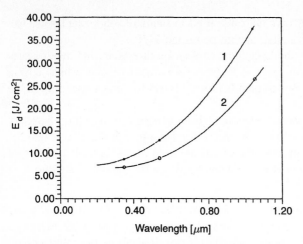

Fig. 11.8. Damage threshold of a multilayer ZrO_2/SiO_2 coating vs. wavelength for a defect density of 280/mm^2 (*curve 1*) and 320/mm^2 (*curve 2*) [11.87]

TiO_2/SiO_2, and Ta_2O_5/SiO_2. In particular, HfO_2/SiO_2 has a very high laser-induced damage threshold; it can be further improved with laser conditioning.

Damage threshold for a multi-layer HfO_2/SiO_2 coating at 1.06 μm as a function of pulse width is depicted in Fig. 11.3. The damage threshold for a multi-layer ZrO_2 /SiO_2 coating measured versus wavelength with a 10 ns long pulse and a Gaussian beam profile is shown in Fig. 11.8. The curves also illustrate the dependence of the damage threshold on defect density. Similar to the observation made with bulk materials (Fig. 11.6), there is a transition region with regard to the pulse width, where the damage morphology changes from heating and melting at the longer pulses, to ablation of the individual dielectric layers at very short pulses. At the short pulse-width regime, pulse-width scaling is less than $t_p^{0.5}$. Figure 11.9 depicts the dependence of the damage threshold on pulse width of very short pulses for two multi-layer dielectric polarizers [11.43]. Damage thresholds for polarizer coatings are typically lower than those of similar mirror coatings. For example, for 3 ns pulses, HfO_2/SiO_2 thin-film polarizers have a damage threshold after laser conditioning of around (15–20) J/cm^2, which is half the value compared to mirror coatings.

In [11.88] damage thresholds were reported for mono-layer and multi-layer coatings produced by different commercial suppliers. The results of the tests performed with 25 ps and 8 ns pulses at 1064 nm are summarized in Table 11.4. The data illustrates the large spread in the damage threshold for coatings from different vendors.

The design and development of eye-safe rangefinders and LIDAR systems requires high-damage-threshold coatings for the wavelength region around 1.5 μm. Measured values of the laser-induced damage threshold at this wavelength at a pulse width of 50 ns are listed in Table 11.5 for different coatings [11.47].

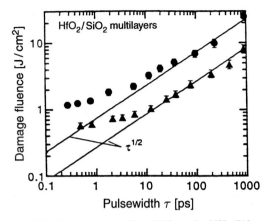

Fig. 11.9. Damage threshold at 1053 nm for HfO$_2$/SiO$_2$ multilayer 45° (▲) and 57° (●) polarizer [11.43]

Table 11.4. Range of damage thresholds at 1064 nm (after preconditioning) for coatings produced by different commercial vendors [11.88]

| | | Damage threshold [J/cm^2] | |
		25 ps	8 ns
Monolayers	Ta$_2$O$_5$	4.6–5.4	25
	TiO$_2$	3.1–3.4	9–18
	SiO$_2$	15–16	140
Multilayer coatings	ZrO$_2$/SiO$_2$	6.5–21.5	44–155
	Al$_2$O$_3$/MgF$_2$	13–15	120–130
Subtrate bare surfaces	Fused silica	17.8–21.5	–
	BK-7	20–21.5	–
	Nd : YLF	10–12.5	–

Table 11.5. Damage threshold of multilayer coatings at 1.54 μm [11.47]

Materials and type of coatings	Number of layers	Laser-damage threshold at 1.54 μm, J/cm^2
ZrO$_2$-SiO$_2$ high reflection	25	65 ÷ 80
ZrO$_2$-SiO$_2$ partially-reflective	10	no less than 100
TiO$_2$-SiO$_2$ high reflection	23	40 ÷ 60
ZrO$_2$-SiO$_2$ antireflection	2	no less than 100

The damage threshold for ZrO_2/SiO_2 coatings were found to be higher as compared to TiO_2/SiO_2 (see also Fig. 11.5). The tests also revealed that the damage threshold for both types of coatings is about (20–30)% higher at 1.5 μm as compared to 1.06 μm.

Mid-Infrared Spectral Range. Interest in high-power coatings for this wavelength regime is driven by the development of solid-state laser pumped OPO's with outputs between 2.5 and 4 μm. The availability of nonlinear crystals, such as KTA and PPLN, has opened the mid-infrared regime to solid-state lasers for a number of important applications such as atmospheric studies, spectroscopy, and IR countermeasures.

Coatings have been well developed in this region as a result of DF and HF laser systems. Dielectric thin films for optical parametric oscillators have special requirements, such as multi-wavelength and dichroic coatings.

Values for the laser damage threshold were reported in [11.89] for a number of dielectric thin films comprised of different materials and deposited by different methods. The stacks were designed around the requirements for OPO's with reflection bands between 2.5 and 4 μm. The highest damage thresholds were found for the ZnS/Al_2O_3 films deposited by e-beam evaporation on either silicon, CaF_2 or ZnS substrates. Damage levels at 3.8 μm for a 55 ns pulse were around 50 J/cm^2, which is close to the damage level of the bare silicon substrate. Stacks of SiO_2/TiO_2 thin films also performed well with damage levels on the order of 45 J/cm^2.

11.5 System Design Considerations

The designer of a laser system although he cannot change the damage thresholds of the materials employed in a laser, can reduce the damage susceptibility and improve the reliability of his system by the right choice of materials and by proper design and operation of the laser. We will enumerate a few design, inspection, and operating procedures which need to be followed to reduce the likelyhood of optical damage to the laser components.

11.5.1 Choice of Materials

In powerful lasers it is important that only high-damage-resistance AR and multilayer coatings are employed. These coatings usually include ZrO_2, SiO_2, TiO_2, Al_2O_3, HfO_2 and MgF_2 compositions which are evaporated with an electron beam gun. Beam splitters, lenses, and prisms located in the beam have to be fabricated from fused silica, sapphire, or highly damage-resistant glass, such as BK-7. The surfaces of these components should be pitch-polished to a 5/10 or 0/0 surface quality. Surface conditioning or "superpolishing", such as chemical etching, ion-beam polishing, laser conditioning, or bowl-feed polishing, are very effective in raising the damage threshold.

All components of the laser system should be inspected for surface defects, such as scratches, fingerprints, "orange peel", and bulk defects, such as inclusions, bubbles, striation, schlieren, and internal stress.

Inclusions in a laser rod can be best checked by illuminating the rod from the side and observing the presence of scatter centers. The fringe pattern obtained from the Twyman-Green interferometer will reveal any density gradients and optical inhomogeneities in the laser rod. Inspection of the laser rod, Q-switch crystal, frequency doubling crystal, etc., between cross-polarizers will reveal internal stresses, striation, and schlieren. Optical surfaces are best inspected under a binocular microscope (magnification of $50\times$) with a strong collimated light beam reflected off the surface.

11.5.2 Design of System

One of the most important considerations in the design of a laser system is the choice of beam size. This usually requires a trade-off between performance, which increases with higher fluence levels, and damage considerations, which favor a larger beam size. Reliable operation requires that the beam diameter and, therefore, the aperture of all optical elements is large enough so that the power and energy densities are well below the damage threshold levels for the particular mode of operation.

The average power density of the beam is only an indicator, but not a reliable measure with regard to damage considerations. Hot spots in a multimode beam can far exceed average values; likewise, temporal fluctuations as a result of mode beating in multi-longitudinal lasers can cause high-intensity temporal spikes. Also, with Gaussian beams the peak intensity at the center rather than the average value has to be considered.

The damage levels listed in this chapter have usually been measured in a carefully controlled laboratory environment, with smooth Gaussian beams, meticulously clean optics and materials of excellent optical quality. The numbers clearly present upper limits. For a typical industrial or military laser employing industrial grade components and coatings the operating fluence and fluxes should be far below the numbers given for damage threshold.

A very important design consideration is the avoidance of damage caused by reflections from optical surfaces in the laser system. For example, very often a telescope is employed between amplifier stages to increase the beam diameter. The design and location of the negative lens employed in such a telescope is very critical, because if reflections from the concave surface are focused into the preceding laser rod, damage will most certainly occur. Therefore, a plano/concave rather than a concave/concave lens should be used and the curved surface should be faced away from the preceding optical components. In this case the reflected light from this surface is diverging rather than converging.

Reflections from the rear surface of an amplifier rod are very dangerous because the reflected radiation is amplified in passing through the amplifier stage, and optical elements at the input side of the amplifier, which are normally operated at much lower energy densities, can be damaged. This is especially the case if a telescope is employed between the oscillator and amplifier stage. If the beam travels backwards through the amplifier chain, the beam diameter decreases for increasing intensities, and damage thresholds are easily exceeded. One way to minimize this kind of damage is to tilt the various amplifier stages and separate them as far as possible from each other. A beam traveling backwards through an amplifier will then miss the aperture of the preceding stage. In order to suppress prelasing, which can cause the same type of damage, the amplifier rod should have wedged ends. Damage to optical surfaces can be minimized by employing highly damage-resistant coatings or by arranging the surfaces at Brewster's angle.

11.5.3 Operation of System

From an operational point of view, the most important factors to consider, if damage is to be avoided, are proper alignment of components and absolute cleanliness of all surfaces. Misalignment of the oscillator components can cause a change of the transverse-mode structure, hot spots and beam uniformities can occur. Misaligned components outside the resonator cavity can cause undesirable secondary reflection leading to damage.

Surfaces which are contaminated by dust, dirt, or fingerprints will rapidly degrade and cause nonuniform beam intensities which, in turn, can cause damage in more expensive optical components. Surfaces should be inspected frequently and cleaned with an airblast or with alcohol. A dental mirror and a flashlight are very convenient for the inspection of optical surfaces in a laser system. Very often components are mounted close together and surfaces are hard to see. For the cleaning of surfaces which are relatively inaccessible, a stream of pressurized clean air or gas should be used. A useful device for cleaning optical surfaces is an ionizing airgun sold by 3M Company. The unit, which contains an α-particle source of polonium isotope 210 in the valve and nozzle sections, can be connected to a gas cylinder. The gas stream containing α-radiation can remove very efficiently statically charged particles from a surface.

The degradation and subsequent damage of optical surfaces by atmospheric dust particles and by cleaning residue has been studied by *Barber* [11.90]. It was found that dust particles consisting of minerals are melted onto the surface, leaving a circular indentation after the particle is dislodged by subsequent pulses.

In another type of degradation, craters are caused by liquids being absorbed a few micrometers into the glass surface during the cleaning operation, then producing miniature explosions when a strong laser pulse strikes the absorbed liquid. Experiments showed that laser rod ends were badly pitted after liquid cleaning without drying, and no damage occurred on rods which were vacuum-dried.

From the foregoing follows that the working area around a laser should be kept as clean as possible, free from dust and contaminants, such as cigarette smoke, approaching clean room conditions as closely as possible.

After a laser is aligned and ready to be operated, the fluence level should be gradually increased starting at a level not higher than half the final output. This provides a chance for the removal of surface defects, and for particles to be evaporated or blown off by the shockwave created by the laser beam. Laser conditioning of coatings and bulk materials is an important procedure in achieving higher damage thresholds, as was discussed earlier in this chapter.

For lasers which are hermetically sealed, a few precautions are necessary to avoid contamination of optical surfaces by the environment created within the enclosure. It has been shown that trace levels of gas-phase contaminants of hermetically sealed Q-switched Nd : lasers can reduce the laser reliability and lifetime, especially for operation at elevated temperature, i.e., above 50°C [11.91]. In these life tests, outgasing from silicone materials used in o-rings, potting materials, and thermally conductive pads were evaluated for their damage potential, along with trace levels of solvents used in cleaning mechanical and optical components.

All silicone materials, except a space-grade encapsulant (Dow Corning 93-500) were found to have a high probability of inducing contamination damage in sealed Nd : YAG lasers. With regard to the cleaning fluids, the aromatic hydrocarbons (benzene derivatives, i.e. toluene) more readily caused damage than nonaromatics, i.e. acetone. Additionally, the presence of oxygen was shown to dramatically inhibit contamination damage due to hydrocarbons. Therefore, purging a hermetically sealed laser with dry air rather than dry nitrogen can significantly reduce optical damage due to outgasing. Implementation of the test results increased the laser operation at 70°C from less than 20 hours to over 500 hours without significant optical damage.

Appendix A

Laser Safety

Hazards associated with solid-state lasers can be separated into two broad categories – those hazards related to the laser beam itself and those hazards related to the electronic equipment.

Radiation Hazards

Although high-power solid-state lasers have the capability of inflicting burns on the skin, the possibility of eye injury is of vastly greater importance. The property of lasers that is of primary concern with regard to eye hazards is their high radiance, i.e. the combination of high power density and directionality. The latter property causes the eye lens to focus the parallel beam emitted from a laser to a tiny spot on the retina. It also makes the laser hazardous over long distances.

Figure A.1 shows a cross section of the human eye. The figure shows a collimated beam of visible light impinging onto the cornea, being focused by the eye lens, passing through the vitreous humor, and impinging onto the retina. At the retina the light is absorbed. Under normal conditions the photon energy is converted by the retina into chemical energy, stimulating optical sensations.

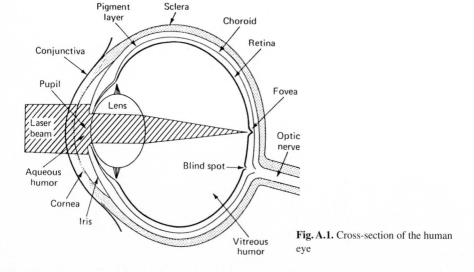

Fig. A.1. Cross-section of the human eye

Figure A.1 illustrates the principal reason for possible eye injury from a laser. The parallel rays of a laser can be focused through the eye lens to a point image of about 10 to 20 μm in diameter. If the pupil is dilated sufficiently to admit the entire beam of a laser, even lasers having power outputs of only a few milliwatts will produce power densities of kilowatts per square centimeter on the retina. Obviously, power densities of this magnitude may cause severe damage to the retina. For worst-case conditions – the eye focused at infinity and the pupil fully dilated (about 0.7 cm diameter) – the ratio of power density at the eye lens to that at the retina will be about 10^5.

By comparison, conventional sources of illumination are extended, they are considerably less bright, and they emit light in all directions. These light sources produce a sizable image on the retina with a corresponding lower power density. For example, looking directly into the sun, the image of it on the retina is approximately 160 μm in diameter which yields a radiation intensity of around 30 W/ cm^2.

The maximum permissible exposure of laser radiation to the eye depends on the wavelength and is dependent on the spectral transmission of the eye. Figure A.2. illustrates the spectral transmission characteristics of the human eye [A1]. Retinal damage is possible from exposure to laser energy in the wavelength region between 400 and 1400 nm [Fig. A.2c]. In particular, radiation between 400 and 700 nm represents the greatest hazard, because the transmission of the anterior parts of the eye (cornea, aqueous humor, lens) is highest for this spectral region. At wavelengths longer than 700 nm, there is some absorption of the radiation before it reaches the retina. At IR wavelengths greater than 1.4 μm absorption of incident radiation occurs in the cornea and aqueous humor, and beyond 1.9 μm absorption is restricted to the cornea alone [Fig. A.2d]. Also ultraviolet radiation with a wavelength shorter than 315 nm is completly absorbed by the cornea [Fig. A.2a]. Between 315 and 400 nm, radiation is absorbed at the lens of the eye [Fig. A.2b.]

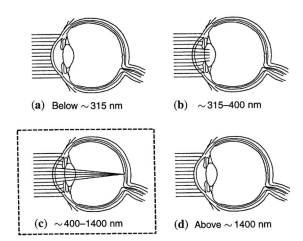

(a) Below ∼315 nm

(b) ∼315–400 nm

(c) ∼400–1400 nm

(d) Above ∼1400 nm

Fig. A.2. Spectral transmission characteristics of the human eye [A.1]

Electrical Hazards

Although the hazards of laser radiation are receiving deserved attention from government agencies, users, and manufacturers, the chief hazard around solid-state lasers is electrical rather than optical. Many solid-state lasers require high-voltage power supplies, and energy-storage capacitors charged to lethal voltages. Furthermore, associated equipment such as Q-switches, optical gates, modulators, etc., are operated at high voltages. The power supply and associated electrical equipment of a laser can produce serious shock and burns and, in extreme cases, can lead to electrocution.

Safety Precautions Applicable to Solid-State Lasers

Enclosure of the beam and target in an opaque housing is the safest way of operating a laser. This level of safety precaution is mandatory for laser materials-processing systems operated in an industrial environment. In these systems interlocked doors, warning signs and lights, key-locked power switches, and emergency circuit breaker, and like precautions are taken to protect operators and passers-by from electrical and radiation hazards of the laser equipment. Also, microscopes and viewing ports are filtered or blocked to prevent the issuance of laser radiation, and laser impact points are surrounded by shields. At points of access for routine maintenance and set-up, warning signs are displayed prominently, and interlocks prevent firing of the laser while doors or ports are open.

In the laboratory it is often not possible to enclose fully a high-power laser. In these situations the following safety precautions should be observed:

 - Wear laser safety glasses at all times! Make sure that their attenuation factor is sufficient at the wavelengths present.
 - Never look directly into the laser light source or at scattered laser light from any reflective surface. Never sight down the beam into the source.
 - Set up the laser and all optical components used with it below eye level. Provide enclosures for the beam. Do not direct the beam outside the table on which the laser is positioned without using protective tubing.
 - The output from Nd lasers and other IR lasers is invisible to the human eye. Use a IR viewing scope to check for specular reflected beams.
 - Avoid operating the laser in a darkened environment. At low light levels the size of the iris is large and can hence accept a large amount of light. Work under as bright light conditions as possible (small iris).
 - Employ a countdown or other audible warning before the laser is fired.
 - Post the appropriate hazard warning signs as specified in ANSI Z136.1 (1993) at the entrance to the laboratory.
 - Control access to the laser area and have a flashing red light on the door when the laser is in operation. Operation of very large laser systems, such as used in inertial confinement fusion experiments, actually require the installation of safety interlocks at the entrance of the laser facility to prevent entry

of unauthorized personnel into the facility while the laser power supply is charged and capable of firing the laser.

– Provide protection against accidentally contacting charged-up capacitors in energy-storage banks, high-voltage power supplies, etc. These components should be installed in cabinets having interlocked doors. Furthermore, capacitor banks should be equipped with gravity-operated dump solenoids.

The key to a successful safety program is the training and familiarization of the personnel involved with laser hazards and subsequent control measures. The appointment of a Laser Safety Officer is essential for companies or laboratories engaged in the manufacture or use of class III and class IV laser systems.

Laser Safety Standards

The potential of lasers for inflicting injury, particularly to the eyes, was recognised early, and as a consequence extensive studies have been undertaken into the biological mechanisms of laser damage, in an attempt to define safe working levels of optical radiation.

A document which is very useful as a guide for the safe use of lasers and systems is the standard published by the American National Standards Institute [A2]. A comprehensive reference work covering every aspect of safety with lasers and other optical sources is the handbook written by *D. Sliney* et al. [A3]. Additional information on laser safety can be found in [A4].

In the United States, regulations for manufacturers are published by the Center for Devices and Radiological Health (formerly Bureau of Radiological Health), under the Food and Drug Administration [A.5.] User requirements in the US are the responsibility of the individual states, but only a few states have such laws.

Under the federal safety standard, four classes of lasers have been created to allow manufacturers to categorize a product based on its potential hazard to a user. These classes are labeled I through IV, from the most safe to the most hazardous. Class I and II are used for lasers of cw outputs of less than $0.39\,\mu\mathrm{W}$ and 1 mW, respectively. Class III and IV include lasers which represent a definite hazard to the eye from direct or scattered radiation. Solid-state lasers generally fall into class III or IV of that standard, which requires that warning signs, labels, and protective housings be installed to prevent human irradiation at levels in excess of a "safe" limit. Safety interlocks must be provided for each portion of the protective housing that can be removed by the user during normal operation or maintenance. In the event that the design of the safety interlock allows the user to defeat the system manually, the manufacturer must incorporate visible or audible warnings of this interlock override during laser operation. In addition, key-lock switches and warning lights are required along with other safety devices. Outside the United States, most countries have adopted the ANSI standard, or a similar standard generated by the International Electrotechnical Commission [A6].

The author found the information presented in Figs. A.3 and 4 quite useful in calculating the safe distance at which accidental direct viewing of the laser

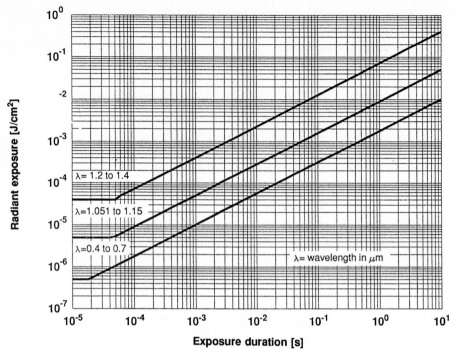

Fig. A.3. Exposure limits for direct viewing of a laser beam as a function of pulse width. Ranges are for visible and near-infrared spectral regions [A.2]

beam does not present a hazard. These situations occur quite frequently in field tests involving rangefinders, LIDAR's and similar equipment. Figure A.3 provides the occupational health and safety limits for pulsed laser radiation in the visible and near infrared spectrum. The curves show the safe fluence limits vs. pulsewidth for the visible and NIR. Compared to the visible regime, the threshold for eye damage is higher in the NIR because the focused spot on the retina is larger, furthermore higher transmission losses are encountered for the longer wavelength. The increase in fluence level with longer pulses reflects the fact that heat produced at the retina can be more effectively dissipated to the surrounding tissue as the exposure time increases. For exposure times shorter than about 50 μs, heat dissipation is insignificant and the damage threshold does not depend on pulsewidth. For exposure levels 10 times as high as indicated in Fig. A.3, there is a 50% probability of the occurence of ophthalmologically visible retinal lesions [A.3].

Figure [A.4] gives the exposure limits for direct viewing of a cw laser beam for wavelengths ranging from 0.4 to 1.4 μm. The damage threshold for the longer wavelengths is higher for the same reasons as stated above. The reader is reminded that the data presented in these figures should only be considered as guidelines.

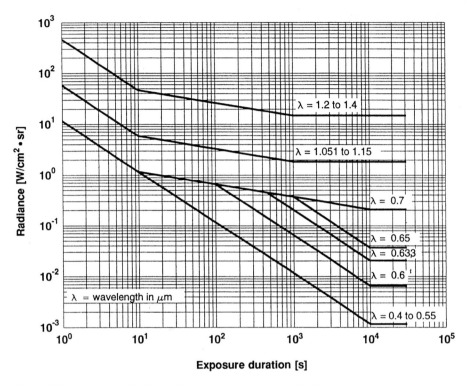

Fig. A.4. Exposure limits for direct viewing of a cw laser beam [A.2]

Appendix B

Conversion Factors and Constants

In this section we have listed some of the most frequently used conversion factors, constants and definitions.

Physical Constants

h	$= 6.626 \times 10^{-34} \, \text{J s}$	Planck constant
e	$= 1.602 \times 10^{-19} \, \text{A s}$	Charge of an electron
k	$= 1.381 \times 10^{-23} \, \text{J K}^{-1}$	Boltzmann constant
c	$= 2.998 \times 10^{8} \, \text{m s}^{-1}$	Speed of light in vacuum
ε_0	$= 8.854 \times 10^{-12} \, \text{A s V}^{-1} \text{m}^{-1}$	Permittivity of free space
μ_0	$= 1.257 \times 10^{-6} \, \text{V s A}^{-1} \text{m}^{-1}$	Permeability of free space
Z_0	$= \sqrt{\mu_0/\varepsilon_0} = 376.7 \, \Omega$	Impedance of free space
g	$= 9.81 \, \text{m s}^{-2}$	Acceleration due to gravity

Force Conversions

$$1 \, \text{dyne} = 1 \, \text{g cm s}^{-2}$$
$$1 \, \text{N} = 1 \, \text{kg m s}^{-2} = 10^5 \, \text{dyne}$$
$$1 \, \text{kp} = 9.81 \, \text{kg m s}^{-2}$$

Energy Conversions

$$1 \, \text{J} = 1 \, \text{Ws}$$
$$1 \, \text{cal} = 4.19 \, \text{J}$$
$$1 \, \text{eV} = 1.60 \times 10^{-19} \, \text{J}$$
$$1 \, \text{erg} = 10^{-7} \text{J}$$

Pressure Conversions

$$1 \, \text{Pa} = 1 \, \text{Nm}^{-2}$$
$$1 \, \text{atm} = 1.013 \, \text{bar} = 1.033 \, \text{kp cm}^{-2}$$
$$1 \, \text{bar} = 10^6 \, \text{dyne cm}^{-2} = 10^5 \, \text{Nm}^{-2}$$
$$1 \, \text{torr} = 133.3 \, \text{Nm}^{-2}$$

Conversion of English Units into the MKS System

1 inch	$= 2.540\,\mathrm{cm}$
1 gal	$= 3.785\,\mathrm{ltr}$
1 lbs	$= 0.453\ \mathrm{kg}$
1 atm	$= 14.7\,\mathrm{psi}$
1 Btu	$= 1055.8\,\mathrm{Ws}$
$T\ [^\circ\mathrm{C}]$	$= \frac{5}{9}(T\ [^\circ\mathrm{F}] - 32)$
$\Delta T\ [^\circ\mathrm{C}]$	$= \frac{5}{9}\Delta T\ [^\circ\mathrm{F}]$
$1\,\mathrm{cal\,cm^{-1}\,C^{-1}\,s^{-1}}$	$= 242\ \mathrm{Btu/hr\,ft\,F}$
$1\,\mathrm{cal\,g^{-1}}$	$= 1.8\,\mathrm{Btu/lb}$

Conversion of Angles

1°	$= 17.45\,\mathrm{mrad};\ 1' = 0.29\,\mathrm{mrad};\ 1'' = 4.85\,\mu\mathrm{rad}$
1 rad	$= 57^\circ\ 17'\ 45'',\ 1\,\mathrm{mrad} = 3'\ 26''$

Conversion of Wavenumber n [cm^{-1}] to Energy

$E = n\,hc$
$hc = 1.986 \times 10^{-23}\ \mathrm{Ws\,cm}$

Conversion of Linewidth Given in Wavelength ($\Delta\lambda$) or Wavenumber (Δn) to Bandwidth ($\Delta\nu$)

$$\Delta\nu/\nu = \Delta\lambda/\lambda = \Delta n/n$$
$$\Delta\nu = c\,\Delta n$$
$$\Delta\lambda = \Delta n\lambda^2 = \Delta\nu\lambda^2/c$$

Amplifier Gain

$g(\mathrm{dB}) = 10\,\log\,(E_2/E_1)$

Optical Units

Wave number	$n\ [1/\mathrm{cm}] \approx 10^4/(\lambda\ [\mu\mathrm{m}])$
Frequency	$\nu\ [\mathrm{Hz}] \approx 3 \times 10^{14}/(\lambda\ [\mu\mathrm{m}])$
Photon energy	$E\ [\mathrm{J}] \approx 1.987 \times 10^{-19}/(\lambda\ [\mu\mathrm{m}])$
Photon energy	$E\ [\mathrm{eV}] \approx 1.24/(\lambda\ [\mu\mathrm{m}])$

Conversion of transmission, T, to optical density, D

$$T = 10^{-D}$$

Conversion of Units for the Nonlinear Index of Refraction n_2

n_2 [esu] $= 238.6 \, n_0 n_2$ [cm^2/W]
n_2[cm^2/W] $= 4.19 \times 10^{-3}(n_2/n_0)$ [esu]

Conversion of Beam Quality Factor M to Beam Diameter-Divergence Product $D\Theta$

Divergence Θ and diameter D of a multimode beam scale with M with respect to TEM$_{00}$ mode (Θ_0, D_0):

$$\Theta = M\Theta_0, \quad D = MD_0$$

Diffraction limited beam divergence:

$$\Theta_0 = 4\lambda/\pi D_0$$

Therefore:

$$
\begin{aligned}
M^2 &= D\Theta/(4\lambda/\pi) \\
D &= \text{beam diameter [mm]} \\
\Theta &= \text{full beam angle [mr]} \\
\lambda &= \text{wavelength [mm]} \\
M &= \text{beam quality factor}
\end{aligned}
$$

Diffraction Limited Beam (TEM$_{00}$ Mode)

$$
\begin{aligned}
M^2 &= 1 \\
\Theta_0 D_0 &= 4\lambda/\pi \\
D_0 &= 2\omega_0, \text{ diameter of beam waist } (1/e^2 \text{ points}) \\
\Theta_0 D_0 &= 1.35 \text{ mm-mr at } 1.064 \, \mu\text{m}
\end{aligned}
$$

References

Chapter 1

1.1   A. Yariv: *Quantum Electronics*, 3rd edn. (Wiley, New York 1988);
      O. Svelto: *Principles of Lasers*, 3rd edn. (Plenum, New York 1989);
      K. Shimoda: *Introduction to Laser Physics*, 2nd edn., Springer Ser. Opt. Sci., Vol. 44
      (Springer, Berlin Heidelberg 1986);
      A.E. Siegman: *Lasers* (University Science Books, Mill Valley, CA 1986)
1.2   M. Garbuny: *Optical Physics* (Academic, New York 1965)
1.3   H. Haken: *Laser Theory* (Springer, Berlin Heidelberg 1984)
1.4   I.I. Sobelman: *Atomic Spectra and Radiative Transitions*, 2nd edn. Springer Ser. Atoms and
      Plasmas, Vol. 12 (Springer, Berlin Heidelberg)
      I.I. Sobelman, L.A. Vainshtein, E.A. Yukov: *Excitation of Atoms and Broadening of Spectral
      Lines*, 2nd edn. Springer Ser. Atoms and Plasmas, (Springer, Berlin Heidelberg 1995)
1.5   H. Statz, G.A. de Mars: In *Quantum Electronics*, ed. by C.H. Townes (Columbia Univ.
      Press, New York 1960) pp. 530–537

Chapter 2

2.1   A.L. Schawlow, C.H. Townes: Phys. Rev. **112**, 1940 (1958)
2.2   T.H. Maiman: Nature **187**, 493 (1960)
2.3   P.P. Sorokin, M.J. Stevenson: Phys. Rev. Lett. **5**, 557 (1960); and in *Advances in Quantum
      Electronics*, ed. by J.R. Singer (Columbia Univ. Press, New York 1961) p.65
2.4   E. Snitzer: Phys. Rev. Lett. **7**, 444 (1961)
2.5   L.F. Johnson, L. Nassau: Proc. IRE **49**, 1704 (1961)
2.6   J.E. Geusic, H.M. Marcos, L.G. Van Uitert: Appl. Phys. Lett. **4**, 182 (1964)
2.7   E. Snitzer, R.F. Woodcock, J. Segre: IEEE J. QE-**4**, 360 (1968)
2.8   S.E. Stokowski: Glass lasers, in *Handbook of Laser Science and Technology*, ed. by M.J.
      Weber (CRC Press, Boca Raton, FL 1982) pp. 215–264
2.9   P.F. Moulton: Paramagnetic ion lasers, in *Handbook of Laser Science and Technology*, ed.
      by M.J. Weber (CRC Press, Boca Raton, FL 1986) Vol. 1, pp. 21–295;
      L.G. DeShazer, S.C. Rund, B.A. Wechsler: Laser crystals, in *Handbook of Laser Science
      and Technology*, ed. by M.J. Weber (CRC Press, Boca Raton, FL 1987) Vol. 5, pp. 281–338;
      J.C. Walling: Tunable paramagnetic-ion solid-state lasers, in *Tunable Lasers*, 2nd edn., ed.
      by L.F. Mollenauer, J.C. White, C.R. Pollock, Topics Appl. Phys., Vol. 59 (Springer, Berlin
      Heidelberg 1992) Chap. 9
2.10 L. DeShazer, M. Bass, U. Ranon, T.K. Guka, E.D. Reed, T.W. Strozyk, L. Rothrock: Laser
      operation of neodymium in YVO_4 and gadolinium gallium garnet (GGG) and of holmium
      in YVO_4. 8th Int'l Quant. Electr. Conf., San Francisco, CA (1974)
2.11 D. Pruss, G. Huber, A. Beimowski: Appl. Phys. B **28**, 355–358 (1982)
2.12 L.F. Johnson, J.E. Geusic, L.G. Van Uitert: Appl. Phys. Lett. **7**, 127 (1965)
2.13 D.P. Devor, B.H. Soffer: IEEE J. QE-**8**, 231 (1972)
2.14 L. Schearer, M. Leduc: IEEE J. QE-**22**, 756 (1986)

708   References

2.15   Kh.S. Bagdasarov, A.A. Kaminskii: JETP Lett. **9**, 303 (1969)
2.16   R.F. Belt, J.R. Latore, R. Uhrin, J. Paxton: Appl. Phys. Lett. **25**, 218 (1974)
2.17   M.J. Weber, M. Bass, K. Andringa, R.R. Monchamp, E. Comperchio: Appl. Phys. Lett. **15**, 342 (1969)
2.18   M.J. Weber, M. Bass, T.E. Varitimos, D.P. Bua: IEEE J. QE-**9**, 1079 (1973)
2.19   R.V. Alves, R.A. Buchanan, K.A. Wickersheim, E.A.C. Yates: J. Appl. Phys. **42**, 3043 (1971)
2.20   R.C. Ohlmann, K.B. Steinbruegge, R. Mazelsky: Appl. Opt. **7**, 905 (1968)
2.21   K.B. Steinbruegge, R.H. Hopkins, G.W. Roland: Increased energy storage neodymium laser material: Silicate oxyapatite. Tech. Report AFAL-TR-72-37, Air Force Avionics Lab., WPAFB (1972)
2.22   K.B. Steinbruegge, G.D. Baldwin: Appl. Phys. Lett. **25**, 220 (1974)
2.23   G.D. Baldwin: Q-switched evaluation of CaLa SOAP : Nd. Tech. Report AFAL-TR-72-334, Air Force Avionics Lab., WPAFB (1972)
2.24   W.W. Krühler, J.P. Jeser, H.G. Danielmeyer: Appl. Phys. **2**, 329 (1973)
2.25   H.P. Weber, T.C. Damen, H.G. Danielmeyer, B.C. Tofield: Appl. Phys. Lett. **22**, 534 (1973)
2.26   H.P. Weber, P.F. Liao, B.C. Tofield: IEEE J. QE-**10**, 563 (1974)
        H.P. Weber, P.F. Liao, B.C. Tofield, P.m. Bridenbaugh: Appl. Phys. Lett. **26**, 692 (1975)
2.27   J.G. Gualtieri, T.R. Aucoin: Appl. Phys. Lett. **28**, 189 (1976)
2.28   H.G. Danielmeyer, G. Huber, W.W. Krühler, J.P. Jeser: Appl. Phys. **2**, 335 (1973)
2.29   K. Nassau, A.M. Broyer: J. Appl. Phys. **33**, 3064 (1962)
2.30   L.L. Harper, J.R. Thornton: Increased energy storage Nd Laser material: Sodium lanthanum molybdate. Tech. Report AFAL-TR-72-38, Air Force Avionics Lab., WPAFB (1972)
2.31   J.R. O'Connor: Appl. Phys. Lett. **9**, 407 (1966)
2.32   A.W. Tucker, M. Birnbaum, C.L. Fincher, J.W. Erler: J. Appl. Phys. **48**, 4907 (1977)
2.33   T. Chin, R.C. Morris, O. Kafri, M. Long, D.F. Heller: *High Power and Solid State Lasers*, ed. by W.W. Simmons. SPIE Proc. **622**, 53 (1986)
2.34   L.F. Johnson: J. Appl. Phys. **33**, 756 (1962); and **34**, 897 (1963)
2.35   E.P. Chicklis: Appl. Phys. Lett. **19**, 119 (1971); and Stimulated emission at 0.85 μm in Er^3 : YLF. 7th Int'l Quant. Electr. Conf., Montreal, Canada (1972)
2.36   D.P. Devor: 2.06 μm laser performance and design options for rangefinder and illuminator applications. IEEE/OSA Conf. Laser Eng. and Appl. (May 1975)
2.37   E.J. Sharp, D.T. Horowitz, T.E. Miller: J. Appl. Phys. **44**, 5399 (1973)
2.38   G. Müller, N. Neuroth: J. Appl. Phys. **44**, 2315 (1973)
2.39   M.J. Weber, M. Bass, G.A. deMars: J. Appl. Phys. **42**, 301 (1971)
2.40   W.F. Krupke, J.B. Gruber: J. Chem. Phys. **41**, 1225 (1964)
2.41   Z.J. Kiss, R.C. Duncan: Proc. IRE **50**, 1531 (1962)
2.42   S.A. Pollack: Proc. IEEE **51**, 1793 (1963)
2.43   E.Snitzer: Appl. Phys. Lett **6**, 45 (1965)
2.44   K.O. White, E.H. Holt: The erbium doped glass laser. Report ECOM-5294, U.S. Army, Fort Monmouth, NJ (1970)
2.45   N.P. Barnes, D.J. Gettemy: IEEE J. QE-**17**, 1303 (1981)
2.46   G.J. Quarles, A. Rosenbaum, C.L. Marquardt, L. Esterowitz: *Solid State Lasers*, ed. by G. Dubé. SPIE Proc. **1223**, 221 (1990)
2.47   T.Y. Fan, G. Huber, R.L. Byer, Mitzscherlich: IEEE J. QE-**24**, 924 (1988)
2.48   R.C. Stoneman, L. Esterowitz: Advanced Solid-State Laser Conf. (Salt Lake City, UT 1990), Techn. digest, p. 176, also *Solid State Lasers*, ed. by G. Dubé. SPIE Proc. **1223**, 231 (1990)
2.49   P.F. Moulton: IEEE J. QE-**21**, 1582 (1985)
2.50   J. Harrison, D. Welford, P.F. Moulton: IEEE J. QE-**25**, 1708 (1989)
2.51   IEEE J. Quantum Electronics. Special issue on solid-state lasers, QE-**24** (June 1988)
2.52   A.A. Kaminskii: *Laser Crystals*, 2nd edn., Springer Ser. Opt. Sci., Vol. 14 (Springer, Berlin Heidelberg 1990)

2.53   R.C. Powell: *Physics of Solid-State Laser Materials* (Springer, New York 1998)

2.54   O.C. Cronemeyer: J. Opt. Soc. Am. **56**, 1703 (1966)

2.55   D.M. Dodd, D.L. Wood, R.L. Barns: J. Appl. Phys. **35**, 1183 (1964)

2.56   T.H. Maiman, R.H. Hoskins, I.J. D'Haenens, C.K. Asawa, V. Evtuhov: Phys. Rev. **123**, 1151 (1961)

2.57   W. Koechner: Rev. Sci. Instr. **41**, 1699 (1970)

2.58   C.A. Burrus, J. Stone: Appl. Phys. Lett. **26**, 318 (1975)

2.59   T. Kushida, J.E. Geusic: Phys. Rev. Lett. **21**, 1172 (1968)

2.60   N.P. Barnes, D.J. Gettemy, L. Esterowitz, R.A. Allen: IEEE J. QE-**23**, 1434 (1987)

2.61   T. Kushida, H.M. Marcos, J.E. Geusic: Phys. Rev. **167**, 289 (1968)

2.62   W.F. Krupke, M.D. Shinn, J.E. Marion, J.A. Caird, S.E. Stokowski: J. Opt. Soc. Am. B **3**, 102 (1986)

2.63   J.K. Neeland, V. Evtuhov: Phys. Rev. **156**, 244 (1967)

2.64   M. Birnbaum, J.A. Gelbwachs: J. Appl. Phys. **43**, 2335 (1972)

2.65   M.J. Weber, T.E. Varitimos: J. Appl. Phys. **42**, 4996 (1971)

2.66   S. Singh, R.G. Smith, L.G. Van Uitert: Phys. Rev. B **10**, 2566 (1974)

2.67   H.G. Danielmeyer, M. Blätte: Appl. Phys. **1**, 269 (1973)

2.68   P.H. Klein, W.J. Croft: J. Appl. Phys. **38**, 1603 (1967)

2.69   H.F. Mahlein: IEEE J. QE-**6**, 529 (1970)

2.70   C.G. Bethea: IEEE J. QE-**9**, 254 (1973)

2.71   R.G. Smith: IEEE J. QE-**4**, 505 (1968);

2.72   S. Singh, R.G. Smith, L.G. Van Uitert: Phys. Rev. B **10**, 2566–2572 (1974)

2.73   A.A. Kaminskii: Sov. Phys. JETP **37**, 388–399 (1968)

2.74   J. Marling: IEEE J. QE-**14**, 56 (1978)

2.75   M. Birnbaum, C.F. Klein: J. Appl. Phys. **44**, 2928 (1973)

2.76   R.W. Wallace, S.E. Harris: Appl. Phys. Lett. **15**, 111 (1969)

2.77   R.W. Wallace: IEEE J. QE-**7**, 203 (1971)

2.78   M.D. Thomas, G.A. Rines, E.P. Chicklis, W. Koechner: High power 1.3 micron Nd: YaG laser. CLEO'86 (San Francisco, CA) paper WM4

2.79   S.M. Yarema, D. Milam: IEEE J. QE-**18**, 1941 (1982)

2.80   M.J. Weber: J. Non-Cryst. Solids **42**, 189 (1980)

2.81   S.E. Stokowski, R.A. Saroyan, M.J. Weber: Nd: doped laser glass spectroscopic and physical properties. Lawrence Livermore Nat'l Lab. Report M-095 (Rev.2) (1981)

2.82   M.J. Weber, D. Milam, W.L. Smith: Opt. Eng. **17**, 463 (1978)

2.83   W.W. Simmons, J.T. Hunt, W.E. Warren: IEEE J. QE-**17**, 1727 (1981)

2.84   D. Duston: IEEE J. QE-**6**, 3 (1970)

2.85   M. Michon: Phys. Lett. **19**, 219 (1965)

2.86   R. Dumanchin: IEEE J. QE-**7**, 53 (1971)

2.87   P.C. Magnante: IEEE J. QE-**8**, 440 (1972)

2.88   A.A. Mak, D.S. Prilezhaev, V.A. Serebryakov, A.D. Starikov: Optics and Spectroscopy **33**, 381 (1972)

2.89   C.F. Rapp: Laser glasses, in *Handbook of Laser Science and Technology*, Vol. V, Pt. 3, ed. by M.J. Weber (CRC Press, Boca Raton, FL 1987) pp. 339–372

2.90   W.F. Krupke: IEEE J. QE-**10**, 450 (1974)

2.91   E.V. Zharikov, N.N. Il'ichev, V.V. Laptev, A.A. Mayutin, V.G. Ostroumov, P.P. Pashinin, I.A. Shcherbakov: Sov. J. Quantum Electr. **12**, 338–341 (1982)

2.92   D. Pruss, G. Huber, A. Beimowski: Appl. Phys. B **28**, 355–358 (1982)

2.93   E.V. Zharikov, N.N. Il'ichev, V.V. Laptev, A.A. Mayutin, V.G. Ostroumov, P.P. Pashinin, A.S. Pimenov, V.A. Smirnov, I.A. Shcherbakov: Sov. J. Quantum Electron. **13**, 82–85 (1983)

2.94   A. Beimowski, G. Huber, D. Pruss, V.V. Laptev, I.A. Shcherbakov, E.V. Zharikov: Appl. Phys. B **28**, 234 (1982)

2.95   E.V. Zharikov, V.A. Zhitnyuk, G.M. Zverev, S.P. Kalitin, I.I. Kuratev, V.V. Laptev, A.M. Onishchenko, V.V. Osiko, V.A. Pashkov, A.S. Pimenov, A.M. Prokhorov, V.A. Smirnov, M.F. Stel'makh, A.V. Shestakov, I.A. Shcherbakov: Sov. J. Quantum Electron., **12**, 1652–1653 (1982)

2.96   D.S. Sumida, D.A. Rockwell: Dependence of Cr : Nd : GSGG pumping on Cr concentration, CLEO'86 (San Francisco, CA) paper WQ3

2.97   P. Fuhrberg, W. Luhs, B. Struve, G. Litfin: Single-mode operation of Cr : doped GSGG and KZnF₃, in *Tunable Solid-State Lasers II*, ed. by A.B. Budgor, L. Esterowitz, L.G. DeShazer, Springer Ser. Opt. Sci., Vol. 52 (Springer, Berlin Heidelberg 1987)

2.98   Lawrence Livermore National Laboratory, medium average power solid-state laser. Technical Information Seminar (October, 1985)

2.99   J.Y. Lice, C. Lice, M.G. Cohen: High average power normal mode Cr : Nd : GSGG lasers. CLEO'86 (San Francisco, CA) paper TuK33

2.100   E. Reed: IEEE J. QE-**21**, 1625 (1985)

2.101   E.P. Chicklis: Private commun. (1991)

2.102   M.G. Knights, G.A. Rines, J. McCarthy, T. Pollak, K.A. Smith, E.P. Chicklis: High power Nd : YLF laser performance. CLEO'84 (Anaheim, CA) paper WM1

2.103   M.G. Knights, M.D. Thomas, E.P. Chicklis, G.A. Rines, W. Seka: IEEE J. QE-**24**, 712 (1988)

2.104   T.M. Pollak, W.F. Wing, R.J. Grasso, E.P. Chicklis, H.P. Jenssen: IEEE J. QE-**18**, 159–163 (1982)

2.105   J.E. Murray: IEEE J. QE-**19**, 488–490 (1983)

2.106   L. De Shazer: Laser Focus World, p. 88 (February 1994)

2.107   A. Brignon, G. Feugnet, J.P. Huignard, J.P. Pocholle: IEEE J. QE-**34**, 577 (1998)

2.108   R.A. Fields, M. Birnbaum, C.L. Fincher: Appl. Phys. Lett. **51**, 1885 (1987)

2.109   A. Agnesi, S. Dell'Acqua, E. Piccinini, G. Reali, G. Piccinno: IEEE J. QE-**34**, 1480 (1998)

2.110   D.C. Brown, R. Nelson, L. Billings: Appl. Opt. **36**, 8611 (1997)

2.111   G. Feugnet, C. Bussac, M. Schwarz, J.P. Pocholle: Opt. Lett. **20**, 157 (1995)

2.112   M. Fuller, D. Matthews, L.R. Marshall: OSA Advanced Solid-State Lasers (Coeur d'Alene, ID 1998) Techn. digest, p. 252

2.113   J.E. Bernard, A.J. Alcock: Opt. Lett. **18**, 968 (1993)

2.114   P. Dekker, J.M. Dawes, J.A. Piper: In *TOPS on Advanced Solid-State Lasers* (Opt. Soc. Am., Washington, DC 1996) Vol. 1, p. 378

2.115   E.V. Zharikov, V.I. Zhekov, L.R. Kulevskii, T.M. Mirina, V.V. Osiko, A.M. Prokhorov, A.D. Savel'ev, V.V. Smirnov, B.P. Starikov, M.I. Timoshenko: Sov. J. Quantum Electron. **4**, 1039 (1975)

2.116   K.S. Bagdesarov, V.I. Zhekov, L.A. Kulevskii, V.A. Lobachev, T.M. Murina, A.M. Prokhorov: P. Sov. J. Quantum Electron. **10**, 1127 (1980)

2.117   K.L. Vodop'vanov, L.A. Kulevskii, A.A. Malyutin, P.P. Pashinin, A.M. Prokhorov: Sov. J. Quantum Electron. **12**, 541 (1982)

2.118   M. Bass, Wei-Quiang Shi, R. Kurtz, M. Kokta, H. Diegl: Room temperature operations of the 50 % doped Er : YAG laser at 2940 nm. CLEO '86 (San Francisco, CA) Postdeadline paper ThT1-1

2.119   C.E. Hamilton, R.J. Beach, S.B. Sutton, L.H. Furu, W.F. Krupke: Opt. Lett. **19**, 1627 (1994)

2.120   H.P. Weber, W. Luthy: Erbium laser for medical applications. CLEO '87 (Baltimore, MD) paper ThJ1

2.121   J.G. Manni, G.A. Rines, P.F. Moulton: Characterization of 2.9 μm laser operation in Er-doped crystals. CLEO'87 (Baltimore, MD) paper ThJ3

2.122   L. Esterowitz: *Growth, Characterisation, and Applications of Laser Host and Nonlinear Crystals*, ed. by J.T. Lin. SPIE Proc. **1104**, 216 (1989)

2.123   E. Snitzer, R. Woodcock: Appl. Phys. Lett. **6**, 45 (1965)

2.124   R. Renari, A. Johnson: *MELIOS: Status report of the U.S. Army's eyesafe laser rangefinder program*, ed. by P.K. Galoff, D.H. Sliney. SPIE Proc. **1207**, 112 (1990)

2.125   Y. Morishige, S. Kishida, K. Washio: Opt. Lett. **9**, 147–149 (1984)
2.126   S.J. Hamlin, J.D. Myers, M.J. Myers: SPIE Symposium on High Power Lasers, Los Angeles (1991), ed. by A.M. Johnson
2.127   L.F. Johnson, R.E. Dietz, H.J. Guggenheim: Phys. Rev. Lett. **11**, 318–320, 1963
2.128   L.F. Johnson, R.E. Dietz, H.J. Guggenheim: Appl. Phys. Lett. **5**, 21–22, 1964
2.129   L.F. Johnson, H.J. Guggenheim, R.A. Thomas: Phys. Rev. **149**, 179–185 (1966)
2.130   L.F. Johnson, H.J. Guggenheim: J. Appl. Phys. **38**, 4837 (1967)
2.131   L.F. Mollenauer, J.C. White, C.R. Pollock (eds.): *Tunable Lasers*, 2nd edn. Topics Appl. Phys., Vol. 59 (Springer, Berlin Heidelberg 1992)
2.132   L.F. Johnson, H.J. Guggenheim: IEEE J. QE-**10**, 442 (1974)
2.133   J.C. Walling, H.P. Jenssen, R.C. Morris, E.W. O'Dell, O.G. Peterson: Annual Meeting of the Optical Society of America (1978)
2.134   J.C. Walling, H.P. Jenssen, R.C. Morris, E.W. O'Dell, O.G. Peterson: Opt. Lett. **4**, 182–183, 1979
2.135   J.C. Walling, O.G. Peterson, H.P. Jenssen, R.C. Morris, E.W. O'Dell: IEEE J. QE-**16**, 1302–1315 (1980)
2.136   P.F. Moulton, A. Mooradian: Tunable transition-metal-doped solid state lasers, in *Laser Spectroscopy IV*, ed. by H. Walther, K.W. Rothe, Springer Ser. Opt. Sci., Vol. 21 (Springer, Berlin Heidelberg 1979) pp. 584–589
2.137   P.F. Moulton: Opt. News **8**, 9 (1982)
2.138   B. Struve, G. Huber: Appl. Phys. B **30**, 117–120 (1983)
2.139   B. Struve, G. Huber: Laser action and broad band fluorescence in Cr^{3+} : GdScGa garnet. 12th Int'l Quantum Electr. Conf. (1982) paper ThR-5
2.140   E.V. Zharikov, V.V. Laptev, E.I. Sidorova, Yu P. Timofeev, I.A. Shcherbakov: Sov. J. Quantum Electron. **12**, 1124 (1982)
2.141   H.P. Christensen, H.P. Jenssen: IEEE J. QE-**18**, 1197–1201 (1982)
2.142   U. Brauch, U. Dürr: Optics Commun. **49**, 61 (1984)
2.143   H.R. Verdun, L.M. Thomas, D.M. Andrauskas, T. McCollum, A. Pinto: Appl. Phys. Lett. **53**, 2593 (1988)
2.144   V. Petricevic, S.K. Gayen, R.R. Alfano: Appl. Phys. Lett. **53**, 2590 (1988)
2.145   V. Petricevic, S.K. Gayen, R.R. Alfano: Appl. Phys. Lett. **52**, 1040 (1988)
2.146   J.G. Daly, C.A. Smith: 2 micron laser applications, ed. by J. Quarles. SPIE Proc. **1627**, 26 (1992)
2.147   B.T. McGuckin, R.T. Menzies: IEEE QE-**28**, 1025 (1992)
2.148   U. Brauch, A. Giesen, M. Karszewski, Chr. Stewen, A. Voss: Opt. Lett. **20**, 713 (1995)
2.149   J.C. Walling, O.G. Peterson, H.P. Jenssen, R.C. Morris, E.W. O'Dell: IEEE J. QE-**16**, 1302 (1980)
2.150   H. Samelson, J.C. Walling, D.F. Heller: SPIE Proc. **335**, 85 (1982)
2.151   M. Shand, H.P. Jenssen: IEEE J. QE-**19**, 480 (1983)
2.152   S. Guch: CLEO '83 (Baltimore, MA) paper ThR3
2.153   R.C. Sam, J.J. Yeh, K.R. Leslie, W.R. Rapoport: IEEE J. QE-**24**, 1151 (1988)
2.154   W.R. Rapoport, J.J. Yeh, R.C. Sam: CLEO '83 (Baltimore, MD) paper ThR4
2.155   P.F. Moulton: J. Opt. Soc. Am. B **3**, 125 (1986); also Laser Focus **14**, 83 (May 1983)
2.156   E.G. Erickson: OSA Proc. on Tunable Solid State Lasers, Vol. 5, North Falmouth, MA (1989) p. 26
2.157   A. Hoffstädt: IEEE J. QE-**33**, 1850 (1997)
2.158   R. Rao, G. Vaillancourt, H.S. Kwok, C.P. Khattak: OSA Proc. on Tunable Solid State Lasers, Vol. 5, North Falmouth, MA (1989) p. 39
2.159   T.R. Steele, D.C. Gerstenberger, A. Drobshoff, R.W. Wallace: Opt. Lett. **16**, 399 (1991)
2.160   G.T. Maker, A.I. Ferguson: Opt. Lett. **15**, 375 (1990)
2.161   J. Harrison, A. Finch, D.M. Rines, G.A. Rines, P.F. Moulton: Opt. Lett. **16**, 581 (1991)
2.162   S.A. Payne, L.L. Chase, L.K. Smith, W.L. Kway, H.W. Newkirk: J. Appl. Phys. **66**, 1051 (1989)

2.163   S.A. Payne, L.L. Chase, H.W. Newkirk, L.K. Smith, W.F. Krupke: IEEE J. QE-**24**, 2243 (1988)

2.164   P.A. Beaud, M. Richardson, E.J. Miesak: IEEE J. QE-**31**, 317 (1995)

2.165   M. Richardson, V. Castillo, P. Beaud, M. Bass, B. Chai, G. Quarles, W. Ignatuk: Photonics Spectra, p. 86 (October 1993)

2.166   R. Scheps, J.F. Myers, H. Serreze, A. Rosenberg, R.C. Morris, M. Long: Opt. Lett. **16**, 820 (1991)

2.167   Q. Zhang, B.H.T. Chai, G.J. Dixon: CLEO '91 (Baltimore, MD) paper CThR6

2.168   S.A. Payne, W.F. Krupke, L.K. Smith, W.L. Kway, L.D. DeLoach, J.B. Tassano: IEEE J. QE-**28**, 1188 (1992)

2.169   M. Stalder, B.H.T. Chai, M. Bass: Appl. Phys. Lett. **58**, 216 (1991)

2.170   P.M.W. French, R. Mellisk, J.R. Taylor, P.J. Delfyett, L.T. Florez: Electron. Lett. **29**, 1262 (1993)

2.171   P.J.M. Suni, S.W. Henderson: Opt. Lett. **16**, 817 (1991)

2.172   I.F. Elder, M.J.P. Payne: OSA Advanced Solid State Lasers, San Francisco, CA (1996) Techn. digest, p. 30; also Appl. Opt. **36**, 8606 (1997)

2.173   E.C. Honea, R.J. Beach, S.B. Sutton, J.A. Speth, S.C. Mitchell, J.A. Skidmore, M.A. Emanuel, S.A. Payne: IEEE J. QE-**33**, 1592 (1997), also *TOPS, Advanced Solid State Lasers* **10**, 307 (Opt. Soc. Am., Washington, DC 1997)

2.174   G. Rustad, H. Hovland, K. Stenersen: OSA Advanced Solid State Lasers, San Francisco, CA (1996), Techn. digest, p. 48

2.175   R.C. Stoneman, L. Esterowitz: Opt. Lett. **15**, 486 (1990)

2.176   T. Yokozawa, H. Hara: Appl. Opt. **35**, 1424 (1996)

2.177   T.S. Kubo, T.J. Kane: IEEE J. QE-**28**, 1033 (1992)

2.178   J. Yu, U.N. Singh, N.P. Barnes, M. Petros: Opt. Lett. **23**, 780 (1998)

2.179   A. Finch, J.H. Flint: OSA Advanced Solid State Lasers, San Francisco, CA (1996) Techn. digest, p. 253

2.180   M.E. Storm: IEEE J. QE-**29**, 440 (1993)

2.181   W.F. Krupke, L.L. Chase: Opt. Quant. Electron. **22**, 1 (1990)

2.182   D.S. Sumida, T.Y. Fan: Proc. Advanced Solid State Lasers **20**, 100 (Opt. Soc. Am., Washington, DC 1994)

2.183   M. Karszewski, U. Brauch, K. Contag, A. Giesen, I. Johannsen, C. Stewen, A. Voss: OSA Advanced Solid State Lasers, Coeur d'Alene, ID (1998), Techn. digest, p. 82

2.184   T.Y. Fan: IEEE J. QE-**29**, 1457 (1993)

2.185   D.S. Sumida, T.Y. Fan: Opt. Lett. **20**, 2384 (1995)

2.186   C. Bibeau, R. Beach, C. Ebbers, M. Emanuel, J. Skidmore: *TOPS Advanced Solid State Lasers* **10**, 276 (Opt. Soc. Am., Washington, DC 1997)

2.187   H. Bruesselbach, D.S. Sumida, R. Reeder, R.W. Byren: *TOPS Advanced Solid State Lasers* **10**, 285 (Opt. Soc. Am., Washington, DC 1997)

2.188   H. Bruesselbach, D.S. Sumida: Opt. Lett. **21**, 480 (1996)

2.189   A. Giesen, M. Karszewski, C. Stewen, A. Voss: *Proc. on Advanced Solid State Lasers* **24**, 330 (Opt. Soc. Am., Washington, DC 1995)

2.190   A. Giesen, M. Brauch, I. Johannsen, M. Karszewski, U. Schiegg, C. Stewen, A. Voss: *TOPS Advanced Solid State Lasers* **10**, 280 (Opt. Soc. Am., Washington, DC 1997)

Chapter 3

3.1   W.W. Rigrod: J. Appl. Phys. **36**, 2487 (1965)

3.2   A.Y. Cabezas, R.P. Treat: J. Appl. Phys. **37**, 3556 (1966)

3.3   A.E. Siegman: *Lasers* (University Science Books, Mill Valley, CA 1986)

3.4   O. Svelto: *Principles of Lasers* (Plenum, New York 1989)

3.5   E. Behrens: Internal Report (Fibertek, Inc. 1991)

3.6   M.J.F. Digonnet, C.J. Gaeta: Appl. Opt. **24**, 333 (1985)

3.7   J.B. Trenholme: Laser Program Annual Report UCRL-50021-86, Livermore National Laboratory, Livermore, CA (1987)

3.8   D. Findlay, R.A. Clay: Phys. Lett. **20**, 277 (1966)

3.9   R. Dunsmuir: J. Electron. Control **10**, 453 (1961)

3.10   H. Statz, G.A. Mars, D.T. Wilson, C.L. Tang: J. Appl. Phys. **36**, 1510 (1965)

3.11   D. Röss: *Lasers, Light Amplifiers and Oscillators* (Academic, New York 1969)

3.12   J.W. Strozyk: IEEE J. QE-**3**, 343 (1967)

3.13   F.R. Marshall, D.L. Roberts: Proc. IRE **50**, 2108 (1962)

3.14   E. Panarella, L. Bradley: IEEE J. QE-**11**, 181 (1975)

3.15   C.H. Thomas, E.V. Price: IEEE J. QE-**2**, 617 (1966)

3.16   R.P. Johnson, N.K. Moncur, L.D. Siebert: CLEO '87 (Baltimore, MD) paper FP2

3.17   T.J. Kane: IEEE Photon. Technol. Lett. **2**, 244 (1960)

3.18   K. Tsubono, S. Moriwaki: Jpn. J. Appl. Phys. **31**, 1241 (1992)

3.19   C.C. Harb, M.B. Gray, H.A. Bachor, R. Schilling, P. Rottengatter, I. Freitag, H. Welling: IEEE J. QE-**30**, 2907 (1994)

3.20   L.W. Casperson: J. Appl. Phys. **47**, 4555 (1976)

3.21   A. Owyoung, G.R. Hadley, R. Esherick, R.L. Schmitt, L.A. Rahn: Opt. Lett. **10**, 484 (1985)

3.22   W. Koechner: IEEE J. QE-**8**, 656 (1972)

3.23   H.G. Danielmeyer: J. Appl. Phys. **41**, 4014 (1970)

3.24   T. Kimura, K. Otsuka: IEEE J. QE-**6**, 764 (1970)

3.25   R. Polloni, O. Svelto: IEEE J. QE-**4**, 481 (1968)

3.26   J.F. Nester: IEEE J. QE-**6**, 97 (1970)

3.27   I.C. Chang, E.G.H. Lean, C.G. Powell: IEEE J. QE-**6**, 436 (1970)

3.28   G.D. Baldwin, I.T. Basil: IEEE J. QE-**7**, 179 (1971)

3.29   R.B. Chesler: Appl. Opt. **9**, 2190 (1970)

3.30   J.M. Eggleston, L. DeShazer, K. Kangas: IEEE J. QE-**24**, 1009 (1988)

3.31   V. Evtuhov, J.K. Neeland: J. Appl. Phys. **38**, 4051 (1967)

3.32   D. Röss: IEEE J. QE-**2**, 208 (1966)

3.33   C.G. Young: Laser Focus **3**, 36 (February 1967)

3.34   R.C. Sam: High average power alexandrite lasers. Allied Corp. (Morristown, NJ). Private commun.

3.35   A.D. Hays, R. Burnham, G.L. Harnagel: CLEO '89 (Baltimore, MA) Postdeadline paper PO 9-1

3.36   W. Koechner, H.R. Verdun: Diode laser pumped solid-state lasers. Final Rept. Contract N00014-85-C-0174, Naval Oceans Systems Command, San Diego, CA (1987)

3.37   N. Barnes, M. Storm, P. Cross, M. Skolaut: IEEE J. QE-**26**, 558 (1990)

3.38   D.L. Sipes: Appl. Phys. Lett. **47**, 74–76 (1985)

3.39   M.J.F. Digonnet, C.J. Gaeta: Appl. Opt. **24**, 333 (1985)

3.40   H.R. Verdun, T. Chuang: Opt. Lett **17**, 1000 (1992)

3.41   E.C. Honea, C.A. Ebbers, R.J. Beach, J.A. Speth, J.A. Skidmore, M.A. Emanuel, S.A. Payne: Opt. Lett. **23**, 1203 (1998)

3.42   C. Bibeau, R. Beach, C. Ebbers, M. Emanuel, J. Skidmore: TOPS, Advanced Solid State Lasers **10**, 276 (Opt. Soc. Am., Washington, DC 1997)

3.43   R.A. Fields, M. Birnbaum, C.L. Fincher: CLEO '87 (Baltimore, MD)

3.44   N. Mermilliod, R. Romero, I. Chartier, C. Garapon, R. Moncorgé: IEEE J. QE-**28**, 1179 (1992)

3.45   K.X. Liu, C.J. Flood, D.R. Walker, H.M. van Driel: Opt. Lett. **17**, 1361 (1992)

3.46   J.R. Lincoln, A.I. Ferguson: Opt. Lett. **19**, 2119 (1994)

3.47   R. Mellisk, P.M.W. French, J.R. Taylor, P.J. Delfyett, L.T. Florez: Proc. Advanced Solid State Lasers **20**, 239 (Opt. Soc. Am., Washington, DC 1994)

3.48   D. Kopf, K.J. Weingarten, L.R. Brovelli, M. Kamp, U. Keller: Opt. Lett. **19**, 2143 (1994)

3.49   E.C. Honea, R.J. Beach, S.B. Sutton, J.A. Speth, S.C. Mitchell, J.A. Skidmore, M.A. Emanuel, S.A. Payne: IEEE J. QE-**33**, 1592 (1997)

3.50   G. Feugnet, C. Bussac, C. Larat, M. Schwarz, J.P. Pocholle: Opt. Lett. **20**, 157 (1995)
3.51   B. Desthieux, R.I. Laming, D.N. Payne: Appl. Phys. Lett. **63**, 586 (1993)
3.52   R.M. Percival, D. Szebesta, J.R. Williams, R.D.T. Lauder, A.C. Tropper, D.C. Hanna: Electron. Lett. **30**, 1598 (1994)
3.53   H. Takara, A. Takada, M. Saruwatari: IEEE Photon. Tech. Lett. **4**, 241 (1992)
3.54   G. Nykolak, S.A. Kramer, J.R. Simpson, D.J. DiGiovanni, C.R. Giles, H.M. Presby: IEEE Photon. Tech. Lett. **3**, 1079 (1991)
3.55   W.H. Cheng, J.H. Bechtel: Electron Lett. **29**, 2055–2057 (1993)
3.56   A.R. Clobes, M.J. Brienza: Appl. Phys. Lett. **21**, 265 (1972)
3.57   A.M. Bonch-Bruevich, V.Yu. Petrun'kin, N.A. Esepkina, S.V. Kruzhalov, L.N. Pakhomov, V.A. Chernov, S.L. Galkin: Sov. Phys.-Tech. Phys. **12**, 1495 (1968)
3.58   T.J. Kane, R.L. Byer: Opt. Lett. **10**, 65 (1985)
3.59   W.R. Trutna, D.K. Donald, M. Nazarathy: Opt. Lett. **12**, 248 (1987)
3.60   A.C. Nilsson, E.K. Gustafson, R.L. Byer: IEEE J. QE-**25**, 767 (1989)
3.61   T.J. Kane, E.A.P. Cheng: Opt. Lett. **13**, 970 (1988)
3.62   D.G. Scerbak: *Solid-State Lasers*, ed. by G. Dubé. SPIE **1223**, 196 (1990)
3.63   J. Harrison, G.A. Rines, P.F. Moulton, J.R. Leger: Opt. Lett. **13**, 111 (1988)
3.64   R. Scheps, J. Myers: *Solid-State Lasers*, ed. by G. Dubé. SPIE Proc. **1223**, 186 (1990)

Chapter 4

4.1   L.M. Frantz, J.S. Nodvik: J. Appl. Phys. **34**, 2346 (1963)
4.2   R. Bellman, G. Birnbaum, W.G. Wagner: J. Appl. Phys. **34**, 780 (1963)
4.3   E.L. Steele: J. Appl. Phys. **36**, 348 (1965)
4.4   E.L. Steele: *Optical Lasers in Electronics* (Wiley, New York 1968)
4.5   P.G. Kriukov, V.S. Letokhov: *Laser Handbook I*, ed. by E.T. Arecchi, E.O. Schulz-DuBois (North-Holland, Amsterdam 1972) pp. 561–595
4.6   P.V. Avizonis, R.L. Grotbeck: J. Appl. Phys. **37**, 687 (1966)
4.7   C.R. Jones, P.V. Avizonis, P. Sivgals: Experimental investigation of the behavior of neodymium-glass laser amplifiers. NBS Spec. Pub. **341**, 28 (1970)
4.8   J.M. McMahon: Glass laser material testing at Naval Research Laboratory. ASTM Report on Damage in Laser Glass STP-**469**, 117 (1969)
4.9   I.F. Balashov, V.A. Berenberg, V.V. Blagoveshchenskii: Sov. Phys. **14**, 692 (1965)
4.10   M. Michon, R. Auffret, R. Dumanchin: J. Appl. Phys. **41**, 2739 (1970)
4.11   J.E. Geusic, H.E.D. Scovil: In *Quantum Electronics III* (Columbia Univ. Press, New York 1964) pp. 1211–1220
4.12   W.R. Sooy, R.S. Congleton, B.E. Dobratz, W.K. Ng: In *Quantum Electronics III* (Columbia Univ. Press, New York 1964) pp. 1103–1112
4.13   J.I. Davis, W.R. Sooy: Appl. Opt. **3**, 715 (1964)
4.14   R. Carman: Laser fusion program. Semiannual Report UCRL-50021-73-1, Lawrence Livermore Lab., Livermore, CA (January–June 1973) p. 154
4.15   D.C. Brown: *High-Peak Power Nd:Glass Laser Systems*, Springer Ser. Opt. Sci. Vol. 25 (Springer, Berlin, Heidelberg 1981)
4.16   Special issue on lasers for fusion, IEEE J. QE-**17** (1981)
4.17   Lawrence Livermore Lab. Laser Program Annual Reports UCRL-50021-74 (1974) to UCRL-50021-98 (1998)
4.18   M.J. Weber: J. Non-Cryst. Solids **42**, 189 (1980)
4.19   G.J. Linford, R.A. Saroyan, J.B. Trenholme, M.J. Weber: IEEE J. QE-**15**, 510 (1979)
4.20   W.E. Martin, D. Milam: IEEE J. QE-**18**, 1155 (1982);
        D.W. Hall, M.J. Weber: IEEE J. QE-**20**, 831 (1984);
        D.W. Hall, R.A. Haas, W.F. Krupke, M.J. Weber: IEEE J. QE-**19**, 1704 (1983)
4.21   S.M. Yarema, D. Milam: IEEE J. QE-**18**, 1941 (1982)

4.22   C. Yamanaka, Y. Kato, Y. Izawa, K. Yoshida, T. Yamanaka, T. Sasaki, M. Nakatsuka, T. Mochizuki, J. Kuroda, S. Nakai: IEEE J. QE-**17**, 1639 (1981)

4.23   C.C. Young, J.W. Kantorski: Optical gain and inversion in Nd : glass lasers. Proc. 1st DOD Conf. on Laser Technology (1964) p. 75

4.24   R.W. Beck: Damage threshold studies of glass laser materials. Ownes-Illinois, Tech. Report ARPA, Contract DAHC-69-C-0303 (January 1970)

4.25   T.G. Crow, T.J. Snyder: Techniques for achieving high-power Q-switched operation in YAG : Nd. Final Tech. Report AFAL-TR-70-69, Air Force, WPAFB (1970); see also Laser J. **18**, (November/December 1970)

4.26   N.P. Barnes, V.J. Corcoran, I.A. Crabbe, L.L. Harper, R.W. Williams, J.W. Wragg: IEEE J. QE-**10**, 195 (1974)

4.27   W. Koechner, R. Burnham, J. Kasinski, P. Bournes, D. DiBiase, K. Le, L. Marshall, A. Hays: High-power diode-pumped solid-state lasers for Optical Space Communications. In Proc. Int'l. Conf. Optical Space Com. (Munich, 10–14 June 1991), ed. by J. Franz. SPIE Proc. **1522**, 169 (1991)

4.28   W.W. Rigrod: J. Appl. Phys. **34**, 2602 (1963)

4.29   E.O. Schulz-DuBois: Bell Systems Techn. J. **43**, 625 (1964)

4.30   A.Y. Cabezas, R.P. Treat: J. Appl. Phys. **37**, 3556 (1966)

4.31   A.Y. Cabezas, G.L. McAllister, W.K. Ng: J. Appl. Phys. **38**, 3487 (1967)

4.32   J. Bunkenberg, J. Boles, D.C. Brown, J. Eastman, J. Hoose, R. Hopkins, L. Iwan, S.D. Jacobs, J.H. Kelly, S. Kumpan, S. Letzring, D. Lonobile, L.D. Lund, G. Mourou, S. Refermat, S. Seka, J.M. Soures, K. Walsh: IEEE J. QE-**17**, 1620 (1981)

4.33   W.B. Bridges: IEEE J. QE-**4**, 820 (1968)

4.34   J.P. Campbell, L.G. DeShazer: J. Opt. Soc. Am. **59**, 1427 (1969)

4.35   J. Trenholme: A user oriented axially symmetric diffraction code. Semiannual Report UCRL-50021-73-1, Lawrence Livermore Lab., Livermore, CA (January–June 1973) p. 46

4.36   D.R. Speck, E.S. Bliss, J.A. Glaze, J.W. Herris, F.W. Holloway, J.T. Hunt, B.C. Johnson, D.J. Kuizenga, R.G. Ozarski, H.G. Patton, P.R. Rupert, G.J. Suski, C.D. Swift, C.E. Thompson: IEEE J. QE-**17**, 1599 (1981)

4.37   J.T. Hunt, J.A. Glaze, W.W. Simmons, P.A. Renard: Appl. Opt. **17**, 2053 (1978)

4.38   S.A. Akhmanov, R.V. Khokhlov, A.P. Sukhorukov: In *Laser Handbook II*, ed. by E.T. Arecchi, E.O. Schulz-DuBois (North-Holland, Amsterdam 1972) p. 1151

4.39   R.Y. Chiao, E. Garmire, C.H. Townes: Phys. Rev. Lett. **13**, 479 (1964)

4.40   V.I. Bespalov, V.I. Talanov: JEPT Lett. **3**, 307 (1966)

4.41   J. Trenholme: Review of small signal theory. Laser Program Annual Report UCRL-50021-74, Lawrence Livermore Lab., Livermore, CA (1974) p. 178

4.42   E.S. Bliss, D.R. Speck, J.F. Holzrichter, J.H. Erkkila, A.J. Glass: Appl. Phys. Lett. **25**, 448 (1974)

4.43   J.F. Holrichter, D.R. Speck: J. Appl. Phys. **47**, 2459 (1976)

4.44   J. Trenholme: Review of small signal theory: Laser Program Annual Report, UCRL-50021-74, Lawrence Livermore Lab., Livermore, CA (1974) p. 178

4.45   J.B. Trenholme: Proc. Soc. Photo-Opt. Instr. Eng. **69**, 158 (1975)

4.46   E.S. Bliss, D.R. Speck, J.F. Holzrichter, J.H. Erkkila, A.J. Glass: Appl. Phys. Lett. **25**, 448 (1974)

4.47   E.S. Bliss, J.T. Hunt, P.A. Renard, G.E. Sommargren, H.J. Weaver: Effects of nonlinear propagation on laser focusing properties. Lawrence Livermore Lab. Preprint UCRL-77557 (1975)

4.48   E.S. Bliss, G.E. Sommargren, H.J. Weaver: Loss of focusable energy due to small scale nonlinear effects. Conf. on Laser Eng. and Appl., Washington, DC (May 1975) paper 8.2

4.49   J.T. Hunt, P.A. Renard, W.W. Simmons: Appl. Opt. **16**, 779 (1977). Also, J.T. Hunt, J.A. Glaze, W.W. Simmons, P.A. Renard: Appl. Opt. **17**, 2053 (1978)

4.50   W.W. Simmons, J.T. Hunt, W.E. Warren: IEEE J. QE-**17**, 1727 (1981)

4.51   W.W. Simmons, S. Guch, F. Rainer, J.E. Murray: A high energy spatial filter for removal of small scale beam instabilities in high power solid state lasers. Conf. on Laser Eng. and Appl., Washington, DC (1975) paper 8.4

4.52   M.A. Duguay, L.E. Hargrove, K.B. Jefferts: Appl. Phys. Lett. **9**, 287 (1966)

4.53   F. Shimuzu: Phys. Rev. Lett. **19**, 1097 (1967)

4.54   R.J. Roenk: Phys. Lett. **24A**, 228 (1967)

4.55   F. DeMartini, C.H. Townes, T.K. Gustafson, P.L. Kelley: Phys. Rev. **164**, 312 (1967)

4.56   M.A. Duguay, J.W. Hansen, S.L. Shapiro: IEEE J. QE-**6**, 725 (1970)

4.57   R.A. Fisher: Picosecond optical pulse nonlinear propagation effects, Ph.D. thesis, Univ. California, Berkeley, CA (1971)

4.58   L. Tonks: J. Appl. Phys. **35**, 1134 (1964)

4.59   J.A. Glaze, S. Guch, J.B. Trenholme: Appl. Opt. **13**, 2808–2811 (1974)

4.60   J.B. Trenholme: Fluorescence amplification and parasitic oscillation limitations in disk lasers, Memorandum Rep. 2480, Naval Research Lab., Washington DC, July 1972

4.61   G.P. Kostometov, N.N. Rozanov: Sov. J. Quantum Electron. **6**, 696–699 (1976)

4.62   D.C. Brown, S.D. Jacobs, N. Nee: Appl. Opt. **17**, 211–224 (1978)

4.63   A. Hardy, D. Treves: IEEE J. QE-**15**, 887–895 (1979)

4.64   S. Guch: Appl. Opt. **15**, 1453 (1976)

4.65   D. Benfey, D. Brown, J. Chernoch: Adv. Solid State Lasers, Salt Lake City, UT (1990) paper WE1-1

4.66   G.J. Linford, E.R. Peressini, W.R. Sooy, M.L. Spaeth: Appl. Opt. **13**, 379 (1974)

4.67   C.G. Young, J.W. Kantorski, E.O. Dixon: Appl. Phys. **37**, 4319 (1966)

4.68   G.J. Linford, L.W. Hill: Appl. Opt. **13**, 1387 (1974)

4.69   J.A. Glaze, S. Guch, J.B. Trenholme: Appl. Opt. **13**, 2808 (1974)

4.70   G.P. Kostometov, N.N. Rozanov: Sov. J. Quantum Electron. **6**, 696 (1976)

4.71   D.C. Brown, S.D. Jacobs, N. Nee: Appl. Opt. **17**, 211 (1978)

4.72   G.J. Linford, R.A. Saroyan, J.B. Trenholme, M.J. Weber: IEEE J. QE-**15**, 510 (1979)

4.73   P. Labudde, W. Seka, H.P. Weber: Appl. Phys. Lett. **29**, 732 (1976)

4.74   A.N. Chester: Appl. Opt. **12**, 2139 (1973)

Chapter 5

5.1   A.G. Fox, T. Li: Bell Syst. Tech. J. **40**, 453 (1961)

5.2   G.D. Boyd, J.P. Gordon: Bell Syst. Tech. J. **40**, 489 (1961)

5.3   G.D. Boyd, H. Kogelnik: Bell Syst. Tech. J. **41**, 1347 (1962)

5.4   K. Kogelnik, T. Li: Appl. Opt. **5**, 1550 (1966); see also Proc. IEEE **54**, 1312 (1966); also H. Kogelnik: In *Lasers I*, ed. by A.K. Levine (Dekker, New York 1966) pp. 295–347

5.5   H.K.V. Lotsch: Optik **28**, 65, 328, 555 (1968/1969); Optik **29**, 130, 662 (1969); and Optik **30**, 1, 181, 217, 563 (1969)

5.6   J.S. Kruger: Electro-Opt. Syst. Designs 12 (September 1972)

5.7   R.J. Freiberg, A.S. Halsted: Appl. Opt. **8**, 335 (1969)

5.8   G. Goubau, F. Schwering: IRE Trans. AP-**9**, 248 (1961)

5.9   J.R. Pierce: Proc. Nat. Acad. Sci. **47**, 1808 (1961)

5.10   T. Li: Bell Syst. Tech. J. **44**, 917 (1965)

5.11   L.A. Vainshtein: Sov. Phys.-JETP **44**, 1050 (1963); Sov. Phys. JETP **17**, 709 (1963)

5.12   D.E. McCumber: Bell Syst. Tech. J. **44**, 333 (1965)

5.13   B.A. See: Laser resonator, properties of laser beams and design of optical systems. N68-16910, Weapons Research Establishment, Salisbury, South Australia (1967)

5.14   J.S. Kruger: Beam divergence for various transverse laser modes. Report AD-729-299, Harry Diamond Lab., Washington, DC (1971)

5.15   G.L. McAllister, M.M. Mann, L.G. DeShazer: Transverse mode distortion in giant-pulse laser oscillators. IEEE Conf. Laser Eng. and Appl., Washington, DC (1969)

5.16   A.G. Fox, T. Li: IEEE J. QE-**2**, 774 (1966)

5.17   H. Kogelnik, Bell Syst. Tech. J. **44**, 455 (1965)

5.18   H. Steffen, J.P. Lörtscher, G. Herziger: IEEE J. QE-**8**, 239 (1972)

5.19   D.R. Whitehouse, C.F. Luck, C. Von Mertens, F.A. Horrigan, M. Bass: Mode control technology for high performance solid state lasers. Rept. TR ECOM-0269-F, US Army Electr. Command, Forth Monmouth, NJ (1963)

5.20   T.J. Gleason: Analysis of complex laser cavities. Rept. HDL-TM-71-5, Harry Diamond Lab., Washington, DC (1971)

5.21   D.C. Hanna: IEEE J. QE-**5**, 483 (1969)

5.22   H.W. Kogelnik: IEEE J. QE-**8**, 373 (1972)

5.23   R.B. Chesler, D. Maydan: J. Appl. Phys. **43**, 2254 (1972)

5.24   A.L. Bloom: Properties of laser resonators giving uniphase wavefronts. Tech. Bulletin No. 7, Spectra Physics, Mountain View, CA (1963)

5.25   R.J. Freiberg, A.S. Halsted: Appl. Opt. **8**, 355 (1969)

5.26   I.M. Belousova, O.B. Danilov: Soc. Phys. **12**, 1104 (1968)

5.27   E.A. Teppo: Nd : YAG laser laboratory experiments, Tech. Note 4051-2 (Feb. 1972); Tech. Note 4051-7, Naval Weapons Center, China Lake, CA (August 1973)

5.28   L.W. Davis: J. Appl. Phys. **39**, 5331 (1968)

5.29   J.E. Geusic, H.J. Levingstein, S. Singh, R.C. Smith, L.G. Van Uitert: Appl. Phys. Lett. **12**, 306 (1968)

5.30   S.C. Tidwell, J.F. Seamens, M.S. Bowers, A. Cousins: IEEE J. QE-**28**, 997–1009 (1992)

5.31   W.C. Fricke: Appl. Opt. **9**, 2045 (1970)

5.32   F.A. Levine: IEEE J. QE-**7**, 170 (1971)

5.33   W.C. Scott, M. DeWit: Appl. Phys. Lett. **18**, 3 (1971)

5.34   Applied Optics Research, 59 Stonington Drive, Pittsford, N.Y. Physical optics analysis, code: GLAD 4.5.

5.35   H.K.V. Lotsch: Jpn. J. Appl. Phys. **4**, 435 (1965)

5.36   D. Ryter, M. Von Allmen: IEEE J. QE-**17**, 2015 (1981)

5.37   J.P. Lörtscher, J. Steffen, G. Herziger: IEEE J. QE-**7**, 505 (1975)

5.38   H.P. Kortz, R. Iffländer, H. Weber: Appl. Opt. **20**, 4124 (1981)

5.39   V. Magni: Appl. Opt. **25**, 107 (1986); also Appl. Opt. **25**, 2039 (1986)

5.40   S. DeSilvestri, P. Laporta, V. Magni: Opt. Commun. **57**, 339 (1986)

5.41   N.K. Berger, N.A. Deryugin, Y.N. Lukyanov, Y.E. Studenikin: Opt. Spectrosc. USSR **43**, 176 (1977)

5.42   R. Hauck, H.P. Kortz, H. Weber: Appl. Opt. **19**, 598 (1980)

5.43   D.C. Hanna, S.G. Sawyers, M.A. Yuratich; Opt. Quantum Electron. **13**, 493 (1981)

5.44   D.C. Hanna, S.G. Sawyers, M.A. Yuratich: Opt. Commun. **37**, 359 (1981)

5.45   P.H. Sarkies: Opt. Commun. **31**, 189 (1979)

5.46   M. Born, E. Wolf: *Principles of Optics* (Pergamon, New York 1964)

5.47   F.A. Jenkins, H.E. White: *Fundamental of Optics* (McGraw-Hill, New York 1957)

5.48   K. Iizuka: *Engineering Optics*, 2nd edn. Springer Ser. Opt. Sci., Vol. 35 (Springer, Berlin, Heidelberg 1986)

5.49   J.K. Watts: Appl. Opt. **7**, 1621 (1968)

5.50   M. Hercher: Appl. Opt. **8**, 1103 (1969)

5.51   B.L. Booth, S.M. Jarrett, G.C. Barker: Appl. Opt. **9**, 107 (1970)

5.52   A. Yariv: *Quantum Electronics*, 3rd edn. (Wiley, New York (1988) Sect. 21.1

5.53   B. Zhou, T.J. Kane, G.J. Dixon, R.L. Byer: Opt. Lett. **10**, 62 (1985)

5.54   S.A. Collins, G.R. White: Appl. Opt. **2**, 448 (1963)

5.55   D.A. Kleinman, P.P. Kisliuk: Bell Syst. Tech. J. **41**, 453 (1962)

5.56   H.K.V. Lotsch: A modified Fabry–Perot interferometer as a discrimination filter and a modulator for longitudinal modes. Sci. Rept. No. 2, US Air Force Contract AFI9(604)-8052 (1962)

5.57   L.A. Rahn: Appl. Opt. **24**, 940 (1985)

5.58   M. Hercher: Appl. Phys. Lett. **7**, 39 (1965)

5.59   D.G. Peterson, A. Yariv: Appl. Opt. **5**, 985 (1966)
5.60   R.M. Schotland: Appl. Opt. **9**, 1211 (1970)
5.61   M. Daehler, G.A. Sawyer, E.L. Zimmermann: J. Appl. Phys. **38**, 1980 (1967)
5.62   G. Magyar: Rev. Sci. Instrum. **38**, 517 (1967)
5.63   M.M. Johnson, A.H. LaGrone: Appl. Opt. **12**, 510 (1973)
5.64   W.B. Tiffany: Appl. Opt. **7**, 67 (1968)
5.65   F.J. McClung, D. Weiner: IEEE J. QE-**1**, 94 (1965)
5.66   P.W. Smith: Proc. IEEE **60**, 421 (1972)
5.67   W. Wiesemann: Appl. Opt. **12**, 2909 (1973)
5.68   E. Snitzer: Appl. Opt. **5**, 121 (1966)
5.69   N.M. Galaktionova, G.A. Garkavi, V.F. Egorova, A.A. Mak, V.A. Fromzel: Opt. Spectrosc. **28**, 404 (1970)
5.70   W. Culshaw, J. Kannelaud: IEEE J. QE-**7**, 381 (1971)
5.71   W. Culshaw, J. Kannelaud, J.E. Peterson: IEEE J. QE-**10**, 253 (1974)
5.72   H.G. Danielmeyer, W.N. Leibolt: Appl. Phys. **3**, 193 (1974)
5.73   D. Roess: Appl. Phys. Lett. **8**, 109 (1966)
5.74   M.P. Vanyukov, V.I. Isaenko, L.A. Luizova, A. Shorokhov: Opt. Spectrosc. **20**, 535 (1966)
5.75   H.G. Danielmeyer: IEEE J. QE-**6**, 101 (1970)
5.76   R.L. Schmitt, L.A. Rahn: Appl. Opt. **25**, 629 (1986)
5.77   B.H. Soffer: J. Appl. Phys. **35**, 2551 (1964)
5.78   B.B. McFarland, R.H. Hoskins, B.H. Soffer: Nature **207**, 1180 (1965)
5.79   J.E. Bjorkholm, R.H. Stolen: J. Appl. Phys. **39**, 4043 (1968)
5.80   K.A. Arunkumar, J.D. Trolinger: Opt. Engineering **27**, 657 (1988)
5.81   .R.F. Wuerker: SPIE J. **12**, 122 (1971)
5.82   W.R. Sooy: Appl. Phys. Lett. **7**, 36 (1965)
5.83   E.S. Fry, Q. Hu, X. Li: Appl. Opt. **30**, 1015 (1991)
5.84   D.C. Hanna, B. Luther-Davies, R.C. Smith: Electron. Lett. **8**, 369 (1972)
5.85   D.C. Hanna, B. Luther-Davies, H.N. Rutt, R.C. Smith: Opto-Electronics **3**, 163 (1971)
5.86   D.C. Hanna, B. Luther-Davies, R.C. Smith: Opto-Electronics **4**, 249 (1972)
5.87   C.L. Tang, H. Statz, G.A. DeMars, D.T. Wilson: Phys. Rev. **136**, A1 (1964)
5.88   A.M. Bonch-Bruevich, V.Yu. Petrun'kin, N.A. Esepkina; S.V. Kruzhalov, L.N. Pakhomov, V.A. Chernov, S.L. Galkin: Sov. Phys. **12**, 1495 (1968)
5.89   A.R. Clobes, M.J. Brienza: Appl. Phys. Lett. **21**, 265 (1972)
5.90   V. Yu. Petrun'kin, L.N. Pakhomov, S.V. Kruzhalov, N.M Kozhevnikov: Sov. Phys. **17**, 1222 (1973)
5.91   F. Biraben: Opt. Commun. **29**, 353 (1979)
5.92   O.E. Nanii, A.N. Shelaev: Sov. J. Quant. Electron. **14**, 638 (1984)
5.93   G.A. Rines, P.F. Moulton, M.G. Knights: CLEO '87 (Baltimore, MD) paper ThQ1
5.94   T.J. Kane, R.J. Byer, Opt. Lett. **10**, 65 (1985)
5.95   L.J. Bromley, D.C. Hanna: Opt. Lett. **16**, 378 (1991)
5.96   D.A. Draegert: IEEE J. QE-**8**, 235 (1972)
5.97   H.G. Danielmeyer, W.G. Nilsen: Appl. Phys. Lett. **16**, 124 (1970)
5.98   H.G. Danielmeyer, E.H. Turner: Appl. Phys. Lett. **17**, 519 (1970)
5.99   W.R. Trutna, D.K. Donald, M. Nazarathy: CLEO '87 (Baltimore, MD) paper WN2
5.100   B. Zhou, T.J. Kane, G.I. Dixon, R.L. Byer: Opt. Lett **10**, 62 (1985)
5.101   Y.K. Park, G. Giulani, R. L. Byer: Opt. Lett. **5**, 96 (1980)
5.102   T. Kedmi, D. Treves: Appl. Opt. **20**, 2108 (1981)
5.103   V. Evtuhov, A.E. Siegman: Appl. Opt. **4**, 142 (1965)
5.104   T.J. Kane, W.J. Kozlovsky, R.L. Byer, C.E. Byvik: Opt. Lett. **12**, 239 (1987)
5.105   J.F. Nester: IEEE J. QE-**6**, 97 (1970)
5.106   W. Koechner: IEEE J. QE-**8**, 656 (1972)
5.107   R.B. Chesler: Appl. Opt. **9**, 2190 (1970)
5.108   G.D. Baldwin: IEEE J. QE-**7**, 179 (1971)

5.109   T. Kimura: IEEE J. QE-**6**, 764 (1970)
5.110   H.G. Danielmeyer: J. Appl. Phys. **41**, 4014 (1970)
5.111   H.W. Kogelnik, E.P. Ippen, A. Dienes, C.V. Shank: : IEEE J. QE-**8**, 373 (1972)
5.112   D. Hanna: IEEE J. QE-**5**, 483 (1969)
5.113   N.P. Barnes, J.C. Barnes: IEEE J. QE-**29**, 2670 (1993)
5.114   J.C. Barnes, N.P. Barnes, L.G. Wang, W. Edwards: IEEE J. QE-**29**, 2684 (1993)
5.115   T.J. Kane: IEEE Photon. Technol. Lett. **2**, 244 (1990)
5.116   K. Tsubono, S. Moriwaki: Jpn. J. Appl. Phys. **31**, 1241 (1992)
5.117   C.C. Harb, M.B. Gray, H.A. Bachor, R. Schilling, P. Rottengatter, I. Freitag, H. Welling:
         IEEE J. QE-**30**, 2907 (1994)
5.118   N.A. Robertson, S. Hoggan, J.B. Mangan, J. Hough: Appl. Phys. B**39**, 149 (1986)
5.119   J. Harrison, A. Finch, J.H. Flint, P.F. Moulton: IEEE J. QE-**28**, 1123 (1992)
5.120   T.J. Kane, A.C. Nilsson, R.L. Byer: Opt Lett. **12**, 175 (1987)
5.121   J.J. Zayhowski, J.A. Keszenheimer: IEEE J. QE-**28**, 1118 (1992)
5.122   E. Ritter: Optical coatings and thin film techniques, in *Laser Handbook*, ed. by F.T. Arrechi,
         E.O. Schulz-DuBois (North-Holland, Amsterdam 1972) pp. 897–921
5.123   P. Baumeister: *Handbook of Optical Design* (US Governement Printing Office, Washington,
         DC 1963)
5.124   D.S. Heavens: *Optical Properties of Thin Solid Films* (Butterworth, London 1955)
5.125   H.A. Macleod: *Thin-Film Optical Filters*, 2nd edn. (MacMillan, New York 1986)
5.126   A. Vasicek: *Optics of Thin Films* (North-Holland, Amsterdam 1960)
5.127   A.E. Siegman: Proc. IEEE **53**, 227 (1965)
5.128   A.E. Siegman, E. Arrathon: IEEE J. QE-**3**, 156 (1967)
5.129   A.E. Siegman: Appl. Opt. **13**, 353 (1974) see also *Lasers* (Univ. Sci. Books, Mill Valley,
         CA 1986)
5.130   W.H. Steier: Unstable resonators, in *Laser Handbook*, ed. by M. Stitch (North-Holland,
         Amsterdam 1979) Vol. 3
5.131   R.L. Herbst, M. Komine, R.L. Byer: Opt. Commun. **21**, 5 (1977)
5.132   Yu.A. Ananev, G.N. Vinokurov, L.V. Kovalchuk, N.A. Sventsitskaya, V.E. Sherstobitov:
         Sov. Phys.-JETP **31**, 420 (1970)
5.133   R.L. Byer, R.L. Herbst: Laser Focus 48-57 (July 1978)
5.134   W.F. Krupke, W.R. Sooy: IEEE J. QE-**5**, 575 (1969)
5.135   G. Vakhimov: Radio Eng. Electron. Phys. **10**, 1439 (1965)
5.136   G. Giulani, Y.K. Park, R.L. Byer: Opt. Lett. **5**, 491 (1980)
5.137   J.M. Eggleston, G. Giulani, R.L. Byer: J. Opt. Soc. Am. **71**, 1264 (1982)
5.138   D.T. Harter, J.C. Walling: Opt. Lett. **11**, 706 (1986)
5.139   E. Armandillo, C. Norrie, A. Cosentino, P. Laporta, P. Wazen, P. Maine: Opt. Lett. **22**,
         1168 (1997)
5.140   S. De Silvestri, P. Laporta, V. Magni, O. Svelto: Opt. Lett. **12**, 84 (1987)
5.141   D. Andreou: Rev. Sci. Instrum. **49**, 586 (1978)
5.142   T.F. Ewanizky, J.M. Craig: Appl. Opt. **15**, 1465 (1976)
5.143   P.G. Gobbi, S. Morosi, G.C. Reali, A.S. Zarkasi: Appl. Opt. **24**, 26 (1985)
5.144   N. McCarthy, P. Lavigne: Appl. Opt. **22**, 2704–2708 (1983)
5.145   N. McCarthy, P. Lavigne: Appl. Opt. **23**, 3845–3850 (1984)
5.146   D.M. Walsh, L.V. Knight: Appl. Opt. **25**, 2947–2954 (1986)
5.147   A. Parent, N. McCarthy, P. Lavigne: IEEE J. QE-**23**, 222–228 (1987)
5.148   P. Lavigne, N. McCarthy, J.G. Demers: Appl. Opt. **24**, 2581–2586 (1985)
5.149   N. McCarthy, P. Lavigne: Opt. Lett. **10**, 553–555 (1985)
5.150   K.J. Snell, N. McCarthy, M. Piché: Opt. Commun. **65**, 377–382 (1988)
5.151   A. Parent, P. Lavigne: Appl. Opt. **28**, 901–903 (1989)
5.152   C. Zizzo, C. Arnone, C. Calì, S. Sciortino: Opt. Lett. **13**, 342–344 (1988)
5.153·   G. Emiliani, A. Piegari, S. De Silvesti, P. Laporta, V. Magni: Appl. Opt. **28**, 2832–2837
         (1989)

5.154   S. De Silvestri, P. Laporta, V. Magni, O. Svelto, B. Majocchi: Opt. Lett. **13**, 201 (1988)
5.155   S. De Silvestri, P. Laporta, V. Magni, O. Svelto: IEEE J. QE-**24**, 1172 (1988)
5.156   S. De Silvestri, V. Magni, O. Svelto, G. Valentini: IEEE J. QE-**26**, 1500 (1990)
5.157   S. De Silvestri, P. Laporta, V. Magni, G. Valentini, G. Cerullo: Opt. Commun. **77**, 179 (1990)
5.158   J.M. Yarborough, J. Hobart: Conf. on Laser Eng. Appl. Washington, DC (1973) postdeadline paper
5.159   A.L. Bloom: J. Opt. Soc. Am. **64**, 447 (1974)
5.160   I.J. Hodgkinson, J.I. Vukusic: Opt. Commun. **24**, 133 (1978)
5.161   J.A. Keszenheimer, J.J. Zayhowski: CLEO '91 (Baltimore, MD) paper CME8
5.162   A. Owyoung, P. Esherick: Opt. Lett. **12**, 999 (1987)
5.163   T.J. Kane, E.A.P. Cheng: Opt. Lett. **13**, 970 (1988)
5.164   W.R. Trutna, D.K. Donald: Opt. Lett. **15**, 369 (1990)
5.165   P.A. Schultz, S.R. Henion: Opt. Lett. **16**, 578 (1991)
5.166   J.A. Keszenheimer, K.F. Wall, S.F. Root: OSA Proc. Adv. Solid-State Lasers **15**, 283 (1992)
5.167   P. Robrish: Opt. Lett. **19**, 813 (1994)
5.168   T. Day, E.K. Gustafson, R.L. Byer: IEEE J. QE-**28**, 1106 (1992)
5.169   N. Uehara, K. Ueda: Opt. Lett. **18**, 505 (1993)
5.170   K. Nakagawa, A.S. Shelkovnikov, T. Katsuda, M. Ohtsu: Appl. Opt. **33**, 6383 (1994)
5.171   N. Hodgson, H. Weber: Opt. Quantum Electron. **22**, 39 (1990)
5.172   N. Hodgson, T. Haase: SPIE Proc. **1277**, 88 (1990)
5.173   V. Magni, S. DeSilvestri, O. Svelto, G.L. Valentini: Lasers and Electro-Optics **10**, 366 (1991)
5.174   V. Magni, S. DeSilvestri, Lie-Jia Qian, O. Svelto: Opt. Commun. **94**, 87 (1992)
5.175   N. Hodgson, G. Bostanjoglo, H. Weber: Opt. Commun. **99**, 75 (1993)
5.176   N. Hodgson, G. Bostanjoglo, H. Weber: Appl. Opt. **32**, 5902 (1993). See also G. Bostan-joglo: Nah- und Fernfeldverteilungen instabiler optischer Laserresonatoren mit apodisierten Reflektoren, Doctoral Thesis, Technische Universität Berlin (1995)
5.177   G. Bostanjoglo, G.A. Witt: Fibertek Report (February 14, 1994)

Chapter 6

6.1   J.H. Goncz, P.B. Newell: J. Opt. Soc. Am. **56**, 87 (1966)
6.2   J.H. Goncz: Instr. Soc. Am. Trans. **5**, 1 (1966)
6.3   G.J. Linford: Appl. Opt. **33**, 8333 (1994)
6.4   J.L. Emmet, A.L. Schawlow: J. Appl. Phys. **35**, 2601 (1964)
6.5   J.R. Oliver, F.S. Barnes: Proc. IEEE **59**, 638 (1971)
6.6   J.H. Goncz, W.J. Mitchell: IEEE J. QE-**3**, 330 (1967)
6.7   J. Oliver, F.S. Barnes: IEEE J. QE-**5**, 225 (1969)
6.8   W.D. Fountain, L.M. Osterink, J.D. Foster: IEEE J. QE-**6**, 232 and 684 (1970)
6.9   M.B. Davies, P. Scharman, J.K. Wright: IEEE J. QE-**4**, 424 (1968)
6.10   E.A. Teppo: Nd : YAG laser laboratory experiments. Tech. Note 4051-2, Naval Weapons Center, China Lake, CA (1972)
6.11   K.R. Lang, F.S. Barnes: J. Appl. Phys. **35**, 107 (1964)
6.12   J.H. Goncz: J. Appl. Phys. **36**, 742 (1965)
6.13   J.P. Markiewicz, J.L. Emmett: IEEE J. QE-**2**, 707 (1966);
J.F. Holzrichter, J.L. Emmett: Appl. Opt. **8**, 1459 (1969)
6.14   D.E. Perlman: Rev. Sci. Instr. **37**, 340 (1966)
6.15   J.H. Rosolowski, R.J. Charles: J. Appl. Phys. **36**, 1792 (1965)
6.16   R.A. Dugdale, R.C. McVickers, S.D. Ford: J. Nuclear Mater. **12**, 1 (1964)
6.17   H.E. Edgerton, J.H. Goncz, J. Jameson: Xenon flashlamps, limits of operation. Proc. 6th Int'l Congr. on High Speed Photography, Haarlem, Netherlands (1963) p. 143
6.18   L. Waszak: Microwaves **130** (May 1969)

6.19   EG&G Linear Xeonon Flash Tubes, Data Sheet 1002-B

6.20   F. Schuda: In *Flashlamp-Pumped Laser Technology*, ed. by F. Schuda. SPIE Proc. **609**, 177 (1986)

6.21   B. Newell, J.D. O'Brian: IEEE J. QE-**4**, 291 (1968)

6.22   J.F. Holzrichter, N. Dobeck, A. Pemberton: Flashlamp development. Laser Progr. Ann. Rept. UCRL-50021-74, Lawrence Livermore Lab. (1974) pp. 107–115

6.23   L. Richter, F. Schuda, J. Degnan: In *Solid State Lasers*, ed. by G. Dubé. SPIE Proc. **1223**, 142 (1990)

6.24   W. Koechner, L. DeBenedictis, E. Matovich, G.E. Mevers: IEEE J. QE-**8**, 310 (1972)

6.25   J. Richards, D. Rees, K. Fueloep, B.A. See: Appl. Opt. **22**, 1325 (1983)

6.26   J.H. Kelly, D.C. Brown, K. Teegarden: Appl. Opt. **19**, 3817 (1980)

6.27   K. Yoshida, Y. Kato, H. Yoshida, C. Yamanaka: Rev. Sci. Instr. **55**, 1415 (1984)

6.28   T.B. Read: Appl. Phys. Lett. **9**, 342 (1966)

6.29   W. Koechner: Laser Focus **5**, 29 (September 1969)

6.30   M. Grasis, L. Reed: Long-life krypton arc lamp for pumping Nd : YAG lasers. Techn. Rept. AFAL-TR-73-156, ILC Technologies, Sunnyvale, CA (1973)

6.31   W.E. Thouret: Illum. Eng. **55**, 295 (1960)

6.32   S. Yoshikawa, K. Iwamoto, K. Washio: Appl. Opt. **10**, 1620 (1971)

6.33   L.M. Osterink, J.D. Foster: Efficient high power Nd : YAG laser characteristics. 1969 IEEE Conf. on Laser Eng. Appl., Washington, DC (1969)

6.34   W. Koechner: *Solid-State Laser Engineering*, 1st edn., Springer Ser. Opt. Sci., Vol. 1 (Springer, Berlin Heidelberg 1976)

6.35   W. Koechner, H. Verdun: Laser-diode array pumped solid state laser systems. Final Rept., Naval Ocean Systems Command, Contract No. N00014-85-C-0174 (1986)

6.36   W. Streifer, D.R. Scifres, G.L. Harnagel, D.F. Welch, J. Berger, M. Sakamoto: IEEE J. QE-**24**, 883 (1988)

6.37   J. Hecht: *The Laser Guidebook*, 2nd. edn. (McGraw-Hill, New York 1991)

6.38   R. Beach, W.J. Benett, B.L. Freitas, D. Mundinger, B.J. Comaskey, R.W. Solarz, M.A. Emanuel: IEEE J. QE-**28**, 966 (1992)

6.39   J.G. Endriz, M. Vakili, G.S. Browder, M. DeVito, J.M. Haden, G.L. Harnagel, W.E. Plano, M.Sakamoto. D.F. Welch, S. Willing, D.P. Worland, H.C. Yao: IEEE J. QE-**28**, 952 (1992)

6.40   A.A. Karpinski: Laser Focus World (October 1994) p. 155; also Nov. 1990 Annual Meeting OSA, Boston Mass. Technical Digest, paper MK10

6.41   K.J. Linden, P.N. McDonnell: NASA Conf. Publ. **3249**, 233 (1993)

6.42   Jae-Hoon Kim, R.J. Lang, A. Larson: Appl. Phys. Lett. **57**, 2048–2050 (1990)

6.43   N.W. Carlson, G. Evans, D. Bour, S. Liew: Appl. Phys. Lett. **56**, 16–18 (1990)

6.44   A. Scherer, J.L. Jewell, Y.H. Lee, J.P. Harbison, L.T. Florez: Appl. Phys. Lett. **55**, 2724 (1989)

6.45   G. Harnagel, D. Welch, P. Cross, D. Scifers: *Lasers and Applications*, June 1986, p. 135

6.46   K.A. Forrest, J.B. Abshire: IEEE J. QE-**23**, 1287 (1987)

6.47   M. Ettenberg: Laser Focus **22**, 86 (May 1985)

6.48   N.P. Barnes, M.E. Storm, P.L. Cross, M.W. Skolaut: IEEE J. QE-**26**, 558 (1990)

6.49   G.C. Osbourn, P.L. Gourley, I.J. Fritz, R.M. Biefield, L.R. Dawson, T.E. Zipperian: Principles and applications of strained-layer superlattices, in *Semiconductors and Semimetals*, Vol. 24, *Applications of Multiple Quantum Wells, Doping, and Superlattices*, ed. by R. Dingle (Academic, San Diego, CA 1987) p. 459

6.50   R.G. Waters: Enhanced longevity strained-layer lasers. LEOS '90 (Boston 1990) paper SDL6.2

6.51   J.J. Coleman: Strained-layer InGaAs-GaAs quantum-well heterostructure lasers. LEOS '90 (Boston 1990) paper SDL6.1

6.52   J.L. Dallas, R.S. Afzal, M.A. Stephen: Appl. Opt. **35**, 1427 (1996)

6.53   G.N. Glascoe, J.V. Lebacqz: *Pulse Generators* (McGraw-Hill, New York 1948)

6.54   L.C. Yang: Laser Focus **9**, 37 (July 1973)

6.55   N.P. DePratti: J. Phys. E. **4**, 1 (1971)

6.56   V.J. Corcoran, R.W. McMillan, S.K. Barnoske: IEEE J. QE-**10**, 618 (1974)

6.57   F. Benjamin: Electro-Optical System Designs, 32 (March 1975)

6.58   W.L. Gagnon, G. Allen: Power conditioning. Laser Progr. Ann. Rept. UCRL-50021-74, Lawrence Livermore Lab. (1974)

6.59   R.H. Dishington, W.R. Hook, R.P. Hilberg: Appl. Opt. **13**, 2300 (1974)

6.60   W.R. Hook, R.H. Dishington, R.P. Hilberg: IEEE Trans. ED-**19**, 308 (1972)

6.61   G.F. Albrecht, J.M. Eggleston, J.J. Ewing: IEEE J. QE-**22**, 2099 (1986)

6.62   J.M. Eggleston, G.F. Albrecht, R.A. Petr, J.F. Zumdiek: IEEE J. QE-**22**, 2092 (1986)

6.63   M.R. Siegrist: Appl. Opt. **15**, 2167 (1976)

6.64   D.A. Huchital: IEEE J. QE-**12**, 1 (1976)

6.65   I.P. Lesnick, C.H. Church: IEEE J. QE-**2**, 16 (1966)

6.66   Yu.A. Kalinin: Sov. J. Opt. Techn. **37**, 1 (1970)

6.67   D. Roess: Microwaves **4**, 29 (1965); see also IEEE J. QE-**2**, 208 (1966)

6.68   C.H. Church, I. Liberman: Appl. Opt. **6**, 1966 (1967)

6.69   G.J. Fan, C.B. Smoyer, J. Nuñez: Appl. Opt. **3**, 1277 (1964)

6.70   T.Y. Fan: IEEE J. QE-**29**, 1457 (1993)

6.71   H.R. Verdun, T. Chuang: Opt. Lett. **17**, 1000 (1992)

6.72   D.C. Shannon, R.W. Wallace: Opt. Lett. **16**, 318 (1991)

6.73   W.A. Clarkson, D.C. Hanna: Opt. Lett. **21**, 869 (1996)

6.74   C. Bibeau, R. Beach, C. Ebbers, M. Emanuel, J. Skidmore: *TOPS Advanced Solid State Lasers* **10**, 276 (Opt. Soc. Am., Washington, DC 1997)

6.75   R.J. Beach, C. Bibeau, E.C. Honea, S.B. Sutton: *ICF Quarterly Report* **7**, 52 (March 1997). Lawrence Livermore National Laboratory Report UCRL-LR-105821-97-2.

6.76   E.C. Honea, R.J. Beach, S.B. Sutton, J.A. Speth, S.C. Mitchell, J.A. Skidmore, M.A. Emanuel, S.A. Payne: IEEE J. QE-**33**, 1592 (1997)

6.77   S.C. Tidwell, J.F. Seamans, M.S. Bowers: Opt. Lett. **18**, 116 (1993); also **16**, 584 (1991)

6.78   H. Zbinden, J.E. Balmer: Opt. Lett. **15**, 1014 (1990)

6.79   Th. Graf, J.E. Balmer: Opt. Lett. **18**, 1317 (1993)

6.80   J.R. Leger, W.C. Goltsos: IEEE J. QE-**28**, 1088 (1992)

6.81   Keming Du, M. Baumann, B. Ehlers, H.G. Treusch, P. Loosen: *TOPS Advanced Solid State Lasers* **10**, 390 (Opt. Soc. Am., Washington, DC 1997)

6.82   A. Babushkin, W. Seka: Advanced Solid State Lasers, Coeur d'Alene, ID (1998), Techn. digest, p. 112

6.83   S. Yamaguchi, T. Kobayashi, Y. Saito, K. Chiba: Appl. Opt. **35**, 1430 (1996)

6.84   W.L. Nighan: Laser Focus World, p. 97 (May 1995)

6.85   E.C. Holnea, C.A. Ebbers, R.J. Beach, J.A. Speth, J.A. Skidmore, M.A. Emanuel, S.A. Payne: Opt. Lett. **23**, 1203 (1998)

6.86   A. Brignon, G. Feugnet, J.P. Huignard, J.P. Pocholle: IEEE J. QE-**34**, 577 (1998)

6.87   A.D. Hays, L.R. Marshall, J. Kasinski, R. Burnham: Proc. Adv. Solid State Lasers, Hilton Head, SC (1991) paper WA 1-1, p. 180

6.88   W. Hughes, A. Hays, D. DiBiase, J. Kasinski, R. Burnham: Adv. Solid State Lasers, Salt Lake City, UT (1990), Techn. digest, p. 102

6.89   A.D. Hays, R. Burnham: CLEO '90 (Anaheim, CA) paper CMA1, p. 4

6.90   R. Burnham: CLEO '90 (Anaheim, CA) paper CMF3, p. 24

6.91   R.A. Utano, D.A. Hyslop, T.A. Allik: Solid State Lasers. SPIE Proc. **1223**, 128 (1990)

6.92   F. Hanson, D. Haddock: Appl. Opt. **27**, 80 (1988)

6.93   L.R. Marshall, A.D. Hays, R.L. Burnham: CLEO '91 (Baltimore, MA) paper CFC7, p. 492

6.94   A.D. Hays, J. Kasinski, L.R. Marshall, R. Burnham: CLEO '91 (Baltimore, MA) paper CFC2, p. 490

6.95   Y. Hiramo, K. Tatsumi, K. Kasahara: CLEO '91 (Baltimore, MA) paper CFC4, p. 490

6.96   J.E. Bernard, A.J. Alcock: Opt. Lett. **19**, 1861 (1994); also **18**, 968 (1993)

6.97   C.E. Hamilton, R.J. Beach, S.B. Sutton, L.H. Furu, W.F. Krupke: Opt. Lett. **19**, 1627 (1994)

6.98   T.M. Baer, D.F. Head, P. Gooding, G. J. Kintz, S. Hutchison: IEEE J. QE-**28**, 1131 (1992)
6.99   J. Richards, A. McInnes: Opt. Lett. **20**, 371 (1995)
6.100   S.B. Schuldt, R.L. Aagard: Appl. Opt. **2**, 509 (1963)
6.101   V. Evtuhov, J.K. Neeland: Appl. Opt. **6**, 437 (1967)
6.102   J.G. Edwards: Appl. Opt. **6**, 837 (1967)
6.103   K. Kamiryo, T. Lano, K. Matsuzawa: Jpn. J. Appl. Phys. **5**, 1217 (1966)
6.104   D. Fekete: Appl. Opt. **5**, 643 (1966)
6.105   C. Bowness: Appl. Opt. **4**, 103 (1965)
6.106   K. Kamiryo: Proc. IEEE **53**, 1750 (1965)
6.107   D.M. Camm: Appl. Opt. **23**, 601 (1984)
6.108   F. Docchio, L. Pallaro, O. Svelto: Appl. Opt. **24**, 3752 (1985)
6.109   F Docchio: Appl. Opt. **24**, 3746 (1985)
6.110   Yu.A. Kalinin, A.A. Mak: Opt. Tech. **37**, 129 (1970)
6.111   D.R. Skinner, J. Tregellas-Williams: Australian J. Phys. **19**, 1 (1966)
6.112   J. Whittle, D.R. Skinner: Appl. Opt. **5**, 1179 (1966)
6.113   D.R. Skinner: Appl. Opt. **8**, 1467 (1969)
6.114   K. Kamiryo, T. Kano, H. Matsuzawa: Proc. IEEE **55**, 1630 (1967)
6.115   W.R. Sooy, M.L. Stitch: J. Appl. Phys. **34**, 1719 (1963)
6.116   C.H. Cooke, J. McKenna, J.G. Skinner: Appl. Opt. **3**, 957 (1964)
6.117   D. Röss: *Lasers, Light Amplifiers and Oscillators* (Academic, New York 1969) p. 426
6.118   V. Daneu, C.C. Sacchi, O. Svelto: Alta Frequenza, 758 (Nov. 1964)
6.119   H.U. Leuenberger, G. Herziger: Appl. Opt. **14**, 1190 (1975)
6.120   W. Koechner: Appl. Opt. **9**, 1429 (1970)
6.121   J. Trenholme: Optimizing the design of a kilojoule laser amplifier chain. Laser Fusion Program, Semiannual. Rept., Lawrence Livermore Lab. (1973) p. 60
6.122   P. Laporta, V. Magni, O. Svelto: IEEE J. QE-**21**, 1211 (1985)
6.123   M.S. Mangir, D.A. Rockwell: IEEE J. QE-**22**, 574 (1986)
6.124   D.D. Bhawalkar, L. Pandit: IEEE J. QE-**9**, 43 (1973)
6.125   W.W. Morey: IEEE J. QE-**8**, 818 (1972)
6.126   A.N. Fletcher: Appl. Phys. B **37**, 31 (1985)

Chapter 7

7.1   H.S. Carslaw, J.C. Jaeger: *Conduction of Heat in Solids* (Oxford Univ. Press, London 1948) p.191
7.2   S.T. Hsu: *Engineering Heat Transfer* (Van Nostrand, Princeton, NJ 1963) p. 274
7.3   W. Koechner: Appl. Opt. **9**, 1429 (1970)
7.4   W. Koechner: J. Appl. Phys. **44**, 3162 (1973)
7.5   S. Timoshenko, J.N. Goodier: *Theory of Elasticity*, 3rd edn. (McGraw-Hill, Singapore 1982)
7.6   W. Koechner: Appl. Phys. **2**, 279 (1973)
7.7   M. Born, E. Wolf: *Principles of Optics* (Pergamon, London 1965)
7.8   J.F. Nye: *Physical Properties of Crystals* (Clarendon, Oxford, UK 1985, reprinted 1993); L.A. Shuvalov (ed.) *Modern Crystallography IV*, Springer Ser. Solid-State Sci., Vol. 37 (Springer, Berlin Heidelberg 1988)
7.9   R.W. Dixon: J. Appl. Phys. **38**, 5149 (1967)
7.10   J.D. Foster, L.M. Osterink: J. Appl. Phys. **41**, 3656 (1970)
7.11   W. Koechner, D.K. Rice: IEEE J. QE-**6**, 557 (1970)
7.12   H. Kogelnik: Bell Syst. Tech. J. **44**, 455 (1965)
7.13   W. Koechner: Appl. Opt. **9**, 2548 (1970)
7.14   K.B. Steinbruegge, T. Henningsen, R.H. Hopkins, R. Mazelsky, N.T. Melamed, E.P. Riedel, G.W. Roland: Appl. Opt. **11**, 999 (1972)
7.15   K.B. Steinbruegge, G.D. Baldwin: Appl. Phys. Lett. **25**, 220 (1972)

7.16   M. Ohmi, M. Akatsuka, K. Ishikawa, K. Naito, Y. Yonezawa, Y. Nishida, M. Yamanaka, Y. Izawa, S. Nakai: Appl. Opt. **33**, 6368 (1994)

7.17   H.J. Eichler, A. Haase, R. Menzel, A. Siemoneit: J. Phys. D **26**, 1884 (1993)

7.18   M.A. Karr: Appl. Opt. **10**, 893 (1971)

7.19   S. Jackel, I. Moshe, R. Lullouz: *OSA Adv. Solid State Lasers*, Coeur d'Alene, ID (1998), Techn. digest, p. 127

7.20   W.C. Scott, M. de Wit: Appl. Phys. Lett. **18**, 3 (1971)

7.21   J.J. Kasinski, R.L. Burnham: Appl. Opt. **35**, 5949 (1996)

7.22   Q. Lü, N. Kugler, H. Weber, S. Dong, N. Müller, U. Wittrock: Opt. Quantum Electr. **28**, 57 (1996)

7.23   N. Kugler, S. Dong, Q. Lü, H. Weber: Appl. Opt. **36**, 9359 (1997)

7.24   S.Z. Kurtev, O.E. Denchev, S.D. Savov: Appl. Opt. **32**, 278 (1993)

7.25   G.D. Baldwin, E.P. Riedel: J. Appl. Phys. **38**, 2726 (1967)

7.26   G. Giuliani, P. Ristori: Optics Commun. **35**, 109 (1980)

7.27   R. Byren: Hughes Aircraft Corp. Electro-Optical and Data Systems Group. Private commun. (1991)

7.28   A.Y. Cabezas, L.G. Komai, R.P. Treat: Appl. Opt. **5**, 647 (1966)

7.29   J.W. Carsons, L.G. Komai: Dynamic optical properties of laser materials. Final Techn. Rept. No.P66-134, Office of Naval Research, Washington, DC (1966)

7.30   D.C. Burnham: Appl. Opt. **9**, 1727 (1970)

7.31   D. White, D. Gregg: Appl. Opt. **4**, 1034 (1965)

7.32   F.W. Quelle: Appl. Opt. **5**, 633 (1966)

7.33   E.P. Riedel, G.D. Baldwin: J. Appl. Phys. **38**, 2720 (1967)

7.34   E. Snitzer, C.G. Young: *Lasers*, ed. by A.K. Levine (Dekker, New York 1968) Vol. 2, p. 191

7.35   E. Matovich: The axial gradient laser. Proc. DOD Conf. on Laser Technology, San Diego, CA (1970) p.311

7.36   M.K. Chun, J.T. Bischoff: IEEE J. QE-**7**, 200 (1971)

7.37   K.R. Richter, W. Koechner: Appl. Phys. **3**, 205 (1974)

7.38   V.I. Danilovskaya, V.N. Zubchaninova: Temperature stresses forming in cylinders under the effect of a luminous flux. US Government Res. & Dev. Rept. 70, No. AD-704-020 (1970)

7.39   B.A. Ermakov, A.V. Lukin: Sov. Phys. **15**, 1097 (1971)

7.40   S.D. Sims, A. Stein, C. Roth: Appl. Opt. **6**, 579 (1967)

7.41   R.F. Hotz: Appl. Opt. **12**, 1834 (1973)

7.42   T.J. Gleason, J.S. Kruger, R.M. Curnutt: Appl. Opt. **12**, 2942 (1973)

7.43   G. Benedetti-Michelangeli, S. Martelluci: Appl. Opt. **8**, 1447 (1969)

7.44   T.S. Chen, V.L. Anderson, O. Kahan: OSA Proc. Tunable Solid State Lasers Conf., Falmouth, MA (1989) p.295;
        M.S. Mangir, D.A. Rockwell: IEEE J. QE-**22**, 574 (1986)

7.45   E.A. Teppo: Nd:YAG laser technology. NWC Techn. Memo 2534, Appendix C (1975); Techn. Note 4051–2 (1972), Naval Weapons Center, China Lake, CA

7.46   J.D. Foster, R.F. Kirk: Rept. NASA-CR-1771, Washington, DC (1971)

7.47   R.A. Kaplan: Conductive cooling of a ruby rod. Technical Note No.109, TRG, Melville, NY (1964)

7.48   U. Wittrock, B. Eppich, O. Holst: CLEO '93 (Baltimore, MD) paper CW17

7.49   W.F. Hagen, C.G. Young, D.W. Cuff, J. Keefe: Segmented Nd:glass lasers. Proc. DOD Conf., San Diego, CA (1970) p.363

7.50   E. Matovich, G.E. Mevers: 1 KW axial gradient Nd:YAG laser. Final Rept. N00014-70-C-0406, ONR, Boston (1971)

7.51   M.M. Heil, D.L. Flannery: A review of axial gradient laser technology, Proc. DOD Conf., San Diego, CA (1970) p.287

7.52   E. Matovich: Segmented ruby oscillator-amplifier. Rept. AFAL-TR-69-317, Air Force Avionics Lab., Wright. Patterson AFB, Ohio (1970)

7.53   J.M. Eggleston, T.J. Kane, K. Kuhn, J. Unternahrer, R.L. Byer: IEEE J. QE-**20**, 289 (1984);

T.J. Kane, J.M. Eggleston, R.L. Byer: IEEE J. QE-**21**, 1195 (1985)

7.54   W.S. Martin, J.P. Chernoch: US Patent 3,633,126 (January 1972)

7.55   J.P. Chernoch, W.S. Martin, J.C. Almasi: Performance characteristics of a face-pumped, face-cooled laser, the mini-FPL. Tech. Rept. AFAL-TR-71-3, Air Force Avionics Lab., Wright Patterson AFB, Ohio (1971)

7.56   Lawrence Livermore National Lab., Medium Average Power Solid State Laser Technical Information Seminar, CA (October 1985)

7.57   W.B. Jones, L.M. Goldman, J.P. Chernoch, W.S. Martin: IEEE J. QE-**8**, 534 (1972); G.J. Hulme, W.B. Jones: Total internal reflection face pumped laser, Proc. Soc., Photo-Optical Instr. Eng. IEEE J. QE-**22**, 2092 (1986)

7.58   G.F. Albrecht, J.M. Eggleston, R.A. Petr: *High Power and Solid State Lasers*, ed. by W.W. Simmons. SPIE Proc. **622**, 18 (1986)

7.59   S. Basu, T.J. Kane, R.L. Byer: IEEE J. QE-**22**, 2052 (1986)

7.60   L.E. Zapata, K.R. Manes, D.J. Christie, J.M. Davin, J.A. Blink, J. Penland, R. Demaret, G. Dallum: Proc. Soc. Photo-Opt. Instrum. Eng. **1223**, 259 (1990)

7.61   N. Hodgson, S. Dong, Q. Lü: Opt. Lett. **18**, 1727 (1993); also CLEO '93 (Baltimore, MD) paper CW16.

7.62   R.J. Shine, R., A.J. Alfrey, R.L. Byer: Opt. Lett. **20**, 459 (1995)

7.63   T. Kanabe, C. Yamanaka, N. Kitagawa, M. Takeda, M. Nakatsuka, S. Nakai: CLEO '92 (Anaheim, CA) paper CTUE5

7.64   A.D. Hays, G. Witt, N. Martin, D. DiBiase, R. Burnham: *UV and Visible Lasers and Laser Crystal Growth*, ed. by R. Scheps, M.R. Kokta. SPIE Proc. **2380**, 88 (1995)

7.65   R. S. Afzal, M.D. Selker: Opt. Lett. **20**, 465 (1995)

7.66   B. Comaskey, G. Albrecht, R. Beach, S. Sutton, S. Mitchell: CLEO '93 (Baltimore, MD) paper CW15

7.67   B.J. Comaskey, R, Beach, G. Albrecht, W.J. Benett, B.L. Freitas, C. Petty, D. VanLue, D. Mundinger, R.W. Solarz: IEEE J. QE-**28**, 992 (1992)

7.68   R. Burnham, G. Moule, J. Unternahrer, M. McLaughlin, M. Kukla, M. Rhoades, D. DiBiase, W. Koechner: Laser '95 (Munich) paper K9

7.69   J. Unternahrer, M. McLaughlin, M. Kukla: General Electric Company, Schenectady, N.Y.: Private commun. (1995)

7.70   J.C. Almasi, W.S. Martin: US Patent .3,631,362 (December 1971)

7.71   D.C. Brown, J.H. Kelly, J.A. Abate: IEEE J. QE-**17**, 1755 (1981)

7.72   J.A. Abate, L. Lund, D. Brown, S. Jacobs, S. Refermat, J. Kelly, M. Gavin, J. Waldbillig, O. Lewis: Appl. Opt. **20**, 351 (1981); D.C. Brown, J.A. Abate, L. Lund, J. Waldbillig: Appl. Opt. **20**, 1588 (1981)

7.73   D.C. Brown: *High-Peak-Power Nd:Glass Laser Systems*, Springer Ser. Opt. Sci., Vol.25 (Springer, Berlin Heidelberg 1981)

7.74   J.H. Kelly, D.L. Smith, J.C. Lee, S.D. Jacobs, D.J. Smith, J.C. Lambropoulos: CLEO '87 (Baltimore, MD), Techn. digest, p. 114

7.75   A. Giesen, M. Karszewski, C. Stewen, A. Voss, L. Berger, U. Brauch: Proc. Adv. Solid State Lasers **24**, 330 (Opt. Soc. Am., Washington, DC 1995)

7.76   M.A. Summers, J.B. Trenholme, W.L. Gagnon, R.J. Gelinas, S.E. Stokowski, J.E. Marion, H.L. Julien, J.A. Blink, D.A. Bender, M.O. Riley, R.F. Steinkraus: *High Power and Solid State Lasers*, ed. by W. Simmons. SPIE Proc. **622**, 2 (1986)

7.77   Lawrence Livermore Laboratory Laser Program Annual Reports UCRL-520021 (1974–1998).

7.78   W.W. Simmons, D.R. Speck, L.J. Hunt: Appl. Opt. **17**, 999 (1987)

7.79   W.E. Martin, J.B. Trenholme, G.T. Linford, S.M. Yarema, C.A. Hurley: IEEE J. QE-**17**, 1744 (1981)

7.80   M.E. Innocenzi, H.T. Yura, C.L. Fincher, R.A. Fields: Appl. Phys. Lett. **56**, 1831 (1990)

7.81   J. Frauchiger, P. Albers, H.P. Weber: IEEE J. QE-**28**, 1046 (1992)

7.82   S.C. Tidwell, J.F. Seamans, M.S. Bowers, A.K. Cousins: IEEE J. QE-**28**, 997 (1992)

7.83   U.O. Farrukh, A.M. Buoncristiani, E.C. Byvik: J. Quantum Electron. **24**, 2253 (1988)
7.84   S.B. Sutton, G. F. Albrecht: Appl. Opt. **32**, 5256 (1983)
7.85   A.K. Cousins: IEEE J. QE-**28**, 1057 (1992)
7.86   C. Pfistner, R. Weber, H.P. Weber, S. Merazzi, R. Gruber: IEEE J. QE-**30**, 1605 (1994)
7.87   R. Weber, B. Neuenschwander, M. MacDonald, M.B. Roos, H.P. Weber: IEEE J. QE-**34**, 1046 (1998)
7.88   B. Neuenschwander, R. Weber, H.P. Weber: IEEE J. QE-**31**, 1082 (1995)

Chapter 8

8.1   R.W. Hellwarth: In *Advances in Quantum Electronics* (Columbia Univ. Press, New York 1961) p. 334
8.2   F.J. McClung, R.W. Hellwarth: Proc. IRE **51**, 46 (1963)
8.3   W.G. Wagner, B.A. Lengyel: J. Appl. Phys. **34**, 2040 (1963)
8.4   R.B. Kay, G.S. Waldmann: J. Appl. Phys. **36**, 1319 (1965)
8.5   J.J. Degnan: IEEE J. QE-**25**, 214 (1989)
8.6   J.E. Midwinter: Brit. J. Appl. Phys. **16**, 1125 (1965)
8.7   A.R. Newberry: Brit. J. Appl. Phys. **1**, 1849 (1968)
8.8   G.D. Baldwin: IEEE J. QE-**7**, 220 (1971)
8.9   R.B. Chesler, M.A. Karr, J.E. Geusic: Proc. IEEE **58**, 1899 (1970)
8.10   R.J. Collins, P. Kisliuk: J. Appl. Phys. **33**, 2009 (1962)
8.11   R.C. Benson, M.R. Mirarchi: IEEE Trans. MIL-**8**, 13 (1964)
8.12   W. Buchman, W. Koechner, D. Rice: IEEE J. QE-**6**, 747 (1970)
8.13   I.W. Mackintosh: Appl. Opt. **8**, 1991 (1969)
8.14   E.L. Steele, W.C. Davis, R.L. Treuthart: Appl. Opt. **5**, 5 (1966)
8.15   E.J. Woodbury: IEEE J. QE-**3**, 509 (1967)
8.16   R. Renari: Proc. Soc. Photo-Opt. Instrum. Eng. **1207**, 112–123 (1990)
8.17   J.G. Manni, G.A. Rines, P.F. Moulton: CLEO '87 (Baltimore, MA) paper THJ3
8.18   J.A. Hutchinson, T.H. Allik: Proc. Soc. Photo-Opt. Instrum. Eng. OE/Laser 92 Los Angeles (1992) paper 1627-01
8.19   M. Marinček, M. Lukač: IEEE J. QE-**29**, 2405 (1993)
8.20   M. Born, E. Wolf: *Principles of Optics,* 2nd edn. (Macmillan, New York 1964)
8.21   I.P. Kaminow, E.H. Turner: Appl. Opt. **54**, 1374 (1966)
8.22   C.L. Hu: J. Appl. Phys. **38**, 3275 (1967)
8.23   B.H. Billings: J. Opt. Soc. Am. **39**, 797 (1949)
8.24   B.H. Billings: J. Opt. Soc. Am. **39**, 802 (1949)
8.25   R.O'B. Carpenter: J. Opt. Soc. Am. **40**, 225 (1950)
8.26   B.H. Billings: J. Opt. Soc. Am. **42**, 12 (1952)
8.27   R. Goldstein: Laser Focus **21** (February 1968)
8.28   J.T. Milek, S.J. Welles: Linear electro-optic modulator materials, Report AD 704-556, Hughes Aircraft Comp., Culver City, CA (January 1970)
8.29   F. Zernike Jr.: J. Opt. Soc. Am. **54**, 1215 (1964)
8.30   J.H. Ott, T.R. Sliker: J. Opt. Soc. Am. **54**, 1442 (1964)
8.31   T.R. Sliker, S.R. Burlage: J. Appl. Phys. **34**, 1837 (1963)
8.32   M. Yamazaki, T. Ogawa: J. Opt. Soc. Am. **56**, 1407 (1966)
8.33   R.A. Phillips: J. Opt. Soc. Am. **56**, 629 (1966)
8.34   D. Milam: Appl. Opt. **12**, 602 (1973)
8.35   W.R. Hook, R.P. Hilberg: Appl. Opt. **10**, 1179 (1971)
8.36   M.G. Vitkov: Opt. Spectroscopy **27**, 185 (1969)
8.37   L.L. Steinmetz, T.W. Pouliot, B.C. Johnson: Appl. Opt. **12**, 1468 (1973)
8.38   M. Dore: IEEE J. QE-**3**, 555 (1967)
8.39   C.H. Clayson: Electron. Lett. **2**, 138 (1966)
8.40   B. Trevelyan, J. Sci. Instr. **2**, 425 (1969)

8.41   J.M. Ley: Electron. Lett. **2**, 12 (1966)

8.42   M. Okada, S. Ieiri: IEEE J. QE-**6**, 526 (1970)

8.43   B. Stádník: ACTA Technica ČSAV **1**, 65 (1970)

8.44   M.B. Davies, P.H. Sarkies, J.K. Wright: IEEE J. QE-**4**, 533 (1968)

8.45   A.S. Bebchuk, L.A. Kulevskiy, V.V. Smirnov, Yu.N. Solov'yeva: Rad. Eng. and Electr. Phys. **14**, 919 (1969)

8.46   R.P. Hilberg, W.R. Hook: Appl. Opt. **9**, 1939 (1970)

8.47   M.K. Chun, J.T. Bischoff: IEEE J. QE-**8**, 715 (1972)

8.48   W.R. Hook, R.P. Hilberg, R.H. Dishington: Proc. IEEE **59**, 1126 (1971)

8.49   J.F. Ney: *Physical Properties of Crystals* (Oxford University Press, London, 1964); L.A. Shuvalov (ed.): *Modern Crystallography IV*, Springer Ser. Solid-State Sci., Vol. 37 (Springer, Berlin Heidelberg 1984)

8.50   A.W. Warner, M. Onoe, G.A. Coquin: J. Acoust. Soc. Am. **42**, 1223 (1967)

8.51   E.A. Teppo: Nd : YAG Laser Lab. Experiments, January 1972 to June 1973, Technical Note 4051-7, Naval Weapons Center, China Lake, CA (August 1973)

8.52   T. Chuang, A.D. Hays, H.R. Verdun: OSA Proc.: Adv. Solid State Lasers **20**, 314 (1994); also Appl. Opt. **33**, 8355 (1994)

8.53   M.J.P. Payne, H.W. Evans: CLEO '84, Anaheim, CA, Techn. digest, paper TUB16

8.54   W.E. Schmid, IEEE J. QE-**16**, 790 (1980)

8.55   M.K. Chun, E.A. Teppo: Appl. Opt. **15**, 1942 (1976)

8.56   M.B. Rankin, G.D. Ferguson: Proc. Soc. Photo-Opt. Instr. Eng. **160**, 67 (1978)

8.57   J. Richards: Appl. Opt. **22**, 1306 (1983)

8.58   S.Z. Kurtev, O.E. Denchev, S.D. Savov: Appl. Opt. **32**, 278 (1993)

8.59   J. Richards: Appl. Opt. **26**, 2514 (1987)

8.60   E.A. Lundstrom: Waveplate for correcting thermally induced stress birefringence in solid state lasers, U.S. patent 4,408,334 (1983)

8.61   A.R. Newberry: Brit. J. Appl. Phys. **1**, 1849 (1968)

8.62   E.I. Gordon: Proc. IEEE **54**, 1391 (1966)

8.63   R. Adler: IEEE Spectrum **4**, 42 (1967)

8.64   C.F. Quate, C.D. Wilkinson, D.K. Winslow: Proc. IEEE **53**, 1604 (1965)

8.65   R.W. Dixon: IEEE J. QE-**3**, 85 (1967)

8.66   N. Uchida, N. Niizeki: Proc. IEEE **61**, 1073 (1973)

8.67   D. Maydan: IEEE J. QE-**6**, 15 (1970)

8.68   T. Nowicki: Electro-Opt. Syst. Design, **24** (January 1974)

8.69   M. Cohen: Electro-Opt. Syst. Design, **23** (April 1972)

8.70   D.E. Flinchbaugh: Electro-Opt. Syst. Design, **24** (January 1974)

8.71   D.A. Pinnow: IEEE J. QE-**6**, 223 (1970)

8.72   R.W. Dixon: J. Appl. Phys. **38**, 5149 (1967)

8.73   Isomet Data Sheet, Acousto-Optic Q-switch, Model 453 (May 1973)

8.74   M.G. Cohen: Optical Spectra, **32** (November 1973)

8.75   M.G. Cohen, R.T. Daly, R.A. Kaplan: IEEE J. QE-**7**, 58 (1971)

8.76   R.B. Chesler, D.A. Pinnow, W.W. Benson: Appl. Opt. **10**, 2562 (1971)

8.77   T. Dascalu, N. Pavel, V. Lupei, G. Philipps, T. Beck, H. Weber: SPIE, Opt. Eng. **35**, 1247 (May 1996)

8.78   T.T. Basiev, A.N. Kravets, A.V. Fedin: Quantum Electron. **23**, 513 (1993)

8.79   M. Hercher: Appl. Opt. **6**, 947 (1967)

8.80   Z. Burshtein, P. Blau, Y. Kalisky, Y. Shimony, M.R. Kokta: IEEE J. QE-**34**, 292 (1998)

8.81   Y. Shimony, Z. Burshtein, Y. Kalisky: IEEE J. QE-**31**, 1738 (1995)

8.82   A.D. Hays, R. Burnham: *Trends in Optics and Photonics* **10**, 129 (Opt. Soc. Am., Washington, DC 1997)

8.83   G. Xiao, M. Bass: IEEE J. QE-**33**, 41 (1997) and QE-**34**, 1142 (1998)

8.84   A. Agnesi, S. Dell'Acqua, E. Piccinini, G. Reali, G. Piccinno: IEEE J. QE-**34**, 1480 (1998)

8.85   J.A. Morris, C.R. Pollock: Opt. Lett. **15**, 440 (1990)

8.86   J.J. Degnan: IEEE J. QE-**31**, 1890 (1995)
8.87   R. Wu, S. Hamlin, J.A. Hutchinson, L.T. Marshall: *Trends in Optics and Photonics* **10**, 145 (Opt. Soc. Am., Washington, DC 1997)
8.88   R.D. Stultz, M.B. Camargo, M. Lawler, D. Rockafellow, M. Birnbaum: *OSA Advanced Solid State Lasers Conference*, Coeur d'Alene, ID (1998) p. 330
8.89   W.R. Hook, R.H. Dishington, R.P. Hilberg: Appl. Phys. Lett. **9**, 125 (1966)
8.90   W.J. Rundle: J. Appl. Phys. **39**, 5338 (1968)
8.91   W.J. Rundle: IEEE J. QE-**5**, 342 (1969)
8.92   W.R. Hook: Proc. IEEE **54**, 1954 (1966)
8.93   W. Rundle, W.K. Pendleton: IEEE J. QE-**12**, (1976)
8.94   A.E. Siegmann: IEEE J. QE-**9**,. 247 (1973)
8.95   D. Milam, R.A. Bradbury, A. Hordvik, H. Schlossberg, A. Szöke: IEEE J. QE-**10**, 20 (1974)
8.96   D. Maydan, R.B. Chesler: J. Appl. Phys. **42**, 1031 (1971)
8.97   R.B. Chesler, D. Maydan: J. Appl. Phys. **42**, 1028 (1971)
8.98   D. Cheng: IEEE J. QE-**9**, 585 (1973)
8.99   D. Maydan: J. Appl. Phys. **41**, 1552 (1970)
8.100   H. A. Kruegle, L. Klein: Appl. Opt. **15**, 466 (1976)
8.101   R.H. Johnson: IEEE J. QE-**9**, 255 (1973)
8.102   C.W. Reno: Appl. Opt. **12**, 883 (1973)
8.103   A. Hays, Fibertek, Inc.: Private commun. (1991)
8.104   J. Trenholme: Extraction of Energy from Laser Amplifiers. Laser Program Annual Report (1986) Chap. 7. Report UCRL 50021-86

Chapter 9

9.1   R. Bracewell: The Fourier Transform and its Applications (McGraw-Hill, New York (1965)
9.2   H.A. Haus, J.G. Fujimoto, E.P. Ippen: IEEE J. QE-**28**, 2086 (1992)
9.3   F. Krausz, M.E. Fermann, T. Brabec, P.F. Curley, M. Hofer, M.H. Ober, C. Spielmann, E. Wintner, A.J. Schmidt: IEEE J. QE-**28**, 2097 (1992)
9.4   E.P. Ippen, H.A. Haus, L.Y. Liu: J. Opt. Soc. Am. B **6**, 1736 (1989)
9.5   S.L. Shapiro (ed.): *Ultrashort Light Pulses*, Topics Appl. Phys., Vol. 18 (Springer, Berlin Heidelberg 1977)
9.6   W. Kaiser (ed.) *Ultrashort Laser Pulses*, 2nd edn., Topics Appl. Phys., Vol. 60 (Springer, Berlin Heidelberg 1993)
9.7   H.W. Mocker, R.J. Collins: Appl. Phys. Lett. **7**, 270 (1965)
9.8   A.J. DeMaria, D.A. Stetser, H. Heynau: Appl. Phys. Lett. **8**, 174 (1966)
9.9   J.A. Fleck: Phys. Rev. B **1**, 84 (1970)
9.10   D.J. Bradley, W. Sibbett: Opt. Commun. **9**, 17 (1973)
9.11   D. von der Linde: Appl. Phys. **2**, 281 (1973)
9.12   K.H. Drexhage: In *Dye Lasers*, 3rd edn., ed. by F.P. Schäfer (Springer, Berlin Heidelberg 1990)
9.13   G. Girard, M. Michon: IEEE J. QE-**9**, 979 (1973)
9.14   D. von der Linde, K.F. Rodgers: IEEE J. QE-**9**, 960 (1973)
9.15   T. Brabec, C.H. Spielmann, F. Krausz: Opt. Lett. **17**, 748 (1992)
9.16   L.F. Mollenauer, R.H. Stolen: Opt. Lett. **9**, 13 (1984)
9.17   J. Goodberlet, J. Wang, J.G. Fujimoto, P.A. Schulz: Opt. Lett. **14**, 1125 (1989)
9.18   J. Goodberlet, J. Jacobson, J.G. Fujimoto, P.A. Schulz, T.Y. Fan: Opt. Lett. **15**, 504 (1990)
9.19   L.Y. Liu, J.M. Huxley, E.P. Ippen, H.A. Haus: Opt. Lett. **15**, 553 (1990)
9.20   M.J. McCarthy, G.T. Maker, D.C. Hanna: OSA Proc. Adv. Solid-State Lasers **10**, 110 (1991)
9.21   G.P.A. Malcolm, P.F. Curley, A.I. Ferguson: Opt. Lett. **15**, 1303 (1990)
9.22   F. Krausz, C. Spielmann, T. Brabec, E. Wintner, A.J. Schmidt: Opt. Lett. **15**, 737 (1990)
9.23   F. Krausz, C. Spielmann, T. Brabec, E. Wintner, A.J. Schmidt: Opt. Lett. **15**, 1082 (1990)
9.24   F.M. Mitschke, L.F. Mollenauer: IEEE J. QE-**22**, 2242, (1986)

9.25   M. Sheik-Bahae, A.A. Said, D.J. Hagan, J.J. Soileau, E.W. Van Stryland: Opt. Eng. **30**, 1228 (1991)

9.26   F. Salin, J. Squier. G. Mourou, M. Piché, N. McCarthy: OSA Proc. Ad. Solid-State Lasers **10**, 125 (1991)

9.27   H. Kogelnik: Imaging of optical modes, resonators with internal lenses. Bell Syst. Tech. J. **44**, 455–494 (1965)

9.28   V. Magni, G. Cerullo, S. DeSilvestri: Opt. Commun. **101**, 365 (1993)

9.29   V. Magni, G. Cerullo, S. DeSilvestri: Opt. Commun. **96**, 348 (1993)

9.30   G. Cerullo, S. DeSilvestri, V. Magni, L. Pallaro: Opt. Lett. **19**, 807 (1994)

9.31   D.E. Spence, P.N. Kean, W. Sibbett: Opt. Lett. **16**, 42–44 (1991)

9.32   D. Huang, M. Ulman, L.H. Acioli, H.A. Haus, J.G. Fujimoto: Opt. Lett. **17**, 511 (1992)

9.33   F. Salin, J. Squier, M. Piché: Opt. Lett. **16**, 1674 (1991)

9.34   M. Piché: Opt. Commun. **86**, 156 (1991)

9.35   D. Kopf, K.J. Weingarten, L.R. Brovelli, M. Kamp, U. Keller: Opt. Lett. **19**, 2143 (1994)

9.36   R. Mellish, P.M.W. French, J.R. Taylor, P.J. Delfyett, L.T. Florez: OSA Proc. Adv. Solid-State Lasers **20**, 239 (1994)

9.37   P.M.W. French, R. Mellish, J.R. Taylor: Opt. Lett. **18**, 1934 (1993)

9.38   S. Tsuda, W.H. Knox, S.T. Cundiff: Appl. Phys. Lett. **69**, 1538 (1996)

9.39   U. Keller, D.A.B. Miller, G.D. Boyd, T.H. Chiu, J.F. Ferguson, M.T. Asom: OSA Proc. Adv. Solid-State Lasers **13**, 98 (1992)

9.40   U. Keller, K.J. Weingarten, F.X. Kärtner, D. Kopf, B. Braun, I.D. Jung, R. Fluck, C. Hönninger, N. Matuschek, J. Aus der Au: IEEE J. Quant. Electr. Selected Topics in Quant. Electr. **2**, 435 (1996)

9.41   I.D. Jung, F.X. Kärtner, N. Matuschek, D.H. Sutter, F. Morier-Genoud, Z. Shi, V. Scheuer, M. Tilsch, T. Tschudi, U. Keller: Appl. Phys. B **65**, 137 (1997)

9.42   J. Aus der Au, D. Kopf, F. Morier-Genoud, M. Moser, U. Keller: Opt. Lett. **22**, 307 (1997)

9.43   C.J. Flood, D.R. Walker, H.M. van Driel: Opt. Lett. **20**, 58 (1995)

9.44   D.J.Kuizenga, A.E. Siegman: IEEE J. QE-**6**, 694 and 709 (1970)

9.45   A.E. Siegman, D.J. Kuizenga: Appl. Phys. Lett. **14**, 181 (1969)

9.46   A.E. Siegman, D.J. Kuizenga: Opto-Electr. **6**, 43 (1974)

9.47   D.J. Kuizenga, D.W. Phillion, T. Lund, A.E. Siegman: Opt. Commun. **9**, 221 (1973)

9.48   D.J. Kuizenga: IEEE QE-**17**, 1694 (1981)

9.49   A. Hays, L.R. Marshall, J.J. Kasinski, R. Burnham: IEEE J. QE-**28**, 1021 (1992)

9.50   R.P. Johnson, N.K. Moncur, L.D. Siebert: CLEO '87 (Baltimore, MD) paper FP2

9.51   T. Sizer, I.N. Duling: IEEE J. QE-**24**, 404 (1988)

9.52   J.L. Dallas: Appl. Opt. **33**, 6373 (1994)

9.53   U. Keller, K.D. Li, B.T. Khuri-Yakub, D.M. Bloom, K.J. Weingarten, D.C. Gerstenberger: Opt. Lett. **15**, 45 (1990)

9.54   S.J. Walker, H. Avramopoulos, T. Sizer II: Opt. Lett. **15**, 1070 (1990)

9.55   J.H. Boyden: Microwaves 58 (March 1971)

9.56   G.T. Maker, A.I. Ferguson: Opt. Lett. **14**, 788 (1989)

9.57   T. Juhasz, S. T. Lai, M. A. Pessot: Opt. Lett. **15**, 1458 (1990)

9.58   G.P.A. Malcolm, J. Ebrahimzadeh, A.I. Ferguson: IEEE J. QE-**28**, 1172 (1992)

9.59   D.W. Hughes, M.W. Phillips, J.R.M. Barr, D.C. Hanna: IEEE J. QE-**28**, 1010 (1992)

9.60   F. Krausz, T. Brabec, E. Wintner, A.J. Schmidt: Appl. Phys. Lett. **55**, 2386 (1989)

9.61   W.H. Lowdermilk, J.E. Murray: J. Appl. Phys. **51**, 2436 (1980)

9.62   J.E. Murray, W.H. Lowdermilk: J. Appl. Phys. **51**, 3548 (1980)

9.63   I.N. Duling, P. Bado, S. Williamson, G. Mourou, T. Baer: CLEO '84 (Anaheim, CA) paper PD3

9.64   I.N. Duling III, T. Norris, T. Sizer II, P. Bado, G.A. Mourou: J. Opt. Soc. Am. B **2**, 616 (1985)

9.65   P. Bado, M. Bouvier, J. Scott: Opt. Lett. **12**, 319 (1987)

9.66   M. Saeed, D. Kim, L.F. DiMauro: Appl. Opt. **29**, 1752 (1990)

9.67   M.D. Selker, R.S. Afzal, J.L. Dallas, A.W. Yu: Opt. Lett. **19**, 551 (1994)

9.68   M. Gifford, K.J. Weingarten: Opt. Lett. **17**, 1788 (1992)

9.69   L. Turi, T. Juhasz: Opt. Lett. **20**, 154 (1995)

9.70   D.R. Walker, C.J. Flood, H.M. Van Driel, U.J. Greiner, H.H. Klingenberg: CLEO/Europe '94 (Amsterdam) paper CThM1

9.71   T.E. Dimmick: Opt. Lett. **15**, 177 (1990)

9.72   M. Rhoades, K. Le, D. DiBiase, P. Bournes, W. Koechner, R. Burnham: Laser '95 (Munich) paper F

9.73   T. Ditmire, M.D. Perry: LLNL Fusion Report Jan-March (1994) Vol. 4, p. 39

9.74   H. Verdun, Fibertek, Inc., V.A. Herndon: Private commun. (1994)

9.75   T. Brabec, C. Spielmann, P.F. Curley, F. Krausz: Opt. Lett. **17**, 1292 (1992)

9.76   T. Brabec, P.F. Curley, C. Spielmann, E. Wintner, A.J. Schmidt: J. Opt. Soc. Am. B **10**, 1029 (1993)

9.77   O. Svelto: Polytechnic Institute of Mian, Italy. Private commun. (1994)

9.78   D.K. Negus, L. Spinelli, N. Goldblatt, G. Feuget: OSA Proc. Adv. Solid-State Lasers **10**, 120 (1991)

9.79   F. Krausz, C. Spielmann, T. Brabec, E. Wintner, A.J. Schmidt: Opt. Lett. **15**, 1082 (1990)

9.80   H.A. Haus, E.P. Ippen: Opt. Lett. **16**, 1331 (1991)

9.81   R.L. Fork, O.E. Martinez, J.P. Gordon: Opt. Lett. **9**, 150 (1984)

9.82   J.P. Gordon, R.L. Fork: Opt. Lett **9**, 153 (1984)

9.83   H.A. Haus, J.G. Fujimoto, E.P. Ippen: J. Opt. Soc. Am. B **8**, 2068 (1991)

9.84   S. Chen, J. Wang: Opt. Lett. **16**, 1689 (1991)

9.85   Y.M. Liu, K.W. Sun, P.R. Prucnal, S.A. Lyon: Opt. Lett. **17**, 1219 (1992)

9.86   G.T. Maker, A.I. Ferguson: Appl. Phys. Lett. **54**, 403 (1989); and Electron. Lett. **25**, 1025 (1989)

9.87   L. Turi, C. Kuti, F. Krausz: IEEE J. QE-**26**, 1234 (1990)

9.88   F. Krausz, L. Turi, C. Kuti, A.J. Schmidt: Appl. Phys. Lett. **56**, 1415 (1990)

9.89   P.F. Curley, C. Spielmann, T. Brabec, F. Krausz, E. Wintner, A.J. Schmidt: Opt. Lett. **18**, 54 (1993)

9.90   D.E. Spence, J.M. Evans, W.E. Sleat, W. Sibbett: Opt. Lett. **16**, 1762 (1991)

9.91   G.R. Huggett: Appl. Phys. Lett. **13**, 186 (1968)

9.92   C. Spielmann, F. Krausz, T. Brabec, E. Wintner, A.J. Schmidt: Opt. Lett, **16**, 1180 (1991)

9.93   P.R. Staver, W.T. Lotshow: OSA Proc. Adv. Solid-State Lasers **20**, 252 (1994);
        Y.M. Liu, P.R. Prucnal: IEEE J. QE-**29**, 2663 (1993);
        M.T. Asaki, C.P. Huang, D. Garvery, J. Zhou, H.C. Kapteyn, M.M. Murnane: Opt. Lett. **18**, 977 (1993)

9.94   P.M.W. French, R. Mellish, J.R. Taylor, P.J. Delfyett, L.T. Florez: Opt Lett. **18**, 1934, (1993)

9.95   P.M. Mellish, P.M.W. French, J.R. Taylor, P.J. Delfyett, LT. Florez: Electron. Lett. **30**, 223 (1994)

9.96   D. Kopf, K.J. Weingarten, L. Brovelli, M. Kamp, U. Keller: CLEO '94 (Anaheim) paper CPD22

9.97   M.J.P. Dymott, A.I. Ferguson: Opt. Lett. **19**, 1988 (1994)

9.98   G. Cerullo, S. DeSilvestri, V. Magni: Opt. Lett. **19**, 1040 (1994)

9.99   A. Seas, V. Petričević, R.R. Alfano: Opt. Lett. **17**, 937 (1992)

9.100   A. Sennaroglu, C.R. Pollock, H. Nathel: Opt. Lett. **18**, 826 (1993)

9.101   Y. Pang, V. Yanovsky, F. Wise, B.I. Minkov: Opt. Lett. **18**, 1168 (1993)

9.102   V. Yanovsky, Y. Pang, F. Wise, B.I. Minkov: Opt. Lett. **18**, 1541 (1993)

9.103   A. Seas, V. Petričević, R.R. Alfano: Opt. Lett. **18**, 891 (1993)

9.104   E. Treacy: IEEE J. QE-**5**, 454 (1969)

9.105   O.E. Martinez: IEEE J. QE-**23**, 59 (1987)

9.106   D. Strickland, G. Mourou: Opt. Commun. **56**, 219 (1985)

9.107   T. Joo, Y. Jia, G. R. Fleming: Opt. Lett. **20**, 389 (1995)

9.108   A. Sullivan, H. Hamster, H.C. Kapteyn, S. Gordon, W. White, H. Nathel, R.J. Blair, R.W. Falcone: Opt. Lett **16**, 1406 (1991)

9.109   J. Zhou, C. Huang, C. Shi, M. Murnane, H.C. Kapteyn: Opt. Lett. **19**, 126 (1994)

9.110   J.D. Kmetec, J.J. Macklin, J.F. Young: Opt. Lett. **16**, 1001 (1991)

9.111   J.V. Rudd, G. Korn, S. Kane, J. Squier, G. Mourou, P. Bado: Opt. Lett. **18**, 2044 (1993)

9.112   P.A. Beaud, M. Richardson, E.J. Miesak: IEEE J. QE-**31**, 317 (1995)

9.113   P. Beaud, M. Richardson, E.J. Miesak, B. Chai: Opt. Lett. **18**, 1550 (1993)

9.114   C. Rouyer, É. Mazataud, I. Allais, A. Pierre, S. Seznec, C. Sauteret, G. Mourou, A. Migus: Opt. Lett. **18**, 214 (1993)

9.115   J. Squier, F. Salin, G. Mourou, D. Harter: OSA Proc. Adv. Solid-State Lasers **10**, 96 (1991)

Chapter 10

10.1   P.A. Franken, A.E. Hill, C.W. Peters, G. Weinreich: Phys. Rev. Lett. **7**, 118 (1961)

10.2   B.Ya. Zel'dovich, Y.I. Popovichev, V.V. Ragul'skii, F.S. Faizullov: Sov. Phys. JETP **15**, 109 (1972)

10.3   N. Bloembergen: *Nonlinear Optics* (Benjamin, New York 1965)

10.4   P.D. Maker, R.W. Terhune, M. Nisenoff, C.M. Savage: Phys. Rev. Lett. **8**, 21 (1962)

10.5   J.A. Giordmaine: Phys. Rev. Lett. **8**, 19 (1962)

10.6   M. Born, E. Wolf: *Principles of Optics* (Macmillan, New York 1964)

10.7   J.E. Midwinter, J. Warner: Brit. J. Appl. Phys. **16**, 1135 (1965)

10.8   H.P. Weber, E. Mathieu, K.P. Meyer: J. Appl. Phys. **37**, 3584 (1966)

10.9   G.D. Boyd, A. Ashkin, J.M. Dziedzic, D.A. Kleinman: Phys. Rev. **137**, 1305 (1965)

10.10   M.V. Hobden: J. Appl. Phys. **38**, 4365 (1967)

10.11   F. Zernike, J.E. Midwinter: *Applied Nonlinear Optics* (Wiley, New York 1973)

10.12   D.A. Kleinman, A. Ashkin, G.D. Boyd: IEEE J. QE-**2**, 425 (1966)

10.13   G.D. Boyd, D.A. Kleinman: J. Appl. Phys. **39**, 3597 (1968)

10.14   R.L. Sutherland: *Handbook of Nonlinear Optics* (Dekker, New York 1996)

10.15   R.W. Boyd: *Nonlinear Optics* (Academic, San Diego, CA 1992)

10.16   D.L. Mills: *Nonlinear Optics*, 2nd edn. (Springer, Berlin Heidelberg 1998)

10.17   D.A. Kleinman: In *Laser Handbook*, ed. by. F.T. Arecchi, E.O. Schulz-DuBois (North-Holland, Amsterdam 1972) Vol. 2, pp. 1229-1258

10.18   D. Hon: In *Laser Handbook*, ed. by M. Stitch (North-Holland, Amsterdam 1979) Vol. 3, pp. 421–456

10.19   F.A. Jenkins, H.E. White: *Fundamentals of Optics* (McGraw-Hill, New York 1957)

10.20   N.F. Nye: *Physical Properties of Crystals* (Clarendon, Oxford 1960)

10.21   W.F. Hagen, P.C. Magnante: J. Appl. Phys. **40**, 219 (1969)

10.22   K. Kato: IEEE J. QE-**10**, 622 (1974)

10.23   K. Kato: IEEE J. QE-**10**, 616 (1974)

10.24   A.V. Smith, D.J. Armstrong, W.J. Alford: J. Opt. Soc. Am. B **15**, 122 (1998)

10.25   R.C. Miller: Phys. Lett. A **26**, 177 (1968)

10.26   A. Ashkin, G.D. Boyd, J.M. Dziedzic: Phys. Rev. Lett. **11**, 14 (1963)

10.27   G.E. Francois: Phys. Rev. **143**, 597 (1966)

10.28   D. Eimerl: Laser Program Annual Report, Lawrence Livermore National Laboratory, Ca., Report UCRL-50021-83, (1983) pp. 6–69

10.29   D. Eimerl: IEEE J. QE-**23**, 1361 (1987)

10.30   F.R. Nash, G.D. Boyd, M. Sargent, P.M. Bridenbaugh: J. Appl. Phys. **41**, 2564 (1970)

10.31   M. Okada, S. Ieiri: IEEE J. QE-**7**, 469 (1971)

10.32   M. Okada, S. Ieiri: J. QE-**7**, 560 (1971)

10.33   R.S. Adhav, R.W. Wallace: IEEE J. QE-**9**, 855 (1973)

10.34   R.G. Smith: J. Appl. Phys. **41**, 3014 (1970)

10.35   D. Eimerl: The potential for efficient frequency conversion at high average power using solid state nonlinear optical materials. Lawrence Livermore National Laboratory Report UCID.20565 (October 1985)

10.36   R. Eckardt, H. Masuda, Y. Fan, R. Byer: IEEE J. QE-**26**, 922 (1990)

10.37   S. Liu: J. Appl. Phys. **67**, 634 (1989)

10.38   K. Kato: Opt. Commun. **9**, 249 (1973)

10.39   R.S. Craxton, S.D. Jacobs, J.E. Rizzo, R. Boni: IEEE J. QE-**17**, 1782 (1981)

10.40   Lasermetrics, Electro-Optics Div., Englewood, New Jersey, Data Sheet 8701, Optical Harmonic Generating Crystals (February 1987)

10.41   J.C. Jacco, G.M. Loiacono: Final Report under contract DAAK20-83-C-0139, Night Vision and Electro-Optics Lab. Fort Belvoir, VA 22060 (1986)

10.42   R.P. Jones: *Solid State Lasers and Nonlinear Crystals*, ed. by G.J. Quarks, L. Esterowitz, L.K. Cheng. SPIE Proc. **2379**, 357 (1995)

10.43   Y.S. Liu, D. Dentz, R. Belt: Opt. Lett. **9**, 76 (1984)

10.44   B.K. Vainshtein: *Fundamentals of Crystals*, Mod. Crystallography, Vol. 1 (Springer, Berlin Heidelberg 1994)

10.45   J.C. Jacco, G.M. Loiacono, M. Jaso, G. Mizell, B. Greenberg: J. Cryst. Growth **70**, 484 (1984)

10.46   J.D. Bierlein: *Growth, Characterization, and Application of Laser Host and Nonlinear Crystals*, ed. by J.T. Liu. SPIE Proc. **1104**, 2 (1989)

10.47   J.W. Yao, T.S. Fahlen: J. Appl. Phys. **55**, 65 (1984)

10.48   K. Kato: IEEE J. QE-**27**, 1137 (1991)

10.49   Castech-Phoenix, Inc., Fuzhou, China, Data Sheet KTP

10.50   R.C. Eckardt, H. Masuda, Y.X. Fan, R.L. Byer: IEEE J. QE-**26**, 922 (1990)

10.51   J.D. Bierlein, H. Vanherzeele: J. Opt. Soc. Am. B **6**, 622 (1989)

10.52   T.Y. Fan, C.E. Huang, B.Q. Hu, R.C. Eckardt, Y.X. Fan, R.L. Byer, R.S. Feigelson: Appl. Opt. **26**, 2390 (1987)

10.53   K. Asaumi: Appl. Opt. **32**, 5983 (1993)

10.54   F.C. Zumsteg, J.D. Bierlein, T.E. Gier: J. Appl. Phys. **47**, 4980 (1976)

10.55   L.K. Cheng, L.T. Cheng, J.D. Bierlein, F.C. Zumsteg: Appl. Phys. Lett. **62**, 346–348 (1993)

10.56   J.D. Bierlein, H. Vanherzeele: Appl. Phys. Lett. **54**, 783–785 (1989)

10.57   S. Zhao, C. Huang, H. Zhang: J. Cryst. Growth **99**, 805 (1990)

10.58   C.L. Tang, W.R. Bosenberg, T. Ukachi, R.J. Lane, L.K. Cheng: Laser Focus World p. 87 (Sept. 1990)

10.59   J.T. Lin, C. Chen: Lasers and Optronics **6**, 59 (November 1987)

10.60   J.T. Lin: Analyses of frequency conversion and application of nonlinear crystals, Proc. Int'l Conf. Lasers (STS, Arlington, VA 1986) p. 262

10.61   C. Chen, Y.X. Fan, R.C. Eckardt, R.L. Byer: CLEO '86 (San Francisco, CA) paper ThQ4

10.62   K.C. Liu, M. Rhoades: CLEO '87 (Baltimore, MD)

10.63   K. Kato: IEEE J. QE-**22**, 1013 (1986)

10.64   M. Bass: IEEE J. QE-**7**, 350 (1971)

10.65   A.M. Glass, D. von der Linde, T.J. Negran: Appl. Phys. Lett. **25**, 233 (1974)

10.66   W.J. Kozlovsky, C.D. Nabors, R.L. Byer: IEEE J. QE-**24**, 913 (1988)

10.67   I. Schutz, R. Wallenstein: Proc. CLEO '90 (Anaheim, CA) paper CWC4

10.68   J.E. Geusic, H.J. Levinstein, S. Singh, R.G. Smith, L.G. Van Uitert: Appl. Phys. Lett. **12**, 306 (1968)

10.69   C.B. Hitz, J. Falk: Frequency doubled neodymium laser. Rept. AFAL-TR-12, Air Force Avionics Lab., Wright-Patterson AFB, Dayton, Ohio (1971)

10.70   J.M. Yarborough, J. Falk, C.B. Hitz: Appl. Phys. Lett. **18**, 70 (1971)

10.71   C.B. Hitz: Final Report, Contract NASA-20967, G.C. Marshall Space Flight Center, Huntsville, Ala. (July 1970)

10.72   W. Culshaw, J. Kannelaud, J.E. Peterson: IEEE J. QE-**10**, 253 (1974)

10.73   R.G. Smith: IEEE J. QE-**6**, 215 (1970)

10.74   T. Baer: J. Opt. Soc. Am. B **3**, 1175 (1986)
10.75   Y.F. Chen, T.M. Huang, C.L. Wang, L.J. Lee: Appl. Opt. **37**, 5727 (1998)
10.76   V. Magni, G. Cerullo, S. DeSilvestri, O. Svelto, L.J. Qian, M. Danailov: Opt. Lett. **18**, 2111 (1993)
10.77   M.D. Selker, T.J. Johnson, G. Frangineas, J.L. Nightingale, D.K. Negus: CLEO '96 (Anaheim, CA), Techn. digest, p. 602
10.78   G.E. James, E.M. Harrell, C. Bracikowski, K. Wiesenfeld, R. Roy: Opt. Lett. **15**, 1141 (1990)
10.79   M. Oka, S. Kubota: Opt. Lett. **13**, 805 (1988)
10.80   J.E. Murray, S.E. Harris: J. Appl. Phys. **41**, 609 (1970)
10.81   D.T. Hon: IEEE J. QE-**12**, 148 (1976)
10.82   B.J. LeGarrec, G.J. Razé, P.Y. Thro, M. Gilbert: Opt. Lett. **21**, 1990 (1996)
10.83   J.J. Chang, E.P. Dragon, C.A. Ebbers, I.L. Bass, C.W. Cochran: *Advanced Solid State Lasers* (Coeur d'Alene, ID) (OSA, Washington, DC 1998) Postdeadline Paper PD15-1
10.84   G.A. Massey, J.M. Yarborough: Appl. Phys. Lett. **18**, 576 (1971)
10.85   M.V. Ortiz, J.H. Fair, D.J. Kuizenga: *Advanced Solid State Lasers*, ed. by L. Chase, A. Pinto **13**, 361 (OSA, Washington, DC 1992)
10.86   T.S. Fahlen, P.E. Perkins: CLEO '84, (Washington, DC) paper ThCl.
10.87   J.Q. Yao, Y. Li, D.P. Zhang: *High Power Solid State Laser*, ed. by H. Weber. SPIE Proc. **1021**, 181 (1988)
10.88   C.J. Kennedy: IEEE J. QE-**10**, 528 (1974)
10.89   C.B. Hitz, L.M. Osterink: Appl. Phys. Lett. **18**, 378 (1971)
10.90   J. Falk: IEEE J. QE-**11**, 21 (1975)
10.91   O. Bernecker: IEEE J. QE-**9**, 897 (1973)
10.92   C.B. Hitz, J. Falk: Frequency doubled neodymium laser. Rept. AFAL-TR-72-12, Air Force Avionics Lab., Wright-Patterson AFB, Dayton, Ohio (1972)
10.93   T.R. Gurski: Appl. Phys. Lett. **15**, 5 (1969)
10.94   J.H. Boyden, E.G. Erickson, R. Webb: Mode-locked frequency doubled neodymium laser. Tech. Report AFAL-TR-70-214, Air Force Avionics Lab., Wright-Patterson, Dayton, Ohio (1970)
10.95   R.R. Rice, G.H. Burkhardt: Appl. Phys. Lett. **19**, 225 (1971)
10.96   J.A. Armstrong, N. Bloembergen, J. Ducuing, P.S. Pershan: Phys. Rev. **127**, 1918 (1962)
10.97   W. Seka, S.D. Jacobs, J.E. Rizzo, R. Boni, R.S. Craxton: Opt. Commun. **34**, 469–473 (1980)
10.98   R.S. Craxton: IEEE J. QE-**17**; 1771 (1981); Opt. Commun. **34**, 474 (1980)
10.99   A.D. Hays, L.R. Marshall, R. Burnham: Advanced Solid State Lasers (Hilton Head, SC) (OSA, Washington 1991) postdeadline paper PdP8
10.100   H. Hemmati, J.R. Lesh: Optics Lett. **19**, 1322 (1994)
10.101   J. Kasinski, P. Bournes, D. DiBiase, R. Burnham: In CLEO '93 (Baltimore, MD) paper CW11; *Technical Digest Series* **11**, 274 (OSA, Washington DC 1993)
10.102   G.J. Linford, B.C. Johnson, J.S. Hildum, W.E. Martin, K. Snyder, R.D. Boyd, W.L. Smith, C.L. Vercimak, D. Eimerl, J.T. Hunt: Appl. Opt. **21**, 3633 (1982)
10.103   B.C. Johnson, T. Marchl, J. Mihoevich, W.L. Smith, J.E. Swain, R. Wilder, J.D. Williams: Int'l Quant. Electr. Conf. Baltimore, MD (1983), Paper TUE3
10.104   Special issue on optical parametric oscillation and amplification. J. Opt. Soc. Am. B **10**, 1659–1791 (1993)
10.105   D.C. Edelstein, E.S. Wachmann, C.L. Tang: Appl. Phys. Lett. **54**, 1728 (1989)
         E.S. Wachmann, D.C. Edelstein, C.L. Tang: Opt. Lett. **15**, 136 (1990)
         E.S. Wachmann, W.S. Pelouch, C.L. Tang: J. Appl. Phys. **70**, 1893 (1991)
10.106   G. Mak, Q. Fu, H.M. van Driel: Appl. Phys. Lett. **60**, 542 (1992)
         Q. Fu, G. Mak, H.M. van Driel: Opt. Lett. **17**, 1006 (1992)
10.107   W.S. Pelouch, P.E. Powers, C.L. Tang: Opt. Lett. **17**, 1070 (1992)
10.108   H.J. Krause, W. Daum: Appl. Phys. Lett. **60**, 2180 (1992)

10.109 R. Danielius, A. Piskarskas, A. Stabinis, G.P. Banfi, P. Di Trapani, R. Righini: J. Opt. Soc. Am. B **10**, 2222 (1993)

10.110 D.R. Walker, C.J. Flood, H.M. van Driel: Opt. Lett. **20**, 145 (1995)

10.111 A. Lauberau: Optical nonlinearities with ultrashort pulses. In *Ultrashort Light Pulses and Applications*, ed. by W. Kaiser, 2nd edn., Topics Appl. Phys., Vol. 60 (Springer, Berlin Heidelberg New York 1993) p. 35

10.112 J.A. Giordmaine, R.C. Miller: Phys. Rev. Lett. **14**, 973 (1965)

10.113 S.E. Harris: IEEE Proc. **57**, 2096 (1969)

10.114 R.G. Smith: In *Laser Handbook I*, ed. by F.T. Arecchi, E.O. Schultz-DuBois (North-Holland, Amsterdam 1972) pp. 837–895, also *Lasers*, ed. by A.K. Levine, A.J. De Maria (Dekker, New York 1976) Vol. 4, pp. 189–307

10.115 S.J. Brosnan, R.L. Byer: IEEE J. QE-**15**, 415–443 (1979)

10.116 R.L. Byer: In *Quantum Electronics: A Treatise*, ed. by H. Rabin, C.L. Tang (Academic, New York 1973) Vol. I, Pt. B, pp. 587–702

10.117 G.D. Boyd, A. Ashkin: Phys. Rev. **146**, 198 (1966)

10.118 L.R. Marshall, A.D. Hays, J.J. Kasinski, R. Burnham: Adv. Solid State Lasers, Salt Lake City, UT (OSA Washington, DC, March 1990) Postdeadline paper WC8PD-2

10.119 J.J. Kasinski, R. Burnham: CLEO'90 (Anaheim, CA) Dig. Tech. Papers, paper CMF5

10.120 L.R. Marshall, A.D. Hays, R. Burnham: CLEO '90 (Anaheim, CA) Dig. Tech. Paper, post-deadline paper CPDP35-1

10.121 J.E. Bjorkholm: IEEE J. QE-**7**, 109 (1971)

10.122 M.J.T. Milton, T.D. Gardiner, G. Chourdakis, P.T. Woods: Opt. Lett. **19**, 281 (1994)

10.123 R.L. Burnham: Final Report Solid-State Mid Infrared Source. Contract: DAAB07-87-C-F074 Fibertek, Inc. (1988)

10.124 J. Yao, W. Sheng, W. Shi: J. Opt. Soc. Am. B **9**, 891 (1992)

10.125 D. Eimerl: J. Appl. Phys. **62**, 1968 (1987)

10.126 C.L. Tang, W.R. Bosenberg, T. Ukachi, R. Lane, L. Cheng: Laser Focus World, p. 107 (Oct. 1990)

10.127 K. Kato: IEEE J. QE-**27**, 1137 (1991)

10.128 J.T. Lin, J.L. Montgomery: Opt. Commun. **75**, 315–320 (1990)

10.129 R. Burnham, R.A. Stolzenberger, A. Pinto: IEEE Photon. Technol. Lett. **1**, 27–28 (1989)

10.130 W.R. Bosenberg, L.K. Cheng, J.D. Bierlein: Adv. Solid State Lasers and Compact Blue-Green Lasers, New Orleans, LA (OSA, Washington, DC 1993) Vol. 2, pp. 134–135

10.131 T. Chuang, J. Kasinski, H.R. Verdun: *TOPS on Advanced Solid State Lasers* **1**, 150 (OSA, Washington, DC 1996)

10.132 S. Chandra, T.H. Allik, J.A. Hutchinson, R. Utano, G. Catella: *TOPS Advanced Solid State Lasers* **10**, 270 (OSA, Washington, DC 1997)

10.133 S. Chandra, T.H. Allik, G. Catella, J.A. Hutchinson: Advanced Solid State Lasers (Coeur d'Alene, ID) (OSA, Washington, DC 1998) p. 102, paper AMD5-1

10.134 M.M. Fejer, G.A. Magel, D.H. Jundt, R.L. Byer: IEEE J. QE-**28**, 2631 (1992)

10.135 L.E. Myers, R.C. Eckardt, M.M. Fejer, R.L. Byer: J. Opt. Soc. Am. B **12**, 2102 (1995)

10.136 L.E. Myers, W.R. Bosenberg: IEEE J. QE-**33**, 1663 (1997)

10.137 G.J. Dixon: Laser Focus World, p. 105 (May 1997)

10.138 D.H. Jundt: Opt. Lett. **22**, 1553 (1997)

10.139 L.E. Myers, W.R. Bosenberg, J.I. Alexander, M.A. Arbore, M.M. Fejer, R.L. Byer: *Advanced Solid State Laser Conference, San Francisco* (OSA, Washington, DC 1996) Post-deadline paper PD5-1

10.140 W.R. Bosenberg, A. Drobshoff, L.E. Meyers: *Advanced Solid State Laser Conference*, San Francisco (OSA, Washington, DC 1996) Techn. digest, p. 68 paper WD1-1

10.141 T. Chuang, R. Burnham: Opt. Lett. **23**, 43 (1998)

10.142 T. Chuang, R. Burnham: *Advanced Solid-State Laser Conference*, Boston, MA (OSA, Washington, DC 1999) Techn. digest p. 118, paper ME3-1

10.143 K.C. Burr, C.L. Tang: Opt. Lett. **22**, 1458 (1997)

10.144   T.P. Grayson, L.E. Myers, M.D. Nelson, V. Dominic: *Advanced Solid State Laser Conference*, San Francisco, CA 1996 (OSA, Washington, DC) Techn. digest, p. 71, paper WD2-1

10.145   G.D. Miller, R.G. Batchko, W.M. Tulloch, D.R. Weise, M.M. Fejer, R.L. Byer: Opt. Lett. **22**, 1834 (1997)

10.146   M. Taya, M.C. Bashaw, M.M. Fejer: Opt. Lett. **21**, 857 (1996)

10.147   J.P. Meyn, M.M. Fejer: Opt. Lett. **22**, 1214 (1997)

10.148   M.E. Klein, D.H. Lee, J.P. Meyn, B. Beier, K.J. Boller, R. Wallenstein: Opt. Lett. **23**, 831 (1998)

10.149   T. Kartaloglu, K.G. Köprülü, O. Aytür, M. Sundheimer, W.P. Risk: Opt. Lett. **23**, 61 (1998)

10.150   V. Pasiskevicius, S. Wang, H. Karlsson, J.A. Tellefsen, F. Laurell: *Advanced Solid State Lasers* (Coeur d'Alene, ID) (OSA, Washington, DC 1998) p. 183

10.151   A.V. Smith: SNLO Nonlinear Optics Code. Sandia National Laboratories, Albuquerque, NM (Software package available from A.V. Smith)

10.152   M.J. Roskar, C.L. Tang: J. Opt. Soc. Am. B **2**, 691 (1985)

10.153   T.Y. Fan, R.C. Eckardt, R.L. Byer, J. Nolting, R. Wallenstein: Appl. Phys. Lett. **53**, 2014 (1988)

10.154   A. Fix, C. Huang, T. Shroder, R. Wallenstein: CLEO '90 (Anaheim, CA) Tech. Digest, p. 248, paper CWE8

10.155   J.G. Haub, M.J. Johnson, B.J. Orr, R. Wallenstein: CLEO '91 (Baltimore, MD) Techn. Digest, paper CFM1

10.156   W.L. Bosenberg, L.K. Cheng, C.L. Tang: Appl. Phys. Lett. **54**, 13 (1989)

10.157   L.R. Marshall, A. Kaz, O. Aytur: Opt. Lett. **18**, 817 (1993)

10.158   E.J. Woodbury, W.K. Ng: Proc. IRE **50**, 2367 (1962);
E.J. Woodbury, G.M. Eckhardt: US Patent 3,371,265 (27 February 1968)

10.159   R.L. Byer: Frequency conversion via stimulated Raman scattering. Electro-Optical Systems Design, p. 24–29 (February 1980)

10.160   Y.R. Shen: Stimulated Raman scattering, in *Light Scattering in Solids I*, 2nd edn., ed. by M. Cardona, Topics Appl. Phys., Vol. 8 (Springer, Berlin Heidelberg 1983) Chap. 7

10.161   G.L. Eesley: *Coherent Raman Spectroscopy* (Pergamon, New York 1981);
A. Owyoung: CW stimulated Raman spectroscopy, in *Chemical Applications of Nonlienar Raman Spectroscopy*, ed. by B. Harvey (Academic, New York 1981) pp. 281–320

10.162   W. Kaiser, M. Maier: Stimulated Rayleigh, Brillouin and Raman spectroscopy, in *Laser Handbook Vol. II*, ed. by F.T. Arecchi, E.O. Schulz-DuBois (North-Holland, Amsterdam 1972)

10.163   F.P. Milanovich: *Handbook of Laser Science and Technology, Vol. 3*, ed. by M.J. Weber (CRC, Boca Raton, Florida 1986) p. 283

10.164   B.E. Perry: Photonics Spectra (1984) p. 45; also Data sheet for Quanta-Ray Model RS-1

10.165   D.G. Bruns, D.A. Rockwell: High energy Raman resonator. Hughes Aircraft Comp. Culver City, California, Final Report, Report FR-81-72-1035 (1981)

10.166   O. Yu Nosach V.I. Popovichev, V.V. Ragul'skii, F.S. Faizullov: JETP Lett. **16**, 435 (1972)

10.167   D.A. Rockwell: IEEE J. QE-**24**, 1124 (1988)

10.168   D.T. Hon: Opt. Eng. **21**, 252–256 (1982)

10.169   I.D. Carr, D.C. Hanna: Appl. Phys. B **36**, 83 (1985)

10.170   D.M. Pepper: Optical Engineering **21**, 156–286 (1982) Special issue on Non-linear optical phase conjugation

10.171   R.A. Fisher (ed.): *Optical Phase Conjugation* (Academic, New York 1983)

10.172   B.Ya. Zel'dovich, N.F. Pilipetsky, V.V. Shkunov: *Principles of Phase Conjugation*, Springer Ser. Opt. Sci., Vol. 42 (Springer, Berlin, Heidelberg 1985)

10.173   D.M. Pepper: Nonlinear optical phase conjugation, in *Laser Handbook, Vol. 4*, pp. 333–485, ed. by M.L. Stitch, M. Bass (North-Holland, Amsterdam 1985)

10.174   N.F. Andreev, M.A. Dvoretskii, A.A. Leshchev, V.G. Manishin, G.A. Pasmanik, T.P. Samarina: Sov. J. Quantum Electron. **15**, 928 (1985)

10.175   A. Kummrow, H. Meng: Opt. Commun. **83**, 342 (1991)

10.176 R. Menzel, H.J. Eichler: Phys. Rev. A **46**, 7139 (1992)

10.177 R. Menzel, D. Schulze: Proc. Int'l Summer School on Applications of Nonlinear Optics, Prague (1993)

10.178 W. Kaiser, M. Maier: Stimulated Rayleigh, Brillouin and Raman spectroscopy, in *The Laser Handbook*, ed. by F.T. Arecchi, E.O. Schulz-DuBois (North Holland, Amsterdam 1972) Vol. 2

10.179 D. Cotter, D.C. Hanna, R. Wyatt: Appl. Phys. **8**, 333 (1975)

10.180 S.A. Ashmanov, Yu.E. D'yakov, L.I. Pavlov: Sov. Phys. JETP 39, 249–256 (1974)

10.181 M. Maier, G. Renner: Phys. Lett. A **34**, 299–300 (1971)

10.182 H.J. Eichler, A. Haase, R. Menzel: IEEE J. QE-**31**, 1265 (1995)

10.183 V.N. Belousov, Yu.K. Nizienko: Opt. Spectroscopy (USSR) **58**, 563 (1985)

10.184 N.G. Basov, V.F. Efimkov, I.G. Zubarev, A.V. Kotov, S.I. Mikhailov, M.G. Smirnov: JETP Lett **28**, 197 (1978)

10.185 H.J. Eichler, A. Haase, R. Menzel: Proc. Int'l Summer School and Top. Meeting on Applications on Nonlinear Optics, Prague (1993)

10.186 J.J. Eichler, R. Menzel, D. Schumann: Appl. Opt. **31**, 5038 (1992)

10.187 M. Ostermeyer, A. Heuer, R. Menzel: IEEE J. QE-**34**, 372 (1998)

Chapter 11

11.1   Test method for the laser-radiation-induced damage threshold of optical surfaces. ISO 11254 (The International Organization for Standardization 1992)

11.2   S.C. Seitel, A. Giesen, J. Becker: Proc. SPIE **1848** 2 (1992)

11.3   J. Becker, A. Bernhardt: Proc. SPIE **2114**, 703 (1993)

11.4   *Laser-Induced Damage in Optical Materials* (1969–1989), National Bureau of Standards (National Institute of Standards and Technology after 1986), Special publication, ed. by H.E. Bennett, L.L. Chase, A.J. Glass, A.H. Guenther, D. Milam, B.E. Newman, M.J. Soileau

11.5   *Laser-Induced Damage in Optical Materials:* SPIE Proc. **1441** (1990), SPIE Proc. **1624** (1991), SPIE Proc. **1848** (1992), SPIE Proc. **2114** (1993), each ed. by H.E. Bennett, A.H. Guenther, L.O. Chase, B.E. Newnam, M.J. Soileau; SPIE Proc. **2428** (1994), SPIE Proc. **2714** (1995), SPIE Proc. **2966** (1996), ed. by H.E. Bennett, A.H. Guenther, M.R. Kozlowski, B.E. Newnam, M.J. Soileau; SPIE Proc. **3244** (1997), ed. by G.J. Exarhos, A.H. Guenther, M.R. Kozlowski, M.J. Soileau.

11.6   R.M. Wood (ed.): Selected papers on laser damage in optical materials. SPIE Milestone Series MS24 (1990)

11.7   N. Bloembergen: Appl. Opt. **12**, 661 (1973)

11.8   N.L. Boling, G. Dubé: Appl. Phys. Lett. **23**, 658 (1973)

11.9   N. Bloembergen: IEEE J. QE-**10**, 375 (1974)

11.10   N.L. Boling, M.D. Crisp, G. Dubé: Appl. Opt. **12**, 650 (1973)

11.11   M.D. Crisp, N.L. Boling, G. Dubé: Appl. Phys. Lett. **21**, 364 (1972)

11.12   M.D. Feit, A.M. Rubenchick, D.R. Faux, R.A. Riddle, A. Shapiro, D.C. Eder, B.M. Penetrante, D. Milam, F.Y. Genin, M.R. Kozlowski: SPIE **2966**, 417 (1996)

11.13   R.W. Hopper, D.R. Uhlmann: J. Appl. Phys. **41**, 4023 (1970)

11.14   S.A. Akhmanov, R.V. Khokhlov, A.P. Sukhorukov: Self-focusing, self-defocusing and self-modulation of laser beams, in *Laser Handbook*, ed. by F.T. Arecchi, E.O. Schulz-DuBois (North-Holland, Amsterdam 1972) p. 1151

11.15   E.S. Bliss: Nonlinear propagation studies, in NBS Spec. Publ. No. 414 (US Gov. Print. Office, Washington, DC 1974) p. 7

11.16   J.H. Marburger: Theory of self-focusing in fast nonlinear response, in NBS Spec. Publ. No. 356 (US Gov. Print. Office, Washington, DC 1971) pp. 51–59

11.17   E.L. Dawes, J.H. Marburger: Phys. Rev. **179**, 862 (1969)

11.18   W.G. Wagner, H.A. Haus, J.H. Marburger: Phys. Rev. **175**, 256 (1968)

11.19   E.L. Kerr: Phys. Rev. A **4**, 1195 (1971)

11.20   M. Sheik-Bahae, A.A. Said, D.J. Hagan, J.J. Soileau, E.W. van Stryland: Opt. Eng. **30**, 1228 (1991)

11.21   P.R. Staver, W.T. Lotshow: Adv. Solid-State Lasers **20**, 252 (OSA, Washington, DC 1994)

11.22   R.E. Bridges, R.W. Boyd, G.P. Agrawal: Opt. Lett. **18**, 2026 (1993)

11.23   J. Davit: Filamentary damage in glasses, in ASTM Spec. Techn. Publ. No. 469 (ASTM Philadelphia, PA 1969) pp. 100–109; In *Damage in Laser Materials*, ed. by A.J. Glass, A.H. Guenther, NBS Spec. Publ. No. 341 (US Gov. Print. Office, Washington, DC 1970) pp. 37–44;

11.24   C.R. Guiliano: Time evolution of damage tracks in sapphire and ruby, in *Damage in Laser Materials*, ed. by A.J. Glass, A.H. Guenther NBS Spec. Publ. No. 356 (US Gov. Print. Office, Washington, DC 1971) pp. 44–50;

11.25   A.J. Campillo, S.L. Shapiro, B.R. Suydam: Appl. Phys.. Lett. **24**, 178 (1974); Appl. Phys. Lett. **23**, 628 (1973)

11.26   E.S. Bliss, D.R. Speck, J.F. Holzrichter, J.H. Erkkila, A.J. Glass: Appl. Phys. Lett. **25**, 448 (1974)

11.27   J.A. Fleck, C. Layne: Appl. Phys. Lett. **22**, 467 (1973)

11.28   J.R. Jokipii, J. Marburger: Appl. Phys. Lett. **23**, 696 (1973)

11.29   M.M.T. Loy, Y.R. Shen: IEEE J. QE-**9**, 409 (1973)

11.30   V.I. Bespalov, V.I. Talanov: JETP Lett. **3**, 307 (1966)

11.31   J.A. Glaze: High energy glass lasers: Proc. Soc. Photo-Opt. Instr. Eng. **69**, 45 (1975)

11.32   R. DeSalvo, A.A. Said, D.J. Hagan, E.W. Van Stryland, M. Sheik-Bahae: IEEE J. QE-**32**, 1324 (1996)

11.33   Data sheets Schott and Hoya laser glasses.

11.34   G. Arisholm: Advanced Solid-State Lasers, Orlando, FL 1997, Techn. digest, p. 355. Also Report No. 96/01807 Norwegian Defense Research Establishment, Kjeller, Norway

11.35   R.W. Boyd: *Nonlinear Optics* (Academic, San Diego, CA 1992)

11.36   J. Herrmann: Opt. Commun. **98**, 111 (1993)

11.37   E.S. Bliss: Opto-electronics **3**, 99 (1971)

11.38   J.R. Bettis, R.A. House II, A.H. Guenther: Natl. Bur. Stand. (U.S.) Spec. Publ. **462**, 338 (1976)

11.39   R.M. Wood: *Laser Damage in Optical Materials* (Hilger, Boston 1986)

11.40   J.H. Campbell, F. Rainer, M.R. Kozlowski, C.R. Wolfe, I.M. Thomas, F.P. Milanovich: Lawrence Livermore National Laboratory, ICF Quarterly Report (January–March 1991) p. 41

11.41   B.C. Stuart, M.D. Feit, S. Herman, A.M. Rubenchik, B.W. Shore, M.D. Perry: Phys. Rev. Lett. **74**, 2248 (1995)

11.42   B.C. Stuart, M.D. Feit, S. Herman, A.M. Rubenchik, B.W. Shore, M.D. Berry: Phys. Rev. B **53**, 1749 (1996)

11.43   B.C. Stuart, M.D. Perry, R.D. Boyd, J.A. Britten, B.W. Shore, M.D. Feit, A.M. Rubenchik: In *Generation, Amplification, and Measurement of Ultrashort Laser Pulses II*, ed. by F.W. Wise, C.P.J. Barty. SPIE Proc. **2377**, 247 (1995);

11.44   S.C. Jones, P. Braunlich, R.T. Casper, X.A. Shen, P. Kelly: Opt. Eng. **28**, 1039 (1989)

11.45   F. Rainer, L.J. Atherton, J.T. DeYoreo: SPIE Proc. **1848**, 46 (1992)

11.46   J.H. Campbell, F. Rainer: Lawrence Livermore Laboratory, Livermore, CA, Report, UCRL-JC-109255, SPIE '92 Conf. Proc., San Diego, CA (1992)

11.47   G.Y. Kolodnyi, E.V. Levchuk, V.V. Novopashin, O.E. Sidoryuk: Laser & Optoelektronik **28**, 61 (1996)

11.48   L. Sheehan, M. Kozlowski, F. Rainer, M. Staggs: SPIE Proc. **2114**, 559 (1994)

11.49   C.R. Wolfe, M.R. Kozlowski, J. Campbell, F. Rainer, A.J. Morgan, R.P. Gonzales: NIST (US) Spec. Pub. **801**, 360 (1990)

11.50   M.R. Kozlowski, M. Staggs, F. Rainer, J. Stathis: SPIE Proc. **1441** 269 (1991)

11.51   M.R. Kozlowski, I.M. Thomas, J.H. Campbell, R. Rainer: In *Thin Films for Optical Systems*, ed. by K.H. Guenther. SPIE Proc. **1782** 105 (1992)

11.52   N. Kaiser, B. Anton, H. Jänchen, K. Mann, E. Eva, C. Fischer, R. Henking, D. Ristau, P. Weissbrodt, D. Mademann, L. Raupach, E. Hacker: SPIE Proc. **2428**, 400 (1994)

11.53   A. Fornier, C. Cordillot, D. Bernardino, D. Ausserre, F. Paris: SPIE Proc. **2714**, 383 (1995)

11.54   M. Richardson, M.J. Soileau, P. Beaud, R. DeSalvo, S. Garnov, D.J. Hagan, S. Klimentov, K. Richardson, M. Sheik-Bahae, A.A. Said, E. Van Stryland, B.H.T. Chai: SPIE Proc. **1848**, 392 (1992)

11.55   R.M. Wood, R.T. Taylor, R.L. Rouse: Optics and Laser Techn. **6**, 105 (1975)

11.56   T.S. Rose, M.S. Hopkins, R.A. Fields: IEEE J. QE-**31**, 1593 (1995)

11.57   N. Neuroth, R. Hasse, A. Knecht: In NBS Spec. Pub. **356**, 3–14 (1971); *Damage in Laser Materials*, ed. by A.J. Glass, A.H. Guenther

11.58   K.E. Montgomery, F.P. Milanovich: J. Appl. Phys. **68**, 15 (1990)

11.59   F. Rainer, L.J. Atherton, J.H. Campbell, F.P. DeMarco, M.R. Kozlowski, A.J. Morgan, M.C. Staggs: SPIE Proc. **1624**, 116 (1991)

11.60   J. Chang, E. Dragon, C. Ebbers, I. Bass, C. Cochran: *Advanced Solid-State Lasers*, ed. by W.R. Bosenberg, M.M. Fejer. OSA Trends in Optics and Photonics, Vol. 19 (OSA, Washington, DC 1998) p. 300

11.61   R.P. Jones, A.M. Floener, J.S. Runkel: Presented at 26th Annual Boulder Damage Symposium (1994)

11.62   R.P. Jones: In *Solid-State Lasers and Nonlinear Crystals*, ed. by G.J. Quarles, L. Esterowitz, Z.K. Cheng. SPIE Proc. **2379**, 357 (1995)

11.63   F.C. Zumsteg, J.D. Bierlein, E.E. Gier: J. Appl. Phys. **47**, 4980 (1976)

11.64   J.K. Tyminski: J. Appl. Phys. **70**, 5570 (1991)

11.65   J.C. Jacco, D.R. Rockafellow, E.A. Teppo: Opt. Lett. **16**, 1307 (1991)

11.66   M.P. Scripsick, D.N. Lolacono, J. Rottenberg, S.H. Goellner, L.E. Halliburton, F.K. Hopkins: Appl. Phys. Lett. **66**, 3428 (1995)

11.67   G.M. Laiacono, D.N. Laiacono, T. McGee, M. Babb: J. Appl. Phys. **72**, 2705 (1992)

11.68   J.P. Fève, B. Boulanger, G. Marnier, H. Albrecht: Appl. Phys. Lett. **70**, 277 (1997)

11.69   M.P. Scripsick, G.J. Edwards, L.E. Halliburton, R.F. Belt, G.M. Loiacono: J. Appl. Phys. **76**, 733 (1994)

11.70   B. Boulanger, M.M. Fejer, R. Blachman, P.F. Bordui: Appl. Phys. Lett. **65**, 2401 (1994)

11.71   P.A. Morris, A. Ferretti, M.G. Roeloffs, J.D. Bierlein, T.M. Baer: SPIE Proc. **1848**, 24 (1992)

11.72   W.D. Fountain, L.M. Osterink, G.A. Massey: In NBS Spec. Pub. **356**, 91–103 (1971); *Damage in Laser Materials*, ed. by A.J. Glass, A.H. Guenther

11.73   A.M. Glass: Opt. Eng. **17**, 470 (1978)

11.74   D.A. Bryan, R.R. Rice, R. Gerson, H.E. Tomaschke, K.L. Sweeney, L.E. Halliburton: Opt. Eng. **24**, 138–143 (1985)

11.75   V. Pruneri, P.G. Kazansky, J. Webjörn, P.St. J. Russell, D.C. Hanna: Appl. Phys. Lett. **67**, 1957 (1995)

11.76   G.A. Magel, M.M. Fejer, R.L. Byer: Appl. Phys. Lett. **56**, 108 (1990)

11.77   M. Taya, M.C. Bashaw, M.M. Fejer: Opt. Lett. **21**, 857 (1996)

11.78   M.A. Acharekar, L.H. Morton, E.W. Van Stryland: SPIE Proc. **2114**, 69 (1993)

11.79   S. Chandra, T.H. Allik, G. Catella, J.A. Hutchinson: Advanced Solid-State Laser Conf. 1998, Coeur d'Alene, ID, Proceedings, p. 102

11.80   M.R. Kozlowski, R. Chow: SPIE Proc. **2114**, 640 (1993)

11.81   Q. Zhao, Z.X. Fan, H. Qiu, Y. Liu, Z.J. Wang: SPIE Proc. **2966**, 238 (1996)

11.82   B. Steiger, H. Brausse: SPIE Proc. **2428**, 559 (1994)

11.83   R.H. Sawicki, C.C. Shang, T.L. Swatloski: SPIE **2428**, 333 (1994)

11.84   A. Bodemann, N. Kaiser, M. Kozlowski, E. Pierce, C. Stolz: SPIE Proc. **2714**, 395 (1995)

11.85   I.M. Thomas: SPIE **2114**, 232 (1993)

11.86   N. Kaiser: Laser & Optoelektronik **28**, 52 (1996)

11.87   C. Li, Z. Li, Y. Sun: SPIE **2114**, 280 (1993)

11.88   S.V. Garnov, S.M. Klimentov, A.A. Said, M.J. Soileau: SPIE **1848**, 162 (1992)

11.89   K.L. Lewis, A.M. Pitt, M. Corbett, R. Blacker, J. Simpson: SPIE **2966**, 166 (1996)

11.90   T.L. Barber: Rev. Sci. Instrum. **40**, 1630 (1969)

11.91   F.E. Hovis, B. Shepherd, C. Radcliffe, H. Maliborski: SPIE Proc. **2428**, 72 (1994)

Appendix

A.1   Courtesy; Coherent, Inc. Laser Group. 5100 Patrick Henry Drive, Santa Clara, CA

A.2   American National Standards Institute, 11 West 42nd Street, New York, NY 10036

A.3   D. Sliney, M. Wolbarsht: Safety with Lasers and Other Optical Sources (Plenum, New York 1980)

A.4   Laser Institute of America, 12424 Research Parkway, Suite 125, Orlando, FL "LIA Laser Safety Guide"

A.5   Center for Devices and Radiological Health (HFZ-300) 8757 Georgia Avenue, Silver Spring, MD 20910. Performance Standards for Laser Products, Regulation 21CFR1040 (latest revision April 1993)

A.6   International Electrotechnical Committee (IEC), Geneva, Switzerland. IEC Standard 825-1 (1993). Radiation Safety of Laser Products, Equipment Classifications, Requirements and User's Guide

Subject Index

Springer Series in Optical Sciences

Editorial Board: A. L. Schawlow† A. E. Siegman T. Tamir

Printing (computer to plate): Mercedes-Druck, Berlin
Binding: Buchbinderei Lüderitz & Bauer, Berlin